DIGITAL COMMUNICATION FOR PRACTICING ENGINEERS

DIGITAL COMMUNICATION FOR PRACTICING ENGINEERS

FENG OUYANG

WILEY

Published by John Wiley & Sons, Inc., Hoboken, New Jersey.
Published simultaneously in Canada.

For general information on our other products and services or for technical support, please contact our Customer Care Department within the United States at (800) 762-2974, outside the United States at (317) 572-3993 or fax (317) 572-4002.

Wiley also publishes its books in a variety of electronic formats. Some content that appears in print may not be available in electronic formats. For more information about Wiley products, visit our web site at www.wiley.com.

Library of Congress Cataloging-in-Publication Data is available.

Hardback: 9781119418009

Printed in the United States of America.

V10013500_082919

CONTENTS

CHAPTER 1

INTRODUCTION

1.1 WHY THIS BOOK?

This book is intended for practicing engineers in the digital communication field. It can be used as a textbook for master's level courses (e.g., for part-time professional education programs) or self-study. As such, the book has some unique characteristics comparing a typical textbook on the same subject.

A typical textbook strives to provide comprehensive and pedagogically sophisticated coverage of the concepts and theories. Such treatment makes it easier for the students to grasp key knowledge points. However, practicing engineers already have good general engineering knowledge and powerful self-learning skills. They need information sources that can be digested quickly. This book selects concepts and technologies that are most relevant to today's communication systems and presents them concisely and intuitively.

Instead of becoming well-versed in the entire field of digital communications, practicing engineers are more interested in getting knowledge on the specific subfields of their work. This book is organized as self-contained chapters. One can choose to read one or several relevant chapters, instead of the entire book.

Practicing engineers are more interested in applying existing techniques to their particular problems, rather than inventing new techniques. This book focuses on the pros and cons of broadly used techniques, rather than detailed mathematical analyses that may lead to discoveries. For example, on adaptive filtering, this book discusses in detail the tradeoff between performance and complexity of various methods and the tradeoff between convergence speed and final accuracy based on parameter choices.

Advanced topics such as orthogonal frequency division multiplexing (OFDM) and multiple-input multiple-output (MIMO), which are enabling technologies for modern communication systems such as WiFi and LTE-Advanced, are covered in more detail than usual (Chapters 10 and 11). This book also briefly describes other emerging technologies, some of which are adopted in the 5G cellular standards. These contents help practicing engineers follow the current technology trend.

Digital Communication for Practicing Engineers, First Edition. Feng Ouyang.
© 2020 by The Institute of Electrical and Electronics Engineers, Inc.
Published 2020 by John Wiley & Sons, Inc.

This book also covers some contents that are usually out of scope for textbooks, such as cyclostationary symbol timing recovery (Chapter 7), adaptive self-interference canceller (Chapter 8), and Tomlinson–Harashima precoder (Chapter 9). These techniques are used in many popular communications systems and are therefore useful to practicing engineers.

In addition to practicing engineers, regular students of digital communications can benefit from this book's unique perspective and treatment, by using it as a primary or supplementary textbook.

1.2   HOW TO USE THIS BOOK

A textbook typically strikes a balance between details and suspense. Omitting some details in derivation and leaving some open questions help to keep the readers engaged and inspired. On the other hand, narrative gaps increase the difficulty in understanding. Since its targeted readers are likely to be self-studying without professors or peers available to answer questions, this book biases to providing more details and leaving fewer gaps. Some of the homework problems provide leads for further exploration and contemplation.

Another balance is between conceptual discourses and mathematical details. Since the book is designed for self-study, it is important to provide detailed derivations to important conclusions. On the other hand, these derivations may distract the readers from the thread of concept development. To address this concern, we mark the important mathematical results with solid-line frames. The readers may focus on the text and framed equations in the first pass. Detailed derivations contained in other equations can be revisited once the conceptual landscape is understood.

This book is based on the author's experience of teaching "Advanced Digital Communication Systems" at the master's level. In general, the material in each chapter is suitable for one 3-hour lecture. The exceptions are Chapters 5 and 9, which are suitable for two lectures each. Overall, this book is suitable for a master's level course of one semester, while some homework problems can be used as class projects.

1.3   SCOPE

1.3.1   The Physical Layer Transceiver

The prevailing Open Systems Interconnection (OSI) model divides a system into seven layers [1]. This book focuses on layer 1, known as the *physical layer*, or the PHY layer.

The PHY layer functions as a "bit pipe" of the system. A PHY transmitter takes bits from the upper layers and sends them through the physical medium (copper wire, fiber optics, electromagnetic waves, etc.) to the receiver. A PHY receiver recovers the bits and passes them to the upper layers. A *transceiver* is a combination of a transmitter and a receiver. The PHY layer is about point-to-point or point-to-multipoint (in the

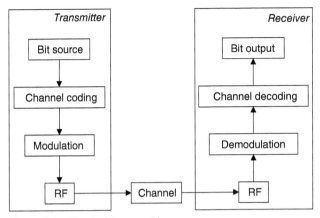

Figure 1.1   Physical layer architecture.

case of broadcast) connections, as opposed to a multi-hop network, which is the concern of the upper layers. The PHY layer transports bits from a transmitter to a receiver with a controlled error probability. The upper layers may perform other functionalities (such as retransmission) to achieve virtually error-free communication.

Figure 1.1 shows the general physical layer architecture of a transmitter/receiver pair. The "Bit Source" block at the transmitter side and the "Bit Output" block at the receiver side are interfaces to the upper layers. At the transmitter side, channel coding is applied to the data bits (Chapter 5) to enhance protection against random errors. Modulation is then performed to convert the bits into signals (i.e., time-varying voltages) (Chapter 3). The signal is then conditioned in various steps and transmitted through the physical medium, known as the channel (Chapters 3 and 4). At the receiver side, the signal is conditioned and demodulated to recover the encoded bits (Chapters 4, 8, and 9). Channel decoding is then performed to recover the data bits (Chapter 5), which is then passed to the upper layer. Chapters 10 and 11 cover more advanced modulation and demodulation techniques.

1.3.2   Prerequisites

This book is for master's level study. It assumes the readers have some training in electrical engineering and beginning digital communications [2]. For example, the readers should have basic knowledge about filters, shift registers, antennas, etc.

As to mathematics, the readers should have knowledge on statistics (Gaussian distribution and the Bayes' theorem), calculus, basic differential equations, linear algebra (especially eigenanalysis and singular value decomposition), and Fourier transform. Some mathematics topics are included in the appendices of the relevant chapters. Notably, Chapter 3 includes the formulation of the Fourier transform, which is also used in other chapters. These appendices intend to clarify conventions and notations, rather than teaching the knowledge from scratch. On the other hand, the required mathematics concepts and properties are very limited in this book. Readers can fill potential knowledge gaps by consulting other textbooks or online tutorials.

1.3.3 Topics Not Covered

This book focuses on advanced and practical topics in digital communications. Some topics such as analog modulation techniques and noncoherent detections are usually covered in the prerequisite courses [2] and are not repeated in this book. While covering basic concepts and techniques of digital communications, the book focuses on the mainstream commercial applications such as mobile cellular systems and wireless local area network (WLAN). Special applications such as underwater communications, satellite communications, military communications, and optical communications are not covered in this book.

This book is on point-to-point communications, that is, a single transmitter–receiver pair. The issue of access, that is, multiple users sharing a medium in coordinated or uncoordinated ways, is only briefly addressed in Chapters 10–13. Since the scope is limited to the PHY-layer, protocols and networking issues are not discussed.

Some of the uncovered topics are briefly discussed in Chapter 13.

1.4 ROADMAP

The book can be roughly divided into four parts.

The first part covers the basic techniques. It starts with Chapter 2, which introduces the foundation of modern digital communication theories: the Shannon theorem. In addition to guiding the development of channel coding, the Shannon theorem also provides some valuable insights into the practical case of white noise channels. Chapter 3 covers several techniques involved in the modulation process, which converts bits to symbols. These techniques include modulation, pulse shaping, and up-converting. They are based on various interesting mathematical concepts discussed in the chapter. Chapter 4 discusses the reverse process, that is, converting received symbols to bits. The chapter focuses on optimal demodulation theories and analyses of error probability. A detailed study of the additive white Gaussian noise (AWGN) channel and various expressions of the signal-to-noise ratio (SNR) is included. We then move on to channel coding in Chapter 5. Channel coding is a technique to improve noise immunity and to ultimately achieve the channel capacity predicted by the Shannon theorem. This book does not provide details on channel code designs and performance evaluations. It only outlines the basic concepts and decoding methodologies of common code types and qualitatively compares the various codes.

The second part continues with more practical matters. Chapter 6 discusses various properties of propagation channels beyond the AWGN. It lays the ground for further discussion of the various receiver techniques that take the channel characteristics into account. This chapter also discusses link budget computation, which is a critical part of the process of system requirement development. Chapter 7 discusses a practical concern: synchronization between the transmitters and receivers. Two issues are addressed in this chapter: how to estimate timing errors and how to correct them. It also points out that the methods of synchronization depend on the magnitude and dynamics of the original timing errors. The next two chapters describe channel equalization, which is a required processing technique for dispersive channels. Chapter 8 provides a general discussion of adaptive filters, which is the basis for equalizers. The chapter also discusses full-duplex radios with self-interference cancellation as another application of the adaptive filters. Chapter 9 describes linear equalizers and decision feedback equalizers, including their structures, training methods, and expected performances.

Chapters 10 and 11 constitute the third part of the book. These two chapters cover the two advanced modulation techniques developed in the 1990s: orthogonal frequency division multiplexing (OFDM) and multiple-in multiple-out (MIMO). These two techniques build onto the basic techniques discussed in part one, with additional gains in performance and simplicity.

The last part of the book provides an overview of the emerging technologies in digital communications. Chapter 12 introduces the 5G cellular technologies currently being developed. The chapter starts with a general overview of the cellular concept and history, as well as the current status of the 5G cellular standard. It then outlines three emerging technologies used in the 3GPP 5G proposal: massive MIMO, millimeter wave communications, and nonorthogonal multiple access. Chapter 13 further outlines several uncovered topics, most of which are current research areas.

The chapters are designed to be self-contained, so that a reader can quickly access a specific topic of interest, instead of reading the entire book. When contents in other chapters are used, explicit cross-references are provided. Each chapter starts with an introduction stating the scope and roadmap, as well as a summary outlining the main points. Figure 1.1 is used in most chapters to orient the readers relative to the overall physical layer architecture. For the sake of narrative flow, nonessential contents are often collected into footnotes and appendices.

1.5 OTHER NOTES

This section contains some notes that are helpful in using this book.

1.5.1 Mathematical Notation

This book follows the common mathematical conventions, with the following particular choices.

- In sections that involve linear algebra, vectors are expressed in boldface lowercase letters, matrices in capital letters, and scalars in normal style lowercase letters. I, in the context of matrix expressions, represents an identity matrix. I_N is an identity matrix of dimension N by N. When there is no confusion, "0" may express a vector or a matrix whose components are all zero.

- Identity matrices may be omitted. For example, if A is a matrix and α is a scalar, then $A + \alpha$ means $A + \alpha I$.
- Letters I to N, both upper case and lowercase, are reserved for integers.
- Exceptions to these practices include traditional electrical engineering notations, such as N_0 for noise power spectrum density, n for noise, j for $\sqrt{-1}$, E_s for total transmit power, t for time, ω for angular frequency, f for frequency, etc.
- $(\cdot)^H$ denotes matrix Hermitian operation. $(\cdot)^T$ denotes transposition operation. $(\cdot)^*$ denotes complex conjugate operation.
- $E(\cdot)$ denotes expectation value.
- $\exp(x)$ is used interchangeably with e^x.
- For summations, we typically use lowercase letters i to n for the summing indices and the corresponding uppercase letters for limits. When the limits are not given explicitly, they are either clear in the context or infinity. If a limit exceeds the range of the summed sequence, the undefined terms are considered as zero. Similar conventions apply to product notations and integrals.
- Notation $\{a_k, k = 1, \ldots, K\}$ denotes the set of a_k, where integer k is in the range of 1 to K. The second part is often omitted when the index range is clear in the context.
- When Fourier transform is involved, the tilde sign is used to indicate the Fourier transform of a function. For example, $\tilde{f}(\omega)$ is the Fourier transform of $f(t)$.
- Following the traditions in electrical engineering, a dimensionless quantity (such as a ratio) can be expressed linearly and in dB, as indicated by the context.

As an engineering textbook, we intentionally overlook some requirements on the rigor of mathematical treatments. For example, we assume all relevant functions are integrable. We also assume all relevant stochastic variables have well-defined probability density functions (pdf).

Mathematical symbols are used consistently within a chapter. Namely, the same symbol always represents the same quantity. However, a symbol may have different meanings in different chapters. Symbols are defined when first used. Such a definition is sometimes repeated later for easy reference.

All equations are numbered for the ease of cross-referencing. Equations representing significant results instead of intermediate steps are marked with shaded boxes.

1.5.2 References

Reference citations in this book are included to provide a historical perspective of the subject, to support statements made in the book without derivation, and to provide leads for further exploration. For the last purpose, newly published papers are preferred, as their introduction sections usually provide an overview of the state-of-the-art with references to the seminal works. When the same content can be found from multiple sources, those available online are preferred.

For current research activities, we often cite review papers published on the *IEEE Communications Surveys & Tutorials*. These papers are not necessarily insightful by themselves. However, they contain an extensive collection of annotated references serving as good starting points for further studies. Special issues of the various IEEE journals are also a good source for overview information. While the collection of papers in a special issue may not reflect the holistic picture of the research field, the combination of their introductions usually provides complete coverage.

Note that some citations (e.g., online articles, technical reports, preprints posted on arxiv.org, and patents) were not peer reviewed. Verification of the derivation and simulation is recommended before adopting the conclusions of these references.

1.5.3 Homework

Practicing engineers typically learn and work at the same time; they do not have much time for homework and practices. On the other hand, they are engaged in work activities related to the knowledge to learn. For this reason, this book contains only a few homework problems for each chapter. The readers are expected to reinforce and evaluate their learning through daily work instead of homework. Therefore, homework problems are more explorative than evaluative.

In addition to the homework, the readers are encouraged to try to reproduce the various simulation results provided in the book. Such exercise ensures that the readers understand the algorithmic details.

1.5.4 Simulation

This book provides quantitative examples through simulations. All simulation plots are generated with MATLAB version R2017a. Some key functions for simulation are mentioned in the text. Parameters used in the simulations are usually stated so that the plots can be reproduced by the readers.

1.5.5 Acronyms

All acronyms are spelled out when first used in a chapter, except for some common terms in electrical engineering such as SNR (signal-to-noise ratio).

ACKNOWLEDGMENTS

This book was made possible by the encouragement and guidance from Mr. Jack Burbank of Johns Hopkins University Applied Physics Laboratory (now at Sabre Systems). The author is grateful to Dr. Nim Cheung, Series Editor of IEEE Press for his support and guidance during the book planning stage. This book was supported in part by the Stuart S. Janney Publication Program of the Johns Hopkins University Applied Physics Laboratory. The author would also like to thank the anonymous reviewers, whose feedbacks were valuable in improving the quality of this book.

REFERENCES

1. International Telecommunication Union (ITU-T), *Information Technology—Open Systems Interconnection—Basic Reference Model: The Basic Model.* ITU-T X.200, 1994.
2. B. P. Lathi and Z. Ding, *Modern Digital and Analog Communication Systems*, 4th ed. New York: Oxford University Press, 2009.

CHAPTER 2

SHANNON THEOREM AND INFORMATION THEORY

2.1 INTRODUCTION

In 1948, Claude Shannon, a 32 years old scientist at the AT&T Bell Labs, published a landmark paper, establishing a mathematical theory of communications known as information theory [1]. His results, known as the Shannon theorem, answered a fundamental question in communications: what is the maximum achievable data rate for a channel? In essence, Shannon showed that there is an upper limit of the data rate for errorless transmission. Such limit can be reached, at least under some limiting conditions.

Today, the information theory focuses on developing various coding and decoding techniques. However, the impact of Shannon theory exceeds coding. In addition to the explicit expression of channel capacity, Shannon theorem provides fundamental insights about the tradeoffs among bandwidth, transmit power, and data rate. Such tradeoffs guide the basic system design considerations. Furthermore, the mathematics framework in the proof of Shannon theorem is beautiful. It reveals the role of randomness in both signal and noise. Therefore, we present Shannon theory as a starting point of our discourse in digital communication.

This chapter introduces Shannon theorem, following the treatment of the original paper. Sections 2.3–2.6 build up the Shannon theorem in a few steps. Sections 2.7 and 2.8 show the two practical applications. Section 2.9 provides a summary. Application of Shannon theorem for multiple antenna channels can be found in Chapter 11. More details on derivation of entropy expression and on compression coding can be found in the appendices, Sections 2.10 and 2.11.

Shannon theorem applies to both discrete and continuous probability distributions. For simplicity of presentation, we focus on discrete distributions at first and switch to continuous distributions when we consider the additive white Gaussian noise (AWGN) channel in Section 2.7. We use capital letters to denote random variables, and lowercase letters to denote their realizations. Furthermore, $p(\cdot)$ denotes probability (for discrete variables) or probability density function (pdf)

Digital Communication for Practicing Engineers, First Edition. Feng Ouyang.
© 2020 by The Institute of Electrical and Electronics Engineers, Inc.
Published 2020 by John Wiley & Sons, Inc.

(for continuous variables). Subscripts are added if there is confusion. For example, $p(x)$ is the pdf of random variable X at value x. $p_x(a)$ is the pdf of X at value a. $p(x \mid y)$ or $p_{(x \mid y)}(x, y)$ denotes the probability density of X taking value x, conditioned on another random variable Y taking the value y.

2.2 RELIABLE TRANSMISSION WITH NOISY CHANNEL

This chapter studies channels that have uncertainty. When a signal is transmitted through such channels, the corresponding received signal carries uncertainty. How do we use such channel to transmit data reliably, that is, without errors?

A typical receiver works as follows. With knowledge about the channel, we know that transmitted symbol A is likely to cause received symbol B. Therefore, when B is received, we decide A is transmitted. However, because of channel uncertainty (noise), it is possible that a different symbol C is transmitted and we still receive B. In this case, an error occurred. A naïve way to mitigate such error is transmitting the same symbol multiple times. Because of the noise, the receiver may reach different decisions (about what was transmitted) among these transmissions. A decision that appears the most times is deemed as the correct one. In such a majority voting approach, an error would occur only when the noise causes many errors in the repeated receptions, and thus is much less likely. The price to pay for such error reduction approach is lower data rate. The channel coding technologies introduced in Chapter 5 are based on such an idea, but with better performance. Nevertheless, such naïve approach shows that it might be possible to achieve errorless communication even with a noisy channel. The next question is the data rate we can achieve through such a channel. We will answer that question in Section 2.6. But first, we will introduce some basic concepts in Sections 2.3 – 2.5.

2.3 ENTROPY AND UNCERTAINTY

2.3.1 Uncertainty and the Value of Communication

Let us start by quantifying uncertainty. Before receiving data, the receiver does not know what is transmitted. In other words, there is great uncertainty. After receiving data, such uncertainty is removed (or significantly reduced, if there are possible errors in receiving).

On the other hand, if an entirely predictable data (e.g., all 1, or bits representing the constant π up to a given precision) is transmitted, the communication has no value, because there is no certainty at the receiver to start with. If the uncertainty is small, the value of communication is limited. For example, since we already know it seldom rains in Sahara, a weather report (rain or no rain) transmitted from there is not very useful.

Therefore, uncertainty is connected to the value or "usefulness" of a communication system. We start our analysis by quantifying uncertainty.

It turns out that under three reasonable assumptions, there is an almost unique expression (up to a constant factor) of uncertainty carried in a random variable X based on the probability distribution of X. The details of derivation will be provided in the appendix (Section 2.10). Here, we only state the result.

Consider a discrete random variable X, characterized by its probability distribution $\{p_n, v_n, n = 1, 2, \ldots, N\}$. p_n is the probability that $X = v_n$. N is the number of possible values X can take. The measure of uncertainty, known as entropy, is given as

$$H(p_1, p_2, \ldots, p_N) \stackrel{\text{def}}{=} -\sum_{n=1}^{N} p_n \log_2(p_n). \tag{2.1}$$

Furthermore, this result can be extended to random variables with continuous probability distribution $p(x)$. In this case, we have

$$H(p) \stackrel{\text{def}}{=} -\int p(x) \log_2[p(x)] dx. \tag{2.2}$$

Equations (2.1) and (2.2) can also be written as

$$H \stackrel{\text{def}}{=} -E\{\log_2[p(x)]\}. \tag{2.3}$$

Here, $E(\cdot)$ is the expectation value over x.

As shown in Section 2.10, the definition of entropy as a measure of uncertainty stems from some "natural" assumptions of its mathematics properties. However, entropy is also very important in the communication theory. As Section 2.4 will show, entropy is related to the number of bits required to represent a random sequence.

2.3.2 Some Properties and Extensions of Entropy

There are many interesting properties of H. The most important ones for our purposes are:

1. $H \geq 0$.
2. $H = 0$ if and only if one of the $\{p_n\}$ is 1 and all others are zero (i.e., X is deterministic).
3. For a given N, $H(p_1, p_2, \ldots, p_N)$ reaches maximum when $p_n = 1/N \, \forall \, n$.

The first two properties are obvious. The third one will be proved later in this section. Furthermore, the definition of $H(X)$ depends only on the probability distribution $\{p_n\}$, not on the actual values X can take $\{v_n\}$. Therefore, if another random variable U has a one-to-one mapping with X, then U has the same entropy as X.

Now consider two random variables X and Y, whose joint distribution is $\{(v_{xn}, v_{ym}), p_{xymn}, n = 1, 2, \ldots, N, m = 1, 2, \ldots, M\}$. p_{xymn} is the probability for $X = v_{xn}$ and $Y = v_{ym}$. We also denote the marginal probability distributions for X and Y as $\{v_{xn}, p_{xn}, n = 1, 2, \ldots, N\}$ and $\{v_{ym}, p_{ym}, m = 1, 2, \ldots, M\}$, respectively.

The conditional probability is noted as $\{(v_{xn}, v_{ym}), p_{(x|y)mn}\}$, where $p_{(x|y)mn}$ is the probability of $X = v_{xn}$ given that $Y = v_{ym}$.

The definition of entropy can be extended to the cases of two variables. The joint entropy is

$$H(X,Y) \overset{\text{def}}{=} -E\left[\log_2\left(p_{xymn}\right)\right] = -\sum_{n,m} p_{xymn} \log_2\left(p_{xymn}\right). \tag{2.4}$$

The conditional entropy is

$$H(X\,|\,Y) \overset{\text{def}}{=} -E\left[\log_2\left(p_{(x|y)mn}\right)\right] = -\sum_{n,m} p_{xynm} \log_2\left(p_{(x|y)mn}\right). \tag{2.5}$$

When X and Y are statistically independent,

$$p_{xymn} = p_{xn}p_{ym}. \tag{2.6}$$

It follows that

$$H(X,Y) = H(X) + H(y). \tag{2.7}$$

What if X and Y are not independent? Intuitively, we expect the entropy to be smaller, as uncertainty is reduced in this case. We will prove this statement mathematically below [2, sec. 1.1].

We start from a well-known inequality:

$$\ln x \le x - 1. \tag{2.8}$$

The equality holds when $x = 1$.

Consider two sets of probabilities $\{p_1, p_2, \ldots, p_N\}$ and $\{q_1, q_2, \ldots, q_N\}$ with normalization

$$\sum_{n=1}^{N} p_n = \sum_{n=1}^{N} q_n = 1. \tag{2.9}$$

From (2.8),

$$\sum_{n=1}^{N} p_n \log_2\left(\frac{q_n}{p_n}\right) = \frac{1}{\ln(2)} \sum_{n=1}^{N} p_n \ln\left(\frac{q_n}{p_n}\right) \le \frac{1}{\ln(2)}\left[\sum_{n=1}^{N} p_n\left(\frac{q_n}{p_n}\right) - 1\right] = 0. \tag{2.10}$$

Therefore,

$$\sum_{n=1}^{N} p_n \log_2 p_n \ge \sum_{n=1}^{N} p_n \log_2 q_n. \tag{2.11}$$

Let $\{p_{xymn}\}$ and $\{p_{xn}p_{ym}\}$ take the roles of $\{p_n\}$ and $\{q_n\}$ in (2.11) and sum over both n and m, we get

$$H(X,Y) \leq H(X) + H(Y). \tag{2.12}$$

Equity holds if and only if x is 1 in (2.8). Namely,

$$p_{xymn} = p_{xn}p_{ym}, \tag{2.13}$$

implying that X and Y are independent.

Inequality (2.11) can also be used to prove the third property above: $H(X)$ reaches its maximum when X is a uniform distribution. In this case, let $q_n = 1/N$ and $\{p_n\}$ be any probability distribution.

2.3.3   Entropy and Gaussian Distribution

In the information theory, random variables with Gaussian distribution occupy a special place. This is not only because Gaussian distribution is a good model for many random processes, but also because Gaussian distribution maximizes entropy under power constraint, as will be shown in this section. Most communication systems subject to power constraints, as we will see throughout this book.

Consider a random variable X with alphabet $\{v_n, n = 1, 2, \ldots, N\}$ with corresponding probability distribution $\{p_n, n = 1, 2, \ldots, N\}$. The power constraint for X can be expressed as

$$E\left(|v_n|^2\right) = \sum_{n=1}^{N} p_n v_n^2 = P. \tag{2.14}$$

In addition, $\{p_n\}$ subjects to the normalization constraint

$$\sum_{n=1}^{N} p_n = 1. \tag{2.15}$$

We wish to find the set of $\{p_n\}$ that maximizes the entropy

$$H(X) = -\sum_{n=1}^{N} p_n \log_2(p_n). \tag{2.16}$$

Following the method of Lagrange multipliers, we construct the Lagrangian

$$L \stackrel{\text{def}}{=} -\sum_{n=1}^{N} p_n \log_2(p_n) - \lambda_1 \sum_{n=1}^{N} p_n - \lambda_2 \sum_{n=1}^{N} v_n^2 p_n. \tag{2.17}$$

L is optimized in terms of $\{p_n\}$ unconditionally. λ_1 and λ_2 are determined afterward from the constraints (2.14) and (2.15). The optimization thus calls for the following derivatives to be zero:

$$\frac{\partial L}{\partial p_n} = -\log_2(p_n) - \frac{1}{\ln(2)} - \lambda_1 - \lambda_2 v_n^2 = 0 \,\forall n = 1, 2, \ldots, N. \tag{2.18}$$

Equation (2.18) leads to

$$p_n = A exp\left(-\frac{v_n^2}{2\,\sigma_x^2}\right);$$

$$\sigma_x^2 \overset{\text{def}}{=} \frac{1}{2\lambda_2} \frac{1}{\ln(2)}, \qquad (2.19)$$

$$A \overset{\text{def}}{=} \exp\{-[1+\lambda_1\ln(2)]\}.$$

Instead of determining λ_1 and λ_2, we can determine parameters A and σ_x^2 directly from the constraints. When $N \to \infty$ and X approaches a continuous random variable, (2.19) becomes the familiar Gaussian distribution. Therefore, the constraints (2.14) and (2.15) lead to

$$\sigma_x^2 = E(X^2) = P,$$

$$A = \left(2\pi\sigma_x^2\right)^{-\frac{1}{2}},$$

$$p_x(x) = \left(2\pi\sigma_x^2\right)^{-\frac{1}{2}} \exp\left(-\frac{x^2}{2\,\sigma_x^2}\right). \qquad (2.20)$$

Therefore, for continuous random variables X with power constraint, to maximize its entropy, X should have a Gaussian distribution. In this case, the entropy is

$$H(X) = -E[\log_2(p_x)] = E\left[\frac{1}{2}\log_2\left(2\pi\sigma_x^2\right) - \frac{1}{\ln 2}\frac{x^2}{2\,\sigma_x^2}\right] = \frac{1}{2}\log_2\left(2\pi\sigma_x^2\right) - \frac{1}{2\ln 2}. \qquad (2.21)$$

2.4 ENTROPY AND BIT LENGTH

Section 2.3 derived an expression, known as the entropy, to measure the uncertainty in a random variable. The next question is: how does such metric help us in characterizing a digital communications system? It turns out that entropy is related to the average number of bits required to represent a random sequence.

2.4.1 The Source Coding Problem

Source coding means representing a data source (a collection of data symbols) with a binary sequence. Consider a random variable X, whose alphabet has size N. Sequence $x_1, x_2, \ldots, x_K$ contains K realizations of X. What is the number of bits needed to represent such sequence?[1] Here, the term "represent" means that one can recover the original sequence from the representing bit sequence. In other words, such coding is "lossless."

A naïve way is representing each member of the alphabet with a bit segment. Since there are N such members, the segment length must be no smaller than $\log_2(N)$. We can then stack the K segments sequentially, to form a binary sequence no shorter than $K \log_2(N)$ bits.

However, this is not necessarily the smartest way. We can try to code the source with binary sequences of shorter length, a process known as "compression." For example, the Morse code for telegraph assigns shorter segments to more frequently used letters, thus reducing the average length, as short segments appear more frequently in the final sequence. However, if all segments do not have the same length, the receiver needs to know the boundary between segments for correct decoding. In Morse code, this is done by inserting extra spaces between letters and words, increasing the total representation length. Therefore, the tradeoff is not that obvious.

In this section, we will find out the minimum length of the binary sequence required to encode a source sequence losslessly. It turns out that such minimum length is related to the entropy of the source.

2.4.2 Typical Sequence (TS)

The length of the binary sequence required to represent a sequence of random variables can be examined using the concept of "typical sequence" (TS).

Consider a very long symbol sequence $\{x_1, x_2, \ldots, x_K\}$, where each symbol x_i is a realization of a random variable X, with probability distribution of $\{v_n, p_n, n = 1, 2, \ldots, N\}$. Namely, X takes possible values $\{v_1, v_2, \ldots, v_N\}$ with corresponding probabilities $\{p_1, p_2, \ldots, p_N\}$. Suppose a particular value v_n appears l_n times in the sequence. A TS is a sequence with the following property[2]:

$$l_n = K p_n \forall n. \tag{2.22}$$

Namely, in a TS, a value v_n appears with a frequency that equals to its probability p_n. According to the law of large numbers, when K approaches infinity, the probability for a sequence to be a TS approaches 1. Namely, practically all observed sequences are TS. For a particular TS, value v_n appears at l_n places. Therefore, the probability for a particular TS to appear is

$$p = \prod_{n=1}^{N} p_n^{l_n} = \prod_{n=1}^{N} p_n^{K p_n}. \tag{2.23}$$

---

[1] In this section, we assume all symbols in the sequence are mutually independent. However, the results can be extended to the cases where the symbols have stationary dependence. See [1] and [5, sec. 6.3] for more details.

[2] In this discussion, K is a very large number. Therefore, we assume l_n expressed in (2.22) is always very close to be an integer.

Note that such probability is the same for all TS. Furthermore,

$$\log_2(p) = \sum_{n=1}^{N} K p_n \log_2(p_n) = -KH(X).$$ (2.24)

On the other hand, the number of such TS can be computed by combinatorics. The number of ways to construct a sequence of length K, with l_1 counts of v_1, l_2 counts of v_2, etc. is

$$M = \binom{K}{l_1} \binom{K-l_1}{l_2} \cdots \binom{K-l_1-l_2-\ldots-l_{N-1}}{l_N} = \frac{K!}{l_1! l_2! \ldots l_N!}.$$ (2.25)

For large numbers, the factorials can be approximated by Stirling's formula [3, sec. 5.11.7]:

$$\ln(n!) \approx n\ln(n).$$ (2.26)

With this approximation, we have

$$\log_2(M) = \frac{1}{\ln(2)} \left[K\ln(K) - \sum_{n=1}^{N} l_n \ln(l_n) \right] = -\frac{K}{\ln(2)} \sum_{n=1}^{N} \frac{l_n}{K} \ln\left(\frac{l_n}{K}\right).$$ (2.27)

For TS, with (2.22), we have

$$\log_2(M) = -K \sum_{n=1}^{N} p_n \log_2(p_n) = KH(X).$$ (2.28)

Since we have M TS (2.28), each appears with a probability of p (2.23), the probability of seeing *any* TS is Mp, which turns out to be 1. This confirms our assertion that TS appears with a probability of 1. Note that all above derivations are predicated on K's approaching infinity.

If every TS is represented by a binary sequence, then we can correctly represent the data with a probability of 1. The original sequences can be recovered from the binary sequence representation (coding). On the other hand, many sequences are not TS and cannot be represented. Errors will occur if a non-TS is to be retrieved. However, as K approaches infinity, the probability for such error is zero. This is the key for source coding. Since there are M TS, the number of bits required to represent these TS is

$$L(K) = \log_2(M) = KH(X).$$ (2.29)

The coding rate (code binary per symbol) is

$$R \overset{\text{def}}{=} \frac{L(K)}{K} = H(X). \qquad (2.30)$$

2.4.3 Source Coding Theorem

Discussion in Section 2.4.2 was formalized as the source coding theorem, first proposed by Shannon and further developed by others [2, sec. 1.1].

The source coding theorem states that for a sequence of length K, each symbol drawn from random variable X, there exists a binary code of average length L, such that

$$L \leq K[H(X) + o(K)], \qquad (2.31)$$

where $o(K)$ is a term, which becomes vanishingly small as K approaches infinity. Conversely, no such code exists, for which

$$L < KH(X). \qquad (2.32)$$

The direct part of the coding theorem (2.31) is supported by the analysis of TS in Section 2.4.2. It can also be proved by specific codes, such as the Huffman code described in the appendix (Section 2.11.1). Detailed proofs of both direct and converse parts can be found in [2, sec. 1.1].

Source coding theorem has broad applications in data compression and data storage. In the context of digital communications, it helps to reveal the significance of entropy: it is related to the number of bits required to represent the information. Practically, with optimal source coding, the average number of bits required to represent (or code) a source of K symbols is approximately

$$L = KH(X). \qquad (2.33)$$

If X has a uniform distribution over N possible values, its entropy is $\log_2 N$. In this case, the minimum coding length from (2.33) is $K \log_2 N$, the same as the naïve coding scheme described in Section 2.4.1. On the other hand, note that a binary sequence of length L has a maximum entropy of L, which is achieved when each bit equally probable between 0 or 1. In fact, one can argue that an optimal binary code of length L must have entropy L. Otherwise, it can be further coded into a shorter sequence. Namely, bits in the coded binary sequence must be equally probable between 0 and 1, and be mutually independent. This observation justifies a common assumption of the input to any communications system: the symbols to transmit have a uniform distribution over the alphabet (i.e., allowed values), and symbols are mutually independent. Furthermore, we see that the entropy of the sequence remains constant in source coding process.

2.5 INFORMATION MEASURED AS REDUCTION OF UNCERTAINTY

Section 2.4 shows the relationship between entropy and the number of bits required for the lossless representation of random sequences. In this section, we continue to introduce another concept: mutual information. Mutual information is the measurement of uncertainty reduction due to communications. Therefore, it is a good metric of channel capacity. As will be shown in Section 2.6, channel capacity based on mutual information is related to the maximum data transmission rate.

2.5.1 Mutual Information

Consider two random variables X and Y, with joint probability distribution $p(x, y)$. If we know the value Y, we would know something about X. Namely, our uncertainty about X is reduced. Such reduction reflects the amount of information that Y carries about X. Thus, mutual information is quantified as

$$I(X;Y) \overset{\text{def}}{=} H(X) - H(X \mid Y). \tag{2.34}$$

Entropies $H(X)$ and $H(X \mid Y)$ are given by (2.3) and (2.5), respectively. For a communications channel, X is the signal transmitted while Y is the signal received. Y is not completely determined by X, because of noise. However, X and Y are correlated, as characterized by the conditional probabilities in (2.5). Thus, we can use $I(X; Y)$ to measure how much received signal tells us about what is transmitted. In Section 2.6, we will link $I(X; Y)$ with achievable data rate.

Let us first consider some salient property of $I(X; Y)$, which are useful for future discussions.

First, the mutual information is symmetric:

$$I(X;Y) = I(Y;X). \tag{2.35}$$

Symmetry is supported by the Bayes theorem, using the same notations for (2.4)–(2.6):

$$\frac{p_{yn}}{P_{(y|x)nm}} = \frac{p_{xm}}{P_{(x|y)mn}} \forall n,m. \tag{2.36}$$

Next, recall again Bayes theorem

$$p_{xymn} = P_{(x|y)mn} p_{yn}, \tag{2.37}$$

which leads to

$$H(X, Y) = H(X \mid Y) + H(Y). \tag{2.38}$$

From (2.12),

$$H(X) + H(Y) \geq H(X, Y) = H(X \mid Y) + H(Y). \tag{2.39}$$

Therefore,

$$I(X; Y) = H(X) - H(X \mid Y) \geq 0. \tag{2.40}$$

Inequality (2.40) shows that the mutual information is nonnegative. This makes sense because knowing Y never increases the uncertainty (and thus the entropy) of X. The mutual information is zero if and only if X and Y are independent. Namely, receiving Y does not tell us anything about X. At the other extreme, if X is completely determined by Y, then $H(X \mid Y) = 0$, which leads to $I(X; Y) = H(X)$.

2.5.2   Channel Capacity

As we can see in Section 2.5.1, mutual information at the receiver side is the difference between $H(X)$ and $H(X \mid Y)$. It depends on the channel property, which is characterized by $p_{(x \mid y)mn}$, and the a priori distribution of X, characterized by p_x.

For example, consider an additive channel:

$$y = x + n, \tag{2.41}$$

where n is a random noise. Such channel can be expressed as

$$p_{(x \mid y)_{mn}} = p_n(n = y - x). \tag{2.42}$$

Here, $p_n(n = a)$ is the probability for n to equal a.

As users of the channel, we cannot control $p_{(x \mid y)mn}$. However, p_x can be controlled to a certain extent through transmission signal design. Therefore, we are interested in the maximum mutual information over all $p_x(x)$ choices. This maximum mutual information is thus considered as the channel capacity, C:

$$C \overset{\text{def}}{=} \max_{p_x} I(Y; X). \tag{2.43}$$

It turns out that channel capacity is closely related to the achievable data rate, as will be shown in Section 2.6. Let us first look at an example.

Consider a binary symmetric channel. In this case, the input X takes two possible values x_1 and x_2. Output Y also takes two possible values y_1 and y_2. The channel is characterized by the following probabilities.

$$P_{(y|x)11} = P_{(y|x)22} = \alpha$$
$$P_{(y|x)12} = P_{(y|x)21} = 1 - \alpha \tag{2.44}$$

If we use a naïve detector at the receiver, which determines the transmit signal from the received ones based on the following mapping:

$$y_1 \rightarrow x_1$$
$$y_2 \rightarrow x_2 \tag{2.45}$$

then the first line in (2.44) shows the probability of correct reception, and the second line shows the probability of error.

Because of (2.35), we can compute $I(Y; X)$, which is easier since we know $p(y_i \mid x_i)$:

$$I(Y; X) = H(Y) - H(Y \mid X). \tag{2.46}$$

Assume that the marginal probabilities of Y are

$$p_{y1} = q$$
$$p_{y2} = 1 - q \tag{2.47}$$

We thus have

$$H(Y) = -\left[p_{y1} \log_2(p_{y1}) + p_{y2} \log_2(p_{y2})\right] = -[q \log_2(q) + (1-q)\log_2(1-q)]. \tag{2.48}$$

Further,

$$H(Y \mid X) = -\sum_{i,j=1}^{2} P_{(y|x)ij} p_{xi} \log_2(P_{(y|x)ij})$$
$$= -\alpha(p_{x1} + p_{x2})\log_2(\alpha) - (1-\alpha)(p_{x1} + p_{x2})\log_2(1-\alpha) \tag{2.49}$$

Here, p_{x1} and p_{x2} are the probabilities for X to take the values of x_1 and x_2, respectively. Since these are the only two choices for X,

$$p_{x1} + p_{x2} = 1. \tag{2.50}$$

Therefore, (2.49) becomes

$$H(Y \mid X) = -[\alpha \log_2 \alpha + (1-\alpha)\log_2(1-\alpha)]. \tag{2.51}$$

Therefore,

$$I(Y;X) = H(Y) - H(Y \mid X)$$
$$= -[q \log_2(q) + (1-q)\log_2(1-q)] + [\alpha \log_2 \alpha + (1-\alpha)\log_2(1-\alpha)] \tag{2.52}$$

To maximize $I(Y; X)$, we only need to maximize the sum in the first square bracket by choosing q. Obviously, maximization is achieved with $q = 0.5$. Therefore, the capacity is

$$C = \max_q I(Y; X) = 1 + \alpha \log_2 \alpha + (1-\alpha)\log_2(1-\alpha). \tag{2.53}$$

Note that we cannot directly control q. However, q depends on our choice of p_{x1} and p_{x2}. $q = 0.5$ implies that $p_{x1} = p_{x2} = 0.5$. Namely, the transmit signals are equiprobable between x_1 and x_2.

From the above example, we can make the following observations.

1. C remains the same when we exchange α with $1 - \alpha$. This is equivalent to exchanging y_1 and y_2 in channel definition. One can change the mapping in (2.45) accordingly and obtain the same channel.

2. C reaches maximum when α is 1 or 0. This corresponds to a channel without error.

3. C reaches minimum (value 0) when α is 0.5. In this case, Y is independent of X.

2.6   SHANNON THEOREM

In Section 2.5, channel capacity was introduced based on mutual information. The Shannon theorem further connects channel capacity with achievable data rates. In this section, we present the statement and supporting argument of the theorem. Rigorous proofs of the Shannon theorem are available in many references and are not provided here.

2.6.1   Shannon Theorem Statement

Shannon proposed and proved the "fundamental theorem for a discrete channel with noise" in his landmark paper [1]. His statement is quoted below.[3]

> Let a discrete channel have the capacity C and a discrete source with the entropy per second R. If $R \leq C$, there exists a coding system such that the output of the source can be transmitted over the channel with an arbitrarily small frequency of errors (or an arbitrarily small equivocation). If $R > C$, it is possible to encode the source so that the equivocation is less than $R - C + \epsilon$, where ϵ is arbitrarily small. There is no method of encoding which gives an equivocation less than $R - C$.

In above statement, "equivocation" is $H(\bar{D} \mid \bar{Y})$, the residual uncertainty about the source after the signal is received. Note that $\bar{D}$ and $\bar{Y}$ are different from X and Y, respectively; they represent the coded sequences, instead of the raw signal. When equivocation is not zero, an error occurs at the receiver. Shannon's discussions are based on "entropy per second," with the corresponding capacity definition. We can understand the theorem based on "entropy per channel use," that is, entropy and capacity per symbol transmission.

2.6.2   Supporting Arguments of the Shannon Theorem

A rigorous proof of the Shannon Theorem can be found in [1], as well as many books on information theory [4]. Here, we only provide some intuitive supporting arguments based on [1].

---

[3] We substituted symbol H in Shannon's statement with R, to be consistent with our convention.

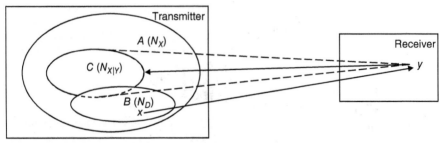

Figure 2.1 Coding and decoding model. On the left side (transmitter): A represents all TSs; B represents TSs that are selected to be codewords; C represents TSs that are considered as possible transmissions when y is received. On the right side (receiver): y represents the sequence received when x is transmitted. It also depends on the noise.

The first part of the theorem says there is a code that can achieve a transmission data rate $R \le C$ with an arbitrarily small frequency of error. This statement can be supported by considering the typical sequences (TSs), which are discussed in Section 2.4.2. Here, we consider only the case where R is strictly less than C.[4]

Figure 2.1 shows the coding and decoding model considered by Shannon. The box "Transmitter" indicates all possible transmitted sequences of length K. Oval A shows all the TSs of length K. There are N_X of them. Of those, N_D of them are chosen as codewords, indicated by oval B. During transmission, one of those codewords x is transmitted, generating y and the receiver. Because of channel uncertainty, the receiver does not know exactly the transmitted sequence. Instead, it can only say that one of the $N_{X|Y}$ sequences (oval C) was transmitted. If this set of possible sequences include legitimate codewords other than x (i.e., the joint of ovals B and C is not empty), then an error occurs. Since the codewords are chosen randomly from oval A, the probability of error is $N_{(X|Y)}N_D/N_X$. Note that for illustration purposes, we express the set of codewords and that of the possible transmitted sequences into ovals B and C. In reality, sequences belonging to these two sets are not concentrated in certain areas. Instead, they are randomly distributed among all TSs in oval A.

Now we work out the details. Suppose we use some of the TS of length K in X to transmit data, and K approaches infinity. The total number of such TSs is N_X, which, according to (2.28), is

$$N_X = 2^{KH(X)}. \tag{2.54}$$

Since we want to achieve a data rate of R bits per channel use, we need to use N_D of the TSs for codewords.

$$N_D = 2^{KR}. \tag{2.55}$$

When one of such codeword is transmitted, the receiver still has uncertainty about the transmission, characterized by entropy $H(X \mid Y)$. As far as the receiver is

---

[4] In fact, some renditions of the Shannon theorem [5, sec. 6.5-2] use strict inequality.

concerned, the transmitted sequence may be one of $N_{X|Y}$ possible TSs, including the correct one. Again, from (2.28),

$$N_{X|Y} = 2^{KH(X|Y)}. \tag{2.56}$$

If one of these TSs is a valid codeword (other than the one actually transmitted), an error occurs. The probability of such error P_e is, assuming a random choice of codewords among the TSs and an optimal choice of p_x

$$P_e = N_{X|Y} \frac{N_D}{N_X} = 2^{K[H(X|Y) + R - H(X)]} = 2^{-K(C-R)}. \tag{2.57}$$

Since $R < C$ as stated in the theorem, P_e can be arbitrarily small as K increases. Namely, with a long enough code, an arbitrarily small probability of error can be achieved. Note this result is for a random selection of TS as codewords. Therefore, the optimal coding will do no worse.

The second part of the statement says when $R > C$, there exists a code that achieves $H(\bar{D}|\bar{Y}) < R - C + \epsilon$, for an arbitrarily small positive ϵ. The meaning of $H(\bar{D}|\bar{Y})$ was explained in Section 2.5: $\bar{D}$ is a binary code for the data. From discussions in Section 2.4, such code has a length of KR, where K is the number of data symbol block to be coded. Such code contains mutually independent bits with an equal probability for values 0 and 1. The entropy of $\bar{D}$ is KR. In the following, we will construct a transmissions scheme so that

$$H(\bar{D}|Y) < R - C + \epsilon. \tag{2.58}$$

For every K channel use, we want to transmit a binary data code stream $\bar{D}$ of length KR, with $R > C$. We can transmit the first segment of $K(C - \eta)$ bits (noted as $\bar{D}_1$) and ignore the remaining segment of KR' bits (noted as $\bar{D}_2$), with

$$R' = R - (C - \eta). \tag{2.59}$$

η is an arbitrarily small positive number. Since the two segments are mutually independent, (2.7) leads to

$$H(\bar{D}|\bar{Y}) = H(\bar{D}_1|\bar{Y}) + H(\bar{D}_2|\bar{Y}). \tag{2.60}$$

According to the first part of the theorem (2.57), $\bar{D}_1$ can be transmitted and received with an arbitrarily small rate of error. This means given $\bar{Y}$, the remaining entropy per bit is arbitrarily small:

$$H(\bar{D}_1|\bar{Y}) < K\nu, \tag{2.61}$$

where ν is an arbitrarily small positive number. Since the second segment is not transmitted at all, its entropy remains the same after receiving.

$$H(\bar{D}_2|\bar{Y}) = H(\bar{D}_2) = KR' = K[R - (C - \eta)]. \tag{2.62}$$

Therefore, (2.60) leads to

$$H(\bar{D} \mid \bar{Y}) < K(R - C + \epsilon)$$
$$\epsilon \stackrel{\text{def}}{=} \eta + \nu \tag{2.63}$$

Obviously, ϵ is an arbitrarily small positive number. Therefore, (2.58) can be achieved. This argument supports the second part of the Shannon theorem.

The third part of the theorem says there can be no code that achieves

$$H(\bar{D} \mid \bar{Y}) < R - C. \tag{2.64}$$

This argument can be supported if we view the combination of the coding and the channel as an effective channel. Such effective channel has a capacity of $C' \le C$.[5] Since capacity is, by definition, the maximum of mutual information, we have

$$C \ge C' \ge H(\bar{D}) - H(\bar{D} \mid \bar{Y}) = R - H(\bar{D} \mid \bar{Y}). \tag{2.65}$$

Therefore, (2.64) is impossible.

2.6.3   Discussions

The Shannon theorem presented in Section 2.6.1 is a very strong statement. It gives the upper limit of the data rate under error constraint and claims that such data rate is achievable. On the other hand, the proof contains a somewhat loose estimate of the bound. It is based on random (as opposed to optimal) coding and ignores sequences other than TS. Such estimate achieves accurate results, due to the property of large numbers. As sequence length K approaches infinity, many random variables take their most probable values with a probability of one. This consideration brings significant simplification to the analysis.

Because of the Shannon theorem, channel capacity is also known as the Shannon bound, as it is the upper bound of the achievable data rate in the absence of error. While Shannon theorem shows that such bound is achievable, it does not provide a specific coding method to achieve it. Achieving Shannon bound, or even approaching it, has been a goal of information theory since the publication of Shannon theorem.

Shannon's analyses show that in order to achieve Shannon bound, very long codes (i.e., codewords containing many symbols) should be used. However, the code length is limited by system latency and complexity constraints. Strategies for code development will be further discussed in Chapter 5. It suffices to say that modern, practical codes can achieve performances very close to the Shannon bound.

Furthermore, in the process of deriving Shannon theorem, Shannon developed an estimate of residue errors as a function of sequence length. This technique is also broadly used in channel coding researches. Details are not covered in this book.

---

[5] It can be proved that if random variables D and X are connected by a channel and so are X and Y, then $H(Y \mid D) = H(Y \mid X) + H(X|D)$. Therefore, $I(Y; D) = H(Y) - H(Y \mid D) \le I(Y; X)$. It follows that $C' \le C$.

The Shannon theorem is commonly used to compute achievable data rate without error. However, Shannon theorem also gives the error rate (in terms of residual entropy at the receiver) when data rate exceeds channel capacity. It turns out, as shown by a special case in Section 2.7.6, the probability of error increases very fast as the data rate exceeds capacity. In other words, even if we allow for reasonable error rate, the data rate cannot be increased significantly. Therefore, the Shannon bound can be viewed as the upper limit of data rate, regardless of allowed error probability.

In Sections 2.7 and 2.8 capacities of two commonly used channels will be examined in more details.

2.7 ADDITIVE WHITE GAUSSIAN NOISE (AWGN) CHANNEL

Additive white Gaussian noise (AWGN) channel is the most common channel model for communications research. While simple in formulation, AWGN is realistic in many situations. AWGN and other channel models are discussed in more details in Chapter 6. In this section, we discuss the application of Shannon theorem to AWGN as an illustration.

2.7.1 Channel Definition

An AWGN channel can be modeled by the following equation:

$$y = x + n. \tag{2.66}$$

x and y are the transmitted and received signals, respectively. n is the additive noise. It obeys Gaussian distribution with zero mean, with the following probability distribution function

$$p_n(n) = \frac{1}{\sqrt{2\pi \sigma_n^2}} \exp\left(-\frac{n^2}{2\sigma_n^2}\right). \tag{2.67}$$

Here, σ_n^2 is the variance and also the expectation value of noise energy per symbol:

$$\sigma_n^2 = E\left(|n|^2\right). \tag{2.68}$$

The term "white" means noise values for each channel use are mutually independent. In the literature, noise energy is also expressed in terms of N_0:

$$N_0 \stackrel{\text{def}}{=} 2\sigma_n^2. \tag{2.69}$$

For more discussion on noise characterization, see Chapter 4.

Usually, operations of such channel are under the constraint of maximum transmit energy per symbol E_t. Namely,

$$E\left(|x|^2\right) \le E_t. \tag{2.70}$$

Note that both signal and noise energies are measured at the receiver. Therefore, channel gain is not included in the model.

In the AWGN model, y, x, n are all continuous random variables. The definitions of entropy and capacity are easily extended from their discrete forms to cover such case. For details concerning continuous random variable cases, see [1].

2.7.2   AWGN Channel Capacity

Given the channel model (2.66), we proceed to compute the channel capacity. Recall from Section 2.5.2, the channel capacity is

$$C = \max_{p_x}[H(y) - H(y\,|\,x)]. \tag{2.71}$$

Here, $H(y)$ and $H(y\,|\,x)$ are unconditioned and conditioned entropies of y. p_x is the probability distribution of x. From (2.66), the conditional probability distribution of y is

$$p_{(y|x)}(y,x) = p_n(y-x) = \frac{1}{\sqrt{2\pi\,\sigma_n^2}}\exp\left(-\frac{x^2}{2\,\sigma_n^2}\right). \tag{2.72}$$

Its entropy is a constant, independent of p_x. $H(y)$, on the other hand, depends on the probability distribution of y, which is determined by the probability distribution of x. From (2.70), the energy of y is constrained:

$$E\left(|y|^2\right) = E\left(|x|^2\right) + E\left(|n|^2\right) \le E_t + \sigma_n^2. \tag{2.73}$$

Obviously, maximum entropy occurs when equality holds in (2.73), as higher energy allows for a "flatter" distribution and thus higher entropy. From Section 2.3.3, $H(y)$ achieves maximum when y has a Gaussian distribution, with the variance

$$\sigma_y^2 = E\left(|y|^2\right) = E_t + \sigma_n^2. \tag{2.74}$$

From the property of Gaussian distributions, (2.66) and (2.74) imply that x also obeys Gaussian distribution, with variance E_t. From (2.21), the entropies are

$$
\begin{aligned}
H(y\,|\,x) &= \frac{1}{2}\log_2\left(2\pi\sigma_n^2\right) - \frac{1}{2\ln(2)} \\
H(y) &= \frac{1}{2}\log_2\left[2\pi\left(\sigma_n^2 + E_t\right)\right] - \frac{1}{2\ln(2)}
\end{aligned}
\tag{2.75}
$$

Equations (2.71) and (2.75) lead to the capacity for AWGN channel:

$$C = H(y) - H(y \mid x) = \frac{1}{2}\log_2\left(\frac{\sigma_n^2 + E_t}{\sigma_n^2}\right) = \frac{1}{2}\log_2(1 + SNR), \tag{2.76}$$

where SNR is the signal-to-noise ratio and is defined as

$$SNR \stackrel{\text{def}}{=} \frac{E_t}{\sigma_n^2}. \tag{2.77}$$

Equation (2.76) is the formula for AWGN channel capacity and is one of the most important formulas in digital communications.

2.7.3   Channel Capacity and Bandwidth

Equation (2.76) gives the channel capacity per channel use. According to the Shannon theorem, this is also the number of bits that can be transmitted without error for every transmitted symbol. The ultimate data rate (in bits per second) also depends on the number of symbols transmitted per second, or symbol rate.

According to the Nyquist theorem (Chapter 3), the symbol rate is limited by the channel bandwidth. If a channel has bandwidth W, the maximum symbol rate is $2W$.[6] For such a channel, the maximum data rate is

$$R = 2CW = W\log_2(1 + SNR). \tag{2.78}$$

If W is in Hz, then R is in bits per second. On the other hand, SNR also depends on bandwidth. Given the symbol rate $2W$, the signal and noise energies E_t and σ_n^2 can be converted to signal power P_t and noise power P_n, respectively:

$$\begin{aligned} P_t &= 2WE_t \\ P_n &= 2W\sigma_n^2 \end{aligned}. \tag{2.79}$$

From (2.77) and (2.79), we can see that SNR is also the power ratio between signal and noise:

$$SNR = \frac{P_t}{P_n}. \tag{2.80}$$

---

[6] We are talking about baseband bandwidth here. See Chapter 3 for more discussions on bandwidths.

Recalling (2.69), the SNR can be expressed in terms of N_0:

$$SNR = \frac{P_t}{WN_0}. \tag{2.81}$$

In the literature, there are three common ways to express channel capacity. The first is (2.76), representing the number of bits per channel use, or the number of bits per symbol transmitted. The second is data rate limit (2.78), in bits per second. The third is

$$\eta \overset{\text{def}}{=} \frac{R}{W}. \tag{2.82}$$

η represents bit rate per Hz, known as spectral efficiency, spectrum efficiency, or bandwidth efficiency. Of the three expressions, the first shows the fundamental relationship between channel capacity and SNR. The second expression shows the actual data rate given channel bandwidth. The third expression provides a fair way to compare different systems by removing the influence of bandwidth.

2.7.4   Limiting Cases at High SNR and Low SNR

From (2.81), the channel capacity (data rate limit) given by (2.78) is

$$R = W \log_2\left(1 + \frac{P_t}{WN_0}\right). \tag{2.83}$$

We can see bandwidth W has two effects on data rate limit. An increase in W increases both symbol rate and noise power. The former causes a linear increase of R, while the latter causes a reduction of R, but within the logarithm function. Therefore, in general, a larger W is always better for data rate. In the following, we analyze two extreme cases: high SNR and low SNR.

At high SNR, that is,

$$\frac{P_t}{WN_0} \gg 1, \tag{2.84}$$

Equation (2.83) can be approximated as

$$R \approx W\log_2\left(\frac{P_t}{WN_0}\right). \tag{2.85}$$

When its variable is large, a logarithm is not sensitive to the variable. Therefore, the data rate is approximately proportional to bandwidth. On the other hand, increasing transmit power does not help much with the data rate, since P_t is contained in the logarithm. In other words, with high SNR, bandwidth is the key for data rate. For this reason, we refer to the high SNR case as "bandwidth-limited."

On the other extreme, at low SNR, that is,

$$\frac{P_t}{WN_0} \ll 1,$$ (2.86)

Equation (2.83) can be approximated as[7]

$$R \approx \frac{1}{\ln(2)} W \cdot \frac{P_t}{WN_0} = \frac{1}{\ln(2)} \frac{P_t}{N_0}.$$ (2.87)

Namely, the data rate is proportional to P_t and independent of W. In this case, increasing bandwidth does not benefit data rate. Therefore, we refer to the low SNR case as "power-limited." In literature, P_t/N_0 is commonly known as the carrier to noise ratio, noted as C/N_0 (more on this in Chapter 4).

Instead of constraining transmit power, sometimes the system is constrained by energy per bit E_b, defined as follows:

$$E_b \overset{\text{def}}{=} \frac{P_t}{R}.$$ (2.88)

The *SNR* in (2.81) can also be expressed in terms of E_b/N_0.

$$SNR = \frac{P_t}{WN_0} = \frac{R\,E_b}{W\,N_0} = \eta \frac{E_b}{N_0},$$ (2.89)

where η is the spectral efficiency defined in (2.82). The channel capacity (2.78) can be expressed in terms of the η:

$$\eta = \log_2(1 + SNR) = \log_2\left(1 + \eta \frac{E_b}{N_0}\right),$$ (2.90)

or

$$\frac{E_b}{N_0} = \frac{2^\eta - 1}{\eta}.$$ (2.91)

Equation (2.91) shows the required energy per bit for a given spectral efficiency. Obviously, E_b/N_0 increases when η increases. This means given the noise level, higher energy per bit is required for higher spectral efficiency. On the other hand, when η approaches 0, E_b/N_0 approaches to a constant $\ln(2)$ (or -1.59 dB).[8] In other words, even when spectral efficiency is sacrificed, we cannot reduce energy per bit below a threshold while transmitting without error. This is in spite of the fact that by reducing the data rate, the transmission *power* can be reduced arbitrarily, as shown in (2.87).

A plot of (2.91) is shown in Figure 2.2, in two scales. We can see that E_b/N_0 is within 0.2 dB of its minimum value when η is 0.1.

Analyses in this section point out two facts that may be counterintuitive. First, significant data rates can be achieved when SNR is less than 1 (or 0 dB), namely, when signal power is less than noise power. Under such condition, the signal can still be

---

[7] Recall that $\ln(1 + x) \approx x$ when $x \ll 1$.
[8] Recall that $\exp(x) \approx 1 + x$ when $x \ll 1$.

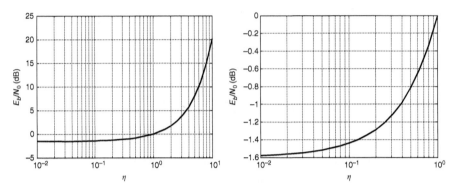

Figure 2.2 Required SNR for various spectral efficiency values. The left and right panels are the same plot with different scales.

recovered through several techniques to be outlined later in this book. Second, while increasing channel bandwidth always increases data rate proportionally for a given E_b/N_0, the latter has a minimum value no matter how much we are willing to increase bandwidth and/or reduce the data rate.

2.7.5 Complex AWGN

In digital communication, analyses are often performed in "baseband," where signals and noises are represented by complex numbers. The AWGN model can be extended to baseband analyses. More details are provided in Chapter 4.

   As discussed in Chapter 4, for AWGN, the noises in the real and imaginary parts of the channel are mutually independent. Therefore, it is optimal to send independent signals through the real and imaginary parts of the channel.[9] In this case, we can view a complex AWGN channel as two independent channels with the same SNR. Therefore, the capacity is doubled. Namely, (2.76) becomes

$$C = \log_2(1 + SNR). \tag{2.92}$$

   It seems that the data rate given by (2.78) should also be doubled for complex AWGN channels. However, in the case of baseband analyses as in Chapters 3 and 4, the bandwidth used usually is the passband bandwidth, which is twice of the bandwidth used in (2.78). Therefore, for complex baseband channels, (2.78) keeps the same form, except that the data rate counts for both real and imaginary parts, and W is the passband bandwidth.[10]

---

[9] There are other reasons to transmit real and imaginary parts of the signal jointly, as shown in Chapter 3. However, channel capacity can be realized by independent signals.

[10] We should keep in mind that noise power also doubles with a complex channel. SNR computations for complex channels are discussed in more details in Chapter 4.

2.7.6 Error Probability and SNR

Shannon theorem, as stated in Section 2.6.1, gives residual entropy at the receiver when data rate exceeds channel capacity:

$$H_y = R_a - R, \tag{2.93}$$

where H_y is the entropy per second at the receiver; R_a is data rate used for transmission, and R is channel capacity given in (2.78). H_y is connected with error probability. If a bit stream is transmitted while each bit has a probability P_e to be wrong, then the residual entropy per bit is

$$H_b = -[P_e \log_2(P_e) + (1 - P_e)\log_2(1 - P_e)]. \tag{2.94}$$

Residual entropy per second is thus

$$H_y = RH_b. \tag{2.95}$$

With (2.78) and (2.93)–(2.95), a relationship between bit error rate P_e and *SNR* can be established. In other words, data rate can be higher than channel capacity, if we accept nonzero P_e.

Figure 2.3 lots P_e against SNR. In this plot, R_a is 1 b/s/Hz. According to (2.78), when the SNR is 0 dB, R is also 1 b/s/Hz, allowing for error-free transmission. When the SNR is lower than 0 dB, we have $R < R_a$, and thus error will occur. Figure 2.3 shows that error increases very sharply as SNR moves below 0 dB. Therefore, we can consider the Shannon bound (i.e., the SNR required for error-free communication) as required SNR even with reasonable bit error rate (e.g., below 10^{-4}).

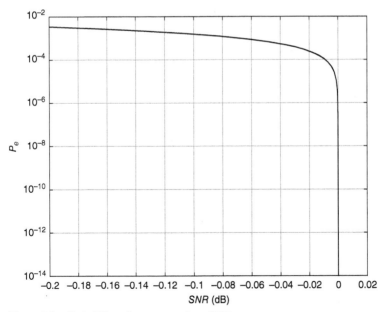

Figure 2.3 Probability of error at various SNR.

2.8   FREQUENCY-SELECTIVE CHANNEL AND WATER FILLING

Equation (2.78) in Section 2.7.3 gives the data rate limit for error-less transmission through a channel with given bandwidth. Within such bandwidth, both noise power spectral density and channel gain are constant (independent of frequency). Therefore, SNR is a constant over the bandwidth. Such a channel is referred to as a "flat channel."

In this section, we consider channels where the SNR depends on frequency, known as frequency-selective channels. For these channels, both channel gain and noise spectral density can be frequency dependent. In this section, we discuss the channel capacity of frequency-selective channels, and capacity optimization through transmit power allocation across frequency. Such optimization is known as "water filling."

In order to leverage capacity results for AWGN channels, we can conceptually divide a frequency-selective channel into many subchannels operating in parallel, each occupying a small frequency band. The total capacity of the channel is the sum of those of the subchannels.[11] Since each subchannel has a very narrow frequency band, its channel gain and noise spectral density can be considered as constant within the subchannel. Therefore, AWGN channel capacity results apply to the subchannels. The channel model for the subchannels is

$$y_k = h_k x_k + n_k,\qquad(2.96)$$

where x_k, y_k, and n_k are transmitted signal, received signal, and noise, respectively, for subchannel k. h_k is the channel gain. Note that in contrast to (2.66), x_k in (2.96) represents the signal at the *transmitter*, while channel gain is expressed explicitly here. The transmit power spectral density allocated to subchannel k, measured at the transmitter, is

$$S_k \overset{\text{def}}{=} \frac{E\left(|x_k|^2\right)R_s}{w},\qquad(2.97)$$

where R_s is the symbol rate, w is the subchannel bandwidth, which is the same for all subchannels. The signal power for subchannel k, measured at the receiver, is

$$P_k = wS_k|h_k|^2.\qquad(2.98)$$

The noise spectral density in subchannel k is

$$N_{0k} \overset{\text{def}}{=} \frac{E\left(|n_k|^2\right)R_s}{w},\qquad(2.99)$$

[11] This statement depends on the assumption that noises in these subchannels are mutually independent. Such assumption is true as long as the noise is wide-sense stationary, which is very common in practice. More discussions are in Chapter 4.

According to (2.83), the total data limit is

$$R = \sum_{k=1}^{K} w \log_2(1 + SNR_k) = w \sum_{k=1}^{K} \log_2 \left(1 + \frac{S_k |h_k|^2}{N_{0k}} \right), \tag{2.100}$$

where K is the number of subchannels. SNR_k is the SNR of subchannel k.

The total transmission power is constrained at the transmitter:

$$\sum_{k=1}^{K} P_k = w \sum_{k=1}^{K} S_k = P. \tag{2.101}$$

Furthermore, we have another constraint that all powers must be nonnegative:

$$P_k \geq 0 \forall k. \tag{2.102}$$

We wish to find the optimal allocation of $\{S_k\}$ that maximizes R in (2.100), while satisfying the constraints (2.101) and (2.102). For this purpose, the Lagrangian multiplier method is used. Construct the Lagrangian

$$L \stackrel{\text{def}}{=} w \sum_{k=1}^{K} \log_2 \left(1 + \frac{S_k |h_k|^2}{N_{0k}} \right) - \lambda \sum_{k=1}^{K} P_k. \tag{2.103}$$

L is to be maximized without constraint. λ can be chosen to satisfy (2.101). The maximization is performed by letting the derivative with regard to $\{S_k\}$ be zero:

$$0 = \frac{\partial L}{\partial S_k} = \frac{w}{\ln(2)} \frac{1}{\dfrac{N_{0k}}{|h_k|^2} + S_k} - \lambda. \tag{2.104}$$

This equation, together with (2.102), implies that the optimal allocation is

$$S_k = \max \left(0, D - \frac{N_{0k}}{|h_k|^2} \right). \tag{2.105}$$

D is a constant, combining the adjustable λ with other parameters, and is determined by (2.101).

The optimal power allocation scheme (2.105) is known as water filling. As shown in Figure 2.4, the chosen transmission power spectral density $\{S_k\}$ "fills" the background

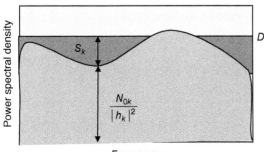

Figure 2.4   Water filling concept according to (2.105).

$(N_{0k}/|h_k|^2)$ to a constant level D. In practice, dividing a channel into subchannels in frequency (known as frequency-division) is not straightforward, as each subchannel needs to occupy a little more bandwidth than required by the Nyquist theorem, reducing total capacity. Chapter 3 will discuss more details on this aspect. However, a modern modulation technique known as orthogonal frequency division multiplexing (OFDM) allows efficient use of bandwidth and flexible allocation of power among various subchannels. OFDM will be described in more details in Chapter 10. The same power allocation scheme can be used to allocate power among spatial subchannels in multiple-in multiple-out (MIMO) systems as well, as will be described in Chapter 11.

2.9 SUMMARY

This chapter reviews Shannon theory and its implications. We started with the question of how to measure uncertainty, which leads to a unique mathematical expression of entropy (Section 2.3). It turned out that entropy is connected to the number of bits required to faithfully represent a source (Section 2.4). Furthermore, mutual information is measured as reduction of entropy (Section 2.5). Such definition of information actually indicates the maximum achievable data rate for error-less data transmission (Section 2.6). Such achievable data rate depends on channel characteristic, especially SNR, as illustrated in the case of AWGN channel (Section 2.7).

As discussed in Section 2.7, for AWGN channel with limited bandwidth, the channel capacity is given by the familiar form of (2.78). Most communications systems operate at SNR levels much higher than 1 (0 dB). For these systems, the channel capacity is approximately proportional to bandwidth. Increasing transmit power also increases channel capacity but to a less extent. However, as bandwidth increases, so does total noise power, resulting in a reduction of SNR. Eventually, we encounter another extreme case where SNR is much less than 1. In this case, there is a minimum level of energy per bit required for any bandwidth and data rate. Section 2.8 extended the results to frequency-selective channels and introduced the water-filling concept.

Although the Shannon bound is not usually achievable in a communications system, as will be shown in Chapter 5, modern channel coding techniques can achieve a performance that is very close to the Shannon bound. On the other hand, sometimes a system has lower performance due to constraints such as complexity and latency. Nonetheless, the Shannon bound always serves as a valuable benchmark in system design.

2.10 APPENDIX: DERIVATION OF ENTROPY AS A MEASURE OF UNCERTAINTY

In this section, we will show that the metric of uncertainty has a unique form (up to a constant factor). Such uniqueness stems from a set of reasonable and straightforward requirements stated in Section 2.10.1.

2.10.1 Properties of an Uncertainty Metric

Consider a discrete random variable X, characterized by its probability distribution $\{p_n, v_n, n = 1, 2, \dots, N\}$. p_n is the probability that $X = v_n$. N is the number of possible values X can take. The collection of possible values $\{v_n\}$ is known as the alphabet of X. Intuitively, X carries more uncertainty if p_n is constant over a large subset of $\{n = 1, 2, \dots, N\}$, which means X can be any value in corresponding subset of the alphabet with equal probability. In contrast, if p_n has a sharp peak at some value m, the uncertainty would be small because we know X is very likely to be equal to v_m.

We would like to have a function $H(p_1, p_2, \dots, p_N)$ to quantify the uncertain in X. We will work out the expression of such function, based on the following three requirements.

1. H is a continuous function of the variables $\{p_n, n = 1, 2, \dots, N\}$.

2. Consider a special case when all $\{p_n\}$ are equal (i.e., equiprobable distribution). In this case, a larger N obviously means higher uncertainty. Therefore, $H(1/N, 1/N, \dots, 1/N)$ monotonically increases with N.

3. If a choice of value is broken down into two cascading choices, H is the weighted sum of the individual values of H for the subchoices. This will be explained in more details below and expressed in (2.108).

The first two requirements are intuitive, while the third one needs some more explanation. Consider a partition of the possible values:

$$(p_1, p_2, \dots, p_K) = q_1(p_{11}, p_{12}, \dots, p_{1N_1}), q_2(p_{21}, p_{22}, \dots, p_{2N_2}), \dots, q_M(p_{M1}, p_{M2}, \dots, p_{MN_M})$$

$$\sum_{m=1}^{M} N_m = K$$

$$\sum_{m=1}^{M} q_m = 1$$

$$\sum_{n=1}^{N_m} p_{mn} = 1 \forall m = 1, 2, \dots, M.$$

$$(2.106)$$

Intuitively, we divide the K possible values of X into M groups. The choice of value for X is performed in two steps. First, the value of X falls into group m with a probability of q_m. Second, given that X is in group m, it takes the value v_{mn} with a probability p_{mn}. Here, v_{mn} is one of the values in $\{v_k\}$ before the partition. Therefore,

$$q_m p_{mn} = p_k. \tag{2.107}$$

Figure 2.5 illustrates the two-step partition concept. The bottom row with grayed squares shows the original possible values, which are mapped to the groups in the row above.

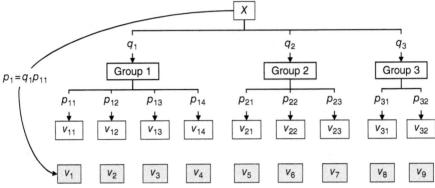

Figure 2.5 Partition of random variable values.

Under such partition, it is reasonable to expect the uncertainty measure obeys the following relationship:

$$H(p_1, p_2, \ldots, p_K) = H(q_1, q_2, \ldots, q_M) + \sum_{m=1}^{M} q_m H(p_{m1}, p_{m2}, \ldots, p_{mN_m}). \qquad (2.108)$$

The first term on the right-hand side represents the uncertainty introduced by the first layer of partition. The second term is a weighed summation of the uncertainties introduced in the second layer of partition. It turns out that based on the above three expected properties, there is a unique expression of H, as will be shown in Section 2.10.2.

2.10.2 The Unique Metric

First, consider the equiprobable case (i.e., $p_n = 1/N \; \forall n$). Define function

$$A(K) \stackrel{\text{def}}{=} H\left(\frac{1}{K}, \frac{1}{K}, \ldots, \frac{1}{K}\right). \qquad (2.109)$$

Now consider the case where there are s groups, each with s members. Namely, in (2.106)

$$\begin{aligned} M &= s \\ N_m &= s \, \forall m \end{aligned} \qquad (2.110)$$

We thus have

$$\begin{aligned} K &= s^2 \\ q_m &= \frac{1}{s} \forall m \\ p_{mn} &= \frac{1}{s} \forall m, n. \end{aligned} \qquad (2.111)$$

Equation (2.108) thus leads to

$$A(s^2) = A(s) + s \cdot \frac{1}{s} A(s) = 2A(s). \tag{2.112}$$

Such process can be repeated, resulting in

$$A(s^l) = lA(s) \quad \forall l = 1, 2, \dots \tag{2.113}$$

Naturally, a logarithm function satisfies this requirement. Now, we will show that this is the only possibility.

For any $1 < s < t$ and any positive integer j, we can always find integer i such that

$$s^i \le t^j < s^{i+1}. \tag{2.114}$$

This equation leads to

$$\frac{i}{j} \le \frac{log_2(t)}{log_2(s)} < \frac{i+1}{j} = \frac{i}{j} + \frac{1}{j}. \tag{2.115}$$

From property 2 stated in Section 2.10.1, $A(s)$ increases with s. Therefore,

$$A(s^i) \le A(t^j) < A(s^{i+1}). \tag{2.116}$$

Combining (2.113) and (2.116):

$$\frac{i}{j} \le \frac{A(t)}{A(s)} < \frac{i+1}{j}. \tag{2.117}$$

Equations (2.115) and (2.117) lead to

$$\left| \frac{log_2(t)}{log_2(s)} - \frac{A(t)}{A(s)} \right| \le \frac{1}{j}. \tag{2.118}$$

Since j can be arbitrarily large, we must have

$$A(s) = F \log_2(s). \tag{2.119}$$

Namely, $A(s)$ is a logarithm function multiplied by an arbitrary constant. Because $A(s)$ increases with s, F should be positive. As shown in Section 2.4, it makes sense to choose $F = 1$. Therefore,

$$A(s) = \log_2(s), \tag{2.120}$$

The measure of uncertainty for an equiprobable distribution is thus

$$H\left(\frac{1}{K}, \frac{1}{K}, \dots, \frac{1}{K}\right) = \log_2(K). \tag{2.121}$$

Next, let us look at the case of unequal probabilities. Let the probabilities be $\{q_1, q_2, \ldots, q_M\}$, where M is very large. Construct the following partition[12]:

$$p_{mn} = \frac{1}{N_m} \quad \forall m = 1, 2, \ldots, M.$$
$$N_m = Mq_m \tag{2.122}$$

Such partition satisfies (2.106). Now we have a two-step partition process. We start with a probability distribution: $\{q_1 p_{11}, q_1 p_{12}, \ldots, q_M p_{MN_M}\}$. This is divided into M groups, with probability $\{q_1, q_2, \ldots, q_M\}$. Each group contains an equal distribution, since $\{p_{mn}\}$ as given by (2.122) is independent of n. Furthermore, through the introduction of $\{p_{mn}\}$, we constructed another equiprobable distribution in the form of (2.106). Indeed, from (2.107):

$$p_k = q_m p_{mn} = \frac{1}{M} \forall m, n. \tag{2.123}$$

With this partition, (2.108) leads to

$$H\left(\frac{1}{M}, \frac{1}{M}, \ldots, \frac{1}{M}\right) = H(q_1, q_2, \ldots, q_M) + \sum_{m=1}^{M} q_m H\left(\frac{1}{Mq_m}, \frac{1}{Mq_m}, \ldots, \frac{1}{Mq_m}\right). \tag{2.124}$$

Using (2.121) for the two equiprobable distributions in (2.124), we get

$$\log_2(M) = H(q_1, q_2, \ldots, q_M) + \sum_{m=1}^{M} q_m \log_2(Mq_m). \tag{2.125}$$

With the normalization of $\{q_m\}$ in (2.106), we finally arrive at

$$H(q_1, q_2, \ldots, q_M) = -\sum_{m=1}^{M} q_m \log_2(q_m). \tag{2.126}$$

Equation (2.126) is the unique expression of uncertainty that satisfies the three requirements listed above, with the choice of a constant factor F in (2.119). H given by (2.126) is known as the entropy.

2.11 APPENDIX: COMPRESSION CODING

The source coding theorem stated in Section 2.4.3 gives the minimum length of lossless binary codes. The next question is how to find such an optimal code. While Section 2.4.2 presented a scheme of coding only the TS, such code becomes lossless only when K approaches infinity, which is not practical in most cases. On the other hand, there are known practical codes that are near-optimal. This section introduces two practical source codes, the Huffman coding and the Lempel–Ziv–Welch coding.

---

[12] Here, we ignore the requirement that N_m must be integers. Such requirement is always met approximately when M is very large.

Detailed descriptions of these algorithms can be easily found in books and literature [5, sec. 6.3], [6]. For the purpose of understanding source coding theorem, only basic principles are described here.

In practice, source coding techniques are often used to reduce the number of bits required to represent a particular content. Therefore, they are also known as data compression methods. An important performance measure for these techniques is the compression ratio, which is the ratio between the bit counts of the source content and its coded version. Obviously, compression ratio depends on the coding techniques, as well as the nature and property of the source content (e.g., its entropy).

Huffman coding and Lempel–Ziv–Welch coding are known as reversible or lossless compression. Source content can be recovered without loss through the decoding process. There are many other compression techniques, some lossless and some lossy. These techniques can also be combined to achieve better results.

2.11.1 Huffman Coding

Huffman coding uses a special kind of variable-length code, known as prefix code. A prefix code has the property that no codeword is a beginning part of another codeword. Therefore, a receiver can tell that a complete codeword has been received as soon as it happens, and proceed to decoding. An example of a prefix code has the following codewords: [0, 10, 110, 1110, 1111]. In this case, the end of codeword is denoted by the bit "0," unless the maximum code length is reached.

Note that a prefix code may have multiple codewords with the same length. For example, [00, 01, 10, 110, 111] is also a prefix code.

Huffman coding uses variable-length codewords to encode "discrete memoryless source (DMS)," which is a source with uncorrelated discrete letters. Huffman coding is based on the idea that shorter codewords are assigned to more frequent source letters. This way, shorter codewords appear more often than the longer ones in the code sequence, reducing the total code sequence length.

A Huffman code is built in two steps. In the first step, the source alphabet is ordered in the sequence of their frequencies of appearance. The two letters with the least frequencies are considered as two branches, which are combined into one node. The node is treated as a new letter, whose frequency is the sum of the two branches. This letter is ordered with the rest in the alphabet according to the updated frequencies of appearance, and the process is repeated. Eventually, a tree is formed, which includes all letters of the alphabet as its branches. In the second step, codewords are built with such tree. At each node, the two branches are labeled as 0 and 1. The final code for each letter is the bits of all the branches leading to that letter.

Figure 2.6 shows two examples of Huffman coding. For each example, the boxes on the left show the source alphabets, each consisting of the five letters A–E. The entry marked as "D (0.1) [0011]" means letter D appears with a frequency 0.1 and is coded by codeword 0011. The traces show the tree structure, with the intermediate nodes marked by circles. The numbers on the traces in parentheses indicate the frequency of appearance associated with the trace. The bold-faced numbers 0 or 1 on each trace are the bit value associated with the trace (branch).

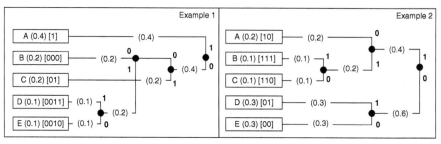

Figure 2.6   Huffman coding examples.

Let us take a closer look at Example 1. The two letters with the lowest frequency: D and E are combined first, to form a node with a frequency of 0.2 (sum of those of D and E). This value is still the lowest frequency among the rest of the letters (A, B, and C). Therefore, it is further combined with B (frequency 0.2), to form a node with frequency 0.4. This node is further combined with the remaining letter with the lowest probability (C, frequency 0.2) to form a node with frequency 0.6, which is finally combined with the remaining node A. After building the tree, bits are assigned to each branch. From each node, the two branches to the left are labeled by 0 or 1. The process continues until all branches are labeled. Finally, the codeword for a letter is constructed by reading the bit labels of the sequence of branches from the root node (the rightmost node) leading to that letter.

Note that the labeling of the two branches of a node is arbitrary. As a result, the codewords are not necessarily in the form of consecutive 1s terminated by 0. Nevertheless, they still qualify for prefix code, since none of the codewords can be a leading part of another codeword.

Example 2 shows a different tree structure. In this case, we have three codewords with length 2 and none with length 1.

Hoffman code is optimal in coding efficiency. It can be shown that the average coding length l, in bits per source letter, is given by [6]:

$$H(S) \le l < H(S) + 1, \tag{2.127}$$

where $H(S)$ is the entropy of the source alphabet. Recall that the source coding theorem says the minimum l is $H(S)$. Therefore, Huffman code is close to optimal. In fact, when the size of the alphabet (and thus its entropy) approaches infinity, l approaches the lower bound on a per-letter basis. Such increase of alphabet size can be achieved by dividing source data sequence into blocks and treating one block as a new "letter." It is easy to see that in this case, the average coding length per original source letter is

$$H(S) \le l < H(S) + \frac{1}{M}, \tag{2.128}$$

where M is the block size [6]. Furthermore, even for small alphabet sizes, Huffman code is optimal among prefix codes, as indicated by the so-called Kraft Inequality [6, sec. 3.6].

2.11.2   Lempel–Ziv–Welch (LZW) Code

Huffman coding is a variable-length code for DMS, which assumes source letters are uncorrelated in the sequence. It also requires the knowledge of frequency (probability) distribution in the source alphabet to construct a codebook. Another method, the Lempel–Ziv–Welch (LZW) algorithm, is quite the opposite. It uses fixed-length codes, does not need prior knowledge of frequency distribution, and takes advantage of correlation among the source symbols. It is not guaranteed to be optimal (in fact, it is possible to achieve no compression at all). However, it typically works quite well for text and image files and is the standard compression method for many computer systems and applications. Huffman coding and LZW coding represent two general classes of compression coding approaches: statistical and dictionary [7].

The LZW algorithm was first proposed by Lempel and Ziv in 1977 and 1978 [8] and was further modified by Welch [9]. There are many variations to the algorithm. Following is the version described by Welch [9].

The basic idea of LZW coding is constructing a coding table (dictionary) to represent frequently appearing letter sequences (known as strings). These strings can be represented by one codeword in the future, thus saving code length. Without prior knowledge about what strings are more frequent, the coding table is built based on our experience with the source sequence so far. The more specific process is described below.

The codebook is initiated with all single letters in the entire alphabet as entries. From then on, it is extended as we encounter new sequences.

The encoder algorithm is as follows. Taking input letters one by one, we accumulate a string in a buffer. Such accumulation continues until the string is not in the codebook. At this point, the string in the buffer is in the form of wK, where w is a string in the codebook. K is a letter. wK is not in the codebook. The following actions are taken:

- The code for w is sent out.
- String wK is added to the codebook.
- The string in the buffer is set to letter K, and the process starts over again.

Such encoding continues until the end of the data stream is reached.

The codeword has a constant length in bits. Therefore, there is a limit on the total number of codewords. When such limit is reached, "older" codewords are purged to make room. The definition of "old" depends on the variations. It could be the earliest entry, or the entry that has not been used for the longest time.

Figure 2.7 shows the flow diagram of the FZW encoder. In the diagram, w is the current string. Code (w) is the codeword corresponding to w. This flow diagram does not show the code table purging process.

Figure 2.8 shows an example of LZW coding [6]. In this case, the input is a binary sequence (i.e., the letters are 0 and 1) listed to the left of the table. Squares around the bits indicate that these bits are coded together as a string. In the table, each row shows the activities during encoding.

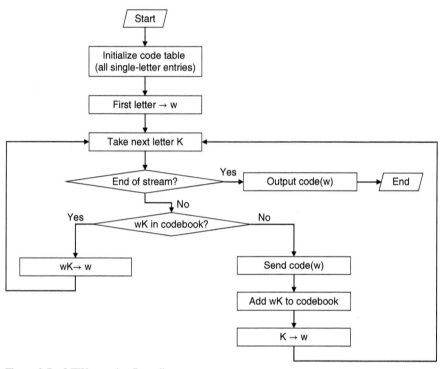

Figure 2.7   LZW encoder flow diagram.

Step	String	Output code	Code table	
			String	Code
			0	0
			1	1
1	1	1	10	2
2	0	0	01	3
3	1	1	11	4
4	10	2	100	5
5	0	0	00	6
6	01	3	010	7
7	010	7	0101	8
8	1	1		

1
0
1
1
0
0
0
1
0
1
0
1

Figure 2.8   LWZ coding example.

At the start, the codebook is initialized with two entries: 0 for bit 0 and 1 for bit 1. At step 1, input 10 was not found in the codebook. Therefore, the code for the previous string (1) is sent out, while string 10 is added to the codebook (code 2). The buffer is set to the last letter (0). At step 2, it was found that string 01 is not in the codebook. So the code for 0 is sent out, and 01 is added to the codebook (code 3). The buffer is set to the last letter (1). The same happens at step 3, where 11 is added to the codebook (code 4). Next (step 4), after finding 10 in the codebook, a new string 100 is encountered. The code for 10 (2) is sent out, and a new entry of the codebook is made for string 100, and so on. Of course, at some point, the code table will be purged. This process is not shown in the example.

In this example, there are nine codewords. Therefore, each codeword requires 4 bits. The eight output codewords consist of 32 bits. This length exceeds the input sequence length (11 bits). Thus, in this example, LZW is not beneficial in data compression. However, with real-word applications, LZW can typically achieve a compression rate of 3–5 for text files [7] and around 2 for image files [10].

Although the codebook is constructed during the encoding process and is dependent on the data content, it does not need to be transmitted to the receiver. The decoder can build a local copy of the same table based on decoded letters so far. Details on the decoding algorithm can be found in [9].

REFERENCES

1. C. E. Shannon, "A Mathematical Theory of Communication," *Bell Syst. Tech. J.*, vol. 27, no. July, pp. 379–423, 1948.
2. A. J. Viterbi and J. K. Omura, *Principles of Digital Communication and Coding—Andrew J. Viterbi, Jim K. Omura—Google Books*. New York: McGraw-Hill, 1979.
3. DLMF, "NIST Digital Library of Mathematical Functions" [Online]. Available at http://dlmf.nist.gov/ [Accessed December 13, 2016].
4. R. G. Gallager, *Information Theory and Reliable Communication*. Wiley, 1968.
5. J. Proakis and M. Salehi, *Digital Communications*, 5th ed. McGraw-Hill Education, 2007.
6. J. D. Gibson, *Information Theory and Rate Distortion Theory for Communications and Compression*. Morgan & Claypool Publishers, 2014.
7. T. Bell, I. H. Witten, and J. G. Cleary, "Modeling for Text Compression," *ACM Comput. Surv.*, vol. 21, no. 4, pp. 557–591, 1989.
8. J. Ziv and A. Lempel, "Compression of Individual Sequences via Variable-Rate Coding," *IEEE Trans. Inf. Theory*, vol. 24, no. 5, pp. 530–536, 1978.
9. T. A. Welch, "A Technique for High-Performance Data Compression," *Computer*, vol. 17, no. 6, pp. 8–19, 1984.
10. G. R. Kuduvalli and R. M. Rangayyan, "Performance Analysis of Reversible Image Compression Techniques for High-Resolution Digital Teleradiology," *IEEE Trans. Med. Imag.*, vol. 11, no. 3, pp. 430–445, 1992.

HOMEWORK

2.1 For a given total power P_x, compute the entropy $H(X)$ with X obeying the following distributions with variance 1:

a. Gaussian distribution

b. Uniform distribution

How much do we lose in spectral efficiency by using the suboptimal uniform distribution?

2.2 Equation (2.73) states that $P_y = P_x + \sigma_n^2$. This comes from the fact that $y = x + n$. However, we can also write $x = y + n'$, with $n' = -n$. Can we then claim that $P_x = P_y + \sigma_{n1}^2 = P_y + \sigma_n^2$? Explain.

2.3 Two binary random variables X and Y are distributed according to the joint distributions $P(X = Y = 0) = P(X = 0, Y = 1) = P(X = Y = 1) = 1/3$. Compute $H(X), H(Y), X(X \mid Y), H(Y \mid X), H(X, Y)$.

2.4 Find the capacity of an additive white Gaussian noise channel with a bandwidth of 1 MHz, received power 10^{-9} W, and noise power spectral density $N_0 = 10^{-19}$ W/Hz. Express the capacity in terms of bit per second. What if the bandwidth is 5 MHz?

2.5 (Optional) Use MATLAB function huffmandict to reproduce the code samples in Figure 2.6. You might not get exactly the same results as Figure 2.6. However, you can show they are equivalent.

2.6 (Optional) Write a flow diagram for an LZW decoder. Compare your algorithm with those presented in [9].

2.7 (Optional) For the Huffman coding examples shown in Figure 2.6, compute the source entropies and the average code lengths. Verify (2.127) with these results.

CHAPTER 3

SINGLE CARRIER MODULATION AND NYQUIST SAMPLING THEORY

3.1 INTRODUCTION

Figure 3.1 shows the high-level physical layer architecture, discussed in Chapter 1. This chapter addresses the Modulation block in the transmitter and the RF blocks on both sides (circled by the ovals).

Digital modulation is the process of converting bits (output of the channel coding) to RF signal (also known as waveform). It contains several steps as shown in Figure 3.2. The Symbol Mapping block transforms bits into symbols. A symbol is a number that can take several possible values. The particular value of a symbol indicates the values of the several bits that it represents. Symbol mapping is covered in Section 3.2. One symbol is produced for each symbol period (or symbol time) T_{Sym}. The inverse of T_{Sym} (i.e., the number of symbols produced per unit time), $R_{Sym} \stackrel{\text{def}}{=} 1/ T_{Sym}$, is known as symbol rate or baud rate. The unit for symbol rate is symbol per second (sps or s/s), also known as baud (Bd). Derivative units such as kilo-symbol per second (ksps or ks/s), also known as kilo-baud (kBd), can also be used.

The Pulse Shaping block in Figure 3.2 converts time-discrete symbols into time-continuous signals, also known as the waveforms. Pulse shaping also converts the signal from digital form to analog form. Section 3.4 covers pulse shaping, while Sections 3.3 and 3.5 provide related concepts and details. The spectrum of the waveform produced by the pulse shaping block has a support that extends to direct current (DC). Therefore, such waveform is referred to as the baseband waveform. The baseband waveform may be further converted to a spectrum whose support extends from a lower frequency bound to a higher frequency bound (i.e., in a frequency band). The waveform after such conversion is referred to as passband waveform. The up-conversion block in Figure 3.2 performs such conversion and is covered in Section 3.6. Section 3.6 also covers the corresponding operation at the receiver, known as down-conversion. The chapter is summarized in Section 3.7. Two appendices, Sections 3.8 and 3.9, discuss formulation and properties of the Fourier transform. Section 3.10 provides details on the Nyquist criterion proof.

Digital Communication for Practicing Engineers, First Edition. Feng Ouyang.
© 2020 by The Institute of Electrical and Electronics Engineers, Inc.
Published 2020 by John Wiley & Sons, Inc.

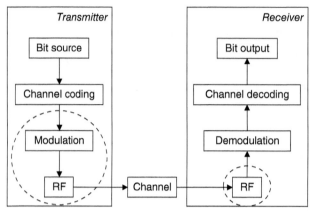

Figure 3.1   Physical layer architecture.

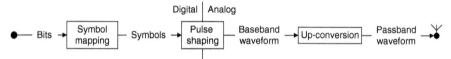

Figure 3.2   Steps of digital modulation.

In a typical communications system, we wish to maximize data transmission rate under the following constraints: signal bandwidth, transmit power, propagation loss, and channel noise. For the portion of the system considered in this chapter (Figure 3.2), such optimization goal includes two separate issues.

1. We want each symbol to carry as many bits as possible. The number of bits is limited by the transmitting power, propagation loss, and channel noise. More bits per symbol mean legitimate symbols are closer to each other in value, and thus cannot be distinguished by the receiver under the influence of noise. This issue is discussed in Section 3.2.

2. We want to transmit with a symbol rate that is as high as possible. The symbol rate is limited by the bandwidth, as discussed in Section 3.3.

Much discussion in this chapter employs the concept of Fourier transform (FT) and inverse Fourier transform (IFT). The readers are assumed to be familiar with FT/IFT and their properties. The basic formulations and properties germane to this chapter are briefly listed in Section 3.8, to clarify notations and conventions. Another basic mathematical results used in this chapter is the Nyquist–Shannon sampling theory, which is introduced in Section 3.3. Also, the Nyquist criterion is an interesting mathematical result applied to pulse shaping. It is presented and applied in Section 3.4 with more mathematical details provided in Section 3.10.

In this chapter, time domain signals (waveforms) are noted in the form of $s(t)$. Their Fourier transforms (also referred to as spectrum) are noted by adding a tilde to the signal names, such as $\tilde{s}(\omega)$. By default, variables t stands for time, f stands for frequency, and $\omega \overset{\text{def}}{=} 2\pi f$ stands for angular frequency. "Support" of a function means the range of the variable where the function value is nonzero. For example, if a spectrum $\tilde{s}(\omega)$ has a support of $[\omega_1, \omega_2]$, then

$$\tilde{s}(\omega) = 0 \forall \omega \notin [\omega_1, \omega_2]. \tag{3.1}$$

3.2 SYMBOL MAPPING

In this section, we discuss how to convert data (bitstream) into symbols. Namely, we focus on the Symbol Mapping block in Figure 3.2. When the literature talks about "modulation schemes," it usually refers to the details of the symbol mapping process.

3.2.1 Constellation and Bit Mapping

Symbol mapping is the process of mapping data bits into symbols in digital domain. For the types of symbol mapping considered in this book, symbols can be expressed as a sequence of complex numbers $\{d_n\}$, which can take a finite set of possible values (known as alphabet). The value of d_n for a particular n depends on the values of the bits it represents. Namely, there is a one-to-one mapping between a value of d_n and a sequence of bits $\{b_1, b_2, \dots, b_K\}$, where K is the number of bits a symbol can represent. It depends on the modulation scheme.

A modulation is said to be M-ary if its alphabet has M possible values. It follows that

$$K = \lfloor \log_2(M) \rfloor. \tag{3.2}$$

Usually, M is a power of 2, that is, $log_2(M)$ is an integer.

It is customary to represent complex-valued symbols as points in the complex plane. With such representation, the alphabet becomes a set of M points in the complex plane. Such pattern is referred to as the constellation of the modulation scheme. As will be shown below, constellation determines the average power and error probability of the modulation. Examples of constellations will be shown as we discuss specific modulation schemes in Sections 3.2.2–3.2.4.

At a given symbol rate R_{Sym}, the bit rate is KR_{Sym}.[1] A larger M results in a larger K, and thus higher bit rate under the same symbol rate. On the other hand, as will be shown in Section 3.2.5, larger M also leads to higher error probability under the same channel condition. Therefore, the choice of M represents a tradeoff between bit rate and error probability. Within the same family of modulations, the one with larger M is said to be a higher order modulation.

Note that constellation is not the only way to represent a modulation scheme. However, for the modulations schemes covered in this book, constellation is a good representation. Furthermore, Figure 3.2 shows that sample mapping is performed in

---

[1] In the presence of channel coding, bit rate is $R_c K R_{Sym}$, where R_c is the coding rate. This will be discussed in Chapter 5.

the digital domain (i.e., symbol values are manipulated digitally). Such modulation schemes are referred to as digital modulation. Alternatively, conversion between bit values and RF signal can be accomplished by manipulating signal in the analog domain, thus bypassing the symbol mapping step. This method is known as analog modulation and is not covered by this book.

3.2.2   Phase Shift Keying (PSK)

3.2.2.1   Basic PSK Modulation

Phase shift keying (PSK) modulation uses phase of the complex numbers to distinguish constellation points. Namely, all constellation points have the same amplitude. This characteristic is advantageous for certain systems since it lowers the requirement on RF circuit linearity.

In PSK, the constellation points can be expressed as[2]

$$P_m = Ae^{j\pi\frac{2m-1}{M}}, m = 1, 2, \ldots, M. \tag{3.3}$$

M is the number of constellation points. P_m is the symbol value for the constellation point with index m. A is the common amplitude. Typically, M has values of 2 (binary PSK or BPSK), 4 (quadrature PSK, quaternary PSK, or QPSK), 8 (8-PSK), and so on.

Figure 3.3 shows the constellation points of the three PSK modulations. The horizontal and vertical axes are the real and imaginary axes of a complex plane.

Plain PSK modulations are easy to understand. There are, on the other hand, many variations for such scheme. Three of the common variations are described in sections 3.2.2.2 – 3.2.2.4.

3.2.2.2   π/4-QPSK

π/4-QPSK is a variation of QPSK, where the constellation points are rotated by π/4 for every successive symbol. For example, if bits 00 is mapped to phase π/4 for symbol n,

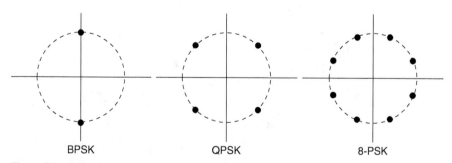

BPSK   QPSK   8-PSK

Figure 3.3   PSK constellations.

[2] Constellation definition may be different from (3.3), see [1, sec. 3.2-2], [2, sec. 7.6].

then the same bits would be mapped to phase $\pi/2$ for symbol $n + 1$. Such constellation rotation guarantees that the symbol values always change from one period to the next, even when data do not. Such change provides a "symbol clock" signature for the receiver to obtain and maintain synchronization. It also avoids a strong DC (i.e., frequency equals zero) spectral component in the waveform. Such DC component may bring difficulties in some transceiver architectures.

3.2.2.3 Offset QPSK

Offset QPSK (OQPSK) is a scheme where the real and imaginary parts of a symbol are staggered in time. Namely, a symbol period T_{Sym} is split into two. At $t = nT_{Sym}$, the real component of the symbol is changed (if necessary). At $t = (n - 1/2)T_{Sym}$, the imaginary component is changed.

As will be shown in Section 3.4, the final waveform produced from a plain QPSK can be expressed as

$$s(t) = \sum_n d_n h(t - nT_{Sym}), \tag{3.4}$$

where $h(t)$ is the (real-valued) pulse shaping function discussed in more details in Section 3.4. Separating the real and imaginary parts, (3.4) can be written as

$$
\begin{aligned}
s_r(t) &= \sum_n d_{nr} h(t - nT_{Sym}) \\
s_i(t) &= \sum_n d_{ni} h(t - nT_{Sym})
\end{aligned}
\tag{3.5}
$$

Here s_r, s_i are the real and imaginary parts of s. d_{nr}, d_{ni} are the real and imaginary parts of d_n. For OQPSK, the real and imaginary parts have a relative time offset:

$$
\begin{aligned}
s_r(t) &= \sum_n d_{nr} h(t - nT_{Sym}) \\
s_i(t) &= \sum_n d_{ni} h\left[t - \left(n - \frac{1}{2} \right) T_{Sym} \right]
\end{aligned}
\tag{3.6}
$$

Such arrangement avoids the cases where a phase is changed by π, and the waveform passes value 0 in the process.

Figure 3.4 shows signals for QPSK and OQPSK, generated from 200 random symbols and a sinc function pulse shaping filter (see Section 3.4 for more explanations on pulse shaping filter). The left column shows the real (thin solid lines) and imaginary (thick dashed lines) parts of the signal over 8-symbol periods. It shows that for OQPSK, the imaginary part has a relative advance of half the symbol period. The right column shows the signal traces plotted with the real part as the x-axis and the imaginary part as the y-axis. It shows that while QPSK signal frequently passes area around the origin, OQPSK signal keeps relatively constant amplitude. Therefore, amplifies with poor linearity may not reproduce QPSK signal faithfully,

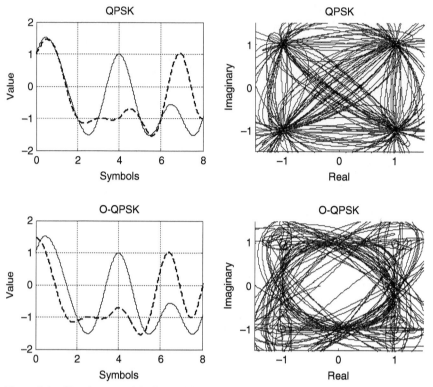

Figure 3.4   Signal traces of QPSK and OQPSK. Left side: signal values as functions of time. Thin solid lines: real parts; thick dashed lines: imaginary parts. Right side: signal traces in complex plains.

while OQPSK signals would fare better in such cases. Obviously, OQPSK and QPSK have the same spectrum.

3.2.2.4   Differential PSK

Another variation is differential PSK (DPSK). In this scheme, the phase difference between successive symbols encodes the data. Namely, (3.3) is replaced by

$$P_m = P_{m-1} e^{j\pi \frac{2m-1}{M}}, m = 1, 2, \ldots, M. \tag{3.7}$$

Equation (3.7) shows that the value of the mth symbol is that of the $(m-1)$th symbol with a phase rotation, which depends on the bit values to be encoded. With such modulation scheme, the receiver does not need to estimate the absolute phase but only the relative one between successive symbols. For example, a receiver can multiply the current symbol value with the complex conjugate of the previous one and estimate the phase of the product. Many channels introduce additional phase shift that are difficult to estimate. Such channel-induced phase shift affects the absolute symbol phase at the receiver. However, such channel-induced phase shift usually remains constant during consecutive symbols. Therefore, the relative phase shift between consecutive symbols is not affected by channel. This makes demodulating DPSK modulation much easier than decoding the basic PSK modulation. On the other hand, since demodulation of DPSK involves two symbols, the noise power involved in demodulation is doubled.

3.2.3   Quadrature Amplitude Modulation (QAM)

While constellation points of PSK are distributed on a circle in the complex plane (see Figure 3.3), Quadrature Amplitude Modulation (QAM) uses a square or rectangular array of constellation points, as shown in Figure 3.5. Note that 4-QAM is the same as QPSK.

The constellation points in QAM scheme can be expressed as

$$P_n = B[(2k_n + 1) + j(2l_n + 1)], \tag{3.8}$$

where B controls the minimum separation between constellation points. Integers k_n and l_n range from $-N$ to $N-1$ and depend on the bits to be represented. The total number of constellation points is $M = 4N^2$. Such modulation scheme is referred to as M-QAM, as labeled in Figure 3.5.[3]

Comparing to PSK, QAM can practically accommodate more constellation points (i.e., larger M), and thus offering a higher data rate under the same symbol rate. However, in order to identify the constellation points, the receiver needs to accurately estimate not only the phase but also the amplitude of the signal. Such task requires not only a high linearity from the analog circuits but also an accurate estimate of channel gain. Currently, a typical wireless system uses up to 64 constellation points in QAM, although some newer technologies (such as 802.11ac, detailed in Chapter 11) use up to 256 constellation points to achieve higher data rates [3, ch. 21].

To further improve system performance, some variations of QAM constellation design produce symbols with Gaussian power distribution, which was identified as optimal according to Shannon theory (see Chapter 2). Figure 3.6 shows two examples from [4, 5].

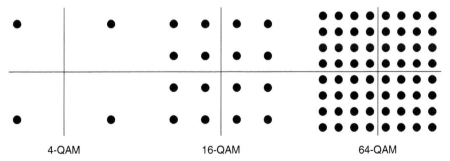

Figure 3.5   QAM constellations.

---

[3] Although less common, k_n and l_n may have different ranges. This results in a rectangular, instead of a square, constellation. For the sake of simplicity, only square QAM constellations are discussed in this chapter.

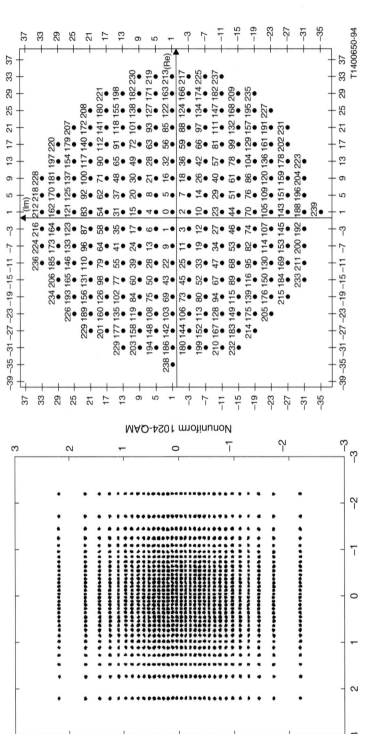

Figure 3.6 Examples of shaped QAM constellations [4, fig. 4], [5, fig. 5]. Reproduced with permission.

3.2.4   A Hybrid between PSK and QAM: Amplitude and Phase Shift Keying (APSK)

As discussed above, PSK has the advantage of constant amplitude among symbols. On the other hand, as will be shown in Section 3.2.6, PSK has the disadvantage of less immunity against noise. The reason is that in PSK, all constellation points are "crowded" onto a circle. Therefore, they are close to each other and are easier to be confused by the receiver under noisy conditions. Amplitude and phase shift keying (APSK) is a compromise between PSK and QAM.

In APSK modulation, constellation points are arranged on multiple circles. It can be viewed as multiple PSK constellations with different amplitudes. With such scheme, more constellation points can be accommodated than in PSK while the amplitude variation is smaller than in QAM. Furthermore, the overall constellation shape is circular as opposed to square (as in QAM). The removal of corners in constellation set reduces average transmit energy.

APSK is used in digital satellite broadcast standard DVB-S2. Figure 3.7 shows the two constellations used in DVB-S2 [6]. The numbers of constellation points are 16 (left) and 32 (right). The power ratios between the circles are variable, depending on the channel coding schemes.

3.2.5   Bit Mapping and Gray Code

In Sections 3.2.2–3.2.4, several modulation schemes are introduced by describing their constellations, that is, the values a symbol can take. In data transmission, each of these values represents a bit sequence of length K. The association (mapping) between bit segments and constellation points is known as bit mapping.

Bit mapping is not arbitrary; it affects the bit error rate (the percentage of bits received in error). As will be shown in Section 3.2.6, most receiving errors happen when one constellation point is mistaken for a *neighboring* one. When such error

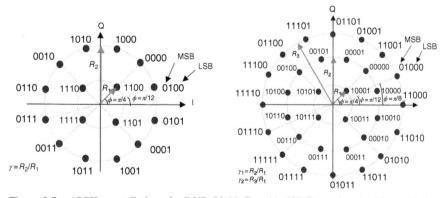

Figure 3.7   APSK constellations for DVB-S2 [6, figs. 11, 12]. Reproduced with permission.

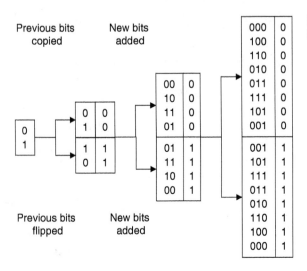

Figure 3.8   Gray code illustration.

occurs, the number of erroneous bits depends on the bit mapping scheme. For example, suppose a symbol with value 1 is mistaken as 2. In one case, symbol values 1 and 2 are mapped to bit sequences 00 and 11. Such symbol error would thus cause two bit errors. If, instead, symbol value 2 is mapped to bit sequence 01, then only one bit error occur for such symbol error.

Therefore, to reduce the probability of bit error (and thus bit error rate), we should reduce the average number of bit errors when a constellation point is mistaken for its neighbor. Ideally, the bit segments mapped to two neighboring constellation points should differ by only one bit.

Gray code, also known as reflected binary code (RBC), is a way to achieve such bit mapping property [7].

Figure 3.8 shows the process of Gray coding for the one-dimensional constellation. It is built successively, from shorter bit segments to longer ones. To add a bit to the mapping (left to right in the figure), the current bit mapping is replicated into two copies, one of them flipped upside down. Next, bits 0 and 1 are appended to the right of the bit sequences for the two copies, respectively. The reader can easily verify that such construction indeed creates a bit mapping that includes all possible bit combinations, and the neighboring entries differ only by one bit. The same concept can be extended to two-dimensional constellations. Figure 3.9 shows an example [3].

3.2.6   Constellation, Power, and Bit Error Rate

Sections 3.2.2–3.2.4 introduced three popular modulation schemes. In this section, we discuss how to assess the performance of PSK and QAM modulations. The same principles discussed here can be applied to other modulations. In this section, we introduce the metrics used to compare the relative performance of various modulation schemes.

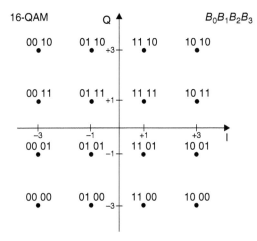

Figure 3.9   Bit mapping for 16-QAM in wireless LAN standard 802.11 [3, fig. 17.10]. Reproduced with Permission.

Discussions in this section are based on the concept of "signal space," which is mathematically defined as the space spanned by the signals. Signal space is discussed in more details in Chapter 4, where it serves as a foundation for the discussion of demodulation techniques. For the purpose of this section, we are only concerned about the signal space distance as defined below. In the presence of noise, two signals are more likely to be confused at the receiver (an instance of error) when their signal space distance is small. Therefore, when designing modulation schemes, we hope to minimize the BER by maximizing the signal space distance between the constellation points.

For the purpose of this section, signal space distance is defined as[4]

$$d \stackrel{\text{def}}{=} |s_m - s_{m'}|^2, \tag{3.9}$$

where s_m and $s_{m'}$ are the complex numbers representing the constellation points. Obviously, we can increase d by simply scaling up the constellation, that is, multiplying all $\{s_m\}$ by a factor $\alpha > 1$. However, with such scaling, the average signal power is also increased by a factor α^2. Therefore, a meaningful metric for comparing modulation designs is the ratio between d^2 and the average power. Such ratio does not change during the scaling. It will be shown in Chapter 4 that the bit error rate depends on this ratio.

Since neighboring constellation points have the smallest mutual distance, mistakes among these points usually dominate the error events. Therefore, we focus on

---

[4] The signal space is defined based on the waveform, instead of the constellation points (symbols). However, as we will see later in the chapter, for the modulation schemes discussed in this book, d defined in (3.9) is indeed the signal space distance, up to a constant factor that is common to all constellation points. Therefore, it suffices to study d.

the signal distance between neighboring constellation points. In the following, we examine PSK and QAM in terms of average energy per symbol and nearest neighbor distance.

For PSK, the constellation points are given by (3.3). For nearest neighbors, the phase angle difference is

$$\theta = \frac{2\pi}{M},\tag{3.10}$$

M being the total number of constellation points. The nearest neighbor distance is thus

$$d_P = 2A\sin\frac{\theta}{2} = 2A\sin\left(\frac{\pi}{M}\right),\tag{3.11}$$

All constellation points have the same energy A^2. Therefore, the average energy is

$$E_P = A^2.\tag{3.12}$$

In other words,

$$\frac{d_P^2}{E_P} = 4\sin^2\frac{\pi}{M}.\tag{3.13}$$

For QAM, the constellation points are given by (3.8). The nearest neighbor distance is

$$d_Q = 2B.\tag{3.14}$$

The average symbol energy is

$$E_Q = \frac{2NB^2\left[\sum_{k_n=-N}^{N-1}(2k_n+1)^2 + \sum_{l_n=-N}^{N-1}(2l_n+1)^2\right]}{4N^2} = \frac{B^2\sum_{k_n=-N}^{N-1}(2k_n+1)^2}{N}.\tag{3.15}$$

Further algebraic manipulation yields

$$E_Q = \frac{2}{3}B^2\left(4N^2-1\right).\tag{3.16}$$

Therefore,

$$\frac{d_Q^2}{E_Q} = \frac{6}{M-1},\tag{3.17}$$

where $M \stackrel{\text{def}}{=} 4N^2$ is the number of constellation points. These results can be extended to nonsquare constellations.

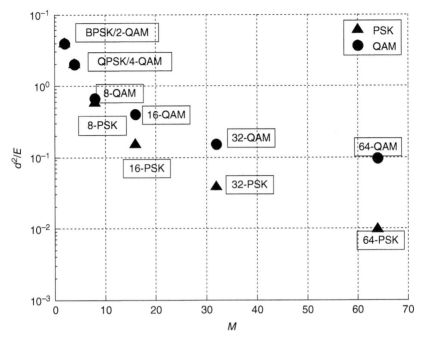

Figure 3.10   Nearest neighbor distances normalized by average power for PSK and QAM.

Note that with reasonably large M, (3.13) shows d_P^2/E_P is proportional to M^{-2}, while (3.17) shows d_Q^2/E_Q is proportional to M^{-1}. Therefore, as M increases, QAM has better noise immunity due to the larger signal distance given the symbol energy.

Figure 3.10 shows the nearest neighbor distances of PSK and QAM for various M values, normalized by average symbol energy.[5] When M has values 2 and 4, QAM and PSK are the same. When M is 8, the two modulation schemes have approximately the same performance. When M increases further, PSK has smaller nearest neighbor distances, and thus is more vulnerable to noise. Therefore, M usually goes only up to 8 for PSK schemes, limiting the data rate. For higher M values, QAM or APSK are usually chosen over PSK.

3.2.7   Summary of Symbol Mapping

Section 3.2 addresses the Symbol Mapping block in Figure 3.2, where bit sequences are mapped to symbol values. As discussed in Section 3.2.1, symbols are described by complex values, and the collection of all valid symbol values is known as the constellation of the modulation scheme. Sections 3.2.2–3.2.4 described three common modulation schemes and their variations. In addition to its constellation, modulation

---

[5] Note that 2-QAM, 8-QAM, and 32-QAM have nonsquare constellations. Therefore, (3.17) does not apply to these cases.

scheme is further defined by bit mapping, which associates the bit segments to be transmitted with the symbol values. The optimal bit mapping should reduce the number of bit errors when a constellation point is mistaken as its neighbors. Section 3.2.5 introduced the Gray code for bit mapping.

Choice of modulation schemes influences many aspects of a communications system. Two of the major impacts are data rate and error probability. As far as modulation is concerned, the data rate is determined by the number of bits represented by each symbol, K, given by (3.2). For common channel conditions, error probability is determined by the signal space distance between neighboring constellation points given by (3.9). Section 3.2.6 compared various PSK and QPSK modulation schemes in terms of error probability. More quantitative evaluations will be given in Chapter 4.

This chapter does not cover all modulation schemes. This chapter includes only single carrier modulation, where the resulting waveform occupies the entire bandwidth allocated to the system. Chapter 10 introduces multicarrier modulation, which contains multiple subcarriers, each being modulated with the schemes discussed in this chapter. There are more modulation schemes than PSK, QAM, and APSK, although these three types are used most frequently in modern digital communications systems, especially when combining with multicarrier modulation. Other modulation schemes include frequency shift keying (FSK), minimum shift keying (MSK), etc.

It should also be noted that the boundary between symbol mapping and pulse shaping (Section 3.4) is not always clear. For example, the OQPSK modulation discussed in Section 3.2.2 includes pulse shaping considerations. The benefit of the constant amplitude in PSK should also be understood in the context of pulse shaping. It will be beneficial if this section is revisited after learning Section 3.4.

3.3   NYQUIST–SHANNON SAMPLING THEORY

In digital communications systems, conversion between analog and digital forms of the signal is an important issue. Analog signal is what is transmitted through the wire or over the air. It is described by voltage as a function of time $s(t)$. Digital signal is the form used for digital processing. It is described as a sequence of digital values $\{d_n\}$. We often say that analog signals are continuous, while digital signals are discrete. Such distinction has two meanings. One is about the time values. While the analog signal is defined for all time points, the digital signal is defined only for discrete time points. The other distinction is on value range. An analog signal can take any value within a range. A digital signal, on the other hand, is represented by a fixed number of bits and, therefore, can only have a limited number of values. The sampling theory in this section studies the connection between the time-continuous signal (waveform) and time-discrete signal (samples or symbols). They are both assumed continuous in the value range. This is a good approximation as long as the number of bits used to represent a sample is large enough.

Sampling theory is used in communications systems in two contexts. At the transmitter, digital samples (symbols) at discrete time points are converted to analog waveforms in continuous time, to be transmitted into the channel. At the receiver, received time-continuous waveform is sampled at discrete time points to form digital data for further processing.

At the transmitter, we wish to have a higher symbol rate so that more data can be transmitted at a given time interval. At the receiver, we wish to have a lower sampling rate, to reduce the amount of data needed for processing. Therefore, the questions are:

A. What is the highest symbol rate one can use to form a waveform within certain transmission constraints?

B. What is the minimum sample rate one can use, while still preserving all the information in a waveform?

These two questions are related and are both answered by the Nyquist–Shannon sampling theory [8, 9]. It turns out that both answers are related to the bandwidth of the waveform. In the rest of this section, we will build up the theoretical foundation necessary to answer these two questions.

Note that although complex symbols are discussed in Section 3.2, in this section all samples, symbols, and signals are real-valued. As will be evident in Section 3.6, a sequence of complex symbols can be considered, within the scope of this section, as two separate sequences of real-valued symbols.

In the following, Section 3.3.1 provides some preliminary mathematical formulations to establish notations for our discussion. Section 3.3.2 discusses a scheme of perfectly reconstructing a waveform from its samples. Such discussion provides the sampling rate that can be used both for converting samples to waveforms and for sampling a waveform with perfect information retention. Sections 3.3.3–3.3.5 show that such rate is indeed the highest symbol rate sought by question A above, as well as the minimum sample rate sought by question B above. Section 3.3.6 provides a summary.

3.3.1 Frequency Domain Analysis of Sampled Waveform

Consider a time-domain function $s(t)$, whose FT is $\tilde{s}(\omega)$. Namely, from (3.100) in Section 3.8.1,

$$s(t) = \frac{1}{2\pi} \int \tilde{s}(\omega) e^{j\omega t} d\omega. \tag{3.18}$$

ω is the angular frequency. $s(t)$ is sampled at a uniform rate:

$$g_n \overset{\text{def}}{=} s(nT_s). \tag{3.19}$$

Here T_s is a constant, representing the sample period (also known as sample time). n is any integer. The invers of T_s is sample rate

$$R_s \overset{\text{def}}{=} \frac{1}{T_s}. \tag{3.20}$$

Now consider a reconstruction of the waveform based on samples $\{g_n\}$:

$$g(t) \overset{\text{def}}{=} T_s \sum_{n=-\infty}^{\infty} g_n \delta(t - nT_s). \tag{3.21}$$

Function $g(t)$ takes samples of $s(t)$ at nT_s and use impulses (i.e., delta functions) to represent these samples. The FT of $g(t)$ is, from (3.98) in Section 3.8.1,

$$\tilde{g}(\omega) = \int g(t)e^{-j\omega t}dt = T_s \sum_{n=-\infty}^{\infty} s(nT)e^{-j\omega nT_s}. \tag{3.22}$$

Combining (3.18) and (3.22),

$$\tilde{g}(\omega) = \sum_{n=-\infty}^{\infty} \frac{T_s}{2\pi} \int \tilde{s}(\omega')e^{-j(\omega-\omega')nT_s}d\omega' = \frac{T_s}{2\pi} \int \tilde{s}(\omega') \left(\sum_{n=-\infty}^{\infty} e^{-j(\omega-\omega')nT_s} \right)d\omega'. \tag{3.23}$$

Equations (3.110) and (3.108) in Section 3.8.2 lead to

$$\sum_{n=-\infty}^{\infty} e^{-j(\omega-\omega')nT_s} = 2\pi \sum_{k=-\infty}^{\infty} \delta[(\omega-\omega')T_s - 2k\pi] = \frac{2\pi}{T_s} \sum_{k=-\infty}^{\infty} \delta\left[(\omega-\omega') - \frac{2k\pi}{T_s}\right]. \tag{3.24}$$

Therefore, (3.23) becomes

$$\tilde{g}(\omega) = \sum_{k=-\infty}^{\infty} \tilde{s}\left(\omega - \frac{2\pi k}{T_s}\right). \tag{3.25}$$

Namely, the spectrum of g is the summation of multiple copies of the spectrum $\tilde{s}$, each shifted by $2\pi k/T_s$ with various integer values of k. This result provides the basis for the sampling theory.

3.3.2   Perfect Reconstruction of Waveform

Suppose waveform $s(t)$ with a limited bandwidth B (in frequency), that is, its spectrum has a support $(-B, B)$[6]:

$$\tilde{s}(\omega) = 0 \; \forall |\omega| \geq 2\pi B. \tag{3.26}$$

First, let us consider only ω in the range of $(-2\pi B, 2\pi B)$. If we have

$$B = \frac{1}{2T_s}, \tag{3.27}$$

then the summation in (3.25) has only one nonzero term: $k = 0$. Under such condition, (3.25) becomes

$$\tilde{g}(\omega) = \tilde{s}(\omega) \forall \omega \in (-2\pi B, 2\pi B). \tag{3.28}$$

---

[6] For simplicity, we exclude frequencies at the edge of the band in our discussion.

Namely, g replicates s in this frequency range.

Now let us consider other angular frequencies. In order to construct a waveform whose spectrum replicates $\tilde{s}(\omega)$ over all frequencies, we need to force its spectrum to be zero outside of the range $(-2\pi B, 2\pi B)$, following (3.26). In other words, we can construct a waveform $c(t)$, such that

$$\tilde{c}(\omega) \stackrel{\text{def}}{=} \tilde{g}(\omega)\tilde{h}(\omega); \; \tilde{h}(\omega) \stackrel{\text{def}}{=} \begin{cases} 1 & \forall \omega \in (-2\pi B, 2\pi B) \\ 0 & \text{otherwise} \end{cases}. \tag{3.29}$$

Since $\tilde{c}$ matches $\tilde{s}$ over the entire frequency range, $c(t)$ is a perfect reconstruction of $s(t)$.

Figure 3.11 shows the reconstruction process. Panel A shows $\tilde{s}(\omega)$, which is band-limited by (3.26). Panel B shows a general from of reconstruction $\tilde{g}(\omega)$ with spectrum replication, according to (3.25). Panel C shows $\tilde{g}(\omega)$ with a special sampling time given by (3.27). In this case, there is no overlap between different copies of $\tilde{c}$. Panel D shows $\tilde{c}(\omega)$, which is the spectrum in panel C, further limited by the band-limiting filter h. The portion of the spectrum cut off by h is shown in dotted lines.

The time domain filter function $h(t)$ is the IFT of $\tilde{h}(\omega)$:

$$h(t) = IFT\left[\tilde{h}(\omega)\right] = \frac{1}{2\pi} \int_{-\infty}^{\infty} \tilde{h}(\omega)e^{j\omega t}d\omega = \frac{1}{2\pi} \int_{-2\pi B}^{2\pi B} e^{j\omega t}d\omega. \tag{3.30}$$

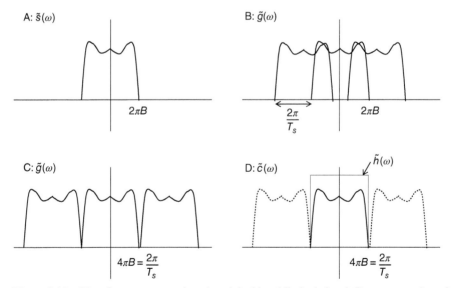

Figure 3.11 Waveform reconstruction. A: original band-limited signal; B: reconstruction of time-continuous signal causing spectrum replications; C: reconstruction with proper sampling rate, avoiding spectrum overlapping; D: low-pass filtering after reconstruction, recovering the original spectrum.

The integration can be performed analytically.

$$\int_{-2\pi B}^{2\pi B} e^{j\omega t}\,d\omega = \frac{2}{t}\sin(2\pi Bt) = 4\pi B\,\mathrm{sinc}(2Bt), \tag{3.31}$$

where the sinc function is defined as [10, sec. 5.2]

$$\mathrm{sinc}(x) \overset{\text{def}}{=} \frac{\sin(\pi x)}{\pi x}. \tag{3.32}$$

Note that for integers n,

$$\mathrm{sinc}(n) = \begin{cases} 1 & \text{if } n = 0 \\ 0 & \text{otherwise} \end{cases}. \tag{3.33}$$

Equations (3.27), (3.30), and (3.31) lead to

$$h(t) = \frac{1}{T_s}\mathrm{sinc}\left(\frac{t}{T_s}\right). \tag{3.34}$$

From the convolution theorem (3.117), (3.21) and (3.29) lead to

$$c(t) = \int g(\tau)h(t-\tau)\,d\tau = \sum_{n=-\infty}^{\infty} s(nT_s)h(t-nT_s). \tag{3.35}$$

To summarize, above discussions and derivations show that when the waveform $s(t)$ has a bandwidth limit of B as shown in (3.26) and the sampling time T_s is related to B as shown in (3.27), the discrete-time sample set $\{g_n\}$, defined by (3.19), contains all the information about $s(t)$. In fact, the original waveform can be reconstructed from this set of samples, using (3.35). Comparing (3.35) with (3.21), we see that instead of using delta functions, we need to use $h(t)$ for each impulse to reconstruct a waveform. $h(t)$ is known as the pulse shaping function in this context.

This result can be understood in another way. Assume for now that the original waveform has a period of NT_s. Later on, we can let N approach infinity to cover the general case. From (3.113) in Section 3.8.3, the spectrum of $s(t)$ consists of discrete lines separated by $2\pi/NT_s$. Since $s(t)$ has a bandwidth limit of $2\pi B$ in angular frequency, the total number of nonzero spectrum values is

$$K_s = 4\pi B \Big/ \left(\frac{2\pi}{NT_s}\right) = 2NBT_s. \tag{3.36}$$

Given the relationship between B and T_s in (3.27), we have

$$K_s = N. \tag{3.37}$$

K_s is the same as the total number of samples if $s(t)$ is sampled over the length of NT_s (the period), with sampling time T_s. Therefore, the N samples and the N spectral points are related by N linear equations. In general, this means the N samples can uniquely determine the N spectral points and thus $s(t)$, and vice versa.

Equation (3.27) means we have two samples in every period based on the maximum frequency. Or, in terms of sampling rate defined in (3.20),

$$R_s = 2B. \tag{3.38}$$

Under such condition, we can perfectly reconstruct the waveform from the samples.

3.3.3   Under-Sampling and Aliasing

All analyses in Section 3.3.2 assume a sampling rate that is connected to the waveform bandwidth through (3.27) and (3.38). With such a sampling rate, a waveform can be perfectly reconstructed using the discrete-time samples. Such set of samples thus captures all information about the waveform.[7]

Suppose we use a lower sampling rate, or, equivalently, a larger sampling time $T_s' > T_s$. This situation is known as under-sampling, that is, we have fewer samples than necessary.

According to discussions in the Section 3.3.2, we can follow (3.35) to construct a waveform $c'(t)$ based on the samples, with the spectrum

$$\tilde{c}'(\omega) = \begin{cases} \sum_{k=-\infty}^{\infty} \tilde{s}\left(\frac{\omega - 2\pi k}{T_s'}\right) & \forall |\omega| < 2\pi B' \\ 0 & \text{otherwise} \end{cases} \tag{3.39}$$

$$B' \stackrel{\text{def}}{=} \frac{1}{2\,T_s'} < B$$

Note that because the bandwidth of $s(t)$ is B, which is larger than B', we cannot remove the summation as in (3.28). Therefore, $\tilde{c}'(\omega)$ represents a waveform $c'(t)$, which is different from $s(t)$. However, these two waveforms result in the same samples:

$$c'(nT) = s(nT), \tag{3.40}$$

for all integer n. Therefore, it is impossible to distinguish $c'(t)$ and $s(t)$ (along with certain other waveforms) based on the samples.

Note that (3.35) is not the only way for waveform reconstruction. The key point of (3.39) is not that with under-sampling, (3.35) will not reconstruct the correct waveform. Instead, (3.39) shows that there can be multiple waveforms, all satisfying the bandwidth limitation and with the same sample values under such sampling rate. This means the samples do not capture all information about the waveform.

---

[7] In this chapter, we use the term "information" not in the sense of information theory as in Chapter 2. Instead, information is used in a more general sense as in an ordinary conversation.

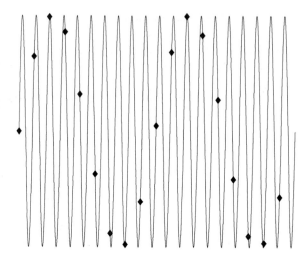

Figure 3.12   Aliasing demonstration. The lines show a sine waveform with frequency *f*. The diamonds show the samples taken at a sampling rate of 0.9*f*.

The summation in (3.39) shows the spectrum of $s(t)$, which has support over $(-2\pi B, 2\pi B)$, is "folded" into a narrower range $(-2\pi B', 2\pi B')$. This is known as aliasing. Simply put, a spectrum component at ω that is outside of the range $(-2\pi B', 2\pi B')$ is "folded" into that range at a new frequency $\omega' = \omega + 4k\pi B'$ with appropriate integer(s) k.

Aliasing can also be expressed in a similar manner with frequencies, instead of angular frequencies. In this case, the frequency shifts during folding is $2kB'$.

Figure 3.12 demonstrates the aliasing effect with a simple example. The solid line shows a sine wave with frequency f. The diamonds show the sampling of the wave, with a sampling rate of 0.9f. We can see that although the samples all fall on the original wave, they also form another sine wave with a lower frequency. Within the plotted time window, there are 20 periods for the original wave and 18 sampling points. The apparent "new wave" has two periods. The readers can verify that this quantitatively agree with the "folding" relationship in (3.39).

Figure 3.13 shows the original data of another, more complicated waveform. It is a combination of three sine waves, with frequencies of 6, 12, and 18 Hz. The upper panel shows the waveform, and the lower one shows its spectrum. Figure 3.14 shows the spectra after sampling and reconstruction. Only spectra within $(-B', B')$ are shown. It can be extended periodically to other ranges. Figures 3.13 and 3.14 are generated using actual simulation. Note that frequencies, instead of angular frequencies, are used in the plots.

In Figure 3.14, titles of each panel show the sampling rate R_s. The middle panel shows the case where the sampling rate is slightly higher than twice the maximum frequency in the waveform, which is 18 Hz. The other two panels show cases with lower sampling rates. The arrows show how spectral lines are moved due to aliasing. For the sake of clarity, only the positive half of the spectrum is traced. The other half is symmetric.

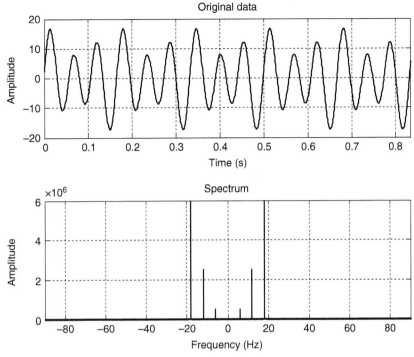

Figure 3.13 Original data for under-sampling demonstration. Upper and lower panels show data in time and frequency domains, respectively.

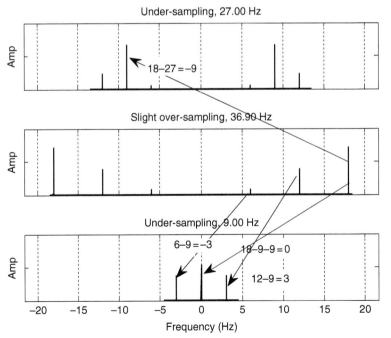

Figure 3.14 Spectra after reconstruction with various sampling rates. The three panels show spectrum after sampling with different sampling rates as indicated in panel titles.

From above derivations and demonstrations, we see that if a waveform is under-sampled, namely, the sampling rate is less than $2B$, then the resulting samples carry ambiguity. Namely, because of frequency folding, we cannot uniquely determine the original waveform based on the samples. Therefore, these samples do not preserve all the information in the original waveform. In other words, to preserve all the information, the minimum sampling rate is $2B$, or twice the frequency bandwidth of the waveform. This is the answer to question B put forward at the beginning of Section 3.3.

3.3.4   Over-Sampling

Section 3.3.3 shows that in order to capture all information from a waveform, the sampling rate cannot be lower than twice the waveform bandwidth. The next question is: what happens if we sample at a higher rate, or over-sampling?

Let $T_s = 1/2B$. Consider a time period between 0 and NT_s. There are N sample values $\{c_n, n = 0, 1, \ldots, (N-1)\}$ with sample period T_s:

$$c_n \overset{\text{def}}{=} c(nT_s), n = 0, 1, \ldots, (N-1). \tag{3.41}$$

When we sample $c(t)$ at a higher rate $1/T_s'$ with $T_s' < T_s$, we get another set of samples $\{d_m, m = 0, 1, \ldots, (M-1)\}$, with

$$d_m \overset{\text{def}}{=} c\left(mT_s'\right), m = 0, 1, \ldots, (M-1)$$

$$M = \left\lfloor \frac{NT_s}{T_s'} \right\rfloor > N \tag{3.42}$$

Since $c(t)$ can be reconstructed from $\{c_n\}$ with Equation (3.35),

$$d_m = \sum_{n=0}^{N-1} c_n h\left(mT_s' - nT_s\right), \forall m = 0, 1, \ldots, M-1. \tag{3.43}$$

Here we ignore the cases when t is near zero or NT_s, so that we can truncate the summation in (3.35) into 0 to $N-1$. This is appropriate as we are interested in the case where both N and M approach infinity. Equation (3.43) shows that the new sample values $\{d_m\}$ are linear combinations of $\{c_n\}$. We can consider (3.43) as a system of linear equations for $\{c_n\}$ given $\{d_m\}$. Since there are M equations and $N < M$ variables, we do not always have solutions. This means $\{d_m\}$ cannot take arbitrary values. They must be linearly dependent to each other, with N degrees of freedom. In other words, some of the sample values in $\{d_m\}$ can be predicted by others. Such constraint is automatically satisfied if $\{d_m\}$ are obtained by sampling the waveform. On the other hand, when such constraint is met, the solution for $\{c_n\}$ from (3.43) is in general unique. This means $\{d_m\}$ also captures all the information in the waveform.

This constraint can also be understood in frequency domain. In general, we can reconstruct the waveform using (3.35) with pulse shaping function given by (3.34), while replacing T_s with T_s' and $s(nT_s)$ with d_n. In general, the bandwidth of such waveform is $1/2T_s'$, which is larger than B, according to (3.29). However, because $\{d_m\}$ are not arbitrary but linearly dependent as discussed above, the waveform has a smaller bandwidth B.

Therefore, sampling at a higher sampling rate produces linearly dependent (and thus redundant) samples. However, such sample set still captures all the information and can be used to perfectly reconstruct the waveform.

3.3.5 Higher Symbol Rate in Transmission

Let us now turn to a different problem: using waveform to transmit data. In this case, we have a set of symbols $\{d_m\}$. Since each symbol represents a different bit sequence (see Section 3.2.1), they are mutually independent. We construct a waveform similar to (3.35):

$$c(t) = \sum_m d_m h\left(t - mT_{Sym}\right). \tag{3.44}$$

$c(t)$ has a limited bandwidth B, as dictated by the pulse shaping function $h(t)$. We hope that the receiver can recover transmitted symbols $\{d_m\}$ based on $c(t)$. Note that unlike (3.35), the pulse shaping function $h(t)$ does not have to be a sinc function. The question is: what is the highest symbol rate $1/T_{Sym}$ that can be used for such transmission, while the bandwidth is limited to B?

This question can be analyzed similarly as above. With any symbol rate, we can always choose a pulse shaping function $h(t)$, so that the resulting waveform has a bandwidth of B as required.[8] However, this waveform can be reconstructed using a set of samples $\{c_n\}$, collected at a sampling rate $1/T_s$, where

$$T_s = \frac{1}{2B}. \tag{3.45}$$

Therefore, similar to (3.43), we have

$$c_n = \sum_n d_m h\left(nT_s - mT_{Sym}\right). \tag{3.46}$$

If $\{d_m\}$ could be recovered from the waveform, it would be recovered from $\{c_n\}$, because the latter captured all information of the waveform. However, if we consider a section of the waveform of duration T with $T \to \infty$, then we have N samples in $\{c_n\}$, and M symbols in $\{d_m\}$ to recover:

$$\begin{aligned} N &= \left\lfloor \frac{T}{T_s} \right\rfloor \\ M &= \left\lfloor \frac{T}{T_{Sym}} \right\rfloor \end{aligned}. \tag{3.47}$$

Namely, (3.46) has N equations and M unknown variables. In order to recover the symbols from the waveform, we must be able to obtain unique solutions for $\{d_m\}$ from (3.46). This requires[9]

---

[8] However, as will be shown in Section 3.4, such filters will not have the property in (3.33) when $T_{Sym} < 1/(2B)$.

[9] When $N > M$, we have more equations than variables. In general, solutions do not exist. However, since $\{c_n\}$ are obtained from sampling, they are not mutually independent. Therefore, as long as the waveform is constructed according to (3.35), (3.46) will have a solution.

$$N \geq M. \tag{3.48}$$

Therefore, the condition for transmitting recoverable symbols in a waveform with bandwidth B is

$$T_{Sym} \geq T_s = \frac{1}{2B}. \tag{3.49}$$

or, in terms of the symbol rate,

$$R_{Sym} \stackrel{\text{def}}{=} \frac{1}{T_{Sym}} \leq 2B. \tag{3.50}$$

This result answers question A at the beginning of Section 3.3.

3.3.6   Summary of the Sampling Theory

With above analysis, we can answer the two questions raised at the beginning of this section.

A. When the transmitted waveform has a bandwidth limitation of B, the maximum symbol rate, in which independent symbols can be sent by the transmitter and recovered by the receiver, is $2B$.

B. For a given waveform with bandwidth limitation of B, the minimum sampling rate required in order to preserve all information is $2B$. With such sampling rate, the waveform can be reconstructed from the samples.

Both conclusions center on an important sampling rate: $2B$. Given waveform bandwidth B, the sampling rate or symbol rate $2B$ is referred to as the Nyquist rate. Conversely, given the sampling/symbol rate R_s, waveform bandwidth $R_s/2$ is referred to as Nyquist frequency or Nyquist bandwidth.

These results can be extended to nonuniform sampling. In this case, for any large enough length of time, the density of the sampling points must be no less than the Nyquist rate in order to sample the waveform properly and must be no more than the Nyquist rate in order to send independent symbols [11].

It should be noted that the Nyquist rate applies to a faithful sampling of a waveform, about which we know nothing except for its bandwidth. Sub-Nyquist sampling, that is, sampling below the Nyquist rate, is feasible if we have more knowledge about the waveform, or we do not need to capture all information. Some sub-Nyquist sampling technologies are applicable to digital communication systems. For example, with a proper sampling rate, passband waveforms can be converted into baseband through proper aliasing [12]. (Passband and baseband are explained in Section 3.6.)

Another example of sub-Nyquist sampling is the detection of signals, where we only care about the power spectral density of the signal, not the actual waveform [13]. Recently, sub-Nyquist sampling has a research thrust known as compressive sensing [14]. However, this is not related to our present topic.

Likewise, the maximum symbol rate given in Section 3.3.5 was derived assuming a symbol can take any value. The symbol rate limit ensures that the receiver can recover the values of all symbols from the waveform. However, in reality, symbol values are drawn from a finite alphabet (see Section 3.2). In this case, the receiver may still be able to recover the symbol values when the symbol rate is slightly higher than the Nyquist rate. Such faster-than-Nyquist signaling techniques can bring limited performance gain [15–17], [18, ch. 2]. Faster-than-Nyquist signaling is not discussed in this book.

3.4 PULSE SHAPING AND NYQUIST CRITERION

This section studies ways of constructing transmitted waveform based on symbols, following (3.44):

$$c(t) = \sum_n d_n h\left(t - nT_{Sym}\right). \tag{3.51}$$

Here, $\{d_n\}$ is the sequence of data-carrying symbols discussed in Section 3.2. T_{Sym} is the time interval between symbols, or symbol period. In most communications systems, $c(t)$ is bandwidth-limited. Namely, its bandwidth cannot exceed B:

$$\tilde{c}(\omega) = 0 \forall \omega \notin (-2\pi B, 2\pi B). \tag{3.52}$$

In this section, we consider some design considerations about the pulse shaping function $h(t)$.

Note that although complex symbols are discussed in Section 3.2, in this section all samples, symbols and signals are real-valued. As will be clear in Section 3.6, a sequence of complex symbols can be considered, in the scope of this section, as two separate sequences of real-valued symbols.

3.4.1 Pulse Shaping Filter

To satisfy the bandwidth constraint, the ideal pulse shaping filter is the sinc function given by (3.34), with a "brick wall" frequency response of (3.29). In addition to the ideal frequency response for limiting waveform bandwidth, the sinc function also has the nice property of (3.33). Equation (3.33) means that the waveform value at nT_{Sym} depends only on the symbol d_n, as contributions from other symbols are all zero at this time point. In other words, there is no intersymbol interference (ISI). The absence of ISI significantly simplifies receiver design: d_n can be recovered by taking the sample of the waveform at nT_{Sym}.

However, as shown in (3.32), a sinc filter is infinitely long. Its envelope decays relatively slowly, on the order of T_{Sym}/t. This causes two practical difficulties [1, sec. 9.2-1]. First, when approximating a sinc filter with a finite impulse response (FIR) filter, a large number of taps are needed in order to reproduce the frequency response of (3.29) with fidelity. This increases implementation cost. Second, if there is a timing error (i.e., sampling time deviates from nT_{sym}), the resulting ISI would be high because it is contributed by many neighboring symbols. Therefore, we wish to design pulse shaping filters with the following properties:

A. The filter tap values decay faster than T_{sym}/t in time domain.

B. Not too much spectral energy beyond the allowed bandwidth B.

C. No ISI.

Requirements A and B involve a tradeoff. In general, if a filter is shorter in the time domain, its spectrum is wider [19, sec. 8.2]. An extreme case is the delta function: it is extremely localized in the time domain and has an infinitely extending spectrum. In fact, it can be shown that a waveform cannot have limited support in both time domain and frequency domain. In other words, if we want the spectrum to be zero for all frequency beyond $(-B, B)$, then the filter must be infinitely long in the time domain. See the appendix (Section 3.9) for more discussions on this matter.

Therefore, to have shorter filters, we need to allow the spectrum to extend beyond the prescribed bandwidth. However, we would like to have the "spill" to be as small as possible. The exact metric for such spill depends on the actual situations and is usually specified as a spectrum mask (maximum power spectral density as a function of frequency). The portion of the spectrum that is outside the bandwidth limits is referred to as excess bandwidth.

Concerning property C above, the Nyquist criterion introduced in Section 3.4.2 provides guidance on filter design.

3.4.2 Nyquist Criterion

The Nyquist criterion, also known as the Nyquist pulse shaping criterion or the Nyquist condition for zero ISI, states that the necessary and sufficient condition for a function $h(t)$ to have the zero ISI property (3.33) is that its FT $\tilde{h}(\omega)$ satisfies

$$\sum_{m=-\infty}^{\infty} \tilde{h}\left(\omega + \frac{2m\pi}{T_{Sym}}\right) = T_{Sym} \forall \omega. \tag{3.53}$$

The Nyquist criterion says the "folded spectrum" of $h(t)$ is a constant over ω, although the actual value of the constant is not important because it is just the overall gain of the system. Typically, the excess bandwidth is not large. Namely, $\tilde{h}(\omega)$ is essentially zero when the frequency is much higher than the Nyquist frequency $1/2T_{sym}$. This "small excess bandwidth" condition can be formally written as[10]

$$\tilde{h}(\omega) = 0 \forall |\omega| > \frac{\pi}{T_{Sym}}(1+\alpha). \tag{3.54}$$

for some $\alpha \ll 1$. In this case, the folding only happens near the Nyquist frequency. Furthermore, when $h(t)$ is real-valued, (3.102) leads to $\tilde{h}(-\omega) = \tilde{h}^*(\omega)$. Under such conditions, (3.53) can be written as

$$\tilde{h}(\omega) + \tilde{h}\left(\omega - \frac{2\pi}{T_{Sym}}\right) = \tilde{h}(\omega) + \tilde{h}^*\left(\frac{2\pi}{T_{Sym}} - \omega\right) = T_{Sym} \forall \left|\omega - \frac{\pi}{T_{Sym}}\right| \le \alpha \frac{\pi}{T_{Sym}}. \tag{3.55}$$

Equation (3.55) exhibits certain symmetry about the Nyquist frequency. In fact, (3.55) leads to

$$\left[\tilde{h}\left(\frac{\pi}{T_{Sym}} - u\right) - \frac{T_{Sym}}{2}\right] = -\left[\tilde{h}^*\left(\frac{\pi}{T_{Sym}} + u\right) - \frac{T_{Sym}}{2}\right] \forall |u| < \alpha \frac{\pi}{T_{Sym}}. \tag{3.56}$$

Similarly, at the negative frequency,

$$\left[\tilde{h}\left(-\frac{\pi}{T_{Sym}} - u\right) - \frac{T_{Sym}}{2}\right] = -\left[\tilde{h}^*\left(-\frac{\pi}{T_{Sym}} + u\right) - \frac{T_{Sym}}{2}\right] \forall |u| < \alpha \frac{\pi}{T_{Sym}}. \tag{3.57}$$

Furthermore, the middle portion of the spectrum is a constant:

$$\tilde{h}(\omega) = T_{Sym} \forall |\omega| \le (1-\alpha)\frac{\pi}{T_{Sym}}. \tag{3.58}$$

From (3.56) to (3.58), we can see that the only Nyquist filter without excess bandwidth (i.e., $\alpha = 0$) is the "brick wall" filter given by (3.29). Furthermore, we cannot even "give up" bandwidth in exchange for the removal of excess bandwidth. In other words, it is impossible to have a filter satisfying Nyquist criterion and

$$\tilde{h}(\omega) = 0 \forall |\omega| \ge B' \text{ for some } B' < \frac{\pi}{T_{Sym}}. \tag{3.59}$$

Figure 3.15 illustrates the characteristics discussed above. In this case, $\tilde{h}(\omega)$ is real-valued.

Proof of the Nyquist criterion is provided in the appendix, Section 3.10.

---

[10] Note that as stated in Section 3.4.2, (3.54) implies that the filter is infinitely long in time domain. It can often be approximated with a finite length filter.

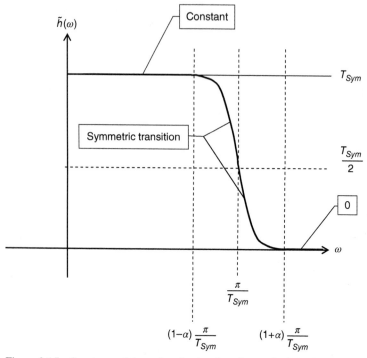

Figure 3.15   Spectrum of the pulse shaping function under Nyquist criterion.

3.4.3   Raised Cosine Filter and Root Raised Cosine Filter

The Nyquist criterion (3.53) is the necessary and sufficient condition for no ISI. A widely used pulse shaping filter satisfying the Nyquist criterion is the raised cosine filter. It is defined in the frequency domain:

$$\tilde{h}(\omega) = \begin{cases} T_{Sym} & 0 \le |\omega| \le \dfrac{\pi(1-\alpha)}{T_{Sym}} \\[2ex] \dfrac{T_{Sym}}{2}\left\{1+\cos\left[\dfrac{T_{Sym}}{2\alpha}\left(|\omega|-\dfrac{\pi(1-\alpha)}{T_{Sym}}\right)\right]\right\} & \dfrac{\pi(1-\alpha)}{T_{Sym}} < |\omega| \le \dfrac{\pi(1+\alpha)}{T_{Sym}} \\[2ex] 0 & |\omega| > \dfrac{\pi(1+\alpha)}{T_{Sym}} \end{cases} . \quad (3.60)$$

The spectrum has three parts. The middle part is a constant. The "transition" part near the Nyquist frequency $\pm\pi/T_{Sym}$ is a cosine function with an offset. The variable for the cosine function changes from 0 to π over the transition region. Therefore, the cosine function is symmetric about the Nyquist frequencies (where the variable for the cosine function is $\pi/2$), satisfying the Nyquist criterion. Beyond the transition frequency, the spectrum is zero. Parameter α (known as the roll-off factor) controls the width of the transition region and the excess bandwidth. When $\alpha = 0$, the raised cosine filter becomes the "brick wall" filter we discussed before.

The time-domain filter can be obtained by taking an inverse Fourier transform of the desired spectrum. Truncation can then be made to obtain a filter with the desired length. There are other optimization methods if a very short filter is required. As will be shown in homework, raised cosine filter offers better performance than sinc filter, when filter length is limited.

In many transceiver architectures, the pulse shaping filter is split between the transmitter and receiver (see Chapter 4). In this case, the desired filter spectrum is the square root of (3.60). Such filter is known as the root raised cosine filter.

Figure 3.16 shows the frequency response (spectrum) of raised cosine filters with various excess bandwidths and filter lengths. The legend marks the α values and filter lengths L (in symbol counts) for the various traces. The x coordinate is the normalized frequency, where the Nyquist frequency $(1/2T_{Sym})$ is set as 1. The y coordinate is the relative power spectrum in dB. Only the positive frequency part is shown.

We can see from Figure 3.16 that while smaller α values result in steeper transition and less excess bandwidth, they are also accompanied by slower decay at higher frequencies. This is especially true for shorter filters. Therefore, there is a tradeoff in design, which is usually addressed through simulation.

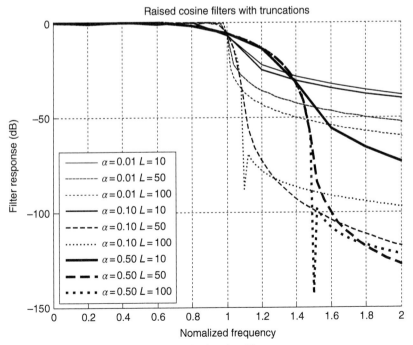

Figure 3.16   Frequency response of raised cosine filters.

3.4.4   Summary and Closing Remarks

Pulse shaping converts symbols in the digital domain to waveforms in the analog domain. This section studies a class of pulse shaping filters, based on the Nyquist criterion. Nyquist criterion guarantees no ISI in the resulting waveform. The receiver can thus recover the symbol values by merely sampling the received waveform at the correct time. A commonly used pulse shaping filter, raised cosine filter, is used in this section as an example to illustrate the tradeoff between filter length and bandwidth accuracy.

Another filter performance metric focused in this section is the excess bandwidth. To reduce interference to neighboring channels, we wish excess bandwidth be small. Usually, system specification provides a spectral mask that the filter response cannot exceed. However, it should be noted that excess bandwidth is not the only performance metric in practical design. Other considerations include sensitivity to timing errors and the peak to root mean squared (RMS) ratio, which affects the linearity requirement of the analog circuitry.

While Nyquist criterion is brilliant mathematically and is often used as a guideline for pulse shaping filter design, we should recognize that "no ISI" is not an absolute requirement, for several reasons as listed below.

- ISI can be corrected by the receiver (see Chapter 9).
- In fact, some modulation schemes intentionally introduce ISI in exchange for other performance gains [1, sec. 9.2-2]. Such techniques are known as partial response.
- If the ISI level is lower than the noise level experienced by the receiver, it can be ignored without performance degradation.
- Many communications channels introduce ISI anyway (Chapter 9). Therefore, it is not necessary to avoid ISI in transmit waveform design.

Similarly, excess bandwidth is not always bad, either. Usually, excess bandwidth requires neighboring channels to be separated by guard bands, wasting spectrum resource. However, if the sidelobe (i.e., spectral energy beyond the bandwidth limit) is lower than the noise level at the receiver, it does not cause additional interference and can be ignored. Furthermore, as discussed in Chapter 7, some synchronization techniques depend on the excess bandwidth.

Pulse shaping filter design can be challenging, as it must consider various tradeoffs discussed above. Other than the Nyquist criterion, most of the optimization is based on trial and error with simulation. Interested readers can search for recent literature on the topic [20]. On the other hand, in most cases, a well-studied pulse shaping filter such as the raised cosine filter is good enough.

3.5   IMPLEMENTATION OF PULSE SHAPING FILTER: UP-SAMPLING

In Section 3.2, we have established that with a waveform with bandwidth limitation of B, we can transmit data-carrying symbols at a rate of $2B$. The waveform is constructed from the symbols according to (3.51) using the pulse shaping filters discussed in

(a)   (b)

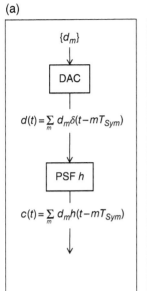
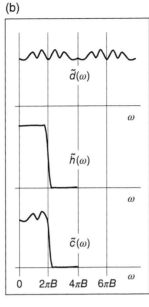

Figure 3.17   Pulse shaping
filter implementation in
analog domain. A: block
diagram. B: spectra
of the various quantities.

Section 3.4. Since a pulse shaping filter is time-continuous, it should be implemented in the analog domain.

Figure 3.17 shows the implementation architecture, where the pulse shaping filter (PSF) is implemented in the analog domain. As shown in panel A, symbols $\{d_m\}$ are generated in the digital domain and are passed to the analog domain through a digital to analog convertor (DAC). The output of the DAC can be approximated as $d(t)$. The pulse shaping filter (PSF) with impulse response h is applied to $d(t)$, yielding the desired output $c(t)$, according to (3.51). Panel B shows the spectrum at the various steps. $d(t)$ has a spectrum with a period of $2\pi/T_{Sym}$, or $4\pi B$, according to (3.25). This extended spectrum is filtered by $\tilde{h}(\omega)$ to form the bandwidth-limited spectrum $\tilde{c}(\omega)$.

Unfortunately, it is generally difficult to design an analog filter with precisely the desired response. In the case of pulse shaping filter, this means the Nyquist criterion is not easy to guarantee. Even without the non-ISI requirement, building an analog filter that has a flat in-band response and a sharp cutoff at the band edge can be difficult. For this reason, we often implement the pulse shaping filter in the digital domain and use over-sampling to approximate the continuous-time.

Figure 3.18 shows the architecture, where the over-sampling rate is 2. Panel A shows the signal flow. Digital symbols $\{d_m\}$ are up-sampled by inserting zeros between the symbols in sequence, emulating the impulse function in (3.21). The new sequence $\{d'_k\}$ is filtered by a digital pulse shaping filter (PSF), whose response is h. The PSF is implemented at the higher sample rate. The output of the filter $\{c'_m\}$ is converted to analog signal $c'(t)$ by DAC. This signal is further filtered by an analog band-limiting filter (BLF), resulting in the final output signal $c(t)$. Panel B shows the spectrum of the various signals. Because of the over sampling rate of 2 (i.e., the sampling rate is $4B$), the spectrum of $c'(t)$ has a period of $8\pi B$, instead of the $4\pi B$ period

(a)   (b)

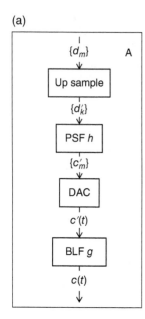   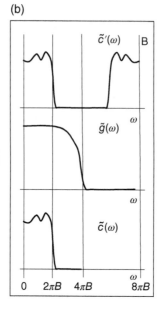

Figure 3.18   Pulse shaping filter implementation in digital domain. A: block diagram. B: spectra of the various quantities

shown in Figure 3.17. Because the useful signals in the range $[-2\pi B, 2\pi B]$ are now well-separated from its "image," a much more gradual filter (as the plot of $\tilde{g}(\omega)$ shows) can be used in the analog domain to reject such image, in contrast with the sharp-edged filter $\tilde{h}(\omega)$ in Figure 3.17. Since the BLF is flat near the original Nyquist frequency $2\pi B$, it would not distort the compliance of Nyquist criteria from h. Through over sampling, we achieve good performance in both ISI reduction and spectral mask compliance, at the cost of a higher sampling rate at the DAC.

Over-sampling is beneficial on the receiver side, as well, as will be discussed in Chapter 4.

3.6   BASEBAND AND PASSBAND

In Sections 3.2 – 3.5, we discussed the methods to transmit data-carrying symbols with waveforms, that is, a signal that is a function of time. In such discussions, the support of the waveform spectrum is from $-B$ to B, that is, around direct current (DC). In other words, the waveform spectrum includes zero frequency. This is referred to as baseband signal.

On the other hand, in wireless communications (and some wired communications systems), the waveform spectrum is centered at some frequency $f_c \gg B$, known as the carrier frequency. Namely, the spectrum support spans between $f_c - B$ to $f_c + B$, with a symmetric part in the negative frequency. Such signal is referred to as passband signal.[11] Passband signal, because of its higher frequency, can be transmitted more

---

[11] In signal processing, the term "passband" also refers to the part of the spectrum that can pass through a filter, as opposed to "stopband."

efficiently through an antenna. Furthermore, by choosing a proper f_c, we can use the most advantageous propagation condition, or share the wired or wireless medium with other users without interference.

In this section, we discuss the conversion between baseband and passband signals, and some properties of the passband signal.

3.6.1 Up-Conversion and Down-Conversion

Figure 3.19 shows the architecture of a typical transceiver. At the transmitter side (the top portion of the figure), digital data are converted into analog signal by digital-to-analog converter (DAC). This analog signal is at baseband. After processing (e.g., filtering) by the baseband circuit, the signal is converted to passband by the up-converter. The passband signal (also referred as radio frequency (RF) signal) is sent to the antenna after some further processing (e.g., amplification). At the receiver (the lower portion of the figure), the process is reversed. The passband signal received from the antenna is converted to baseband by the down-converter, and further converted to digital form by the analog-to-digital converter (ADC). Note that in some systems, the up-conversion and down-conversion are performed in two or more steps, resulting in one or more intermediate frequency (IF) paths. For simplicity, such cases are not discussed here.

Figure 3.20 shows the details of up-converter and down-converter. $\omega_c = 2\pi f_c$ is the center carrier angular frequency. The left panel shows the up-converter. Two

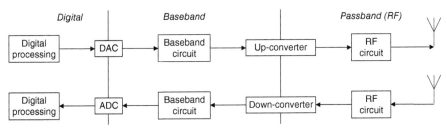

Figure 3.19   Transceiver analog architecture. Top portion: transmitter. Lower portion: receiver.

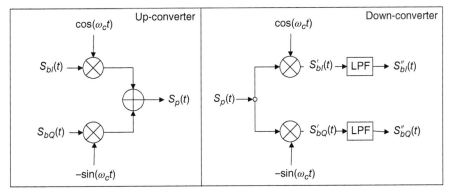

Figure 3.20   Up-converter and down-converter.

streams of real-valued baseband signal $S_{bI}(t)$ and $S_{bQ}(t)$ are fed into the up-converter. They are multiplied by $\cos(\omega_c t)$ and $-\sin(\omega_c t)$, respectively, before being summed to form the passband signal $S_p(t)$. S_{bI} and S_{bQ} are referred to as in-phase (I) and quadrature (Q) signals, respectively. The right panel shows the down-converter. The passband signal $S_p(t)$ is split into two streams and multiplied by $\cos(\omega_c t)$ and $-\sin(\omega_c t)$, respectively. The products pass their low-pass filters (LPF) to form the baseband signals $S''_{bI}(t)$ and $S''_{bQ}(t)$. In the following, we will study the spectra of the passband signals and show that the output of the down-converter is the same as the input of the up-converter. Typically, the local signals are generated from one close with a $90°$ phase shifter to form $\cos(\omega_c t)$ and $\sin(\omega_c t)$.

Recall that sine and cosine functions can be expressed as exponential functions:

$$\cos(\omega_c t) = \frac{1}{2}\left(e^{j\omega_c t} + e^{-j\omega_c t}\right)$$
$$\sin(\omega_c t) = \frac{1}{2j}\left(e^{j\omega_c t} - e^{-j\omega_c t}\right) \qquad (3.61)$$

From the up-conversion process, we have

$$S_p(t) = S_{bI}(t)\cos(\omega_c t) - S_{bQ}(t)\sin(\omega_c t). \qquad (3.62)$$

Its FT, from (3.98) in Section 3.8.1 and (3.61), is

$$\tilde{S}_p(\omega) = \frac{1}{2}\left\{\int [S_{bI}(t) + jS_{bQ}(t)]e^{-j(\omega - \omega_c)t}dt + \int [S_{bI}(t) - jS_{qQ}(t)]e^{-j(\omega + \omega_c)t}dt\right\}. \qquad (3.63)$$

Therefore,

$$\tilde{S}_p(\omega) = \frac{1}{2}\left\{\left[\tilde{S}_{bI}(\omega - \omega_c) + j\tilde{S}_{bQ}(\omega - \omega_c)\right] + \left[\tilde{S}_{bI}(\omega + \omega_c) - j\tilde{S}_{bQ}(\omega + \omega_c)\right]\right\}. \qquad (3.64)$$

In (3.64), the first square bracket is the spectra of S_{bI} and S_{bQ} shifted to center at ω_c. The second square bracket is the same spectra shifted to center at $-\omega_c$.

Figure 3.21 shows conceptually the spectrum shifting between baseband and passband. For simplicity, only one baseband signal is shown.

At the down-converter, the multiplication of sine and cosine functions produces the results:

$$S'_{bI}(t) = S_p(t)\cos(\omega_c t)$$
$$S'_{bQ}(t) = -S_p(t)\sin(\omega_c t) \qquad (3.65)$$

Combining with (3.62), we have

$$S'_{bI}(t) = S_{bI}(t)\cos^2(\omega_c t) - S_{bQ}(t)\sin(\omega_c t)\cos(\omega_c t)$$
$$S'_{bQ}(t) = -S_{bI}(t)\cos(\omega_c t)\sin(\omega_c t) + S_{bQ}(t)\sin^2(\omega_c t) \qquad (3.66)$$

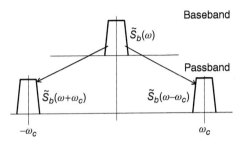

Figure 3.21   Spectra in up-conversion.

Simple trigonometry manipulations lead to

$$S'_{bI}(t) = \frac{1}{2}S_{bI}(t) + \frac{1}{2}[S_{bI}(t)\cos(2\omega_c t) - S_{bQ}(t)\sin(2\omega_c t)]$$
$$S'_{bQ}(t) = \frac{1}{2}S_{bQ}(t) + \frac{1}{2}[-S_{bI}(t)\sin(2\omega_c t) - S_{bQ}(t)\cos(2\omega_c t)]$$
(3.67)

As discussed above, the terms in the square brackets in (3.67) have spectra centered at $\pm 2\omega_c$. Therefore, they can be removed by the subsequent low pass filters, resulting in

$$S''_{bI}(t) = \frac{1}{2}S_{bI}(t)$$
$$S''_{bQ}(t) = \frac{1}{2}S_{bQ}(t)$$
(3.68)

Therefore, the down-converter can recover the baseband signals sent to the up-converter, with a factor of 1/2.

Thus, we see that up-converter (3.62) generated the desired spectrum for passband signal. Down-converter (3.65) recovers the baseband signal at the receiver.

3.6.2 Complex Signal Expression

Equation (3.63) is often expressed in complex form:

$$S_p(t) = Re\left(\bar{S}_p(t)\right)$$
$$\bar{S}_P(t) \overset{\text{def}}{=} \bar{S}_b(t)e^{j\omega_c t}$$
(3.69)

$\bar{S}_b(t)$ is the complex form of the baseband signal:

$$\bar{S}_b(t) \overset{\text{def}}{=} S_{bI}(t) + jS_{bQ}(t).$$
(3.70)

The first line of (3.69) can also be written as

$$S_p(t) = \frac{1}{2}\left[\bar{S}_p(t) + \bar{S}_p^*(t)\right]. \tag{3.71}$$

From (3.69), we see that $\bar{S}_p$ has a spectrum concentrated around ω_c. Therefore, $\bar{S}_p^*$ has a spectrum concentrated around $-\omega_c$. Equation (3.71) shows that S_p has a symmetric spectrum, as shown in (3.63) and Figure 3.21.

For the down conversion, (3.65) can be written as

$$\bar{S}_b'(t) = e^{-j\omega_c t} S_p(t), \tag{3.72}$$

where $\bar{S}_b'$ is the complex form of S_{bI}' and S_{bQ}':

$$\bar{S}_b'(t) \overset{\text{def}}{=} \bar{S}_{bI}'(t) + j\bar{S}_{bQ}'(t). \tag{3.73}$$

With (3.71), (3.72) leads to

$$\bar{S}_b'(t) = \frac{1}{2}e^{-j\omega_c t}\bar{S}_p(t) + \frac{1}{2}e^{-j\omega_c t}\bar{S}_p^*(t). \tag{3.74}$$

From (3.69), we see that the spectrum of $\bar{S}_p^*(t)$ concentrates around $-\omega_c$. Therefore, the second term in (3.74) has a spectrum concentrating around $-2\omega_c$ and thus is rejected by the low pass filter. If we define the complex-valued baseband signal

$$\bar{S}_b''(t) \overset{\text{def}}{=} S_{bI}''(t) + jS_{bQ}''(t), \tag{3.75}$$

then

$$\bar{S}_b''(t) = \frac{1}{2}e^{-j\omega_c t}\bar{S}_p(t). \tag{3.76}$$

Combining (3.69) with (3.76), we have

$$\bar{S}_b''(t) = \frac{1}{2}\bar{S}_b(t). \tag{3.77}$$

This result agrees with (3.68).

For most analyses, we can stay in the complex form. Therefore, we often combine the in-phase and quadrature into a complex baseband signal. This approach was used in Section 3.2 and will be the approach for the rest of this book. In fact, complex baseband signals are the object of a large portion of the signal analysis.

Figure 3.22 shows the complex signal version of the converters. The left panel shows the up-converter based on (3.69); the right panel shows two versions of the down-converter, based on (3.76) (upper path) and (3.74) (lower path).

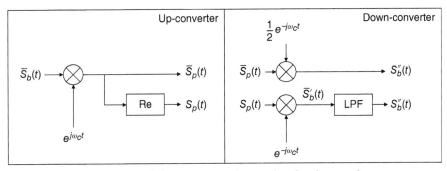

Figure 3.22   Up-converter and down-converter in complex signal expression.

3.6.3   Bandwidths and Symbol Rates in Baseband and Passband

In this section, for the sake of clarity, we use W_B and W_P to note the bandwidths of baseband and passband signals.

$$W_B \stackrel{\text{def}}{=} B. \tag{3.78}$$

In Sections 3.6.1 and 3.6.2, we see that if the support of the baseband signal spectrum extends between $-W_B$ and W_B, then that of the passband signal extends between

$f_c - W_B$ to $f_c + W_B$ and $-f_c - W_B$ to $-f_c + W_B$, where f_c is the carrier frequency. For such a signal, we say the baseband bandwidth is W_B, while the passband bandwidth is W_P, with

$$W_P = 2W_B. \tag{3.79}$$

Figure 3.23 illustrates the two bandwidth definitions.

On the other hand, from (3.63) we see that the spectrum amplitude of passband signal is 1/2 of that of the baseband signal at the transmitter sides, for both in-phase and quadrature components. Of importance is any possible change in signal to noise ratio. This issue will be discussed in more details in Chapter 4.

As discussed in Section 3.2, the maximum symbol rate for a waveform with baseband bandwidth W_B is $2W_B$. From the passband point of view, the maximum symbol rate is W_P. Namely, the minimum symbol period is given by

$$\frac{1}{T_{Sym}} = W_P. \tag{3.80}$$

Remember that we are actually talking about two streams of symbols, in-phase and quadrature. They can be considered as separate channels based on (3.68).

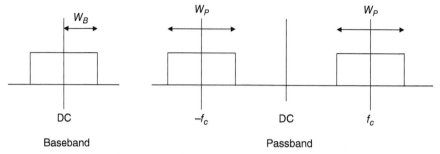

Figure 3.23   Baseband and passband bandwidth definitions.

Therefore, the total capacity of the channel is twice of that given in Chapter 2. With white noise, the channel capacity C is given as

$$C = 2W_B\log_2(1 + SNR) = W_P\log_2(1 + SNR). \tag{3.81}$$

The fact that a passband channel can transmit two separate baseband channels can be understood intuitively in both time domain and frequency domain.

In the time domain, consider a function that is a summation of cosine and sine function:

$$s(t) = A\cos(\omega t) + B\sin(\omega t). \tag{3.82}$$

We have

$$s(t)\cos(\omega t) = A\cos^2(\omega t) + B\sin(\omega t)\cos(\omega t). \tag{3.83}$$

In the first term, $\cos^2(\omega t)$ is always larger than 0, while $\sin(\omega t)\cos(\omega t)$ in the second term oscillates between positive and negative. If we integrate over a time period much larger than the oscillation period $2\pi/\omega$, only the first term yields a significant nonzero quantity. Therefore, A is "extracted from the mix" by "projecting" $s(t)$ toward the cosine function. Same wise, we can extract B by integrating the product $(t)\sin(\omega t)$. In this sense, we can say that the two functions $\sin(\omega t)$ and $\cos(\omega t)$ are the orthogonal basis. A and B are signal components of these basis and can be recovered independently from $s(t)$. When A and B are time-varying, above analysis is still valid when A and B changes much slower than the sine and cosine function. Therefore, $s(t)$ given by (3.82) carries two independent components, the in-phase and quadrature. In fact, the lower pass filtering in Figure 3.20 is equivalent to the integration of time discussed here.

In the frequency domain, (3.102) in Section 3.8.1 stats that if a function is real-valued, then its spectrum at the negative frequency is the complex conjugate of that at the positive frequency. Namely,

$$\tilde{S}_b(-\omega) = \tilde{S}_b^*(\omega). \tag{3.84}$$

Therefore, for the real-valued baseband signal, only the portion with positive frequency carries information. The portion with negative frequency is completely determined by the other portion, and therefore does not carry additional information. For passband signal, however, such constraint no longer applies. Namely, $\tilde{S}_p(\omega_c + \omega)$ and $\tilde{S}_p(\omega_c - \omega)$ are independent and both carry information.[12] Therefore, an additional stream can be carried by the passband signal. To be more specific, let us return to (3.64) and look at only the first term (i.e., the portion around ω_c).

$$\tilde{S}_{p+}(\omega) = \frac{1}{2}\left[\tilde{S}_{bi}(\omega - \omega_c) + j\tilde{S}_{bq}(\omega - \omega_c)\right]. \tag{3.85}$$

Introduce

$$\nu \overset{\text{def}}{=} \omega - \omega_c. \tag{3.86}$$

Equation (3.85) becomes

$$\tilde{S}_{p+}(\omega_c + \nu) = \frac{1}{2}\left[\tilde{S}_{bi}(\nu) + j\tilde{S}_{bq}(\nu)\right]. \tag{3.87}$$

Changing the sign of ν and using (3.84):

$$\tilde{S}_{p+}(\omega_c - \nu) = \frac{1}{2}\left[\tilde{S}_{bi}(-\nu) + j\tilde{S}_{bq}(-\nu)\right] = \frac{1}{2}\left[\tilde{S}_{bi}^*(\nu) + j\tilde{S}_{bq}^*(\nu)\right], \tag{3.88}$$

or

$$\tilde{S}_{p+}^*(\omega_c - \nu) = \frac{1}{2}\left[\tilde{S}_{bi}(\nu) - j\tilde{S}_{bq}(\nu)\right]. \tag{3.89}$$

Therefore, based on $\tilde{S}_{p+}(\omega_c + \nu)$ and $\tilde{S}_{p+}^*(\omega_c - \nu)$, we can solve for $\tilde{S}_{bi}(\nu)$ and $\tilde{S}_{bq}(\nu)$. Namely, $\tilde{S}_{p+}$ carries information from both $\tilde{S}_{bi}(\nu)$ and $\tilde{S}_{bq}(\nu)$.

Of course, the second term in (3.64),

$$\tilde{S}_{p-}(\omega) = \frac{1}{2}\left[\tilde{S}_{bi}(\omega + \omega_c) - j\tilde{S}_{bq}(\omega + \omega_c)\right], \tag{3.90}$$

is the same as $\tilde{S}_{p+}^*(-\omega)$. It does not carry additional information.

From above discussions, we can understand that by moving from baseband to passband, both the bandwidth and the amount of data carried (measured in bits per symbol) are doubled, while the symbol rate remains the same.

3.6.4 Hilbert Transform

The last part of the analysis in Section 3.6.3 shows that S_{p+}, whose spectrum is defined in (3.87), carries all the information about S_{bi} and S_{bq}. If we want to build a down-converter to recover S_{bi} and S_{bq} based on (3.87), we would need to separate S_{p+} from

---

[12] For passband, the constraint.

S_p. This can be done through the Hilbert transform [1, sec. 2.1-1]. Hilbert transform is important because it has many interesting and useful mathematical properties, and has practical implementations [21].

Let $\hat{s}(t)$ be the Hilbert transform of waveform $s(t)$. The spectrum of $\hat{s}$ is

$$\tilde{\hat{s}}(\omega) \overset{\text{def}}{=} \begin{cases} -j\tilde{s}(\omega) & \omega > 0 \\ 0 & \omega = 0. \\ j\tilde{s}(\omega) & \omega < 0 \end{cases} \tag{3.91}$$

It is easy to verify that $\tilde{\hat{s}}(-\omega) = \tilde{\hat{s}}^*(\omega)$. Therefore, $\hat{s}(t)$ is real-valued. With Hilbert transform, we can construct S_{p+} easily:

$$S_{p+}(t) = \frac{1}{2}\left[S_p(t) + j\hat{S}_p(t)\right]. \tag{3.92}$$

$S_{p+}(t)$ is a complex function. It can be frequency-shifted into baseband:

$$S_{b+}(t) \overset{\text{def}}{=} S_{p+}(t)e^{-j\omega_c t}. \tag{3.93}$$

Equations (3.93) and (3.87) leads to

$$\tilde{S}_{b+}(\omega) = \tilde{S}_{p+}(\omega_c + \omega) = \frac{1}{2}\left[\tilde{S}_{bi}(\omega) + j\tilde{S}_{bq}(\omega)\right]. \tag{3.94}$$

In the time domain, such relationship translates to:

$$S_{b+}(t) = \frac{1}{2}\left[S_{bi}(t) + jS_{bq}(t)\right]. \tag{3.95}$$

Since both $S_{bi}(t)$ and $S_{bq}(t)$ are real-valued,

$$\begin{aligned} S_{bi}(t) &= S_{b+}(t) + S_{b+}^*(t) \\ S_{bq}(t) &= \frac{1}{j}\left[S_{b+}(t) - S_{b+}^*(t)\right] \end{aligned} \tag{3.96}$$

Combining (3.96) with (3.93) and (3.92) and remembering both $S_p(t)$ and $\hat{S}_p(t)$ are real-valued, we get

$$\begin{aligned} S_{bi}(t) &= \frac{1}{2}\left\{\left[S_p(t) + j\hat{S}_p(t)\right]e^{-j\omega_c t} + \left[S_p(t) - j\hat{S}_p(t)\right]e^{j\omega_c t}\right\} \\ &= S_p(t)\cos(\omega_c t) + \hat{S}_p(t)\sin(\omega_c t) \\ S_{bq}(t) &= \frac{1}{2j}\left\{\left[S_p(t) + j\hat{S}_p(t)\right]e^{-j\omega_c t} - \left[S_p(t) - j\hat{S}_p(t)\right]e^{j\omega_c t}\right\} \\ &= \hat{S}_p(t)\cos(\omega_c t) - S_p(t)\sin(\omega_c t) \end{aligned} \tag{3.97}$$

Figure 3.24 shows the block diagram of down-converter according to (3.97) [1, sec. 2.1-2]. From the left side, passband signal $S_p(t)$ is split into two branches.

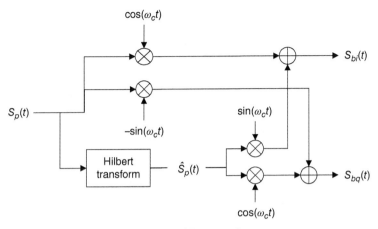

Figure 3.24   Down-converter with Hilbert transform.

The bottom branch undergoes Hilbert transform and forms $\hat{S}_p(t)$. S_p and $\hat{S}_p$ are then multiplied to sine and cosine functions of time, according to (3.97), yielding S_{bi} and S_{bq}. Such implementation does not require the LPFs shown in Figure 3.20. Outputs of such down-convertor are the baseband waveforms, without the factor 1/2 in (3.68).

3.6.5   Summary of Baseband and Passband

As shown in Figure 3.20, the baseband signal can be converted into the passband for transmission, and be converted back to the baseband at the receiver. A passband waveform with a bandwidth of W_P carries two baseband waveforms, each with a bandwidth of $W_B = 1/2 W_P$. See Figure 3.23 for bandwidth definitions. The two baseband waveforms are often expressed as the real and imaginary parts of a complex-valued waveform. Therefore, the up-conversion and down-conversion can also be expressed as complex operations as described in Section 3.6.2. The down conversion process can also be achieved with the Hilbert transform, as shown in Figure 3.24.

3.7   SUMMARY

This chapter discusses the process of converting bits to RF signals. Such process involves several steps, as shown in Figure 3.2.

First, the bits are converted to symbols through modulation (Section 3.2). A symbol has a set of M legitimate values (known as the alphabet, or the constellation set when expressed in the complex plane), M being an integer depending on the modulation scheme. Such modulation is said to be M-ary. All possible bits sequences of length $K = \lfloor log_2(M) \rfloor$ are mapped to these constellation points. Once the receiver determines which value of a symbol was transmitted, it recovers the bit sequence the symbol carries. The mapping of bit sequences to constellation points can be chosen to lower the bit error probability, as described in Section 3.2.5. The choice of M is a balance between noise immunity and data rate, as discussed in Section 3.2.6. A larger M means more bits can be carried by a symbol, thus increasing transmit data rate. On the other hand, a larger M means more possible values for a symbol, which means these values are closer to each other and are more prone to confusion in the presence of noise. We will return to this topic in Chapter 4 when we talk about receiver design and performance. This chapter covers two common modulations schemes, the PSK and the QAM, as well as their hybrid APSK (Sections 3.2.2–3.2.4). PSK is typically used for $M \leq 8$. All constellation points have the same amplitude. Therefore, such modulation can tolerate higher nonlinearity in the RF circuit. QAM can support much higher M (up to thousands). APSK offers a compromise between PSK and QAM.

Next, the symbols are converted into a time-continues signal (waveform) through pulse shaping (Sections 3.3 and 3.4). Pulse shaping is performed according to (3.51). The pulse shaping function must meet the following constraints:

1. The total power is limited. This constraint can be met by setting the overall gain.
2. The bandwidth is limited. This constraint can be met by choosing the pulse shaping function with a proper power density spectrum.
3. The constituting symbols can be uniquely resolved from the waveform. Namely, not two distinct symbol sequences should yield the same waveform. Consequently, the symbol rate cannot exceed twice the baseband bandwidth, as dictated by the Nyquist–Shannon sampling theorem, described and discussed in Section 3.3.

Furthermore, it is desirable that the waveform value at some point in time depend only on the value of one particular symbol. Such property makes it much easier to resolve symbols from received waveform. Namely, we want the intersymbol interference (ISI) to be zero. One necessary and sufficient condition for zero ISI is the Nyquist criterion, introduced in Section 3.3 and proved in Section 3.10. A commonly used pulse shaping function based on the Nyquist criterion is the raised cosine filter described in Section 3.4.3. Section 3.5 further discussed some implementation details: pulse shaping filtering with over-sampling in the digital domain.

The waveform then needs to be "moved" to a higher frequency for wireless transmission and media sharing. Such effect is achieved by the process of up-converting (with the corresponding process of down-converting at the receiver). Up-converting and down-converting are covered in Section 3.6. These conversions are performed by multiplying the baseband signal with sine and cosine functions. It should be noted that the term "bandwidth" is defined differently in baseband and passband, as discussed in Section 3.6.3.

This chapter covers only the basics of the subject of converting bits to waveforms. While essential concepts and knowledge are provided, there are a number of topics not covered, as briefly listed below.

1. There are more modulation schemes than PSK, QAM, and APSK, although these three types are used most frequently in modern digital communications

systems, especially when combining with multicarrier modulation (see below). Other modulation schemes include frequency shift keying (FSK), minimum shift keying (MSK), and so on.

2. This chapter covers only single carrier modulation, where the resulting waveform occupies the entire bandwidth allocated to the system. Chapter 10 introduces multicarrier modulation, which contains multiple subcarriers, each being modulated with the schemes discussed in this chapter.

3. Zero ISI is not the only criterion for pulse shaping function design. There are other considerations, outlined in Section 3.4.4.

4. Nyquist–Shannon theorem, which connects sampling rate and symbol rate to waveform bandwidth, is valid only when we know nothing about the waveform other than its bandwidth. In some cases, lower sampling rates or higher symbol rates can be used, as briefly described in Section 3.3.6.

5. Transceivers may have different architectures for up-converting and down-converting. For example, some receivers (known as superheterodyne structure) perform down-conversion in two steps, with an intermediate frequency (IF). Another example is the software-defined radio (SDR), who typically performs analog to digital conversion and digital to analog conversion at an intermediate frequency. Conversions between baseband and passband signals are thus performed in the digital domain.

3.8 APPENDIX: FOURIER TRANSFORM

It is assumed that the readers are well-versed in Fourier transform. This section does not provide a comprehensive tutorial. It only summarizes a few properties that are important to the rest of this chapter. The primary purpose is to clarify the notations and traditions related to Fourier transform operations.

3.8.1 Fourier Transform

Let $s(t)$ be a function of real variable t. Its Fourier transform (FT) is defined as

$$\tilde{s}(\omega) \stackrel{\text{def}}{=} FT[s(t)] \stackrel{\text{def}}{=} \int_{-\infty}^{\infty} s(t)e^{-j\omega t}\,dt. \tag{3.98}$$

j is the unit imaginary number $\sqrt{-1}$. ω is a real variable. In the context of digital communications, t represents time. Therefore, $s(t)$ is said to be a function in the time domain. ω represents the angular frequency. Its relationship with the conventional frequency f is

$$\omega \stackrel{\text{def}}{=} 2\pi f. \tag{3.99}$$

$\tilde{s}(\omega)$ is thus said to be a function in the frequency domain. $\tilde{s}(\omega)$ is also known as the spectrum of $s(t)$, while $s(t)$ is referred to as the waveform of $\tilde{s}(\omega)$.

The inverse Fourier transform (IFT) is

$$s(t) \overset{\text{def}}{=} IFT[\tilde{s}(\omega)] = \frac{1}{2\pi} \int_{-\infty}^{\infty} \tilde{s}(\omega)e^{j\omega t} d\omega. \tag{3.100}$$

FT and IFT are unique. Namely, $s(t) = g(t)$ if and only if $\tilde{s}(\omega) = \tilde{g}(\omega)$. Frequently used properties of the FT include the following.

$$FT[s(t+a)] = e^{j\omega a} FT[s(t)]. \tag{3.101}$$

Namely, an offset in the time domain is translated to a phase ramp (i.e., a phase shift proportional to frequency) in the frequency domain. Furthermore, if $s(t)$ is real-valued, then

$$\tilde{s}(-\omega) = \tilde{s}^*(\omega). \tag{3.102}$$

Namely, the spectrum of a real-valued function has a symmetry. In this case, the part of the spectrum with negative frequency is entirely determined by the part with positive frequency.

There is another form of symmetry across the time and frequency domain:

$$FT[s^*(-t)] = \tilde{s}^*(\omega). \tag{3.103}$$

Another commonly used FT convention is using frequency instead of angular frequency:

$$\hat{s}(f) \overset{\text{def}}{=} FT[s(t)] \overset{\text{def}}{=} \int s(t)e^{-2\pi jft} dt$$

$$s(t) \overset{\text{def}}{=} IFT[\hat{s}(f)] = \int \hat{s}(f)e^{2\pi jft} df \tag{3.104}$$

This definition leads to

$$\hat{s}(f) = \tilde{s}(2\pi f). \tag{3.105}$$

3.8.2   Dirac Delta Function

The FT and IFT are connected by the mathematical property [22, sec. 1.17]:

$$\int e^{j\omega(t-t')} d\omega = 2\pi\delta(t-t'). \tag{3.106}$$

Here, $\delta(t-t')$ is the Dirac delta function, or simply delta function. It has the following property. For any function $s(t)$,

$$\int s(t)\delta(t-t_0)dt = s(t_0). \tag{3.107}$$

Another property of the delta function follows:

$$\delta(ax) = \frac{1}{a}\delta(x).$$

(3.108)

Equation (3.106) has a discrete form [22, sec. 1.17]:

$$\frac{1}{2\pi}\sum_{n=-\infty}^{\infty} e^{jnx} = \delta(x), \forall x \in (-\pi, \pi].$$

(3.109)

For x values beyond this range, the left-hand side of (3.109) is a periodic function of x with a period of 2π. Therefore, (3.109) can be written as

$$\frac{1}{2\pi}\sum_{n=-\infty}^{\infty} e^{jnx} = \sum_{k=-\infty}^{\infty} \delta(x-2k\pi), \forall x \in (-\infty, \infty).$$

(3.110)

When the frequency, instead of the angular frequency, is used, (3.106) becomes

$$\int e^{2\pi f(t-t')}df = \delta(t-t').$$

(3.111)

3.8.3 Fourier Transform of Periodic Functions

If $s(t)$ has a period of T_p, that is, $s(t+T_p) = s(t)$, its FT becomes, from (3.98):

$$\tilde{s}(\omega) = \sum_{n=-\infty}^{\infty} \int_0^{T_p} s(t)e^{-j\omega(t-nT_p)}dt = \sum_{n=\infty}^{-\infty} e^{j\omega(nT_p)} \int_0^{T_p} s(t)e^{-j\omega t}dt.$$

(3.112)

The value of the summation is given by (3.110). Therefore, recalling (3.108),

$$\tilde{s}(\omega) = 2\pi \sum_{k=-\infty}^{\infty} \delta(\omega T_p - 2k\pi) \int_0^{T_p} s(t)e^{-j\omega t}dt$$
$$= \frac{2\pi}{T_p} \sum_{k=-\infty}^{\infty} \delta\left(\omega - \frac{2k\pi}{T_p}\right) \int_0^{T_p} s(t)e^{-j\omega t}dt$$

(3.113)

Equation (3.113) shows that when $s(t)$ is periodic, its spectrum has discrete tones separated by $2\pi/T_p$, where T_p is the period. This treatment is also useful when we deal with a waveform with limited length. We can extend the waveform definition by duplicating it beyond the original length, resulting in a periodic function.

Performing IFT of (3.113), we have from (3.100):

$$
s(t) = \frac{1}{2\pi} \int_{-\infty}^{\infty} \tilde{s}(\omega) e^{j\omega t} d\omega = \sum_{k=-\infty}^{\infty} s_k e^{j\frac{2k\pi t}{T_p}}
$$

$$
s_k \stackrel{\text{def}}{=} \frac{1}{T_p} \int_0^{T_p} s(t) e^{-j\frac{2k\pi t}{T_p}} dt
$$

(3.114)

Namely, a periodic function can be expressed as a series.

3.8.4 Convolution

Convolution of two time domain functions $s(t)$ and $g(t)$ is a time domain function $h(t)$:

$$
h(t) \stackrel{\text{def}}{=} \int s(t-\tau)g(\tau)d\tau.
$$

(3.115)

Convolution is also noted as

$$
h = s \otimes g = g \otimes s.
$$

(3.116)

To look at convolution in the frequency domain, perform Fourier transform of $h(t)$:

$$
\begin{aligned}
\tilde{h}(\omega) &= \int h(t) e^{-j\omega t} dt \\
&= \int s(t-\tau)g(\tau) e^{-j\omega t} d\tau dt \\
&= \int s(t-\tau) e^{-j\omega(t-\tau)} g(\tau) e^{-j\omega \tau} d\tau dt \\
&= \tilde{s}(\omega)\tilde{g}(\omega)
\end{aligned}
$$

(3.117)

Namely, a convolution in the time domain corresponds to multiplication in the frequency domain. This property is known as the convolution theorem.

3.8.5 Parseval's Theorem

In (3.115), let $g(t) = s^*(-t)$. We then get

$$
h(0) = \int s(-\tau)g(\tau)d\tau = \int |s(-\tau)|^2 d\tau = \int |s(\tau)|^2 d\tau.
$$

(3.118)

$h(0)$ in this case is the total "energy" of waveform $s(t)$. From (3.117) and (3.100),

$$h(0) = \frac{1}{2\pi} \int \tilde{h}(\omega)d\omega = \frac{1}{2\pi} \int \tilde{s}(\omega)\tilde{g}(\omega)d\omega = \frac{1}{2\pi} \int |s(\omega)|^2 d\omega. \qquad (3.119)$$

The last step used the FT property in (3.103). Therefore, we can connect the total energy in both domains:

$$\int |s(\tau)|^2 d\tau = \frac{1}{2\pi} \int |\tilde{s}(\omega)|^2 d\omega. \qquad (3.120)$$

This result is known as the Parseval's theorem or the Parseval's formula.

3.9 APPENDIX: FUNCTION LOCALIZATION IN FREQUENCY AND TIME DOMAINS

3.9.1 Motivation

When we send a data-carrying signal as a function of time $s(t)$, we naturally wish it to be "localized" in frequency and time. Namely, we wish the signal to occupy a small "phase space," which is spanned by time and frequency. Such localization allows the signal to coexist with other signals without mutual interference, thus allowing more data to be transmitted by the system.

Ideally, we would like to have a signal to be limited in both time and frequency domains with a duration of T and a bandwidth B:

$$\begin{aligned} s(t) &= 0 \quad \forall t \notin [-T/2, T/2] \\ \tilde{s}(\omega) &= 0 \quad \forall \omega \notin [-2\pi B, 2\pi B] \end{aligned}. \qquad (3.121)$$

Unfortunately, it is well-known that a signal cannot have finite supports in both frequency and time domains. As discussed in Section 3.3.2, if the signal $c(t)$ is strictly band-limited in the frequency domain between B and $-B$, its spectrum can be written as in (3.29), that is, as a spectrum multiplied by a "brick wall" filter $\tilde{h}(\omega)$. According to Section 3.8.4, a multiplication in the frequency domain is equivalent to a convolution in the time domain. Therefore, $c(t)$ is some function convolved with $h(t)$, given by (3.34). Since $h(t)$ is a sinc function, which extends to infinite times (delaying on the order of $|t|^{-1}$), $c(t)$ also extends to infinite times. Therefore, it is impossible to have a nonzero function satisfying (3.121).

We can relax the requirement and hope a function is "concentrated" in a limited range. A commonly used "concentrated" function is the Gaussian function, which extends to positive and negative infinities yet has a predominant peak in the middle. The variance σ^2 is proportional to the "width" of the peak, and can be considered as a measure of localization. The Fourier transform of a Gaussian function with variance σ^2 is also a Gaussian function, whose variance is σ^{-2} [22, sec. 1.14 (vii)]. Therefore, if a function is more "localized" in the time domain (i.e., the variance is smaller), it is less localized in the frequency domain.

The tradeoff between localizations in the two domains can also be understood with scaling properties. Consider another function $d(t)$:

$$d(t) \overset{\text{def}}{=} c(at). \tag{3.122}$$

It is easy to see its FT is

$$\tilde{d}(\omega) = \tilde{c}\left(\frac{\omega}{a}\right). \tag{3.123}$$

Therefore, dilation in the time domain results in compression in the frequency domain, and vice versa.

Despite the tradeoff shown above, it is still possible that one function is more localized than another in both time and frequency domains. In Sections 3.9.2 and 3.9.3, we try to find the "maximum localization" from various angles.

3.9.2   The Weyl–Heisenberg Inequality

The Heisenberg uncertainty principle is known as one of the most fundamental laws in quantum mechanics. It is also a general constraint concerning a function and its FT. In this context, it is known as the Weyl–Heisenberg inequality [23], [24, sec. 4].[13] This inequality is the foundation of the Heisenberg Uncertainty Principle in quantum physics.

Consider a function $s(t)$ and its Fourier transform $\hat{s}(\omega)$. Without losing generality and considering the Parseval's theorem (3.120), we can assume the following properties of s:

$$\int |s(t)|^2 dt = \frac{1}{2\pi} \int |\tilde{s}(\omega)|^2 d\omega = 1$$

$$\bar{t} \overset{\text{def}}{=} \int t|s(t)|^2 dt \qquad . \tag{3.124}$$

$$\bar{\omega} \overset{\text{def}}{=} \int \omega|\tilde{s}(\omega)|^2 d\omega$$

Both $\bar{t}$ and $\bar{\omega}$ are assumed to be finite numbers. The Weyl–Heisenberg inequality says that for any function $s(t)$ and its FT $\tilde{s}(\omega)$ given by (3.98) in Section 3.8.1:

$$\int (t-\bar{t})^2 |s(t)|^2 dt \left[\frac{1}{2\pi} \int (\omega-\bar{\omega})^2 |\tilde{s}(\omega)|^2 d\omega \right] \geq \frac{1}{4}. \tag{3.125}$$

Namely, the product of the variances in the time domain and frequency domains has a minimum value, when we consider $|s(t)|^2 / \int |s(t)|^2 dt$ and $|\tilde{s}(\omega)|^2 / \int |\tilde{s}(\omega)|^2 d\omega$ as the probability distribution functions in the two domains.

---

[13] The Weyl–Heisenberg inequality was expressed in terms of linear frequency f in the cited references. Here we use angular frequency to be consistent with the rest of the book.

As an example, let us revisit the Gaussian function mentioned in Section 3.9.1. Let $g(t)$ be a Gaussian function with variance σ^2. $\tilde{g}(\omega)$ is also a Gaussian function with variance σ^{-2}. Furthermore, $|g(t)|^2$ and $|\tilde{g}(\omega)|^2$ are Gaussian functions with variances $\sigma^2/2$ and $\sigma^{-2}/2$, respectively. Therefore, the equality holds in (3.125). In other words, A Gaussian function achieves "maximum localization" in the sense of variances.

We provide a proof of the inequality (3.125) in the following. First, let us point out that $\bar{t}$ and $\bar{\omega}$ are not important. Consider a function

$$r(t) \overset{\text{def}}{=} e^{-j\bar{\omega}t} s(t+\bar{t}). \tag{3.126}$$

Its FT is

$$\begin{aligned}\tilde{r}(\omega) &= \int r(t)e^{-j\omega t}dt = e^{j(\omega+\bar{\omega})\bar{t}}\int s(t+\bar{t})e^{-j(\omega+\bar{\omega})(t+\bar{t})}dt \\ &= e^{j(\omega+\bar{\omega})\bar{t}}\tilde{s}(\omega+\bar{\omega})\end{aligned}. \tag{3.127}$$

It can be verified that

$$\begin{aligned}\int(t-\bar{t})^2|s(t)|^2dt &= \int(t-\bar{t})^2|r(t-\bar{t})|^2dt = \int t'|r(t')|^2dt' \\ \int(\omega-\bar{\omega})|\tilde{s}(\omega)|^2d\omega &= \int(\omega-\bar{\omega})^2|\tilde{r}(\omega-\bar{\omega})|^2d\omega = \int\omega'|\tilde{r}(\omega')|^2d\omega'\end{aligned}. \tag{3.128}$$

Therefore, the Weyl–Heisenberg inequality (3.125) is proven if

$$\int t^2|r(t)|^2dt\frac{1}{2\pi}\int\omega^2|\tilde{r}(\omega)|^2d\omega \geq \frac{1}{4} \tag{3.129}$$

is proven, where $r(t)$ is any function with the following property:

$$\int|r(t)|^2dt = \frac{1}{2\pi}\int|\tilde{r}(\omega)|^2d\omega = 1. \tag{3.130}$$

Namely, we can disregard the nonzero average values $\bar{x}$ and $\bar{\omega}$. Inequality (3.129) can be proved by the Cauchy–Schwarz inequality in its integration form [22, sec. 1.7(ii)]. For any functions $u(x)$ and $v(x)$, the Cauchy–Schwarz inequality states

$$\int|u(t)|^2dt\int|v(t)|^2dt \geq \left|\int u(t)v^*(t)dt\right|^2. \tag{3.131}$$

The equality holds when $u(t) = av(t)$ for some constant a. Now choose our functions as follows:

$$\begin{aligned}u(t) &\overset{\text{def}}{=} t\,r(t) \\ v(t) &\overset{\text{def}}{=} \frac{d}{dt}r(t) = \frac{1}{2\pi}j\int\omega\tilde{r}(\omega)e^{j\omega t}d\omega\end{aligned}. \tag{3.132}$$

The last step in the second line is based on the inversed Fourier transform from $\tilde{r}(\omega)$ to $r(t)$, given by (3.100) in Section 3.8.1. Now let us compute the various factors in (3.131).

$$\int |u(t)|^2 dt = \int t^2 |r(t)|^2 dt, \tag{3.133}$$

$$\int |v(t)|^2 dt = \left(\frac{1}{2\pi}\right)^2 \int \omega_1 \omega_2 \tilde{r}(\omega_1) \tilde{r}^*(\omega_2) e^{j(\omega_1-\omega_2)t} d\omega_1 d\omega_2 dt$$
$$= \frac{1}{2\pi} \int \omega^2 |\tilde{r}(\omega)|^2 d\omega \tag{3.134}$$

The last step was obtained by conducting integration over t, using the property of the delta function (3.106) in Section 3.8.2. Therefore, the left side of (3.131) is the same of that of (3.129) except for a constant factor. Regarding the right side of (3.131), we recognize that

$$\left| \int u(t) v^*(t) dt \right|^2 \geq \left\{ Re\left[\int u(t) v^*(t) dt \right] \right\}^2. \tag{3.135}$$

The real part can be evaluated as follows:

$$Re\left[\int u(t) v^*(t) dt \right] = \frac{1}{2} \left[\int u(t) v^*(t) dt + \int u^*(t) v(t) dt \right]$$
$$= \frac{1}{2} \int t \left[r(t) \frac{d}{dt} r^*(t) + r^*(t) \frac{d}{dt} r(t) \right] dt = \frac{1}{2} \int t \frac{d}{dt} |r(t)|^2 dt. \tag{3.136}$$
$$= -\frac{1}{2} \int |r(t)|^2 dt = -\frac{1}{2}$$

The second to the last step was obtained through integration by parts, and considering that $t|r(t)|^2$ vanishes at the plus and minus infinity. The last step above considered the normalization of $r(t)$ given by (3.130).

Combining (3.133)–(3.136) and using (3.131), we arrive at

$$\int t^2 |r(t)|^2 dt \frac{1}{2\pi} \int \omega^2 |\tilde{r}(\omega)|^2 d\omega \geq \frac{1}{4}. \tag{3.137}$$

This result proves (3.129) and thus the Weyl–Heisenberg inequality (3.125).

3.9.3   Imperfect Bandwidth and Time Limitation

Section 3.9.2 uses variances as a measure of localization. In this section, we deal with imperfect bandwidth and time limitation with a "leakage" measurement [25, 26].

Instead of using the variances, here we use "leakage" to characterize localization. Let B and T be the limits in frequency domain and time domain. Namely, we wish to have the function $s(t)$ and its Fourier transform $\tilde{s}(\omega)$ satisfying (3.121). Of course, this is impossible as discussed in Section 3.9.1. However, we can measure the degree of "imperfectness" from such limits. Let

$$\alpha^2(T) \stackrel{\text{def}}{=} \frac{\displaystyle\int_{-T/2}^{T/2} |s(t)|^2 dt}{\displaystyle\int_{-\infty}^{\infty} |s(t)|^2 dt}$$

$$\beta^2(B) \stackrel{\text{def}}{=} \frac{\displaystyle\int_{-2\pi B}^{2\pi B} |\tilde{s}(\omega)|^2 d\omega}{\displaystyle\int_{-\infty}^{\infty} |\tilde{s}(\omega)|^2 d\omega} \tag{3.138}$$

The deviations of α^2 and β^2 from 1 indicate the "leakage" from time and frequency limits, respectively.

Therefore, it is interesting to seek the function $s(t)$ that maximizes α^2 and β^2 given T and B. Because of the scaling relationship between the time and frequency domains, the only parameter that matters is the product BT. For a given BT, the maximum values of α^2 and β^2 form a pair of tradeoff. The detailed results are given in [26, fig. 3]. For example, with $BT = 1$, when $\alpha^2(T) = 1$, the maximum value for $\beta^2(B)$ is approximately 0.35. With $BT = 4$ and the same $\alpha^2(T)$ value, the maximum value for $\beta^2(B)$ is is approximately 0.9.

Another interesting result from this approach is the degrees of freedom in signal space. The Nyquist–Shannon sampling theory discussed in Section 3.3 states that the maximum symbol rate is $2B$, where B is the bandwidth. In other words, during time T, we can send $2BT$ independent symbols that can be recovered at the receiver. On the other hand, as discussed in Section 3.9.1, the signal of each symbol extends to infinite times, due to the strict bandwidth limitation.

What if the bandwidth limitation is relaxed to allow for some leakage? Consider the signal space F limited by the bandwidth B and time window T, with the level of leakage of ϵ. Namely, for any function $s(t) \in F$, we have

$$\int_{|t| > T/2} |s(t)|^2 dt < \epsilon$$

$$\int_{|\omega| > 2\pi B} |\tilde{s}(\omega)|^2 d\omega < \epsilon \tag{3.139}$$

We say the dimension of F is N at level ϵ', if there exists N functions $\{\phi_k(t) \in F,$ $k = 1, 2, \ldots, N\}$, such that for any $s(t) \in F$ there is a set of coefficients $\{c_i\}$ for the approximation

$$\int_{-T/2}^{T/2} \left| s(t) - \sum_{k=1}^{N} c_k \phi_k(t) \right|^2 dt < \epsilon'. \tag{3.140}$$

Furthermore, such approximation cannot be made for all functions $s(t)$ if the number of the members in $\{\phi_k(t)\}$ is limited to $N - 1$. In other words,

$\{\phi_k, k = 1, 2, \ldots, N\}$ forms a set of linearly independent basis for the function space F in the approximation (3.140). This dimension N is also the number of independent symbols that can be transmitted (and recovered at the receiver) with the bandwidth and time window limits with an error proportional to ϵ'.

It turns out that the dimension is approximately $2BT$. More precisely, if $N(B, T, \epsilon, \epsilon')$ is the dimension of the signal space as discussed above and $\epsilon' > \epsilon$, then [25, 26]

$$\begin{aligned} lim_{T \to \infty} N(W, T, \epsilon, \epsilon')/T &= 2B \\ lim_{B \to \infty} N(W, T, \epsilon, \epsilon')/B &= 2T \end{aligned} \tag{3.141}$$

Namely, one can transmit $2BT$ independent symbols when band-limited and time-limited with arbitrarily small leakage. The catch is that either the bandwidth or the time window needs to approach infinity.

3.10   APPENDIX: PROOF OF THE NYQUIST CRITERION

The Nyquist criterion, introduced in Section 3.4.2, states that the necessary and sufficient condition for a function $h(t)$ to have the zero ISI property (3.33) is that its FT $\tilde{h}(\omega)$ satisfies

$$\sum_{m=-\infty}^{\infty} \tilde{h}\left(\omega + \frac{2m\pi}{T_{Sym}}\right) = T_{Sym} \forall \omega. \tag{3.142}$$

Zero ISI means that $h(t)$ is zero at all symbol times except the origin:

$$h(nT_{Sym}) = \delta_{n,0}. \tag{3.143}$$

The proof of the Nyquist criterion is as follows. For sufficient condition, we need to show that (3.142) leads to (3.143).

Starting with the IFT (3.100) in Section 3.8.1, we break the integration over frequency into intervals of $2\pi/T_{Sym}$:

$$h(t) = \frac{1}{2\pi} \int_{-\infty}^{\infty} \tilde{h}(\omega)e^{j\omega t} d\omega = \frac{1}{2\pi} \sum_{m=-\infty}^{\infty} \int_{(2m-1)\pi/T_{Sym}}^{(2m+1)\pi/T_{Sym}} \tilde{h}(\omega)e^{j\omega t} d\omega. \tag{3.144}$$

With substitute of variables $\omega \to u + (2m\pi)/T_{sym}$ for each period, the above equation becomes

$$h(t) = \frac{1}{2\pi} \sum_{m=-\infty}^{\infty} \int_{-\pi/T_{Sym}}^{\pi/T_{Sym}} \tilde{h}\left(u + \frac{2m\pi}{T_{Sym}}\right) e^{j\left(u + \frac{2m\pi}{T_{Sym}}\right)t} du. \tag{3.145}$$

For any integer n,

$$h(nT_{Sym}) = \frac{1}{2\pi} \sum_{m=-\infty}^{\infty} \int_{-\pi/T_{Sym}}^{\pi/T_{Sym}} \tilde{h}\left(u + \frac{2m\pi}{T_{Sym}}\right) e^{junT_{Sym}} e^{j2nm\pi} du. \tag{3.146}$$

The last factor $e^{j2nm\pi}$ is always 1 and can be ignored. Exchanging the order of integration and summation, we get:

$$h(nT_{Sym}) = \frac{1}{2\pi}\int_{-\pi/T_{Sym}}^{\pi/T_{Sym}} \left[\sum_{m=-\infty}^{\infty} \tilde{h}\left(u + \frac{2m\pi}{T_{Sym}}\right)\right] e^{junT_{Sym}} du. \tag{3.147}$$

The Nyquist criterion (3.142) thus leads to

$$h(nT_{Sym}) = \frac{1}{2\pi}\int_{-\pi/T_{Sym}}^{\pi/T_{Sym}} T_{T_{Sym}} e^{junT_{Sym}} du. \tag{3.148}$$

Similar to (3.31) and (3.33), we can work out the integration and get, based on Equations (3.32) and (3.33),

$$h(nT_{Sym}) = \frac{1}{2\pi}\frac{2\pi}{T_{Sym}} T_{T_{Sym}} \operatorname{sinc}(n) = \delta_{n,0}. \tag{3.149}$$

Therefore, the Nyquist criterion is sufficient for zero ISI. Now, we continue to prove the necessity. Namely, we will show that (3.143) leads to (3.142). Starting with (3.147):

$$h(nT_{Sym}) = \frac{1}{2\pi}\int_{-\pi/T_{Sym}}^{\pi/T_{Sym}} \tilde{b}(u) e^{junT_{Sym}} du$$

$$\tilde{b}(u) \overset{\text{def}}{=} \sum_{m=-\infty}^{\infty} \tilde{h}\left(u + \frac{2m\pi}{T_{Sym}}\right) \tag{3.150}$$

We wish to show that $\tilde{b}(u)$ is a constant.

Obviously, $\tilde{b}(u)$ has a period of $2\pi/T_{Sym}$, because of the summation. Therefore, according to (3.114) in Section 3.8.3,[14]

$$\tilde{b}(u) = \sum_{k=-\infty}^{\infty} b_k e^{-jkT_{Sym}u}. \tag{3.151}$$

From (3.143), (3.150), and (3.151), we have

$$h(nT_{Sym}) = \frac{1}{2\pi}\sum_{k=-\infty}^{\infty} b_k \int_{-\pi/T_{Sym}}^{\pi/T_{Sym}} e^{ju(n-k)T_{Sym}} du = \delta_{n,0}. \tag{3.152}$$

The integral in (3.152) is

$$\int_{-\pi/T_{Sym}}^{\pi/T_{Sym}} e^{ju(n-k)T_{Sym}} du = \frac{2\pi}{T_{Sym}}\delta_{n-k,0}. \tag{3.153}$$

Therefore,

$$b_n = T_{Sym}\delta_{n,0}. \tag{3.154}$$

---

[14] Note that the sign in the exponent is opposite to that in the first equation of (3.114). This sign change is equivalent to replacing k with $-k$ in the derivation.

Combining (3.150), (3.154), and (3.151),

$$\sum_{m=-\infty}^{\infty} \tilde{h}\left(u + \frac{2m\pi}{T_{Sym}}\right) = \tilde{b}(u) = T_{Sym}. \tag{3.155}$$

This result is the same as (3.142). Thus, we proved the necessity. QED.

REFERENCES

1. J. Proakis and M. Salehi, *Digital Communications*, 5th ed. McGraw-Hill Education, 2007.
2. S. Haykin, *Digital Communication Systems*. Wiley, 2014.
3. IEEE, *802.11-2016—IEEE Standard for Information Technology—Telecommunications and Information Exchange between Systems Local and Metropolitan Area Networks—Specific Requirements—Part 11: Wireless LAN Medium Access Control (MAC) and Physical Layer (PHY)*. IEEE, 2016.
4. W. Zhou and L. Zou, "An Adaptive QAM Transmission Scheme for Improving Performance on an AWGN Channel," U.S. Patent No. 9,036,694 B2, 2015.
5. ITU-T, *A Modem Operating at Data Signalling Rates of up to 33 600 bit/s for Use on the General Switched Telephone Network and on Leased Point-to-Point 2-Wire Telephone-Type Circuits*. ITU-T Recommendation V.34, 1998.
6. ETSI—European Telecommunications Standards Institute and ETSI, *Digital Video Broadcasting (DVB); Second Generation Framing Structure, Channel Coding and Modulation Systems for Broadcasting, Interactive Services, News Gathering and Other Broadband Satellite Applications; Part 1: DVB-S2*. V1.4.1. ETSI EN 302 307-1, 2014.
7. F. Gray, "Pulse Code Communication," U.S. Patent No. 2,632,058 A, 1953.
8. H. Nyquist, "Certain Topics in Telegraph Transmission Theory," *Trans. Am. Inst. Electr. Eng.*, vol. 47, no. 2, pp. 617–644, 1928.
9. C. E. Shannon, "Communication in the Presence of Noise," *Proc. IRE*, vol. 37, no. 1, pp. 10–21, 1949.
10. E. W. Hansen, *Fourier Transforms: Principles and Applications*. Wiley, 2014.
11. H. J. Landau, "Sampling, Data Transmission, and the Nyquist Rate," *Proc. IEEE*, vol. 55, no. 10, pp. 1701–1706, 1967.
12. M. Mishali and Y. C. Eldar, "Sub-Nyquist sampling," *IEEE Signal Process. Mag.*, vol. 28, no. 6, pp. 98–124, 2011.
13. C.-P. Yen, Y. Tsai, and X. Wang, "Wideband Spectrum Sensing Based on Sub-Nyquist Sampling," *Signal Process. IEEE Trans.*, vol. 61, no. 12, pp. 3028–3040, 2013.
14. R. Baraniuk, "Compressive Sensing [Lecture Notes]," *IEEE Signal Process. Mag.*, vol. 24, no. 4, pp. 118–121, 2007.
15. F. Rusek and J. B. Anderson, "Constrained Capacities for Faster-Than-Nyquist Signaling," *IEEE Trans. Inf. Theory*, vol. 55, no. 2, pp. 764–775, 2009.
16. J. B. Anderson, F. Rusek, and V. Owall, "Faster-Than-Nyquist Signaling," *Proc. IEEE*, vol. 101, no. 8, pp. 1817–1830, 2013.
17. Y. J. D. Kim, J. Bajcsy, and D. Vargas, "Faster-Than-Nyquist Broadcasting in Gaussian Channels: Achievable Rate Regions and Coding," *IEEE Trans. Commun.*, vol. 64, no. 3, pp. 1016–1030, 2016.
18. F.-L. Luo and C. Zhang, *Signal Processing for 5G: Algorithms And Implementations*. Wiley, 2016.
19. E. Stade, *Fourier Analysis*. Hoboken, NJ: Wiley, 2005.
20. S. Traverso, "A Family of Square-Root Nyquist Filter with Low Group Delay and High Stopband Attenuation," *IEEE Commun. Lett.*, vol. 20, no. 6, pp. 1136–1139, 2016.
21. S. L. Hahn, *Hilbert Transforms in Signal Processing*. Boston, MA: Artech House, 1996.
22. DLMF, "NIST Digital Library of Mathematical Functions" [Online]. Available at http://dlmf.nist.gov/ [Accessed December 13, 2016].

23. I. I. Hirschman Jr., "A Note on Entropy," *Am. J. Math.*, vol. 79, no. 1, p. 152, 1957.
24. W. Beckner, "Inequalities in Fourier Analysis," *Ann. Math.*, vol. 102, no. 1, p. 159, 1975.
25. D. Slepian, "On Bandwidth," *Proc. IEEE*, vol. 64, no. 3, pp. 292–300, 1976.
26. D. Slepian, "Some Comments on Fourier Analysis, Uncertainty and Modeling," *SIAM Rev.*, vol. 25, no. 3, pp. 379–393, 1983.

HOMEWORK

3.1 Study MATLAB functions fft and ifft, which perform FT and IFT with discrete-time samples.

   a. Based on MATLAB documentation (i.e., the mathematical relationships between input and output parameters). Consider only the most basic form of the function calls (i.e., ignore all optional parameters). Verify the equations with some simple numerical tests.

   b. How do we use fft to perform FT as defined in (3.98)?

   i. Given a sequence $\{s_k, k = 1, 2, \dots, K\}$ representing samples of a time-continuous function $s(t)$: $s_k = s(kT_s)$, where T_s is a constant sampling time. How is the sequence generated by fft(s) related to the spectrum $\tilde{s}(\omega)$?

   ii. Consider a time domain waveform $\bar{s}(t)$: it has a period of $K\Delta t$. When $t \in [0, KT_s)$, $\bar{s}(t)$ is constructed from $\{s_k\}$ in the form of (3.21). What is the FT of $\bar{s}$? How is it related to the output of function fft?

   c. Use functions fft and ifft to verify properties of FT presented in this chapter, that is, Equations (3.101), (3.102), (3.103), (3.113), (3.117), and (3.120).

3.2 Supposed you are asked to design a communication system with bit rate of 20 Mbps without channel coding. You should use QPSK based on channel conditions. In addition, the hardware designer tells you that the passband bandwidth should be no more than 1/100 of the center frequency. What is the lowest center frequency you can use for the system?

3.3 Verify the descriptions of Hilbert transform in Section 3.6.4. Consider a complex function $s_l(t)$, whose bandwidth is much smaller than ω_c. Given $s(t) = Re(s_l(t)e^{j\omega_c t})$, show that the

$$\tilde{s}_+(\omega) = \begin{cases} \tilde{s}(\omega) & \omega > 0 \\ \dfrac{1}{2}\tilde{s}(0) & \omega = 0 \\ 0 & \omega < 0 \end{cases} \text{ is } \frac{1}{2}\tilde{s}_l(\omega) \text{ shifted up by } \omega_c. \text{ (Hint: you can separate } s_l \text{ into the real}$$

and imaginary parts for consideration, or express $s(t)$ as $\frac{1}{2}\left[s_l(t)e^{j\omega_c t} + s_l^*(t)e^{-j\omega_c t}\right]$).

3.4 Use MATLAB simulation (could be Monte Carlo method or other methods) to study a signal, in the form of $s_b(t) = \sum a_n h(t - nT)$, where $\{a_n\}$ is a sequence of random complex numbers whose real and imaginary parts obey uniform distribution between -1 and 1. $h(t) = \text{sinc}(t/T)$, $\text{sinc}(x) = \sin(\pi x)/\pi x$. Study the following. Submit resulting plot as well as the MATLAB code generating the plots.

   a. Ideally, $h(t)$ extends from $t = -\infty$ to $t = \infty$. However, in practice, we have to truncate it as an approximation. Generate a plot showing the relationship between excess bandwidth (defined as the percentage of power outside of the theoretical band) and filter length. Suggestion: plot the y-axis in log scale. Note: you can use MATLAB function fft to generate a spectrum. You can also explore the use of MATLAB function pwelch.

   b. Plot the mean squared error (MSE) of $s_b(nT - \delta t)$ relative to $\delta t/T$. Suggestion: plot the MSE in log scale. Do that for three different truncation lengths.

  c. Repeat the last problem with raised cosine filters, with $\beta = 0.1, 0.2$, and 0.5. You can use the built-in MATLAB function rcosine. However, convince yourself that you get the filter you expected.

3.5 Verify analytically that the raised cosine filter satisfies Nyquist criterion.

3.6 Consider a baseband BPSK modulation: $s(t) = a_n$, $nT - (1/2)T < t \leq nT + (1/2)T$, where $a_n = \pm 1$. Note that instead of a modulation constructed by impulse functions, the signal here is constructed by square waves.

  a. What spectrum would you expect from such signal?

  b. Make a simple simulation to evaluate the spectrum. Compare with your expectation.

  c. Reformulate the modulation scheme in terms of pulse shaping (3.51). What is the pulse shaping function in this case? Does that predict the same spectrum?

  d. Check that this pulse shaping function satisfies the Nyquist criterion, using either analytical or numerical methods.

3.7 Use MATLAB to code two baseband modulator/demodulator pair for 16-QAM and 8-PSK. Several built-in functions can serve the purpose. For example, you may use functions qammod, qamdemod, pskmod, pskdemod.

  a. Verify that the bitstream can be recovered after the modulation and demodulation.

  b. Generate constellation plots for both cases.

  c. Do these functions use the optimal bit mapping as discussed in Section 3.2.5?

  d. Code modulators from scratch, and compare your results with those generated by MATLAB built-in functions.

3.8 In the time domain, verify analytically that baseband signal $s_b(t) = s_i(t) + js_q(t)$ has twice the power comparing to a passband signal $s_P = s_i(t) \cos \omega_c t - s_q(t) \sin \omega_c t$. Use the formula: $P = \lim_{t \to \infty} (1/T) \int_{-T/2}^{T/2} E\left(|s(t)|^2 \right) dt$. Note the following assumptions:

  a. $s_i(t)$ and $s_q(t)$ are independent. Therefore, $E\left(s_i(t)s_q^*(t) \right) = 0$.

  b. $s_i(t)$ changes much slower (i.e., with much lower frequency) than $\cos \omega_c t$. Therefore, $\int_{T/2}^{T/2} E\left(|s_i(t)|^2 \right) \cos (2\omega_c t) dt = 0$. There are similar results for other quantities.

3.9 The signal space distance defined in Chapter 4 is in the form of

$$d \stackrel{\text{def}}{=} \int_{-\infty}^{\infty} c_m(t)c_{m'}(t)dt$$

  where $c_m(t)$ and $c_{m'}(t)$ are waveforms constructed in the form of (3.51) with symbols d_m and $d_{m'}$. Compare this definition with what was used in this chapter, (3.9).

3.10 Suppose a pulse shaping function $g(t)$ satisfies the Nyquist criterion. Show that for any "well-behaved" function $h(t)$, $h(t)g(t)$ also satisfies the Nyquist criterion. Provide the proofs in both time domain and frequency domain.

CHAPTER 4

STATISTICAL DETECTION
AND ERROR PROBABILITY

4.1 INTRODUCTION

Figure 4.1 shows the high-level physical layer architecture, discussed in Chapter 1. This chapter addresses the Demodulation block at the receiver. As discussed in Chapter 3, in a digital transmitter, the bitstream is converted into a waveform, which is transmitted through the channel. At the receiver, the function of the Demodulation block is recovering the bitstream from the received waveform.

The key issue in demodulation is noise. Because of the noise, the received waveform is not the same as the transmitted one. A demodulator needs to determine what was transmitted based on received waveform, which is corrupted by noise. This chapter discusses the best way to achieve this goal.

This chapter can be divided into four parts. The first part (Sections 4.2 and 4.3) provides preliminary knowledge on the treatment of noise in general and that of the additive white Gaussian noise (AWGN) in particular. The focus is the relationship between the time domain behavior (autocorrelation) and the frequency domain behavior (power spectral density, PSD). The second part (Section 4.4) presents optimal detection algorithms (the maximum a posteriori [MAP] and maximum likelihood [ML] detectors) in a general sense. The third part (Sections 4.5 and 4.6) applies such algorithms to the AWGN channel. In particular, we discuss the principle and implementation of the matched filter (MF) in more details. MF is used not only for AWGN channels but also many other applications. The fourth part (Section 4.7) discusses the performance of receivers regarding error probabilities, focusing on the phase shift keying (PSK) and the quadrature amplitude modulation (QAM) that were described in Chapter 3. After that, Section 4.8 provides a summary. Finally, more mathematical details are provided in Sections 4.9–4.11 as appendices.

This chapter addresses two key concepts: optimal detection methods and error probability. These concepts form the foundation of receiver design. In Chapter 5, these concepts will be applied to code design and analysis. The optimal demodulator

Digital Communication for Practicing Engineers, First Edition. Feng Ouyang.
© 2020 by The Institute of Electrical and Electronics Engineers, Inc.
Published 2020 by John Wiley & Sons, Inc.

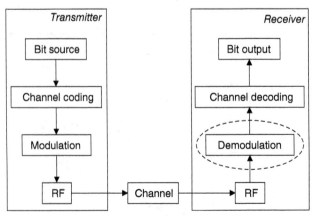

Figure 4.1   Physical layer architecture.

(matched filter) discussed in this chapter is also practical for implementation. It can be further integrated into the equalizer design (Chapter 9).

The mathematical analyses in the chapter center on the stochastic process. For the purpose of our discussion, stochastic processes are viewed as random variables, which are functions of continuous or discrete time. Many statements and derivations are not mathematically rigorous, especially when dealing with the white noise. The reader should consult a textbook specialized on stochastic processes if more precise and systematic mathematical background is desired. On the other hand, the treatment in this chapter is adequate for the analyses of communication systems.

For a communication system, signal and noise are functions of time. However, we are often interested in not the instantaneous values of signal and noise, but their integrals of some form. For this purpose, it is more convenient to express signal and noise in vector forms so that the integrals can be expressed as inner products. This concept is introduced in Section 4.5.1, while more details are provided in an appendix, Section 4.11. The "functions of time" representation (also referred to as waveform) can be considered as a special case of the vector presentation. Since the vector presentation of signal and noise is used in several parts of this chapter, it is recommended that readers who are not already familiar with the subject read Section 4.11 before moving on to Section 4.5 and later.

This chapter is closely related to Chapter 3. Concepts, results, and notations concerning the Fourier transform, up-conversion and down-conversion, and modulation schemes introduced in Chapter 3 will be used in this chapter.

4.2   WIDE-SENSE STATIONARY (WSS) PROCESS

This section provides some mathematical background on the wide-sense stationary (WSS) process. WSS is a class of stochastic process, which can be used to describe many types of signal and noise in a digital communications system. WSS processes can be described in the time domain by autocorrelation, and in the frequency domain by power spectral density (PSD).

For communication systems, we would like to carry out our analysis in baseband, where both signals and noises are complex-valued, as described in Chapter 3. For discussions about the autocorrelations of complex-valued processes, the concept of "properness" is introduced in Section 4.2.2.

Discussions about the power spectral density (PSD) in this section use the definition and properties of Fourier transform. Preliminary information on these topics is provided in an appendix in Chapter 3.

4.2.1 Definition of WSS Process

Let X be a time-dependent stochastic process. At any given time t, the corresponding value of X is a random variable, noted as $x(t)$. We are interested in how $x(t)$ depends on its values at other time intervals [1].

Given a set of time offsets $\tau_1, \tau_2, \ldots \tau_n$, we can look at the joint probability distribution $p[x(t + \tau_1), x(t + \tau_2), \ldots, x(t + \tau_n)]$. If this probability distribution is independent of t for all n values, such process X is said to be stationary (also known as strongly stationary or strictly stationary). If such probability distribution is independent of t only for $n = 1$ and $n = 2$, then process X is said to be wide-sense stationary, or WSS. Another common definition of WSS is that the expectation value and autocorrelation functions:

$$m(t) \overset{\text{def}}{=} E[x(t)]$$
$$c(\tau, t) \overset{\text{def}}{=} E[x(t + \tau)x(t)]$$

(4.1)

are independent of t [2, sec. 15.4]. This is a slightly weaker definition than the previous one. However, it is enough for our discussion in this chapter. We will use the latter definition from now on. Furthermore, all stochastic processes considered in this chapter have zero mean. Therefore, unless otherwise noted, we assume $m(t) = 0$ for the rest of this chapter.

If $\{x(t_1), x(t_2), \ldots, x(t_n)\}$ obeys a joint Gaussian distribution for any values of n and any set of $\{t_i, i = 1, 2, \ldots, n\}$, the process is said to be a Gaussian process. If a Gaussian process is WSS, it is also strictly stationary [2, sec. 16.8]. Usually, the noise of a communications system is a Gaussian process, and the signal of such system can sometimes be approximated by a Gaussian process.

For a WSS process, the autocorrelation function of x is defined as

$$c(\tau) \overset{\text{def}}{=} E[x(t + \tau)x(t)].$$

(4.2)

As discussed above, $c(\tau)$ is independent of t.

Now consider what happens when WSS process $x(t)$ passes a filter $h(t)$. The result is a stochastic process Y:

$$y(t) = \int x(t - u)h(u)du.$$

(4.3)

The autocorrelation of y is

$$E[y(t+\tau)y(t)] = \iint E[x(t+\tau-u_1)x(t-u_2)]h(u_1)h(u_2)du_1du_2$$
$$= \iint c(\tau-u_1+u_2)h(u_1)h(u_2)du_1du_2 \qquad (4.4)$$

Since its autocorrelation is independent of t, y is also a WSS process.

4.2.2 Complex-Valued Stochastic Processes: The Proper Processes

4.2.2.1 Autocorrelation of Complex-Valued Radom Process
The definition in the Section 4.2.1 can be extended to complex-valued processes [3]. If $x(t)$ is a complex-valued random process, it has a real part $x_r(t)$ and an imaginary part $x_i(t)$:

$$x(t) = x_r(t) + jx_i(t). \qquad (4.5)$$

Therefore, its autocorrelation contains four parts:

$$c_{rr}(t,\tau) \stackrel{\text{def}}{=} E[x_r(t+\tau)x_r(t)]$$
$$c_{ri}(t,\tau) \stackrel{\text{def}}{=} E[x_r(t+\tau)x_i(t)]$$
$$c_{ir}(t,\tau) \stackrel{\text{def}}{=} E[x_i(t+\tau)x_r(t)] \qquad (4.6)$$
$$c_{ii}(t,\tau) \stackrel{\text{def}}{=} E[x_i(t+\tau)x_i(t)]$$

Equivalently (see homework), we can define the following two complex-valued autocorrelation functions:

$$c(t,\tau) \stackrel{\text{def}}{=} E[x(t+\tau)x^*(t)] = [c_{rr}(t,\tau) + c_{ii}(t,\tau)] + j[c_{ir}(t,\tau) - c_{ri}(t,\tau)]$$
$$\bar{c}(t,\tau) \stackrel{\text{def}}{=} E[x(t+\tau)x(t)] = [c_{rr}(t,\tau) - c_{ii}(t,\tau)] + j[c_{ir}(t,\tau) + c_{ri}(t,\tau)] \qquad (4.7)$$

Here, $c(t, \tau)$ is known as the autocorrelation function, while $\bar{c}(t,\tau)$ is known as the pseudo-autocorrelation function. As in the real-valued case, we can also define the mean value

$$m(t) = E[x(t)]. \qquad (4.8)$$

A complex-valued process $x(t)$ is WSS if and only if c, $\bar{c}$, and m are all independent of t.

4.2.2.2 Proper Processes

A complex-valued stochastic process $x(t)$ is called proper, if its pseudo-autocorrelation $\bar{c}$, as defined in (4.7), is zero. This implies

$$
\begin{aligned}
c_{rr}(t,\tau) &= c_{ii}(t,\tau) \\
c_{ri}(t,\tau) &= -c_{ir}(t,\tau)
\end{aligned}
\tag{4.9}
$$

It is easy to see that the result of linear transformation or filtering of a proper process is still proper.

If $x(t)$ is proper, (4.9) shows that its real and imaginary parts have the same auto-correlation. Furthermore, if $c(t, \tau)$ is real-valued, then $c_{ri}(t, \tau) = c_{ir}(t, \tau) = 0$. Namely, the real and imaginary parts of $x(t)$ are mutually independent.

Proper WSS processes are important in communication systems. As will be shown in Section 4.3.2, typical baseband noise is a proper WSS process. Some commonly used baseband channel models are also proper WSS processes (see Chapter 6).

A typical baseband signal is also proper because the real and imaginary parts of the symbols are independent and identically distributed (iid). However, such signals are not WSS, but cyclostationary (see Chapter 7). There are, however, some works in the literature on using improper signals to improve performance in multiuser situations [4].

4.2.3 Power Spectral Density (PSD)

When considering a stochastic process in the frequency domain, it should be noted that the Fourier transform (FT) of its waveform does not necessarily converge, because its waveform value does not approach to zero when time t approaches infinity. Therefore, some special treatments are required. In particular, we can consider the signal in terms of power, that is, energy divided by time.

Let us consider a real-valued stochastic process $x(t)$, which is not necessarily WSS. Let $\bar{x}(t,T)$ be the windowed version of $x(t)$. Namely,

$$
\bar{x}(t,T) =
\begin{cases}
x(t) & \text{if } t \in [-T,T] \\
0 & \text{otherwise}
\end{cases}
,
\tag{4.10}
$$

where T specifies a window size. Later on, we will let $T \to \infty$, thus making $\bar{x}(t,T)$ the same as $x(t)$. With such windowing, $\bar{x}$ has legitimate FT for any T. The power of $\bar{x}(t,T)$ is

$$
P_{\bar{x}}(T) = E\left[\frac{1}{2T}\int |\bar{x}(t,T)|^2 dt\right].
\tag{4.11}
$$

From the Parseval's theorem (see appendix of Chapter 3),

$$\int |\bar{x}(t,T)|^2 dt = \frac{1}{2\pi} \int |\tilde{\bar{x}}(\omega,T)|^2 d\omega, \tag{4.12}$$

where $\tilde{\bar{x}}(\omega,T)$ is the FT of $\bar{x}(t,T)$:

$$\tilde{\bar{x}}(\omega,T) \overset{\text{def}}{=} \int \bar{x}(t,T) e^{-j\omega t} dt. \tag{4.13}$$

Therefore, the total power can also be expressed as

$$P_{\bar{x}}(T) = E\left[\frac{1}{2T}\frac{1}{2\pi}\int |\tilde{\bar{x}}(\omega,T)|^2 d\omega\right] = E\left[\frac{1}{2T}\int |\tilde{\bar{x}}(2\pi f,T)|^2 df\right] = \frac{1}{2T}\int E\left[|\tilde{\bar{x}}(2\pi f,T)|^2\right] df. \tag{4.14}$$

Here, $f = \omega/2\pi$ is the frequency, while ω is the angular frequency.

We can interpret the right-hand side of (4.14) as the "power density" integrated over the frequency. Therefore, we can define the power spectral density (PSD) of a stochastic process:

$$P_W(f) \overset{\text{def}}{=} \lim_{T\to\infty} \frac{1}{2T} E\left[|\tilde{\bar{x}}(2\pi f,T)|^2\right] \tag{4.15}$$

This definition can be justified as follows. We can "extract" a portion of the power in $\bar{x}$ that falls within a frequency band, by filtering $\bar{x}$ with a bandpass filter:

$$\bar{y}(t,T) = \int \bar{x}(t-\tau,T) g(\tau) d\tau, \tag{4.16}$$

where the spectrum of g is given by

$$\tilde{g}(\omega) = \begin{cases} 1 & \text{if } 2\pi W_1 \le |\omega| < 2\pi W_2 \\ 0 & \text{otherwise} \end{cases}. \tag{4.17}$$

$\tilde{g}(\omega)$ is the FT of $g(t)$. W_1 and W_2 are the lower and upper edges of the frequency band. From the property of convolution (see Chapter 3), the FT of $\bar{y}$ is

$$\tilde{\bar{y}}(\omega,T) = \tilde{\bar{x}}(\omega,T)\tilde{g}(\omega) = \begin{cases} \tilde{\bar{x}}(\omega,T) & \text{if } 2\pi W_1 \le |\omega| < 2\pi W_2 \\ 0 & \text{otherwise} \end{cases}. \tag{4.18}$$

According to (4.14) and (4.18), the extracted power is

$$P_{\bar{y}}(T) = E\left[\frac{1}{2T}\int |\tilde{\bar{y}}(2\pi f,T)|^2 df\right] = \frac{1}{2T}\left[\int_{-W_2}^{-W_1} |\tilde{\bar{x}}(2\pi f,T)|^2 df + \int_{W_1}^{W_2} |\tilde{\bar{x}}(2\pi f,T)|^2 df\right] \tag{4.19}$$

When $T \to \infty$, (4.15) and (4.19) lead to

$$E\left[\lim_{T\to\infty} P_{\bar{y}}(T)\right] = \int_{-W_2}^{-W_1} P_W(f) df + \int_{W_1}^{W_2} P_W(f) df. \tag{4.20}$$

Therefore, $P_W(f)$ is a proper definition of spectral power density when $T \to \infty$.

In above definition, the total power is the integration of PSD over the frequency range of $(-\infty, \infty)$, as shown in (4.14). Such PSD is known as the *two-sided PSD*. For real-valued stochastic processes, on the other hand, $\tilde{\bar{x}}(-\omega, T) = \tilde{\bar{x}}^*(\omega, T)$. Namely, the PSD in negative frequency is the same as that for positive frequency. Therefore, we can combine $P_W(f)$ with $P_W(-f)$.

$$P_{WS}(f) \stackrel{\text{def}}{=} P_W(f) + P_W(-f) = 2P_W(f) \tag{4.21}$$

$P_{WS}(f)$ is known as the *one-sided PSD*. With P_{WS}, we only consider positive frequency range when computing power. Thus, (4.20) becomes

$$\lim_{T \to \infty} P_{\bar{y}}(T) = \int_{W_1}^{W_2} P_{WS}(f) df. \tag{4.22}$$

When we measure PSD with instruments such as a spectrum analyzer, the result is given in terms of the one-sided PSD.

4.2.4   Wiener–Khinchin Theorem

Now, we try to connect PSD with the autocorrelation function. Suppose $x(t)$ discussed in the Section 4.2.3 is a real-valued WSS process.

From (4.10) and (4.13),

$$\begin{aligned} E\left[|\tilde{\bar{x}}(\omega, T)|^2\right] &= \iint_{-\infty}^{\infty} E[\bar{x}(t_1, T)\bar{x}(t_2, T)] e^{-j\omega(t_1 - t_2)} dt_1 dt_2 \\ &= \iint_{-T}^{T} E[x(t + \tau)x(t)] e^{-j\omega\tau} d\tau dt \end{aligned} \tag{4.23}$$

Recalling the definition of autocorrelation (4.2), for a WSS process, the expectation value in the integrant is no other than the autocorrelation $c(\tau)$. When $T \to \infty$, the integration over τ converges to the FT of $c(\tau)$, noted as $\tilde{c}(\omega)$. The integration over t is $2T$, which cancels out the denominator in (4.15). Therefore, (4.15) becomes

$$P_W(f) = \tilde{c}(2\pi f). \tag{4.24}$$

Equation (4.24) is known as the Wiener–Khinchin theorem. Through this derivation, we also see that although the FT of a WSS process does not exist, the power spectral density of a WSS process is well-defined.

4.3 AWGN CHANNEL

Section 4.2 discussed a general class of stochastic process, known as WSS process. Most type of the noise in communication systems can be described by WSS process. In this section, we apply such concept to the simplest channel model for communications: the additive white Gaussian noise (AWGN). AWGN is used for basic analyses of communication techniques, such as demodulation in this chapter and error correction coding in Chapter 5. Many other channel models are built on top of the AWGN model (Chapter 6).

4.3.1 AWGN Channel Model

The AWGN channel model has three characteristics: additive, white, and Gaussian. They are discussed individually below.

4.3.1.1 Additive Noise

In AWGN channel model, the noise is additive, that is, the received signal is the sum of the transmitted signal and noise[1]:

$$r(t) = s(t) + n(t),\tag{4.25}$$

where $r(t)$, $s(t)$, and $n(t)$ are the received signal, the transmitted signal, and the noise, respectively. Furthermore, it is implicitly assumed that the noise and signal are uncorrelated.

4.3.1.2 White Noise

Furthermore, the noise is "white," which means its autocorrelation is a delta function:

$$c_n(\tau) \stackrel{\text{def}}{=} E(n(t+\tau)n(t)) = \frac{N_0}{2}\delta(\tau)\tag{4.26}$$

N_0 is a constant characterizing the power level of the noise. In fact, it will be shown below that N_0 is the one-sided PSD of the noise.

According to the Wiener–Khinchin theorem (Section 4.2.4), the PSD of the noise is

$$P_{Wnp}(f) = \tilde{c}_n(2\pi f) = \int c_n(\tau)e^{-j2\pi f\tau}d\tau\tag{4.27}$$

---

[1] Here, we assume that the gain of the channel is 1. This can be achieved by applying an appropriate gain at the receiver amplifier.

From (4.26), we get

$$P_{Wnp}(f) = \frac{N_0}{2} \tag{4.28}$$

Therefore, the two-sided PSD of a white noise specified by (4.26) is $N_0/2$, independent of frequency (this is the meaning of whiteness). The one-sided PSD of the same noise is N_0.

Note that since the noise PSD is a constant across frequency, the total power of the noise process, as given by (4.14), is infinite. Therefore, the truly white noise does not exist in practice. However, white noise, that is, noise with a constant PSD, is a good approximation within limited bandwidth in many cases.

4.3.1.3 Gaussian Noise

Gaussian noise is a stochastic process $n(t)$, where all joint distributions $\{n(t_1), n(t_2), \ldots, n(t_K)\}$ is a joint Gaussian distribution for any K and $\{t_k, k = 1, 2, \ldots, K\}$. Especially, its probability density function for any given t is

$$p[n(t)] = \left(2\pi\sigma_n^2\right)^{-1/2} \exp\left[-\frac{n^2(t)}{2\sigma_n^2}\right], \tag{4.29}$$

where σ_n^2 is the variance:

$$E\left(|n(t)|^2\right) = \sigma_n^2 \tag{4.30}$$

Special attention is needed when it comes to the *white* Gaussian noise. Because the power of white noise is infinite as pointed out in Section 4.3.1.2, σ_n^2 in (4.30) is not well-defined. In reality, we only consider band-limited Gaussian noise, of which white Gaussian noise is an approximation under certain conditions. More details on dealing with white Gaussian noise are given in Sections 4.5.1 and 4.11.

Since the sum of Gaussian variables with zero mean is still a Gaussian variable, any Gaussian noise passing a filter is still Gaussian noise. In fact, a WSS Gaussian noise with given autocorrelation can be viewed as a filtered version of white Gaussian noise.

4.3.1.4 Summary

AWGN is a commonly used channel model for communication systems. Under this model, the relationship between transmitted and received signals is given by (4.25) (additive). The noise autocorrelation is given by (4.26) with PDF given by (4.28) (white). Furthermore, the noise is a Gaussian process.

4.3.2   Baseband AWGN Channel

As described in Chapter 3, at the receiver, the passband signal is down-converted to the baseband. Many analyses of signal and noise are performed in baseband. Therefore, it is important to understand the baseband noise.

Conversion from passband to baseband (down-conversion) is summarized by (4.177) in Section 4.9.1. For noise, we cannot assume it is contained in the signal passband. Therefore, we need to express the low pass filter at the down converter explicitly. Namely,

$$n_b(t) = \int e^{-j\omega_c u} n_p(u) g(t-u) du. \tag{4.31}$$

Here, ω_c is the angular frequency for carrier. $n_p(t)$ is the passband noise. $g(t)$ is an ideal low pass filter, whose frequency response is given as

$$\tilde{g}(\omega) = \begin{cases} 1 & \text{if } |\omega| \le 2\pi W_b, \\ 0 & \text{otherwise} \end{cases}, \tag{4.32}$$

W_b is the baseband bandwidth; $2\pi W_b < \omega_c$.

As shown in an appendix (Section 4.10), if n_p is WSS, then n_b is a proper WSS process. Therefore, we are concerned only about its autocorrelation

$$c_b(\tau) \overset{\text{def}}{=} E\big[n_b(t+\tau) n_b^*(t)\big]. \tag{4.33}$$

Equation (4.204) in Section 4.10 gives the spectrum of $c_b(\tau)$:

$$\tilde{c}_b(\omega) = |\tilde{g}(\omega)|^2 \tilde{c}_p(\omega + \omega_c), \tag{4.34}$$

where $\tilde{c}_p(\omega)$ is the FT of the passband autocorrelation function $c_p(\tau)$:

$$c_p(\tau) \overset{\text{def}}{=} E\big[n_p(t+\tau) n_p(t)\big]. \tag{4.35}$$

$c_b(\tau)$ can thus be obtained through inverse Fourier transform:

$$c_b(\tau) = (2\pi)^{-1} \int |\tilde{g}(\omega)|^2 \tilde{c}_p(\omega + \omega_c) e^{j\omega\tau} d\omega, \tag{4.36}$$

Therefore, the PSDs of baseband and passband have the relationship, according to (4.24):

$$P_{Wnb}(f) = |\tilde{g}(2\pi f)|^2 P_{Wnp}(f + f_c), \tag{4.37}$$

where P_{Wnb} and P_{Wnp} are baseband and passband noise PSD, respectively.

For AWGN in passband, P_{Wnp} is given by (4.28). With the ideal filter whose response is given by (4.32),

$$P_{Wnb}(f) = \begin{cases} \dfrac{N_0}{2} & \text{if } |f| \le W_b \\ 0 & \text{otherwise} \end{cases}. \tag{4.38}$$

Therefore, within the baseband bandwidth, the baseband noise is also white (i.e., with a flat PSD). One can easily see that if passband noise is Gaussian with zero mean, so is baseband noise. Furthermore, for AWGN, $c_b(\tau)$ is real-valued. Therefore, the real and imaginary parts of the noise are mutually independent, as discussed in Section 4.2.2.1.

Within the baseband bandwidth, we can consider baseband noise as white noise. Its autocorrelation is thus a delta function, similar to that in passband:

$$c_b(\tau) = \frac{N_0}{2}\delta(\tau). \tag{4.39}$$

4.3.3   Specifying Signal Quality

For AWGN channel, signal quality is characterized by comparing the power levels of signal and noise. This section describes several ways of specifying signal quality in this sense. Such signal quality metrics are used in Section 4.7 to predict error probability at the receiver.

4.3.3.1   Signal-to-Noise Ratio (SNR)

The basic definition of signal quality is signal-to-noise ratio (SNR). SNR is the ratio between the signal power and the noise power. SNR is dimensionless and is usually specified in dB. Signal power can be specified in many ways. In this section, it is specified in terms of energy per symbol E_{Sym} (in passband) and symbol interval T_{Sym}. Noise power is usually specified in terms of N_0 and bandwidth. Therefore, it is useful to express SNR using these parameters. In this section, we will examine such expression in passband and baseband.

In passband, this definition is straightforward:

$$SNR_p \overset{\text{def}}{=} \frac{P_{sp}}{P_{np}} = \frac{\int P_{Wp}(f)\left|\tilde{g}_p(2\pi f)\right|^2 df}{\int P_{Wnp}(f)\left|\tilde{g}_p(2\pi f)\right|^2 df}. \tag{4.40}$$

Here, P_{sp} and P_{np} are signal and noise powers in the passband. $P_{Wp}(f)$ is the passband signal PSD, discussed in an appendix (Section 4.9.2). $P_{Wnp}(f)$ is the passband noise PSD. $\tilde{g}_p(\omega)$ is the spectrum of the equivalent receiver filter in the passband $g_p(t)$. We often approximate g_p with an ideal bandpass filter:

$$\tilde{g}_p(f) = \begin{cases} 1 & \text{if } |f-f_c| \le \dfrac{W_p}{2} \text{ or } |f+f_c| \le \dfrac{W_p}{2}, \\ 0 & \text{otherwise} \end{cases} \tag{4.41}$$

where f_c is the carrier frequency, and W_p is the passband bandwidth.

For an AWGN channel, $P_{Wnp}(f)$ is given by (4.28) as a constant $N_0/2$. The total bandwidth (including positive and negative frequency ranges) is $2W_p$. Therefore,

$$P_{np} = N_0 W_p. \tag{4.42}$$

Furthermore, we usually take the approximation that all signal power falls within the passband bandwidth (i.e., the excessive bandwidth discussed in Chapter 3 does not represent a significant portion of the signal power). The passband PDF of the signal, as given by (4.187) and (4.192) in an appendix (Section 4.9.2), is

$$P_{Wp}(f) = \frac{1}{2} E_{Sym} \left[T_{Sym} \int |\tilde{h}(2\pi f)|^2 df \right]^{-1} \left[|\tilde{h}(2\pi f - 2\pi f_c)|^2 + |\tilde{h}(2\pi f + 2\pi f_c)|^2 \right], \tag{4.43}$$

where E_{Sym} is the energy per symbol in the passband. T_{Sym} is the symbol interval. $\tilde{h}(\omega)$ is the spectrum of the pulse shaping filter. Therefore,

$$P_{sp} = \int P_{Wp}(f) df = \frac{E_{Sym}}{T_{Sym}}. \tag{4.44}$$

Equation (4.40) thus leads to the passband SNR:

$$SNR_p = \frac{E_{Sym}}{N_0 W_p T_{Sym}}. \tag{4.45}$$

In baseband, SNR is defined similarly.

$$SNR_b \stackrel{\text{def}}{=} \frac{P_{sb}}{P_{nb}} = \frac{\int P_{Wb}(f) |\tilde{g}_b(2\pi f)|^2 df}{\int P_{Wnb}(f) |\tilde{g}_b(2\pi f)|^2 df}. \tag{4.46}$$

Here, $P_{Wb}(f)$ and $P_{Wnb}(f)$ are the signal and noise PSD in baseband, respectively. $\tilde{g}_b(\omega)$ is the spectrum of the receiver baseband filter. P_{sb} and P_{nb} are the signal and noise powers in baseband, respectively. We can further compute SNR_b under the similar approximation as in passband. Namely, we assume the receiver filter is ideal, as characterized by (4.32). Also, all signal energy falls within the bandwidth of such filter.

The signal PSD in baseband is given by (4.192) in Section 4.9.2:

$$P''_{Wb}(f) = \frac{1}{2}E_{Sym}\left[T_{Sym}\int\left|\tilde{h}(2\pi f)\right|^2 df\right]^{-1}\left|\tilde{h}(2\pi f)\right|^2. \tag{4.47}$$

Therefore,

$$P_{sb} = \frac{1}{2}\frac{E_{Sym}}{T_{Sym}} \tag{4.48}$$

For AWGN channel, the noise PSD in baseband is $N_0/2$ within the bandwidth, as given by (4.38) in Section 4.3.2. Therefore, the total noise power is

$$P_{nb} = \frac{N_0}{2}2W_b = N_0 W_b. \tag{4.49}$$

Note that, as discussed in Chapter 3, $W_b = W_p/2$. Therefore,

$$SNR_b = \frac{1}{2}\frac{E_{Sym}}{N_0 W_b T_{Sym}} = \frac{E_{Sym}}{N_0 W_p T_{Sym}}. \tag{4.50}$$

Therefore, the baseband and passband SNR values are the same, as expected.

Note that given the passband PSDs, the baseband PSDs may differ by a constant factor, depending on how down-converting is described. For example, some literature [3, 5] includes an additional factor of $\sqrt{2}$ in the down-conversion formulation (4.177) in Section 4.9.1. However, such factor affects both signal and noise powers. Therefore, the SNR value remains the same.

4.3.3.2 Energy per Bit over Noise Density

Another way to specify signal quality is the ratio between energy per bit (commonly noted as E_b) and N_0. This ratio is noted as E_b/N_0, and colloquially referred to as "eb-naught" or "eb-no." E_b/N_0 is dimensionless and is usually specified in dB. It can be converted to SNR, as shown below. E_b is related to symbol energy E_{Sym}:

$$E_b = \frac{E_{Sym}}{N_b}, \tag{4.51}$$

where N_b is the number of bits per symbol. For uncoded modulation,

$$N_b = \log_2 M, \tag{4.52}$$

where M is the number of constellation points in modulation. Usually, M is chosen such that N_b is an integer. In the presence of channel coding (Chapter 5), N_b is the number of "data bits" per symbol. It depends on both modulation order (Chapter 3) and coding rate. In this case, N_b can be a fraction and can be smaller than 1.

The SNR (4.45) and (4.50) can be expressed as

$$SNR = \frac{N_b}{W_p T_{Sym}} \frac{E_b}{N_0} \tag{4.53}$$

Furthermore, communication systems commonly use Nyquist symbol rate (Chapter 3), where

$$T_{Sym} W_p = 1. \tag{4.54}$$

In this case, we have

$$SNR = N_b \frac{E_b}{N_0}. \tag{4.55}$$

E_b/N_0 is commonly used for characterizing the performance of communication techniques [6, sec. 3.1.5]. It provides a fairer comparison among the various modulation and coding techniques, as discussed in Chapters 2 and 5.

4.3.3.3 Carrier-to-Noise Density Ratio

The third way to express signal quality is the carrier-to-noise PSD ratio, noted as C/N_0. It is the ratio between signal power (P_{sp} in our notation) and noise PSD (N_0 in our notation). C/N_0 is commonly specified in dB-Hz. It can be converted to SNR:

$$SNR = \frac{1}{W_p} \frac{C}{N_0}, \tag{4.56}$$

where W_p is the passband bandwidth.

4.3.4 Summary and Notes

The AWGN channel model is widely used in communication system analysis. This section discussed the mathematical property of the model, its baseband representation, and several ways to characterize signal quality.

It should be noted that while AWGN model applies to a wide variety of communication systems, there are important cases beyond this model.

1. Additiveness is a good description of noises that are independent of the signals, such as thermal noise, background interference, and noise from amplifiers. However, there are other sources of noises that are not additive. Examples include phase noise, amplifier nonlinearity and clipping, synchronization error (see Chapter 7), residual inter-symbol interferences, and inaccuracy in

transmission. For a well-designed system in most applications, additive noise should be dominant. In fact, other noise sources mentioned above are often ignored in analyses. When they must be included, they can be treated as additive noise while signal power level is kept constant. Therefore, additive noise is often a good model. An important exception is optical communication systems. In such systems, because the signal energy arrives in discrete units (photons), there is a fluctuation in signal power, known as the shot noise. Shot noise is not additive.

2. Whiteness is a good description of the thermal noise, amplifier noise, and background interference when the bandwidth is relatively narrow. However, there are non-white noises in many applications. A common example is narrow-band interferences, unintentional or intentional. Non-white noises are also known as colored noise. It will be discussed more in Chapter 9.

3. Gaussianity can be the consequence of the central limit theorem when the noise is contributed by many independent sources. There are many other reasons to expect natural noises to be Gaussian. However, noises from the system itself, such as the examples listed item 1 above, are not necessarily Gaussian. On the other hand, Gaussian noise often remains to be an excellent approximation that is mathematically tractable. As will be shown in the following chapters, except for error probability computations, most of the conclusions in this chapter do not depend on Gaussianity of noise.

In Chapter 9, the AWGN channel model will be extended to include frequency-dependent gain, or equivalently, inter-symbol interference (ISI). However, most concepts and notions presented in this chapter also apply there. Chapter 6 will present a broader picture of channel characteristics, especially channels with time-variant gains.

4.4 DETECTION PROBLEM AND MAXIMUM LIKELIHOOD DETECTION

The Sections 4.2 and 4.3 focus on channel and noise characterization. In this section, we turn our attention to the optimal receiver (or detector). We will first state the problem and present the solution in terms of conditional probabilities. Then we will analyze error probability under optimal detector. In this section, we discuss the problem of detection in general terms. Later in this chapter, we will apply these results to the special case of AWGN channel.

4.4.1 Problem Statement

The task of the receiver (detector) is recovering transmitted data in the presence of noise and other channel effects. Mathematically, such task can be stated as the following detection problem.

The transmission alphabet $\{s_m, m = 1, 2, \ldots, M\}$, and conditional probability density $p(r \mid s)$ are known. Here, r is the received signal. $s \in \{s_m\}$ is the transmitted signal. The detection problem is to determine the transmitted symbol identity $\hat{m}$, based on r.

Note that channel effects (such as noise) introduce uncertainty in the received signal, as reflected in $p(r \mid s)$. For example, if the channel has additive noise, the received signal is expressed as

$$r = s + n, \tag{4.57}$$

where s is the transmitted symbol and n is the noise. In this case, we have

$$p(r \mid s) = p_n(r - s), \tag{4.58}$$

where $p_n(x)$ is the probability density function for noise n.

Note that at the front end of the receiver, r, s, and n are all functions of t. Mathematically, these functions can be treated as entities, based on which the probabilities can be defined. We will not discuss the details in this section. In Section 4.6, we will discuss the conversion from functions of t (i.e., waveforms) to scalars (i.e., samples or symbols), for the special case of additive white Gaussian noise (AWGN) channel. More mathematical background can be found in an appendix, Section 4.11.

4.4.2 Optimal Detection

We wish to find an optimal way of detection, to minimize the probability of error. Given r, if we decide that the transmitted symbol identity is $\hat{m}$, then the probability of making an error is $1 - p(s_{\hat{m}} \mid r)$. Therefore, to minimize the error probability, we should choose the decision

$$\hat{m} = \arg \max_m p(s_m \mid r). \tag{4.59}$$

$p(s_m \mid r)$ is known as the a posteriori probability, because it is the probability of s_m after the fact that r is received. Equation (4.59) shows that the optimal detection strategy is selecting $\hat{m}$ so that the a posteriori probability for $s_{\hat{m}}$ is maximized. Therefore, such detection method is known as maximum a posteriori, or MAP.

However, in our detection problem stated in Section 4.4.1 $p(r \mid s)$, instead of $p(s_m \mid r)$, is known. According to the Bayes theorem,

$$p(s_m \mid r) = \frac{p(r \mid s_m) p_s(s_m)}{p_r(r)}, \tag{4.60}$$

where $p_s(s_m)$ is the a priori probability of s_m, and $p_r(r)$ is the a priori probability of r. For the purpose of MAP detection, we can ignore $p_r(r)$ because it is independent of m. Therefore, (4.59) leads to

$$\hat{m} = \arg\max{}_{m} p(r \mid s_m) p_s(s_m).\tag{4.61}$$

Furthermore, in most communications systems, all symbols are equiprobable. In other words, $p_s(s_m)$ is a constant, independent of m. In this case, the MAP detection becomes

$$\hat{m} = \arg\max{}_{m} p(r \mid s_m).\tag{4.62}$$

Namely, we look for the m that is most likely to result in received signal r. Equation (4.62) is known as the maximum likelihood (ML) detector. When the transmitted symbols are equiprobable, ML detector is equivalent to the MAP and is thus optimal.

4.4.3 Error Probability

Because of the noise, it is possible to make errors in detection. An error occurs when the detector determines that $\hat{m} = m'$ while symbol $m \neq m'$ is transmitted in fact. With ML detector, this means the received signal r meets the following condition:

$$p(r \mid s_{m'}) > p(r \mid s_m), m' \neq m.\tag{4.63}$$

When s_m is transmitted, the probability of error is thus

$$P_e = p\left(r \in \cup_{m' \neq m} R_{m',m} \mid s_m\right),\tag{4.64}$$

where $R_{m',m}$ is a subset of r, where (4.63) is true. The symbol $\cup_{m' \neq m}$ means union of all subsets $R_{m',m}$ with $m' \neq m$. Such subsets and their unions are often difficult to handle. Therefore, some approximations are usually used.

If we ignore the intersects (overlaps) among the subsets $\{R_{m',m}\}$, (4.64) can be written as a summation of probabilities. Such approximation yields an upper bound of the error probability:

$$P_e \leq \sum_{m' \neq m} p(r \in R_{m',m} \mid s_m).\tag{4.65}$$

Equation (4.65) is known as the union bound. The probability $p(r \in R_{m',m} \mid s_m)$ is the error probability when s_m is mistaken as $s_{m'}$. This is known as the pairwise error probability, noted as $P_{e,(m,m')}$.

On the other hand, we recognize that (4.63) is most likely to be true when $s_{m'}$ is close to s_m. Therefore, we may need to only consider the "neighborhood" of m when we compute error probability. Mathematically, we can define some form of distance between any pair of transmitted signals $d(m, m')$. For example, we may have the Euclidean distance

$$d(m,m') = \int [s_m(t) - s_{m'}(t)]^2 dt. \tag{4.66}$$

It is a reasonable assumption that the pairwise error probability $P_{e,(m,m')}$ decreases sharply when $d(m, m')$ increases. Section 4.7 will provide some examples.[2] In this case, when we compute error probability, we only need to consider the subset of m', where $d(m, m')$ takes the minimum value. This subset is referred to as the "nearest neighbor" of m, noted as N_m. The error probability among this "neighborhood" provides an approximation of the overall error probability:

$$P_e \approx K_n P_{e,(m,m_n)}, \tag{4.67}$$

where K_n is the number of members in the nearest neighbor. m_n represents a transmitted symbol within this neighborhood. m_n may have multiple values, but the pairwise error probability $P_{e,(m,m_n)}$ is the same for all such values. Equation (4.67) is known as the nearest neighbor approximation.[3]

Both the union bound and the nearest neighbor approximation make the computation of error probability much more tractable. These two approximations are widely used in the analysis of communication systems. When error probabilities are low (as in the case of most practical systems), these approximations provide very good agreements with the actual values. Section 4.7 will show some comparisons.

Note that above discussions are based on a specific m. To compute the overall error probability, results (4.64), (4.65), and (4.67) should be averaged over all m values.

Furthermore, P_e is the probability of symbol error. Another common metric is bit error rate P_b, which is the probability of bit error. As an approximation, we can assume that all errors happen among the nearest neighbors. With Gray coding (Chapter 3), such a symbol errors causes 1 bit to be wrong. Therefore,

$$P_b = \frac{1}{N_b} P_e. \tag{4.68}$$

N_b is the number of bits per symbol. Bit error rate P_b is often noted as BER, while symbol error rate P_e is noted as SER. Another common error metric is block error rate, noted as BLER. It is the probability for a block of bits to have an error in transmission. It is often used in the presence of channel coding (Chapter 5). Computation of BLER may be more complicated, depending on the details of modulation and coding.

---

[2] Chapter 5 will provide other examples where the distance is defined differently.

[3] By accounting only for the nearest neighbor points, we underestimate the error probability. However, union bound is also used in (4.67), resulting in an over estimation. Therefore, it is difficult to say whether the approximation in (4.67) is higher or lower than the actual value.

4.5 MAP AND ML DETECTION WITH AWGN CHANNEL

In this section, we apply the MAP and ML detectors obtained in Section 4.4 to a special case: the AWGN channel discussed in Section 4.3. First, we need some mathematical treatment of the white noise, to remove the infinity in its power. We will then derive the specific forms of MAP and ML detectors under AWGN channel model. Both of these forms utilize the matched filter (MF), which will be discussed in more details in Section 4.6.

4.5.1 MAP and ML Detection for AWGN Channel

MAP detection is formulated in (4.61) in Section 4.4.2:

$$\hat{m} = arg\,\max_m p(r\,|\,s_m)p_s(s_m). \tag{4.69}$$

For AWGN channel and with the vector representation described in Section 4.11 (an appendix), $p(r\,|\,s_m)$ is the probability of receiving $\vec{r}$ given that $\vec{s}_m$ is transmitted. $p_s(s_m)$ is the a priori probability of $\vec{s}_m$ being transmitted. Let us first examine $p(r\,|\,s_m)$. From (4.80), we can see that

$$p\big(\vec{r}\,|\,\vec{s}_m\big) = p_n\big(\vec{r}-\vec{s}_m\big), \tag{4.70}$$

where $p_n\big(\vec{x}\big)$ is the probability density function for noise $\vec{n}$. As stated in Section 4.5.1, $\vec{n}$ has K independent components, each obeying Gaussian distribution with variance $N_0/2$. From (4.231) in Section 4.11.3, we have

$$p_n\big(\vec{n}\big) = \prod_{k=1}^{K}\left[(\pi N_0)^{-1/2}e^{-\frac{|n_k|^2}{N_0}}\right] \tag{4.71}$$

$$= (\pi N_0)^{-\frac{K}{2}}e^{-\frac{1}{N_0}\sum_{k=1}^{K}|n_k|^2} = (\pi N_0)^{-\frac{K}{2}}e^{-\frac{|\vec{n}|^2}{N_0}}$$

Here, $|\vec{n}|$ is the amplitude of the vector $\vec{n}$. Equations (4.70) and (4.71) thus lead to

$$p\big(\vec{r}\,|\,\vec{s}_m\big) = (\pi N_0)^{-\frac{K}{2}}e^{-\frac{|\vec{r}-\vec{s}_m|^2}{N_0}}. \tag{4.72}$$

The arguments of maxima in (4.69) remain the same if we replace the function with its logarithm. With such replacement and ignoring some constant terms and factors, we have

$$\hat{m} = \arg\max_m \ln p(r \mid s_m) p_s(s_m) = \arg\max_m \left\{ -\frac{|\vec{r} - \vec{s}_m|^2}{N_0} + \ln \left[p_s(s_m) \right] \right\}.$$ (4.73)

$$= \arg\max_m \left\{ Re\,(\vec{r} \cdot \vec{s}_m) - \frac{1}{2} |s_m|^2 + \frac{1}{2} N_0 \ln \left[p_s(s_m) \right] \right\}$$

Equation (4.73) can be implemented directly, where the vectors can be computed from (4.77), as shown in Section 4.6.3. Alternatively, we can convert the vector operations in the last expression back to waveform operations, based on (4.79)[4]:

$$\hat{m} = \arg\max_m \left\{ Re\,\left(\int r(t) s_m^*(t) dt \right) - \frac{1}{2} \int |s_m(t)|^2 dt + \frac{1}{2} N_0 [\ln p_s(s_m)] \right\}.$$ (4.74)

Equation (4.74) is the MAP detection algorithm for an AWGN channel. When all symbols are equally probable, the last term in (4.74) becomes a constant that can be dropped from consideration. In this case, we arrive at the ML detection:

$$\hat{m} = \arg\max_m \left[Re\,\left(\int r(t) s_m^*(t) dt \right) - \frac{1}{2} \int |s_m(t)|^2 dt \right].$$ (4.75)

A typical communication system transmits all possible symbols with equal probability, to maximize source entropy (see Chapter 2). Therefore, ML detection is widely applicable. However, sometimes additional information on a priori symbol probability is available to the receiver, necessitating MAP detection. One such example is coded symbols, as will be discussed in Chapter 5. Also note that MAP algorithm needs the knowledge of N_0 at the receiver, while the ML algorithm does not. The term $\int r(t) s_m^*(t) dt$ is discussed in more details in the Section 4.6.

4.5.2   Signal Space and Vector Expression of AWGN Channel

In the AWGN channel described by Section 4.3.1, we have, as given in (4.25),

$$r(t) = s(t) + n(t).$$ (4.76)

---

[4] In (4.74) and thereafter, $r(t)$ is assumed to fall within a subspace that contains all $\{s_m\}$ and has finite dimension. This can be done by common filtering, and thus is true in practice. For the sake of simplicity, we keep the same notation $r(t)$ with such limitation.

Here, r and s are received and transmitted signals, respectively. n is the noise. In AWGN channel, $n(t)$ is white and Gaussian, as described in Section 4.3.1. To further analyze the system, we face three difficulties.

1. Usually, the detection decision is not based on the received signal at one instant t. Instead, we look at the received waveform over time in entirety.

2. As pointed out in Section 4.3.1.2, the power of white Gaussian noise in infinite, because its bandwidth is infinite.

3. In reality, the passband bandwidth is finite and known. However, (4.76) does not capture that fact.

These difficulties are related and can be addressed together. The approach we take here is representing waveforms as vectors [7, sec. 4.2]. More mathematics details are provided in an appendix (Section 4.11). Main conclusions are listed here for future references.

As shown in (4.215) in an appendix (Section 4.11), a function $f(t)$ can be expressed as a linear combination of some orthonormal basis functions $\{\phi_k, k = 0, 1, 2, \ldots\}$:

$$f(t) = \sum_{k=0}^{\infty} f_k \phi_k(t)$$
$$f_k = \int f(t)\phi_k^*(t)dt]$$

(4.77)

Therefore, we can establish a mapping between function $f(t)$ and vector (with infinite dimension) $\vec{f}$, whose components are $\{f_k, k = 1, 2, \ldots\}$. The orthonormal basis function set has the following property:

$$\langle \phi_k, \phi_l \rangle \overset{\text{def}}{=} \int \phi_k(t)\phi_l^*(t)dt = \delta_{k,l}.$$

(4.78)

It can be shown (see homework) that $\vec{f}$ indeed behaves as a vector. In particular, the inner product of the vectors can be expressed as

$$\vec{f} \cdot \vec{g} \overset{\text{def}}{=} \sum_k f_k g_k^* = \int f(t)g^*(t)dt.$$

(4.79)

Our detection algorithm can be formulated in terms of the vectors corresponding to $r(t)$, $s(t)$, and $n(t)$. By using vectors, the whole waveforms, instead of their values at one instant, are captured in the algorithm. This addresses point 1 above.

Furthermore, it is shown in Section 4.11 that for white Gaussian noise vector $\vec{n}$, its components are mutually independent. Each component obeys Gaussian distribution with variance $N_0/2$. Therefore, the infinite power problem (point 2 above) is addressed.

Although these vectors are usually of infinite dimension, we only need to consider the subspace spanned by all possible signal waveforms. Since the number of possible signal waveforms is finite, such subspace is of finite dimension. The choice of such subspace reflects our prior knowledge of the signal characteristics, such as its bandwidth mentioned in point 3 above. Such subspace is referred to as the *signal space*.

Therefore, the vector representation of the signal and noise resolves all of the difficulties listed above. As will be shown below, the details of the basis function and decomposition are not important. We will use the vector representation only as an intermediate tool and return to waveform representation later.

With the vector representation, (4.76) can be written as

$$\vec{r} = \vec{s} + \vec{n}, \tag{4.80}$$

where $\vec{r}, \vec{s}$, and $\vec{n}$ are the vectors corresponding to $r(t)$, $s(t)$, and $n(t)$. Furthermore, we limit our consideration to the signal space, that is, a subspace containing all $\{\vec{s}_m\}$ and with finite dimension. Therefore, all vectors here have a finite dimension, noted as K. Let M be the number of possible signal waveforms $\{s_m, m = 1, 2, \ldots, M\}$, K is upper-bounded by M.

$$K \leq M. \tag{4.81}$$

In fact, in some familiar cases discussed in the Section 4.6, K is much smaller than M.

In (4.80), $\vec{n}$, and thus $\vec{r}$, are not necessarily inside the signal subspace. However, as will be discussed in Section 4.6, the application of the matched filter rejects any components outside of the signal subspace. Therefore, the assumption made above is valid.

4.6   MATCHED FILTER (MF)

As shown in the Section 4.5, in both MAP detection (4.74) and ML detection (4.75), the only metric depending on the received signal is its inner product with the signal "templates" $\{s_m\}$:

$$d_m \stackrel{\text{def}}{=} \int r(t) s_m^*(t) dt. \tag{4.82}$$

In this section, we further discuss the implementation of these detectors.

4.6.1   Matched Filter for Symbol Sequence

Discussions in Sections 4.4 and 4.5 are based on a single symbol. Namely, $s(t)$ represents one of the M possible symbols $\{s_m(t)\}$. In this case, (4.25) and (4.82) lead to

$$d_m = \int s(t) s_m^*(t) dt + \int n(t) s_m^*(t) dt. \tag{4.83}$$

In reality, symbols are transmitted in a sequence. Therefore, the received signal is

$$\bar{r}(t) = \sum_l s^{(l)} \left(t - lT_{Sym} \right) + n(t), \tag{4.84}$$

where $s^l(t)$ is the signal for the lth symbol. $s^l(t)$ is one of the legitimate symbols $\{s_m(t), m = 1, 2, \ldots, M\}$. $n(t)$ is the noise. T_{Sym} is the symbol interval. To detect the lth symbol, we desired the following metric to be extracted from $\bar{r}(t)$:

$$d_m^{(l)} = \int s^{(l)}(t) s_m^*(t) dt + \int n(t) s_m^* \left(t - lT_{Sym} \right) dt. \tag{4.85}$$

The time shift of lT_{sym} introduced in the second term is for the convenience of expression later. It does not affect the results because the statistics of $n(t)$ does not change over time.

Consider the following filtering on the received signal:

$$\bar{r}_{fm}(t) \overset{\text{def}}{=} \bar{r}(t) \otimes s_m^*(-t) = \int \bar{r}(\tau) s_m^*(\tau - t) dt. \tag{4.86}$$

From (4.84),

$$\begin{aligned} \bar{r}_{fm}(t) &= \sum_l \int s^{(l)} \left(\tau - lT_{Sym} \right) s_m^*(\tau - t) d\tau + \int n(\tau) s_m^*(\tau - t) dt \\ &= \sum_l \int s^{(l)}(\tau') s_m^* \left[\tau' - \left(t - lT_{Sym} \right) \right] d\tau' + \int n(\tau) s_m^*(\tau - t) dt \end{aligned} \tag{4.87}$$

Obviously, if

$$\int s^{(l)}(\tau') s_m^* \left[\tau' - kT_{Sym} \right] d\tau' = P^{(l, m)} \delta_{k,0} \tag{4.88}$$

where $P^{(l, m)}$ is a nonzero constant depending on l and m, then

$$\bar{r}_{fm} \left(lT_{Sym} \right) = d_m^{(l)}. \tag{4.89}$$

Therefore, if we sample the output of the filter (4.86) at instances lT_{sym}, we get the metric used for MAP or ML detection of the lth symbol. Because of (4.86) and (4.89), $s_m^*(-t)$ is known as the matched filter (MF) corresponding to transmitted symbol m. The metric $d_m^{(l)}$ is obtained by taking the received signal, filtering it with the MF, and sampling at time lT_{Sym}.

With MF, the MAP and ML detection algorithms (4.74) and (4.75) can be expressed for a symbol sequence. The detector's estimation of the transmitted symbol at lT_{Sym} is noted as $\hat{m}_l$, which is given by:

$$\hat{m}_l = \arg\max_m \left\{ Re\left[\bar{r}_{fm}\left(lT_{Sym} \right) \right] + \eta_m \right\}, \tag{4.90}$$

where $\bar{r}_{fm}(t)$ is given in (4.86).

$$\eta_m \overset{\text{def}}{=} \begin{cases} -\dfrac{1}{2}\displaystyle\int |s_m(t)|^2 dt + \dfrac{1}{2}N_0 & \ln\left[p_s(s_m) \right] \quad MAP \\[3mm] -\dfrac{1}{2}\displaystyle\int |s_m(t)|^2 dt & ML \end{cases}, \tag{4.91}$$

where N_0 is the noise power spectral density. $p_s(s_m)$ is the probability for s_m to be transmitted.

Condition (4.88) is related to the Nyquist criteria discussed in Chapter 3. If the pulse shaping function is $h(t)$, then the left-hand side of (4.88) is proportional to

$$\check{h}(t) \overset{\text{def}}{=} h(t) \otimes h^*(-t) = \int h(\tau)h^*(\tau - t)d\tau \tag{4.92}$$

sampled at kT_{sym}. Therefore, condition (4.88) says $\check{h}(t)$ should satisfy the Nyquist criterion. Namely,

$$\check{h}\left(kT_{Sym} \right) = C\delta_{k,0}, \tag{4.93}$$

where C is a constant.

In the frequency domain, the spectra of $\check{h}(t)$ and $h(t)$, $\tilde{\check{h}}(\omega)$ and $\tilde{h}(\omega)$, are related by

$$\tilde{\check{h}}(\omega) = \left| \tilde{h}(\omega) \right|^2. \tag{4.94}$$

Therefore, for this type of receiver, the pulse shaping filter should be the "square root" of a Nyquist filter. For example, the root-raised-cosine filter described in Chapter 3 can be used as a pulse shaping filter.

4.6.2 SNR with MF

In this section, we examine the input and output SNR of the baseband MF. As shown in an appendix (Section 4.9.2), the baseband signal PSD is given by (4.192):

$$P_{Wb}''(f) = \frac{1}{2}E_{Sym}\left[T_{Sym}\int \left| \tilde{h}(2\pi f) \right|^2 df \right]^{-1} \left| \tilde{h}(2\pi f) \right|^2, \tag{4.95}$$

where T_{Sym} is the symbol interval. $\tilde{h}(\omega)$ is the spectrum of the pulse shaping function. E_{Sym} is the passband symbol energy at the receiver input. The baseband noise PSD is given by (4.38) in Section 4.3.2:

$$P_{Wnb}(f) = \begin{cases} \dfrac{N_0}{2} & \text{if } |f| \le W_B \\ 0 & \text{otherwise} \end{cases}, \tag{4.96}$$

where W_B is the baseband frequency bandwidth. At the input of the MF, the baseband SNR is given by (4.50) in Section 4.3.3.1:

$$SNR_b = \frac{E_{Sym}}{N_0 W_p T_{Sym}}, \tag{4.97}$$

where W_p is the passband frequency bandwidth. With Nyquist symbol rate (Chapter 3),

$$W_p T_{Sym} = 1. \tag{4.98}$$

Equation (4.97) thus leads to

$$SNR_b = \frac{E_{Sym}}{N_0} \tag{4.99}$$

At the output of MF, we get a sample, which is the sum of a signal component and noise component. Therefore, instead of comparing the powers of signal and noise, we compare the energy contained in the signal and noise components of the output sample.

Let us first consider the case for a given m. Assume that $s_m(t)$ is transmitted and the MF $s_m(-t)$ is used. After the matched filtering (4.86), the signal energy is

$$E_{Symf}^{(m)} = \left[\int |s_m(t)|^2 dt \right]^2. \tag{4.100}$$

The noise energy after the MF is

$$E_{nf}^{(m)} = E\left[\left| \int n(t)s_m(t)dt \right|^2 \right] = \int E[n(t_1)n^*(t_2)]s_m^*(t_1)s_m(t_2)dt_1 dt_2. \tag{4.101}$$

With the noise autocorrelation (4.39) in Section 4.3.2, we have

$$E_{nf}^m = \frac{N_0}{2} \int |s_m(t)|^2 dt. \tag{4.102}$$

Therefore, SNR at the output of MF, when symbol m is transmitted, is

$$SNR_{fm} \stackrel{\text{def}}{=} \frac{E_{Symf}^{(m)}}{E_{nf}^{(m)}} = \frac{2}{N_0} \int |s_m(t)|^2 dt. \tag{4.103}$$

Averaging over all possible symbols, the expected output SNR is

$$SNR_f = \frac{2}{N_0} E\left[\int |s_m(t)|^2 dt\right]. \tag{4.104}$$

Recognize that $E[\int |s_m(t)|^2 dt]$ is the symbol energy in baseband E_{Symb}. According to (4.191) in Section 4.9.2,

$$E_{Symb} = \frac{1}{2} E_{Sym}. \tag{4.105}$$

Therefore,

$$SNR_f = \frac{E_{Sym}}{N_0}. \tag{4.106}$$

This result is the same as the input SNR in (4.99), when the Nyquist symbol rate given by (4.98) is used. Note that a larger bandwidth would allow more noise to enter the detector based on (4.97), reducing SNR_b. However, the matched filter rejects such excess noise automatically, leaving SNR_f in (4.106) unchanged. Therefore, when MF is used, any preceding low pass filters do not need to be very tight.

Equation (4.106) can be written as

$$SNR_f = \frac{E_b}{N_0} N_b. \tag{4.107}$$

Here E_b/N_0 is discussed in Section 4.3.3.2. N_b is the number of bit per symbol.

4.6.3   Implementation of MAP and ML Detectors

This section discusses some implementation considerations of MAP and ML detectors based on MF. The first two subsections discuss two alternative architectures. Section 4.6.3.3 provides an example with QAM modulation. Section 4.6.3.4 discusses the choice of the sampling rate.

4.6.3.1   Direct Matched Filters

MFs, as given in (4.86), can be implemented directly for the detectors specified in (4.90). In this case, there is one filter for each m. Figure 4.2 shows an example of such implementation.

In Figure 4.2, the input $\bar{r}(t)$ on the left side is the baseband received signal, modeled by (4.84). It is filtered by a bank of matched filters $s_m^*(-t)$ to form $\bar{r}_{fm}$, as shown in (4.86). The real part of $\bar{r}_{fm}$ is sampled at time lT_{Sym}, as shown in (4.89). η_m, given in (4.91),

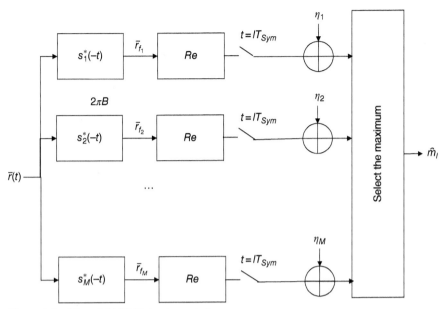

Figure 4.2   Direct MAP/ML implementation.

is added to the samples, to form M inputs to the selector. M is the number of possible symbols. The selector outputs $\hat{m}_l$, corresponding to the index with the maximum value, realizing the last step in (4.90).

4.6.3.2   Projection onto Basis Functions

Implementation in Section 4.6.3.1 is conceptually simple. However, in practice, M can be large (up to 256 or even higher in modern communication systems), requiring a large number of filters and thus a high complexity. On the other hand, in many cases, the signal space as described in Section 4.5.2 has very low dimension. It is thus advantageous to use the basis for signal space to perform matched filtering (4.73).

Figure 4.3 shows the MAP/ML detection implementation with the basis functions. Suppose the signal space has K basis functions $\{\phi_k(t), k = 1, 2, \ldots, K\}$. The input of the detector is baseband received signal $\bar{r}(t)$, modeled by (4.84). It is filtered by a bank of K filters $\{\phi_k(-t)\}$ and sampled at lT_{Sym}. The output of the filtering and sampling are the components of $\vec{r}$ for each symbol, as expressed in (4.77). The inner product (indicated in Figure 4.3 by a dot inside a circle) between $\vec{r}$ and vectors $\{\vec{s}_m\}$, as expressed in (4.79) in Section 4.5.2, is performed at the next stage. The vector $\{\vec{s}_m\}$ is obtained by projecting function $s_m(t)$ onto the basis. Namely, the kth component of the vector $\vec{s}_m$ is

$$(\vec{s}_m)_k = \int s_m(t)\phi_k^*(t)dt. \tag{4.108}$$

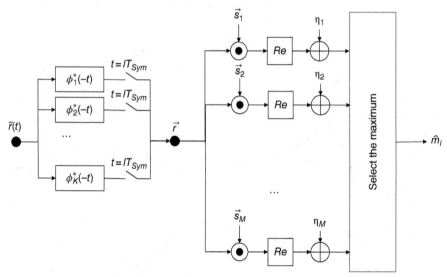

Figure 4.3   Basis projection MAP/ML implementation.

The real parts of the inner products between $\vec{r}$ and $\{\vec{s}_m\}$ are added to $\{\eta_m\}$, which are given by (4.91). The resulting values are fed to the selector, which finds the $\hat{m}_l$ that yields the maximum output, as expressed in (4.73).

An especially interesting case is the PSK and QAM modulation schemes described in Chapter 3. The baseband signal for these modulations can be expressed as follows:

$$s(t) = \sum_{l} S_b^{(l)} h\left(t - lT_{Sym}\right), \tag{4.109}$$

where $S_b^{(l)}$ is the lth complex-valued baseband symbol. T_{Sym} is the symbol duration. In this case, the only basis function in signal space is $h(t)$, the pulse shaping function. Therefore, only one (complex-valued) filter is required in Figure 4.3.

4.6.3.3   ML Detector for PSK and QAM Modulations
As described in Section 4.6.3.2, for PSK and QAM modulations, the signal space has only one basis function, which is the pulse shaping function. Furthermore, for QAM, ML detection decisions can be determined based on the real and imaginary parts of signal separately, as explained below. In this case, not only the MF can be implemented as a single filter, but the selector in Figure 4.3 also has a simple implementation. For QAM and PAM, we have

$$s_m(t) = S_m h(t), \tag{4.110}$$

where S_m is the complex-valued symbol. $h(t)$ is the pulse shaping function, which is real-valued. Equation (4.86) thus becomes

$$\bar{r}_{fm}(t) = S_m^* \int \bar{r}(\tau) h(\tau-t) d\tau. \tag{4.111}$$

where $\bar{r}(t)$ is the baseband received signal. Equation (4.111) shows that, independent of m, the only filtering needed is

$$\bar{r}_f(t) \stackrel{\text{def}}{=} \int \bar{r}(\tau) h(\tau-t) d\tau$$
$$\bar{r}_{fm}(t) = S_m^* \bar{r}_f(t) \tag{4.112}$$

This operation is a special case of the projection concept described in Section 4.5.2. Equations (4.90) and (4.91) (for ML) thus become

$$\hat{m}_l = \arg \max_m \left\{ Re\left[\bar{r}_{fm}\left(lT_{Sym} \right) \right] + \eta_m \right\}$$
$$= \arg \max_m \left\{ Re\left[S_m^* \bar{r}_f\left(lT_{Sym} \right) \right] - \frac{1}{2} |S_m|^2 \right\} \tag{4.113}$$

Therefore, the only information needed for detection is the sequence $\{R_l\}$:

$$R_l \stackrel{\text{def}}{=} \bar{r}_f\left(lT_{Sym} \right). \tag{4.114}$$

Above discussions apply to both PSK and QAM, as well as any other modulations in the form of (4.110). In the following, we focus on QAM modulation to show how to make detection decision. The same approach, with proper modifications, applies to PSK, as well.

Let us express the I and Q parts explicitly. Express the complex-valued quantities in real and imaginary parts:

$$S_m = S_{mI} + jS_{mQ}$$
$$R_l = R_{lI} + jR_{Ql} \tag{4.115}$$

Since $h(t)$ is real-valued, The MF can operate separately on I and Q parts:

$$R_{lI} = \int \bar{r}_I(\tau) h\left(\tau - lT_{Sym} \right) d\tau$$
$$R_{Ql} = \int \bar{r}_Q(\tau) h\left(\tau - lT_{Sym} \right) d\tau \tag{4.116}$$

Recall that in QAM modulation (Chapter 3), constellation points can be expressed as

$$P_m = B[(2m_I + 1) + j(2m_Q + 1)], \tag{4.117}$$

where B determines the spacing between constellation points. Integers m_I and m_Q determines the constellation point position in real and imaginary dimensions, respectively. The pair (m_I, m_Q) is mapped to symbol index m as defined in modulation (e.g., through Gray coding).

Therefore, in the context of (4.115), S_{mI} depends only on m_I, and S_{mQ} only on m_Q:

$$S_{mI} = S_{m_I I} \stackrel{\text{def}}{=} B(2m_I + 1)$$
$$S_{mQ} = S_{m_Q Q} \stackrel{\text{def}}{=} B(2m_Q + 1)$$

(4.118)

We thus separate the I and Q parts in (4.113):

$$\hat{m}_l = \arg\max_m \left\{ \left[S_{m_I I} R_{Il} - \frac{1}{2} S_{m_I I}^2 \right] + \left[S_{m_Q Q} R_{Ql} - \frac{1}{2} S_{m_Q Q}^2 \right] \right\}.$$

(4.119)

The first and second terms in (4.119) depend only on m_I and m_Q, respectively. We can thus defined the two terms as $d_{m_I I}$ and $d_{m_Q Q}$:

$$d_{m_I I}(l) \stackrel{\text{def}}{=} S_{m_I I} R_{Il} - \frac{1}{2} S_{m_I I}^2$$
$$d_{m_Q Q}(l) \stackrel{\text{def}}{=} S_{m_Q Q} R_{Ql} - \frac{1}{2} S_{m_Q Q}^2$$

(4.120)

Thus, (4.119) can be written as

$$\hat{m}_l = \arg\max_{m_I, m_Q} \left[d_{m_I I}(l) + d_{m_Q Q}(l) \right].$$

(4.121)

Since $d_{m_I I}(l)$ depends only on m_I and $\bar{r}_{fl}(lT_{Sym})$ while $d_{m_Q Q}(l)$ depends only on m_Q and $\bar{r}_{fQ}(lT_{Sym})$, $d_{m_I I}$ and $d_{m_Q Q}$ can be independently maximized by choosing m_I and m_Q:

$$\hat{m}_{Il} = \arg\max_m d_{mI}(l)$$
$$\hat{m}_{Ql} = \arg\max_m d_{mQ}(l)$$

(4.122)

Focus on d_{mI} for the moment. Consider two symbol candidates m_1 and m_2, with $S_{m_1 I} > S_{m_2 I}$. From (4.120),

$$d_{m_1 I} - d_{m_2 I} = (S_{m_1 I} - S_{m_2 I}) \left[R_{Il} - \frac{1}{2}(S_{m_1 I} + S_{m_2 I}) \right].$$

(4.123)

Therefore, whether m_1 or m_2 is a better choice depends on whether R_{Il} is larger or smaller than the middle point between $S_{m_1 I}$ and $S_{m_2 I}$. Namely, the decision on $\hat{m}_{Il}$ can be made by a "slicer," which chooses the $\hat{m}$ so that $S_{\hat{m}I}$ is the closest to R_{Il}. Similarly, $\hat{m}_Q$ can be determined by examining d_{mQ}, that is, by comparing R_{Ql} with S_{mQ}. The final decision $\hat{m}$ is a combination of $\hat{m}_I$ and $\hat{m}_Q$.

Figure 4.4 shows the structure of a slicer. The solid lines indicate the values of $S_{m_I I}$ or $S_{m_Q Q}$, with indices m_I or m_Q noted on the left. The dashed lines, which lay in the middle of spaces between the solid lines, demarcate the "decision regions." If the input falls within a region, then the corresponding index value is the decision.

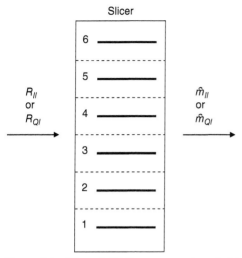

Figure 4.4   Slicer. Solid lines: assumed symbol levels; dashed lines: decision region boundaries.

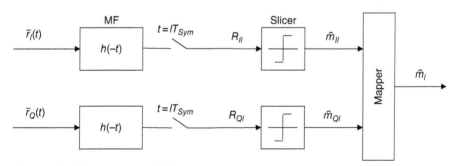

Figure 4.5   ML detector for QAM.

Figure 4.5 shows the QAM ML detector architecture based on the above discussion. The input data are filtered by MF, with I and Q signals following separate paths. The sampled outputs of MF are fed into the slicers, which determine $\hat{m}_I$ and $\hat{m}_Q$ based on constellation definition. These decisions are combined by the mapper to form a decision of $\hat{m}$. This figure also shows the standard symbol for slicers.

The concept of slicers can be used for PSK detection, as well. In this case, instead of separating the signal into real and imaginary parts, we focus on the phase of the signal and compare that with the expected constellation values.

4.6.3.4   Sampling Rates and Timing
Conceptually, an MF, as expressed in (4.86), operates in continuous time. Its output is sampled at multiples of the symbol interval T_{Sym}, forming a sequence with one sample

per symbol. In practice, since MF is implemented in the digital domain, its input is also a time-discrete sequence.

The sampling rate at the input of the MF depends on the bandwidth of pre-filtering. As discussed in Section 4.6.2, the output SNR of MF is insensitive to the signal bandwidth at the input. Therefore, one can use a loose low pass filter to reject higher frequency components after down-conversion, while allowing more out-of-band noise to enter MF. However, the input sampling rate of MF must be high enough to avoid "folding" the out-of-band noise into the signal band. To achieve this, the sampling rate R_s should satisfy

$$R_s \geq 2W_{BF}, \tag{4.124}$$

where W_{BF} is the baseband bandwidth of the prefiltering (not that of the signal). See Chapter 3 for more explanations.

Usually, the noise is injected before the MF in simulation. Care should be taken to match sampling rate and noise power per sample so that the output SNR of the MF is at the desired level (see Section 4.6.2 for more discussions on SNR).

MF can also be used to adjust for timing synchronization. If there is a timing offset t_o at the receiver, then instead of receiving $r(t)$, $r(t+t_o)$ is received. In this case, the corresponding MF in (4.86) is changed to $s_m^*(-t-t_o)$. Sampling after the MF is still the same. Such change of MF can be implemented by changing the tap values of the filters. No additional timing correction processing is necessary. Of course, how to determine the tap values in response to timing error is a different issue. In addition to dedicated synchronization techniques discussed in Chapter 7, adaptive filters can be used for MF, to respond to timing offsets. Chapter 9 will provide more detailed discussions.

4.6.4 Discrete Channel Model

In this section, we build onto the MF formulated by (4.111). This formulation is a special case in the theory of MF, applicable only when the modulated signals are generated in the form of (4.110). However, the most popular modulations, QAM and PSK, use such form. Therefore, people often mean (4.111) when they use the term "matched filter."

As discussed in Chapter 3 and Section 4.3.2, at the input of the receiver (baseband), the received signal $y_b(t)$ from an AWGN channel can be modeled as

$$\bar{r}(t) = s_b(t) + n_b(t)$$
$$s_b(t) = \sum_m S_m h(t - mT_{Sym})$$
$$E[n_b(t_1)n_b(t_2)] = \frac{N_0}{2}\delta(t_1 - t_2) \tag{4.125}$$

Here $\{S_m\}$ is the sequence of transmitted symbols. $s_b(t)$ is the transmitted signal in baseband. $n_b(t)$ is the added noise in baseband, which is white. $h(t)$ is the pulse shaping function in the baseband.

After *MF*, the signals are sampled at the symbol rate, to form a discrete-time sample sequence. As shown in Section 4.6.3.3, such sample sequence can be separated into the signal component $\{s_l\}$ and noise component $\{n_l\}$:

$$R_l = \int \bar{r}(\tau)h(\tau - lT_{Sym})d\tau = s_l + n_l$$
$$s_l \overset{\text{def}}{=} \int \sum_m S_m h(\tau - mT_{Sym})h(\tau - lT_{Sym})d\tau. \qquad (4.126)$$
$$n_l \overset{\text{def}}{=} \int n_b(\tau)h(\tau - lT_{Sym})d\tau$$

As discussed in Section 4.6.1, the square of the pulse shaping function $h(t)$ satisfies the Nyquist criterion, which leads to:

$$\int h(\tau - mT_{Sym})h(\tau - lT_{Sym})d\tau = A\delta_{m,l}, \qquad (4.127)$$

where A is a constant. With such property,

$$s_l = AS_l. \qquad (4.128)$$

As to noise, we have

$$E(n_m n_l) = \int E[n_b(\tau_1)n_b(\tau_2)]h(\tau - mT_{Sym})h(\tau - lT_{Sym})d\tau_1 d\tau_2. \qquad (4.129)$$

From the correlation of $n_b(t)$ in (4.125) and the property (4.127), (4.129) leads to

$$E(n_m n_l) = \frac{N_0}{2}A\delta_{m,l}. \qquad (4.130)$$

Namely, the noise is white. Furthermore, we can always scale the receiver so that A is 1. Therefore, at the output of MF, we have a time-discrete channel model:

$$R_l = S_l + n_l \qquad (4.131)$$

Here, S_l is the lth transmitted symbol, and n_l is a sample from a *white Gaussian* noise.

Naturally, one would like to know the SNR in R_l, namely, the ratio between $E(|S_l|^2)$ and $E(|n_l|^2)$. SNR at the output of MF is discussed in Section 4.6.2. From the expected SNR in (4.106), we have (with proper scaling at the receiver)

$$E\left(|S_l|^2\right) = E_{Sym}$$
$$E\left(|n_l|^2\right) = N_0 \qquad (4.132)$$

Equations (4.131) and (4.132) form the discrete channel model with an MF.

4.6.5 More Discussions on ML Detector and Matched Filter

As shown in Section 4.6.3, the MF is a major part in MAP or ML detectors for AWGN channel. In fact, as far as these detectors are concerned, the output of MF is the only information about the received signal. In this section, we will provide some more discussions on MF in the context of ML detection, to help understanding why MF leads to optimal detection. Most of the conclusions can easily be extended to MAP detection. These discussions also show that noise Gaussianity is not required for ML to be a building block of ML and MAP detectors.

4.6.5.1 ML Detection in Signal Space

As shown in the derivation from (4.69) to (4.75) in Section 4.5.1, the concept of MF was introduced based on the need to minimize $|\vec{r} - \vec{s}_m|^2$, which can also be expressed in terms of functions: $\int |r(t) - s_m(t)|^2 dt$. Both $|\vec{r} - \vec{s}_m|$ and $\int |r(t) - s_m(t)|^2 dt$ are distances in signal space. In this context, noise can also be expressed as a vector $\vec{n}$.

In the signal space, $\{s_m, m = 1, 2, \ldots M\}$ represents M points in the space (in the case of PSK and QAM, these are actually constellation points in a two-dimensional space). One of such points is transmitted as a signal. Ideally, received signal r should fall onto the same point. However, because of the noise, received signal is "kicked" away from the transmitted point, as shown in the AWGN channel model (4.25) or (4.80). As discussed in Section 4.5, ML detection picks the closest point to r among $\{s_m, m = 1, 2, \ldots M\}$ as the transmitted signal.

For such determination to be consistent with the ML principle (4.62), the noise probability density function $p_n(\vec{n})$ distribution needs to have the following properties:

1. $p_n(\vec{n})$ depends on $|\vec{n}|^2$ only. Namely, it is "isotropic" in signal space.

2. $p_n(\vec{n})$ decreases when $|\vec{n}|^2$ increases.

In fact, the entire derivation in Section 4.5.1 depends only on the above properties, although Gaussian noise is used there. With such property, the ML detection has a geometric interpretation. It picks the legitimate transmitted signal point that has the shortest distance from the received signal in signal space. Such strategy works for a large family of noises. MF, which computes the projection of the received signal onto the transmitted signal candidates, is a tool for calculating the distances.

4.6.5.2 MF as SNR Maximization

In this section, we will show that with white noise, among all choices of its tap values, the MF given by (4.86) produces maximum SNR at the output.

Let $f(-t)$ be the receiver filter. With the channel model (4.25) and assume that $s_m(t)$ is transmitted, received signal $r(t)$ is expressed as

$$r(t) = s_m(t) + n(t). \tag{4.133}$$

The second term is the noise. After filtering and sampling at $t = 0$, we have

$$r_f = \int s_m(t)f(t)dt + \int n(t)f(t)dt. \tag{4.134}$$

The first term is the signal, and the second the noise. From this and the property of white noise (4.26), we can follow the approach in Section 4.6.2 and get the output SNR as

$$SNR_f = \frac{\left|\int s_m(t)f(t)dt\right|^2}{(N_0/2)\int |f(t)|^2 dt}. \tag{4.135}$$

Maximization of SNR_f can be achieved with the Cauchy–Schwarz inequality [8, sec. 1.7(ii)], which states that for any functions $a(t)$ and $b(t)$,

$$\left|\int a(t)b^*(t)dt\right|^2 \le \int |a(t)|^2 dt \int |b(t)|^2 dt. \tag{4.136}$$

The equality holds if and only if $b(t)/a(t)$ is a constant. Therefore, maximum SNR_f is obtained with

$$f(t) = c s_m^*(t), \tag{4.137}$$

where c is any constant. For convenience, we can choose $c = 1$. We have thus shown that the MF given by (4.86) indeed maximizes SNR_f. Again, such conclusion is based on the assumption that the noise is white, but not necessarily Gaussian.

4.6.5.3 Matched Filter and Sufficient Statistics

As shown in Section 4.5, for AWGN channels, the optimal MAP and ML detection lead to MF. Namely, the received signal can be "reduced" to its filtered and sampled version as shown in (4.90) in Section 4.6.1. Apparently, information is discarded in the process.[5] In other words, received signal $r(t)$ cannot be reconstructed from $\{\bar{r}_{fm}(lT_{Sym}), m = 1, 2, ..., M, l = ..., -1, 0, 1, ...\}$ as defined by (4.86). Yet, such reduced version is enough for optimal detection. The question is: can such "sufficiency" of MF be extended to more general cases? It turns out that the "sufficiency" of MF relies on weaker conditions. Therefore, it is more general than the AWGN channel model.

There is a mathematical tool to address this issue, known as sufficient statistics (see, e.g., [9]). In this section, we provide an intuitive explanation following [5].

As outlined in Section 4.5.1 and detailed in an appendix (Section 4.11), any waveform (i.e., functions of t) can be expressed as a vector, with orthonormal basis functions $\{\phi_k(t), k = 1, 2, ...\}$. All transmitted signals lay in a subspace $\mathcal{S}$ (known as the signal space), which is of finite dimension $K_\mathcal{S}$. On the other hand, the entire space

---

[5] The term "information" here is in general sense, not in the sense of information theory as discussed in Chapter 2.

can be "divided" into $\mathcal{S}$ and its "orthogonal complement" $\mathcal{S}_\perp$. $\mathcal{S}_\perp$ is the set of vectors that are orthogonal to all vectors in $\mathcal{S}$. Such division is "complete," in the sense that any vector $\vec{u}$ can be decomposed into

$$\vec{u} = \vec{u}_S + \vec{u}_{S_\perp}, \tag{4.138}$$

with $\vec{u}_S \in \mathcal{S}$ and $\vec{u}_{S_\perp} \in \mathcal{S}_\perp$. Thus, we can decompose received signal $r(t)$ and noise $n(t)$ into these two subspaces:

$$\begin{aligned} r(t) &= r_S(t) + r_{S_\perp}(t) \\ r_S(t) &\in \mathcal{S} \\ r_{S_\perp}(t) &\in \mathcal{S}_\perp \end{aligned} \quad . \tag{4.139}$$

n_S and $n_{S_\perp}$ are defined similarly. Because transmit signal $s(t)$ is in subspace $\mathcal{S}$ only, the additive noise channel model (4.25) can be written as

$$\begin{aligned} r_S(t) &= s(t) + n_S(t) \\ r_{S_\perp}(t) &= n_{S_\perp}(t) \end{aligned} \quad . \tag{4.140}$$

Furthermore, for white noise, it is shown in an appendix (Section 4.11.3) that noise projections onto different basis functions are mutually independent. Therefore, n_S and $n_{S_\perp}$ are also mutually independent. Consequently, (4.140) implies that $r_{S_\perp}$ is independent of r_S and s.

Through the analysis in Section 4.6.1, we see that the sampled MF output $\{\bar{r}_{fm}(lT_{Sym}), m = 1, 2, \ldots, M, l = \ldots, -1, 0, 1, \ldots\}$ can be used to reconstruct $r_S(t)$.[6] Therefore, the only information we discarded in the process is $r_{S_\perp}(t)$. Since it is independent of s and r_S, namely,[7]

$$p(s \mid r_S, r_{S_\perp}) = p(s \mid r_S), \tag{4.141}$$

$r_{S_\perp}(t)$ has no impact to the detection result. Therefore, MF outputs are sufficient. Namely, they capture all information needed for detection.

It can be seen from above discussion that even if the channel is not AWGN, MF can still be used to reduce the input signal $r(t)$ to a simpler form $\{\bar{r}_{fm}(lT_{Sym}), m = 1, 2, \ldots, M, l = \ldots, -1, 0, 1, \ldots\}$, in preparation for future processing. Such approach will be used for channels with inter-symbol interference, treated in Chapter 9. The only condition for such conclusion is that additive noise n_S is independent of $n_{S_\perp}$. White noise is a sufficient but not necessary condition for such independence.

---

[6] $\bar{r}_{fm}(lT_{Sym})$ is the projection of $r(t - lT_{Sym})$ onto $s_m(t - lT_{Sym})$. Since $\{s_m(t), m = 1, 2, \ldots, M\}$ is a basis (it may not even be linearly independent) in the signal space $\mathcal{S}$, $\{\bar{r}_{fm}(lT_{Sym}), m = 1, 2, \ldots, M\}$ captures all components of $r_S(t - lT_{Sym})$.

[7] Note that just saying $r_{S_\perp}$ is independent of s is not enough. $r_{S_\perp}$ must be independent of s *given* r_S, or equivalently, it must be independent of r_S, as well.

4.6.6   Summary

Section 4.6 Introduces MF as an implementation of MAP/ML detection for AWGN channel. Section 4.6.1 shows that for symbol sequences, MF realizes the MAP/ML detection algorithms, given that the pulse shaping function satisfies the Nyquist criterion. Section 4.6.2 derives the output SNR of MF. Such result will be used in Section 4.7 for error probability evaluation. Section 4.6.3 discusses MF implementation approaches. For low-dimensional signals such as (4.109), the MF structure can be quite simple. Slicer is further introduced in Section 4.6.3.3 for QAM and PSK demodulation. Section 4.6.5 provides some more understanding of MF from various perspectives.

Although MF was introduced as a tool for MAP/ML detection for AWGN channel, its application is more general. In fact, most discussions in this section only require the additive noise to be white (as a sufficient but not necessary condition). Gaussianity is in general not needed for the optimality of MF-based detectors and associated SNR analysis. Furthermore, MF output is a sufficient statistics of the transmitted signal. Namely, it can be used in receivers that require further signal processing without affecting optimality.

4.7   ERROR PROBABILITY OF UNCODED MODULATIONS UNDER AWGN MODEL

Sections 4.4–4.6 introduced the optimal detectors, and their application to the AWGN channels. In this section, we move on to examine the error probability of PSK and QAM signals with AWGN channels. It is assumed that all constellation points are equiprobable. Therefore, ML detection is used. In this section, we consider the time-discrete channel model presented in Section 4.6.4.

4.7.1   Pairwise Error Probability

Recall the channel model at the output of the MF, (4.131) and (4.132) in Section 4.6.4. The received symbol R is expressed as

$$R = S + n$$

$$E\left(|S|^2\right) = E_{Sym} .$$ (4.142)

$$E\left(|n|^2\right) = N_0$$

Here S is the transmitted symbol. n is a white Gaussian noise. E_{Sym} is the signal energy per symbol at the receiver input. N_0 is the noise power spectrum density. Since we consider single symbol detection here, we drop the indices in (4.131) and (4.132).

We start by studying the error probability between two constellation points m and m'. As discussed in Section 4.4.3, the pairwise error probability is given by (4.63). When transmit signal is S_m and the received signal is R, an error occurs during ML detection when

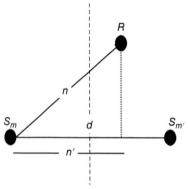

Figure 4.6   Pairwise errors. The various quantities are given from (4.142) to (4.145).

$$p(R \mid S_{m'}) > p(R \mid S_m), m' \neq m. \tag{4.143}$$

Of course, the probability of this happening depends on the noise level.

Now, for the two constellation points S_m and $S_{m'}$. Figure 4.6 illustrates the relationship among these quantities. The two bottom dots represents signals S_m and $S_{m'}$ in the complex plane. The distance between them is d. Received signal R is represented by the dot on top. R is the sum of the transmitted signal S_m and noise n (represented by the line connecting S_m and R). For ML detector under AWGN model,[8] an error happens (i.e., the transmitted signal is mistaken as $s_{m'}$) when

$$|R - S_{m'}| < |R - S_m|. \tag{4.144}$$

Such situation happens when R is located to the right of the vertical dashed line, which is the perpendicular bisector of the line segment connecting m and m' (the solid horizontal line). Let n' be the projection of n onto the line segment connecting s_m and $s_{m'}$, as shown in the figure. Error happens when

$$n' > \frac{d}{2}. \tag{4.145}$$

Since n is white Gaussian, any dimension of n, and thus n', obeys Gaussian distribution with the same variance. Since this is a two-dimensional space (expressed as real and imaginary parts) and the total variance is N_0 as stated before, the variance for each dimension is $N_0/2$. Namely, the probability density function for n' is, from (4.29) in Section 4.3.1.3,

$$p_n(n') = \frac{1}{\sqrt{\pi N_0}} e^{-\frac{n'^2}{N_0}}. \tag{4.146}$$

---

[8] Actually (4.144) is valid as long as $p_n(n)$ decreases with $|n|$. It does not need to be Gaussian.

Therefore, the pairwise error probability is

$$P_{mm'} = p\left(n' > \frac{d}{2}\right) = \frac{1}{\sqrt{\pi N_0}} \int_{d/2}^{\infty} e^{-\frac{n'^2}{N_0}} dn' = Q\left(\sqrt{\frac{d^2}{2N_0}}\right), \qquad (4.147)$$

where

$$Q(x) \overset{\text{def}}{=} \frac{1}{\sqrt{2\pi}} \int_x^{\infty} e^{-\frac{u^2}{2}} du \qquad (4.148)$$

is the well-known "Q-function." Q-function and its inverse function are supported by MATLAB functions qfunc and qfuncinv. Q-function is also related to the "error function," commonly known as erf:

$$erf(x) \overset{\text{def}}{=} \frac{2}{\sqrt{\pi}} \int_0^x e^{-t^2} dt. \qquad (4.149)$$

Various properties (including approximation formulas) for the Q-function and error function have been reported in literature and mathematical handbooks [8, ch. 7].

If S_m and $S_{m'}$ are the two constellation points in a BPSK modulation, we have $d = 2\sqrt{E_{Sym}}$. Therefore, the pairwise error probability for BPSK is

$$P_{eBPSK} = Q\left(\sqrt{\frac{2E_{Sym}}{N_0}}\right). \qquad (4.150)$$

Equation (4.147) is the starting point of error probability computation in Sections 4.7.2 and 4.7.3.

4.7.2 Error Probability of PSK

For PSK, the minimum distance between constellation points is given in Chapter 3:

$$d_{PSK} = 2A\sin\left(\frac{\pi}{M}\right), \qquad (4.151)$$

where A is the distance between the constellation points and origin:

$$A = \sqrt{E_{Sym}}. \qquad (4.152)$$

M is the number of constellation points (i.e., the order of modulation).

Therefore, according to (4.147), the symbol error probability between neighboring points is

$$P_{mm'} = Q\left[\sqrt{\frac{2E_{Sym}}{N_0}}\sin\left(\frac{\pi}{M}\right)\right].$$ (4.153)

For PSK, each point has two nearest neighbors and $M-3$ other points. As discussed in Section 4.4.3, an upper bound for symbol error probability is the union bound (UB), where all pairwise error probability are further upper-bounded by the nearest neighbor case (4.153):

$$P_{UB,PSK} = (M-1)P_{mm'} = (M-1)Q\left[\sqrt{\frac{2E_{Sym}}{N_0}}\sin\left(\frac{\pi}{M}\right)\right].$$ (4.154)

Another approximation, also discussed in Section 4.4.3, is the nearest neighbor (NN) approximation, where only errors among the nearest neighbors are counted:

$$P_{NN,PSK} = 2P_{mm'} = 2Q\left[\sqrt{\frac{2E_{Sym}}{N_0}}\sin\left(\frac{\pi}{M}\right)\right].$$ (4.155)

Equations (4.154) and (4.155) can be converted to bit error rate by replacing symbol error probability with BER through (4.68) and replacing E_{Sym} with E_b through

$$E_{Sym} = N_b E_b,$$ (4.156)

where N_b is the number of bits per symbol. The approximations of the probability of bit error (under Gray coding) for PSK are thus:

$$\begin{aligned} P_{bUB,PSK} &= \frac{(M-1)}{N_b}Q\left[\sqrt{2N_b\frac{E_b}{N_0}}\sin\left(\frac{\pi}{M}\right)\right] \\ P_{bNN,PSK} &= \frac{2}{N_b}Q\left[\sqrt{2N_b\frac{E_b}{N_0}}\sin\left(\frac{\pi}{M}\right)\right] \end{aligned}.$$ (4.157)

For low-order PSK (BPSK and QPSK), more accurate estimates can be obtained. For BPSK, the result is given by (4.150) in Section 4.7.1, and $N_b = 1$.

$$P_{b,BPSK} = Q\left(\sqrt{2\frac{E_b}{N_0}}\right).$$ (4.158)

QPSK can be viewed as two orthogonal BPSK modulations. The energy per bit is still the same as energy per dimension. Therefore, the total symbol error rate is

$$P_{QPSK} = 1 - (1 - P_{BPSK})^2 = 1 - \left[1 - Q\left(\sqrt{2\frac{E_b}{N_0}}\right)\right]^2. \quad (4.159)$$

The bit error rate is

$$P_{b,QPSK} = \frac{1}{2}\left\{1 - \left[1 - Q\left(\sqrt{2\frac{E_b}{N_0}}\right)\right]^2\right\}. \quad (4.160)$$

When the pairwise error probability (i.e., the value of the Q-function) is small, (4.160) can be approximated by dropping the square of the Q-function:

$$P_{b,QPSK} \approx Q\left(\sqrt{2\frac{E_b}{N_0}}\right). \quad (4.161)$$

This result is the same as (4.158). Moreover, it is the same as $P_{bNN,PSK}$ in (4.157) with $N_b = 2$, $M = 4$.

Figure 4.7 shows the BER versus E_b/N_0 curves for QPSK, 8-PSK, and 32-PSK. For each modulation, the union bound $P_{bUB,PSK}$ and nearest neighbor approximation $P_{bNN,PSK}$, both given by (4.157), as well as the exact error rate, are plotted. The nearest neighbor approximation almost coincides with the exact rate, except for very high BER. The union bound has a vertical offset, representing an over estimation. However, usually, the accuracy of an approximation is evaluated by the difference in E_b/N_0 for a given BER. From that point of view, the union bound is still a fairly good approximation (at low BER, the error in E_b/N_0 is less than 1 dB).

Plots similar to Figure 4.7 are commonly used in digital communications to characterize system performance. They are often referred to as the "waterfall plot" in the trade.[9] We will see more of such plots in Section 4.7.3 and Chapter 5.

4.7.3 Error Probability of QAM

In this section, we continue to study the error probability of QAM. We limit ourselves to square QAMs (i.e., M is an integer power of 4). The basic approach can be extended to nonsquared QAM modulations, although details will change. As shown in Chapter 3, if the distance between nearest neighbors for a square QAM is $2B$, then the average symbol energy is

---

[9] Outside of communications, waterfall plot refers to a type of plots for three-dimensional data.

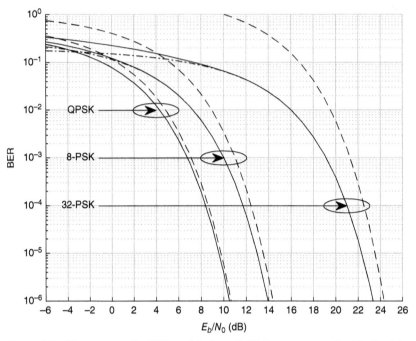

Figure 4.7   Bit error rates for PSK modulations. Solid lines: exact results (obtained from MATLAB function berawgn); dashed lines: union bounds; dot-dashed lines: nearest neighbor approximations.

$$E_{Sym} = \frac{2}{3}B^2(M-1) \tag{4.162}$$

where M is the number of constellation points. Assuming M is always an integer power of 2, the number of bits per symbol is

$$N_b = \log_2(M). \tag{4.163}$$

Therefore, the minimum distance d_m can be expressed in terms of E_{Sym}:

$$d^2 = (2B)^2 = \frac{6E_{Sym}}{M-1}. \tag{4.164}$$

The pair-wise symbol error probability (4.147) becomes

$$P_{mm'} = Q\left[\sqrt{\frac{3E_{Sym}}{(M-1)N_0}}\right]. \tag{4.165}$$

Similar to the PSK case in Section 4.7.2, we can have the union bound symbol error probability:

$$P_{UB,QAM} = (M-1)P_{mm'} = (M-1)Q\left[\sqrt{\frac{3E_{Sym}}{(M-1)N_0}}\right]. \tag{4.166}$$

For the nearest neighbor approximation, we need to know the number of nearest neighbors of the constellation points. For square constellations with M constellation points, there are three types of points in terms of the number of nearest neighbors (assuming $M \geq 4$):

$\left(\sqrt{M}-2\right)^2$ interior points, each with four neighbors.

4 corner points, each with two neighbors.

$4\left(\sqrt{M}-2\right)$ edge points, each with three neighbors.

Therefore, the average number of neighbors is

$$K_N = \frac{1}{M}\left[4\left(\sqrt{M}-2\right)^2 + 2\times4 + 3\times4\left(\sqrt{M}-2\right)\right] = \frac{4}{\sqrt{M}}\left(\sqrt{M}-1\right). \tag{4.167}$$

When the error probabilities are much smaller than 1 (as in most practical cases), the error probability with nearest neighbor approximation is

$$P_{NN,QAM} = \frac{4}{\sqrt{M}}\left(\sqrt{M}-1\right)P_{mm'} = \frac{4}{\sqrt{M}}\left(\sqrt{M}-1\right)Q\left[\sqrt{\frac{3E_{Sym}}{(M-1)N_0}}\right]. \tag{4.168}$$

Or, in terms of bit errors and E_b/N_0,

$$\begin{aligned} P_{bUB,QAM} &= \frac{M-1}{N_b}Q\left[\sqrt{\frac{3N_b}{(M-1)}\frac{E_b}{N_0}}\right] \\ P_{bNN,QAM} &= \frac{4\left(\sqrt{M}-1\right)}{N_b\sqrt{M}}Q\left[\sqrt{\frac{3N_b}{(M-1)}\frac{E_b}{N_0}}\right] \end{aligned}. \tag{4.169}$$

For 4-QAM ($M = 4$), $P_{bNN,\,QAM}$ in (4.169) becomes the same as (4.161) for QPSK, as expected.

Figure 4.8 shows BER for QAM with various orders of modulation, organized in the same way as Figure 4.7. Figure 4.8 shows the same trends as for PSK. Namely, the union bound overestimates BER; however, in terms of offset in E_B/N_0, the error is not very big. On the other hand, the nearest neighbor approximation matches with the exact results very well, expect for very high BER values.

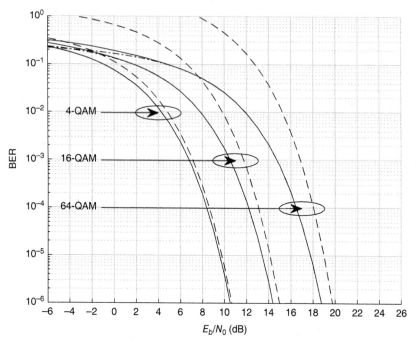

Figure 4.8 Bit error rates for QAM modulations. Solid lines: exact results (obtained from MATLAB function berawgn); dashed lines: union bounds $P_{bUB,QAM}$ given by (4.169); dot-dashed lines: nearest neighbor approximations $P_{bNN,QAM}$ given by (4.169).

From (4.169) and (4.106), we can further express the BER (under nearest neighbor approximation) in terms of SNR:

$$P_{bNN,QAM} = \frac{4\left(\sqrt{M}-1\right)}{\sqrt{M}N_b} Q\left[\sqrt{\frac{3}{(M-1)}SNR}\right]. \qquad (4.170)$$

Consider the case when N_b is increased by 2, that is, when M is multiplied by 4. What does SNR need to change in order to keep $P_{bNN,\ QAM}$ approximately the same? As M changes, the Q-function changes much faster than the factor $4\left(\sqrt{M}-1\right)/\sqrt{M}N_b$. Therefore, we wish to keep the argument of the Q-function constant. Namely, SNR needs to increase by a factor of 4, or approximately 6 dB. Therefore, the "rule of thumb" is that when adding two more bits per symbol in QAM, the required SNR for a given BER is increased by 6 dB. This is known in the trade as the "6 dB rule."

Table 4.1 illustrates the "6 dB rule" estimation. The top row lists the M values in QAM modulations. For each BER value (listed in the first column), the table lists the required SNR (in dB). For each BER value, the exact SNR is computed from (4.170).

TABLE 4.1 Exact and estimated SNR requirements (dB).

BER		4-QAM	16-QAM	64-QAM	256-QAM
1e–05	Exact	12.6	19.5	25.6	31.5
	Estimate	13.5	19.5	25.5	31.5
	Difference	0.9	0.0	−0.1	−0.1
1e–07	Exact	14.3	21.2	27.4	33.4
	Estimate	15.2	21.2	27.2	33.2
	Difference	0.9	0.0	−0.2	−0.2

The estimate SNR is obtained from the "6 dB rule" stated above, using the exact SNR for 16-QAM as a reference. The differences between the two SNR evaluations are also listed in the third row. As shown in Table 4.1, when M is large (64 and 256), the "6 dB rule" estimate yields accurate results. However, when M is 4, the error becomes larger.

In practice, we can memorize the SNR requirements for one of the M values and easily estimate the requirements for other M values using the "6 dB rule."

4.7.4 About Channel Coding

Sections 4.7.2 and 4.7.3 focus on PSK and QAM modulations without channel coding. Channel coding, discussed in Chapter 5, can significantly improve system performance (i.e., achieving lower BER at the same SNR). Although channel coding is ubiquities in modern communication systems, results for uncoded systems are still useful. The reason is that the benefit of channel coding can often be characterized as "coding gain," which is the difference in required SNR values (in dB) between the coded and uncoded systems when they achieve the same BER. Such SNR differences are considered to be insensitive to the BER values of concern and the type of modulations.

Figure 4.9 compares performance between coded and uncoded modulations. These data were obtained from MATLAB application "bertool." The thick lines are BER versus E_b/N_0 for uncoded modulations, as marked in the legend. The thin lines show the performance of the same modulations with convolutional coding with coding rate 1/2. We can see from these plots that in the BER range of 10^{-8}–10^{-5}, coding gain is very close to 4 dB across all modulations. Therefore, in this case, performance for coded modulations can be estimated based on that of the uncoded ones with a constant coding gain.

Note that while such behavior is typical for convolutional codes and block codes, things are different for other modern codes such as turbo codes and low-density parity check (LDPC) codes. With these codes, the slopes of the BER versus E_b/N_0 curve are typically steeper than those of the uncoded modulations. This means coding gain changes with the BER values of interest. More detailed discussions will be provided in Chapter 5.

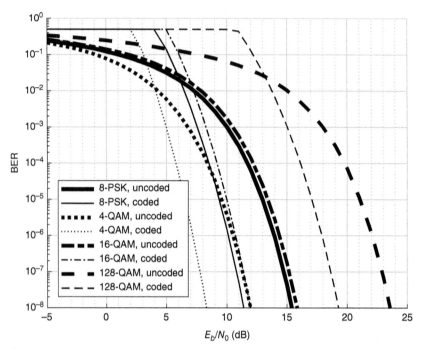

Figure 4.9 Performance of coded and uncoded modulations. Thick lines: uncoded modulations; thin lines: convolutional-coded modulations.

4.8 SUMMARY

This chapter examines the demodulation part of the receiver, as indicated in Figure 4.1. To understand receiver algorithms, we first need to understand the nature of noise. In particular, we are interested in the time domain behavior (autocorrelation) and frequency domain behavior (PSD). Such understanding is based on the mathematical concept of WSS and proper random variables, as summarized in Sections 4.2 and 4.3.

As pointed out in Section 4.4, under fairly general assumptions, the optimal receiver for signals corrupted by noise is MAP, and, if all signal alphabets are equiprobable, ML. These detectors are based on the idea of finding the most probable transmitted symbol, conditioned on the received signal.

In Section 4.5, MAP and ML detectors are applied to the special case of AWGN channels, where these detectors are implemented by MF. It turns out that MF has some more general properties that make it applicable beyond AWGN. More details on MF is provided in Section 4.6.

Given the detection algorithms, we move on to the error probabilities in Section 4.7. Although we focus on the two modulations PSK and QAM, the basic

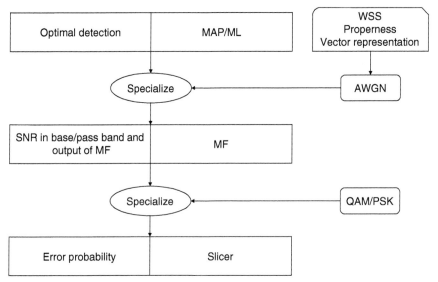

Figure 4.10   Concept architecture of this chapter.

concept of error probability computations introduced there (i.e., using minimum signal distance) is widely applicable. For example, in Chapter 5 will show another application.

Figure 4.10 shows the concept architecture of this chapter. On the left side, three groups of boxes show the progression of specialization, whose steps are indicated by the rounded rectangles on the right side. For each step, the left box notes the theoretical concepts and the right box notes the implementation techniques. In addition, three mathematical concepts are introduced to support the AWGN model, as indicated by the block on the upper right.

Although most of the analyses in this chapter are in the realm of AWGN, the conclusions are actually less limiting. Some of the conclusions require only that the noise is WSS, while most of the others need the noise to be white. The optimality of ML depends on the fact that the probability of noise n is a decreasing function of $|n|$. This requirement is more general than noise Gaussianity. However, the error probability computations in Section 4.7, especially the use of Q-function, are based on Gaussianity of noise, although the same methodology is applicable to other noise statistics.

This chapter gives explicit detector formulation only for QAM modulation (Section 4.6.3.3). The same approach can be extended to PSK modulations with relative ease. For other modulations (including some variations of PSK modulations), the general principles are still applicable, yet the details in design and implementation may be different. More modulation schemes are analyzed in other textbooks [7].

In this book, some other chapters are built on the concepts and principles discussed in this chapter. For example, Chapter 5 extends performance analysis to systems with channel coding. Chapter 9 extends channel model to include inter-symbol interference. Chapter 11 extends the concept of MF to the spatial dimension.

4.9   APPENDIX: PSD OF MODULATED SIGNALS

In this section, we consider the PSDs of modulated signals. A modulated signal is a stochastic process because it depends on the random modulating symbols. However, such signal is not WSS. In this section, we will establish the relationship between the passband and baseband PSDs at the receiver and connect them with symbol energy in the passband.

4.9.1   Up-Conversion and Down-Conversion Formulation

The formulation of up-conversion and down-conversion is detailed in Chapter 3. For simplicity, we consider one form of modulation, where the complex baseband waveform is

$$\bar{S}_b(t) = \sum_{n=1}^{N} \bar{S}_n h(t - nT_{Sym}), \qquad (4.171)$$

where $h(t)$ is the pulse shaping function, which has a baseband bandwidth of W_b.[10] T_{Sym} is the symbol interval. Following the same time windowing idea as in Section 4.2.3, we are limiting the summation over symbols to between 1 and N. Later on, we will let N go to infinity.

This form covers the modulation schemes discussed in Chapter 3, PSK and QAM. Results in this section can be extended to more general modulations, where $\{s_m(t)\}$ takes different forms depending on m.

The complex modulating symbols can be expressed as

$$\bar{S}_n \stackrel{\text{def}}{=} S_{In} + jS_{Qn}, \qquad (4.172)$$

where S_{In} and S_{Qn} are the real and imaginary parts of $\bar{S}_n$. In other words, the modulating symbols form two discrete stochastic processes $\{S_{In}, n = 1, 2, \dots\}$ and $\{S_{Qn}, n = 1, 2, \dots\}$. They have the following statistical properties:

$$E(S_{In}) = E(S_{Qn}) = 0, \forall n$$
$$E(S_{In}S_{Im}) = E(S_{Qn}S_{Qm}) = G\delta_{n,m}. \qquad (4.173)$$
$$E(S_{In}S_{Qm}) = 0 \forall n, m$$

Namely, all modulating symbols are independent and identically distributed (iid) with zero mean. G is the energy per symbol per dimension.

Based on the statistics of the symbols (4.173), it is easy to verify that $\bar{S}_b(t)$ is a proper stochastic process with zero mean (see Section 4.2.2 for more details on proper stochastic processes):

$$E\left(\bar{S}_n \bar{S}_m^*\right) = 2G\delta_{n,m}$$
$$E(\bar{S}_n \bar{S}_m) = 0 \qquad (4.174)$$

---

[10] h is real-valued. However, our discussions bellow applies to complex-valued filters, as well.

The complex form of the passband signal is given by

$$\bar{S}_p(t) \overset{\text{def}}{=} \bar{S}_b(t)e^{j\omega_c t}, \tag{4.175}$$

where ω_c is the carrier angular frequency. The real-valued passband signal $S_p(t)$ is

$$S_p(t) = Re\left[\bar{S}_p(t)\right] = \frac{1}{2}\left[\bar{S}_p(t) + \bar{S}_p^*(t)\right], \tag{4.176}$$

The down-conversion result $\bar{S}_b''$ is given by two versions:

$$\bar{S}_b''(t) = \frac{1}{2}e^{-j\omega_c t}\bar{S}_p(t) = \int e^{-j\omega_c u}S_p(u)g(t-u)du, \tag{4.177}$$

where $g(t)$ is a low pass filter whose cutoff frequency is $W_b \ll \omega_c/2\pi$. The result of down-conversion recovers the transmitted signal, with a factor of $1/2$:

$$\bar{S}_b''(t) = \frac{1}{2}\bar{S}_b(t). \tag{4.178}$$

4.9.2 Baseband and Passband PSD

Since we focus on receiver, we are interested in the relationship between passband and baseband PSD *at the receiver*. Now, let us start with the passband. As explained in Chapter 3, the two terms in (4.176) represent the two parts of the signal concentrating around ω_c and $-\omega_c$, respectively. The spectra of the two parts are not overlapped. Therefore, the contribution to the PSD of $S_p(t)$ by the two parts can be separated.

$$P_{Wp}(f) = P_{Wp}^+(f) + P_{Wp}^-(f). \tag{4.179}$$

Here, $P_{Wp}(f)$ is the PSD of $S_p(t)$. $P_{Wp}^+(f)$ and $P_{Wp}^-(f)$ are the PSDs of $1/2\bar{S}_p(t)$ and $1/2\bar{S}_p^*(t)$, respectively.

Let us first study $\bar{S}_p(t)$. From (4.175) and (4.171):

$$\bar{S}_p(t) = \bar{S}_b(t)e^{j\omega_c t} = \sum_{n=1}^{N}\bar{S}_n h\left(t-nT_{Sym}\right)e^{j\omega_c t} = \sum_{n=1}^{N}\bar{S}_{pn}h_p\left(t-nT_{Sym}\right)$$

$$\bar{S}_{pn} \overset{\text{def}}{=} \bar{S}_n e^{j\omega_c nT_{Sym}} \tag{4.180}$$

$$h_p(t) \overset{\text{def}}{=} h(t)e^{j\omega_c t}$$

As explained in Chapter 3, h_p has the same spectrum as h but shifted up by ω_c:

$$\tilde{h}_p(\omega) = \tilde{h}(\omega - \omega_c). \tag{4.181}$$

Furthermore, based on (4.174), $\{\bar{S}_{pn}\}$ is also a proper stochastic process:

$$E\left(\bar{S}_{pn}\bar{S}_{pm}^*\right) = E\left(\bar{S}_n\bar{S}_m^*\right)e^{j\omega_c(n-m)T_{Sym}} = 2G\delta_{n,m}$$

$$E\left(\bar{S}_{pn}\bar{S}_{pm}\right) = E(\bar{S}_n\bar{S}_m)e^{j\omega_c(n+m)T_{Sym}} = 0 \tag{4.182}$$

Therefore, the spectrum of $\bar{S}_p(t)$ from N transmitted symbols is

$$\tilde{\bar{S}}_p(\omega) = \sum_{n=1}^{N} \bar{S}_{pn}\tilde{h}_p(\omega) = \tilde{h}(\omega - \omega_c)\sum_{n=1}^{N}\bar{S}_{pn}. \tag{4.183}$$

Note that based on (4.182):

$$E\left(\left|\sum_{n=1}^{N}\bar{S}_{pn}\right|^2\right) = \sum_{n=1}^{N}\sum_{m=1}^{N}E\left(\bar{S}_{pn}\bar{S}_{pm}^*\right) = \sum_{n=1}^{N}\sum_{m=1}^{N}2G\delta_{n,m} = N. \tag{4.184}$$

Therefore, (4.15) in Section 4.2.3 leads to

$$P_{Wp}^+(f) = \lim_{N\to\infty}\frac{1}{NT_{Sym}}E\left[\frac{1}{2}\tilde{\bar{S}}_p(2\pi f)\right]^2 = \frac{G}{2T_{Sym}}\left|\tilde{h}(2\pi f - 2\pi f_c)\right|^2, \tag{4.185}$$

where $f_c = \omega_c/2\pi$ is the carrier frequency. Likewise, the PSD around $-\omega_c$ is

$$P_{Wp}^-(f) = \frac{G}{2T_{Sym}}\left|\tilde{h}(2\pi f + 2\pi f_c)\right|^2. \tag{4.186}$$

Therefore, the passband PDF for the modulated signal is

$$P_{Wp}(f) = P_{Wp}^+(f) + P_{Wp}^-(f) = \frac{G}{2T_{Sym}}\left[\left|\tilde{h}(2\pi f - 2\pi f_c)\right|^2 + \left|\tilde{h}(2\pi f + 2\pi f_c)\right|^2\right]. \tag{4.187}$$

Let E_{Sym} be the total symbol energy at the *passband*. We wish to connect E_{Sym} with G given in (4.173). Equation (4.187) leads to

$$E_{Sym} = T_{Sym}\int P_{Wp}(f)df = G\int\left|\tilde{h}(2\pi f)\right|^2 df. \tag{4.188}$$

For baseband on the receiver side, from (4.177) we have

$$\tilde{\bar{S}}_b''(\omega) = \frac{1}{2}\tilde{\bar{S}}_p(\omega + \omega_c). \tag{4.189}$$

Therefore, the baseband PSD is

$$P_{Wb}''(f) = P_{Wp}^+(f + 2\pi f_c) \tag{4.190}$$

The total symbol energy in baseband is

$$E_{Symb} = T_{Sym}\int P_{Wb}''(f)df = \frac{1}{2}T_{Sym}\int P_{Wp}(f)df = \frac{1}{2}E_{Sym}. \tag{4.191}$$

We can then use (4.188) to express the passband and baseband PSDs in terms of E_{Sym}:

$$P''_{Wb}(f) = \frac{1}{2}E_{Sym}\left[T_{Sym}\int\left|\tilde{h}(2\pi f)\right|^2 df\right]^{-1}\left|\tilde{h}(2\pi f)\right|^2$$

$$P^+_{Wp}(f) = \frac{1}{2}E_{Sym}\left[T_{Sym}\int\left|\tilde{h}(2\pi f)\right|^2 df\right]^{-1}\left|\tilde{h}(2\pi f - 2\pi f_c)\right|^2 . \qquad (4.192)$$

$$P^-_{Wp}(f) = \frac{1}{2}E_{Sym}\left[T_{Sym}\int\left|\tilde{h}(2\pi f)\right|^2 df\right]^{-1}\left|\tilde{h}(2\pi f + 2\pi f_c)\right|^2$$

Note that the baseband PSD and symbol energy given in (4.191) and (4.192) are based on our particular down-conversion formulation (4.177).

4.10 APPENDIX: BASEBAND NOISE

In this section, we will show that when passband noise is a real-valued WSS white Gaussian stochastic process, baseband noise is a complex-valued proper WSS white Gaussian stochastic process.

4.10.1 WSS Baseband Noise

In this section, we will show that if the passband noise is a WSS process, then the baseband noise is a proper WSS process, following Wozencraft and Jacobs [5].

We start with the passband noise $n_p(t)$, which is a real-valued WSS stochastic process with autocorrelation $c_p(\tau)$:

$$E\left[n_p(t+\tau)n_p(t)\right] = c_p(\tau)\forall t \in (-\infty, \infty). \qquad (4.193)$$

The complex baseband noise n_b is described by (4.31) in Section 4.3.2:

$$n_b(t) = \int n_p(u)e^{-j\omega_c u}g(t-u)du. \qquad (4.194)$$

Here, ω_c is the angular carrier frequency. $g(t)$ is a real-valued low pass filter, whose bandwidth is less than ω_c.

We continue to examine the autocorrelation property of baseband noise c'_b:

$$c'_b(t_1, t_2) \stackrel{\text{def}}{=} E\left[n_b(t_1)n^*_b(t_2)\right]. \qquad (4.195)$$

Note that here the autocorrelation is a function of t_1 and t_2 instead of $\tau = t_1 - t_2$ as in (4.193), since we do not know whether n_b is a WSS process at this point. From (4.194),

$$c_b'(t_1,t_2) = \iint E\left[n_p(u_1)n_p(u_2)\right]e^{-j\omega_c(u_1-u_2)}g(t_1-u_1)g(t_2-u_2)du_1du_2. \qquad (4.196)$$

From (4.193) and substituting u_1 with $v \overset{\text{def}}{=} u_1 - u_2$, (4.196) becomes

$$c_b'(t_1,t_2) = \iint c_p(v)e^{-j\omega_c v}g(t_1-u_2-v)g(t_2-u_2)dvdu_2. \qquad (4.197)$$

Let us separate the two integrals over v and u_2:

$$c_b'(t_1,t_2) = \int d(t_1,u_2)g(t_2-u_2)du_2$$

$$d(t_1,u_2) \overset{\text{def}}{=} \int c_p(v)g(t_1-u_2-v)e^{-j\omega_c v}dv \qquad (4.198)$$

Express the filter in terms of its frequency responses $\tilde{g}(\omega)$:

$$g(t) = IFT[\tilde{g}(\omega)] = (2\pi)^{-1}\int \tilde{g}(\omega)e^{j\omega t}d\omega. \qquad (4.199)$$

We thus have

$$
\begin{aligned}
d(t_1,u_2) &= (2\pi)^{-1}\iint dvd\omega c_p(v)\tilde{g}(\omega)e^{j\omega(t_1-u_2-v)}e^{-j\omega_c v} \\
&= (2\pi)^{-1}\int d\omega\tilde{g}(\omega)e^{j\omega(t_1-u_2)}\int c_p(v)e^{-j(\omega_c+\omega)v} . \qquad (4.200) \\
&= (2\pi)^{-1}\int d\omega\tilde{g}(\omega)\tilde{c}_p(\omega_c+\omega)e^{j\omega(t_1-u_2)}
\end{aligned}
$$

The last step introduced the FT of c_p:

$$\tilde{c}_p(\omega) - \int c_p(t)e^{-j\omega t}dt. \qquad (4.201)$$

Now go back to the first line of (4.198) and use the IFT of g in its complex conjugate form:

$$
\begin{aligned}
c_b'(t_1,t_2) &= (2\pi)^{-2}\int du_2\iint d\omega_1 d\omega_2 \tilde{g}(\omega_1)\tilde{c}_p(\omega_c+\omega_1)e^{j\omega_1(t_1-u_2)}\tilde{g}^*(\omega_2)e^{-j\omega_2(t_2-u_2)} \\
&= (2\pi)^{-2}\iint d\omega_1 d\omega_2 \tilde{g}(\omega_1)\tilde{c}_p(\omega_c+\omega_1)e^{j\omega_1(t_1)}\tilde{g}^*(\omega_2)e^{-j\omega_2(t_2)}\int du_2 e^{ju_2(\omega_2-\omega_1)} \\
&= (2\pi)^{-1}\int d\omega_1 \tilde{c}_p(\omega_c+\omega_1)|\tilde{g}(\omega_1)|^2 e^{j\omega_1(t_1-t_2)}
\end{aligned}
$$
$$(4.202)$$

In the process of simplifying (4.202), we recognized that integral over u_2 yields a Dirac delta function, as pointed out in Chapter 3. We can see that $c_b'(t_1,t_2)$ depends only on $\tau \overset{\text{def}}{=} t_1 - t_2$ instead of depending on t_1 and t_2 separately. Therefore, the correlation can be noted as $c_b(\tau)$:

$$c_b(\tau) \overset{\text{def}}{=} c_b'(t_1+\tau,t_2) = (2\pi)^{-1}\int |\tilde{g}(\omega_1)|^2\tilde{c}_p(\omega_1+\omega_c)e^{j\omega_1\tau}d\omega_1. \qquad (4.203)$$

Equation (4.203) shows that the autocorrelation at the baseband is the IFT of $|\tilde{g}(\omega)|^2 \tilde{c}_p(\omega + \omega_c)$. In other words, the FT of $c_b(\tau)$ is

$$\tilde{c}_b(\omega) = |\tilde{g}(\omega)|^2 \tilde{c}_p(\omega + \omega_c). \tag{4.204}$$

Since c_b is complex-valued, we also need to examine the pseudo-correlation

$$\bar{c}'_b(t_1, t_2) \stackrel{\text{def}}{=} E[n_b(t_1)n_b(t_2)]. \tag{4.205}$$

Similar to (4.196), we have

$$\bar{c}'_b(t_1, t_2) = \iint E\left[n_p(u_1)n_p(u_2)\right] e^{-j\omega_c(u_1 + u_2)} g(t_1 - u_1) g(t_2 - u_2) du_1 du_2. \tag{4.206}$$

In parallel with (4.198)–(4.202), we get

$$\bar{c}'_b(t_1, t_2) = (2\pi)^{-1} e^{-j2\omega_c t_2} \int \tilde{g}(\omega_1)\tilde{g}^*(\omega_1 + 2\omega_c) e^{j\omega_1(t_1 - t_2)} d\omega_1 \tilde{c}_p(\omega_c + \omega_1) d\omega_1. \tag{4.207}$$

Because $g(t)$ is a low pass filter whose bandwidth is much less than ω_c, the supports of $\tilde{g}(\omega_1)$ and $\tilde{g}^*(\omega_1 + 2\omega_c)$ do not overlap. Namely,

$$\tilde{g}(\omega)\tilde{g}^*(\omega + 2\omega_c) \equiv 0 \forall \omega. \tag{4.208}$$

Therefore,

$$\bar{c}_b(t_1, t_2) = 0. \tag{4.209}$$

Equations (4.203) and (4.209) show that the baseband noise $n_b(t)$ is a proper WSS process. This conclusion is based on two conditions:

1. The passband noise $n_p(t)$ is WSS. It does not need to be AWGN.
2. The low pass filter $g(t)$ has the property that supports of $\tilde{g}(\omega)$ and $\tilde{g}^*(\omega + 2\omega_c)$ do not overlap. Namely, the bandwidth of the filter is much less than ω_c. $g(t)$ does not need to be an ideal bandpass filter.

4.10.2 Gaussian Noise

An interesting special case is the Gaussian process [3]. If $x(t)$ is a complex-valued proper Gaussian stochastic process with zero mean, then its probability density function is

$$p[x(t)] = \frac{1}{\pi\sigma^2}e^{-\frac{|x(t)|^2}{\sigma^2}}, \tag{4.210}$$

where σ^2 is the variance:

$$E\left[|x(t)|^2\right] = \sigma^2 \tag{4.211}$$

A proper stochastic process with zero mean is also known as a *circular process*. Such process has the property that when $x(t)$ is rotated by a constant phase, that is, $x(t) \rightarrow x(t)e^{j\theta}$, its statistical properties (autocorrelation and mean) do not change. Therefore, we often say $x(t)$ is *circularly symmetric*. Circular Gaussian variables are discussed more in Chapter 11.

Note that (4.210) is different from the familiar Gaussian distribution probability density function for real-valued processes. The probability density function of a real-valued Gaussian variable with zero mean and variance of σ^2 is given by (4.29):

$$p_G(x,\sigma^2) = \frac{1}{\sqrt{2\pi\sigma^2}}e^{-\frac{x^2}{2\sigma^2}}. \tag{4.212}$$

The relationship between (4.210) and (4.212) can be understood as follows. As stated in Section 4.2.2.1, since $x(t)$ is proper and its autocorrelation is real-valued when $\tau = 0$, as shown in (4.211), the real and imaginary parts of $x(t)$, namely $x_r(t)$ and $x_i(t)$, are mutually independent and have the same variance. Therefore, $x_r(t)$ and $x_i(t)$ should be Gaussian processes with a variance of $\sigma^2/2$. Therefore, the total probability density function is

$$p(x) = p[x_r(t), x_i(t)] = p_G\left[x_r(t), \frac{\sigma^2}{2}\right]p_G\left[x_i(t), \frac{\sigma^2}{2}\right] = \frac{1}{\pi\sigma^2}\exp\left[\frac{|x_r(t)|^2}{\sigma^2} + \frac{|x_i(t)|^2}{\sigma^2}\right]. \tag{4.213}$$

Equation (4.213) is the same as (4.210).

4.11 APPENDIX: REPRESENTING SIGNALS AND NOISES WITH VECTORS

This section provides more mathematical details concerning the statements made in Section 4.5.1. We will first introduce the concept of functions treated as vectors in general terms, and then study the behavior of white Gaussian noise in such representation. More rigorous treatment of the subject can be found in [5]. Vector representation of functions is a subject in functional analysis. Some mathematical details are ignored here; however, the results and conclusions are applicable to the problems treated in this Chapter.

4.11.1 Functions Treated as Vectors

Functions can be treated as vectors, whereas the inner product is defined as integral over time:

$$<a(t),b(t)> \overset{\text{def}}{=} \int a(t)b^*(t)dt, \qquad (4.214)$$

where $a(t)$ and $b(t)$ are two functions.[11] As in "regular vectors," we can introduce a basis $\{\phi_k(t), k = 1, 2, \ldots\}$, so that any function $f(t)$ can be expressed as a linear combination of the basis[12]:

$$f(t) = \sum_k f_k \phi_k(t), \qquad (4.215)$$

where $\{f_k, k = 1, 2, \ldots\}$ is a set of constants. Comparing with linear algebra, here the basis consists of functions. More importantly, the number of basis functions (i.e., the dimension of the vector space) is in general infinite.[13] The basis is known as orthonormal if all basis functions are mutually orthogonal and have unit norms. Namely,

$$\langle \phi_i(t), \phi_j(t) \rangle = \int \phi_i(t)\phi_j^*(t)dt = \delta_{i,j}. \qquad (4.216)$$

In this case, the coefficients in (4.215) are given by

$$f_k = \langle f(t), \phi_k(t) \rangle = \int f(t)\phi_k^*(t)dt. \qquad (4.217)$$

If any function can be expressed by (4.215) with an orthonormal basis, we call such basis complete. In this case, we have

$$f(t) = \sum_k f_k \phi_k(t) = \sum_k \left[\int f(\tau)\phi_k^*(\tau)d\tau \right] \phi_k(t) = \int f(\tau) \left[\sum_k \phi_k^*(\tau)\phi_k(t) \right] d\tau \qquad (4.218)$$

Equation (4.218) implies the necessary and sufficient condition for completeness:

$$\sum_k \phi_k^*(\tau)\phi_k(t) = \delta(t-\tau). \qquad (4.219)$$

---

[11] These functions must meet certain conditions to ensure the integral converges. Here, we ignore these conditions and assume that such integrals always converge.

[12] Here, we assume the functions have countable basis. See footnote 13 for further clarification.

[13] In some cases, the basis is even uncountable. However, for our purposes we consider only countable bases. In other words, we assume the space is *separable*.

For our discussion, we are not concerned about how to find basis functions. It suffices to know that orthonormal basis functions exist for the type of functions we are interested in.

In the case where the basis functions are complete and orthonormal, two equations are equal if and only if all corresponding projection coefficients are equal[14]:

$$f(t) = g(t)\forall t \Leftrightarrow f_k = g_k \forall k. \tag{4.220}$$

Therefore, given an orthonormal basis $\{\phi_k, k = 1, 2, \ldots\}$, we can treat the project coefficients $\{f_k, k = 1, 2, \ldots\}$ as a vector, noted as $\vec{f}$ thereafter. The mapping between $f(t)$ and $\vec{f}$ (noted as $f(t) \leftrightarrow \vec{f}$) is given by (4.215) and (4.217). One can be easily verify that $\vec{f}$ indeed behaves as a vector (see homework). Especially, we have

$$\langle f(t), g(t) \rangle = \int f(t)g^*(t)dt = \sum_{j,k} f_j g_k^* \int \phi_j(t)\phi_k^*(t)dt = \sum_k f_k g_k^* = \vec{f} \cdot \vec{g}. \tag{4.221}$$

The linear space (mathematically known as a separable Hilbert space) is known as the signal space, when $f(t)$ represents a waveform or other types of signal and noise. Signal space is a concept used in Sections 4.5 and 4.6.

4.11.2   Example: Fourier Series and Fourier Transform

The vector expression in Section 4.11.1 seems to be abstract. However, there is a familiar example: the Fourier transform (FT). For a function $f(t)$ with support in $[0, T_p)$, we can extend it periodically to $(-\infty, \infty)$. As shown in Chapter 3, such a periodic function $f(t)$ can be expressed as

$$f(t) = \sum_{k=-\infty}^{\infty} f_k e^{j\frac{2k\pi t}{T_p}} \qquad \text{for } t \in [0, T_p). \tag{4.222}$$

$$f_k \overset{\text{def}}{=} \frac{1}{T_p}\int_0^{T_p} f(t)e^{-j\frac{2k\pi t}{T_p}} dt$$

Equation (4.222) is the same expression as (4.215) and (4.217) (with the difference of a constant factor $\sqrt{T_p}$, with $\{1/\sqrt{T_p}e^{j(2k\pi t)/T_p}, k = \ldots, -1, 0, 1, \ldots\}$ as the basis function.

We can further let $T_p \to \infty$ to extend the arguments to general functions. In this case, note that

---

[14] Actually, this is true under a more general condition: when the basis functions are linearly independent.

$$lim_{T_p \to \infty} f_k = \frac{1}{T_p} \tilde{f}\left(\frac{2\pi k}{T_p}\right),$$ (4.223)

where $\tilde{f}(\omega)$ is the Fourier transform (FT) of $f(t)$ as formulated in Chapter 3. The first equation of (4.222) becomes

$$\begin{aligned}
\lim_{T_p \to \infty} f(t) &= \lim_{T_p \to \infty} \sum_{k=-\infty}^{\infty} \frac{1}{T_p} \tilde{f}\left(\frac{2\pi k}{T_p}\right) e^{j\frac{2k\pi t}{T_p}} \\
&= \frac{1}{2\pi} \lim_{T_p \to \infty} \sum_{k=-\infty}^{\infty} \frac{2\pi}{T_p} \tilde{f}\left(\frac{2\pi k}{T_p}\right) e^{j\frac{2k\pi t}{T_p}} = \frac{1}{2\pi} \int \tilde{f}(\omega) e^{j\omega t} dt
\end{aligned}$$ (4.224)

The last step represents the transition from summation to integral:

$$\lim_{\Delta x \to 0} \sum_{k=-\infty}^{\infty} \Delta x g(k\Delta x) = \int_{-\infty}^{\infty} g(x) dx.$$ (4.225)

Equation (4.224) is the same as the inverse Fourier transform (IFT) described in Chapter 3. Therefore, the frequency analysis we are familiar with can be viewed as a projection of waveforms onto the basis function set $\{e^{j\omega t}\}$ with varying ω. Of course, there is a difference between (4.224) and (4.215). Although the basis function set in both expressions are infinite, such set is countable in (4.215) and uncountable in (4.224). However, for our purposes, we will ignore such difference.

4.11.3   White Gaussian Noise as Vectors

Now apply the techniques introduced in Section 4.11.1 to baseband white Gaussian noise, as described in Section 4.3.2. Based on (4.215) and (4.217) and with an complete orthonormal basis $\{\phi_k\}$, we can express $n(t)$ in terms of vector $\vec{n}$, whose components are $\{n_k, k = 1, 2, \ldots\}$:

$$n(t) = \sum_k n_k \phi_k(t)$$ (4.226)

$$n_k = \int n(t) \phi_k^*(t) dt.$$ (4.227)

The cross correlation of the projection coefficients can be computed based on (4.227):

$$E\left(n_j n_k^*\right) = \int E[n(t_1) n^*(t_2)] \phi_j^*(t_1) \phi_k(t_2) dt_1 dt_2$$ (4.228)

With the autocorrelation of white noise (4.39) in Section 4.3.2 and the orthonormal property (4.216), (4.228) leads to

$$E\left(n_j n_k^*\right) = \frac{N_0}{2}\int \delta(t_1 - t_2)\phi_j^*(t_1)\phi_k(t_2)dt_1 dt_2 = \frac{N_0}{2}\delta_{j,k}, \tag{4.229}$$

where N_0 characterizes the autocorrelation of $n(t)$ in (4.39). Therefore, noise in different dimensions are mutually independent. Noise variance for each dimension is $N_0/2$. Furthermore, since a summation of inde
pendent Gaussian variables with the same mean is still Gaussian, the projection of $n(t)$ on each dimension is Gaussian with zero mean. Furthermore, in parallel to (4.228), it can be shown that n_k is proper (see Section 4.2.2). Namely,

$$E\left(n_j n_k\right) = 0. \tag{4.230}$$

Therefore, n_k has the probability density function of (4.29) in Section 4.3.1.3 with variance $N_0/2$.

$$p(n_k) = (\pi N_0)^{-1/2} e^{-\frac{|n_k|^2}{N_0}}. \tag{4.231}$$

4.11.4 Signal and Noise in a Subspace

We have seen so far that given an complete orthonormal basis, a function $f(t)$ can be expressed as a vector with infinite dimension $\vec{f}$, with the mapping given by (4.215) and (4.217). Therefore, the AWGN channel model (4.25) can be expressed in terms of vectors:

$$\vec{r} = \vec{s} + \vec{n}. \tag{4.232}$$

Here, $\vec{r}, \vec{s}$, and $\vec{n}$ are vectors corresponding to $r(t)$, $s(t)$, and $n(t)$.

It should be noted that although these vectors, in general, have infinite dimension, the transmitted signal $\vec{s}$ is usually limited to a subspace with a finite dimension. For example, in the QAM and PSK modulations schemes discussed in Chapter 3, passband transmitted signal can be written as (for one symbol only)

$$s(t) = S_I h(t)\cos(\omega_c t) - S_Q h(t)\sin(\omega_c t), \tag{4.233}$$

where S_I and S_Q are baseband I and Q symbols; $h(t)$ is baseband pulse shaping function; ω_c is the angular carrier frequency. Therefore, no matter what the modulating data are, $s(t)$ belongs to a subspace spanned by two functions $h(t)\cos(\omega_c t)$ and

$h(t) \sin (\omega_c t)$.[15] More generally, if there are M possible signals $\{s_m(t), m = 1, 2, \ldots, M\}$, then the dimension of such subspace is no more than M.

Let the subspace containing s be $\mathcal{S}$ with finite dimension $K_{\mathcal{S}}$:

$$s_m \in \mathcal{S} \forall m = 1, 2, \ldots, M. \tag{4.234}$$

Now choose an orthonormal basis $\{\phi_k(t), k = 1, 2, \ldots\}$ so that the first $K_{\mathcal{S}}$ basis functions are inside $\mathcal{S}$ and the rest are outside. Namely,

$$\begin{aligned} \phi_k(t) \in \mathcal{S} \quad \forall k \leq K_{\mathcal{S}} \\ \phi_k(t) \notin \mathcal{S} \quad \forall k > K_{\mathcal{S}} \end{aligned} \tag{4.235}$$

With such choice, any signal components are zero for $k > K_{\mathcal{S}}$. Equation (4.232) thus leads to

$$r_k = \begin{cases} s_k + n_k & k \leq K_{\mathcal{S}} \\ n_k & k > K_{\mathcal{S}} \end{cases} . \tag{4.236}$$

The second line in (4.236) contains only noise that is outside of $\mathcal{S}$. Since noises in deferent dimensions are mutually independent as shown in (4.229), the "out of space" noise does not contain any information about the signal and noise in $\mathcal{S}$. It can thus be disregarded in detection. Therefore, $r(t)$ can be preprocessed to remove the part outside of $\mathcal{S}$. Namely, the vectors in (4.232) can be viewed as having finite dimension. This idea is further exploited in Section 4.6.5.3.

For example, $s(t)$ usually has limited bandwidth. In this case, a simple filtering can limit $r(t)$ to the same bandwidth. The part of $r(t)$ outside of signal bandwidth contains only noise and can be ignored in receiver processing. The exact choice of $\mathcal{S}$ is not important as long as (4.234) is satisfied. In other words, $\mathcal{S}$ can be chosen to be larger than necessary. The important point is it is of finite dimension, so that noise power $\left|\vec{n}\right|^2$ is finite.

REFERENCES

1. J. Cacko, M. Bily, and J. Bukoveczky, *Random Processes: Measurement, Analysis and Simulation*. Burlington: Elsevier Science, 1987.
2. P. Kousalya, *Probability, Statistics and Random Processes*. Dorling Kindersley (India), 2013.
3. F. D. Neeser and J. L. Massey, "Proper Complex Random Processes with Applications to Information Theory," *IEEE Trans. Inf. Theory*, vol. 39, no. 4, pp. 1293–1302, 1993.
4. Z. K. M. Ho and E. Jorswieck, "Improper Gaussian Signaling on the Two-User SISO Interference Channel," *IEEE Trans. Wirel. Commun.*, vol. 11, no. 9, pp. 3194–3203, 2012.
5. J. M. Wozencraft and I. M. Jacobs, *Principles of Communication Engineering*. New York: Wiley, 1965.
6. B. Sklar, *Digital Communications: Fundamentals and Applications*. Prentice-Hall PTR, 2001.
7. J. Proakis and M. Salehi, *Digital Communications*, 5th ed. McGraw-Hill Education, 2007.

---

[15] Although these two basis functions are not strictly orthonormal, an orthonormal basis for the same two-dimensional subspace can be found in ways similar to those in linear algebra.

8. DLMF, "NIST Digital Library of Mathematical Functions" [Online]. Available at http://dlmf.nist.gov/ [Accessed December 13, 2016].

9. L. L. Scharf and C. Demeure, *Statistical Signal Processing: Detection, Estimation, and Time Series Analysis*. Addison-Wesley Pub. Co, 1991.

HOMEWORK

The following MATLAB functions may be useful for some problems: pskmod, pskdemod, qammod, qamdemod, qfunc, qfuncinv, randi, randn, de2bi, bi2de, var. You can look up the documentation in MATLAB.

4.1 Show that if a WSS with PSD $P_W(f)$ is filtered with a filter with a frequency response of $\tilde{h}(2\pi f)$, then the result is still WSS, with PSD $P_W(f)|h(2\pi f)|^2$.

4.2 Express the expectation values given in (4.6) with the autocorrelation and pseudo-autocorrelation defined in (4.7).

4.3 Try to extend the definitions and properties in Sections 4.2.3 and 4.2.4 to complex-valued random processes. You may need to introduce additional constraints such as properness.

4.4 Show that the first stage of Figure 4.3, when applied to the symbol sequence (4.84), produces the same vector decomposition as given by the second line of (4.77). You can follow the general approach in Section 4.6.1 and assume (4.88) is valid.

4.5 Consider optimal detection when the noise is not white. Results in this chapter can be reused by introducing a noise-whitening filter before the MF, to make the noise appear to be white to the MF.

    a. Given baseband noise PSD $P_{nb}(f)$, derive MF (either in the frequency domain or in the time domain).

    b. Derive an expression of SNR in this case.

    c. Does MF still maximize SNR, as discussed for the case of white noise in Section 4.6.5.2?

4.6 Can you find an example in engineering or in real life where the difference between MAP and ML is very important?

4.7 Using frequency domain formulations, show that a matched filter maximizes output SNR (see Section 4.6.5.2).

4.8 Look up "maximum ratio combining" and compare it to the matched filter. This technique is another way to show that matched filter maximizes SNR.

4.9 A binary digital communication system employs the signals $s_0(t) = 0$, $s_1(t) = A$ for $0 \leq t \leq T_{Sym}$ for transmitting bits 1 and 0. This is called on-off signaling.

    a. Determine the optimum detector for an AWGN channel and the optimum threshold for the slicer, assuming that the signals are equally probable.

    b. Determine the probability of error as a function of the SNR. How does on–off signaling compare with antipodal signaling with the same transmission power? Antipodal signaling uses two possible signals $s_0(t) = -B$, $s_1(t) = B$ for $0 \leq t \leq T_{Sym}$.

4.10 Suppose that binary PSK is used for transmitting information over an AWGN channel with a noise power spectral density of $1/2N_0 = 10^{-10}$ W/Hz. The transmitted signal energy at the receiver is $E_b = 1/2A^2T_{Sym}$, where T_{Sym} is the symbol interval and A is the signal amplitude. Using the Shannon theorem (Chapter 2), determine the signal amplitude required to achieve an error probability of 10^{-6} when the data rate is

    a. 10 kbps (kilobits per second)

    b. 100 kbps

    c. 1 Mbps (megabits per second)

4.11 In the homework of Chapter 3, you constructed MATLAB codes that modulate and demodulate the baseband signals. In the current assignment, you are asked to add noise to the modulated signal, before demodulation. You can first measure the signal power that you generated (proportional to the variance of the signal), and compute E_b. Then you can figure out the noise power from the desired E_b/N_0. You can then use MATLAB function randn to generate a sequence of complex noise, and scale them to get the correct power level. When the noise is added to the signal, the demodulator may produce bit errors, which you can detect by comparing the demodulator output with the modulator input. By counting the bit errors and the number of bits passed through, you can compute the bit error rate BER.

    Construct a MATLAB program that computes the BER for BPSK, 4-QAM, and 16-QAM through simulation as described above, and compare the results with those shown in Section 4.7. Plot the resulting BER vs. E_b/N_0 curve in the range of $BER = 10^0$ to $BER = 10^{-7}$. Be careful to select proper symbol numbers in simulation to get significant statistics at low BER.

4.12 In the context of Section 4.11.1, given an orthonormal basis $\{\phi_k, k = 1, 2, \ldots\}$, we can treat the project coefficients $\{f_k, k = 1, 2, \ldots\}$ as a vector noted as $\vec{f}$ thereafter. The mapping between $f(t)$ and $\vec{f}$ (noted as $f(t) \leftrightarrow \vec{f}$) is given by (4.215) and (4.217). Verify that $\vec{f}$ indeed behave as a vector. Namely, show the following:

    a. For any constant α, $af(t) \leftrightarrow a\vec{f}$

    b. For any two functions $f(t)$ and $g(t)$, $f(t) + g(t) \leftrightarrow \vec{f} + \vec{g}$

    c. For any two functions $f(t)$ and $g(t)$, $\langle f(t), g(t) \rangle \leftrightarrow \vec{f} \cdot \vec{g}$

4.13 In the context of Section 4.11.2, verify that the basis $\{1/\sqrt{T_p}e^{j(2k\pi t)/T_p},$ $k = \ldots, -1, 0, 1, \ldots\}$ is indeed a complete orthonormal basis with $t \in [0, T_p)$, satisfying (4.216) and (4.219). See appendices in Chapter 3 for the relevant mathematical properties.

CHAPTER 5

CHANNEL CODING

5.1 INTRODUCTION

Figure 5.1 shows the architecture of a typical transmitter and receiver pair for digital communication.[1] This chapter covers the blocks of channel coding channel decoding.

In the following, Section 5.2 provides a general discussion of the concepts related to channel coding. As explained therein, channel coding schemes can be divided into two families: classical codes and modern codes. Sections 5.3–5.6 describe some classical codes, while Sections 5.7 and 5.8 describe some modern codes. Section 5.9 provides a summary. Some mathematical details are provided in the appendices, Sections 5.10 and 5.11.

Channel coding is a big subject. There are many books dedicated to the channel coding, as well as voluminous current researches. This chapter provides only an introduction. More specifically, the intended audience for this chapter is not those who design new codes or implement encoders and decoders. Instead, this chapter aims to help the communications engineers who need to select a proper channel coding scheme for a particular application. We will focus on the basic principles behind each coding class. Some details on encoding and decoding will be discussed to illustrate these principles. These encoding and decoding methods are the simplest in concept but not necessarily prevalent in practice. Furthermore, code performance evaluations are not discussed quantitatively. Only some examples from the literature are given to provide a general impression of the pros and cons of each code.

Channel coding is related to demodulation, especially in performance analysis (i.e., estimation of error probability). Some basic concepts in performance analysis are covered in Chapter 4. This chapter assumes that the reader is familiar with those concepts such as union bound, nearest neighbor approximation, maximum a posteriori (MAP) and maximum likelihood (ML) detection techniques, and noise characterization.

---

[1] While transceiver structures shown in Figure 5.1 are widely used, channel coding and modulation can be further integrated. At the receiver side, "soft decision" decoding (Section 5.2.4) combines modulation and decoding functionalities. At the transmitter side, trellis-coded modulation (TCM) (Section 5.5.1) combines channel coding and modulation.

Digital Communication for Practicing Engineers, First Edition. Feng Ouyang.
© 2020 by The Institute of Electrical and Electronics Engineers, Inc.
Published 2020 by John Wiley & Sons, Inc.

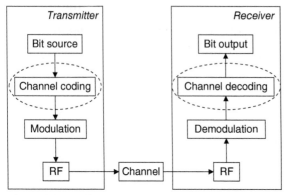

Figure 5.1   Physical layer architecture.

Note that unless otherwise noted, all additions in this chapter that involve binary numbers are modulo-2. Namely:

$$0 + 0 = 1 + 1 = 0$$
$$0 + 1 = 1 + 0 = 1$$

(5.1)

5.2   CHANNEL CODING OR FORWARD ERROR CORRECTION (FEC)

5.2.1   Error Control and Channel Coding

In Chapter 4, we see that even in the presence of noise, a receiver can still determine the transmitted symbol with high reliability, because not all possible symbol values are legit in a modulation scheme. There are only M possible transmitted signal waveforms $\{s_m, m = 1, 2, \ldots, M\}$. The receiver can pick the $\hat{m}$ with $s_{\hat{m}}$ being the closest alternative from the received waveform. As long as the noise is not too strong, such a decision is most likely to be correct. In other words, because only a limited set of waveforms are transmitted, the system has some noise immunity. For a given symbol rate (dictated by allowed bandwidth through Nyquist symbol rate) and transmit power, there is a tradeoff between data rate (determined by the number of allowed signal waveforms M) and noise immunity (determined by the minimum distance between constellation points).

This idea can be extended from symbols to symbol sequences. In uncoded systems studied in Chapter 4, the symbols are mutually independent. Therefore, all sequences of symbols are possible. If, instead, only a subset of the sequences is transmitted, then even if some symbols are received wrong, the receiver would not confuse one transmitted sequence with another. Additional immunity against noise is thus obtained. As in the case of modulation design, there is a tradeoff between data rate and noise immunity, both determined by the selection of the legit symbol sequences.

Channel coding, also known as forward error correction (FEC), is based on such an idea. The legit sequences used for transmission are known as codewords. A code is defined by the collection of all codewords and their mapping to user data.

In addition to FEC, there is another common way to ensure error-free communication: automatic repeat request (ARQ). In ARQ schemes, the receiver notifies the transmitter if an error in transmission is detected. The transmitter then resends the information. Coding can also be used for error detection in ARQ, as will be discussed in Section 5.3.5.1. FEC and ARQ can be further integrated into a more efficient error control scheme, as will be described briefly in Section 5.9.

5.2.2 Shannon Theorem and Channel Coding

Chapter 2 introduces Shannon theorem. Simply put, Shannon showed that error-free communications could be achieved through a noisy channel if the data rate is below a specific limit. The way to realize noise immunity is transmitting only a subset of the "typical sequences." Such a subset is the codewords in our language. At the receiver, the probability of mistaking one codeword with another, even at the presence of noise, approaches zero as the length of the codewords approaches infinite.

Although elegant, Shannon's codeword choice is not easy to implement in practice. One difficulty is that very long codewords cause a significant delay in communication because the receiver must finish receiving the entire codeword before decoding for the data. The other difficulty is the complexity involved. Suppose the data rate (measured as the number of bits per symbol) is N_b, which can be a fraction number or less than 1. If the codeword length is K symbols, then the number of codewords will be

$$M = 2^{N_b K}. \tag{5.2}$$

At the receiver, one needs to compare the received signal with M possible transmitted waveform, in order to find the most likely transmitted data. The complexity of such tasks normally increases exponentially with K.

Furthermore, Shannon's claim is about the *average* performance of a set of the randomly chosen codewords. It follows that there must be a particular code whose performance is no worse than the average. In fact, there should be many such codes. However, we do not know which one. Therefore, Shannon theorem does not tell us how to construct a "good performing" code.

Although Shannon theorem points out channel coding as an approach for achieving robust data communication in the presence of noise, the actual search efforts for channel code designs have taken multiple directions, resulting in several families of codes, as will be discussed in the rest of this chapter.

For a given data rate, Shannon theorem gives a required signal-to-noise ratio (SNR) level for error-free communication, known as the Shannon limit.[2] As pointed out in Chapter 2, required SNR is not sensitive to allowed error probability under

---

[2] "Shannon limit" has various, yet inter-related definitions in the literature. For example, sometimes it refers to the achievable data rate for a given SNR.

Shannon theory. On the other hand, for each channel coding scheme, there is a relationship between the bit error rate (BER) and the required SNR. For a given BER and with a particular code, the difference between the required SNR under a particular code and the Shannon limit is known as the Shannon gap (or "gap to the capacity") for such code. Shannon gap is an important measure of code performance. One should be careful, however, about the definition of Shannon limit (and thus Shannon gap). In the literature, Shannon limit is often defined with some modulation and coding rate constraints, and different from what is given in Chapter 2 [1, sec. 1.5].

5.2.3 Classical Codes and Modern Codes

There are many ways to classify the various codes [2]. From an application point of view, it makes sense to divide the current mainstream codes into two classes: the classical codes and the modern codes. This classification is based on the time of discovery of these codes, but also shed light on the characteristics of them [1].

Classical codes were discovered in the 1950s and 1960s (with some additions in later decades). These codes take advantage of certain mathematical structures for good performances and low complexities in optimal decoding. Three predominant types of the classical codes are block codes, convolutional codes and trellis-coded modulation (TCM) codes, as discussed in Sections 5.3–5.5. Because of the constraint on complexity, classical codes typically have limited length (up to hundreds of bits), leading to significant Shannon gaps.

Modern codes are newer members of the code family, discovered or rediscovered in the 1990s. These codes use very long codewords (thousands or tens of thousands of bits) without fine-tuned structures. It relies on the same principle as in Shannon's analysis: since the average of all codes is good, a random code is likely to be good. Since the optimal decoding of long codes is impractical, suboptimal decoding methods are introduced. Nonetheless, these codes can achieve performances very close to Shannon bounds. Two of the most popular modern codes, the turbo code and the low-density parity-check (LDPC) code, are described in Sections 5.7 and 5.8. Modern codes are still under active research.

5.2.4 Hard Decision and Soft Decision

Most channel encoding methods work at the bitstream level. Namely, the data bits are encoded by the channel encoder, whose output is a stream of coded bits. This coded bitstream is sent to the modulator to form symbols. The process is shown in the left part of Figure 5.1.

The most straightforward decoding method is shown in the right part of Figure 5.1. The received signal is first demodulated into a coded bitstream, which is then sent to the channel decoder to recover the data bits. In this method, the demodulator makes "hard decisions," that is, the output of the demodulator is either 1 or 0. This method is known as "hard decision decoding." Hard decision decoding does not take advantage of all information available in the received signal. For example, consider a binary phase shift keying (BPSK) modulation with constellation points at ± 1 V for bits

1 and 0, respectively. If the received signal is 0.999 V in case 1 and 0.1 V in case 2, the optimal hard decision for the demodulator is 1 in both cases. However, the certainty of such a decision is much higher in case 1. Therefore, in case 1, the channel decoder should be more "willing" to pick a codeword that is consistent with such decision. On the other hand, in case 2, the channel decoder may defer to implications from other received symbols in the codeword because the current demodulate decision is not reliable.

To improve performance, "soft decision decoding" can be adopted. In this scheme, the demodulator does not output bits, but the "likelihood" for a bit being 1 or 0. The channel decoder chooses codewords based on such "soft" input. In general, soft decision decoding can improve noise immunity by 2–3 dBs comparing to hard decision decoding [2]. However, complexity is also higher for soft decision decoding.

5.2.5   Code Parameters and Performances

A code has three most important types of parameters: code rate, code length, and noise immunity. In this section, we discuss the general principles only. The actual parameters depend on the code type.

Code rate is the ratio between the numbers of data bits and coded bits, that is, the numbers of bits at the input and output of the encoder. In essence, coding introduces redundancy in exchange for better noise immunity. Therefore, the code rate is, in general, less than 1.

Code length is the number of bits or symbols for a code unit. A code unit can be decoded independently. For example, for block codes, a code unit is a block. For convolutional codes, on the other hand, the code length is characterized by the constraint length.

Noise immunity is in general characterized by the relationship between bit error rate (BER) and SNR. Such a performance metric is also used for uncoded modulations in Chapter 4. SNR is often measured in E_b/N_0. For uncoded modulations, noise immunity depends on the distance between constellation points in signal space. For channel codes, noise immunity depends also on the "distance" between codewords. The definition of such distance depends on the context. It could be the "Hamming distance," that is, the number of bits that are different between two codewords. It can also be the "Euclidean distance," which is the distance in signal space:

$$d_e \overset{\text{def}}{=} \int |s_m(t) - s_n(t)|^2 dt, \tag{5.3}$$

where $s_m(t)$ and $s_n(t)$ are signal waveforms of the two codewords m and n. This book will not discuss performance assessments. However, the terminology of distance will be used for qualitative comparison among codes.

The question of noise immunity for classical codes is usually mathematically tractable because optimal decoders are typically used, and the distance between codewords can be computed explicitly. At least, in most cases, a tight upper bound of error probability can be obtained. For modern codes, the problem is more complicated. Noise immunity depends on not only code design but also choices of the decoder. Simulations are often used for performance assessment.

When comparing codes, noise immunity is often quantified as "coding gain" and "Shannon gap." The Shannon gap was explained in Section 5.2.2. Coding gain is the difference in required E_b/N_0 for a given BER, when compared with uncoded modulation. Naturally, these two measures are equivalent. In other words, if codes A and B have a difference of X dB in Shannon gap, then their coding gains also differ by $-X$ dB. However, one must be careful in such comparison, when two codes have different code rates. A fair comparison should maintain the same user data rate.

As explained in Chapter 2, there are two communications situations: power-limited and bandwidth-limited. In power-limited case, the SNR is so low that additional bandwidth does not bring significant benefit to performance. In other words, we can use as much bandwidth as we wish, while constrained by total power. In this case, we can keep the modulation scheme the same between codes, and adjust the symbol rate (and thus the Nyquist bandwidth) to compensate for code rate difference. Remember that E_b is defined as energy per *user data* bit. Since user data rate is kept constant, the relationship between E_b and signal power is also constant. As an example, let us compare a rate 1/2 code with the uncoded case. To keep the user data rate constant, the coded signal needs to double the symbol rate (and thus double the bandwidth) comparing to the uncoded signal. For the same E_b and thus the same signal power, this means the coded signal has less energy per symbol and a lower PSD level. Such changes bring disadvantage in terms of noise immunity. Therefore, to result in a positive coding gain, the benefit of channel coding must exceed such disadvantage. Chapter 4 shows a comparison of error probabilities of the same modulation with and without channel coding. Such comparison applies to the power-limited case.

On the other hand, in bandwidth-limited case, symbol rate is limited by the Nyquist bandwidth and thus cannot be changed. For a fair comparison, we need to keep both bandwidth and user data rate the same. In other words, the spectral efficiency (the ratio of user data rate and bandwidth) needs to be kept constant. Therefore, different coding rates lead to different modulation orders. With the same example above where we compare a rate 1/2 code with uncoded modulation, if the uncoded modulation uses BPSK, the coded modulation should use QPSK to achieve the same spectral efficiency. Using higher order modulation results in noise immunity degradation, which needs to be compensated by the benefit from channel coding. Therefore, when computing coding gain, we need to compare BER–SNR performances of coded and uncoded schemes with the same spectral efficiency (and thus difference modulation orders).

In order to take into account both spectral efficiency and noise immunity, a normalized SNR can be used to measure code performance. As described in Chapter 2, for additive white Gaussian noise (AWGN) channels, the upper bound of spectral efficiency R (bits per second per Hz in passband) can be expressed in terms of SNR:

$$R = \log_2(1 + SNR), \tag{5.4}$$

Inversely,

$$SNR = 2^R - 1 \tag{5.5}$$

Therefore, we can define the normalized SNR as

$$SNR_{norm} \stackrel{\text{def}}{=} \frac{SNR}{2^R - 1}. \tag{5.6}$$

SNR_{norm} has the property that for any spectral efficiency, the Shannon bound is $SNR_{norm} = 0$ dB.

Of course, noise immunity is not the only performance metric when comparing coding. As discussed before, code length affects transmission latency and implementation complexity and is an important consideration. Demodulation complexity also depends on code family and implementation technology, for example, whether the decoder uses a general-purpose processor or an application-specific integrated circuit (ASIC). Detailed discussions of these issues are beyond the scope of this book.

5.2.6 Interleaver

Typically, an error correction code works best when the input bit errors are genuinely random. Namely, there are no burst errors (i.e., the concentration of many errors in a short time period). Such is the case with the AWGN channels. However, in practice, burst errors can happen as a result of channel fading or burst noise (see Chapter 6 for more details). Burst errors can also be produced by an accompanying decoder (see Sections 5.6 and 5.7).

An interleaver converts burst errors to random errors by rearranging the bit order. An interleaver takes a collection of bits and shuffles them into a pseudorandom order. At the transmitter, encoded bits are "scrambled" by an interleaver before being sent through the channel. At the receiver, received bits are "unscrambled" by a deinterleaver before entering the decoder. This way, a burst error caused by the channel is spread over the sequence at pseudorandom positions.

Common challenges in interleaver design include tradeoffs between performance (i.e., how far and how random the bits are spread), additional latency, and memory requirements. Details of interleaver design are beyond the scope of this book [3].

There are two classical types of interleavers: block interleavers [4] and convolutional interleavers [5]. They are different in structure but similar in functionality. Interleavers are often used as a supplement to or as an integral part of channel coding (see Section 5.7).

5.3 BLOCK CODE

5.3.1 Block Code in General

Block codes are the first family of error correction codes in history. Early block codes predated Shannon's seminal paper on information theory and were mentioned in the latter [6]. A block code takes a "block" of k data bits, to form a codeword of n bits. The minimum Hamming distance among the codewords is d. Such a block code is noted as (n, k, d). The code rate is k/n.

The earliest block codes include the Hamming code [7], the Golay code [8], and the Reed–Muller (RM) code [9, 10]. Codes with more intricate algebraic structures include the cyclic codes, the Bose–Chaudhuri–Hocquenghem (BCH) codes [11, 12], and the Reed–Solomon (RS) codes [13]. It was found later that the RS and BCH codes are closely related.

5.3.2   Linear Code and Systematic Code

In this book, we consider only a special class of block code, known as the linear code, which includes the most popular block codes. For linear codes, it is convenient to express the data blocks and codewords as row vectors. Operations on them can be expressed by matrices. Note that all linear algebra operations are defined in the field F_2. Namely, all numbers are 0 or 1, and all additions are modulo-2, as described in (5.1). Moreover, for the sake of clarity, in this section, we do not denote vectors with bold-faced letters as in the rest of this book.

A codeword of a linear code is a linear transform of the input data:

$$c_m = s_m G, m = 1, 2, \ldots, 2^k. \tag{5.7}$$

Here, s_m is a row vector of size k, representing the input bit block. G is a coding matrix of size $k \times n$ and of rank k, specifying the code. c_m is the resulting codeword, which is a row vector of size n. Obviously, rows of G represent the "basis" of codewords. All codewords are linear combinations of the basis codewords. Conversely, any linear combination of the basis codewords is a valid codeword.

The weight of a codeword is the number of 1s it contains. Obviously, a linear code always contains a codeword with weight 0 (i.e., all bits are 0). The minimum Hamming distance between codewords is the minimum weight of nonzero codewords.

Furthermore, a special type of linear code is the systematic code. For a systematic code, the first k bits of a codeword are the same as input bits. The rest $n - k$ bits are known as parity check bits. The simplest parity-check adds one bit after the input bits, which is the (modulo-2) sum of all input bits. More generally, a systematic code can have multiple parity check bits. Its codewords are generated by (5.7), where the first k columns of G form an identity matrix:

$$G = [I_k, P]. \tag{5.8}$$

Here, I_k is an identity matrix of size $k \times k$. P has the size of $k \times (n - k)$ and is known as the parity matrix (not to be confused with the party-check matrix in Sections 5.3.3.2 and 5.8). All linear codes are equivalent to systematic codes in some sense.[3] Therefore, we can limit our consideration to systematic codes without sacrificing performance.

5.3.3   Hard Decision Decoding of Systematic Code

In this section, we discuss a hard decision decoding method for systematic code. Given the received bit vector y of size n, we wish to find the optimal codeword c_m. We start by stating the ML detection problem. Then, we introduce the concepts of parity-check matrix, syndrome and coset. These concepts lead to a decoding method.

5.3.3.1   Maximum Likelihood (ML) Detection

With maximum likelihood (ML) detection (see Chapter 4), the transmitted codeword is identified by

$$\hat{m} = \arg \max_m p(y \mid c_m). \tag{5.9}$$

Here, $\{c_m\}$ are all possible codewords, and y is the received bit sequence. If c_m is transmitted and y is received, the number of bit errors happened in the n bits is $d_H(y, c_m)$:

$$d_H(y, c_m) \overset{\text{def}}{=} \sum_{l=1}^{n} |y_l - c_{ml}|, \tag{5.10}$$

where y_l and c_{ml} are the lth bits of y and c_m, respectively. $d_H(y, c_m)$ is the number of bits that are different between y_l and c_{ml}. It is known as the Hamming distance between y and c_m. If the probability of bit error is p and the bit errors in a codeword are mutually independent, then

$$p(y \mid c_m) = p^{d_H(y, c_m)}(1-p)^{[n - d_H(y, c_m)]} = \left(\frac{p}{1-p}\right)^{d_H(y, c_m)} (1-p)^n. \tag{5.11}$$

Usually, we have $p < 0.5$. Therefore, to maximize $p(y \mid c_m)$ we should minimize $d_H(y, c_m)$. Above assumptions apply to a large collection of channels (including AWGN). For these channels, the decoding scheme

$$\hat{m} = \arg \min_m d_H(y, c_m) \tag{5.12}$$

is an ML detection algorithm. In Sections 5.3.3.2 to 5.3.3.5, we will see how to choose the codeword that minimizes $d_H(y, c_m)$.

---

[3] Coding matrices $G' = FG$ generate the same set of codewords as G, where F is a $k \times k$ matrix. Therefore, codes based on G' and G have the same performance. Given that G is of rank k, one can always choose F so that the first k column of G' is an identity matrix. In other words, G' represents a systematic code.

5.3.3.2   Parity-Check Matrix

As described in Section 5.3.2, a systematic code is generated by (5.7), where generation matrix G is given by (5.8). All codewords are linear combinations of the k row vectors in G.

Consider a matrix H of size $(n-k) \times n$:

$$H \stackrel{\text{def}}{=} \left(-P^T, I_{n-k} \right). \tag{5.13}$$

Here, $(\cdot)^T$ denotes transpose operation. It can be easily verified that

$$GH^T = 0. \tag{5.14}$$

The right-hand side of (5.14) is a matrix of size $k \times (n-k)$, whose elements are all zero. From (5.14) and (5.7), we see that

$$c_m H^T = 0 \, \forall \, m. \tag{5.15}$$

Namely, any codeword, when right-multiplied by H^T, leads to a zero vector of size k. It can also be shown that all vectors satisfying (5.15) are codewords.[4]

Matrices satisfying (5.14) are known as the parity-check matrix.[5] They can be used to check whether a vector is a codeword based on (5.15). Note that H is not unique given a code matrix G. In fact, for any full-rank matrix F of size $(n-k) \times (n-k)$,

$$H' \stackrel{\text{def}}{=} FH \tag{5.16}$$

is also a parity-check matrix.

5.3.3.3   Syndrome and Coset

Given the received vector y, we define the *syndrome* as vector u:

$$u \stackrel{\text{def}}{=} yH^T \tag{5.17}$$

Consider y as a sum of a codeword c_m and an error vector e:

$$y = c_m + e. \tag{5.18}$$

Considering (5.15), (5.17) leads to

$$u = eH^T. \tag{5.19}$$

---

[4] H has $n-k$ rows. Therefore, in an n-dimensional space, there are k linearly independent vectors $\{d_l, l = 1, 2, \ldots, k\}$ that are orthogonal to H, as expressed in (5.15). Since all rows in G satisfy (5.15) and are linearly independent, they must be linear combinations of $\{d_l\}$. Furthermore, any vector c_m satisfying (5.15) is a linear combination of $\{d_l\}$. Therefore, c_m is also a linear combination of the rows in G. Namely, it can be expressed by (5.7). Therefore, c_m is a codeword.

[5] Parity matrix can be defined for linear codes that are not systematic, by (5.14), although in this case (5.13) does not apply.

Therefore, a syndrome is connected to the error vector. There are many e vectors corresponding to the same syndrome u. In fact, if both e_1 and e_2 satisfy (5.19), then the difference between e_1 and e_2 is a codeword. Therefore, for each syndrome u, there are 2^k corresponding e vectors. Such collection of e are referred to as the *coset* associated with u. Within a coset, the member that has the least weight (i.e., has the least bits equal to 1) is known as the coset leader.

Note that e is of size n, and u is of size $n - k$. In other words, there are 2^n possible e vectors. They form 2^{n-k} cosets, each containing 2^k members.

5.3.3.4 Standard Array

The standard array is a table listing all cosets. In this section, we will describe the construction steps with an example, which is a systematic block code with $n = 5$, $k = 2$. The G matrix is

$$G = \begin{bmatrix} 1 & 0 & 1 & 0 & 1 \\ 0 & 1 & 0 & 1 & 1 \end{bmatrix}. \tag{5.20}$$

We will take the following step to construct the standard array.

Step 1: Put all codewords on the top row. Codewords can be found based on (5.7), where s_m takes all possible values. s_m is of size k. In our example, s_m can be (0, 0), (0, 1), (1, 0), (1, 1). The corresponding codewords are listed in the first row of Table 5.1. Make sure codeword with zero weight (i.e., all bits are zero) is the left-most one. Obviously, this row is also a coset, corresponding to $u = (0, 0, 0)$.

Step 2: Starting with vectors with least weight (i.e., with the least number of bits equal to 1), find a vector that is not already in the table. In our example, we can choose (0, 0, 0, 0, 1), becomes the coset leader for the next coset.

Step 3: Generate the coset containing this vector by adding (with modulo-2) codewords to the coset leader. In our example, the coset members are (0, 0, 0, 0, 1), (0, 1, 0, 1, 0), (1, 0, 1, 0, 0), (1, 1, 1, 1, 1). This forms the second row in Table 5.2.

Step 4: Repeat Steps 2 and 3, until all 2^{n-k} cosets and corresponding coset leaders are found. In our example, the eight coset results are shown in Table 5.3.

TABLE 5.1   Standard array step 1.

00000	01011	10101	11110

TABLE 5.2   Standard array step 3.

00000	01011	10101	11110
00001	01010	10100	11111

TABLE 5.3   Completed standard array.

00000	01011	10101	11110
00001	01010	10100	11111
00010	01001	10111	11100
00100	01111	10001	11010
01000	00011	11101	10110
10000	11011	00101	01110
00110	01101	10011	11000
01100	00111	11001	10010

TABLE 5.4   Standard array with mapped syndrome.

00000	01011	10101	11110	000
00001	01010	10100	11111	001
00010	01001	10111	11100	010
00100	01111	10001	11010	100
01000	00011	11101	10110	011
10000	11011	00101	01110	101
00110	01101	10011	11000	110
01100	00111	11001	10010	111

After the standard array is obtained, each coset can also be mapped to a syndrome based on (5.19), as shown in the last column of Table 5.4.

Note that the choice of coset leaders can be somewhat arbitrary. For example, in Table 5.4 row 7, both 00110 and 11000 have the same weight. Either of them can be the coset leader. Changing the coset leader results in a rearrangement of members of coset in the row. However, the coset still has the same members, and the corresponding u value does not change. The implication of such ambiguity to decoding will be discussed in Section 5.3.3.5.

Also, note that the standard array and its mapping to syndromes are both independent of the received signal. Therefore, they can be computed during the design process and stored in the receiver.

5.3.3.5 Decoding

Armed with the results so far, we can proceed to decoding. The receiver model is described in (5.18). Given u, ML detection says we should choose the e that satisfies (5.19) with minimum weight. Such an error is the most likely one under our assumptions. Therefore, decoding can be done as follows:

1. Given received bit block y, compute u from (5.17).
2. Look up the corresponding row in the standard array, as in Table 5.4. Let e be the coset leader of that row.
3. The optimal codeword is $y - e$ (with modulo-2).

As shown in Section 5.3.3.4, the selection of coset leader is ambiguous when there are multiple members of a coset with minimum weight. Such a situation is similar to the case of uncoded demodulation when the received signal is in the middle of two constellation points. In both cases, neither of the alternative decisions is reliable. Therefore, the choice of coset leader is not important in such case. See Section 5.3.5.1 for more discussions. In our example in Section 5.3.3.4, such ambiguity happens when e has weight 2. This means decoding is reliable only when there is one-bit error in the block of five bits.

5.3.3.6 Summary

Section 5.3.3 described a hard decision decoding method for linear codes. An example of a $(5, 2, 3)$ systematic code is used to illustrate the construction of the standard array. We start by computing the syndrome caused by the received bit vector, using the parity-check matrix. Next, the syndrome is mapped to the estimated error vector (the coset leader), based on the standard array. The decoded codeword is the input bit vector minus the error vector. This decoding method is quite practical if the added redundancy (i.e., $n - k$) is not too big.

The decoding method described in this section is referred to as syndrome decoding and is used to illustrate the principles behind block code decoding. There are other decoding techniques not covered by this book.

5.3.4 Soft Decision Decoding of Systematic Codes

As discussed in Section 5.2.4, soft decision decoding, that is, decoding based on the received signal instead of demodulated bits, can improve noise immunity. For block codes, the most straightforward soft decoding method is correlator selection.

Figure 5.2 shows a generic soft decision decoder for block codes. The basic idea is ML detection described in Chapter 4: finding the codeword that has a maximum correlation with the input signal. Correlation computation is divided into two parts. First, a matched filter reduces the time-continuous signal to samples at symbol times (see Chapter 4 for more details). Next, such a sample sequence is correlated with the expected sample sequences based on the 2^k codewords. The outputs of the correlators are sent to the selector. The codeword corresponding to the maximum output is selected as the decoder decision.

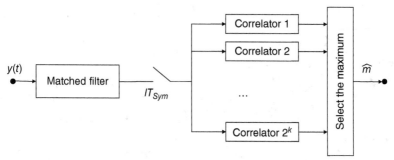

Figure 5.2   Soft decision decoder for block code.

While Figure 5.2 illustrates the basic concept, there may be ways for simplification. For example, for linear codes, we can use the basis of the codewords to reduce the number of correlators needed. Similar idea for in the case of uncoded modulation was discussed in Chapter 4. One may also be able to take advantage of the mathematical structures of specific codes [14]. Another example of simplification is the Wagner algorithm [15].

5.3.5   Performance of Block Codes

In this section, we discuss several topics on block code performance. As stated in Section 5.1, a detailed performance analysis is out of the scope of this book.

5.3.5.1   Error Detection and Error Correction with Hard Decision Decoding

Let us consider hard decision decoding, that is, decoding based on the received bitstream produced by the demodulator at the receiver. The decoder can perform two functions: error detection and error correction. For error detection, we know we *cannot* decode correctly based on the received signal. Error correction, on the other hand, means correct data can be recovered by the decoder.

While the value of error correction is obvious, error detection provides additional protection of data integrity. In practical communication systems, error detection is often provided independent of, and in addition to, channel coding. For example, an Ethernet physical layer frame includes "frame check sequence (FCS)" field, which contains 32 bits. These bits are used to detect errors in the frame. In the familiar transmission control protocol (TCP)/internet protocol (IP) for internet, for example, a TCP packet provides a 16-bit checksum field in its header, to protect data integrity of the packet. When a data error is detected, it can be addressed by other mechanisms. For example, the transmitter and receiver can coordinate to resend the corrupted packet or frame.

The common error detection scheme is the cyclic redundancy check (CRC), which can be viewed as a form of block code. Both Ethernet and TCP use CRC for error detection. With CRC protection, additional bits (checksum bits) are appended to the data. Checksum bits are computed from the values of data bits. At the receiver, errors can be detected by checking the consistency between checksum bits and data bits (similar to parity checking in Section 5.3.3).

In general, any block code can be used for error detection. If a code has a minimum Hamming distance d_{min} between codewords, a transmitted codeword can have a maximum of $d_{min} - 1$ bit errors, without turning into a different codeword. Therefore, if the number of bit errors within a block is between 1 and $d_{min} - 1$, the receive will see an illegal codeword (i.e., the syndrome in Section 5.3.3.3 is nonzero), indicating the presence of errors. In other words, a block code can detect up to $d_{min} - 1$ bit errors in a block:

$$m_d \le d_{\min} - 1, \tag{5.21}$$

where m_d is the number of bit errors in a block that we can detect. If the number of bit errors is more than $d_{min} - 1$, the received sequence may be a legit codeword but is different from the transmitted one. In this case, the error cannot be reliably detected by the receiver.[6]

Now let us come back to error correction. As stated in Section 5.3.3.1, an ML decoder selects the codeword closest to the received bit block (in the sense of Hamming distance) as the decoder decision. Obviously, such a decision is always correct when the number of bit error is no more than $(d_{min} - 1)/2$. Namely,

$$m_c \le \frac{d_{\min} - 1}{2}, \tag{5.22}$$

where m_c is the number of bit errors we can reliably correct. Let us look at the example shown in Section 5.3.3.4. For this particular example, $d_{min} = 3$. Therefore, according to (5.22), the code can correct all errors that involve one bit. When an error involves more than one bit in error, the decoding result may choose the wrong codeword.

Recall that the decoding process in Section 5.3.3.5 is as follows. We compute the syndrome from the received bit block. The syndrome indicates that the error vector must be in a specific coset (i.e., a specific row in Table 5.4). Moreover, the coset leader is picked to construct the output codeword.

Let us examine rows 2–6 of Table 5.4. For these rows, the coset leaders have weight 1, and there are no other members of weight 1 in the cosets. Therefore, if the corresponding syndromes are received and there is at most one-bit error, the error vector must be the coset leaders. Namely, the decoder decision described in Section 5.3.3.5 is always correct.

However, when there are two-bit errors, we may end up with rows 7 and 8 in Table 5.4. In both cases, the coset leaders have weight 2. However, there is another

---

[6] On the other hand, for a (n, k) code, there are 2^k possible codewords and 2^n possible received sequences. Therefore, with high bit error rates, the probability of a codeword turning into another one due to random bit errors is approximately 2^{n-k}, which can be a small number. Therefore, even with a high number of bit errors in a block, a block code can still detect most of the errors.

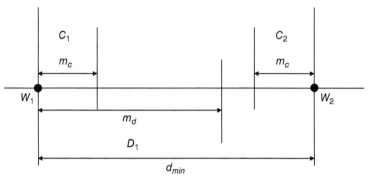

Figure 5.3   Error detection and correction for block codes. W_1 and W_2 denotes two codewords separated by the minimum distance d_{min}. m_c, distance for error correction; m_d, distance for error detection.

member with weight 2 in each coset. Namely, if there are two-bit errors, we would not know which member in the coset is the correct one to choose. Furthermore, two-bit errors may bring us to the wrong coset. For example, if the error vector is 00011, we would end up with row 5 in Table 5.4, which leads the decoder to believe the error vector is 01000. In other words, in these cases, the error cannot be corrected.

From the above discussions, we see that a block code with a minimum distance of d_{min} can *detect* up to $d_{min} - 1$ bit errors and *correct* up to $(d_{min} - 1)/2$ bit errors in a block. Furthermore, a code can perform both error detection and correction. This is done by limiting the weight of error vectors to correct. A code may choose to correct up to m_c bits and detect up to m_d bits of errors. This means that if a block as no more than m_c bit errors, these errors can be corrected. If the number of bit errors is between $m_c + 1$ and m_d, the decoder will declare error but cannot correct it.

Consider a pair of codewords with Hamming distance d_{min}. The regions for error detection and error correction are shown in Figure 5.3. The two codewords are marked as W_1 and W_2, separated at minimum Hamming distance d_{min}. C_1 and C_2 are correction regions. If the received vector falls into C_1 or C_2, it would be "corrected" into W_1 or W_2, respectively. The size of C_1 and C_2 is m_c. Region D_1 is the error detection region. If the received vector falls into D_1 but not C_1, it is determined to have an error, although we do not know whether W_1 or W_2 was transmitted.

On the other hand, if W_1 is transmitted and too many bit errors happens, the received vector can be in C_2. In this case, the detector determines that W_2 was transmitted. This is therefore an undetected error. Therefore, D_1 and C_2 can have adjacent boundary:

$$m_d = (d_{min} - m_c) - 1. \qquad (5.23)$$

In other words,

$$m_c + m_d = d_{\min} - 1. \tag{5.24}$$

Equation (5.24) shows the tradeoff between m_c and m_d, or the powers of error correction and error detection.

Let us go back to the example in Section 5.3.3.4. We can see that if we choose not to perform error correction (i.e., $m_c = 0$), we can detect two-bit errors per block (i.e., $m_d = 2$). This is because, as shown in the first row of Table 5.4, all codewords are differ by at least three bits from each other (i.e., $d_{\min} = 3$). Therefore, if the error vector has a weight no more than 2, the result will not be a valid codeword and the error can be detected. This is consistent with (5.21). On the other hand, suppose we want to perform error correction. From (5.22), m_c can only be 1. In Table 5.4, we can see that some error vectors with weight 2 are listed in rows 2–5. These cases will be "corrected" using the corresponding coset leader. The result is still not correct. This means that when there are two or more bit errors, the receiver cannot detect the error. Namely, m_d can only be 1. This relationship is consistent to (5.24).

5.3.5.2 Error Probability of Hard Decision Decoding

For hard decision decoding, the decoder error probability depends on the bit error rate (BER) at the demodulator output. Suppose such BER is p. As described in Section 5.3.5.1, the hard decision decoding will not make an error when the number of bit errors in the block is m_c or less. When the number of bit errors exceeds m_c, a decoding error may occur. Therefore, the total probability of error, with the union bound, is[7]

$$P_e \leq \sum_{m=m_c+1}^{n} \binom{n}{m} p^m (1-p)^{n-m}. \tag{5.25}$$

When $p \ll 1$, only the first term is important. Therefore,

$$P_e \approx \binom{n}{m_c+1} p^{m_c+1} (1-p)^{n-m_c-1} \tag{5.26}$$

This result is equivalent to the nearest neighbor approximation in Chapter 4, except that we assume here that any error vector with a weigh of m_c reaches to a nearest neighbor.

Equations (5.25) and (5.26), together with (5.22), establish the relationship between the approximate hard decision decoding error rate and the minimum distance.

---

[7] As an upper bound estimation, (5.25) assumes that if any $m > m_c$ bits in the block are in error, we will get a legitimate codeword.

In the example shown in Section 5.3.3.4, $m_c = 1$, $n = 5$. When p is very small, (5.26) leads to

$$P_e \approx 10p^2. \tag{5.27}$$

Therefore, in order to reach an error probability of 10^{-7}, which is typical for data transmission, we need p to be 10^{-4}.

Note that P_e is the error probability for the whole block, not BER. To convert P_e to BER, we need to estimate the number of data bits in error for each error event. The exact conversion can be quite involved. However, since we are usually only interested in the order of magnitude when dealing with BER, accurate conversion between P_e and BER is not critical.

5.3.5.3 Error Probability of Soft Decision Decoding

Performance of soft decision decoding is much more complicated, and in many cases, only bounds are known. In this section, we provide a simplified example [2]. Consider a block code (n, k, d) with binary phase shift keying (BPSK) modulation, transmitted through an additive white Gaussian noise (AWGN) channel. It is shown in Chapter 4 that bit error rate for BPSK is

$$P_{eBPSK} = Q\left(\sqrt{\frac{2E_{Sym}}{N_0}}\right), \tag{5.28}$$

where E_{Sym} is the symbol energy. N_0 is the one-sided noise spectral density. The block code in this example has a block length of n bits, with k data bits. Let d be the minimum Hamming distance between codewords. In other words, the two closest codewords A and B have d bits in difference. If the received signals corresponding to A and B are run through the correlator for code A, the resulting energy difference would be $2dE_{Sym}$. This energy difference corresponds to the $2E_{Sym}$ in (5.28) for BPSK. Furthermore, because of the code rate difference, the number of bits per symbol is k/n. Therefore,

$$E_b = \frac{n}{k}E_{Sym}. \tag{5.29}$$

Block error rate under nearest neighbor approximation is thus [16, sec. 3.2.1]

$$P_{eBlock} \approx N_d Q\left(\sqrt{\frac{dk}{n}\frac{2E_b}{N_0}}\right), \tag{5.30}$$

where N_d is the number of the nearest codewords. To convert block error probability to bit error rate, assume that each block error causes one bit error (similar to the Gray coding in Chapter 4). In this case, the bit error probability is

$$P_b = \frac{P_{eBlock}}{k} = \frac{N_d}{k}Q\left(\sqrt{\frac{dk}{n}\frac{2E_b}{N_0}}\right). \tag{5.31}$$

Again, this is a rough estimation to illustrate the concept. For more detailed analyses, consult specialized books on coding.

5.3.5.4 Performance Examples

Figure 5.4 shows an example of a block code performance computed using the MATLAB tool bertool. In the figure, BER versus E_b/N_0 are plotted for a block code with a code rate of 1/2, QPSK modulation.[8] Theoretical performances for both hard decision decoding and soft decision decoding are plotted. These performances are consistent with those of the Golay code (24, 12, 8) [17, fig. 2.19]. For comparison, performance for uncoded modulation is also plotted. The uncoded modulation is BPSK, to keep the same spectral efficiency (i.e., number of data bit per symbol) as the coded one.

We can examine the relative coding gains (i.e., the difference in required E_b/N_0 for a given BER level) among the various schemes. This example shows that in low BER region (BER between 10^{-6} and 10^{-8}), the soft decision gains about 2 dB in E_b/N_0 comparing to the hard decision performance. The coding gain for the hard decision decoding is about 2.5 dB for this particular code. Note that although Figure 5.4 shows coding gain being insensitive to the reference BER level, this is not always the case. For example, a Reed–Solomon code with a block size of 255 bits

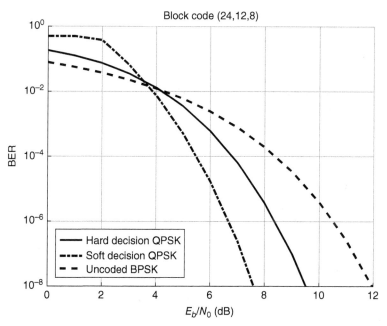

Figure 5.4 Block code performance example.

[8] A higher order modulation is used for coded transmissions, so that the user data rates for coded and uncoded modulations are the same.

shows much faster decreases in BER than the uncoded modulation when E_b/N_0 increases [1, sec. 3.4]. More examples of block code performance can be found in the literature, for example, [2].

5.3.6 Block Code Summary

Section 5.3 provides a brief introduction to block codes. A block code introduces redundancy to enhance noise immunity. The key performance figure is the minimum Hamming distance d between codewords. Such distance determines the code noise immunity. Other specifications include block length n and the number of data bits per block k. Therefore, a block code is often expressed as three integers (n, k, d). In addition to seeking for a large d for a given set of n and k, block codes are typically designed with some algebraic structure to make decoding easier.

A common subclass of block code is the linear code, which also includes the popular systematic code. Because of its structure, linear codes are easier to analyze and implement.

Besides being a relatively simple channel coding method, block code can also be combined with other codes. Section 5.6 provides some examples.

Decoding of block codes can be in either hard decision or soft decision. The latter method has better performance and higher complexity. This section described hard decision decoding using the syndrome and the standard array, as well as soft decision decoding based on correlators. There are many other decoding methods, the choice among which depends on the codes and implementation situations.

Block codes can also be used for error detection. While a code with minimum Hamming distance d_{min} can correct up to $(d_{min} - 1)/2$ bit errors, it can detect up to $d_{min} - 1$ bit errors. A combination of these two functionalities can also be performed by a block code.

5.4 CONVOLUTIONAL CODE

Another class of classical codes, the Convolutional code, was invented by Peter Elias in 1955 [18]. Incidentally, Elias was also the thesis advisor of Robert Gallager, who invented the low-density parity-check (LDPC) codes in his PhD work (Section 5.8). In 1967, Andrew Viterbi introduced the Viterbi algorithm (VA) as an optimal and practical decoding method for convolutional codes [19]. Today, most channel coding systems in practice use convolutional codes, if not modern codes [2]. Convolutional codes are also a building block for many combined codes and many modern turbo codes.

5.4.1 Several Ways to Specify a Convolutional Code

The block codes discussed in Section 5.3 maps a block of input data to a codeword with a fixed length. On the other hand, the underlying structure of the convolutional code is a continuous stream of coded bits, generated by the input data bitstream. Each input bit influences only a range of output bits (i.e., with an impulse response).

With the addition of initialization and tailing bits, a convolution code can be applied to a block of data bits, forming a (usually very long) block code. However, performance and decoding method are all based on the streams of input and output instead of the block structure and are determined by the impulse response, as will be shown in more details in this section.

Sections 5.4.1.1 to 5.4.1.3 discuss several ways to represent the convolutional codes, providing different perspectives. We will use a relatively simple example to illustrate the methods, while extensions to more general cases are discussed. Section 5.4.1.1 provides the basic description of convolutional codes based on the shift registers and generators. This description is a direct and intuitive way to specify the code. Shift registers are also a common way to implement convolutional encoders. Sections 5.4.1.2 and 5.4.1.3 provide two alternative representations: the state machine and the trellis. These representations are useful in analyzing code performance and designing decoders.

5.4.1.1 The Shift Register Representation and the Generator

Figure 5.5 shows a convolutional code represented by the shift registers. In this example, the input data bits are fed to a bank of shift registers (also known as a tapped delay line), which is depicted as the row of squares. Each square represents one register. At every time step in operation, one data bit is put into the leftmost register, while the content in each register is transferred to the one to the right, with the content in the rightmost register discarded. In this example, there are three registers. In general, the number of the registers is K, known as the constraint length of the code.

There are also two summers in Figure 5.5, tapping into the data in the registers selectively. The summations are performed with modulo-2 as described by (5.1). The outputs of the summers are multiplexed into a single sequence of coded bits. In this case, two coded bits are produced at each step (i.e., for each input data bit). In general, there can be n summers, producing n coded bits at each step.

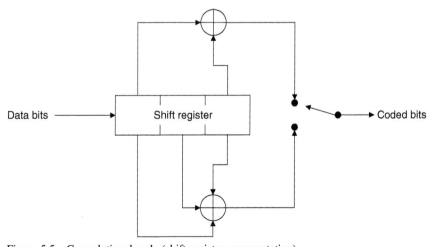

Figure 5.5 Convolutional code (shift register representation).

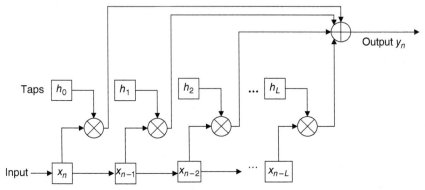

Figure 5.6   Filter implementation with shift registers.

The registers are typically initialized to zero at the beginning. At the end of the transmission block, trailing bits (typically all zeros) can be added to the data bits, so that the last bits in data travel through all the registers before transmission ends.

Convolution codes and the conventional filters (i.e., the convolution operation) are closely related. Figure 5.6 shows a structure of a filter of length $L + 1$ using the shift registers. The input data $\{x_n\}$ propagate in the shift register bank shown in the bottom row of the squares in Figure 5.6. The top row of the square shows the static memory for tap values $\{h_n\}$. At each step, data values and tap values at the corresponding positions are multiplied, and the products are summed to form output $\{y_n\}$. This realizes the filter function:

$$y_n = \sum_k x_{n-k} h_k. \tag{5.32}$$

The convolutional encoder in Figure 5.5 is a special case of the filter structure Figure 5.6, where both data and filter tap values are binary. Therefore, a summer i in the convolutional encoder can be specified by a binary sequence $\{g_{il}, i = 1, 2, \ldots, n, l = 1, 2, \ldots, K\}$. For each position l of the shift register bank, g_{il} is 1 if there is a tap to the ith summer, and is 0 otherwise. g is known as the *generator* for the convolutional code. Traditionally, the sequence $\{g_{il}, l = 1, 2, \ldots, K\}$ is expressed as an octal number $\bar{g}_i$. Such octal number is understood as a binary sequence, where the least significant bit corresponds to the oldest position (i.e., the rightmost position in Figure 5.5) of the shift register bank. Therefore, if the code as n bits per step, it is specified by n octal numbers $\{\bar{g}_i, i = 1, 2, \ldots, n\}$.

In the example shown in Figure 5.5, the tap sequences can be expressed as (101) (the top row) and (111) (the bottom row). Therefore, the generator for this code is (5,7).

In general, k bits, instead of 1 bit, can be fed into the encoder at each step, leading to a code rate of k/n. In this case, there are two ways to represent the encoder. In the first way, the shift register bank has kK registers. At each step, k input bits are taken, and data are shifted by k registers at each step. In the second way, there are k banks of shift registers. The results are summed to form output. The second representation allows the flexibility that different register banks can have different constraint lengths.

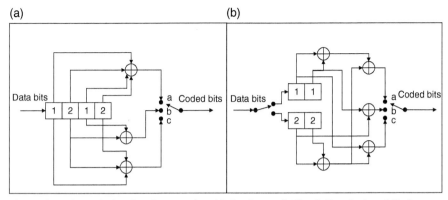

Figure 5.7 Convolutional code example with $k = 2$, $n = 3$, $K = 2$. Panels A and B show two equivalent realizations.

Figure 5.7 shows the two representations with $k = 2$, $n = 3$, $K = 2$. In Panel A, the shift register bank hosts all input data bits. It is advanced by k positions at each step. The data bits are grouped into "1" and "2," whose positions in the shift registers are marked in the squares. The n output bits are formed by tapping into the register bank, similar to Figure 5.5. The three output bits in each step are marked as a, b and c. They are multiplexed into one bitstream. Panel B shows the second way of representation. The input bits are de-multiplexed into groups 1 and 2, and fed to two shift register banks. The two banks are encoded separately as in Figure 5.5, while their results are further summed to form final output bits.

For the cases of $k > 1$, the generators can be defined in either long sequences (length of kK), or k separate ones (cf., Figure 5.7 for the two representations). In this case, the generator is a k by n array.

Furthermore, feedback can be introduced in convolutional encoding. Figure 5.8 shows an example. The encoder structure is similar to that in Figure 5.5, except

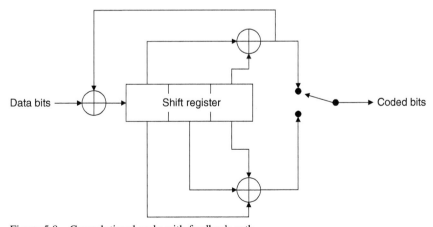

Figure 5.8 Convolutional code with feedback path.

that one of the coded bits is fed back to the input. Such codes are also referred to as recursive convolutional code. Convolution codes with feedbacks can be described in the form of a ratio between two polynomials, similar to the Z transform of IIR filters. Recursive codes are used in turbo code, described in Section 5.7.

To summarize, this section describes the convolutional code encoding process, based on the shift register representation. A convolutional code can be denoted by the triplet (n, k, K). The three parameters are:

- n: the number of coded bits produced at each step;
- k: the number of data bits used at each step;
- K: the constraint length.

The code rate is

$$R_c = \frac{k}{n}. \tag{5.33}$$

In addition, the code is specified by its generator g, which can be expressed as a matrix of size $k \times n$. An element of g is an octal number, whose binary form describes the taps used by the corresponding summation.

Note that the above scheme does not specify the order in which the k input bits are distributed to the k groups, nor the order in which the n output bits are arranged into a sequence. Obviously, such orderings do not affect the performance of the code.[9] As long as the encoder and decoder are kept consistent about the ordering, there is no need to concern.

On the other hand, MATLAB functions convenc and poly2trellis use the following convention (using the representation with separate shift register banks for each input bit, as shown in panel B of Figure 5.7). For input, within a group of k bits, the first bit (i.e., the earliest bit) is assigned to the branch expressed by the first row of g. For output, the first coded bit in the group is produced by the summation expressed by the first column of g.

5.4.1.2   State Machine Representation

An alternative way to describe a convolutional code is the state machine. In fact, any shift register bank can be described by a state machine. Let us first consider the case where $k = 1$. Intuitively, the state of the encoder is determined by the contents of the shift registers. Therefore, for an encoder with constraint length K, there are K shift registers, and thus 2^K states. One state can transition to another when the contents

---

[9] Since the input data bits are considered as mutually independent, their ordering does not affect performance. With regard to the coded bits, it will be shown in Section 5.4.3 that the n bits are processed as a group. Therefore, their ordering does not matter.

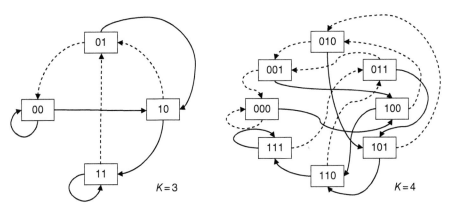

Figure 5.9   State diagrams for shift registers. Solid lines: the input is 1; dashed lines: the input is 0.

in the registers are shifted to the right. As described in Section 5.4.1.1, each state corresponds to one set of output coded bits (n bits).

However, we can reduce the number of states by associating the coded bit output to transitions between states, instead of the states themselves. In this new scheme, the states are defined by the bit values held by the shift registers, excluding the first one. Thus the number of states is reduced to 2^{K-1}. However, the content of the first register can be recovered by looking at transition. For example, assume $K = 4$. If we have a state specified as (101), we know that register 2, 3, and 4 has values 1, 0, and 1, respectively. The value of the first register remains unknown. On the other hand, when we see this state transition to (110), we would know the first register held 1 in the previous state. Therefore, we can determine the output coded bits for the previous state. In this scheme, we consider the output coded bits as associated with the transition. In the following, the values of the first register is referred to as the input to the state machine.

Figure 5.9 shows two state diagrams. The left side has four states ($K = 3$), and the right side has eight states ($K = 4$). Each box represents a state, marked with bit values contained in the shift registers excluding the first one. A transition of states indicates the shift between the registers. Solid lines indicate that the content in register 1 (i.e., the input) was 1 before the transition. Dashed lines indicate such content was 0.

Let us go back to the code described in Figure 5.5. We can see that when the shift register has values (100), the coded bits are (11). To express such code with the "reduced state" scheme, we associate the output with the transition between state (00) and (10).

Figure 5.10 shows the state diagram for the convolutional code described in Figure 5.5. The state diagram is the same as the left part of Figure 5.9. However, the output coded bits associated with transitions are also noted in this Figure.

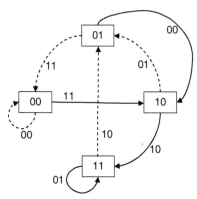

Figure 5.10   State diagrams and output coded bits. Solid lines: the input is 1; dashed lines: the input is 0. Bits next to the lines are output bits. Bits in the boxes indicate states.

State diagrams shown in Figures 5.9 5.10 can be extended to $k > 1$ cases in various ways. For example, based on the right part of Figure 5.7, a state can be defined as a collection of the k banks of shift registers. Therefore, the number of states is

$$N_s = 2^{\sum_{i=1}^{k}(K_i-1)}, \tag{5.34}$$

where K_i is the constraint length for the ith bank of shift registers. Each state has 2^k transition paths to other states, corresponding to the k input bits.

Note that the state machine representation is more general than the shift register representation. While a general state diagram can connect any states with transitions, only the connection configurations showed in Figure 5.9 represent shift register structures. On output values, the convolutional code is a linear code. Such linearity is natural in the shift register representation. For the state machine representation, however, linearity constraints the choice of output values associated to the state transitions.

5.4.1.3   Trellis Representation

Armed with the state diagram, we can trace the transitions through the states. As will be shown in Section 5.4.2, for performance analysis, it is important to know that if we start from state 0 (i.e., all registers have value 0) and transition to a different state, how many steps it would take to return to state 0. In other words, we need to view state transition in the time dimension. For this purpose, trellis representation is a handy tool.

A trellis can be graphically represented on a grid. The horizontal axis is time (in steps), while the vertical one denotes states. Line segments are drawn between grid points indicating state transitions. Connected line segments form a "trace," enabling us to track time evolution of the encoder state.

Figure 5.11 shows the trellis representation of the state machine shown in Figure 5.10. The numbers on the left side indicate the states. The numbers on the top indicate the steps. The thin lines serve as grid guides. The thick lines are the traces showing state transitions, with solid lines showing transitions with input 1, and dashed

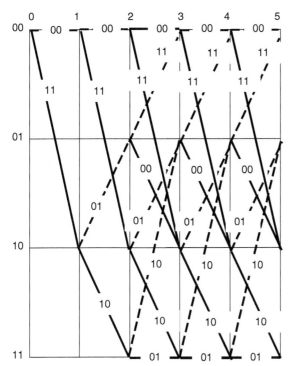

Figure 5.11   Trellis of convolutional code. Solid lines: input is 1; dashed lines: input is 0. Bits on the lines indicate output bits. Bits on the left side indicate states. Numbers on the top indicate time steps.

lines, 0. The numbers on the traces show the output coded bits associated with these transitions.

The traces all start with the initial state 00. At each subsequent state, a trace bifurcates into two, corresponding to inputs 1 or 0. In step 2, all states are reached. Therefore, traces in all subsequent steps are the same.

From the trellis diagram, we can also see that the shortest path returning to state 00 takes three steps, passing states 10, 01, and 00.

The trellis representation can be extended to cases where $k > 1$. In this case, the number of states is given by (5.34), and the number of branches stemming from each state is 2^k.

5.4.1.4   Initial State, End State, and Codewords

As described above, a convolution encoder continues to take in data bits and to put out coded bits. Such encoding process can theoretically run indefinitely. In practice, however, data are always broken down to blocks (packets or frames) in transmission. It is more convenient to have an encoding process contained in one data block.

When an encoding process starts, the encoder and decoder need to agree on the initial values of the shift registers (or, equivalently, the starting state of the trellis or state diagram). Typically, the encoder is initialized so that all shift registers contain 0 except for the first one, which contains the first data bit. Therefore, the states described in Sections 5.4.1.2 and 5.4.1.3 are initialized to 0.

At the end of the block, a predetermined bit sequence (typically all zero) is appended to the data bits, to bring the encoder to a known end state. These bits are known as the trailing bits. They provide the last data bits with the same coding protection as the rest of the data bits.

With the initial state and the trailing bits, we can consider convolutional code as a special case of block code, except that the "block" is much longer than a typical block code. In this sense, we can view a particular sequence of output coded bits as a codeword. It corresponds to a trace across the trellis diagram in Figure 5.11.

5.4.1.5 *Punctured Convolutional Code*

As can be seen in Section 5.4.1.1, the simplest convolutional code is those with $k = 1$. Such codes have the code rate of $1/n$, which is 1/2 or lower. To get a higher code rate, we need to have $k > 1$. This increases the complexity of both encoder and decoder. Furthermore, the code rate may need to be fine-tuned frequently to match the current channel condition. Changing code rate by choosing different k and n values would be too crude and too cumbersome in these cases.

Alternatively, convolutional codes can be "punctured" to increase code rate. In the puncturing process, some of the coded bits are deleted at the transmitter, forming a shorter codeword, and thus achieving higher code rate. The choice of the bits to be deleted is important to code performance. Therefore, designing punctured convolutional code is complicated and usually involves both analytical considerations and computer search and simulation. Some known puncturing structures are given in the literature [20, sec. 8.4].

5.4.2 Performance and Free Distance

5.4.2.1 *Error Probability and Trace Deviation*

Section 5.4.1 shows how a convolutional code is specified and represented in different ways. In this section, we analyze the noise immunity of such code. We limit the discussion to hard decision decoding. In this case, the question becomes the relationship between "raw" bit errors and decoding error. Since the convolutional code is a linear code, all codewords have the same performance. Therefore, for convenience, we consider the probability of error when the transmitted codeword is all zero. Namely, the states follow the top trellis trace in Figure 5.11. (See Section 5.4.1.4 for the concept of codeword in convolutional code.)

For a convolutional code, all legitimate state transition sequences are shown as traces in the trellis diagram in Figure 5.11. Each trace corresponds to a coded bit sequence, or codeword. When a bit sequence is received at the decoder and under ML hard decision decoding, we look for the codeword that has the minimum Hamming distance from the received sequence.

Therefore, a decoding error occurs when there are enough bit errors in the received bit sequence so that it is closer to an alternative codeword (in terms of Hamming distance) than the correct one. Naturally, such an error is more likely to happen when the Hamming distance between the two codewords is small. For convolutional code, the codewords are as long as the entire transmission sequence.

A "close" alternative trace breaks away from the correct trace for a few steps and merges back. For example, in Figure 5.11, suppose the correct trace is 00, 00, 00, 00, 00, 00. A close alternative trace can be 00, 00, 10, 01, 00, 00. Namely, it deviates for two steps and comes back to the correct one. The three wrong transitions [00 → 10, 10 → 01, 01 → 00] generate coded bits 11, 01, 11. Comparing to the correct coded bits 00, 00, 00, the Hamming distance in such deviation is 5. More generally, the Hamming distance between an all-zero trace and an alternative trace is the number of coded bits with value 1 generated by the alternative trace. Such bit count is called the weight of the bit sequence, or the weight of the corresponding trace.

Given that two codewords A and B have a Hamming distance of d between them, what would be the pairwise error probability P_{eAB}? Obviously, only bit errors in the d bits that are different in A and B would affect the decoder decision between the two codewords. A decoding error would occur if and only if no less than $(d+1)/2$ of them is wrong.[10] Furthermore, there are $\binom{d}{m}$ ways to have m bit errors among the d bits. Therefore, the total pairwise probability of error is

$$P_{eP}(d) = \sum_{m=(d+1)/2}^{d} \binom{d}{m} p_b^m (1-p_b)^{d-m}, \tag{5.35}$$

where p_b is the raw bit error probability.

Therefore, in order to evaluate the error probability of a convolutional code, we need to know the range of Hamming distances between the codewords, and the number of codewords that have Hamming distance d from the reference codeword (commonly all zero). In Section 5.4.2.2, we introduce a tool known as the transfer function [20, sec. 8.1-2] to answer these two questions. The key is associating a state with a polynomial that tracks the weight of the traces leading to that state.

The overall error probability, with the union bound, is the summation of all pairwise errors:

$$P_e \leq \sum_{d=d_{free}}^{\infty} a_d P_{eP}(d), \tag{5.36}$$

where d_{free}, known as the free distance, is the minimum Hamming distance among codewords.[11] a_d is the number of codewords that has Hamming distance d from the reference codeword. Furthermore, we can take the nearest neighbor approximation and consider only the first term in summation.

$$P_e \approx a_{d_{free}} P_{eP}(d_{free}). \tag{5.37}$$

---

[10] This is obviously true when d is an odd number. If d is even, $d/2$ wrong bits would produce a tie between the two codewords. Special treatment (e.g., randomly select one codeword) is needed for this special situation. For the sake of simplicity, we ignore this complication.

[11] In the literature, the term "minimum distance" is also used. There is a subtle difference between minimum distance and free distance [24]. However, in this book, these two terms are used interchangeably.

Note that P_e is the probability of error for a codeword, not the probability of bit error. When a codeword error occur, the number of bit errors depends on the code structure, which can be analyzed in a way similar to the transfer function described in Section 5.4.2.2 [20, sec. 8.2-2], [21]. Detailed analysis in this aspect is not discussed in this book.

In practice, we are usually only concerned about the order of magnitude of bit error probability (or bit error rate). For such purpose, P_e given above could be a good estimate.

5.4.2.2   Transfer Function

In this section, we describe the transfer function as a tool for estimating the Hamming distances among the convolutional codewords.

Introduce a dummy variable Z. We define the transfer function $T_{AB}(Z)$ for a pair of states A and B as a polynomial:

$$T_{AB}(Z) \overset{\text{def}}{=} \sum_l a_l^{[A,B]} Z^l. \tag{5.38}$$

Here, l denotes the weight of coded bit sequence, when we follow a particular trace from A to B. $a_l^{[A,B]}$ is the number of traces between A and B with weight l. Since the summation limit can be infinite, we assume $|Z|$ is small enough to ensure convergence.

The transfer function can be constructed using the following observations, which can be easily justified by the readers.

First, if states C and B are connected with a single transition that generates a coded bit sequence with weight m, then

$$T_{CB}(Z) = Z^m. \tag{5.39}$$

Second, let $T_{AB}^{[C]}(Z)$ be the transfer function between A and B where only traces passing through C are included, then

$$T_{AB}^{[C]}(Z) = P_{AC}(Z)P_{CB}(Z). \tag{5.40}$$

Third, let $T_{AB}^{[CD]}$ be the transfer function between A and B, where traces passing through either D or D are included, then[12]

$$T_{AB}^{[CD]} = T_{AB}^{[C]}(Z) + T_{AB}^{[D]}(Z). \tag{5.41}$$

With these three observations, we can construct transfer functions for any pairs of states in a state machine. While the general procedure can be found in the literature [1, sec. 4.6], [22], an example will be shown in Section 5.4.2.3.

---

[12] We do not consider traces passing both C and D, because in application of this property, we only care about states C and D that happen at the same time point.

Note that depending on the definition of exponent m in the first step above (5.39), the same approach can be used to collect statistics of other metrics, as long as these metrics of a trace is the sum of those of each segment. Examples of such metrics include the number of transitions (where m is 1 for each transition), the weight of input data bits (where m is the weight of the input data bits driving the transition), etc.

5.4.2.3 Transfer Function Example

Let us now compute the transfer function of A back to A, after going over other states. When A is the state of 0, a trace starting from A and back to A represents a deviation from the correct path, if it passes any other states then A. The weight of such trace is a part of the error probability computation discussed in Section 5.4.2.1.

We use the example described in Figure 5.10 to illustrate the basic concept. In our purpose, the self-loops around state 00 are ignored. Furthermore, we split state 00 into two states, $00'$ and $00''$, and evaluate the transfer function between them. For that purpose, Figure 5.10 is redrawn as Figure 5.12.

In Figure 5.12, the power of Z are marked next to the coded bits for each transition. Let x_s be the transfer function between state $00'$ and state s. Using the three properties observed in Section 5.4.2.2, transitions in Figure 5.12 translate to the following equations:

$$
\begin{cases}
x_{01} = x_{11}Z + x_{10}Z \\
x_{10} = x_{01} + Z^2 \\
x_{11} = x_{11}Z + x_{10}Z \\
x_{00''} = x_{01}Z^2
\end{cases}
\tag{5.42}
$$

For example, the first equation in (5.42) is obtained by observing that two transitions lead to state 01: one from state 11 with output coded bits 10 (weight 1), and the other from state 10 with output coded bits 01 (weight 1).

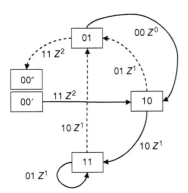

Figure 5.12 State diagram for transfer functions. Solid lines: the input is 1; dashed lines: the input is 0. Bits next to the lines are output bits, whose total weights are indicated by the exponents of Z. Bits in the boxes indicate states.

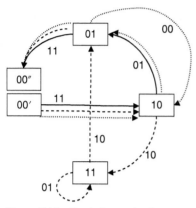

Figure 5.13   Deviation paths from state 00′ to state 00″. Solid, dashed, and dotted lines show three loops among the states. Other notations are the same as previous figures.

Equations in (5.42) lead to the solution[13]

$$x_{00''} = \frac{Z^5}{1-2Z} = \sum_{l=0}^{\infty} 2^l Z^{5+l}. \tag{5.43}$$

The last equation in (5.43) comes from the Taylor expansion of $(1 - 2Z)^{-1}$ when $|Z| < 1/2$. We can see that for this code, the lowest power of Z is 5, that is, $d_{free} = 5$. Namely, for a trace to start from state 00 and to return to state 00, at least five bits of value 1 need to occur in the coded bit sequence.

We can look at the first terms in (5.43) based on Figure 5.10, as shown in Figure 5.13.

In Figure 5.13, the solid transition lines form the path 00′ to 10 to 01 to 00″, representing the first term in (5.43) (i.e., $l = 0$). The total weight of coded bits along this path is 5. The dashed transition lines form a path with a total weight of 6: 00′ to 10 to 11 to 01 to 00″. The dotted transition lines form a path with the same weight: 00′ to 10 to 01 to 10 to 01 to 00′. These two paths represent the second term in (5.43) (i.e., $l = 1$). From then on, paths with weight $l + 5$ can be constructed from any path with weight $(l-1) + 5$ by adding either a loop 11 to 11, or a loop 10 to 01 to 10, all with the additional weight of 1 in output coded bits. Therefore, the number of paths with weight $l + 5$ is twice that with weight $(l-1) + 5$. This agrees with (5.43).

5.4.2.4   Error Probability and Free Distance

Now let us return to the error probability given by (5.35), (5.36), and (5.37). In order to compute error probability, we need to know d_{free} and $\{a_d, d = d_{free}, d_{free} + 1, \ldots \}$. Both parameters are provided by the transfer function (5.38), which can be evaluated following the example in Section 5.4.2.3.

---

[13] Note that there are several loops in state transition diagram. The existence of loops causes Z to appear in the denominator in (5.43), leading to infinite summation. This approach is similar to the handling of infinite impulse response (IIR) filters in the Z-transformation. Z-transformation is discussed briefly in Chapter 9.

These results are for hard decision decoding. For soft decision decoding, a good estimate can also be provided based on the transfer function [20, sec. 8.2-2].

Note that the transfer function, as evaluated in Section 5.4.2.3, considers only one "trip" of deviating from the reference trace and returning to it later. In reality, if the transmission sequence lasts long enough, multiple deviations may happen. However, since the Hamming distance between these traces and the reference trace is at least multiples of d_{free}, multiple deviations happen in small probabilities and can be ignored.

Therefore, given the code parameters n, k, and K (defined in Section 5.4.1.1), we wish to find a code with the maximum d_{free} for optimal noise immunity. It has been found that free distance can be bounded as follows [21]:

$$d_{free} \le \min_{l \le I} \left\lfloor \frac{2^{l-1}}{2^l - 1}(K + l - k)\frac{n}{k} \right\rfloor, \tag{5.44}$$

where

$$I = \begin{cases} 1 & \text{if } K < 2k - 1 \\ k & \text{if } K \ge 2k - 1 \end{cases}. \tag{5.45}$$

This bound turns out to be rather tight. Usually, convolutional codes can be found, whose free distance is within 1 from the bound [20, sec. 8.3]. Equation (5.44) shows that in general, d_{free} increases as the code rate k/n becomes lower. For the same code rate, d_{free} increases with the constraint length K.

5.4.3 Viterbi Algorithm (VA)

Viterbi algorithm (VA) can be used for both hard decision and soft decision decoding. It is an optimal decoding technique based on the ML principle. The invention of VA played an important role in the broad adoption of the convolutional code. The same principle can also be used in some other estimation problems.

In this section, we describe the Viterbi decoding concept and algorithm. We start by examining the decoding problem more closely in Section 5.4.3.1. Such discussion leads to the Viterbi algorithm, which is described in Section 5.4.3.2. After that, Section 5.4.3.3 briefly discusses soft output VA.

5.4.3.1 The ML Decoding Problem in AWGN Channel
As discussed in Section 5.3.3.1, for AWGN channel, the ML decoding looks for the codeword that is the closest to the received signal:

$$\hat{m} = \arg \min_m d(c_m, y). \tag{5.46}$$

Here, the minimization is over all possible codewords $\{c_m\}$. As discussed in Section 5.4.1.4, for convolutional codes, a codeword corresponds to a trace in the trellis diagram that starts at the given initial state and ends at the given ending state (usually both being the zero state). Conversely, any such trace represents a legitimate codeword. y is the received signal. $d(c_m, y)$ is the signal distance between the codeword c_m and the received signal y. It is also known as the cost function in this context. For hard decision, $d(c_m, y)$ is the Hamming distance:

$$d_H(c_m, y) = \sum_j \left| c_{mj} - \hat{y}_j \right|. \tag{5.47}$$

Here, c_{mj} and $\hat{y}_j$ are the jth bit of the codeword and received bitstream (output of the demodulator, derived from y), respectively. For the soft decision, the demodulator is not used. Instead, the output of the matched filter is taken as the input to the decoder. The distance to minimize is the Euclidian distance:

$$d_E(c_m, y) = \sum_j \left| \bar{c}_{mj} - y_j \right|^2 \tag{5.48}$$

Here, $\bar{c}_{mj}$ and y_j are the jth symbol of the codeword and received signal, respectively.

For the ease of discussion, we can group both the bits and the symbols into units of "code letter." The ith code letter is the n bits generated by the state transition *leading* to the ith state. Such a transition is also referred to as the ith step. In this sense, we can write a common expression of ML decoding algorithm:

$$d(c_m, y) = \sum_i \mu\left({}^m s_i, {}^m s_{i-1}, y_i \right), \tag{5.49}$$

where ${}^m s_i, {}^m s_{i-1}$ are the states of codeword m in steps i and $i-1$, respectively. y_i is the part of the received signal corresponding to the ith code letter. The stepwise cost function is the signal distance for one code letter, and is defined as

$$\mu(s_1, s_2, y) \overset{\text{def}}{=} \begin{cases} \sum_j \left| c_j^{[s_1, s_2]} - b_j \right| & \text{hard decision} \\ \sum_j \left| \bar{c}_j^{[s_1, s_2]} - a_j \right|^2 & \text{soft decision} \end{cases} \tag{5.50}$$

Here, $c_j^{[s_1, s_2]}$ and $\bar{c}_j^{[s_1, s_2]}$ are the jth bit or symbol, respectively, caused by the transition from state s_2 to state s_1. b_j and a_j are the bit or symbol, respectively, in the jth position as derived from received signal y. Summation of j runs over all bits (hard decision) or symbols (hard decision) contained in the ith code letter.

Unlike block codes, which usually have small block sizes, a convolutional codeword can span over thousands or tens of thousands of bits, depending on the application. Let N be the number of code letters included in a codeword. The total number of data bits transmitted is kN. The total number of possible traces (i.e., possible codewords) is thus 2^{Nk}, which can be a huge number. Therefore, it is not practical to implement ML decoding in (5.46) by exhaustive search.

Fortunately, because of the special form of the cost function shown in (5.49), the decoding complexity can be reduced dramatically with the Viterbi algorithm (VA), to be described in Section 5.4.3.2.

5.4.3.2 Viterbi Algorithm

The cost function for the convolutional code decoding is shown in (5.49). Given input y, the ith term in the summation depends on the codeword output at the particular state transition at the ith step. The total cost function is an accumulation of contributions from each step. Such property allows us to find the optimal trace through local search. Let

$$^m d[i,j](y) \overset{\text{def}}{=} \sum_{l=i+1}^{j} \mu\left(^m sl, {}^m sl-1, y_l\right) \tag{5.51}$$

where all other parameters are defined in the same way as in (5.49). $^m d^{[i,j]}(y)$ is the cost function accumulated between steps i and j, given the received signal y and the codeword m, We then have

$$d(c_m, y) = {}^m d[1, N](y), \tag{5.52}$$

where N is the total number of steps in the codeword. Definition (5.51) leads to the property

$$^m d[i,j](y) = {}^m d[i,l](y) + {}^m d[l+1,j](y) \tag{5.53}$$

For any $i \le l < j$. When the state in step l is fixed as s_l, $^m d^{[i,\,l]}$ and $^m d^{[l+1,\,j]}$ (y) can be minimized independently, by choosing traces extending backward and forward from s_l, respectively. Therefore, the minimization of $^m d^{[i,\,l]}$ is broken down to two minimization tasks. Especially, for each state s_l, we only need to keep one trace going backward, which carries the minimum cost. This trace will be a part of any optimal end-to-end trace that passes through s_l. Such property makes the complexity of decoding much lower than an exhaustive search.

This concept can be illustrated by an analogy. Suppose we want to find the shortest route from Rutgers University in New Jersey to New York University in New York City. Moreover, we want to use the Lincoln Tunnel to cross the Hudson River. There are N candidate routes from Rutgers to Lincoln Tunnel, and M candidate routes from Lincoln Tunnel to New York University. In this case, a naive exhaustive search consists of evaluations of distance over the NM route combinations. On the other hand, since distance along a route is additive as in (5.49), we can first find the shortest route from Rutgers to Lincoln Tunnel (N evaluations) and then find the shortest route from Lincoln Tunnel to New York University (M evaluations). This approach leads to a total of $M + N$ evaluations, typically much less than the NM evaluations in the naive approach. Such simplification can be extended to multiple middle points, for example, in case we want to pass either Lincoln Tunnel or Holland Tunnel.

Viterbi Algorithm (VA) is based on such a concept. Instead of searching globally for the most likely codeword as suggested by (5.46), VA constructs the optimal codeword through local optimizations. The algorithm can be summarized in the following operations. Remember that received signal or bits are given. Therefore, they are not explicitly referenced in most of the expressions.

1. Start with step count $l = 1$.

2. Note all eligible states in step l as $s_l^{[i]}$. For $l = 1$, the only eligible state is zero (initial state). As l grows, so does the number of eligible states. Eventually, i runs from 0 to $2^{k(K-1)} - 1$ for the $2^{k(K-1)}$ states (see Sections 5.4.1.2 and 5.4.1.3).

3. In step l, each state $s_l^{[i]}$ is associated with one trace extending back to step 1. This trace carries the minimum cost function, noted as $d_i^{[1,l]}(y)$, among all traces leading to $s_l^{[i]}$. Here, y is the received signal or bits so far. Obviously, the "one trace" assumption is valid when $l = 1$ and $l = 2$. We will show later how this trace is built beyond that point.

4. Now move from step l to step $l + 1$. Consider all states at this point $s_{l+1}^{[j]}$. Each state is connected to a subset of the states in step l, $\left\{ s_l^{[i]}, i \in \mathcal{S}_j \right\}$ on the trellis diagram. $\mathcal{S}_j$ is the subset of states in step l that can transition to state $s_{l+1}^{[j]}$. Typically, $\mathcal{S}_j$ has 2^k members.

5. Each of such transition incurs additional cost $\mu\left(s_{l+1}^{[j]}, s_l^{[i]}, y_l \right)$. This cost contributes to $d_j^{[1,l+1]}\Big|_i$, which is the cost function for the trace extending from step 1 to step $l + 1$ with the last two states being $s_l^{[i]}$ and $s_{l+1}^{[j]}$.

$$d_j^{[1,l+1]}\Big|_i = d_i^{[1,l]}(y) + \mu\left(s_{l+1}^{[j]}, s_l^{[i]}, y_l \right). \tag{5.54}$$

6. Among all $i \in \mathcal{S}_j$, choose the one that yields minimum cost function:

$$\hat{i} = \arg \min_i d_j^{[1,l+1]}\Big|_i. \tag{5.55}$$

7. The trace connecting $s_l^{[\hat{i}]}$ back to step 1 is extended forward to include the transition to $s_{l+1}^{[j]}$. Other connections to $s_{l+1}^{[j]}$ are removed. The cost function associated with state $s_{l+1}^{[j]}$ is

$$d_j^{[1,l+1]} = d_j^{[1,l+1]}\Big|_{\hat{i}} \tag{5.56}$$

8. Operations 5–7 are repeated over all states in step $l + 1$. When this is completed, each state in step $l + 1$ has one trace reaching back to step 1, validating the assumption made in operation 3.

9. Operations 4–8 are repeated for further steps until the end of the data block is reached.

10. In the end, the final state is known based on the trailing bits. Therefore, the decoding result is the trace extending from the known final state back to step 1.

Figure 5.14 shows an example, using the trellis diagram in Figure 5.11 ($k = 1$, $K = 3$). The traces are constructed in several stages from panel A to panel F. The dashed lines show the state-step grids (thin lines) and trellis (thicker lines), as references. The thick solid lines show the candidate traces.

- Panel A shows the trellis building of the first four steps (operation 4 above). All traces start with the initial state 00. In steps 1–3, each state has only one ancestor (i.e., states in the previous step that can transit to the current state). However, in step 4, each step has two ancestors. At this point, the traces are to be trimmed based on the associated cost functions.

- Panel B shows the result of trimming in step 4 (operations 5–7 above), leaving only one trace extending back from each state. Furthermore, traces are extended forward to step 5 (operation 4 above). Note that there is an orphan link (i.e., a terminated trace) as indicated.

- Panel C shows the result of trimming in step 5, the removal of the orphan link, and the extension of the traces to step 6. Note that trimming in step 5 results in another orphan link as indicated.

- Panel D shows the result of trimming in step 6, removal of the previous orphan links and extension to step 7 (the final step).

- Panel E shows the result of trimming in step 7 and the removal of previous orphan links. Note that all remaining traces are merged in step 4. Such merging is not guaranteed. However, it is most likely to happen in practice. Since step 7 is the final step, its valid state is known. In this case, the final state is 00. Only the trace going to the final state is valid.

- Panel F shows the final trace as the result of decoding.

Note that the cost functions are not shown in Figure 5.14. Therefore, the choices of the traces to keep are not derived from Figure 5.14 but depend on the received signal.

For a convolutional code with parameter (n, k, K), there are $2^{k(K-1)}$ states at each step. For each state, the decoder needs to keep track of a trace. Also, at each step, the decoder needs to compute 2^k transition cost (5.50) for each state. These considerations can be used to estimate the decoder complexity.

Ideally, a VA decoder maintains $2^{k(K-1)}$ traces until the end of the data block, when the final decision is made. In addition to the large memory requirement, such method also introduces significant delay in output (latency), since the decoding result is not available until all transmitted symbols are received and processed. Alternatively, a "delayed decision" decoding can be used. Namely, in step l, we select, as our final choice, the state in step $l - \delta$ that is associated with the minimum cost function at present. This way, only one trace is maintained before step $l - \delta$, and all user data up to that point are deemed as decoded. As shown in the above example, traces tend to merge after a while. In such case, the optimal trace has already emerged. Otherwise, our decision may not be optimal. It has been found experimentally that δ should be chosen to be $4K–5K$ to ensure good performance [1, sec. 4.5.1],

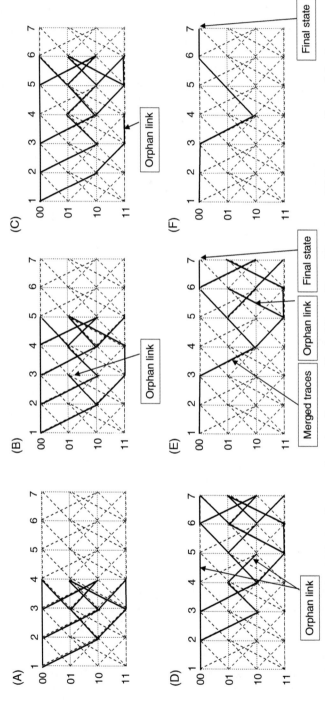

Figure 5.14 Example of VA decoding at various decoding steps. Dashed lines: all possible traces; solid lines: currently surviving traces.

[20, sec. 8.2-1]. Note that these values may depend on characteristics such as coding rates. δ can be further fine-tuned for particular codes through computer simulation.

5.4.3.3 Soft Output Viterbi

As shown in Section 5.4.3.2, a VA decoder naturally supports both hard decision and soft decision decoding. The only difference is the cost function definition in (5.50). On the other hand, a VA decoder can also produce soft output. Namely, it provides both the selected trace (with corresponding data bits) and the reliability metric. Such soft output can be further used in the next layer of decoding (see Sections 5.6 and 5.7). There are many design choices for soft output VA decoder. Interested readers may consult the literature, for example, [23].

5.4.4 Performance Figures and Code Examples

Good convolutional codes can be found with a combination of computer search and bound computations. Known good codes with various parameters are compiled by [20, sec. 8.3]. More examples can be found in [21, 24, 25] and references therein. A convolutional code with the very low code rate of 1/32 is presented in [26].

Figure 5.15 shows some sample performance of convolutional codes. The rate 1/2 code has $K = 7$, $k = 1$, $n = 2$. The generator is (171, 133). The rate 1/4 code has $K = 8$, $k = 1$, $n = 4$. The generator is (235, 275, 313, 357). The free distances are 15 and 22 for the two codes, respectively [20, sec. 8.2-2]. The BER values are the theoretical upper bounds computed by MATLAB tool bertool. Both hard and soft decision decoding for the rate 1/2 code are evaluated. For comparison, BER from uncoded modulation is also plotted. For power limited case (left panel), each case have the same modulation of four-point quadrature-amplitude modulation (4-QAM) (i.e., for the same data rate, the coded transmissions use twice the symbol rate, and thus twice the bandwidth). For bandwidth limited case (right panel), modulation schemes are adjusted from BPSK to 16-QAM, so that each coding scheme yields the same spectral efficiency (i.e., number of user data bits per symbol).

As shown in the figure, significant coding gains are obtained relative to uncoded modulation. Soft decision decoding brings about 2.5 dB additional coding gain comparing to hard decision decoding. When we go to a lower code rate, there is an advantage of about 1 dB coding gain in the power limit case. However, such an advantage is lost when higher order modulation must be used in the bandwidth limit case.

More illustrations of convolutional code performance can be found in [20, sec. 8.6].

5.4.5 Convolutional Code Summary

A convolutional code has three characteristics. First, it has a set of internal states that are connected by transition paths. Namely, the output codeword is not only influenced by the current input data but also the previous ones, which are "memorized" in the states. Second, the input data causes state transitions. Third, the output codewords are mapped to the state transitions. Therefore, to specify a convolutional code, we

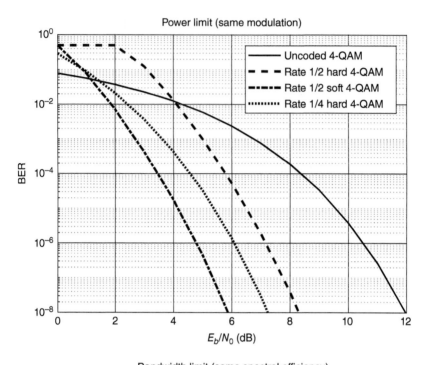

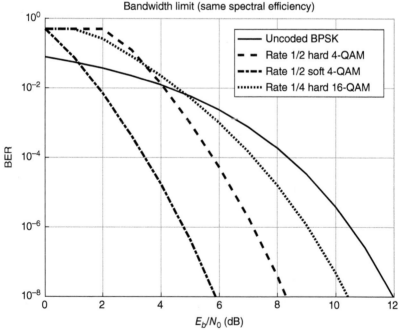

Figure 5.15   Convolutional code performances. Rate 1/2 code: $K = 7$, $k = 1$, $n = 2$; rate 1/4 code: $K = 8$, $k = 1$, $n = 4$.

need to specify its states and transitions among them. We also need to specify how input data trigger state transitions and how these transitions are mapped to the output data. Section 5.4.1 shows several ways to represent a convolutional code. The shift register representation and the associated generator provide a concise and intuitive description of the codes. They also provide a practical implementation of encoders. The state machine and trellis representations focus more on the intrinsic code structure and are more useful in analyzing code performance. The trellis representation is also closely related to the Viterbi algorithm (VA) for decoding, which is described in more details in Section 5.4.3. Although VA is the most widely used decoding technique for convolutional code and the only one covered in this book, other decoding methods are advantageous in certain cases [17]. Noise immunity of convolutional code can be analyzed in terms of free distance, which is the minimum weight of any nonzero codeword. Section 5.4.2 discusses the concept and computation of free space. Usually, analyzing convolutional code performance is complicated when constraint length is long. We often rely on various types of bounds. On the other hand, with the increase in computer power, direct simulations become more practical. Therefore, this book does not cover related topics beyond the basic concept of free distance. For the intended readers, the most important task is selecting well-studied convolutional codes for particular applications. Section 5.4.4 provided sample performances to set expectations. That section also includes some references that list practical codes.

5.5 CODING FOR BANDWIDTH-LIMITED CHANNELS AND TRELLIS-CODED MODULATION (TCM)

For bandwidth-limited channels, symbol rate is limited by the Nyquist theorem, as discussed in Chapter 3. Higher order modulation (i.e., modulations with more than one bit per symbol) is typically used to increase data rate (at the cost of higher transmission power for the same bit error rate). Furthermore, as discussed in Section 5.2.5, channel coding results in even higher order modulation, in order to provide the same user data rate. On the other hand, the block code (Section 5.3) and convolutional code (Section 5.4) are designed at the bit level, to maximize the Hamming distance between codewords. While they can be used for higher order modulation, which applies to the coded bits as shown in Figure 5.1, they are not necessarily optimal in this case. This section discusses coding techniques specifically designed for higher-order modulations.

5.5.1   Trellis-Coded Modulation (TCM)

Trellis-coded modulation (TCM) was developed in the 1970s and 1980s [27, 28]. It is based on the following observations. In a high order modulation, an error is most likely to happen between a constellation point and its neighbor. Therefore, channel coding protection can be applied to such pairs of constellation points, while not worrying about protecting a constellation point from other far-way points. Such an

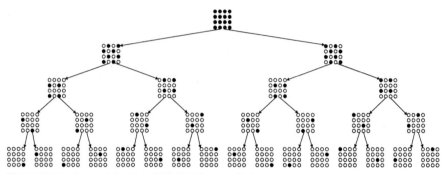

Figure 5.16 Set partitioning of 16-QAM constellation.

idea is realized by the concept of set partition (also referred to as coset in some literature, but not to be confused with the coset discussed in Section 5.3.3).

For TCM, constellation points in a modulation scheme are divided into sets iteratively, with nearest neighbors belonging to different sets.

Figure 5.16 shows an example of set partitioning, starting with a 16-QAM constellation [27]. The constellation points used in the sets are shown as solid circles. At each step, the current set is partitioned into two smaller sets, separating the nearest neighbor points. Each partition increases the minimum distance between constellation points by $\sqrt{2}$. Similar partitions can be performed to PSK modulations, as well.

In TCM coding, input data bits are separated into two groups. Group 1 is coded for error protection. The resulting coding bits are used to select the sets. Group 2 is not coded and is used to select constellation points inside a set. Since the nearest neighbors in the original constellation are assigned to different sets, the most likely errors lead to the wrong set identity. Since such identity is protected by coding, noise immunity is improved.

For example, consider the transmission of three bits per symbol by QAM. Without coding, we can use 8-QAM, with one of the constellations in the second row of Figure 5.16. With TCM coding, we can send one bit through a rate 1/2 convolutional coder, to produce two coded bits. These bits are used to select one of the four sets in row 3 of Figure 5.16. The other two-uncoded bits are used to select one of the four points in the set. The scheme is shown in Figure 5.17.

Figure 5.18 shows the constellation points in the sets and the total constellation. As shown in the figure, the top-left points of the four sets are mapped to the four top-left points in the 16-QAM. Therefore, the interset error probability is the same as that among the nearest neighbors in 16-QAM. The intraset error probability is the same as that in 4-QAM. With high enough coding gain, the interset error probability can be reduced enough so that the intra-set errors are dominant. In such case, comparing with the reference modulation (8-QAM), there is a 3 dB coding gain, because the nearest neighbor distance of 4-QAM is $\sqrt{2}$ of that of 8-QAM.

More coding gain can be obtained by dedicating more bits to the coding group. In the above example, if we allocate two bits for coding and one bit uncoded, we would be using the sets in the fourth row of Figure 5.16, and potentially achieving

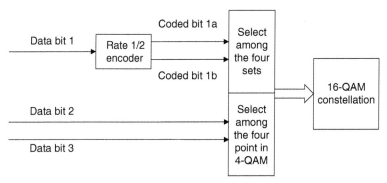

Figure 5.17   Sample TCM for 3 bit per symbol QAM.

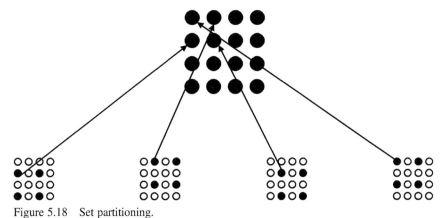

Figure 5.18   Set partitioning.

6 dB coding gain. However, in such cases, the coding gain is more likely to be limited by the channel code.[14] Namely, the interset error probability would be dominant. In this case, the performance of the TCM code depends on the underlining channel code.

Design and analysis of the channel code used for TCM employ the same approach as the convolutional code, as discussed in Section 5.4. However, the output of the code is not coded bits that are associated with the state transition, but the set identities that are associated with the states. Therefore, the number of states is an essential parameter for TCM code, as it also specifies the set partition scheme. For example, a four-state TCM code partitions the original constellation into four sets.

As with the convolutional code, interset noise immunity of TCM is also determined by the free distance f_{free}, which is the distance accumulated along a path that deviate from the all-zero path and returning to it. However, the "distance" for each segment (i.e., state transition) is not the Hamming distance (i.e., the weight of the output coded bits associated with the segment) as with the convolutional code, but the

---

[14] It is possible to achieve more than 6 dB coding gain with TCM, although some practical codes provide coding gains around 4 dB [2].

Euclidian distance (i.e., the distance in signal space between the two constellation points associated with the state transition). Therefore, designs of the trellis and the mapping between states and constellation sets are both important for TCM performance.

Since the advent of TCM, some good codes with a various number of states and modulation types were compiled in [29]. Comparing to uncoded QAM, TCM can achieve a coding gain up to 6 dB. Furthermore, in 1984 and 1987, Wei published two TCM codes that employed nonlinear convolutional code, which can accommodate phase rotations due to timing synchronization errors. The first code is 8-state [30]. The second code is 16-state and in four-dimensional (i.e., coding two consecutive symbols together) [31]. These two codes are used in the voiceband modem standards V.32 and V.34, respectively. The second code is also used in the asymmetric digital subscription line (ADSL) standard of ITU-992.1 and other DSL standards based on it.

Performances of the Wei codes are shown in Figure 5.19 [2]. The horizontal axis is SNR_{norm} defined by (5.6) in Section 5.2.5. The vertical axis is the symbol error probability. At an error probability of 10^{-5}, the coding gains (comparing to uncoded QAM) for the 2D and 4D Wei codes are 4.0 and 4.2 dB, respectively. The small difference in coding gain between the two standardized codes shows how valuable coding gain improvement was to the modem industry at the time. The meaning of "shaping" mentioned in the figure will be discussed in Section 5.5.2. Note that the uncoded QAM error probability in Figure 5.19 was approximated as [2, sec. V.A]:

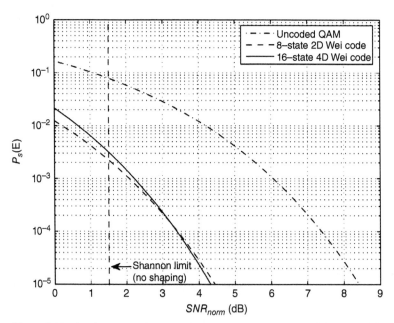

Figure 5.19   Performance of Wei codes comparing with uncoded QAM and Shannon limit [2, fig. 9]. Reproduced with permission.

$$P_s \approx 4Q\left(\sqrt{3SNR_{norm}}\right). \tag{5.57}$$

Such value agrees with those given in Chapter 4 when $M \gg 1$.

TCM codes can be decoded based on VA decoder, described in Section 5.4.3. First, in each set, the closest constellation point from the received signal is found. Its Euclidian distance from the received signal, that is, the "minimum distance" between the received signal and the set, serves as the basis of cost function computation in VA. Once the set identifications are decoded, the uncoded bits can be determined by a demodulator described in Chapter 4.

Theoretically, the underlying channel coding in TCM does not need to be convolutional coding. It can be block coding, for example [32]. However, in practice, convolutional coding is advantageous for various reasons. Therefore, TCM is the prevailing code used for high-order modulations.

5.5.2 Lattice Code and Shaping Gain

In this section, we briefly describe two other concepts related to bandwidth-limited channel coding, or coding with high order modulations: lattice code and shaping gain [2]. Although the lattice code is not widely used at present, its analysis illustrates the concept of shaping gain, which is important in coding theory.

A *lattice* is a regular array of points, generated by moving the origin by an integer number of steps towards various directions. A lattice constellation is the set of lattice points that are confined in a given *bounding region*.

Figure 5.20 shows two examples of lattice constellation. The left-hand side shows a square lattice confined to a square. The right-hand side shows a hexagonal lattice confined to a circle. More formally, an n-dimensional lattice can be defined as a set of points

$$\Lambda \stackrel{\text{def}}{=} \left\{ aG = \sum_{i=1}^{n} a_i g_i, a \in \mathbb{Z}^n \right\}. \tag{5.58}$$

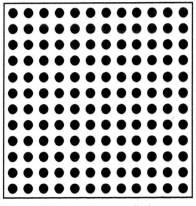
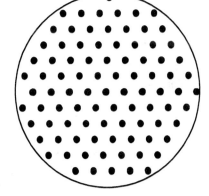

Figure 5.20   Lattice constellation examples.

Here, $\{a\}$ are arbitrary row vectors, whose components $\{a_j, j = 1, 2, \ldots, n\}$ are integers (i.e., they belong to the integer set $\mathbb{Z}$). G is the generator matrix of size $n \times n$ that defines the lattice. g_i is the vector formed by the ith row of G. Λ is a set of row vectors, each specifying a lattice point in the n-dimensional real space. A lattice constellation is defined as

$$C(\Lambda, \mathcal{R}) \overset{\text{def}}{=} \Lambda \cap \mathcal{R}, \tag{5.59}$$

where $\mathcal{R} \subset \mathbb{R}^n$ is the bounding region in the n-dimensional real space.

In the context of coding, the n-dimensional space is constructed by grouping $n/2$ symbols together, each contributing two dimensions (in-phase and quadrature, see Chapter 3). In that space, a lattice constellation is constructed. Each point in the constellation represents a codeword. Therefore, the lattice code is conceptually a block code with a length of $n/2$ symbols, except that it codes on symbols, instead of bits.

For a lattice, the important parameters are the minimum squared distances between the lattice points $d_{min}^2(\Lambda)$, the number of nearest neighbors to any lattice point $K_{min}(\Lambda)$, and the volume of the unit cell $V_\Lambda(\Lambda)$. A unit cell is defined as the parallelotope formed by the following points: $\{bG, b \in [0, 1)^n\}$. Here, b is any n-dimensional vector within the parallelotope. Obviously, $d_{min}^2(\Lambda)$ and $V_\Lambda(\Lambda)$ both depend on scale. The *Hermite parameter* is a normalized density parameter, which is independent of scale:

$$\gamma_c(\Lambda) \overset{\text{def}}{=} \frac{d_{min}^2(\Lambda)}{V(\Lambda)^{2/n}}. \tag{5.60}$$

For the constraining shape $\mathcal{R}$, we can further define its volume

$$V_R(\mathcal{R}) \overset{\text{def}}{=} \int_\mathcal{R} dx, \tag{5.61}$$

and its average energy density

$$P(\mathcal{R}) \overset{\text{def}}{=} \frac{1}{n V_R(\mathcal{R})} \int_\mathcal{R} \|x\|^2 dx. \tag{5.62}$$

Remember x here is an n-dimentional vector. Again, we can normalize the energy density into a scale-independent parameter

$$Q(\mathcal{R}) \overset{\text{def}}{=} \frac{P(\mathcal{R})}{V_R(\mathcal{R})^{2/n}}. \tag{5.63}$$

Let us return to the examples in Figure 5.20. For the square lattice and square bounding region shown on the left side, we have

$$G = \begin{pmatrix} c & 0 \\ 0 & c \end{pmatrix}$$

$$\begin{aligned} d_{min}^2(\Lambda) &= c^2 \\ V_\Lambda(\Lambda) &= c^2 \\ \gamma_c(\Lambda) &= 1 \end{aligned} \tag{5.64}$$

where c is the grid size, which determines the chosen scale. For the bounding region, assuming the half-side length of the square is a (another chosen scale),

$$V_R(\mathcal{R}) = 4a^2$$

$$P(\mathcal{R}) = \frac{a^2}{3} \quad . \tag{5.65}$$

$$Q(\mathcal{R}) = \frac{1}{12}$$

For the hexagonal lattice shown on the right side, we have

$$G = \begin{pmatrix} c & 0 \\ \frac{1}{2}c & \frac{\sqrt{3}}{2}c \end{pmatrix}$$

$$d_{\min}^2(\Lambda) = c^2 \quad , \tag{5.66}$$

$$V_\Lambda(\Lambda) = \frac{\sqrt{3}}{2}c^2$$

$$\gamma_c(\Lambda) = \frac{2}{3}\sqrt{3}$$

where c is the grid size. For the circular bounding region, assuming the circle diameter is r (another chosen scale),

$$V_R(\mathcal{R}) = \pi r^2$$

$$P(\mathcal{R}) = \frac{r^2}{4} \quad . \tag{5.67}$$

$$Q(\mathcal{R}) = \frac{1}{4\pi}$$

When the number of lattices points is large, we can make the following approximations. The average symbol energy per dimension is $P(\mathcal{R})$. The number of bits per codeword is

$$N_b = \log_2\left(\frac{V_R(\mathcal{R})}{V_\Lambda(\Lambda)}\right) \tag{5.68}$$

Since each codeword takes $n/2$ symbols, the spectral efficiency at the Nyquist rate (i.e., number of bits per symbol) is

$$R = \frac{N_b}{n/2} = \log_2\left(\frac{V_R(\mathcal{R})}{V_\Lambda(\Lambda)}\right)^{\frac{2}{n}} \tag{5.69}$$

Therefore, the normalized SNR given by (5.6), when $R \gg 1$, is

$$SNR_{norm} = \frac{P_R}{P_N}\frac{1}{2^R - 1} \approx \frac{P_R}{P_N}\frac{1}{2^R} = \frac{P(\mathcal{R})}{P_N}\left(\frac{V_R(\mathcal{R})}{V_\Lambda(\Lambda)}\right)^{-\frac{2}{n}}, \tag{5.70}$$

where P_N is the noise power per dimension. $V_R(\mathcal{R})$ and $V_\Lambda(\Lambda)$ can be expressed in terms of scale-independent parameters defined in (5.60) and (5.63):

$$SNR_{norm} = \frac{1}{P_N} Q(\mathcal{R}) \frac{d_{min}^2}{\gamma_c(\Lambda)}. \tag{5.71}$$

Therefore,

$$d_{min}^2 = 12 SNR_{norm} P_N \gamma_c(\Lambda) \gamma_s(\mathcal{R}), \tag{5.72}$$

where

$$\gamma_s(\mathcal{R}) \overset{\text{def}}{=} \frac{1}{12Q(\mathcal{R})}. \tag{5.73}$$

The reason for the factor 12 will be made clear later.

As discussed previously, the error probability of a code depends on d_{min}^2 (or d_{free}^2, we ignore the difference). Therefore, for higher performance, we wish to increase d_{min}^2 under a given SNR_{norm}. This means we wish to increase γ_c and γ_s.

γ_c is given in (5.60). It is the ratio between the minimum distance and the unit cell size per dimension. γ_c depends only on the lattice (specified by the generation matrix G). A larger γ_c means a smaller unit cell volume V_Λ with the same d_{min}^2. Geometrically, we can think of each lattice point as an n-dimensional sphere with radius $d_{min}/2$. To maximize γ_c, we need to design a lattice, such that as many spheres as possible can be packed into a given volume. Such geometric problem is known as the sphere-packing problem [33]. Note that for the square lattice, $\gamma_c = 1$, or 0 dB, as shown in (5.64) for two dimensions. In fact, this γ_c value is true for any dimension. Therefore, γ_c reflects the performance improvement of any lattice relative to the square lattice. It is known as the coding gain.[15] For two-dimensional hexagonal lattice, γ_c given by (5.66) is 1.155, or 0.6 dB. Higher coding gains can be obtained by going to higher dimensional space, namely, larger code block size.

γ_s is given by (5.73) and (5.63) and is known as the shaping gain. When γ_s is large, the bounding region average energy is small for the same bounding region volume. A larger γ_s reduces the SNR while keeping the same d_{min} and the number of constellation points (i.e., the number of codewords). In other words, for the same data rate and error probably, the required SNR level is reduced. γ_s depends only on the bounding region $\mathcal{R}$ and not the lattice design. Because of the factor 12 in (5.73), γ_s is 1, or 0 dB, for a square bounding region. Obviously, γ_s is the largest when the bounding region is an n-dimensional sphere, and n goes to infinity. It is shown in an appendix (Section 5.10) that the shaping gain is upper-bounded to $\pi e/6$, or 1.53 dB.

---

[15] Note that the term "coding gain" in this section is different from what is used in the rest of the chapter, where it refers to the difference in noise immunity compared to an uncoded modulation.

r_c and r_s can be optimized independently, through the choices of lattice and bounding region, respectively. Shaping of the constellation can be applied to other codes (such as TCM codes), as well. In order to achieve the Shannon bound, both gains must be optimized. In other words, without shaping (i.e., when the square bounding region is used), the best a code can do is achieving error-free transmission at $SNR_{norm} = 1.53$ dB. This level is marked as "Shannon Limit (no shaping)" in Figure 5.19.

A practical lattice code was published and used for voiceband modems [34, 35]. The sphere-packing problem has been an active research area [33], although the alternative TCM codes have much broader applications. On the other hand today, shaping has been used together with TCM coding in the voiceband modem standard V.34 [36, sec. 9]. Shaping gain is also used with other coding schemes such as turbo code, to be introduced in Section 5.7 [37]. For a good overview of the coding gain concept and a collection of classical references on the topic, see [38].

5.6 COMBINED CODES

As discussed in Section 5.2.2, it requires infinitely long codes to achieve Shannon limit. In general, longer codes are expected to have better noise immunity. However, decoder complexity also increases rapidly with code length. In order to solve this problem, shorter codes are combined to form long codes. These component codes (also known as constituent codes) can be decoded somewhat separately, limiting complexity. Such encoding and decoding method is not necessarily optimal. Nonetheless, combined codes serve as an approach to improve code performance.

One form of combined codes is the product code, where two block codes are combined. The structure is shown in Figure 5.21. In this case, there are two component

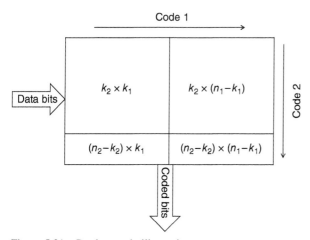

Figure 5.21   Product code illustration.

codes, with sizes (n_1, k_1) and (n_2, k_2). For the purpose of illustration, we assume the codes are systematic. For a size (n, k) systematic code, the first k bits of the codeword is the input data. The rest $n - k$ bits are referred to as parity check bits (Section 5.3.2).

For each block of product coding, $k_2 \times k_1$ data bits are taken to fill the upper-left rectangle. Code 1 is used to encode the rows of bits, generating the parity check bits that are filled to the upper-right rectangle of Figure 5.21. After such coding, the first k_2 rows of the entire area are the resulting codewords of code 1. Next, code 2 is applied to each column of the data, generating the parity check bits, which are filled to the lower rectangles. When encoding is completed, each column of the entire area in Figure 5.21 is a codeword of code 2. These codewords are output as the coded bits. The overall code length is $n_1 n_2$. The code rate is $k_1 k_2 / n_1 n_2$.

One can easily see that a two-step decoder can be constructed, where code 2 is decoded first, with results serving as input to the code 1 decoder.

Another form of combined code is concatenated codes, first introduced in the early 1960s [39]. Figure 5.22 shows a configuration [39]. In this scheme, there is an outer code, whose encoding results are fed to an inner code for secondary coding. The outer code operates on "symbols" (i.e., a group of bits), instead of on bits. Consider the case with an outer code with size (N, K) and an inner code of size (n, k). Input bits are grouped into "symbols," each containing k bits. The outer coder codes K of such symbols into a codeword containing N symbols. The inner coder then operates on each "symbol," converting the k bits into an n-bit codeword. Overall, one coding "block" consists of Kk input data bits and Nn output coded bits. At the decoder, the inner code is decoded first, generating input for the outer decoder.

Typically, the outer code is the Reed–Solomon (RS) code, which works naturally in symbols. The RS code has an advantage in correcting multiple bit errors within a symbol. Bit errors generated from the inner code tend to be concentrated in a codeword (i.e., symbols to the outer code) instead of uniformly distributed. Therefore, the RS code is a good fit.

There are variations of the above scheme. For example, in asymmetric digital subscriber line (ADSL) related systems [40, sec. 7.6, 7.8], an RS outer code is used, which operates on bytes (8-bit symbols). The output of the outer encoder goes through a byte-level interleaver, before being fed into the inner coder, which is a Wei 4-D TCM (see Section 5.5.1). The interleaver is used to counter burst errors caused by the inner coder that extends beyond byte boundaries. Burst errors within a byte are not spread by the interleaver since the RS code can deal with them effectively.

Combined codes are used in many practical systems. It is also a prelude to one of the modern capacity-approaching codes: the turbo code (Section 5.7).

Figure 5.22   Concatenated code structure.

5.7 TURBO CODE

So far, we discussed three code families: the block code (Section 5.3), the convolutional code (Section 5.4), and the TCM (Section 5.5.1). These codes are all invented before 1990 and are considered classical codes. In this section and Section 5.8, we describe two more code families emerged after 1990, classified as modern codes.

Turbo code was first published in 1993 and was claimed to be "near Shannon limit" from the start [41]. The term "turbo" refers to the iterative nature of its decoder, as will be explained briefly in this section. The same iterative method is also used for joint sequence estimations (not necessarily decoding) in equalization (Chapter 9).

5.7.1 Turbo Encoder

The primary concept of a turbo code is connecting multiple codes (known as constituent codes) with interleavers, resulting in long and random-like codewords. Because of such structure, a turbo code can be decoded by jointly decoding its constituent codes.

Figure 5.23 shows two types of turbo codes. Panel A shows the parallel concatenated convolutional code (PCCC). PCCC is the code proposed in the original paper [41]. In this configuration, the input data is fed to $n + 1$ parallel paths. One is the direct path, one directly to an encoder EC 1, and the rest of the $n - 1$ paths are fed to their respective encoders EC 2 to EC n, with different interleavers IL 2 to IL n. All of the encoders are rate 1. Namely, they generate one output bit for each input bit. The $n + 1$ outputs are multiplexed to form the output stream. Therefore, the overall code rate is $1/(n + 1)$. Panel B shows a serial concatenated convolutional code (SCCC). The input bits are coded by the first encoder EC 1, which generates multiple output bits depending on the code rate. In this particular case, two coded bits are generated for each data bit. These output bits pass through an interleaver, and enter the second encoder EC 2. Output bits of EC 2 are multiplexed into the coded bitstream.

In both cases, the encoders are convolutional codes. They typically include feedback paths, as in the example shown in Figure 5.8 in Section 5.4.1.1. These codes also include the direct path (as shown by the top path in the case of PCCC), known as the systematic path.[16] Such codes are known as the recursive systematic convolutional code (RSC). The n constituent codes in PCCC shown in Figure 5.23 can be considered individually as rate 1/2 systematic codes, while the systematic path is shared among the codes.

Other forms of turbo code are also possible. For example, block codes can be used as constituent codes [42]. The two structures (PCCC and SCCC) can be mixed to form a hybrid structure. The product codes described in Section 5.6 can also be viewed as turbo codes if they are decoded iteratively, known as the turbo product code (TPC).

Turbo codes are used in modern communication systems including cellular systems, space data systems and digital video broadcast systems [43, ch. 9]. Figure 5.24

---

[16] The encoder shown in Figure 5.8 does not have a systematic path.

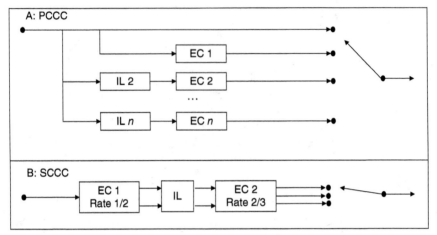

Figure 5.23   Turbo encoders. A: parallel concatenated convolutional code (PCCC); B: serial concatenated convolutional code (SCCC).

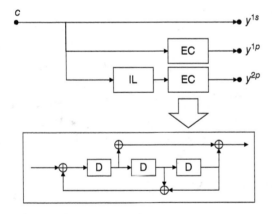

Figure 5.24   Turbo encoder used in 3GPP cellular systems.

shows the PCCC encoder used in the cellular systems as specified by the 3rd Generation Partnership Project (3GPP). Such code is used in the Long-Term Evolution (LTE) and its predecessor, the 3G cellular systems [44, sec. 5.1.3]. The same code is also used for the digital video broadcast services for the satellite to handheld devices (DVB-SH) [45, sec. 5.3].

The upper part of Figure 5.24 shows the overall structure, similar to panel A in Figure 5.23. Each data bit c goes through three parallel paths. The first generates the systematic bit y^{1s}, which simply duplicates the input bit. The second path contains a convolutional encoder EC, leading to the parity bit y^{1p}. The third path contains the same encoder following an interleaver (IL). The output of that path is another parity bit y^{2p}. The three output bits are further selected for transmission, based on the intended code rate adjustment (similar to the punctured code described in

Section 5.4.1.5). The constituent encoder used in the EC blocks is an RSC as shown in the lower part of Figure 5.24. The blocks marked as "D" are the shift registers for convolutional coding (see Section 5.4.1.1).

5.7.2 Symbol-Based Maximum A Posteriori Decoding (BCJR Algorithm)

Before discussing the decoding of turbo codes, we first introduce another way of decoding convolutional codes. Unlike the VA introduced in Section 5.4.3, this method estimates the probability distribution of bits, instead of the sequences. The probability of a bit to be 1 or 0 also depends on other bits, because of the coding structure. The method was first published by Bahl, Cocke, Jelinek, and Raviv in 1974 [46], and is known as the BCJR algorithm. For an overview of BCJR and other turbo decoding algorithms, see [47].

5.7.2.1 Problem Definition

Consider a convolutional code sequence transmitted through a channel. Let i denote the time steps for state transitions. The user data sequence is $\{u_i\}$. Such data drives a state sequence $\{\sigma_i\}$ at the encoder. The state sequence generates a coded symbol sequence $\{c_i\}$, which, with noise from the channel, becomes a received sequence $\{y_i\}$. The tasks is deriving $\{u_i\}$ from $\{y_i\}$. Actually, we are interested in "soft output." Namely, instead of determining whether u_i is 0 or 1, we estimate the probability for u_i to be 1.

For simplicity, we consider the special case of a rate $1/n$ convolutional code. Therefore, c_i is a bit vector of size n. y_i is also a vector, whose size depends on the modulation scheme.[17]

The BCJR algorithm is based on the maximum a posteriori (MAP) principle, which is described in more details in Chapter 4. The MAP detection method looks for the decision that yields the maximum a posteriori probability, given the received signal. Namely, the decision is given by

$$\hat{l} = \arg \max_l p\left(u_i = l \mid \{y_j\}\right). \tag{5.74}$$

Here, l can take the values of 0 and 1. $p(u_i = l \mid \{y_j\})$ is the probability of $u_i = l$ given the received *sequence* $\{y_j\}$. Note that because of the coding, u_i cannot be determined by y_i alone as was the case in Chapter 4. Instead, the entire sequence of the received signal needs to be considered. Fortunately, as will be shown in Sections 5.7.2.2 to 5.7.2.6, the problem is more tractable than it may seem.

Furthermore, Bayes' law states that

$$p\left(u_i = l \mid \{y_j\}\right) = \frac{p\left(u_i = l, \{y_j\}\right)}{p\left(\{y_j\}\right)}. \tag{5.75}$$

---

[17] For convenience, we assume that the n bits in c_i fit into integer number of symbols.

The denominator is common to all ls. Therefore, it can be ignored in the decision consideration (5.74). Furthermore, as mentioned before, our goal is not the make a decision as shown in (5.74) but to evaluate probabilities in support of such a decision. Therefore, we focus on the evaluation of $p(u_i = l, \{y_j\})$.

5.7.2.2   Decomposition of Probability

To compute $p(u_i = l, \{y_j\})$, we focus on the state transitions that are caused by u_i. For convolutional codes described in Section 5.4.1.2,

$$p(u_i = l, \{y_j\}) = \sum_{\sigma_i, \sigma_{i-1} \in S_l} p(\sigma_i, \sigma_{i-1}, \{y_j\}). \tag{5.76}$$

Here, S_l is the set of all state pairs (σ_i, σ_{i-1}) whose transitions are triggered by $u = l$. It depends on the code design.

Let us first focus on $p(\sigma_i, \sigma_{i-1}, \{y_j\})$. To further study the dependency of states on $\{y_j\}$, the sequence is broken into three parts. For a sequence of $\{y_j\}$ with a total of N terms, define:

$$\{y_j\} = \{y_p, y_i, y_f\}$$
$$y_p \overset{\text{def}}{=} \{y_1, \ldots, y_{i-1}\} . \tag{5.77}$$
$$y_f \overset{\text{def}}{=} \{y_{i+1}, \ldots, y_N\}$$

y_p and y_f are two sequences collecting the "past" and "future" parts of the received signal. y_i is the current received symbol in step i. With such partition and applying Bayes' law repeatedly, we have

$$\begin{aligned} p(\sigma_i, \sigma_{i-1}, \{y_j\}) &= p(\sigma_i, \sigma_{i-1}, y_p, y_i, y_f) \\ &= p(y_f \mid \sigma_i, \sigma_{i-1}, y_p, y_i) p(\sigma_i, \sigma_{i-1}, y_p, y_i) \\ &= p(y_f \mid \sigma_i, \sigma_{i-1}, y_p, y_i) p(\sigma_i, y_i \mid \sigma_{i-1}, y_p) p(\sigma_{i-1}, y_p) \end{aligned} \tag{5.78}$$

The reason for such seemingly convoluted and arbitrary decomposition will be apparent once we consider the true dependence of the various quantities.

First, y_f depends only on σ_i and the data input after that. Namely, it is independent of σ_{i-1}, y_p, and y_i. Therefore,

$$p(y_f \mid \sigma_i, \sigma_{i-1}, y_p, y_i) = p(y_f \mid \sigma_i). \tag{5.79}$$

Furthermore, once σ_{i-1} is known, y_p does not provide additional information to σ_i and y_i. Therefore,

$$p(\sigma_i, y_i \mid \sigma_{i-1}, y_p) = p(\sigma_i, y_i \mid \sigma_{i-1}). \tag{5.80}$$

Therefore, (5.78) leads to

$$\begin{aligned} p(\sigma_i, \sigma_{i-1}, y_p, y_i, y_f) &= \beta_i(\sigma_i) \gamma_i(\sigma_i, \sigma_{i-1}) \alpha_{i-1}(\sigma_{i-1}) \\ \beta_i(\sigma_i) &\overset{\text{def}}{=} p(y_f \mid \sigma_i) \\ \gamma_i(\sigma_i, \sigma_{i-1}) &\overset{\text{def}}{=} p(\sigma_i, y_i \mid \sigma_{i-1}) \\ \alpha_{i-1}(\sigma_{i-1}) &\overset{\text{def}}{=} p(\sigma_{i-1}, y_p) \end{aligned} \tag{5.81}$$

The newly introduced functions α_i, β_i, and γ_i describe the dependency between states and signals for the past, future, and present, respectively. Note that these functions all depend on i in addition to σ_i, because they implicitly depend on y_p, y_f, and y_i, which all depend on index i.

5.7.2.3 Iterative Relationships

In this section, we discuss how to compute α_i, β_i, and γ_i defined in (5.81). They are all done through iteration.

For function α, let us take one step forward, from $i-1$ to i. Remember that y_p is the y sequence up to $i-1$. Therefore,

$$\alpha_i(\sigma_i) = p(\sigma_i, y_p, y_i) = \sum_{\sigma_{i-1}} p(\sigma_i, \sigma_{i-1}, y_i, y_p). \tag{5.82}$$

Here, the summation runs over all possible states σ_{i-1}.[18] Applying Bayes' law, the probabilities on the right-hand side are

$$p(\sigma_i, \sigma_{i-1}, y_i, y_p) = p(\sigma_i, y_i \mid \sigma_{i-1}, y_p) p(\sigma_{i-1}, y_p). \tag{5.83}$$

The first factor on the right-hand side is $\gamma_i(\sigma_i, \sigma_{i-1})$ (note that its dependency on y_p is superfluous). The second fact is $\alpha_{i-1}(\sigma_{i-1})$. Therefore,

$$\alpha_i(\sigma_i) = \sum_{\sigma_{i-1}} \gamma(\sigma_i, \sigma_{i-1}) \alpha_{i-1}(\sigma_{i-1}). \tag{5.84}$$

For function β, let us take one step backwards from $i+1$ to i.

$$\beta_i(\sigma_i) = p(y_f \mid \sigma_i) = \sum_{\sigma_{i+1}} p(y_{f(i+1)}, y_{i+1}, \sigma_{i+1} \mid \sigma_i). \tag{5.85}$$

Here, $y_{f(i+1)}$ is the y_f for step $i+1$: $\{y_{i+2}, \ldots, y_N\}$. Applying Bayes' law, we have

$$p(y_{f(i+1)}, y_{i+1}, \sigma_{i+1} \mid \sigma_i) = p(y_{f(i+1)} \mid \sigma_{i+1}, y_{i+1}, \sigma_i) p(\sigma_{i+1}, y_{i+1} \mid \sigma_i). \tag{5.86}$$

The first fact on the right-hand side is $\beta_{i+1}(\sigma_{i+1})$, whose dependency on y_{i+1} and σ_i is superfluous. The second factor is $\gamma_{i+1}(\sigma_{i+1}, \sigma_i)$. We thus obtained a backward iterative relationship:

$$\beta_i(\sigma_i) = \sum_{\sigma_{i+1}} \beta_{i+1}(\sigma_{i+1}) \gamma_{i+1}(\sigma_{i+1}, \sigma_i). \tag{5.87}$$

Equations (5.84) and (5.87) shows the iterative approaches for computing functions α and β. To start such process, the first and last state of the code σ_1 and σ_N are known, as in the case of a well-terminated convolutional code. At each iteration step, α and β need to be computed for all possible states, in order to facilitate the computation for the next iteration. α and β are known as the forward and backward metrics, respectively, reflecting the direction of their iteration directions.

---

[18] For a given σ_i, there are some states in step $i-1$ that cannot transition to σ_i due to the trellis definition. For these states σ_{i-1}, the corresponding probability will be zero. Therefore, they can still be included in the summation (5.87)

Equations (5.84) and (5.87) are quite general for convolutional codes. They do not show how code structure and received signal influence the probabilities. Such dependency is contained in the computation of $\gamma_i(\sigma_i, \sigma_{i-1})$, as will be shown in Section 5.7.2.4.

5.7.2.4　γ for AWGN Channel

Now let us focus on γ. According to (5.81) and applying Bayes' law,

$$\gamma_i(\sigma_i, \sigma_{i-1}) = p(\sigma_i, y_i \mid \sigma_{i-1}) = p(y_i \mid \sigma_{i-1}, \sigma_i) p(\sigma_i \mid \sigma_{i-1}). \tag{5.88}$$

Remember in (5.76) we consider only state transitions that are produced by $u_i = l$. Therefore, $p(\sigma_i \mid \sigma_{i-1})$ is the a priori probability $p(u_i = l)$. $p(y_i \mid \sigma_{i-1}, \sigma_i)$ is the probability that such transition produces signal y_i. Such a probability depends on the code and channel.

For the AWGN channel, we have (see Chapter 4)

$$p(y_i \mid \sigma_{i-1}, \sigma_i) = p(y_i \mid c_i) = \frac{1}{(\pi N_0)^{\frac{m}{2}}} \exp\left(-\frac{|y_i - c_i|^2}{N_0}\right), \tag{5.89}$$

where N_0 is the noise spectral density, c_i is the transmitted symbol corresponding to the state transition $\sigma_{i-1} \rightarrow \sigma_i$. m is the dimension of symbols c_i and y_i.

More specifically, with BPSK modulation (and code rate $1/n$), both y_i and c_i are vectors of size $m = n$. c_i takes the value of $\pm\sqrt{E_c}$, where E_c is the energy per symbol. In this case,

$$|y_i - c_i|^2 = |y_i|^2 + E_c - 2y_i^H \cdot c_i. \tag{5.90}$$

In summary, for AWGN channel and BPSK modulation, we have

$$\gamma_i(\sigma_i, \sigma_{i-1}) = p(u_i = l) \frac{1}{(\pi N_0)^{\frac{n}{2}}} \exp\left[-\frac{1}{N_0}\left(|y_i|^2 + E_c - 2y_i^H \cdot c_i\right)\right]. \tag{5.91}$$

5.7.2.5　The Log-Likelihood Ratio (LLR)

Sections 5.7.2.3 and 5.7.2.4 show that a posteriori probability of a symbol can be computed with the following steps. First, functions α_i, β_i, and γ_i can be computed for each symbol, and for all possible states, using (5.84), (5.87), and (5.91). Next, (5.81), (5.78), and (5.76) can be used to compute $p(u_i = l, \{y_j\})$. In this section, we further compute the log-likelihood ratio (LLR), defined as

$$L(u_i) \overset{\text{def}}{=} \ln \frac{P(u_i = 1 \mid \{y_j\})}{P(u_i = 0 \mid \{y_j\})}. \tag{5.92}$$

LLR is used in soft decision decoding and turbo decoding. When $L(u_i) > 0$, u_i is more likely to be 1. Otherwise, it is more likely to be 0. Based on the results in Sections 5.7.2.3 and 5.7.2.4,

$$L(u_i) = \ln \frac{P(u_i = 1 \mid \{y_j\})}{P(u_i = 0 \mid \{y_j\})} = \ln \frac{P(u_i = 1, \{y_j\})}{P(u_i = 0, \{y_j\})} = \ln \frac{\sum_{\sigma_i, \sigma_{i-1} \in S_1} [\alpha_{i-1}(\sigma_{i-1})\gamma_i(\sigma_i, \sigma_{i-1})\beta_i(\sigma_i)]}{\sum_{\sigma_i, \sigma_{i-1} \in S_0} [\alpha_{i-1}(\sigma_{i-1})\gamma_i(\sigma_i, \sigma_{i-1})\beta_i(\sigma_i)]} \tag{5.93}$$

Note that the forms in the numerator and denominator are the same, except that the summation domains are S_1 for the numerator and S_0 for the denominator. As defined in Section 5.7.2.2, S_1 and S_0 are the sets of states $\{\sigma_i, \sigma_{i-1}\}$ that are driven by $u_i = 1$ or $u_i = 0$, respectively. Although (5.92) is the customary notation, L actually depends on i, and u_i is a dummy variable.

In (5.93), we can separate some common factors in a special case. In addition to AWGN channel and BPSK, consider a rate 1/2 systematic code. Namely, $c_i = \left(c_i^s, c_i^p\right)$. The first is the systematic symbol, whose sign is a direct mapping from u_i:

$$c_i^s = \begin{cases} \sqrt{E_c} & \text{if } u_i = 1 \\ -\sqrt{E_c} & \text{if } u_i = 0 \end{cases}. \tag{5.94}$$

Correspondingly, the received signal is $y_i = \left(y_i^s, y_i^p\right)$. In this case, (5.91) becomes

$$\gamma_i(\sigma_i, \sigma_{i-1}) = \left[\frac{1}{(\pi N_0)^{\frac{n}{2}}} \exp\left(-\frac{|y_i^s|^2}{N_0}\right) \exp\left(-\frac{|y_i^p|^2}{N_0}\right) \exp\left(-\frac{E_c}{N_0}\right) \right] \cdot$$

$$\left[p(u_i = 1) \exp\left(\frac{2 y_i^s c_i^s}{N_0}\right) \right] \exp\left(\frac{2 y_i^p c_i^p}{N_0}\right) \tag{5.95}$$

The factors in the first square bracket are independent of the assumed values of u_i and independent of the states. Therefore, they cancel out between the numerator and denominator in (5.93). Those in the second square bracket are independent of the states, but dependent on the assumed value of u_i. They can be moved out of the summations as common factors. The remaining factor needs to remain in place. Equation (5.93) thus becomes

$$L(u_i) = \ln\left[\frac{p(u_i = 1)}{p(u_i = 0)}\right] + 4 \frac{y_i^s \sqrt{E_c}}{N_0} + \ln \frac{\sum_{\sigma_i, \sigma_{i-1} \in S_1} \left[\alpha_{i-1}(\sigma_{i-1})\beta_i(\sigma_i) \quad \exp\left(\frac{2 y_i^p c_i^p}{N_0}\right)\right]}{\sum_{\sigma_i, \sigma_{i-1} \in S_0} \left[\alpha_{i-1}(\sigma_{i-1})\beta_i(\sigma_i) \quad \exp\left(\frac{2 y_i^p c_i^p}{N_0}\right)\right]}. \tag{5.96}$$

Namely, the LLR is separated into three terms:

$$L(u_i) = L_a(u_i) + L_c y_i^s + L^e(u_i)$$

$$L_a(u_i) \overset{\text{def}}{=} \ln\left[\frac{p(u_i = 1)}{p(u_i = 0)}\right]$$

$$L_c \overset{\text{def}}{=} \frac{4\sqrt{E_c}}{N_0} \tag{5.97}$$

$$L^{(e)}(u_i) \overset{\text{def}}{=} \ln \frac{\sum_{\sigma_i, \sigma_{i-1} \in S_1} \left[\alpha_{i-1}(\sigma_{i-1})\beta_i(\sigma_i)\exp\left(-\frac{2 y_i^p c_i^p}{N_0}\right)\right]}{\sum_{\sigma_i, \sigma_{i-1} \in S_0} \left[\alpha_{i-1}(\sigma_{i-1})\beta_i(\sigma_i)\exp\left(-\frac{2 y_i^p c_i^p}{N_0}\right)\right]}$$

L_a reflects the a priori probability distribution of bit u_i. L_c reflects channel condition. $L^{(e)}$ is known as the extrinsic information, which reflects the biases in probability estimation imposed by other symbols in the sequence. Extrinsic information is the additional information provided by channel coding.[19] Equation (5.97) can be easily extended to lower rate systematic codes, that is, codes with more parity bits.

5.7.2.6   Summary of the BCJR Algorithm

BCJR algorithm computes the LLR of a convolutional code, given the received sequence $\{y_j\}$. The algorithm consists of the following operations.

1. For each transition step and all possible states, compute $\gamma_i(\sigma_i, \sigma_{i-1})$ from (5.88) and (5.89) (for AWSN channels). In the special case of AWGN channel and BPSK modulation, (5.91) can be used.

2. Starting from the known beginning and end states, α_i and β_i for all steps and all states can be computed through the iteration Equations (5.84) and (5.87).

3. Then, LLR can be computed for any step using (5.93) and, in the special case of AWGN channel, BPSK modulation, and rate 1/2 systematic code, (5.97).

Note that the computation of γ_i uses the received signal $\{y_j\}$, whose influence to LLR propagates from there. Furthermore, the BCJR algorithm also needs information on noise (N_0 in the case of AWGN), which may be obtained as a part of channel estimation. However, depending on channel types and noise level, decoder performance may not be very sensitive to noise estimation [48].

5.7.3   Turbo Decoding

Section 5.7.2 describes the BCJR algorithm, which estimates LLR for a convolutional code. Especially in the case of AWGN channel, BPSK modulation, and systematic coding, as shown in (5.97), LLR can be decomposed into contributions from a priori probability, channel conditions, and extrinsic information. In this section, we apply BCJR to the decoding of turbo code.

Turbo code has very long correlations because of the use of interleavers. Effectively, it has a large number of very long codewords. Therefore, it is impractical to use sequence estimators such as VA. Symbol-based MAP decoding is a suboptimal method, but still with very good performance.

Figure 5.25 shows a turbo decoder, corresponding to the turbo encoder shown in Figure 5.24. DEC 1 and DEC 2 are the two decoders corresponding to the ECs in Figure 5.24. y^{1s}, y^{1p}, and y^{2p} are coded bits generaged by the encoder as shown in Figure 5.24. They serve as input to the decoder. IL is the interleaver, which is the same as that in the encoder. IL^{-1} is the deinterleaver, which reverses the effect of the interleaver. The "Final Estimate" block produces LLR and bit values u for output, as described in operation 6 below.

---

[19] Note that although γ is separated into several factors in (5.97), it is still used in its original form in computation of α and β in (5.84). Therefore, the influence of a priori probability distribution is not limited to L_a.

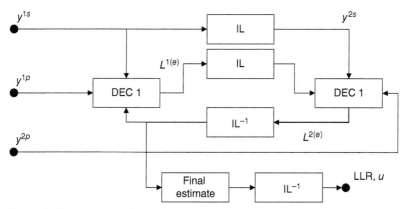

Figure 5.25 Turbo decoder.

The iteration operations are as follows.

1. Start with $p(u_k = 1) = p(u_k = 0) = 0.5$.

2. For decoder 1 (DEC 1), compute γ, α, β using (5.91), (5.84), and (5.87), based on received signals y^{1s} and y^{1p}. Compute $L^{1(e)}$ using (5.97).

3. For decoder 2 (DEC 2), compute γ, α, β using (5.91), (5.84), and (5.87), based on received signals y^{2s} and y^{2p}. y^{2s} is an interleaved version of y^{1s}. In this case, use the interleaved version of $L^{1(e)}$ as the a priori probability ratio $\ln[p(u_k = 1)/p(u_k = 0)]$. Compute $L^{2(e)}$ using (5.97).

4. Send $L^{2(e)}$ through the deinterleaver, and feed the result back to DEC 1, to serve as $\ln[p(u_k = 1)/p(u_k = 0)]$.

5. Repeat operations 2–4 until convergence or for a preset number of cycles.

6. Compute LLR based on (5.97) as the final output. If necessary, determine all the bits depending on whether LLR is positive or negative.

Note that only the extrinsic information $L^{(e)}$, instead of the entire LLR, is used as a priori estimation. This is because extrinsic information is the only additional information from the companion encoder.

5.7.4 Turbo Code Performances

Turbo code performance can be characterized in two aspects: the convergence speed and the final bit error rate.

Convergence speed can be studied analytically and with simulation, with a tool known as the extrinsic information transfer (EXIT) chart [49]. In general, the number of iterations needed is manageable.

Figure 5.26 shows an example [50]. This plot is based on a PCCC with two rate 1/2 constituent codes, and an interleaver of length 2^{16} (65,536). We can see that the bit error performance improves with more iterations. However, after six iterations, the

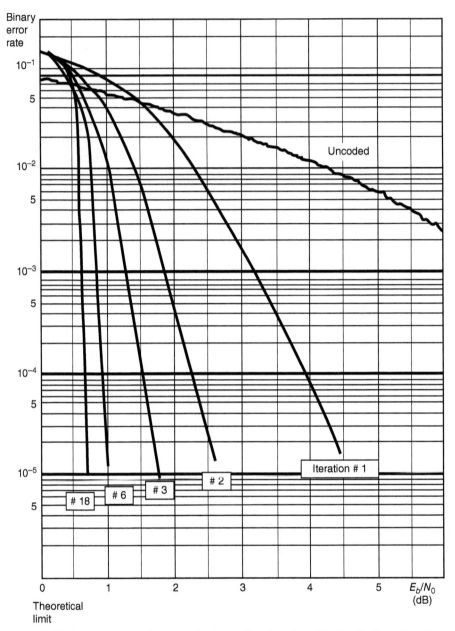

Figure 5.26 Bit error rates of turbo code after various iterations [50, fig. 9]. Reproduced with permission.

Bit error probability (BER)

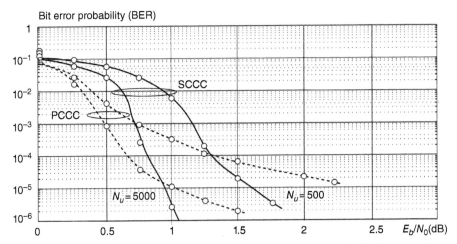

Figure 5.27   Performance comparison of several turbo codes [43, fig. 8.22]. Reproduced with permission.

performance is pretty close to the ultimate one. The ultimate performance is worse than the Shannon bound by only a fraction of dB. (With BPSK, Shannon theory predicts error-less communication at $E_b/N_0 = 0$ dB.) Furthermore, Figure 5.26 shows that the slope of the BER versus E_b/N_0 curve for turbo code is much steeper than those for uncoded modulation. Steeper slot in such "waterfall" plots is a common character of long codes.

Figure 5.27 compares PCCC with SCCC, and also with block sizes of 500 and 5000 bits [43]. The data is obtained through simulation with seven iterations. Both codes are rate 1/3 and use the same constituent codes. The result shows the benefit of longer codes. At low SNR, PCCC codes are advantageous. SCCC codes, on the other hand, perform better at high SNR.

Figure 5.28 shows the performance of three turbo codes of rate 1/2 [43, fig. 8.7]. Two of them (marked with "random") use random interleavers, with different constituent codes, whose generators are marked on the graph. The other code (marked as "spread") uses an interleaver designed to achieve maximum spread (i.e., maximum free distance). At medium SNR, where BER drops rapidly as E_b/N_0 increases, the random interleaver results in better performance. This shows that turbo code depends on randomness, instead of code structure, to achieve good performance. On the other hand, at high SNR region, the random codes display a less steep slope in their BER versus E_b/N_0 curve. This is known as the *error floor*. The code with the maximum spread, on the other hand, behaves well at high SNR, achieving lower BER.

Error floor is a common phenomenon for all random-based codes, including turbo code as well as the LDPC code to be described in Section 5.8. The main reason is that these codes have a small portion of codewords with small mutual

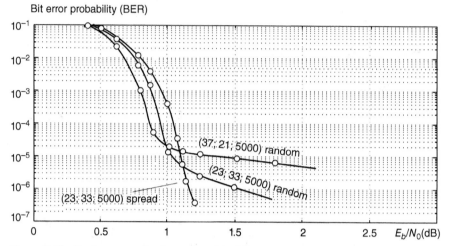

Figure 5.28   Performance of turbo codes with different constituents and interleavers [43, fig. 8.7]. Reproduced with permission.

distance. Their contribution to BER becomes dominant at high SNR, limiting the overall BER reduction potential. Error floor can be alleviated by improving code design methodology [51, 52], or by using other means to counter bit errors [53].

5.7.5   Turbo Code Summary

Section 5.7 provides a brief overview of the turbo code. Turbo code uses multiple constituent codes, connected in parallel or in serial through interleavers that randomize the codewords. It achieves noise immunity performance very close to the Shannon bound (within a fraction of dB) with a code block length of tens of thousands bits. Turbo code is widely used in modern communications systems, most notably 3G and 4G cellular systems.

Because of the very long codewords, optimal decoding of turbo code such as VA is impractical. Instead, iterative decoding methods are commonly used. Sections 5.7.2 and 5.7.3 describe one algorithm, known as the BCJR algorithm. BCJR was the algorithm used in the early turbo code works [41]. However, other coding methods have been developed, and approximations are widely used to reduce complexity [43, 54].

Although the turbo code achieves good performance in general, an issue that requires consideration is the error floor. Careful designs of the constituent codes and interleavers can improve performance regarding error floor.

Turbo code is still an active research area. This section provides only the very basic concepts that aid the readers to get started in the field. There are numerous books and online tutorials providing more information on the design and implementation of turbo codes. It is worth noting that in digital communication research, the term "turbo" refers to the method of iterative detection, where estimates of a posteriori probabilities are refined through iteration between different analysis functions. Therefore, in addition to turbo coding, there are also turbo equalization, turbo MIMO receiving, etc.

5.8 LOW-DENSITY PARITY-CHECK (LDPC) CODE

Low-density parity-check (LDPC) code was first invented in the 1960s [55]. Despite its promise of excellent performance, LDPC was not practical at the time due to its complexity. In the 1990s, the advent of turbo code sparked interest in codes with long and random codewords. As a code that shares many traits with turbo code, LDPC was brought to attention in the community [56, 57], with improved decoding techniques. The LDPC code is currently used in several modern communications systems including wireless local area network (WLAN) standard IEEE 802.11 [58] and digital video broadcast for satellites (DVB-S2) [59]. It is also proposed for the 5G cellular systems (Chapter 12) [60]. The 802.11 standard supports codeword lengths of 648, 1269, and 1944 in the high throughput (HT) mode (also known as 802.11n) and very high throughput (VHT) mode (also known as 802.11ac), and codeword length of 672 in the directional multi-gigabit (DMG) mode (also known as 802.11ad). DVB-S2 uses LDPC as inner code in a concatenated code structure (Section 5.6). The LDPC codeword lengths are 64,800 for normal frames and 16,200 for short frames.

5.8.1 LDPC Structure and Encoding

LDPC is a linear block code. As described in Section 5.3.2, a linear code can be defined by its parity-check matrix H. A bit sequence c (viewed as a row vector of length n) is a codeword if and only if

$$cH^T = 0. \tag{5.98}$$

H is a matrix of size $(n-k) \times k$ for a code with k data bits and n coded bits per block. However, LDPC is different from the traditional block code, because H is usually of very large size (n can be in thousands or tens of thousands). H is also very sparse (or low-density) for LDPC. Namely, relatively few elements in H are nonzero.

From (5.98), the parity check process can be expressed as follows. c is a codeword if and only if

$$\sum_{\{k_l\}} c_{k_l} = 0 \forall l. \tag{5.99}$$

Here, c_n is the nth bit of c. $\{k_l\}$ are the positions of "1" elements in the lth row of H:

$$H_{l,m} = \begin{cases} 1 & \text{if } m \in \{k_l\} \\ 0 & \text{otherwise} \end{cases}. \qquad (5.100)$$

In other words, bit m is included in the lth summation in (5.99), if and only if $H_{l,m}$ is 1. Since H is "low density," relatively few terms are included in each summation.

A *regular* LDPC code has the same number of 1s in each row of its parity-check matrix, and the same number of 1s in each column of the matrix. If the parity-check matrix of a code has n_r 1s in each row and n_c 1s in each column, then each bit in a codeword is checked n_c times. Each parity check includes n_r bits. A regular code can be expresses as (n, n_c, n_r). Figure 5.29 shows an example of the parity-check matrix for (20, 3, 4) [55]. The 15 rows are grouped into three 5 row submatrices, to show the structure of this particular code.

Regular LDPC codes are easier to generate because it has some structure. However, irregular codes may provide better performances [61, 62].

For regular LDPC codes, the parity matrices can be designed in several ways. They can be generated randomly, constrained on n_c and n_r and some other guidelines. They can also be generated with algebraic methods, to provide easier encoding and more optimization [1, sec. 5.1.2], [63]. Alternatively, a hybrid method can be used, which starts with a small algebraic construction and then randomize the result to form a larger parity-check matrix.

Conceptually, given H, LDPC can be encoded using (5.7) in Section 5.3.2, where the coding matrix G can be constructed from the vectors in the null space of H, using (5.14) in Section 5.3.3.2. Generating G requires some computation, but it needs to be done only once for a code. Execution of (5.7) requires kn multiplications

```
1 1 1 1 0 0 0 0 0 0 0 0 0 0 0 0 0 0 0 0
0 0 0 0 1 1 1 1 0 0 0 0 0 0 0 0 0 0 0 0
0 0 0 0 0 0 0 0 1 1 1 1 0 0 0 0 0 0 0 0
0 0 0 0 0 0 0 0 0 0 0 0 1 1 1 1 0 0 0 0
0 0 0 0 0 0 0 0 0 0 0 0 0 0 0 0 1 1 1 1
1 0 0 0 1 0 0 0 1 0 0 0 1 0 0 0 0 0 0 0
0 1 0 0 0 1 0 0 0 1 0 0 0 0 0 1 0 0 0 0
0 0 1 0 0 0 1 0 0 0 0 0 1 0 0 0 1 0 0 0
0 0 0 1 0 0 0 0 0 0 1 0 0 0 1 0 0 0 1 0
0 0 0 0 0 0 0 1 0 0 0 1 0 0 0 1 0 0 0 1
1 0 0 0 0 1 0 0 0 0 1 0 0 0 0 0 1 0 0 0
0 1 0 0 0 0 1 0 0 0 1 0 0 0 1 0 0 0 0 0
0 0 1 0 0 0 0 1 0 0 0 0 1 0 0 0 0 0 1 0
0 0 0 1 0 0 0 0 1 0 0 0 0 1 0 0 1 0 0 0
0 0 0 0 1 0 0 0 0 1 0 0 0 0 1 0 0 0 0 1
```

Figure 5.29   Parity check matrix of a (30, 3, 4) regular code.

per block, where k and n are the number of bits in user data and in the codeword for a block, respectively. Note that

$$kn = R_c n^2, \tag{5.101}$$

where R_c is the code rate. Therefore, encoding complexity is proportional to n^2. However, taking advantage of the "low-density" of the parity-check matrix, the encoding complexity can be reduced to be linear with n [64].

5.8.2 Graphic Representation of LDPC: The Tanner Graph

The parity checking relationship in LDPC can be expressed with a type of graphs known as the Tanner graph. We will introduce the Tanner graph in this section, and describe a decoding algorithm based on the Tanner graph in Section 5.8.3.

With Tanner graph, a linear block code is expressed with two types of the nodes. One type is "variable nodes," each representing a bit in the codeword. The other type is the "check nodes," each representing a parity check summation in (5.99) (i.e., a row in H).

Figure 5.30 shows an example of the Tanner graph. The top box shows the parity-check matrix of a (7,4) block code. The bottom part shows two equivalent graph representation of the code. The form on the right is commonly used to describe a code, while the form on the left may be more intuitive in examining the structure of the graph. In these graphs, the solid circles (variable nodes) represent the bits in codewords. The numbers next to the circles indicate the bit positions. The boxes (parity nodes) represent the summations prescribed by the rows in the parity-check matrix. The numbers next to them indicate the corresponding row numbers of the parity-check

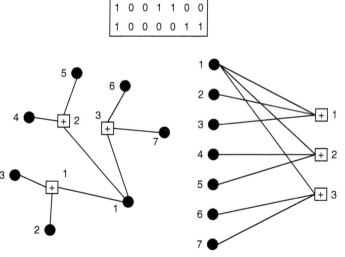

Figure 5.30   Tanner graph for block code. Top box: parity-check matrix; bottom-left and bottom-right: two equivalent Tanner graphs.

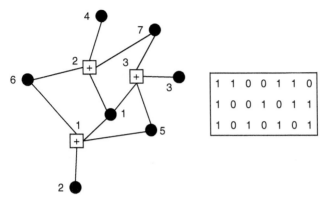

Figure 5.31   Block code with cycles. Left: Tanner graph; right: parity-check matrix.

matrix. The lines indicate that certain bits are used in the summation. For example, the first row of the parity-check is (1110000). It means adding the first three bits. Therefore, bits 1, 2, and 3 have lines linking to the parity node 1.

Figure 5.31 shows the Tanner graph of another block code. The parity-check matrix is shown on the right side. In this case, the Tanner graph shows several cycles. Cycles in Tanner graphs, especially those with a small number of edges, tend to degrade the performance and decoder convergence behavior [1, sec. 5.1.2].

5.8.3   LDPC Decoding with Tanner Graph

LDPC codes can be decoded iteratively, based on the Tanner graph introduced in Section 5.8.2. The basic concept is "belief propagation," also known as "message passing." Essentially, all parity nodes "negotiate" through the variable nodes to find a codeword that is considered to be most likely based on all parity checks. Such negotiation is performed at the bit level, as described in the following.

A variable node k passes the tentative estimation about the corresponding bit k to the connected parity nodes. It then collects feedback from these parity nodes to form the estimation for the next iteration. These two steps are referred to as the horizontal step and the vertical step, respectively.

In hard decision case, a variable node sends its estimation of the bit value x_k to all connected parity nodes. The initial estimation is based on received signal demodulation (same as the uncoded case). A parity node determines the "suggested" value for a particular bit x'_k based on input values of *other* bits $x_{i \neq k}$. Such suggestion is based on the parity condition (5.99). Namely,

$$x'_k = \sum_{i \neq k} x_i. \tag{5.102}$$

The summation runs over all bits that are connected to the parity node. Again, the summations here are modulo-2 as described in (5.1). This computation is the horizontal step. In the vertical step, a variable node collects such suggestions from all linked parity nodes. If there are more than d suggestions that are opposite to the current bit estimation value, then the bit is flipped to form the next estimation. Otherwise, the current estimation is kept for the next round. Threshold d can be adjusted in the design process and vary among variable nodes (e.g., depending on the number of the linked parity nodes). It can also change as iteration progresses.

In soft decision case, a variable node k estimates the probability of the corresponding bit being 1, p_k. Initial estimation is made based on the received signal, for example, by (5.89) in Section 5.7.2.4 for AWGN channel. In the horizontal step, the variable nodes send such estimations to the linked parity nodes. A parity node suggests the probability of a particular bit being 1, by considering the probability estimates reported by *other* bits. As shown in an appendix (Section 5.11), the suggested probability is[20]

$$P_{j,k} = \frac{1}{2} - \frac{1}{2}\prod_{i \neq k}(1 - 2p_i).$$

(5.103)

Here, $P_{j,\,k}$ is the suggested probability of bit k being 1, from parity node j. p_i is the estimated probability of bit i being 1, as reported by variable node i. The product in (5.103) is over all bits connected to this parity node, except for node k.

In the vertical step, a variable node combines suggested probabilities from all linked parity nodes to form a new estimate for the bit. The following equations can be used for such combination at variable node k:

$$p_k = P(x_k = 1) = F_k p_{k0} \prod_j P_{j,k}$$

$$P(x_k = 0) = F_k(1 - p_{k0}) \prod_j (1 - P_{j,k})$$

(5.104)

where F_k is chosen so that $P(x_k = 1) + P(x_k = 0) = 1$. p_{k0} is the initial estimation of $P(x_k = 1)$, based on the received signal. $P(x_k = 1)$ forms the probability estimation p_k for the next iteration. At the end of the iteration process, (5.104) is again used to estimate the final log-likelihood ratio (LLR)

$$LLR_k = \ln\left(\frac{P(x_k = 1)}{P(x_k = 0)}\right).$$

(5.105)

Equation (5.105) is the soft output of the decoder. A hard output can also be formed based on the resulting LLR_k.

In both hard decision and soft decision cases, the iterative process can terminate under a variety of conditions such as convergence (i.e., further iteration does not result in significant change) or a maximum number of iteration.

---

[20] Note that most literatures use the probability for bit k to be 0. The expression is therefore different from (5.103).

Note that in both hard decision and soft decision cases, the parity nodes determine feedback to variable node k, based solely on information from *other* nodes. This forms the extrinsic information to bit k, similar to the case for the turbo code (Section 5.7.3). The variable nodes combine the received extrinsic information with its intrinsic information (current value of x_k in the case hard decision and p_{k0} in the case of soft decision) to form the next estimation, also similar to the case for the turbo code.

Convergence of such decoding iteration is guaranteed if there is no cycle in the Tanner graph for the code. With cycles, convergence is not always possible. However, many practical codes do achieve convergence.

The decoding method described above have many variations for performance improvement and complexity reduction. More information can be found in the literature [1]. Decoding through "nodes" and "messages" is a natural fit for parallel processing architectures.

5.8.4 LDPC Performances

Figure 5.32 shows a comparison between LDPC (solid lines) and turbo codes (dashed lines) (all rate 1/2) with various block sizes as marked on the graph [62]. The three rows of numbers at the bottom are different scales for SNR. The top row is the familiar E_b/N_0. The second row is noise standard deviation σ, with bit energy set a 1. Namely,

$$\frac{E_b}{N_0} = 20\log_{10}\sigma. \tag{5.106}$$

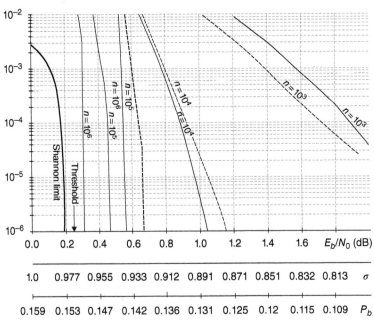

Figure 5.32   LDPC and turbo codes performances [62, fig. 3]. Reproduced with permission.

The third row is the bit error rate at the input of the decoder (i.e., that of the uncoded modulation). The vertical axis is the bit error rate. The "threshold" is the theoretical value of SNR for LDPC to achieve an error rate of 0 when block size approaches infinity. We can see that the performance of both codes are sensitive to block size (as expected) and can be close to Shannon limit (within a dB) with reasonable block sizes (thousands and tens of thousands of bits per block).

With even larger block sizes (on the order of 10^7 bits), LDPC code has been shown to achieve performances that are less than 0.01 dB from Shannon limit [65]. Furthermore, LDPC also provides very attractive performance on erasure channels [2].

As turbo codes, LDPC also faces the challenge of error floor [43, ch. 7]. Careful analysis and optimization of the code help in mitigating this problem.

5.8.5   LDPC Summary

LDPC is a modern code similar to the turbo code in many ways. It is a type of linear block code. With very large block sizes and controllable complexity, LDPC code can achieve performances within a fraction of dB from the Shannon bound. LDPC code can be classified into regular and irregular codes. The latter is more complex in design but may provide better performance. LDPC is typically specified by its parity-check matrix, which can also be expressed in the form of a Tanner graph.

Decoding of LDPC is based on iteration. One method uses "belief propagation," also known as "message passing," between the variable nodes (representing codeword bits) and parity nodes (representing parity-checking summations). One such method is described in Section 5.8.3. There are other decoding methods in the literature and practice.

As turbo codes, LDPC is still an active research area. This section introduces the basic concepts, based on the early works in the 1990s and the early 2000s. More information on the latest advances and implementation details can be found in the literature and from online resources.

5.9   SUMMARY

Channel coding is a ubiquitous component of modern communication systems. Its function is reducing receiver errors through redundancy in transmission. According to Shannon theorem, optimal performance is achieved only by some codes with infinite lengths. In practice, however, the code length is limited by complexity and latency constraints.

Channel codes can be classified as classical and modern. Prominent members of the classical code family include block codes (Section 5.3) and convolutional codes (Section 5.4). For higher order modulations, TCM (Section 5.5) integrates coding with modulation design. Classical codes usually have relatively small block sizes (or constraint lengths for the convolutional code) due to decoding complexity constraints. However, one can combine two codes to form a longer code (Section 5.6) with moderate complexity.

Modern codes are those attracted attention since 1990. Unlike classical codes who regularly use optimal decoders, modern codes typically use suboptimal iterative decoding techniques, which can still achieve near-optimal performances. This chapter describes two most common modern codes: the turbo codes (Section 5.7) and the LDPC codes (Section 5.8). While still researched actively, these modern codes have been applied to modern communication systems such as cellular, WLAN, and satellite communications.

The most important code performance metric is noise immunity, measured by bit error rate (BER) under given SNR (usually in E_b/N_0). If the channel is bandwidth limited, code performance needs to be evaluated under the constraint of spectral efficiency. Code noise immunity can also be expressed in terms of coding gain (at given BER levels) and Shannon gap, as detailed in Section 5.2.5. Typically and with practical implementations, classical codes have Shannon gaps around 3 dB, while modern codes have Shannon gaps near 1 dB.

There are a number of topics on channel coding not covered by this chapter; some of them are mentioned briefly below.

Turbo and LDPC codes both belong to the "graph-based codes." The iterative decoding methods for the two code families can be understood from the view of factor graph [66, 67]. Such a unified view explains many common traits of the two code families (such as error floor and convergence behavior).

Some less common codes are not covered in this chapter. Notable examples include erasure codes, which are used for channels where some of the bits are known to be lost (e.g., in magnetic media). The polar code is another modern code not covered here [68, 69]. Invented in the 2000s, the polar code attracts attention recently as a candidate for the 5G cellular systems (Chapter 12).

A significant part of the channel coding theory is code performance evaluation. A number of theories and techniques have been developed for this purpose. In most cases, performance *bounds* can be obtained through theoretical analysis. For practicing engineers who are not specialized in code design, however, code performance can usually be obtained from the literature or through simulation. Therefore, this chapter does not discuss performance evaluation, except for the basic concept of the free distance. Furthermore, only AWGN channels are discussed. Code performance analysis for other channels (such as fading channels) can be found in the literature [16].

In addition to channel coding, another common way to combat transmission error is automatic repeat request (ARQ). In this scheme, the receiver checks the integrity of data. Upon discovering an error, the receiver sends a request to the transmitter for retransmission. ARQ is often used in combination of channel coding. For example, for the internet, ARQ is used in the transmission control protocol (TCP), which is a transport layer protocol. The underlying physical layer services for internet systems typically employ channel coding. ARQ relies on coding to detect (but not correct) errors, as described in Section 5.3.5.1. A more sophisticated, yet now widely used, ARQ method is the hybrid ARQ (HARQ). In this scheme, instead of retransmitting the entire data, the transmitter sends additional bits of a codeword in response to previous transmission errors, to increase error correction power at the receiver. Such a technique is also referred to as "incremental redundancy." It requires a channel code that can be decoded with a portion of the codeword (similar to the case of erasure channel). One such code is the fountain code [70].

This chapter aims to provide a basic understanding of channel coding and to reduce the barrier in choosing and learning about specific codes needed for an application. For each code family, construction, encoding, and decoding examples are provided. For a more in-depth understanding of the theories and design options behind channel coding and implementation issues, the readers are referred to specialized books and literature, such as [1, 16, 17, 43, 71]. Costello and Forney provided an excellent overview of channel coding techniques with a wealth of additional references [2]. There are also many online introductions and tutorials on various coding topics. Especially, complex to real is a practical tutorial collection worth exploring [72].

5.10 APPENDIX: UPPER BOUND OF SHAPING GAIN

In this section, we show that the upper bound of shaping gain introduced in Section 5.5.2 is $e\pi/6$. Intuitively, the upper bound is the coding gain for a sphere of infinite dimension comparing with a hypercube of the same dimension. We start by considering the case of n-dimension and later let n go to infinity.

We start with $P(\mathcal{R})$ for a sphere of n-dimension. According to (5.62) in Section 5.5.2,

$$P(\mathcal{R}) \stackrel{\text{def}}{=} \frac{1}{n V_R(\mathcal{R})} \int_{\mathcal{R}} \|x\|^2 dx. \tag{5.107}$$

For a sphere with radius R, the volume is [73, sec. 5.19 (iii)]:

$$V_R(R) = \frac{\pi^{\frac{n}{2}}}{\Gamma\left(\frac{n}{2}+1\right)} R^n \tag{5.108}$$

The power is

$$\int_{\mathcal{R}} \|\vec{x}\|^2 d\vec{x} = \int_0^R r^2 d\frac{\pi^{\frac{n}{2}}}{\Gamma\left(\frac{n}{2}+1\right)} r^n = \frac{n\pi^{\frac{n}{2}}}{\Gamma\left(\frac{n}{2}+1\right)} \int_0^R r^{(n+1)} dr = \frac{n\pi^{\frac{n}{2}}}{(n+2)\Gamma\left(\frac{n}{2}+1\right)} R^{n+2}. \tag{5.109}$$

Therefore,

$$P(\mathcal{R}) = \frac{1}{n+2} R^2. \tag{5.110}$$

From (5.63) in Section 5.5.2, we have

$$Q(\mathcal{R}) = \frac{P(\mathcal{R})}{V_R(\mathcal{R})^{2/n}} = \frac{1}{n+2} \Gamma\left(\frac{n}{2}+1\right)^{\frac{2}{n}} \pi^{-1}. \tag{5.111}$$

When $z \to \infty$, we have [73, sec. 5.11 (i)]

$$\Gamma(z) \sim e^{-z} z^z \sqrt{\frac{2\pi}{z}}. \tag{5.112}$$

Therefore, when $n \to \infty$,

$$\Gamma\left(\frac{n}{2}+1\right) \sim \left(\frac{n+2}{2e}\right)^{\frac{n+2}{2}} \sqrt{\frac{4\pi}{n+2}}. \tag{5.113}$$

As a result,

$$\lim_{n\to\infty} [Q(\mathcal{R})] = \lim_{n\to\infty} \left[\frac{1}{n+2}\left(\frac{n+2}{2e}\right)^{\frac{n+2}{n}} \left(\frac{4\pi}{n+2}\right)^{\frac{1}{n}} \pi^{-1}\right] = \frac{1}{2e\pi}. \tag{5.114}$$

From (5.73) in Section 5.5.2,

$$\lim_{n\to\infty} [\gamma_s(\mathcal{R})] = \lim_{n\to\infty} \left[\frac{1}{12Q(\mathcal{R})}\right] = \frac{e\pi}{6}. \tag{5.115}$$

That completes the proof.

5.11 APPENDIX: PROBABILITY UPDATE AT PARITY NODE

In this appendix, we prove (5.103) in Section 5.8.3. The problem is stated as follows.

Given that the probability of bit $i \neq k$ to be 1 is p_i, and the parity check condition that the sum (with modulo-2) of all bits must be 0, then the probability of bit k to be 1 is given by (5.103):

$$P_{e,k} = \frac{1}{2} - \frac{1}{2}\prod_{i\neq k}(1-2p_i). \tag{5.116}$$

Let us consider the set of N bits, which includes all of the bits in the product in (5.116). Let P_N be the probability that the parity is even. Namely, the number of bits with value 1 is even. We use mathematical induction to show that

$$P_N = \frac{1}{2} + \frac{1}{2}\prod_{n=1}^{N}(1-2p_n), \tag{5.117}$$

where p_n is the probability for bit n to be 1.

When $N = 2$, the parity can be even if and only if both bits are 1 (with probability $p_1 p_2$), or no bit is 1 (with probability $(1-p_1)(1-p_2)$). Therefore,

$$P_2 = p_1 p_2 + (1-p_1)(1-p_2) = \frac{1}{2} + \frac{1}{2}(1-2p_1)(1-2p_2). \tag{5.118}$$

Equation (5.117) is thus valid when $N = 2$. Under mathematical induction, we assume that (5.117) is valid when $N = K$. Namely,

$$P_K = \frac{1}{2} + \frac{1}{2}\prod_{n=1}^{K}(1-2p_n) \tag{5.119}$$

Now consider the case when $N = K + 1$. In this case, the parity can be even in two mutually exclusive cases: either the K bits form even parity and the $(K + 1)$th bit is zero (with a probability of $P_K(1 - p_{K+1})$), or the K bits form odd parity and the $(K + 1)$th bit is 1 (with a probability of $(1 - P_K)p_{K+1}$). Therefore,

$$P_{K+1} = P_K(1 - p_{K+1}) + (1 - P_K)p_{K+1} = P_K(1 - 2p_{K+1}) + p_{K+1}. \qquad (5.120)$$

Combining (5.119) and (5.120), we have

$$P_{K+1} = \frac{1}{2} + \frac{1}{2}\prod_{n=1}^{K+1}(1 - 2p_n). \qquad (5.121)$$

Namely, (5.117) is valid for $N = K + 1$. Therefore, by mathematical induction, (5.117) is valid for any integer $N > 2$.

Now return to (5.116) and look at bit k. Bit k is 1 if and only if the parity of the rest of the bits is odd since the total parity must be even. Therefore,

$$P_{e,k} = 1 - P_N = \frac{1}{2} - \frac{1}{2}\prod_{n=1}^{N}(1 - 2p_n) = \frac{1}{2} - \frac{1}{2}\prod_{i \neq k}(1 - 2p_i). \qquad (5.122)$$

This result proves (5.116) and thus (5.103).

REFERENCES

1. W. Ryan and S. Lin, *Channel Codes: Classical and Modern*. Cambridge University Press, 2009.
2. D. J. Costello and G. D. Forney, "Channel Coding: The Road to Channel Capacity," *Proc. IEEE*, vol. 95, no. 6, pp. 1150–1177, 2007.
3. K. Andrews, C. Heegard, and D. Kozen, "A Theory of Interleavers," 1997 [Online]. Available at https://ecommons.cornell.edu/handle/1813/7289 [Accessed July 19, 2017].
4. I. Richer, "A Simple Interleaver for Use with Viterbi Decoding," *IEEE Trans. Commun.*, vol. 26, no. 3, pp. 406–408, 1978.
5. J. Ramsey, "Realization of Optimum Interleavers," *IEEE Trans. Inf. Theory*, vol. 16, no. 3, pp. 338–345, 1970.
6. C. E. Shannon, "A Mathematical Theory of Communication," *Bell Syst. Tech. J.*, vol. 27, no. July, pp. 379–423, 1948.
7. R. W. Hamming, "Error Detecting and Error Correcting Codes," *Bell Syst. Tech. J.*, vol. 29, no. 2, pp. 147–160, 1950.
8. M. J. E. Golay, "Notes on Digital Coding," *Proc. Inst. Radio Eng.*, vol. 37, no. 6, p. 657, 1949.
9. D. E. Muller, "Application of Boolean Algebra to Switching Circuit Design and to Error Detection," *Trans. IRE Prof. Gr. Electron. Comput.*, vol. EC-3, no. 3, pp. 6–12, 1954.
10. I. Reed, "A Class of Multiple-Error-Correcting Codes and the Decoding Scheme," *Trans. IRE Prof. Gr. Inf. Theory*, vol. 4, no. 4, pp. 38–49, 1954.
11. A. Hocquenghem, "Codes Correcteurs D'erreurs," *Chiffres*, vol. 2, pp. 147–156, 1959.
12. R. C. Bose and D. K. Ray-Chaudhuri, "On a Class of Error Correcting Binary Group Codes," *Inf. Control*, vol. 3, no. 1, pp. 68–79, 1960.
13. I. S. Reed and G. Solomon, "Polynomial Codes over Certain Finite Fields," *J. Soc. Ind. Appl. Math.*, vol. 8, no. 2, pp. 300–304, 1960.

14. M. P. C. Fossorier and S. Lin, "A Unified Method for Evaluating the Error-Correction Radius of Reliability-Based Soft-Decision Algorithms for Linear Block Codes," *IEEE Trans. Inf. Theory*, vol. 44, no. 2, pp. 691–700, 1998.

15. R. Silverman and M. Balser, "Coding for Constant-Data-Rate Systems," *Trans. IRE Prof. Gr. Inf. Theory*, vol. 4, no. 4, pp. 50–63, 1954.

16. E. Biglieri, *Coding for Wireless Channels*. Boston: Springer, 2005.

17. A. J. Viterbi and J. K. Omura, *Principles of Digital Communication and Coding–Andrew J. Viterbi, Jim K. Omura–Google Books*. New York: McGraw-Hill, 1979.

18. P. Elias, "Coding for Noisy Channels," *IRE Conv. Rec.*, vol. 4, pp. 37–46, 1955.

19. A. Viterbi, "Error Bounds for Convolutional Codes and an Asymptotically Optimum Decoding Algorithm," *IEEE Trans. Inf. Theory*, vol. 13, no. 2, pp. 260–269, 1967.

20. J. Proakis and M. Salehi, *Digital Communications*, 5th ed. McGraw-Hill Education, 2007.

21. D. Daut, J. Modestino, and L. Wismer, "New Short Constraint Length Convolutional Code Constructions for Selected Rational Rates (Corresp.)," *IEEE Trans. Inf. Theory*, vol. 28, no. 5, pp. 794–800, 1982.

22. S. Lin and D. J. Costello, *Error Control Coding: Fundamentals and Applications*. Englewood Cliffs, NJ: Prentice-Hall, 1983.

23. J. Hagenauer and P. Hoeher, "A Viterbi Algorithm with Soft-Decision Outputs and Its Applications," in *IEEE Global Telecommunications Conference, 1989, and Exhibition "Communications Technology for the 1990s and Beyond,"* 27–30 November, Dallas, TX, 1989, pp. 1680–1686.

24. D. J. Costello, "A Construction Technique for Random-Error-Correcting Convolutional Codes," *IEEE Trans. Inf. Theory*, vol. 15, no. 5, pp. 631–636, 1969.

25. J. Justesen, "New Convolutional Code Constructions and a Class of Asymptotically Good Time-Varying Codes," *IEEE Trans. Inf. Theory*, vol. 19, no. 2, pp. 220–225, 1973.

26. P. Shaft, "Low-Rate Convolutional Code Applications in Spread-Spectrum Communications," *IEEE Trans. Commun.*, vol. 25, no. 8, pp. 815–822, 1977.

27. G. Ungerboeck, "Channel Coding with Multilevel/Phase Signals," *IEEE Trans. Inf. Theory*, vol. 28, no. 1, pp. 55–67, 1982.

28. G. Ungerboeck and I. Csajka, "On Improving Data-Link Performance by Increasing the Channel Alphabet and Introducing Sequence Coding," in *1976 International Symposium on Information Theory*, 21–24 June, Ronneby Brunn, Sweden, 1976.

29. G. Ungerboeck, "Trellis-Coded Modulation with Redundant Signal Sets Part II: State of the Art," *IEEE Commun. Mag.*, vol. 25, no. 2, pp. 12–21, 1987.

30. Lee-Fang Wei, "Rotationally Invariant Convolutional Channel Coding with Expanded Signal Space—Part II: Nonlinear Codes," *IEEE J. Sel. Areas Commun.*, vol. 2, no. 5, pp. 672–686, 1984.

31. Lee-Fang Wei, "Trellis-Coded Modulation with Multidimensional Constellations," *IEEE Trans. Inf. Theory*, vol. 33, no. 4, pp. 483–501, 1987.

32. M. Ardakani, T. Esmailian, and F. R. Kschischang, "Near-Capacity Coding in Multicarrier Modulation Systems," *IEEE Trans. Commun.*, vol. 52, no. 11, pp. 1880–1889, 2004.

33. J. H. Conway and N. J. A. Sloane, *Sphere Packings, Lattices and Groups*. New York: Springer, 1988.

34. R. de Buda, "The Upper Error Bound of a New Near-Optimal Code," *IEEE Trans. Inf. Theory*, vol. 21, no. 4, pp. 441–445, 1975.

35. G. R. Lang and F. M. Longstaff, "A Leech Lattice Modem," *IEEE J. Sel. Areas Commun.*, vol. 7, no. 6, pp. 968–973, 1989.

36. ITU-T, *A Modem Operating at Data Signalling Rates of up to 33 600 Bit/s for Use on the General Switched Telephone Network and on Leased Point-to-Point 2-Wire Telephone-Type Circuits*. ITU-T Recommendation V.34, 1998.

37. D. Raphaeli and A. Gurevitz, "Constellation Shaping for Pragmatic Turbo-Coded Modulation with High Spectral Efficiency," *IEEE Trans. Commun.*, vol. 52, no. 3, pp. 341–345, 2004.

38. R. Laroia, N. Farvardin, and S. A. Tretter, "On Optimal Shaping of Multidimensional Constellations," *IEEE Trans. Inf. Theory*, vol. 40, no. 4, pp. 1044–1056, 1994.

39. G. D. Forney, *Concatenated Codes*. Cambridge: MIT Research Laboratory of Electronics, Technical Report 440, 1965.

40. ITU-T, *Asymmetric Digital Subscriber Line (ADSL) Transceivers*. ITU-T Recommendation G.992.1, 1999.

41. C. Berrou, A. Glavieux, and P. Thitimajshima, "Near Shannon Limit Error-Correcting Coding and Decoding: Turbo-Codes. 1," in *Proceedings of ICC '93—IEEE International Conference on Communications*, 1993, vol. 2, pp. 1064–1070.

42. S. Benedetto, D. Divsalar, G. Montorsi, and F. Pollara, "Serial Concatenation of Interleaved Codes: Performance Analysis, Design, and Iterative Decoding," *IEEE Trans. Inf. Theory*, vol. 44, no. 3, pp. 909–926, 1998.

43. C. B. Schlegel and L. C. Perez, *Trellis and Turbo Coding: Iterative and Graph-Based Error Control Coding*. Wiley-IEEE Press, 2015.

44. 3GPP, *3rd Generation Partnership Project; Technical Specification Group Radio Access Network; Evolved Universal Terrestrial Radio Access (E-UTRA); Multiplexing and Channel Coding*. 3GPP Technical Specifications 36.212 V13.0.0, 2015.

45. ETSI, *Digital Video Broadcasting (DVB); Framing Structure, Channel Coding and Modulation for Satellite Services to Handheld Devices (SH) below 3 GHz*. V1.2.1. ETSI EN 302 583, 2011.

46. L. Bahl, J. Cocke, F. Jelinek, and J. Raviv, "Optimal Decoding of Linear Codes for Minimizing Symbol Error Rate (Corresp.)," *IEEE Trans. Inf. Theory*, vol. 20, no. 2, pp. 284–287, 1974.

47. P. Robertson, P. Hoeher, and E. Villebrun, "Optimal and Sub-optimal Maximum A Posteriori Algorithms Suitable for Turbo Decoding," *Eur. Trans. Telecommun.*, vol. 8, no. 2, pp. 119–125, 1997.

48. M. A. Khalighi, "Effect of Mismatched SNR on the Performance of Log-MAP Turbo Detector," *IEEE Trans. Veh. Technol.*, vol. 52, no. 5. pp. 1386–1397, 2003.

49. S. ten Brink, "Convergence Behavior of Iteratively Decoded Parallel Concatenated Codes," *IEEE Trans. Commun.*, vol. 49, no. 10, pp. 1727–1737, 2001.

50. C. Berrou and A. Glavieux, "Near Optimum Error Correcting Coding and Decoding: Turbo-Codes," *IEEE Trans. Commun.*, vol. 44, no. 10, pp. 1261–1271, 1996.

51. R. Garello, F. Chiaraluce, P. Pierleoni, M. Scaloni, and S. Benedetto, "On Error Floor and Free Distance of Turbo Codes," in *ICC 2001. IEEE International Conference on Communications. Conference Record (Cat. No.01CH37240)*, 11–14 June, Helsinki, Finland, 2001, vol. 1, pp. 45–49.

52. L. C. Perez, J. Seghers, and D. J. Costello, "A Distance Spectrum Interpretation of Turbo Codes," *IEEE Trans. Inf. Theory*, vol. 42, no. 6, pp. 1698–1709, 1996.

53. T. Tonnellier, C. Leroux, B. Le Gal, B. Gadat, C. Jego, and N. Van Wambeke, "Lowering the Error Floor of Turbo Codes with CRC Verification," *IEEE Wirel. Commun. Lett.*, vol. 5, no. 4, pp. 404–407, 2016.

54. Hong Chen, R. G. Maunder, and L. Hanzo, "A Survey and Tutorial on Low-Complexity Turbo Coding Techniques and a Holistic Hybrid ARQ Design Example," *IEEE Commun. Surv. Tutorials*, vol. 15, no. 4, pp. 1546–1566, 2013.

55. R. Gallager, "Low-Density Parity-Check Codes," *IEEE Trans. Inf. Theory*, vol. 8, no. 1, pp. 21–28, 1962.

56. D. A. Spielman, "Linear-Time Encodable and Decodable Error-Correcting Codes," *IEEE Trans. Inf. Theory*, vol. 42, no. 6, pp. 1723–1731, 1996.

57. D. J. C. MacKay and R. M. Neal, "Near Shannon Limit Performance of Low Density Parity Check Codes," *Electron. Lett.*, vol. 32, no. 18, p. 1645, 1996.

58. IEEE, *802.11-2016—IEEE Standard for Information Technology—Telecommunications and Information Exchange between Systems Local and Metropolitan Area Networks—Specific Requirements—Part 11: Wireless LAN Medium Access Control (MAC) and Physical Layer (PHY)*. IEEE, 2016.

59. European Telecommunications Standards Institute (ETSI), *Digital Video Broadcasting (DVB); Second Generation Framing Structure, Channel Coding and Modulation Systems for Broadcasting, Interactive Services, News Gathering and Other Broadband Satellite Applications; Part 1: DVB-S2*. V1.4.1. ETSI EN 302 307-1, 2014.

60. T. Richardson and S. Kudekar, "Design of Low-Density Parity Check Codes for 5G New Radio," *IEEE Commun. Mag.*, vol. 56, no. 3, pp. 28–34, 2018.

61. M. Luby, M. Mitzenmacher, A. Shokrollah, and D. Spielman, "Analysis of Low Density Codes and Improved Designs Using Irregular Graphs," in *Proceedings of the Thirtieth Annual ACM Symposium on Theory of Computing—STOC '98*, 24–26 November, Dallas, TX, 1998, pp. 249–258.

62. T. J. Richardson, M. A. Shokrollahi, and R. L. Urbanke, "Design of Capacity-Approaching Irregular Low-Density Parity-Check Codes," *IEEE Trans. Inf. Theory*, vol. 47, no. 2, pp. 619–637, 2001.

63. R. Tanner, "A Recursive Approach to Low Complexity Codes," *IEEE Trans. Inf. Theory*, vol. 27, no. 5, pp. 533–547, 1981.

64. T. J. Richardson and R. L. Urbanke, "Efficient Encoding of Low-Density Parity-Check Codes," *IEEE Trans. Inf. Theory*, vol. 47, no. 2, pp. 638–656, 2001.

65. Sae-Young Chung, G. D. Forney, T. J. Richardson, and R. Urbanke, "On the Design of Low-Density Parity-Check Codes within 0.0045 dB of the Shannon Limit," *IEEE Commun. Lett.*, vol. 5, no. 2, pp. 58–60, 2001.

66. H. Loeliger, "An Introduction to Factor Graphs," *IEEE Signal Process. Mag.*, vol. 21, no. 1, pp. 28–41, 2004.

67. T. R. Halford, K. M. Chugg, and A. J. Grant, "Which Codes Have 4-Cycle-Free Tanner Graphs?," in *IEEE International Symposium on Information Theory—Proceedings*, 9–14 July, Seattle, WA, 2006, vol. 45, no. 6, pp. 871–875.

68. K. Niu, K. Chen, J. Lin, and Q. T. Zhang, "Polar Codes: Primary Concepts and Practical Decoding Algorithms," *IEEE Commun. Mag.*, vol. 52, no. 7, pp. 192–203, 2014.

69. B. Zhang et al., "A 5G Trial of Polar Code," in *2016 IEEE Globecom Workshops (GC Wkshps)*, 4–8 December, Washington, DC 2016, pp. 1–6.

70. D. J. C. MacKay, "Fountain Codes," *IEE Proc. Commun.*, vol. 152, no. 6, p. 1062, 2005.

71. B. Sklar and F. J. Harris, "The ABCs of Linear Block Codes," *IEEE Signal Process. Mag.*, vol. 21, no. 4, pp. 14–35, 2004.

72. E. Rate, "Complex to Real" [Online]. Available at http://complextoreal.com/ [Accessed July 30, 2017].

73. DLMF, "NIST Digital Library of Mathematical Functions" [Online]. Available at http://dlmf.nist.gov/ [Accessed December 13, 2016].

HOMEWORK

5.1 Write a MATLAB program to reproduce Table 5.4 in Section 5.3.3.4. The input is the G matrix in (5.20). Note that due to the ambiguity discussed at the end of Section 5.3.3.4, your results may be different from what is in the text.

5.2 Based on (5.30) in Section 5.3.5.3, what is the coding gain for a block code (n, k, d) in a power-limited case (i.e., modulation is the same for coded and uncoded transmission)? Can you extend the results to other modulation types? You may need to consult Chapter 4 for the expression of BER under various modulations.

5.3 Study how to use MATLAB functions convenc and ploy2trellis to perform convolutional code encoding, given generation matrix g. Write your own MATLAB code for an encoder, and verify that the encoding results are the same as that from convenc. In your code, you do not have to use a trellis as an intermediate result.

5.4 Justify the statement about the cases with $k > 1$ at the end of Section 5.4.1.3.

5.5 Convince yourself that the properties described in Section 5.4.2.2 are consistent with the definition of the transfer function.

5.6 Justify the factor 2^l in (5.43) in Section 5.4.2.3.

5.7 Consider the set partitioning described in Section 5.5.1.

   a. Draw a partition diagram similar to Figure 5.16 for 8-PSK modulation (see [27] for an answer).

   b. Discuss the connection between the set partition and the one-dimensional Gray code construction rule discussed in Chapter 3.

   c. Construct a two-dimensional Gray code for 16-QAM based on the partition in Figure 5.16.

5.8 Use MATLAB "bertool," with the same modulation (e.g., 4-PSK) and various coding schemes (convolutional code and block code) with various constraint lengths and coding rates, plot BER over E_b/N_0 for theoretical bounds.

    a. How do block code and convolutional code compare, in low BER and high BER regions?

    b. What is the difference in coding gain for soft-decision and hard-decision methods with a convolutional code?

    c. Compare Hamming code with $N = 7$ and $N = 511$. Comment on the performance difference.

    d. Explore any other way you would like. (You may reference discussions in [20, sec. 8.6]).

5.9 A block code that was not covered in detail in this book yet widely used is the Reed–Solomon code (RS). Comparing with the convolutional code (CC), RS has better performance under impulse noise. In this problem, let us verify this. (Read the whole assignment before starting to work.)

    a. MATLAB has two objects: comm.RSEncoder and comm.RSDecoder to support the RS codes. Read the documentation to find out how to use them, and write a test code so that a bitstream can be encoded and decoded back to the same stream. Use the following parameters (referring to MATLAB documentation for comm.RSEncoder): BitInput=true, CodewordLength=255, MessageLength=129. All other parameters can be left to default. These parameters specify an RS code whose polynomial order is 8, with a code rate of approximately 1/2.

    b. MATLAB has two objects supporting convolutional coding: comm.ConvolutionalEncoder and comm.ViterbiDecoder. Read the documentation to find out how to use them, and write a test code so that a bitstream can be encoded and decoded back to the same stream. Use the following code: $k = 14$, $g = (21675, 27123)$ (in octal). Set TerminationMethod to "Terminated." For the decoder, use hard input and set OutputDataType to "int32."

    c. We can treat the output bits (0 and 1) as BPSK signals. We can add noise to the bits, and slice them to get the output bits. Think about how much noise you need to add to each sample for a given value of E_b/N_0. Figure out the noise power for continuous noise. Write the simulation for the RS code (with noise added), and compare the resulting BER versus E_b/N_0 curve with what you can get from bertool. They will not match exactly because bertool gives approximate results. Just make sure your results are not off by more than 1 dB. Otherwise, your calculation of noise power might be wrong.

    d. Implement impulse noise as regular noise pulses. Such pulses can be characterized by duty cycle (the portion of time when the noise is on) and duration (the width of the noise pulse measured by the number of bits). Use the following parameters: duty cycles (0.2, 0.5, 1), duration (5, 10, 50). Note that the noise power needs to be scaled according to the duty cycle to obtain desired average noise power level. Figure out how to implement a stream of impulse noise.

    e. Put everything together. Plot BER versus E_b/N_0 curves for both RS and CC codes, under impulse noise with parameters specified in (d) above. Take the E_b/N_0 range between 0 and 10 dB.

f. Programming notes:

    i. You will need a large number of bits (about 10 million) to get a reasonable estimate of BER. It would be easier to invoke encoder and decoder multiple times, each time passing a fraction of the bits. Otherwise, you might run into a memory problem.

    ii. This code will run for a long time (possibly days). It would be a good idea to print out progress information, so that you are sure the program is running normally. It is also a good idea to build in a mechanism for saving intermediate results and resume running from such results.

5.10 In Chapter 4, we showed that when input symbols are equiprobable, MAP detection is equivalent to ML detection. Explain why in turbo decoding we must use MAP instead of ML detection.

5.11 Write a MATLAB code to reproduce the turbo code example in the following file: http://complextoreal.com/wp-content/uploads/2013/01/turbo2.pdf. For more information about the method and notations, refer to part 1 of the tutorial at http://complextoreal.com/wp-content/uploads/2013/01/turbo1.pdf. More intermediate data (for checking) can be found at http://complextoreal.com/wp-content/uploads/2013/01/turbo_example.xls. After repeating the tutorial results, use the code to compute BER for some more SNR values. Use your own judgment on the number of points to compute, depending on the computation time and the shape of the curve. Note that the example data contains some mistakes and inconsistencies. It is a part of the assignment to sort out and overcome these problems.

5.12 Given a parity-check matrix for LDPC code, can the code always be systematic?

CHAPTER 6

CHANNEL CHARACTERISTICS

6.1 INTRODUCTION

Figure 6.1 shows the high-level physical layer architecture, discussed in Chapter 1. This chapter addresses the channel, as circled by a dashed oval. The channel is not a part of the transceiver, and we usually do not have much control over it. However, channel properties profoundly affect communication system design.

- Channel attenuation of the communication signal (path loss) determines the signal power at the receiver. This power level and the noise power at the receiver lead to the expected signal-to-noise ratio (SNR) at the receiver. As discussed in Chapter 2, SNR sets the data rate limit. It also affects many aspects of the receiver design, as discussed throughout this book.

- Channel time dispersion causes mutual interference between subsequent symbols. The requirement of removing such interference affects both waveform design and receiver implementation. This issue will be further discussed in Chapters 9 and 10.

- Channel response may change with time (fading). The speed of fading affects how well the transmitter and receiver know about the channel state information (CSI) and how often CSI estimation should be performed. Channel estimation is not a focus of this book, but it will be touched upon in Chapters 9 and 10. Chapter 11 also discusses spatial diversity as a means to counter channel fading.

As discussed in Section 6.4, there are various mechanisms causing channel gain changes. On the other hand, from the system point of view, it is often desirable to separate average path loss and other effects such as fast fading and time dispersion. Therefore, in the literature, channels are often "normalized" so that the average path loss is 1 (i.e., 0 dB), while the average SNR is introduced as a separate parameter. Average path loss is covered in Sections 6.3 and 6.4.1.2, while Sections 6.4.1.3 and 6.5 mainly deal with normalized channels.

This chapter focuses on the mobile wireless channels because it is more complicated in general. Characterization of wired channels uses the same framework

Digital Communication for Practicing Engineers, First Edition. Feng Ouyang.
© 2020 by The Institute of Electrical and Electronics Engineers, Inc.
Published 2020 by John Wiley & Sons, Inc.

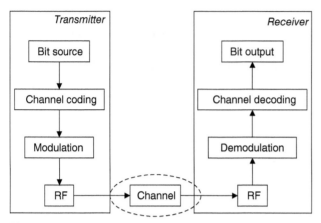

Figure 6.1 Physical layer architecture.

while focusing on different aspects. Details can be found in the literature. There are other channels outside of the mainstream consumer applications, such as ground-to-space, air-to-air, and special frequency bands such as the high frequency (HF) band. These channels are mentioned only briefly in this chapter.

Furthermore, channels for multiantenna systems are briefly discussed in Chapter 11. Channels for the emerging millimeter-wave (30–300 GHz) communications are briefly discussed in Chapter 12.

This chapter considers the channel between the transmitting and receiving antennas. On the other hand, it sometimes makes sense to include a portion of the transceiver circuitry, such as the filters in the RF chain, into the channel. In this case, the basic concepts and methodologies in this chapter still apply, although details may be different.

In keeping with conventions in the literature, in this chapter, the terms "radio" and "wireless" are considered interchangeable. In the literature, many channel properties are expressed in either linear or dB forms (see Chapter 1). To avoid confusion, in this chapter we use linear forms for all equations. However, typical values are usually expressed in dB forms.

For the rest of this chapter, Section 6.2 provides an overview and a basic formulation of channel description. Section 6.3 considers the simplest channels, which remain constant over time and do not have time dispersion. These channels are usually used to describe rural and open space areas. Section 6.4 adds time dependency experienced by mobile communications, especially in urban areas. Section 6.5 studies time dispersion. Time dependency and time dispersion are expressed in time and frequency domains in Section 6.6. Section 6.7 provides some guidelines on channel modeling practice. Section 6.8 covers a relatively independent topic: link budget computation, where other factors such as noise are combined with channel path loss to compute the SNR at the receiver. Section 6.9 provides summary and closing remarks. An appendix linking the channel gains in the passband and baseband is provided in Section 6.10.

6.2 CHANNEL GAIN AND CHANNEL CLASSIFICATION

6.2.1 Mathematical Representation of Wireless Communication Channels

In general, a *channel* is a linear transform between the input (transmitted) signal $x(t)$ and the output (received) signal $y(t)$:

$$y(t) = \int h(t,\tau)x(t-\tau)d\tau, \tag{6.1}$$

where $h(t, \tau)$ is known as the channel gain.[1] For wireless channels, by default, $x(t)$ and $y(t)$ are signals transmitted and received by *ideal isotropic antennas*.

For transmission, an ideal isotropic antenna radiates equally in all directions, and the total radiation power equals input power. For receiving, the ideal isotropic antenna collects all energy flowing through an isotropic aperture, whose area is [1]

$$A_{iso} = \frac{\lambda^2}{4\pi}, \tag{6.2}$$

where λ is the wavelength of the signal concerned.[2] In the literature, the isotropic antennas are also known as omnidirectional antennas.

As explained in Chapter 3, signal analysis can be performed in passband and baseband. Equation (6.1) applies to both cases, as shown in the appendix (Section 6.10). For the rest of the chapter, we stay in baseband, where channel gain is in general complex-valued. The absolute phase of the channel is not important, as it can be easily corrected at the receiver. However, the phase change with time is important.

As stated in Section 6.1, in many analyses the channel gain is "normalized" so that its average value is 1. Such normalization is acceptable because the absolute amplitude of the received signal is usually not important. What we care more is the ratio between the signal and noise, that is, the SNR. While the SNR depends on channel gain, it is often taken as an input parameter to analysis, thus allowing the channel gain to be normalized. Section 6.3 focuses on the "raw" channel gain and its connection with SNR. The rest of this chapter uses normalized channel gain. Section 6.8 discusses the computation of SNR from channel gain.

---

[1] Note that in this chapter, as in the literature, another term "channel power gain" is often used. Channel power gain is $|h(t, \tau)|^2$. The path loss, another term used frequently in this chapter, is also in terms of power.
[2] From electromagnetic theory point of view, channel propagation is better described as electrical and magnetic field strength at given points. In vacuum, which is usually a very good approximation for air, we have the relationship $S = E^2/120\pi$ [2]. Here E is the root-mean-squared value of electric field strength (in V/m). S is power flux density in W/m^2. 120π is the characteristic impedance, in Ohm.

6.2.2   Channel Classification

Equation (6.1) can be simplified in some cases. If $h(t, \tau)$ is independent of t, the channel is said to be time-invariant, or constant. Furthermore, if

$$h(t,\tau) = g(t)\delta(t-\tau), \tag{6.3}$$

such channel is said to be memoryless, or without time dispersion. For these channels, the output at time t depends only on the input at the same time[3]:

$$y(t) = g(t)x(t). \tag{6.4}$$

For reasons that will become apparent later, channels described by (6.4) are also known as flat channels.

6.2.3   Near-Field and Far-Field

For radio channels, it is important to divide the space into the "near-field" and the "far-field" [3]. Near-field is the region close to an antenna. In near-field, the electromagnetic field has complex patterns. In the far-field, on the other hand, there is only the radiating electromagnetic wave. The commonly used criterion for the boundary between the two zones is [3]

$$R_f = \frac{2D^2}{\lambda}, \tag{6.5}$$

where D is the size of the largest antenna structure. λ is the radiation wavelength. A point whose distance from an antenna is larger than R_f belongs to the far-field. In communication systems, a receiver is typically in the far-field region of the transmitter, and vice versa. This chapter studies only the far-field channels.

6.2.4   Frequency Bands and Special Channels

The frequency is a critical parameter in wireless communications. As will be shown in the rest of the chapter, many channel characteristics are frequency-dependent. Furthermore, frequency also affects antenna size and gain. Therefore, different applications use different frequency ranges.

In communications engineering, the entire spectrum of radio frequencies is divided into bands for easy reference. Table 6.1 is the band definition from the International Telecommunication Union (ITU), which is a primary international standards body for communications [4].

As shown in Table 6.1, we start with the medium frequency (MF) at 300 kHz to 3 MHz (1 km to 100 m in wavelength). For the rest of the frequencies, each decade is

---

[3] Physically, there is a delay due to propagation the speed of light. Since such delay is a constant, it is inconsequential to most physical layer analysis and thus can be ignored. However, propagation delay may be important when we consider two-way signal exchanges and their latencies.

TABLE 6.1 Nomenclature of frequency and wavelength [4].

Band number	Symbols	Frequency range (Hz)	Wavelength range (m)	Metric subdivision	Metric abbreviations
3	ULF	$3 \times 10^2 - 3 \times 10^3$	$10^6 - 10^5$	Hectokilometric	B.hkm
4	VLF	$3 \times 10^3 - 3 \times 10^4$	$10^5 - 10^4$	Myriametric	B.Mam
5	LF	$3 \times 10^4 - 3 \times 10^5$	$10^4 - 10^3$	Kilometric	B.km
6	MF	$3 \times 10^5 - 3 \times 10^6$	$10^3 - 10^2$	Hectometric	B.hm
7	HF	$3 \times 10^6 - 3 \times 10^7$	$10^2 - 10^1$	Decametric	B.dam
8	VHF	$3 \times 10^7 - 3 \times 10^8$	$10^1 - 10^0$	Metric	B.m
9	UHF	$3 \times 10^8 - 3 \times 10^9$	$10^0 - 10^{-1}$	Decimetric	B.dm
10	SHF	$3 \times 10^9 - 3 \times 10^{10}$	$10^{-1} - 10^{-2}$	Centimetric	B.cm
11	EHF	$3 \times 10^{10} - 3 \times 10^{11}$	$10^{-2} - 10^{-3}$	Millimetric	B.mm
12		$3 \times 10^{11} - 3 \times 10^{12}$	$10^{-3} - 10^{-4}$	Decimillimetric	B.dmm
13		$3 \times 10^{12} - 3 \times 10^{13}$	$10^{-4} - 10^{-5}$	Centimillimetric	B.cmm
14		$3 \times 10^{13} - 3 \times 10^{14}$	$10^{-5} - 10^{-6}$	Micrometric	B.μm
15		$3 \times 10^{14} - 3 \times 10^{15}$	$10^{-6} - 10^{-7}$	Decimicrometric	B.dμm

assigned to a band. The lower frequency bands are named as low frequency (LF), very-low frequency (VLF), and ultra-low frequency (ULF). Even lower frequencies (not shown in Table 6.1) are named extremely low frequency (ELF), but they are not typically used for communications. The higher frequencies relative to the MF are named high frequency (HF), very-high frequency (VHF), ultra-high frequency (UHF), super-high frequency (SHF), and extremely high frequency (EHF).

In addition to the band definitions in Table 6.1, there are other band nomenclature conventions for specific industries. For example, in satellite communications, the spectrum bands are named L, S, C, X, Ku, Ka, and so on. The exact frequency allocations depend on regions and regulatory bodies. For example, the Federal Communications Commission (FCC) in the United States (as Region I in the ITU framework) defines the C-band as 3.9–8.5 GHz [5, ch. 2]. On the other hand, ITU specifies the C-band as 3.4–7.1 GHz for space communications and 4–8 GHz for radar [4].

In general, lower frequency bands (below 1 GHz) are better in terrestrial communications because they can go around obstacles and have better penetration through building walls. Higher frequency bands (above 1 GHz) are able to support larger bandwidths (and thus higher data rate) and help to reduce antenna sizes (see Section 6.8.3). Multiantenna techniques are advantageous or even necessary for higher frequency communications (Chapters 11 and 12). For space communications, dish antennas are often used. Higher frequencies (above 2 GHz) increase antenna gain for a given dish size (see Section 6.8.3). Attenuation from atmosphere also varies with frequency.

LF, VLF, and ULF bands have special propagation modes. They have global reach either through the earth or in multiple reflections between the earth's surface and the ionosphere. They can also penetrate the water body to some extent. HF band has global reach as well, due to ionosphere reflection. These bands are not covered in this book.

At present, modern commercial communication systems are concentrated in the frequency range between 700 MHz and 5 GHz, with some emerging applications at higher frequencies (up to 100 GHz). Unless otherwise noted, our discussions of wireless channels focus on the frequency range of 1–5 GHz.

6.3 CONSTANT FLAT CHANNELS

In this section, we consider the constant flat channels, that is, channels that are constant over time and memoryless. For such channels, (6.1) is simplified to

$$y(t) = gx(t). \tag{6.6}$$

The constant channel gain g is the single parameter that characterizes the channel. $1/|g|^2$ represents the power loss through the channel, known as channel loss, channel attenuation, path loss, and so on. In this book, we use the term path loss. Path loss is affected by a number of factors; some of the major ones are discussed in the following.

6.3.1 Free Space Loss

The simplest channel is free space, that is, the transmitter and receivers are separated by vacuum, without any surrounding objects.

Let the input power be P_T, and the distance between the transmitter and receiver be d. As stated in Section 6.2, the ideal isotropic antenna radiates the power over all directions equally. At distance d, such power is evenly distributed over the spherical surface with an area of $4\pi d^2$. At the receiver, the ideal isotropic antenna collects power through an aperture with the area of A_{iso} given by (6.2). Therefore, the received power P_R is given by

$$P_R = P_T \frac{A_{iso}}{4\pi d^2} = P_T \left(\frac{\lambda}{4\pi d}\right)^2, \tag{6.7}$$

where λ is the wavelength of the carrier. This leads to the path loss

$$T_{free} \overset{\text{def}}{=} \frac{P_T}{P_R} = \left(\frac{4\pi d}{\lambda}\right)^2 = \left(\frac{4\pi fd}{c}\right)^2, \tag{6.8}$$

where f is the carrier frequency. c is the speed of light.

Although free space transmission is an abstract, path loss (6.8) applies accurately to many situations, such as ground-to-ground line-of-sight (with high-rise antennas), ground-to-air, and ground-to-space transmissions. The key to free space is that reflected and scattered energies are unimportant. This condition can happen when the transmitted power is constrained in space free of other objects (such as situations with directional antennas), or reflection or scattering paths are much longer than the direct path (such as situations with high-rise antennas and line of sight). Furthermore, additional attenuations by the media (air, cloud, moisture, and so on) can be added as corrections to the free space path loss, as will be discussed in Section 6.3.4.

The frequency dependency of free space path loss needs some clarification. Equation (6.8) may lead to the impression that for the same transmitted and received power, higher frequency leads to a shorter communication distance. This is not necessarily the case.

As shown in the derivation above, frequency dependency of the path loss originates from the definition of the isotropic aperture. According to (6.2), aperture size reduces as frequency increases, resulting in lower received power. However, (6.2) is based on the ideal isotropic antenna and may not reflect reality.

As will be detailed in Section 6.8.3, there are two types of antennas. One type is represented by a dipole antenna, whose size is scaled with wavelength. With higher frequencies (and thus shorter wavelength), the antenna size becomes smaller, and thus collecting less power, as (6.8) predicts. For this type of antennas, a higher frequency does lead to less communication distance due to the higher path loss given by (6.8). On the other hand, a smaller antenna reduces the size and weight of the communications device.

Another type of antennas is represented by a dish antenna, whose size is independent of wavelength. It turns out, as will be shown in Section 6.8.3, that these antennas have gains (i.e., the ratio between its received power and that from an ideal isotropic antenna) that increase with the square of the frequency. Therefore, the total received power, which is proportional to the ratio between receiving antenna gain and path loss, remains constant over frequency change. For this type of antennas, higher frequency does not reduce communication distance. In fact, as Section 6.8.3 will show, the transmitting antenna gain also increases with frequency for this type of antennas. Therefore, the overall link condition *improves* with frequency increase. This fact has been recognized as millimeter wave finds application in the latest mobile communications [6, sec. I-A], [7, sec. II-B]. See Chapter 12 for more details on millimeter-wave communications.

6.3.2 Flat Ground

Let us move from free space to a more complicated case: flat ground. In this case, the transmitting and receiving antennas are close to the ground in the absence of terrain features such as hills. We have some issues to consider.

6.3.2.1 Radio Horizon
Because the earth's surface has a curvature, a radio link may be blocked by earth. The maximum transmission distance without blockage is known as the radio horizon.

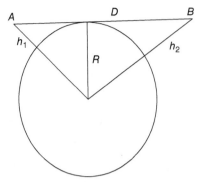

Figure 6.2   Radio horizon. Circle: the earth with radius R. A and B: the two antennas with heights h_1 and h_2, respectively. D: the maximum distance between the antennas without earth blockage.

Figure 6.2 shows the geometry for radio horizon. R is the earth's radius (nominally 6,371 km). h_1 and h_2 are antenna heights for the two parties. D is the maximum radio distance without blockage. From the figure, it is easy to see

$$D = \sqrt{(R+h_1)^2 - R^2} + \sqrt{(R+h_2)^2 - R^2} \approx \sqrt{2R}\left(\sqrt{h_1} + \sqrt{h_2}\right). \qquad (6.9)$$

The approximation holds when h_1 and h_2 are much smaller than R.

There are other effects in propagation such as atmosphere bending the propagation toward earth, increasing radio horizon. A simple way to account for such effect is using "equivalent earth's radius" instead of the actual one. A typical choice is

$$R_e = \frac{4}{3}R, \qquad (6.10)$$

where R_e is the equivalent earth's radius [8, sec. 2.7.4], [9, sec. C15]. This value can be adjusted depending on the weather condition and transmitting frequency.

Note that radio transmissions are not necessarily limited by the radio horizon. Low-frequency radio waves can get over the horizon to some extent through diffraction. Some frequency bands can achieve global reach through reflection from the ionosphere (Section 6.2.4).

6.3.2.2   Two-Ray Model

When the antennas are not high enough to approximate free space, we need to consider reflection from the ground surface. Figure 6.3 shows the geometry.

In Figure 6.3, A and B are the transmitter and receiver. They are, respectively, h_1 and h_2 above ground, which is represented by the thick horizontal line. The lateral distance between A and B is d. There are two rays reaching from A to B. One is the direct ray AB, the other is the reflected ray ACB. Both rays follow free space propagation. However, in addition to the path loss in (6.8), we also need to consider the phase of each ray. The complex *amplitude* of a ray can be written as

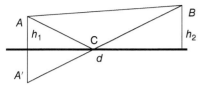

Figure 6.3   Two-ray model. The thick line represents the ground. A and B are the two antennas with heights h_1 and h_2. C is the reflection point on the ground. d is the horizontal distance between A and B.

$$A(D) = \frac{\lambda}{4\pi D} \exp\left(j\frac{2\pi}{\lambda}D\right), \tag{6.11}$$

where D is the length of the ray. λ is the wavelength. The first factor in (6.11) reflects the free space loss in amplitude. For the reflection path, we can consider the equivalent ray coming from the image of A, marked as A' in Figure 6.3. Therefore, the lengths of the rays can be computed from Pythagorean theorem. The sum amplitude at B is

$$\begin{aligned}
A_B = A_0 &\frac{\lambda}{4\pi\sqrt{d^2 + (h_1 - h_2)^2}} \exp\left(j\frac{2\pi}{\lambda}\sqrt{d^2 + (h_1 - h_2)^2}\right) \\
&+ \eta A_0 \frac{\lambda}{4\pi\sqrt{d^2 + (h_1 + h_2)^2}} \exp\left(j\frac{2\pi}{\lambda}\sqrt{d^2 + (h_1 + h_2)^2}\right)
\end{aligned}, \tag{6.12}$$

where η is the complex-valued reflection coefficient on the ground surface. A_0 is the amplitude at the transmitter. For glazing reflections, we can take η to be -1 [10, sec. 4.2.1]. When the two antennas are well separated,

$$\begin{aligned} h_1 &\ll d \\ h_2 &\ll d \end{aligned}, \tag{6.13}$$

Under these conditions,

$$\begin{aligned}
A_B = A_0 \frac{\lambda}{4\pi d} \exp&\left(j\frac{2\pi}{\lambda}\sqrt{d^2 + (h_1 - h_2)^2}\right) \\
&\left[1 - \exp\left\{j\frac{2\pi}{\lambda}\left[\sqrt{d^2 + (h_1 + h_2)^2} - \sqrt{d^2 + (h_1 - h_2)^2}\right]\right\}\right]
\end{aligned}. \tag{6.14}$$

Expanding the exponent part using (6.13), the received power is

$$P_B = |A_B|^2 = |A_0|^2 \left(\frac{\lambda}{4\pi d}\right)^2 \left|1 - \exp\left(j\frac{2\pi}{\lambda}\frac{2h_1 h_2}{d}\right)\right|^2. \tag{6.15}$$

This result leads to the path loss

$$T_{2ray} \overset{\text{def}}{=} \frac{|A_0|^2}{P_B} = \left(\frac{4\pi d}{\lambda}\right)^2 \left|1 - \exp\left(j\frac{2\pi}{\lambda}\frac{2h_1 h_2}{d}\right)\right|^{-2}. \tag{6.16}$$

In general, $|1 - \exp(j(2\pi/\lambda)(2h_1h_2/d))|^2$ oscillates between 0 and 4, with the last minimum point occurring at

$$d = \frac{2h_1h_2}{\lambda}.$$

(6.17)

Furthermore, when

$$d \gg \frac{4\pi h_1 h_2}{\lambda},$$

(6.18)

Equation (6.16) can be expanded to the first order of $4\pi h_1 h_2/\lambda d$:

$$T_{2ray} \approx \left(\frac{4\pi d}{\lambda}\right)^2 \left(\frac{4\pi h_1 h_2}{\lambda d}\right)^{-2} = \frac{d^4}{(h_1 h_2)^2}.$$

(6.19)

Namely, the path loss is proportional to d^4, in contrast with the free space path loss, which is proportional to d^2 as shown in (6.8).

Figure 6.4 shows an example of (6.16). The thin solid line shows the path loss given by (6.16). The thick dotted line is the free space path loss given by (6.8). The

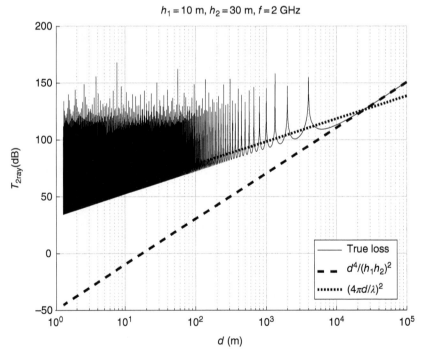

Figure 6.4 Two-ray model path loss.

thick dashed line shows the approximation (6.19). At smaller distances, path loss can get very large due to the destructive interference between the direct and reflected rays. The lower bound of the two-ray path loss is 3 dB lower than the free space loss due to construction interference, as shown in (6.16). At larger distances, path loss approaches the approximation given by (6.19).

Two-ray model is frequently used to estimate real-world path losses. Path loss characteristics demonstrated in the two-ray model, that is, the oscillation at smaller distances and the faster than–free space increase at larger distances, are common to other flat ground models.

6.3.3   Terrain Effects

Going beyond the flat ground model, we need to consider terrain effects. Terrain features impact radio propagation in various ways. The most important ones include the following.

1. Obstruction. Terrain objects such as hills may block the line of sight of a radio link. However, radio wave may propagate around the obstacles due to diffraction. Such effect depends strongly on frequency. The size of the object that a radio wave can "get around" is on the same order of magnitude as the wavelength. More detailed computation methods are provided in [11].

2. Reflection. Terrain features modify the ground reflection by adding more paths, changing the existing paths, or changing reflection coefficients. With terrain, reflection is not limited by the ground immediately underneath the direct path. Lateral reflections may come from nearby hills, for example.

3. Scattering. With rough ground surfaces and undulating surfaces, reflected waves are not concentrated at the specular direction but spread over all angles.

Path loss can be estimated given the terrain classification. Some engineering rules and approximate equations are given in the literature [10, ch. 4, 5]. More accurate pass loss predictions can be obtained by the various propagation modeling tools. For example, the "Terrain Integrated Rough Earth Model" (TIREM) is widely used by the United States Government for ground-to-ground propagation modeling [12]. It integrates with 3D map data sets to provide accurate modeling of specific locations. IF-77 is another tool focusing on air/ground propagation [13]. When choosing a tool, a user should pay attention to its modeled effects, frequency range, and validation credential.

6.3.4   Other Effects

Other than terrain effects, atmosphere, rain, moisture, vegetation may also significantly affect path loss. Usually, these effects are more pronounced at higher frequencies (above 10 GHz). There are numerous research works in this area. For practical guides, ITU publishes various recommendations concerning the excess attenuation effects of cloud and fog [14], vegetation [15], atmosphere [16], precipitation [17], and so on. More recommendations can be found in ITU-R recommendation P series [18].

6.4   FLAT FADING CHANNEL

Section 6.3 discusses channels that are constant in time and memoryless. For such channels, the constant path loss is the only parameter for channel characteristics. This section examines channels that change with time during communication. There are two main reasons for the time dependency of wireless channels, also known as fading. The first is the motion of the transmitter and/or the receiver. Their instantaneous positions determine the channel between them. The second reason is time variation of the communication media, such as the motion of various reflectors or change of atmospheric conditions. We will focus on the first reason for fading, which is usually the dominating factor in mobile communication.

As far as mobile channel fading is concerned, it is important to identify the three factors that affect channel gain, because they act at three different scales. These factors are discussed in some details in Section 6.4.1. Next, we study fading in two different dimensions. Section 6.4.2 studies the statistical distribution of channel gains, introducing Rayleigh fading and Ricean fading models. Section 6.4.3 studies how a channel gain evolves with time. One popular model for such behavior is the Jakes model, which provides the autocorrelation function of channel gain.

In this section, we limit our discussion to flat channel fading. Namely, (6.3) and (6.4) are valid. A channel gain can be expressed as $g(t)$.

6.4.1   Channel Gain Components

Figure 6.5 illustrates the three mechanisms for fading through space. The thick solid line is the long-range path loss discussed in Section 6.3. It is a smooth decrease in channel gain as distance increases. The dashed line includes the effect of shadowing, which is caused by large objects such as buildings and hills blocking the line of sight.

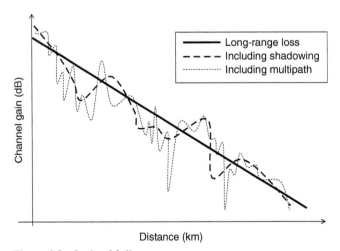

Figure 6.5   Scale of fading.

Shadowing results in channel gain fluctuations at the scale of these objects (10s or 100s of meters). The dotted line further includes the effect of multipath, or scattering. Radio waves through multiple reflections and scattering paths combine with mutual interference, forming a standing wave pattern in space. Such effect results in channel gain fluctuations at the scale of the wavelength (e.g., 0.1 m for 3 GHz). With the motion of the transmitter and/or receiver, these spatial scales translate to the fading time scales.

As discussed in Section 6.1, spatial and time scales of channel variation are important to the operation of communication systems. Therefore, it is important to study long-range path loss, shadowing, and multipath separately. In Sections 6.4.1.1 – 6.4.1.3, these three mechanisms are modeled in some more details.

6.4.1.1 Long-Range Path Loss
The long-range path loss is based on models described in Section 6.3, including the free space model and the two-ray model. Other models have been developed for specific situations. For example, attenuation due to foliage can be modeled in several ways [15, 19].

Some additional models are proposed in the literature. A widely used one is the Hata model for urban area land mobile radio links [20]. The Hata model is similar to the free space model, except that the exponent of the distance is different. In the Hata model, the path loss is given by

$$T_{Hata} \stackrel{\text{def}}{=} Ad^{\beta/10}, \tag{6.20}$$

where d is the distance. A depends on frequency, base station and mobile antenna heights, and propagation setting (e.g., small city, large city, and so on). β changes between 30 and 36, depending on base station antenna height. (For free space model, β would be 20.) The Hata model is adopted by several standard bodies [21].

6.4.1.2 Shadowing
The shadowing effect can be modeled by the log-normal distribution [22], [23, sec. 13.1-2]. The probability density function (pdf) of the path loss x due to shadowing is given by

$$p(x) = \begin{cases} \dfrac{1}{\sqrt{2\pi}\sigma x} \exp\left[\dfrac{\ln(x) - \ln(x_0)}{2\sigma^2}\right] & x > 0 \\ 0 & x \le 0 \end{cases}. \tag{6.21}$$

Here, $\ln(x_0)$ and σ^2 are the mean and variance of $\ln(x)$, respectively. They are usually determined by measurements. σ^2 is 5–12 dB for typical cellular and microcellular environments at 900 MHz, and slightly higher at 1800 MHz [24].

As described before, x here should be considered as the mean value of the path loss, with faster fading due to multipath averaged out. Such path loss due to shadowing is to be multiplied to the long-range path loss to form the dashed curve shown in Figure 6.5.

Equation (6.21) describes the distribution of path loss at given points. Furthermore, path losses in neighboring points are correlated. Let us define "decorrelation distance" as the distance at which the correlation coefficient is 0.3. Measurements show that such decorrelation distance is about 500 m for suburban areas at 900 MHz, and 10 m for urban areas at 1700 MHz [25]. For normal vehicular speeds and cellular systems, such results imply that the shadowing effect is almost constant for a transmission frame (typically in tens of milliseconds). Therefore, time dependency of shadowing effect is not usually a major concern.

6.4.1.3   Multipath Fading: Rayleigh and Ricean Channels

Figure 6.6 illustrates a multipath scattering environment. The transmitted signal arrives at the receiver in many "copies," or paths, each being scattered or reflected from a distinct object. There may also be a direct path between the transmitter and receiver. Signals through various paths superimpose at the receiver. For the purpose of Section 6.4, these signals are assumed to arrive at the same time, that is, no time dispersion. Therefore, the overall channel response obeys (6.4). However, there are phase differences among the signals, due to scattering path length differences. In other words, there is interference among the paths.[4] Signal intensity at the receiver is the result of such interference, which can change from constructive to destructive when the receiver moves by a wavelength. Therefore, the multipath scattering causes short-scale channel fading, which changes at a spatial scale of the carrier wavelength. Channel fading due to multipath is further described in Sections 6.4.2 and 6.4.3.

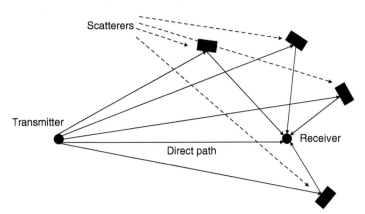

Figure 6.6   Multipath scattering environment.

---

[4] Such assumption is valid when the difference in time of arrival is comparable or larger than the carrier frequency period, yet much smaller than the symbol period.

6.4.2 Statistic Distribution of Channel Gain

In many situations such as mobile communications, the positions and properties of the scatterers are considered random. Therefore, the resulting channel gain is a random variable. It is typically modeled by Rayleigh distribution or Ricean distribution, as explained below. This section lists only the results. More details in derivation can be found in the literature [26, sec. 2.1.3].

First, let us consider the case without direct paths. The channel due to scattering can be expressed as

$$g = \sum_i u_i, \tag{6.22}$$

where g is the total channel gain, and u_i is the channel gain contributed by the scattering path i. They are considered as complex-valued random numbers with independent and identically distributed (iid). With a large number of scatterers and from the theorem of large numbers, both the real and imaginary parts of g obey independent and identical normal distribution. In other words

$$\begin{gathered} g = g_R + jg_I \\ g_R, g_I \sim N(0, \sigma^2) \end{gathered}, \tag{6.23}$$

where g_R and g_I are real numbers representing the real and imaginary parts of $g \cdot N(0, \sigma^2)$ denote normal distribution with mean 0 and variance σ^2. We can write g in terms of its modulus and phase:

$$g = re^{j\theta}. \tag{6.24}$$

It can be shown that in this case, θ obeys a uniform distribution:

$$p(\theta) = \frac{1}{2\pi}, \theta \in [0, 2\pi). \tag{6.25}$$

r obeys the Rayleigh distribution with the following pdf:

$$p(r) = \frac{r}{\sigma^2} \exp\left(-\frac{r^2}{2\sigma^2}\right) = \frac{2r}{\Omega_s} \exp\left(-\frac{r^2}{\Omega_s}\right) \tag{6.26}$$

for any value $r \geq 0$. Here, σ^2 and $\Omega_s \overset{\text{def}}{=} 2\sigma^2$ are related to the average channel power gain:

$$E(r^2) = 2\sigma^2 = \Omega_s. \tag{6.27}$$

Channels whose gains obey (6.25) and (6.26) are known as the Rayleigh fading channels, which is a good approximation for scattering-rich channels without a direct (i.e., line of sight) path.

When the direct path is present, (6.23) should be modified so that g_R and g_I have nonzero means reflecting the contribution from the direct path. In terms of modulus r and phase θ, the joint distribution can be written as [26, sec. A.3.2.5]

$$p(r,\theta) = \frac{r}{2\pi\sigma^2}\exp\left(-\frac{r^2+u_d^2}{2\sigma^2}\right)\exp\left(-r\frac{m_1\cos(\theta)+m_2\sin(\theta)}{\sigma^2}\right), \tag{6.28}$$

where $2\sigma^2$ is the average scattering channel power gain in (6.27). u_d^2 is the channel power gain of the direct path; m_1 and m_2 are the real and imaginary parts of the direct path channel gain.

$$u_d^2 = m_1^2 + m_2^2. \tag{6.29}$$

The marginal pdf for r is the Ricean distribution:

$$p(r) = \frac{2r(K+1)}{\Omega}\exp\left[-K-\frac{(K+1)r^2}{\Omega}\right]I_0\left(2x\sqrt{\frac{K(K+1)r}{\Omega}}\right), r \geq 0. \tag{6.30}$$

I_0 is the zero-order modified Bessel function of the first kind:

$$I_0(x) \overset{\text{def}}{=} \frac{1}{2\pi}\int_0^{2\pi}\exp[x\cos(\theta)]\,d\theta. \tag{6.31}$$

Ω specifies the channel gain. K specifies the relative strength of the direct path. More specifically,

$$K \overset{\text{def}}{=} \frac{u_d^2}{2\sigma^2}. \tag{6.32}$$
$$\Omega \overset{\text{def}}{=} u_d^2 + 2\sigma^2$$

The average channel power gain is

$$E\left(r^2\right) = \Omega. \tag{6.33}$$

K can change from 0 (no direct path) to infinity (no scattering). Rayleigh fading is a special case of Ricean fading with $K = 0$.

The Rayleigh fading and Ricean fading are broadly used in the literature to represent scattering-rich channel conditions. They are mathematically tractable (especially the Rayleigh fading), easy to generate in simulations, and realistic.

Another commonly used channel model is the Nakagami fading [26, sec. 2.1.3.3]. The channel modulus pdf takes the form

$$p(x) = 2\left(\frac{m}{\Omega}\right)^m\frac{x^{2m-1}}{\Gamma(m)}\exp\left(-\frac{mx^2}{\Omega}\right), m \geq \frac{1}{2}, r \geq 0. \tag{6.34}$$

With the additional parameter m, Nakagami model can be used to better fit imperial (i.e., measured) data. Equation (6.33) still holds for Nakagami fading. Ricean distribution (including its special case Rayleigh distribution) can be well approximated by Nakagami distribution with the following choice of m:

$$m \approx \frac{(K+1)^2}{2K+1}. \tag{6.35}$$

Approximating Ricean fading with (6.34) may be advantageous because it avoids the computation of the Bessel function in the simulation.

Multipath fading models discussed here address short-range fading caused by scatterers (see Section 6.4.1.3). They need to be combined with other fading mechanisms to form total channel gain. Such a combination can be achieved through standard statistics techniques [22].

On the other hand, since the three fading mechanisms operate at different scales, we often need to consider only one of them in a particular problem. A large volume of wireless communications research work considers only the Rayleigh or Ricean fading, with the average channel power gain normalized to 1.

6.4.3 Time Dependency

Section 6.4.2 discussed channel fading in terms of channel gain at a given point in time. In this section, we study how channel gain changes with time. As stated at the beginning of Section 6.4, we focus on the effect of transmitter and receiver motion. Such motion converts spatial variation of the channel gain to time variation. Conventionally, such motion is captured by the Doppler frequency spread.

To understand the relationship between channel fading and Doppler frequency, let us look at a simple case as shown in Figure 6.7. A scatterer is located at the origin, while the transmitter (Tx) and receiver (Rx) are located at the same y coordinate. The receiver moves to the right with speed v, while the transmitter is stationary. The angle between v and the connection between scatterer and Rx is θ. Since the receiver is in

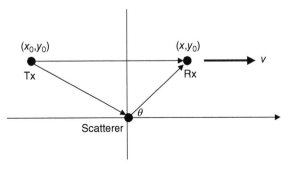

Figure 6.7   Receiver in motion.

motion, its received signal changes with time. Such change can be viewed in two ways. First, we can look at the result of interference between the direct and scattered paths, which varies with position. Therefore, as the receiver moves through various positions, it experiences different channel gains. Alternatively, we can use the receiver as a reference frame. From this point of view, signals from the two paths have different Doppler shifts. Such relative frequency difference creates a "beat" at the receiver. In the following, we formulate both views in more details.

Let us first consider spatial interference. The path length difference s between the direct and scattering paths can be expressed as

$$s = (x - x_0) - \left[\sqrt{x_0^2 + y_0^2} + \sqrt{x^2 + y_0^2} \right]. \tag{6.36}$$

When the receiver moves by a small displacement Δx, s changes by

$$\Delta s = \Delta x \frac{ds}{dx} = \Delta x \left(1 - \frac{x}{\sqrt{x^2 + y_0^2}} \right) = \Delta x \left[1 - \cos(\theta) \right]. \tag{6.37}$$

Channel fading goes through a cycle when

$$\Delta s = \lambda, \tag{6.38}$$

where λ is the signal wavelength. Therefore, the fading period is

$$t_{c1} = \left. \frac{\Delta x}{v} \right|_{\Delta s = \lambda} = \frac{\lambda}{v[1 - \cos(\theta)]}. \tag{6.39}$$

Now, let us consider the alternative view, using Doppler shift. From Doppler shift theory, the frequency shift Δf is given by

$$\Delta f = f \frac{v}{c} \cos(\alpha), \tag{6.40}$$

where f is the signal frequency. c is the speed of light. v is the object motion speed. α is the angle between the motion speed and the direction of the signal. Let the Doppler frequency shift for the direct path and scattering path be Δf_d and Δf_s, respectively. We thus have

$$\Delta f_d = f \frac{v}{c}$$
$$\Delta f_s = f \frac{v}{c} \cos(\theta) \tag{6.41}$$

The beating frequency when the two paths meet is $\Delta f_d - \Delta f_s$. The fading period is thus

$$t_{c2} = \frac{1}{\Delta f_d - \Delta f_s} = \frac{c}{f} \frac{1}{v[1-\cos(\theta)]} = \frac{\lambda}{v[1-\cos(\theta)]}. \qquad (6.42)$$

Therefore, the two methods yield the same results. Namely, the difference in Doppler frequencies is the frequency of fading.

Doppler frequency spread is a commonly used parameter to characterize motion because it combines carrier frequency and motion speed into a single parameter.[5] In the presence of multiple scatterers, the multiple paths have different Doppler frequencies. The average of them can be absorbed into the carrier frequency offset and be compensated at the receiver (see Chapter 7). Channel fading depends on the *spread* of Doppler frequencies, referred to as the *Doppler spread* of a channel.

Channel fading can be described in terms of channel gain autocorrelation over time, or by its Fourier transform. A commonly used model is the Jakes model, which was first proposed by Clarke [24, 27, 28]. This model is based on an idealized scenario, where many scatterers are uniformly distributed around a circle centered at the receiver, who has an isotropic antenna. It can be shown that the autocorrelation of channel gain for such a model is given by

$$r_{gg}(\tau) = 2\sigma_0^2 J_0(2\pi f_{max}\tau). \qquad (6.43)$$

Here, J_0 is the zeroth-order Bessel function of the first kind. σ_0^2 is proportional to the channel power gain. f_{max} is the maximum Doppler frequency shift among the scatters:

$$f_{max} = f\frac{v}{c}, \qquad (6.44)$$

where f is the carrier frequency. v is the speed of movement. c is the speed of light.

In the frequency domain, the power spectral density of the channel gain, that is, the Fourier transform of r_{gg}, is

$$S_{gg}(f) = \begin{cases} \dfrac{2\sigma_0^2}{\pi f_{max}\sqrt{1-(f/f_{max})^2}} & |f| \le f_{max} \\ 0 & |f| \ge f_{max} \end{cases}. \qquad (6.45)$$

---

[5] This does not mean the models are frequency independent, because scattering properties are usually frequency-dependent. However, sensitivity to frequency is greatly reduced after Doppler frequency is used to capture the scaling relationship between frequency and motion speed.

$S_{gg}(f)$ is also the power spectrum density as seen by the receiver when a single tone is transmitted, as shown in homework. Section 6.6 discusses more on the relationship between time domain and frequency domain expressions.

Figure 6.8 shows an example of Jakes model, with $f_{max} = 91\,\text{Hz}, \sigma_0^2 = 1$ [24]. The left side is the power spectral density. The right side is the autocorrelation function in the time domain. Note these plots are for correlations among the real parts of the channel gain. Therefore, the values are one half of those given by (6.43) and (6.45). Correlations among imaginary parts of the channel gain are the same. At 2 GHz carrier frequency, such Doppler shift corresponds to a speed of 49 km/h, or 30 mph.

(A)

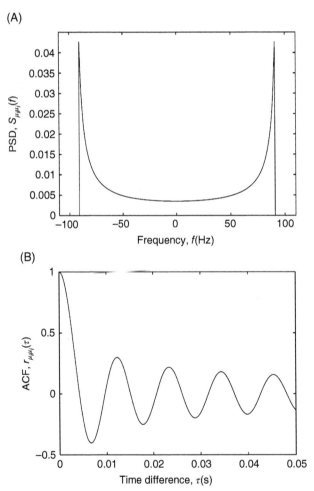

(B)

Figure 6.8   Channel gain autocorrelation in Jakes model: (A) frequency domain and (B) time domain [24, fig. 3.2]. Reproduced with permission.

Jakes model has been experimentally validated in scattering-rich outdoor mobile communication situations and is broadly used in the literature. However, if a direct path exists, corrections are needed.[6] Jakes model is quite practical in simulation, as well [29].

There are other well-accepted fading models. For example, aeronautical and some mobile channels can be modeled by a Gaussian model, whose autocorrelation is a Gaussian function [24].

As discussed in Section 6.1, in communication system design, it is essential to know how fast the channel changes. The speed of channel change can be qualitatively characterized by coherence time t_c. A channel is considered constant within a duration of t_c. The definition of coherence time varies in the literature, although many authors use the time at which the autocorrelation is 1/2 of its maximum value. Under such definition, the coherence time for the Jake's model given by (6.43) is [30, sec. 15.4]

$$t_c = \frac{9}{16\pi f_{max}}.$$ (6.46)

More generally, people often take

$$t_c = \frac{a}{f_{max}},$$ (6.47)

where a can take the value of 0.5 of 1. The relationship between coherence time and Doppler shift frequency can be understood based on Figure 6.7 and the accompanying discussions.

Let us consider an example in mobile communications. Suppose the maximum vehicle speed is 100 miles per hour. The carrier frequency is 1 GHz. In this case, the maximum Doppler shift is 149 Hz. Taking $\alpha = 1$ in (6.47), t_c is approximately 7 ms. These numbers can be compared with the design of LTE, the 4G cellular system indented for such mobile channels. In LTE, modulation and coding schemes (MCSs) are adaptive to channel condition. The MCS can be updated every subframe (1 ms), based on channel condition feedback. LTE transmission includes predetermined reference signals to help with channel estimation. Reference signals are transmitted in every slot, whose duration is 0.5 ms.

In addition to coherence time, other interesting channel characteristics include the duration of fades and level-crossing rates, which can also be derived from the statistical channel model [24]. Furthermore, some research works attempt to predict future channel gains based on the statistical models.

---

[6] When studying time correlation, people often talk about Ricean and Rayleigh cases, actually meaning with and without direct path, respectively. The time autocorrelation functions are not tied to specific statistics discussed in Section 6.4.2.

6.5 TIME DISPERSION AND FREQUENCY-SELECTIVE FADING

Sections 6.3 and 6.4 cover flat fading channels. Although there may be multiple scattering paths, signals through these paths arrive at the receiver at the same time. Therefore, the overall channel is "instantaneous," as described by (6.4). However, this is not always the case. When the signals from the multiple paths do not arrive at the same time, there is time dispersion in the channel. Namely, if an infinitely short pulse is transmitted, the receiver will get a signal with a finite duration, as described in (6.1), where the channel gain $h(t, \tau)$ is not a delta function of τ. Such time-dispersive channel is the focus of this section.

Strictly speaking, except for free space with infinite bandwidth, no channel can be described by (6.4). However, if the maximum delay (i.e., support of $h(t, \tau)$ with respect to τ) is within a symbol period, the receiver still experiences instantaneous channel. On the other hand, if the maximum delay is larger than the symbol period, (6.1) shows that signals belonging to the neighboring symbols are combined. Such a phenomenon is known as inter-symbol interference (ISI). In this case, channel time dispersion would be important. Therefore, time dispersion needs to be studied in the context of particular symbol periods. In addition to multiple scattering, time dispersion can also be caused by the transceiver RF circuit elements, such as filters.

6.5.1 Time Dispersion due to Multipath

In general, a multipath channel can be expressed as

$$h(t, \tau) = \sum_n s_n(t) \delta(\tau - \tau_n).$$ (6.48)

Here, the summation is over all paths. $s_n(t)$ is the strength at time t for path n. It is a complex number for the baseband model. τ_n is the time delay for path n, relative to the reference path. Typically, the reference path is the one with the least delay (e.g., it can be the direct path). In this case, all τ_n are nonnegative. Furthermore, there is a maximum delay t_p (also known as channel length or arrival time spread), so that

$$t_n \le t_p \forall n.$$ (6.49)

There are two extreme cases for time dispersion modeling. In the case of the scattering-rich environment, (6.1) can be approximated by a continuous function of τ, instead of a collection of delta functions. In the other extreme case, there are very few scatterers over a relatively long channel length t_p. In this case (known as a sparse channel), (6.48) remains an appropriate expression.

Some examples of dispersive channels can be found in [31–34].

Path strengths $\{s_n(t), n = 0, 1, \ldots\}$ are random processes. It is usually reasonable to assume $\{s_n(t)\}$ to be mutually independent among the different n values, because they represent different scatters. Namely, the autocorrelation of channel gain can be expressed as:

$$E[h(t_1,\tau_1)h^*(t_2,\tau_2)] = r_{hh}(t_1,t_2,\tau_1)\delta(\tau_1-\tau_2).\qquad(6.50)$$

Such a condition is known as uncorrelated scattering (US). Furthermore, in the short term, (i.e., no significant change in long-range path loss and shadowing), the channel can often be considered as wide-sense stationary (WSS)[7]:

$$E[h(t_1,\tau_1),h^*(t_2,\tau_2)] = \hat{r}_{hh}(t_1-t_2,\tau_1,\tau_2).\qquad(6.51)$$

A channel model that is both US and WSS is known as the WSSUS model [35], which is used almost exclusively for mobile radio channels [24]. For WSSUS channels, (6.50) and (6.51) lead to

$$E[h(t_1,\tau_1),h^*(t_2,\tau_2)] = \bar{r}_{hh}(t_1-t_2,\tau_1)\delta(\tau_1-\tau_2).\qquad(6.52)$$

Here, $\hat{r}$ and $\bar{r}$ are the reduced forms of the autocorrelation. We will return to this assumption in Section 6.6.

6.5.2   Frequency-Selective Fading and Coherence Bandwidth

In this section, we examine time dispersion in the frequency domain. For simplicity, we assume that the channel gain is independent of time. Namely, (6.1) becomes

$$y(t) = \int h(\tau)x(t-\tau)d\tau.\qquad(6.53)$$

This assumption is valid in our context when the channel coherence time (Section 6.4.3) is much larger than the channel length t_p and the passband signal period $1/f_c$. Time dependency will be examined together with time dispersion in Section 6.6.

Based on the property of convolution (see Chapter 3), (6.53) leads to

$$\tilde{y}(\omega) = \tilde{h}(\omega)\tilde{x}(\omega).\qquad(6.54)$$

Here, $\tilde{y}(\omega)$ and $\tilde{x}(\omega)$ are the Fourier transforms of $y(t)$ and $x(t)$, respectively. $\tilde{h}(\omega)$ is the Fourier transform of $h(\tau)$ with respect to τ. Fourier transform is specified in Chapter 3:

$$\tilde{f}(\omega) = \int f(t)\exp(-j\omega t)dt.\qquad(6.55)$$

---

[7] For more discussions on WSS, see Chapter 4.

Now let us consider the average frequency autocorrelation of $\tilde{h}(\omega)$:

$$C_{HH}(\omega) \overset{\text{def}}{=} E[h(\omega')h^*(\omega'-\omega)]. \tag{6.56}$$

As will be shown below, $C_{HH}(\omega)$ as defined in (6.56) is independent of ω'. Equations (6.53) and (6.55) lead to

$$
\begin{aligned}
C_{HH}(\omega) &= \int E[h(\tau_1)h^*(\tau_2)]\exp\{-j[\omega'(\tau_1-\tau_2)+\omega\tau_2]\}d\tau_1 d\tau_2 \\
&= \int E\left[|h(\tau_2)|^2\right]\exp[-j\omega\tau_2]\delta(\tau_1-\tau_2)d\tau_1 d\tau_2 \\
&= \int E\left[|h(\tau_2)|^2\right]\exp(-j\omega\tau_2)d\tau_2
\end{aligned}
\tag{6.57}
$$

The second equality above used the property of US channels (6.52).

In other words, $C_{HH}(\omega)$ is proportional to the Fourier transform of the delay profile $E[|h(\tau)|^2]$. Especially, if $C_{HH}(\omega)$ is a constant, then $E[|h(\tau)|^2]$ is a delta function, meaning there is no time dispersion. This is the reason why the channels discussed in Sections 6.3 and 6.4 are known as "*flat fading*" channels. The frequency autocorrelation of these channels is "flat," that is, constant across the frequency band of interest.

As an autocorrelation function, $C_{HH}(\omega)$ indicates how $\tilde{h}(\omega)$ changes with ω. When $r_{HH}(\omega)$ drops significantly at some ω, it means that channel frequency responses separated by ω tend to change independently. A "coherence bandwidth" can be introduced to indicate a frequency range where channel response correlate strongly. The coherence bandwidth B_C is defined such that [24, sec. 7.3.2.3]

$$|C_{HH}(B_C)| = \frac{1}{2}|C_{HH}(0)|. \tag{6.58}$$

Because of the property of Fourier transform, B_C can be used to estimate channel length t_p given by (6.49):

$$t_p \approx \frac{1}{B_C}. \tag{6.59}$$

As stated at the beginning of Section 6.5, a channel can be considered as flat fading if the channel length t_p is smaller than the symbol period. Considering the Nyquist theory (see Chapter 3), (6.59) shows that such a requirement for flat fading is equivalent to the requirement that the coherence bandwidth B_C is larger than the passband bandwidth. Therefore, the same physical channel may be flat for a narrow-band waveform, while being time dispersive for a broadband waveform.

6.6   CHANNEL FORMULATION IN FREQUENCY AND TIME DOMAINS

In Sections 6.3 and 6.4, the issues of time dependency and time dispersion of channel fading are examined in both time and frequency domains. Channel time dependency is described in the time domain by channel gain autocorrelation and in the frequency domain by Doppler frequency and the power spectral density (PSD) of channel gain (Section 6.4). Time dispersion is described in the time domain as channel delay profile, and in the frequency domain as autocorrelation of channel frequency response (Section 6.5). In this section, we introduce more general ways to describe channels in both domains [24]. These various forms are used in the literature. Therefore, it is helpful to understand how they are defined and how they relate to each other.

The expressions for Fourier transform (FT) and inverse Fourier transform (IFT) used in this section follow the convention in Chapter 3. ω and ν are *angular* frequencies corresponding to the time domain variables t and τ. Subscripts to FT and IFT notations are used to indicate the variables involved. For any function $g(t, \tau)$, we have the following definitions:

$$FT_t[g(t,\tau)] \overset{\text{def}}{=} \int g(t,\tau)\exp(-j\omega t)dt$$
$$FT_\tau[g(t,\tau)] \overset{\text{def}}{=} \int g(t,\tau)\exp(-j\nu\tau)d\tau$$
$$IFT_\omega[g(\omega,\nu)] \overset{\text{def}}{=} \frac{1}{2\pi}\int g(\omega,\nu)\exp(j\omega t)d\omega \qquad (6.60)$$
$$IFT_\nu[g(\omega,\nu)] \overset{\text{def}}{=} \frac{1}{2\pi}\int g(\omega,\nu)\exp(j\nu\tau)d\nu$$

Note that when FT or IFT is applied to one variable, the other one can be in either the frequency or the time domain. For example, based on the definition of FT_t above, we also have

$$FT_t[g(t,\nu)] \overset{\text{def}}{=} \int g(t,\nu)\exp(-j\omega t)dt. \qquad (6.61)$$

In the following expressions, the reader should pay attention to the distinction between the definitions and the derived equations.

Section 6.6.1 outlines the four versions of channel gain, based on the two pair of variables. Variables t and ω indicates channel time dependency in time and frequency domains. The other variable pair, τ and ν, indicates time dispersion in the two domains. Section 6.6.2 extends such duality to autocorrelation functions. In general, an autocorrelation function has four variables. However, for WSSUS channels, autocorrelation functions can take a reduced variable form, as shown in Section 6.6.3.

6.6.1 Channel Gain Expressions in Time and Frequency Domains

As stated in Section 6.2.1, the channel gain can be expressed in general as $h(t, \tau)$. Here, t indicates time variance, and τ is related to time dispersion. FT can be performed relative to t and τ separately:

$$
\begin{aligned}
s(\omega,\tau) &\stackrel{\text{def}}{=} FT_t[h(t,\tau)] \\
H(t,\nu) &\stackrel{\text{def}}{=} FT_\tau[h(t,\tau)]
\end{aligned}
\tag{6.62}
$$

which leads to the IFT relationship:

$$
h(t,\tau) = IFT_\omega[s(\omega,\tau)] = IFT_\nu[H(t,\nu)]. \tag{6.63}
$$

FT can also be performed to both t and τ:

$$
\begin{aligned}
T(\omega,\nu) &\stackrel{\text{def}}{=} FT_t\{FT_\tau[h(t,\tau)]\} \\
&= FT_\tau[s(\omega,\tau)] \\
&= FT_t[H(t,\nu)]
\end{aligned}
\tag{6.64}
$$

Consequently,

$$
\begin{aligned}
h(t,\tau) &= IFT_\omega\{IFT_\nu[T(\omega,\nu)]\} \\
s(\omega,\tau) &= IFT_\nu[T(\omega,\nu)] \\
H(t,\nu) &= IFT_\omega[T(\omega,\nu)]
\end{aligned}
\tag{6.65}
$$

The four quantities h, s, H, and T are equivalent ways to express the channel gain. Their relationships can be shown in Figure 6.9. For example, the upper-right corner shows that $h(t, \tau)$ and $H(t, \nu)$ can be transformed to each other through FT_τ and IFT_ν. The middle column shows that $h(t, \tau)$ and $T(\omega, \nu)$ can be transformed into each other through two consecutive FTs and two consecutive IFTs.

The four expressions of channel gains h, s, H, and T are referred to as *time-variant impulse response*, *Doppler-variant impulse response*, *time-variant transfer function*, and *Doppler-variant transfer function*, respectively.

6.6.2 Autocorrelation Functions

As discussed in Sections 6.4 and 6.5, the most interesting statistical property for channel gains is the autocorrelation across time or frequency. Therefore, with the four

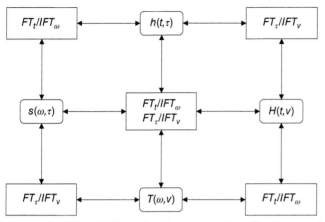

Figure 6.9   FT and IFT relationships among channel gain expressions. The quantities in the rounded boxes are given by (6.62)–(6.66). FT_x and IFT_x in the boxes mean Fourier transform and inverse Fourier transform with regard to variable x.

forms of channel gain expressions in time and frequency domains, four forms of auto-correlation can be defined:

$$r_{hh}(t_1,t_2,\tau_1,\tau_2) \overset{\text{def}}{=} E[h^*(t_1,\tau_1)h(t_2,\tau_2)]$$

$$r_{HH}(t_1,t_2,\nu_1,\nu_2) \overset{\text{def}}{=} E[H^*(t_1,\nu_1)H(t_2,\nu_2)]$$

$$r_{ss}(\omega_1,\omega_2,\tau_1,\tau_2) \overset{\text{def}}{=} E[s^*(\omega_1,\tau_1)s(\omega_2,\tau_2)]$$

$$r_{TT}(\omega_1,\omega_2,\nu_1,\nu_2) \overset{\text{def}}{=} E[T^*(\omega_1,\nu_1)T(\omega_2,\nu_2)]$$

(6.66)

Since autocorrelation functions involve two time or frequency variables for each category, we defined the "double FT" and "double IFT" as follows:

$$FT_{t_1 t_2}[g(t_1,t_2,\tau_1,\tau_2)] \overset{\text{def}}{=} \iint g(t_1,t_2,\tau_1,\tau_2)\exp(-j\omega_1 t_1 - j\omega_2 t_2)dt_1 dt_2$$

$$FT_{\tau_1 \tau_2}[g(t_1,t_2,\tau_1,\tau_2)] \overset{\text{def}}{=} \iint g(t_1,t_2,\tau_1,\tau_2)\exp(-j\nu_1 \tau_1 - j\nu_2 \tau_2)d\tau_1 d\tau_2$$

$$IFT_{\omega_1 \omega_2}[g(\omega_1,\omega_2,\nu_1,\nu_2)] \overset{\text{def}}{=} \frac{1}{(2\pi)^2}\iint g(\omega_1,\omega_2,\nu_1,\nu_2)\exp(j\omega_1 t_1 + j\omega_2 t_2)d\omega_1 d\omega_2$$

$$IFT_{\nu_1 \nu_2}[g(\omega_1,\omega_2,\nu_1,\nu_2)] \overset{\text{def}}{=} \frac{1}{(2\pi)^2}\iint g(\omega_1,\omega_2,\nu_1,\nu_2)\exp(j\nu_1 \tau_1 + j\nu_2 \tau_2)d\nu_1 d\nu_2$$

(6.67)

Autocorrelation functions defined in (6.66) are actually connected through these transforms. For example,

$$r_{HH}(t_1,t_2,\nu_1,\nu_2) = E\left[\iint h^*(t_1,\tau_1)h(t_2,\tau_2)\exp(j\nu_1\tau_1 - j\nu_2\tau_2)d\tau_1 d\tau_2\right]$$
$$= \iint E[h^*(t_1,\tau_1)h(t_2,\tau_2)]\exp(j\nu_1\tau_1 - j\nu_2\tau_2)d\tau_1 d\tau_2$$

(6.68)

Therefore,

$$r_{HH}(t_1,t_2,-\nu_1,\nu_2) = FT_{\tau_1\tau_2}\left[r_{hh}(t_1,t_2,\tau_1,\tau_2)\right].$$

(6.69)

Similarly,

$$r_{ss}(-\omega_1,\omega_2,\tau_1,\tau_2) = FT_{t_1t_2}\left[r_{hh}(t_1,t_2,\tau_1,\tau_2)\right]$$
$$r_{TT}(-\omega_1,\omega_2,\nu_1,\nu_2) = FT_{t_1t_2}\left[r_{HH}(t_1,t_2,\nu_1,\nu_2)\right]$$
$$r_{TT}(-\omega_1,\omega_2,-\nu_1,\nu_2) = FT_{\tau_1\tau_2}\left[r_{ss}(-\omega_1,\omega_2,\tau_1,\tau_2)\right]$$

(6.70)

Note that some frequency domain variables have negative signs, because of the complex conjugates in definitions (6.66). These relationships are summarized in Figure 6.10.

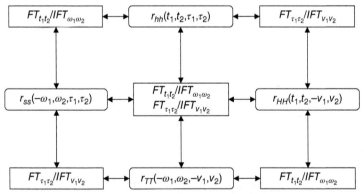

Figure 6.10 FT and IFT relationships among autocorrelation functions. The quantities in the rounded boxes are given by (6.67). FT_x and $IFT_{x,y}$ in the boxes mean Fourier transform and inverse Fourier transform with regard to variables x and y, as described in (6.69) and (6.70).

6.6.3   Autocorrelation Functions for WSSUS Channels

The time domain autocorrelation function for WSSUS channels is given by (6.52) in Section 6.5.1, which is rewritten with the current notations[8]:

$$r_{hh}(t_1,t_2,\tau_1,\tau_2) = \bar{r}_{hh}(t_2-t_1,\tau_1)\delta(\tau_1-\tau_2) \qquad (6.71)$$

From such form and (6.69)–(6.70), other forms of the autocorrelation function can be expressed.

$$
\begin{aligned}
r_{ss}(\omega_1,\omega_2,\tau_1,\tau_2) &= \iint r_{hh}(t_1,t_2,\tau_1,\tau_2)\exp(j\omega_1 t_1 - j\omega_2 t_2)dt_1 dt_2 \\
&= \int \bar{r}_{hh}(t,\tau_1)\exp(-j\omega_2 t)dt \int \exp[-j(\omega_2-\omega_1)t_1]dt_1 \delta(\tau_1-\tau_2), \\
&= 2\pi\bar{r}_{ss}(\omega_2,\tau_1)\delta(\omega_1-\omega_2)\delta(\tau_1-\tau_2)
\end{aligned}
$$
$$(6.72)$$

where $\bar{r}_{ss}$ is the FT of $\bar{r}_{hh}$ as defined in (6.60):

$$\bar{r}_{ss}(\omega,\tau) \overset{\text{def}}{=} FT_t[\bar{r}_{hh}(t,\tau)]. \qquad (6.73)$$

In the second equation in (6.72), a new variable $t \overset{\text{def}}{=} t_2 - t_1$ was introduced to replace t_2 in integration. Similarly,

$$
\begin{aligned}
r_{HH}(t_1,t_2,\nu_1,\nu_2) &= \iint r_{hh}(t_1,t_2,\tau_1,\tau_2)\exp(j\nu_1\tau_1 - j\nu_2\tau_2)d\tau_1 d\tau_2 \\
&= \iint \bar{r}_{hh}(t_2-t_1,\tau_1)\delta(\tau_1-\tau_2)\exp(j\nu_1\tau_1 - j\nu_2\tau_2)d\tau_1 d\tau_2, \\
&= \bar{r}_{HH}(t_2-t_1,\nu_2-\nu_1)
\end{aligned}
$$
$$(6.74)$$

where $\bar{r}_{HH}$ is the FT of $\bar{r}_{hh}$ as defined in (6.60):

$$\bar{r}_{HH}(t,\nu) \overset{\text{def}}{=} FT_\tau[\bar{r}_{hh}(t,\tau)]. \qquad (6.75)$$

Continuing to the all-frequency domain:

$$
\begin{aligned}
r_{TT}(\omega_1,\omega_2,\nu_1,\nu_2) &= \iint r_{HH}(t_1,t_2,\nu_1,\nu_2)\exp(j\omega_1 t_1 - j\omega_2 t_2) \\
&= \iint \bar{r}_{HH}(t_2-t_1,\nu_2-\nu_1)\exp(j\omega_1 t_1 - j\omega_2 t_2)dt_1 dt_2, \\
&= 2\pi\bar{r}_{TT}(\omega_2,\nu_2-\nu_1)\delta(\omega_1-\omega_2)
\end{aligned}
$$
$$(6.76)$$

where $\bar{r}_{TT}$ is the FT of $\bar{r}_{SS}$ and $\bar{r}_{ss}$:

$$\bar{r}_{TT}(\omega,\nu) \overset{\text{def}}{=} FT_t[\bar{r}_{HH}(t,\nu)] = FT_\tau[\bar{r}_{ss}(\omega,\tau)]. \qquad (6.77)$$

---

[8] r_{hh} and $\bar{r}_{hh}$ were introduced in (6.50) and (6.52) in Section 6.5.1.

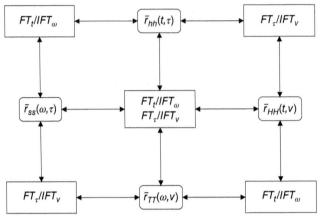

Figure 6.11   FT and IFT relationships among reduced-variable autocorrelation functions. The quantities in the rounded boxes are given by (6.71)–(6.77). FT_x and IFT_x in the boxes mean Fourier transform and Inverse Fourier transform with regard to variables x.

In summary, for WSSUS channels, the autocorrelation functions can be expressed with reduced variable sets:

$$r_{hh}(t_1,t_2,\tau_1,\tau_2) = \bar{r}_{hh}(t_2-t_1,\tau_1)\delta(\tau_1-\tau_2)$$
$$r_{ss}(\omega_1,\omega_2,\tau_1,\tau_2) = 2\pi\bar{r}_{ss}(\omega_2,\tau_1)\delta(\omega_1-\omega_2)\delta(\tau_1-\tau_2)$$
$$r_{HH}(t_1,t_2,\nu_1,\nu_2) = \bar{r}_{HH}(t_2-t_1,\nu_2-\nu_1)$$
$$r_{TT}(\omega_1,\omega_2,\nu_1,\nu_2) = 2\pi\bar{r}_{TT}(\omega_2,\nu_2-\nu_1)\delta(\omega_1-\omega_2)$$

$$(6.78)$$

The reduced variable functions $\bar{r}_{hh}$, $\bar{r}_{ss}$, $\bar{r}_{HH}$, and $\bar{r}_{TT}$ are connected by FT and IFT, as shown in Figure 6.11. They are, respectively, referred to as delay cross-power spectral density, scattering function, time-frequency correlation function, and Doppler cross-power spectral density [24, sec. 7.3.2.3].

These reduced forms of autocorrelation functions are often used in the literature for characterizing the channel. Therefore, it is important to know their connections with each other (Figure 6.11) and their connections with the full-form autocorrelation functions (6.78).

6.7   CHANNEL MODELING METHODS

As shown in Sections 6.3 – 6.6, channel gain $h(t, \tau)$ depends on many factors. In communications system design, we need to have some information about the expected channel gain. Such information is commonly referred to as channel models. Channel models have varying levels of fidelity. For example, the various models discussed in

Sections 6.3 and 6.4 usually have general forms with a few parameters, whose values depend on the general characteristics of the environment such as urban/rural/hilly, or indoor/outdoor, as well the frequency band. A higher fidelity can be achieved when more specifics of the environment, such as the building layout in an urban area, is known. Depending on the specific concerns in system design, various approaches with various fidelity levels can be chosen.

6.7.1   Two Approaches to Channel Modeling

From the first principle, channel gain can be obtained by electromagnetic computations. If the wavelength is long comparing to the obstacles, reflectors, and scatterers, we need to consider diffraction. In this case, solving the wave equations is advisable. Such an approach is often referred to as physics-based channel models. For shorter wavelengths, ray tracing is often used. In ray-tracing models, individual scattering paths are evaluated, and the results are added together at the receiver's position. Both methods are based on specific positions and properties (such as dielectric constants) of all objects in the propagation path. Such information can be collected for specific scenarios or generated with some randomness to represent a generic situation.

Another approach is empirical, based on actual measurements. Statistical properties such as mean, standard deviation, and regression fitting can be obtained from measurement results. Alternatively, parameters in simple channel models introduced in Sections 6.3 and 6.4 can be fitted based on measured data.

The two approaches can be further combined in channel model construction. For example, in ray-tracing methods, path loss along each ray can be computed using empirical-based formulas.

6.7.2   Channel Modeling Resources

Channel construction and characterization can be a substantial undertaking. Existing results in the literature can often be leveraged. In addition to research papers reporting channel measurement campaigns and channel models, various standard bodies often provide detailed recommendations on channel models used for specific systems and applications. ITU has a series of channel model recommendations [18, 36]. Especially, there is an ITU recommendation on 3G cellular system channel models in various scenarios [37]. Another standard organization the 3rd Generation Partnership Project (3GPP) recommends a ray tracing–based spatial channel model for cellular systems [38, 39]. The Institute of Electrical and Electronics Engineers (IEEE) also publishes channel models for the various wireless local area network (WLAN) systems. For example, several channel models for 802.11n (high throughput mode WLAN) are outlined in [40]. A MATLAB-based simulation package for these models is available for free [41, 42]. Most modern standards recommend MIMO channel models for transmitters and receivers with multiple antennas, of which the point-to-point channel discussed in this chapter can be a special case. MIMO technologies and channels are further discussed in Chapter 11. For wired channels, telephone wires have been studied in much detail [43]. Relevant cable models for Ethernet can also be found in IEEE 802 standards committee documents.

Also, there are many commercial or free channel modeling and simulation tools available. As practicing communications engineers, it is rarely necessary or feasible to develop a channel model from scratch. However, to utilize available channel modeling resources, we need to develop the following skills.

- Have a sense of what aspects of the channel model are important to the task. Is the mean path loss enough? Do we need to know channel time dependency (or equivalently Doppler spread)? Do we need to know channel delay spread (or equivalently frequency response)?

- Determine the level of fidelity required for the channel models. Are generic models such as free space, Rayleigh fading, or Ricean fading good enough? Are scenario-specific detailed channel models necessary or helpful?

- Determine the various factors (such as forage, moisture, building blocking) to be included in the model. Such a choice depends on the surrounding, the frequency used, and the accuracy desired.

- Select channel models based on considerations of applicability, level of verification demonstrated, ease of use, and so on.

- Have an intuition on the expected results (e.g., the path loss for a particular link should be around X dB). When using channel models, one often encounters a large set of input parameters and many nuances on selecting their values. Therefore, an intuitive "sanity check" of the results is indispensable.

These skills are best cultivated through experience, and through consultation of subject matter experts.

Usually, antenna characteristics are not a part of the channel model. As to be discussed in Section 6.8, antenna gains are combined with path loss to compute received signal power. However, in scattering channels, the number of scatters that the radio wave "sees" depends on antenna beam width or directivity. Therefore, when selecting channel models, one should pay attention to the underlying assumption of antenna type.

There are many good channel model reviews in the literature [44]. They can serve as a starting point when searching for a suitable channel model.

6.7.3   Channel Emulation

When developing a communication system, we often need to perform lab tests that include the effects of a channel. In this case, channel gain needs to be applied to the RF signal path in real time. Such operation is known as channel emulation. There are various channel emulator products on the market, covering various frequency bands. Typically, a channel emulator supports several channel models specified by standard groups and allows the user to set certain parameters such as Doppler shift. Some include additional impairments such as noise and timing errors. More advanced channel emulators allow the user to load captured or simulated channel fading sequences.

Channel emulators are usually expensive. An alternative is building one with standard digital signal processing (DSP) platform (often called "testbed" in the literature). A software-defined radio (SDR) can be used as the RF front end, to convert RF signals to baseband digital samples. Channel effects can be applied to these samples through DSP computations. The results are converted back to RF with an SDR. Such a solution is usually much cheaper than an off-the-shelf channel emulator is. However, depending on the quality of modules, unintended degradations such as timing jitter and noise may be introduced. Also, considerable development time is needed, especially if sophisticated standard channel models are to be implemented. There are many such works reported in the literature. Open source DSP codes are also available online.

6.8 LINK BUDGET COMPUTATION

In this section, we cover the basics of link budget computation. Link budget involves computing the expected SNR at the receiver and comparing it with what is required for proper receiver operations. The link is said to be "closed" when the expected SNR can satisfy receiver operational requirements.

6.8.1 Link Budget Overview

Figure 6.12 shows a typical signal chain of wireless communication systems. The transmitted signal has power P_t. Such signal is transmitted through an antenna, which has a gain of G_t. The transmitted signal passes the channel and suffers a loss L. At the receiver, the signal is received by the antenna with a gain of G_r and has power P_r at antenna output. Noise is equivalently injected at this point, influencing the signal-to-noise ratio (SNR). This SNR is approximately kept constant as the signal moves further into the receiver processing modules (more about this in Section 6.8.4.5). Figure 6.12 is arguably the simplest model for the signal chain. Some variations will be shown later. Furthermore, additional signal losses, such as cable losses, may be included in the model. For wired systems, a similar architecture applies without the antennas.

It is easy to see the relationship

$$P_r = P_t \frac{G_t G_r}{L}. \tag{6.79}$$

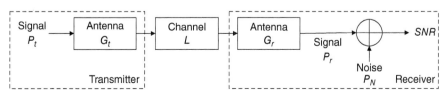

Figure 6.12   Wireless signal chain.

The receiver SNR is

$$SNR = \frac{P_r}{P_N},\qquad(6.80)$$

where P_N is the noise power. The main goal of the link budget computation is to assess SNR at the receiver. Channel path loss L has been discussed in Section 6.3. Other variables in (6.79) and (6.80) will be discussed in Sections 6.8.2 – 6.8.4.

6.8.2   Transmitted Power

Transmitted power P_t is the power at the input of the transmitting antenna. It can be expressed in units of W or mW. In communications, transmitted power is also expressed in dBW or dBm. For the same transmitted power, the value in dBW is 10 times the base-10 logarithm of the value in W. For example, a transmitted power of 100 W can be expressed as 20 dBW. Units dBm and mW have the same relationship.

Furthermore, transmitted power is sometimes expressed in power spectral density (PSD) in units such as W/Hz or dBW/Hz.

Transmitted power can be combined with the antenna gain to form equivalent isotropic radiation power (EIRP).

$$EIRP \overset{\text{def}}{=} P_t G_t.\qquad(6.81)$$

G_t, and therefore EIRP, depends on the spatial orientation of the link unless the antenna is isotropic.

Obviously, a higher transmitted power improves signal quality. However, in practice, transmitted power is limited by a number of factors.

On the hardware side, transmitted power is limited by the maximum power output of the high-power amplifier (HPA) at the transmitter. HPA is the last-stage amplifier before the transmitting antenna. An HPA capable of higher output power usually has larger size, weight, and power consumption, and is more expensive. Furthermore, as actual transmitted power approaches the limit of an HPA, the HPA tends to exhibit higher levels of nonlinearity. Nonlinearity causes higher out-of-band transmission, which interferes with communications in the neighboring frequency channels. Nonlinearity also introduces distortion that degrades signal quality at the receiver end. Different types of modulation have different levels of tolerance of nonlinearity, resulting in different maximum transmitted power using the same HPA. A rule of thumb is that modulations with less peak to average ratio (PAR), such as phase shift keying (PSK) modulations, work better under HPA nonlinearity (see Chapter 3).

Transmitted power is also limited by regulations, in order to control mutual interference among users. Limits of transmitted power in various frequency bands are a part of the spectrum management framework. In the United States, the Federal

Communications Commission (FCC) is in charge of spectrum management. FCC regulations are contained in Title 47 of the Federal Regulation [45].

In the FCC regulation framework, the frequency bands are divided into licensed bands and unlicensed bands. Users of a licensed band must obtain authorization from FCC for the geographical region of operation and adhere to the corresponding power limitation. Cellular systems and TV broadcast systems operate in licensed bands. Unlicensed bands are open to anyone without requiring authorization. However, the devices still need to meet certain FCC requirements, most importantly transmitted power and bandwidth. The wireless local area network (WLAN), also known as WiFi, operates in unlicensed bands. Other systems in unlicensed bands include Bluetooth and cordless phones.

Typically, much higher transmitted power levels are allowed in licensed bands. Two examples are provided below.[9] The Personal Communications Services (PCS) band between 1850 and 1990 MHz is a licensed band allocated for 2G and 3G cellular systems (47 C.F.R. §24.229). Maximum transmitted power (in EIRP) for broadband base stations in PCS band is limited to 1640 W/MHz for antennas less than 300 m in height. Mobile units are allowed for 2 W of maximum transmitted power (47 C.F.R. §24.232). On the other hand, the Unlicensed National Information Infrastructure (U-NII) is an unlicensed band. The frequency allocation is 5.15–5.35 GHz and 5.47–5.85 GHz (47 C.F.R. §15.401). The power limits for U-NII bands are specified in 47 C.F.R. §15.407. For devices with 20 MHz bandwidth, transmitted power is limited to 1 W for 5.15–5.25 GHz and 5.725–5.85 GHz, and 0.25 W for 5.25–5.35 GHz and 5.47–5.725 GHz.

Furthermore, transmitted power may be limited by total power consumption, especially with battery-powered devices. To save power consumption, many systems use adaptive power setting. The transmission is turned off when there is no data to send. When data traffic is low, transmitted power and data rate may be reduced, as well.

6.8.3 Antenna Gains

Antenna gains are commonly specified as dB (isotropic), or dBi, which is the gain compared with the hypothetical ideal isotropic antenna. As described in Section 6.2.1, an ideal isotropic antenna converts all transmitted power to radiation power, which is uniformly distributed over all directions. As a receiver antenna, the ideal isotropic antenna collects power through an aperture, whose area is given by (6.2). Antenna gain is usually a function of the propagation direction. Such a function is known as the gain pattern. The direction with maximum gain is often referred to as the main lobe or the boresight, depending on the type of the antenna. If an antenna gain is given without specifying the direction, it usually means the maximum gain (i.e., gain at the main lobe). A normalized gain pattern is often used as well, where the antenna gain at various directions is divided by the maximum gain.

---

[9] Note that these descriptions are highly simplified to highlight the difference in power limit. For more details, consult the regulations cited.

Because of the reciprocal principle, an antenna has the same transmitting and receiving gains and gain patterns at a given frequency [46, sec. 2.13].

Based on the conservation of energy, if the antenna gain is larger than 1 at some directions, it must be smaller than 1 in some other directions. Namely, the antenna gain is connected with directivity, which is a measure of how radiation is "focused." High gain antennas usually have high directivity. Namely, radiation energy is concentrated within a small solid angle in space. On the other hand, there are antennas with gain less than 1 in all directions, due to inefficient coupling and radiation.

A common high gain antenna is the dish antenna, or parabolic antenna, which is commonly used for satellite and microwave communications. If the wavelength is λ and the antenna diameter is D, then the boresight gain of a dish antenna is [47, sec. 3.4]

$$G = \frac{\pi^2 D^2}{\lambda^2} \eta, \tag{6.82}$$

where η is an efficiency factor, typically between 0.5 and 0.7. Note that the first factor in (6.82) is the ratio between the antenna area $\pi D^2/4$ and that of the isotropic aperture given by (6.2) [48]. The half-beam width (defined as the angle off the boresight, where the gain is 3 dB below the maximum), in degrees, is approximately [47, sec. 3.4]

$$BW = \frac{70\lambda}{D}. \tag{6.83}$$

The horn antennas are another type of high gain antennas with similar gain values, given the aperture area [49, ch. 7]. Dish and horn antennas usually have large sizes and can achieve high gain values. For example, a commercial direct TV satellite dish has a diameter of 1 m, with receiving frequency of 12.2 GHz. Assuming the efficiency η to be 0.5, the antenna gain is 39 dBi according to (6.82).

For dish and horn antennas, one is free to choose antenna size for a given frequency. As shown above, a larger antenna size leads to a higher gain and a smaller beam width.

Another category of antennas includes dipole and monopole antennas, among others [48, ch. 5]. For these antennas, the gain and the beam pattern are independent of frequency, while the antenna size scales with wavelength. These antennas consist of one or two pieces of wires, sometimes adjacent to a ground plane. The radiation is slightly concentrated at the plane perpendicular to the wire. Therefore, a gain of a few dB can be obtained, as shown in Table 6.2 [50].

The third category, the electrically small antennas, has overall sizes less than one-quarter wavelength [51, ch. 6]. They are broadly used in miniature electronic devices. These antennas may have gains significantly smaller than 1 (or 0 dB).

TABLE 6.2   Gain of typical wire antennas [50].

Type	Hertz dipole	Half-wave dipole	Monopole on ground plane	Quarter wave monopole on ground plane
Gain (dBi)	1.75	2.15	4.8	5.2

Sometimes, antenna gains are defined relative to other references such as dipole antennas. Some examples are listed in [50]. However, dBi is by far the common metric.

6.8.4   Noise and SNR

In Sections 6.8.2 and 6.8.3, transmitted power and antenna gain are considered. These results, together with the path loss discussed in Section 6.3, enable us to compute the received power from (6.79). In this section, we consider the other factor in (6.80): the noise.

Figure 6.13 shows the model of noise insertion. The upper part shows a relatively detailed model. The signal comes from the left end. N_e is the external noise picked up by the antenna (Section 6.8.4.4). N_T is the thermal noise at the receiver load (Section 6.8.4.1). The boxes represent processing stages. Stage m introduces additional noise, which is represented by the noise power spectral density N_m at its input (Section 6.8.4.2). State m also has a gain G_m. A stage can represent various objects, such amplifier (with $G_m > 1$), connecting cable (with $G_m < 1$, $N_m = 0$), mixer (with $G_m = 1$), and so on. As will be shown in Section 6.8.4.5, typically only the front stages have a significant impact to the final signal-to-noise ratio (SNR). The lower part of Figure 6.13 shows an equivalent circuit, where all noise contributions are aggregated into one source with power N_S, also known as system noise (Section 6.8.4.5).

From the receiver point of view, the achievable SNR is affected by both noise level and antenna gain. In some cases, it is more convenient to combine the two quantities, as described in Section 6.8.4.3.

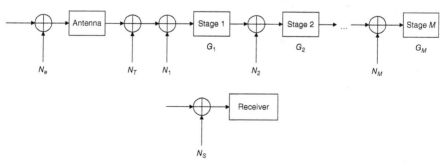

Figure 6.13   Receiver chain with noise insertions.

We would like to lower the noise level for better performance. Some noises (such as external noise and thermal noise) are difficult to change. Some (e.g., noises introduced by analog components) can be reduced to some extent, usually as a trade-off with other system parameters such as cost and size. Other noises (such as quantization noise) can be practically eliminated if higher costs are acceptable. On the other hand, the total noise power is a linear sum of all contributors. Therefore, the reduction of noise power from major contributors is the focus when system performance is concerned.

It should be noted that this section does not cover all possible noise sources. While only additive noises are considered in this section, some receiver noises such as phase noise and synchronization errors are multiplicative. Even for added noise, there are nuances not covered by this section. For more information on noise analysis, the reader can consult a textbook on analog circuit design. We further limit our discussion to white noises, that is, those with power spectrum density (PSD) values independent of frequency. Such discussion covers most noise sources when the system bandwidth is not too wide.

In the discussions below, we use the symbols in Figure 6.13 to represent the one-sided PSD of the corresponding noise. The total noise power of a system with passband bandwidth W is[10]

$$P_N = WN_S. \tag{6.84}$$

6.8.4.1   Thermal Noise

Thermal noise is caused by random electron movements in analog components, especially in the load resister at the input. It is also known as the Johnson–Nyquist noise [52, 53]. Simply speaking, the one-sided PSD is [54, sec. 9.3]

$$N_T = k_B T_T. \tag{6.85}$$

Here, k_B is the Boltzmann constant (1.38×10^{-23} J/K). T_T is the temperature of the component, commonly chosen as 290 K. Under such condition, the noise PSD is −174 dBm/Hz, or −114 dBm/MHz.

It is customary to express other noise power levels in terms of temperature, following (6.85). Although without physical meanings, such notation allows for easy comparison between these noises and the thermal noise. For example, the system noise temperature T_S and system noise PSD N_S (shown in Figure 6.13) are related by

---

[10] In (6.84), W is known as the noise equivalent bandwidth. It is approximately the same as the system bandwidth but may differ from the latter due to filter details.

$$N_S = k_B T_S. \tag{6.86}$$

6.8.4.2 Additional Noise: Noise Figure

Additional noise at stage m are typically specified by the noise figure F_{Nm} [55]. Noise figure is the ratio between the total noise power and the thermal noise power (at 290 K). Namely, the additional noise PSD is

$$N_m = (F_{Nm} - 1)k_B T_0, \tag{6.87}$$

where $T_0 = 290$ K is the reference temperature. N_m and F_{Nm} are the additional noise and noise figure of the mth stage, respectively. Noise figure is often expressed in dB.

As will be shown in Section 6.8.4.5, the noise figure of the first stage is most critical to overall noise performance. Therefore, a low noise amplifier (LNA) is often used for the first stage. Typical noise figures for LNA range from 0.5 to 5 dB [56]. Passive elements such as cable do not introduce additional noise, and thus have a noise figure of 1 (0 dB).

6.8.4.3 G/T

In some cases, especially in satellite communications, receiver antenna gain and noise level are combined into one figure of merit: G/T, sometimes written as GT or G-T [49, sec. 1.15]. It is simply the ratio between antenna gain and system noise temperature given by (6.86):

$$G/T \overset{\text{def}}{=} \frac{G_r}{T_S} \tag{6.88}$$

G/T has the unit of K^{-1}. However, it is typically expressed in dB form, whose unite is taken as $dB \cdot K^{-1}$. With such figure of merit, (6.79) and (6.80) can be combined as

$$SNR = \frac{1}{k_B W} \frac{P_t G_t}{L} G/T. \tag{6.89}$$

Here, W is the system (passband) bandwidth. All other quantities are the same as in (6.79). In (6.89), G/T is the only parameter depending on the receiver. Therefore, it can be used to compare performances among various receivers. Section 6.8.4.5 discusses system noise in more detail.

6.8.4.4 External Noises

In addition to the internal noise, a receiver also picks up noise through the antenna. Such noise may come from atmosphere radiation, radiation from other objects with finite temperatures (such as Earth or celestial bodies), and human activities. There are various ways to estimate external noise levels [49, 55]. However, in most cases, external noise is lower than thermal noise and thus can be ignored. One notable exception is the 4G cellular systems, where neighboring cells are designed to have a certain level of mutual interference. Such interference may even dominate the system noise computation.

6.8.4.5   System Noise and SNR

Sections 6.8.4.1–6.8.4.4 discuss the various components of noises in a receiver. These components combine to form the total system noise. However, since noises are generated at different stages of processing (see Figure 6.13), they need to be "reflected" to the receiver input, where the signal power is measured.

Consider noise N_m, which is generated at the input of the mth stage. The same noise level can be obtained if an equivalent noise N'_m was inserted at the receiver input, with the PSD level of

$$N'_m = \frac{N_m}{\prod_{l=1}^{m-1} G_l}. \tag{6.90}$$

Note that these gains may be smaller than 1 (e.g., for cable loss). Correspondingly, we can also define the noise temperature T_m at the mth stage and the equivalent T'_m at the receiver input:

$$T_m \overset{\text{def}}{=} \frac{N_m}{k_B} = (F_{Nm} - 1)T_0$$

$$T'_m \overset{\text{def}}{=} \frac{N'_m}{k_B} = \frac{T_m}{\prod_{l=1}^{m-1} G_l} \tag{6.91}$$

Here, F_{Nm} and T_0 are defined in Section 6.8.4.2. These equivalent noises can be summed together to form effective noise.

$$N_S = N_e + N_T + \sum_{m=1}^{M} N'_m = N_e + N_T + \sum_{m=1}^{M} \frac{N_m}{\prod_{l=1}^{m-1} G_l}. \tag{6.92}$$

The second term in (6.92) is the thermal noise given by (6.85). Considering Figure 6.13 and (6.87), Equation (6.92) can also be expressed in terms of noise temperatures (c.f. Figure 6.13):

$$T_S = T_e + T_T + \sum_{m=1}^{M} T'_m = T_e + T_T + \sum_{m=1}^{M} \frac{(F_{Nm} - 1)T_0}{\prod_{l=1}^{m-1} G_m}. \tag{6.93}$$

Note that although the noise figures and gains are usually expressed in dB, the summations in (6.92) and (6.93) are in linear scale. From (6.92) and (6.93), we can see that noise at the mth stage is "discounted" by a factor $\prod_{l=1}^{m-1} G_m$, which is usually much larger than 1. This is the result of converting N_m to N_m' based on (6.90). Such discount can also be understood as follows. By the time the signal arrives at the mth stage, it is already amplified by the previous stages. Therefore, the newly added noise is less important, compared to the amplified signal level. Because of such "discount," noise generated by the first state of amply is typically the most important noise component.

With these results, the final SNR from the link can be computed by (6.80)[11]:

$$SNR_L = \frac{P_r}{P_N} = \frac{P_r}{N_S W} = \frac{P_r}{k_B T_S W}. \tag{6.94}$$

6.8.5 Required SNR and Margin

Sections 6.8.1–6.8.4 show how to compute the SNR at the receiver. Such a result is to be compared to the required SNR for the communications system.

For a given modulation and coding scheme (MCS) and a desired bit error rate (BER), required SNR can be computed from theory or simulation, as described in Chapters 3 and 4. Such required SNR is noted as SNR_r. On the other hand, implementation details, such as finite resolution and approximation in algorithms may introduce additional performance degradation, which can be aggregated into an additional "implementation loss" L_I. L_I typically varies from 0.5 to 2 dB. SNR_r and L_I are to be compared with the link-provided SNR_L from (6.94) to arrive at the link margin M_L:

$$M_L \overset{\text{def}}{=} \frac{SNR_L}{L_I SNR_r} \tag{6.95}$$

M_L indicates the excess SNR from the link, after satisfying the requirement imposed by the desired communication performance. If $M_L < 1$, the system cannot achieve the desired BER. On the other hand, if $M_L > 1$, communication link can still function properly, if available SNR suffers an additional drop of M_L. Typically, it is desirable to have an M_L of 6 dB to account for uncertainties in link calculation and unexpected channel variations. SNR_r is usually computed for static channels. If such value is applied to fading channels, a higher margin is advisable.

---

[11] There are other forms of SNR expression, as detailed in Chapter 4.

6.8.6 Link Budget Summary

Link budget computation evaluates the signal and noise levels at the receiver and determines whether the intended radio link can work properly (or "whether the link can be closed" in trade jargon). The basic equations are (6.79), (6.80), and (6.95). The result is the link margin. If the link margin is larger than 1 (or 0 dB), the link is closed. Various components in these equations can be evaluated following the guidelines provided in this section.

The required or desired fidelity in link computation varies among applications. When the margin is large, lower fidelity is acceptable. Fidelity depends on the uncertainty in the estimation of the various components in link computation. Usually, path loss estimation carries the largest uncertainty if it is based on generalized models. Furthermore, antenna gain may deviate from theoretical values due to manufacturing tolerances, and there are miscellaneous losses such as cable and connector losses and mismatches that may or may not be included in the computation. For example, link budget computations can be very coarse in mobile communication systems, because path loss is very uncertain (it can be off by 10 dB or more). On the other hand, satellite communications usually require very accurate link budget computations (to a fraction of dB), because the path loss is often well known, and significant cost savings can be realized if accurate link budget computation leads to a reduction of the required margin.

This section provides a general approach to link budget computation and some common values. Actual practice often varies among the industries and may include factors not considered in this section. Adaptation to specific situations is crucial in link budget computations. There are many different formulations for link budget computation, some are equivalent (e.g., in linear form and dB form), some include different factors. Instead of memorizing formulas, it is better to understand the principle and devise solutions suitable for specific situations. There are also many link budget calculators available online, with varying fidelities. For example, the National Institute of Standards and Technology (NIST) authored a spreadsheet tool for link budget computation [57]. When using these tools, it is advisable to ascertain the formulas behind them or test-run with some simple cases for verification.

6.9 SUMMARY

Channel characteristics are a complicated and broad field, especially for wireless channels. This chapter provides a general framework for understanding these characteristics.

In general, a channel is described by channel gain $h(t, \tau)$. The variable t describes the channel time dependency, while the variable τ describes channel time dispersion. On the other hand, various aspects of these characteristics can be considered separately. This chapter focuses on the three most relevant characteristics: path loss, fading, and time dispersion.

In the basic sense, channel gain depends on the distance between the transmitter and receiver, and any obstacles between them. In addition to the line-of-sight (i.e., direct) path, signals can also propagate through reflection, diffraction, and scattering. These phenomena are discussed in Section 6.3.

Taking a step further, we can study time dependency of channel gain, that is, fading, in mobile communication scenarios. Fading occurs when standing wave zones are formed due to multiple scattering, and the transmitter and/or receiver move through these zones. Equivalently, we can also consider fading as interference among signals through different scattering paths. These signals have slightly different frequencies due to Doppler shift. Therefore, their interference results in a time-varying channel gain. The speed for time variation can be characterized by the Doppler frequency spread. Fading can be characterized statistically with the prevalent Rayleigh or Ricean models. The speed of fading can be captured in the autocorrelation function. A widely used model for time dependency is the Jakes model. These models are covered in Section 6.4.

Section 6.5 continues to discuss time-dispersive channels. Time disperse is equivalent to nonflat frequency response. The dispersion length is the inverse of the coherence bandwidth. Channel time dispersion causes inter-symbol interference (ISI), which is the primary subject of Chapters 9 and 10.

Both time dependency and time dispersion can be described in either the time domain or the frequency domain. Section 6.6 discusses the relationship between the various forms of expressions. Especially, in most cases, the channel can be considered as wide-sense stationary in time dependency and uncorrelated scattering in time-dispersive behavior. The so-called WWSUS channel allows for further simplification in autocorrelation expressions.

Channel path loss can be combined with other system parameters such as transmitted power, antenna gains, and noise power level, to estimate expected system performance measured by the SNR margin. This process is known as link budget computation and is covered in Section 6.8. Link budget computation is usually the first step in system design. Besides link budget computation, channel characteristics affect other aspects of system design, as will be discussed in Chapters 9, 10, and 12.

There are some related topics not covered in this chapter. Interested readers can leverage the basic concepts and framework provided here for further studies. Some of the uncovered topics are briefly mentioned below.

- Channel for multiple antenna systems. It is briefly discussed in Chapter 11.
- Channel simulation techniques. Interested users can consult a specialized textbook (e.g., [24]) and the recent literature on the subject.
- Verification of channel models by field measurement. There are voluminous works reported in the literature on model verification. Selecting the results applicable to specific applications is essential.
- Situation-specific channel models. Besides the general channel models discussed in this chapter, many models are more specific to situations such as indoor, urban, over water, ground to air, and so on. These models are also

specific to some frequency range. Situation-specific channel models are reported in the literature and publications from standard bodies and industrial consortiums.

- Special propagation effects. Radio waves in certain frequency ranges have special properties when propagating around the earth. They are mentioned in Section 6.2.4. Usually, there are specialized books or conference proceedings addressing these frequency bands.

- Channel prediction. The possibility of predicting future channel states from past measurements attracts continued research effort. However, the current performance of channel prediction is not good enough for widespread applications.

Overall, this chapter focuses more on mobile communications channels, especially those for the modern cellular systems. However, the principles and framework provided herein are valuable to other applications such as wired communications. For further exploitation of wireless channel characteristics, one can start with some textbooks that have a focus on channel models [26, 58, 59].

6.10 APPENDIX: CHANNEL GAIN IN PASSBAND AND BASEBAND

In this section, we show that the channel model (6.1) applies to both baseband and passband channels, and establish the relationship between channel gains in the two bands.

First, let us restate the up-conversion and down-conversion relationships given in Chapter 3. Let the complex-valued baseband signal be $\bar{S}_b(t)$ and carrier angular frequency by ω_c. After up-conversion, the passband signal can be expressed as

$$S_p(t) = Re\left[\bar{S}_p(t)\right] = \frac{1}{2}\left[\bar{S}_p(t) + \bar{S}_p^*(t)\right] \tag{6.96}$$

where $\bar{S}_p(t)$ is the complex-valued passband signal:

$$\bar{S}_p(t) \overset{\text{def}}{=} \bar{S}_b(t)\exp(j\omega_c t). \tag{6.97}$$

In down-conversion, we have

$$\bar{S}_b''(t) = LPF\left[S_p(t)\exp(-j\omega_c t)\right]. \tag{6.98}$$

Here $LPF(\cdot)$ denotes the low pass filtering operation. Given that $\bar{S}_p^*$ has a spectrum concentrated around $-\omega_c$, the low pass filtering removes its contribution. Namely, (6.96) and (6.98) lead to

$$\bar{S}_b'' = \frac{1}{2}\bar{S}_p(t)\exp(-j\omega t). \tag{6.99}$$

Now, let us consider the channel. The physical channel applies to passband. Therefore, (6.1) leads to

$$y_p(t) = \int h_p(t,\tau) x_p(t-\tau) d\tau, \tag{6.100}$$

where y_p and x_p are received and transmit signals in the passband, respectively. h_p is the passband channel gain and is real-valued. Based on the up-conversion Equations (6.96) and (6.97), (6.100) can be rewritten as

$$
\begin{aligned}
y_p(t) &= \int h_p(t,\tau) \frac{1}{2} \left[\bar{x}_p(t-\tau) + \bar{x}_p^*(t-\tau) \right] d\tau \\
&= \frac{1}{2} \left[\int h_p(t,\tau) \bar{x}_p(t-\tau) d\tau + \int h_p(t,\tau) \bar{x}_p^*(t-\tau) d\tau \right],
\end{aligned}
\tag{6.101}
$$

where

$$\bar{x}_p(t) \stackrel{\text{def}}{=} \bar{x}_b(t) \exp(j\omega_c t). \tag{6.102}$$

$\bar{x}_b(t)$ is the complex-valued transmit signal in the baseband. Again, the second term in (6.101) has a spectrum concentrating around $-\omega_c$. Its contribution after down converting is removed by the low pass filter. Therefore, similar to (6.99), the baseband complex-valued received signal y_b is given by

$$\bar{y}_b(t) = \frac{1}{2} \int h_p(t,\tau) \bar{x}_p(t-\tau) d\tau \exp(-j\omega_c t). \tag{6.103}$$

The factor 1/2 can be absorbed into receiver gain.[12] Therefore, we can consider the following received signal:

$$\bar{y}_b'(t) \stackrel{\text{def}}{=} 2\bar{y}_b(t) = \int h_p(t,\tau) \bar{x}_p(t-\tau) d\tau \exp(-j\omega_c t). \tag{6.104}$$

Equations (6.102) and (6.104) thus lead to

$$
\begin{aligned}
\bar{y}_b'(t) &= \int h_p(t,\tau) \{ \bar{x}_b(t-\tau) \exp[j\omega_c(t-\tau)] \} d\tau \exp(-j\omega_c t) \\
&= \int \left[h_p(t,\tau) \exp(-j\omega_c \tau) \right] \bar{x}_b(t-\tau) d\tau
\end{aligned}
\tag{6.105}
$$

Therefore, the baseband channel model is:

$$
\begin{aligned}
\bar{y}_b'(t) &= \int h_b(t,\tau) \bar{x}_b(t-\tau) d\tau \\
h_b(t,\tau) &\stackrel{\text{def}}{=} h_p(t,\tau) \exp(-j\omega_c \tau)
\end{aligned}
\tag{6.106}
$$

---

[12] When performing such scaling, we must ensure that both signal and noise are scaled the same way so that SNR remains the same Chapter 4 provides detailed discussion on passband and baseband SNR conversions.

Equation (6.106) shows that the baseband signal has the same channel model (6.1), and the baseband channel gain $h_b(t, \tau)$ is related to the passband channel gain $h_p(t, \tau)$ through (6.106).

REFERENCES

1. H. T. Friis, "A Note on a Simple Transmission Formula," *Proc. IRE*, vol. 34, no. 5, pp. 254–256, 1946.
2. ITU-R, *Calculation of Free-Space Attenuation*. Recommendation ITU-R P.525-3, 2016.
3. R. C. Johnson, H. A. Ecker, and J. S. Hollis, "Determination of Far-Field Antenna Patterns from Near-Field Measurements," *Proc. IEEE*, vol. 61, no. 12, pp. 1668–1694, 1973.
4. ITU-R, *Nomenclature of the Frequency and Wavelength Bands Used in Telecommunications*. Recommendation ITU-R V.431-8, 2015.
5. R. Cochetti, *Mobile Satellite Communications Handbook*, 2nd ed. Hoboken, NJ: Wiley, 2014.
6. J. G. Andrews, T. Bai, M. Kulkarni, A. Alkhateeb, A. Gupta, and R. W. Heath, "Modeling and Analyzing Millimeter Wave Cellular Systems," *IEEE Trans. Commun.*, vol. 65, no. 1, pp. 403–430, 2017.
7. F. Khan, Z. Pi, and S. Rajagopal, "Millimeter-Wave Mobile Broadband with Large Scale Spatial Processing for 5G Mobile Communication," in *2012 50th Annual Allerton Conference on Communication, Control, and Computing (Allerton)*, 1–5 October, Monticello, IL, 2012, pp. 1517–1523.
8. S. Kumar and S. Shukla, *Wave Propagation and Antenna Engineering*. PHI Learning, 2016.
9. ITU-R, *Definitions of Terms Relating to Propagation in Non-ionized Media*. Recommendation ITU-R, P.310-9, 1994.
10. W. C. Lee, *Mobile Communications Engineering, Theory and Applications*, 2nd ed. New York: McGraw-Hill, 1982.
11. ITU-R, *Propagation by Diffraction*. Recommendation ITU-R P.526-13, 2015.
12. Alion, "TIREM™—Terrain Integrated Rough Earth Model™" [Online]. Available at https://www.alionscience.com/terrain-integrated-rough-earth-model-tirem/ [Accessed September 10, 2017].
13. ITS, "IF-77 Electromagnetic Wave Propagation Model (Gierhart-Johnson)" [Online]. Available at https://www.its.bldrdoc.gov/resources/radio-propagation-software/if77/if-77-electromagnetic-wave-propagation-model-gierhart-johnson.aspx [Accessed September 12, 2017].
14. ITU-R, *Attenuation due to Clouds and Fog*. Recommendation ITU-R P.840-6, 2013.
15. ITU-R, *Attenuation in Vegetation*. Recommendation ITU-R P.833-9, 2016.
16. ITU-R, *Attenuation by Atmospheric Gases*. Recommendation ITU-R P.676-11, 2016.
17. ITU-R, *Characteristics of Precipitation for Propagation Modelling*. Recommendation ITU-R P.837-7, 2017.
18. ITU-R, "ITU-R Recommendations P Series Radiowave Propagation" [Online]. Available at http://www.itu.int/rec/R-REC-P/en [Accessed August 18, 2017].
19. ITU-R, *Method for Point-to-Area Predictions for Terrestrial Services in the Frequency Range 30 MHz to 3000 MHz*. Recommendation ITU-R P.1546-5, 2013.
20. M. Hata, "Empirical Formula for Propagation Loss in Land Mobile Radio Services," *IEEE Trans. Veh. Technol.*, vol. 29, no. 3, pp. 317–325, 1980.
21. European Commission, *Digital Mobile Radio Towards Future Generation Systems*. COST Action 231, 1999.
22. F. Hansen and F. I. Meno, "Mobile Fading—Rayleigh and Lognormal Superimposed," *IEEE Trans. Veh. Technol.*, vol. 26, no. 4, pp. 332–335, 1977.
23. J. Proakis and M. Salehi, *Digital Communications*, 5th ed. New York: McGraw-Hill Education, 2007.
24. M. Pätzold, *Mobile Radio Channels*, 2nd ed. Chichester, UK: Wiley, 2012.
25. M. Gudmundson, "Correlation Model for Shadow Fading in Mobile Radio Systems," *Electron. Lett.*, vol. 27, no. 23, p. 2145, 1991.
26. G. L. Stuber, *Principles of Mobile Communication*. New York: Springer, 2011.
27. W. C. Jakes and D. C. Cox, *Microwave Mobile Communications*. Wiley, 1974.

28. R. H. Clarke, "A Statistical Theory of Mobile-Radio Reception," *Bell Syst. Tech. J.*, vol. 47, no. 6, pp. 957–1000, 1968.

29. C. S. Patel, G. L. Stuber, and T. G. Pratt, "Comparative Analysis of Statistical Models for the Simulation of Rayleigh Faded Cellular Channels," *IEEE Trans. Commun.*, vol. 53, no. 6, pp. 1017–1026, 2005.

30. B. Sklar, *Digital Communications: Fundamentals and Applications*. Prentice Hall PTR, 2001.

31. Chia-Chin Chong and Su Khiong Yong, "A Generic Statistical-Based UWB Channel Model for High-Rise Apartments," *IEEE Trans. Antennas Propag.*, vol. 53, no. 8, pp. 2389–2399, 2005.

32. J. B. Andersen, T. S. Rappaport, and S. Yoshida, "Propagation Measurements and Models for Wireless Communications Channels," *IEEE Commun. Mag.*, vol. 33, no. 1, pp. 42–49, 1995.

33. H.-J. Zepernick and T. A. Wysocki, "Multipath Channel Parameters for the Indoor Radio at 2.4 GHz ISM Band," in *1999 IEEE 49th Vehicular Technology Conference (Cat. No.99CH36363)*, 16–20 May, Houston, TX, vol. 1, pp. 190–193.

34. P. Bisaglia, R. Castle, and S. H. Baynham, "Channel Modeling and System Performance for HomePNA 2.0," *IEEE J. Sel. Areas Commun.*, vol. 20, no. 5, pp. 913–922, 2002.

35. P. Bello, "Characterization of Randomly Time-Variant Linear Channels," *IEEE Trans. Commun.*, vol. 11, no. 4, pp. 360–393, 1963.

36. ITU-R, "Mobile, Radiodetermination, Amateur and Related Satellite Services" [Online]. Available at https://www.itu.int/rec/R-REC-M/en [Accessed: September 16, 2017].

37. ITU-R, *Guidelines for Evaluation of Radio Transmission Technologies for IMT-2000*. Recommendation ITU-R M.1225, 1997.

38. 3GPP, *Universal Mobile Telecommunications System (UMTS); Spatial Channel Model for Multiple Input Multiple Output (MIMO) Simulations*. 3GPP Technical Report TR 25.996 V14.0.0, 2017.

39. 3GPP, *Study on Channel Model for Frequencies from 0.5 to 100 GHz (Release 14)*. 3GPP Technical Report TR 38.901 V14.3.0, 2017.

40. V. Erceg, *TGn Channel Models*. IEEE 802.11 Doc. 03/940r4, 2004.

41. FUNDP-INFO, "Distribution Terms of MATLAB Implementation of IEEE 802.11 HTSG Channel Model Proposal" [Online]. Available at https://staff.info.unamur.be/lsc/Research/IEEE_80211_HTSG_CMSC/distribution_terms.html [Accessed September 11, 2017].

42. J. P. Kermoal, L. Schumacher, K. I. Pedersen, P. E. Mogensen, and F. Frederiksen, "A Stochastic MIMO Radio Channel Model with Experimental Validation," *IEEE J. Sel. Areas Commun.*, vol. 20, no. 6, pp. 1211–1226, 2002.

43. P. Golden, H. Dedieu, and K. Jacobsen, *Fundamentals of DSL technology*. Auerbach Publications, 2006.

44. C. Phillips, D. Sicker, and D. Grunwald, "A Survey of Wireless Path Loss Prediction and Coverage Mapping Methods," *IEEE Commun. Surv. Tutorials*, vol. 15, no. 1, pp. 255–270, 2013.

45. eCFR, "Code of Federal Regulations" [Online]. Available at https://www.ecfr.gov/cgi-bin/text-idx?&c=ecfr&tpl=/ecfrbrowse/Title47/47tab_02.tpl [Accessed September 13, 2017].

46. S. Silver and H. M. James, Eds., *Microwave Antenna Theory and Design*. New York: McGraw-Hill Books Company, 1949.

47. D. Minoli, *Satellite Systems Engineering in an IPv6 Environment*. Auerbach Publications, 2009.

48. H. J. Visser, *Antenna Theory and Applications*. Chichester, UK: Wiley, 2012.

49. T. A. Milligan, *Modern Antenna Design*, 2nd ed. Hoboken, NJ: Wiley-Interscience IEEE Press, 2005.

50. ITU-R, *The Concept of Transmission Loss for Radio Links*. Recommendation ITU-R P.341-6, 2016.

51. J. L. Volakis, *Antenna Engineering Handbook*, 4th ed. New York: McGraw-Hill, 2007.

52. J. B. Johnson, "Thermal Agitation of Electricity in Conductors," *Phys. Rev.*, vol. 32, no. 1, pp. 97–109, 1928.

53. H. Nyquist, "Thermal Agitation of Electric Charge in Conductors," *Phys. Rev.*, vol. 32, no. 1, pp. 110–113, 1928.

54. A. B. Carlson and P. B. Crilly, *Communication Systems*. McGraw-Hill, 2010.

55. ITU-R, *Radio Noise*. Recommendation ITU-R P.372-13, 2016.

56. Broadcom, "MGA-631P8 Low Noise, High Linearity, Active Bias Low Noise Amplifier Data Sheet" [Online], 2011. Available at https://docs.broadcom.com/docs/AV02-0174EN [Accessed June 28, 2017].

57. NIST, "LinkCalc: NIST Link Budget Calculator" [Online]. Available at rfic.eecs.berkeley.edu/~niknejad /
ee242/pdf-lock/NIST_LinkBudgetCalc_2_4_konglk.xls [Accessed September 15, 2017].
58. T. S. Rappaport, *Wireless Communications*, 2nd ed. Prentice Hall PTR, 2002.
59. A. F. Molisch, *Wireless Communications*. Hoboken, NJ: Wiley, 2011.

HOMEWORK

6.1 Consider a flat fading channel whose gain is $g(t)$, which is a wide-sense stationary (WSS) stochastic process. A tone at frequency ω_0: $s(t) = Ae^{j\omega_0 t}$ is transmitted through the channel. If necessary, review Chapter 4 for the concepts of PSD, autocorrelation, and Fourier transform.

   a. Evaluate the power spectral density (PSD) of the received signal $P_s(\omega)$.

   b. Show that the $P_S(\omega - \omega_0)$ is proportional to the Fourier transform of the autocorrelation function of the channel gain.

6.2 Establish a connection between an autocorrelation function introduced in Section 6.6.3 and the function C_{HH} introduced in Section 6.5.2. Try to extend the definition of C_{HH} to time-varying channels (see [24, sec. 7.3.2.3]).

6.3 In Section 6.5.2, it is said that if the frequency autocorrelation of a channel C_{HH} is flat, then the channel has no time dispersion. It seems we can make a similar statement from (6.54): if $\tilde{h}(\omega)$ is flat, then $h(\tau)$ is a delta function, meaning the channel has no time dispersion. Compare these two statements.

6.4 Section 6.8.4 states "reduction of noise power from major contributors is the focus when system performance is concerned." In this context, consider the following example. Consider two noise components, with noise powers of $P_{n1} = 10$ dBm and $P_{n2} = 1$ dBm.

   a. What is the total noise power when the two noises are added together?

   b. If we manage to reduce P_{n1} by 1 dB, what would be the total noise power?

   c. If we manage to reduce P_{n2} by 1 dB instead, what would be the total noise power?

6.5 Consider the thermal noise definition (6.85), verify that we get the correct unit for the PSD.

6.6 Suppose you are writing a simulation program for a receiver. From link computation, you know the received power is −120 dBm. The symbol rate is 10 kHz. The receiver noise figure is 5 dB.

   a. What is N_0 for the receiver at room temperature 290 K (including unit)?

   b. In the simulation, you make eight complex-valued samples per symbol. These samples have a mean of 0 and a variance of 1. What should be the variance of noises added to these samples, to reflect the correct SNR?

   c. Suppose the modulation is QPSK, what is E_b/N_0 in dB?

6.7 Consider a space link between a geostationary satellite (altitude 35,786 km) and a receiver directly under it. Here are some parameters for the downlink (from satellite to earth terminal). Transmitted power 10 W, transmit waveguide losses 1.5 dB, transmitting antenna gain 27.0 dBi, carrier frequency 3.95 GHz, atmospheric absorption 0.1 dB, receiver station G/T 18.2 dB/K, bandwidth 25 MHz. Compute the signal-to-noise ratio at the receiver.

6.8 Consider three Rayleigh fading channels with maximum Doppler shifts of 20, 50, and 100 Hz. Plot the Doppler spectrum of each channel. Hint: look up MATLAB document for "Channel Visualization" for an example. You need to choose proper values of other parameters (sampling time, number of samples, and so on).

6.9 Suppose you are assigned a task to explore the possibility of using the LTE cellular system for ground-to-air communications to support control and data links for unmanned aerial vehicles. What are the considerations about channel characteristics that should be a part of your study?

CHAPTER 7

SYNCHRONIZATION

7.1 INTRODUCTION

Synchronization is the process of aligning receiver clocks with the corresponding transmitter clocks. At first glance, synchronization may appear to be trivial. Indeed, most theoretical works and even simulations assume perfect synchronization as a starting point. Even hardware prototyping often bypasses the synchronization issue by driving the transmitter and receiver with the same clock (referred to as "under the table timing"). Such a practice can be justified because, when designed and implemented properly, synchronization usually does not impact system performance. Therefore, perfect synchronization can be assumed when focusing on other parts of the system. On the other hand, synchronization is a critical part of a communications system and needs careful consideration during various design and implementation stages.

Figure 7.1 shows the physical layer architecture. This chapter concerns the radio frequency (RF) and demodulation blocks at the receiver, as shown by the dashed line circle.

To choose a synchronization technique for a particular system, we need to first consider some factors. First, we need to know how timings in the system are regulated. Namely, we need to know the clock architecture. Next, the size and stability of possible timing errors is important, as they determine the search space and frequency of update of the synchronization algorithm to be used. A particular algorithm is effective only when the timing errors are within some limits. Of equal importance is the target synchronization accuracy. The accuracy is chosen so that the received signal is not degraded significantly due to timing errors. All these issues are discussed in Section 7.2.

The task of synchronization can be broken into two parts. One is to estimate the current timing errors. The other is correcting such errors by manipulating captured signal or the system clocks. Such conceptual separation is helpful, although these two tasks are often closely related in practice. Section 7.3 discusses the easier part of synchronization (correcting timing errors), while Section 7.4 covers the more difficult part (detecting timing errors).

Digital Communication for Practicing Engineers, First Edition. Feng Ouyang.
© 2020 by The Institute of Electrical and Electronics Engineers, Inc.
Published 2020 by John Wiley & Sons, Inc.

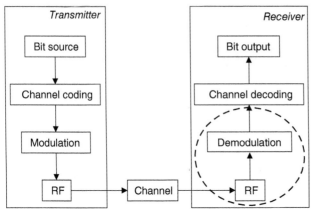

Figure 7.1   Physical layer architecture.

Lastly, Section 7.5 discusses the initial acquisition, namely, to obtain timing alignment without any prior timing knowledge and with a large initial error. Section 7.6 summarizes the chapter.

This chapter discusses synchronization techniques that do not use a separate clocking channel; the receiver measures timing error from analysis of received data-carrying signals. In contrast, a system may use a dedicated channel for synchronization (e.g., some satellite communication systems use dedicated tones for clock synchronization [1]). Other systems may lock transmitter and receiver clocks to a third standard, such as a global positioning system (GPS) time reference. These techniques are not covered in this chapter.

In keeping with traditions in the literature, the terms "timing recovery" or "frequency recovery" are used in this chapter to refer to the process of obtaining synchronization between the transmitter and receiver. The term "symbol period" refers to the time duration of a symbol, that is, the inverse of symbol rate (also known as baud rate). "Sample period" is the inverse of sample frequency.

In addition to clock frequencies, timing and synchronization also concern the relative positions of the clock pulses. The relative time delay between transmitter and receiver "clock ticks" is known as "relative phase," or "phase" for short.[1] Phase depends on the clock operations as well as propagation delay. The goal of synchronization is to keep clock frequencies at the transmitter and receiver identical. Agreement in frequencies results in a constant relative phase, which is treated as a part of the channel effect at the receiver. On the other hand, as will become apparent later, the phase is often easier to monitor than frequency. Therefore, many synchronization algorithms use phase as timing error indicator and adjust the clocks to achieve the

---

[1] Usually, "phase" is used in the context of harmonic oscillations and expressed in angles. However, here the term refers to the relative time difference between clock waveforms, which can be a sinusoid, a square wave or in other forms. As we can see later, sometimes "angle" is still a convenient measure of such time difference.

desired phase (e.g., zero). Such "phase locking" effectively ensures that the transmitter and receiver are running at the same frequency.

In this chapter, several commercial products are mentioned to illustrate the practical values of some important parameters. Such mentioning does not imply the endorsements of those products.

7.2   SYNCHRONIZATION OVERVIEW

Synchronization is usually performed at the receiver. However, it is not a receiver-only issue. Supports for synchronization operations need to be built into protocol and waveform designs. To choose the proper synchronization techniques, it is important to understand the problem. Namely, we need to know which clocks need synchronizing, the size of the possible timing errors, and the required accuracy of the synchronization. This section discusses and answers those questions.

7.2.1   Typical Clock Architectures

Figure 7.2 shows the general clocking architecture of a typical transmitter or receiver. The top row shows the functional blocks. For the transmitter (arrows pointing to the right), the source data (from upper layers) are divided into frames (for our purpose, a frame is a block of bits). Frames are sent under the control of frame timing. Consecutive frames (and possibly fill-in bits between frames) form a bit stream, which are sent to baseband processing (e.g., channel coding and modulation) to form symbols. The symbols form the baseband signal and are sent to the radio frequency (RF) block, driven by symbol timing. The RF block generates the carrier frequency signal. The carrier frequency is controlled by carrier timing. At the receiver, the reverse of such processes happens. For the communications system to work properly, the corresponding clocks on both sides must run at the same rate and have a known phase relationship. Frame timing is an issue of upper layers and is not discussed in this chapter. Therefore, going forward we focus on symbol timing and carrier timing.

For many communications systems, a master clock (e.g., a crystal oscillator) controls all timing. In this case, symbol timing and carrier timing are locked together. Only one of them needs to be recovered; the other can be derived. However, this is not always the case. For example, a cellular base station may have baseband processing, which generates the baseband symbols, located separately from the RF unit, which is

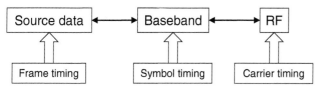
Figure 7.2   Transceiver clocking architecture.

usually placed close to the antenna. In this case, symbol timing and carrier timing are driven by separate clocks and need to be recovered separately.

Furthermore, in many cases, for bidirectional communication systems, clocks are locked together between the transmitting and receiving portions of a transceiver. Such clock locking can be done even when the two directions use different RF frequencies, that is, in a frequency division duplex (FDD) system. In this case, timing recovery is needed only in one of the directions. The party (station) that performs timing recovery is referred to as the "slave," who is responsible for locking its timing to the other party, the "master."

7.2.2   Clock Accuracy

To choose a proper timing recovery technique, it is essential to know the characteristics of timing errors. In particular, we would like to know the maximum amount of timing errors, which determines the search space of initial synchronization and the range for correction mechanism, as well as the speed of timing change, which determines how frequent synchronization needs to be updated.

A major source of timing error is from the system clock, which is usually a quartz crystal oscillator, commonly referred to as a crystal. A crystal generates an oscillating signal (sinusoidal wave or square wave) at a specified frequency. However, it is not perfect. There are four main sources of timing errors from a crystal.

1. Because of the variance in the manufacturing process, the crystal intrinsic frequency is different from its specified value. Such an error is known as calibration error, calibration tolerance, or initial accuracy.
2. Temperature change causes frequency drift (temperature stability).
3. There is a frequency drift over the device lifetime (aging effect).
4. Clock frequency fluctuates with the driving voltage.

In addition, there are very short-term frequency variations, referred to as phase noise or jitter. These quick and random phase changes affect system performance but are not corrected by synchronization. Therefore, they are not considered in this chapter. A crystal is also sensitive to mechanical shock and acceleration. These effects are important in some cases (such as for devices working on airplanes or under other extreme conditions). Clock errors can occur in both transmitters and receivers. The total timing error is the sum from both sides.

A clock signal from the system clock is distributed to the various modules of the transmitter or receiver, and its frequency may be converted to match the needs of the various processing operations (see Section 7.3.2.1 for methods of converting clock frequency). However, the relative frequency accuracy, defined as the ratio between the frequency error and the frequency itself, does not change in an ideal frequency conversion process. Therefore, clock accuracies are usually measured in relative frequency error, in the units of part per million (ppm) or part per billion (ppb).

To illustrate the magnitude of clock timing errors, two examples of crystals are presented below. The first product example [2] has a calibration tolerance of ±15 ppm, a temperature stability of ±15 ppm over the whole operating temperature range, and a typical aging of 2 ppm per year. The second product example [3] has a total stability of ±30 to ±100 ppm, depending on grades. The total stability includes initial accuracy, temperature stability (−40 to +85 °C), and 10-year aging.

Higher frequency accuracy can be obtained by installing the crystal in an oven to control its temperature. Such a device is referred to as an oven–controlled crystal oscillator (OCXO). An OCXO also has tighter tolerance on other aspects. Reference [4] presents the specification of a typical OCXO. This device has a total power consumption of 180 mW, warming up to operational condition in 90 seconds. It has an initial tolerance of 0.1 ppm, a temperature stability (−40 to +85 °C) of 0.5 ppb, and an aging of 20 ppb per year. These performances are several orders of magnitude better than the normal crystal oscillators.

When designing transceivers for industrial standards, sometimes clock accuracy is specified by the standard. For example, IEEE WLAN standard 802.11 high throughput mode (commonly known as 802.11n) specifies a transmitter clock tolerance of ±20 ppm for the 5-GHz band and ±25 ppm for the 2.4-GHz band, with the symbol and RF timings derived from the same clock [5, sec. 19.3.18.4].

Another source of timing error is the Doppler effect. When there is relative motion between the transmitter and receiver, the receiver experiences a frequency shift relative to the transmitted frequency, referred to as the Doppler shift. Its magnitude Δf is given by

$$\Delta f = f_c \frac{v}{c}, \tag{7.1}$$

where v is the relative radial speed between the transmitter and receiver, and c is the speed of light. As an example, a mobile phone operating at 2 GHz with a speed of 100 miles per hour has a Doppler shift of 0.15 ppm, or 280 Hz. Although Doppler shift has relatively small amplitude, it may change quickly as motion conditions change.

In summary, under normal conditions (i.e., no extreme temperature and acceleration or vibration, no extremely high speed), if ordinary crystal clocks are used, the timing errors are on the order of tens to hundreds of ppm. If the OCXO is used, the timing errors are on the order of 0.1 ppm for stationary platforms, and higher for moving platforms due to Doppler shift.

7.2.3 Synchronization Requirement

Another question critical to synchronization design is the desired accuracy. This section gives some very rough estimation. The effect of timing error is expressed in terms of the resulting additional noise. Such an estimate is usually enough for system design purposes. More detailed estimates can be found in [6].

To analyze the synchronization requirement, consider a frame with N symbols. We assume that the first symbol has perfect timing, as guaranteed by the initial acquisition process (Section 7.5). If there is an error in carrier frequency or symbol frequency, timing error will build up as time progresses, causing a degradation of signal-to-noise ratio (SNR). Therefore, the worst timing error happens at the end of the frame.

As will be discussed in more detail in Section 7.3.1.1, a carrier frequency offset (CFO) causes a phase rotation for the baseband symbols:

$$\Delta\Phi_n = 2\pi n T_{Sym}\Delta f_c, \tag{7.2}$$

where $\Delta\Phi_n$ is the baseband phase rotation for the nth symbol after initial phase alignment. T_{Sym} is the symbol period, and Δf_c is the CFO. For a symbol with voltage V, such rotation causes an error (noise) of $V(e^{j\Delta\Phi_n} - 1)$, which is approximated by $V\Delta\Phi_n$ when $\Delta\Phi_n \ll 1$. Therefore, the resulting signal-to-noise ratio (SNR) for the nth symbol is

$$SNR_n = \left(\frac{V}{V\Delta\Phi_n}\right)^2 = (\Delta\Phi_n)^{-2} = \left(2\pi n T_{Sym}\Delta f_c\right)^{-2}. \tag{7.3}$$

Namely, the resulting SNR becomes worse as n becomes larger. For the worst case, we should consider the last symbol in a frame. Let N be the number of symbols in a frame:

$$SNR_N = \left(2\pi N T_{Sym}\Delta f_c\right)^{-2} = \left(2\pi T_F \Delta f_c\right)^{-2}, \tag{7.4}$$

where $T_F \stackrel{\text{def}}{=} NT_{Sym}$ is the frame duration. Given the parameters for the system, required accuracy Δf_c can be estimated from (7.4). The target SNR is chosen so that the symbol can be correctly demodulated in the presence of phase error. Its value depends on the modulation and coding scheme. If a system has a higher order modulation (requiring higher target SNR) and longer frames, its requirement for synchronization accuracy is higher.

As an example, consider a system with a carrier frequency of 2 GHz. The symbol period T_{Sym} is 0.05 microsecond (a Nyquist bandwidth of 20 MHz). The frame length is 10,000 symbols. The target SNR is 20 dB (suitable for binary phase shift keying [BPSK] with convolutional coding and a margin of 15 dB). The required carrier frequency accuracy would be 32 Hz, or 16 ppb.

The previous discussion assumes coherent detection, and the first symbol has the correct phase. If incoherent detection is used, for example, in the case of differential phase shift keying (DPSK) modulation (Chapter 3), required carrier frequency accuracy would be much lower. In this case, the factor n in (7.3) and the factor N in (7.4) should be dropped.

For symbol timing, the time (i.e., phase) error is

$$\Delta T_n = n\left(\frac{1}{f_{Sym}} - \frac{1}{f_{Sym} + \Delta f_{Sym}}\right) \approx n\frac{\Delta f_{Sym}}{f_{Sym}^2}, \tag{7.5}$$

where ΔT_n is the time error for the nth symbol ($n = 0$ at initial acquisition). f_{Sym} is the correct symbol frequency. $\Delta f_{Sym} \ll f_{Sym}$ is the symbol frequency error. Again, timing error becomes larger as n increases.

Symbol timing error has two impacts. First, it introduces an error ΔV in the voltage value received:

$$\Delta V_n = S \Delta T_n \approx V f_M \Delta T_n, \tag{7.6}$$

where V is the signal peak voltage, and S is the average slope (i.e., dV/dt) of the signal. Such slope is approximated by $V f_M$, where f_M is the maximum frequency component in the signal (i.e., signal bandwidth). Therefore, the resulting SNR is

$$SNR_n = \left(\frac{V}{\Delta V_n}\right)^2 = \frac{1}{f_M^2 \Delta T_n^2}. \tag{7.7}$$

Assuming Nyquist symbol rate, we have $f_{Sym} = f_M$. Therefore,

$$SNR_n = \frac{1}{f_M^2 \Delta T_n^2} = \frac{1}{f_{Sym}^2} \left(n \frac{\Delta f_{Sym}}{f_{Sym}^2}\right)^{-2} = \left(\frac{n \Delta f_{Sym}}{f_{Sym}}\right)^{-2}. \tag{7.8}$$

Namely, SNR is inversely proportional to the square product of relative symbol timing error and symbol count.

In addition to the waveform distortion discussed above, timing error also introduces inter-symbol interference (ISI), as discussed in Chapter 3. The extent of ISI depends on the pulse shaping function (see homework of Chapter 3).[2] Therefore, it is difficult to make a general statement on the impact.

Remembering $f_{Sym} = T_{Sym}^{-1}$, we can see that (7.8) is very similar to (7.4), except for the factor of 2π and the difference between Δf_{Sym} and Δf_c. Therefore, the same *SNR* requirement leads to limits for Δf_c and Δf_{Sym} that are of the same order of magnitude. Because usually $f_{Sym} \ll f_c$, the required relative timing accuracy for the symbol frequency is much lower than that of the carrier frequency. Take the previous example, *without considering the impact of ISI*. With target SNR of 20 dB for the last symbol ($n = 10,000$), we get $\Delta f_{Sym}/f_{Sym} = 10^{-5}$ from (7.8). Namely, symbol frequency accuracy required is 10 ppm, as opposed to the 16 ppb required for carrier frequency accuracy.

It should also be noted that "noise contributions" introduced by timing error are to be combined with other noise sources to yield the final SNR at the demodulator. Therefore, the target SNR levels in (7.4) and (7.7) should be higher than what is required by demodulation. Usually, because high-accuracy synchronization is relatively easy to achieve, we want to design the system so that the synchronization error's

---

[2] Theoretically, ISI can be removed or reduced by an equalizer, as discussed in Chapter 9. However, as shown in (7.5), the timing error changes from symbol to symbol, as does ISI. An equalizer cannot adapt fast enough to handle this type of ISI.

contribution to performance degradation is minimal. Therefore, it is usually appropriate to have the target SNR for synchronization be 6–10 dB higher than the final target SNR.

Furthermore, some systems are not frame-based. The receiver continues operation indefinitely after initial acquisition. In this case, we can use "effective frame size" based on how fast clock frequency errors drift between positive and negative values, under the control of synchronization. However, more often than not, these systems achieve synchronization by directly controlling total phase errors $\Delta\Phi_n$ and ΔT_n given in (7.2) and (7.5). In this case, synchronization performance requirements can be derived in terms of phase errors instead of frequency errors, as in (7.3) and (7.7).

This section presents very rough estimations on the expected signal degradation caused by timing errors. It is important to keep in mind that the actual degradation depends on many factors such as channel coding and the rest of the receiver design because noise caused by timing errors is, in general, not Gaussian. Especially, the choice of pulse shaping function has a significant impact on performance under timing errors. More detailed analyses can be found in the literature for specific modulation schemes and in [7, sec. 2.2]. On the other hand, as stated previously, it is common to aim for timing accuracy that is much higher than the minimum requirement. In this case, the accuracy of SNR values is not so important because of the large margin. Therefore, the requirement estimate presented in this section is often good enough for system design purposes. Furthermore, many synchronization methods aim to maintain a constant phase relation, instead of controlling frequency error. For such systems, accuracy requirement estimations in this section need to be modified accordingly.

7.2.4 Separation of Carrier and Symbol Timing

In general, the received signal is distorted by both carrier and symbol timing errors. For example, as shown in (7.2), the CFO introduces a phase rotation among successive baseband samples. Such phase rotation is affected by the CFO Δf_c as well as the baseband symbol period T_{Sym}. Therefore, the two timing errors need to be estimated and corrected jointly.

However, in practice, both carrier and symbol timing errors are small. To state more quantitatively, we assume throughout the chapter that

$$\Delta f_c T_{Sym} \ll 1, \frac{\Delta T_{Sym}}{T_{Sym}} \ll 1, \tag{7.9}$$

where Δf_c is the CFO. T_{Sym} is symbol period, and ΔT_{Sym} is symbol period error. Equation (7.9) means that the phase change caused by CFO given in (7.2) is approximately constant during a symbol period. Therefore, symbol timing errors, which are usually much smaller than T_{Sym}, do not affect the observed phase.

Let us consider a practical example to justify the approximation (7.9). Consider a system with parameters similar to the Long-Term Evolution (LTE) cellular system or WiFi wireless local area network (WLAN) system[3]: carrier frequency at 2 GHz and bandwidth at 20 MHz. Assume both carrier clock and symbol clock have an accuracy of 100 ppm. Based on these numbers, T_{Sym} is 0.05 μs with Nyquist symbol rate, and $\Delta f_c = 200$ kHz. Therefore, $\Delta f_c T_{Sym} = 0.01$, and $\Delta T_{Sym}/T_{Sym} = 10^{-4}$.

In this and many other practical cases, the "coupling effect" between the CFO and symbol timing error is negligible. Therefore, CFO and symbol timing error can be estimated and corrected separately, which is the approach used in this chapter. Namely, when we estimate and correct one of the timing errors, we assume the other one is zero. One should bear in mind that such method may fail when very large timing errors are involved, such as during initial acquisition, discussed in Section 7.5.

Another implication of (7.9) is that the performance of the matched filter (discussed in Chapter 4) is not sensitive to CFO. Because matched filter typically lasts only a few symbols, the relative phase rotation caused by CFO is not important for such a period. Therefore, CFO correction can be placed after the matched filter.

7.3   TIMING CONTROL AND CORRECTION

Assuming timing errors are known, they can be corrected in two ways:

1. Adjusting the corresponding clock so that future samples will be captured with correct timing.
2. Digitally manipulating the baseband samples already captured. CFO is corrected by the phase rotation of the samples. Symbol timing error is corrected by interpolation. In this method, estimated timing error can be used to correct past or future samples.

Based on these two approaches, a timing correction architecture can be classified as "feedforward" or "feedback." In a feedforward, or "open loop," architecture, timing error estimation is used to modify the same block of samples used for estimation, as discussed in Section 7.3.1. In a feedback, or "closed loop," architecture, timing error estimation is used to modify the timing of future samples, which are then used for the next round of timing error estimation, as discussed in Section 7.3.2. Section 7.3.3 compares the two architectures and offers further comments.

Timing control and correction rely on timing error estimate, which is discussed in Section 7.4. However, there are two ways to describe timing error: frequency offset and phase difference. These two quantities are connected:

$$\Delta\phi = 2\pi\Delta f t + \Delta\phi_0, \tag{7.10}$$

where $\Delta\phi$ and Δf are phase difference and frequency offset, respectively. t is the time. $\Delta\Phi_0$ is the initial phase difference (at $t = 0$). One needs to choose either frequency or phase to describe timing error. In many cases, the timing error estimate output and the

---

[3] LTE and WiFi use orthogonal frequency division multiplexing (OFDM) modulations; therefore, the symbol rate here is actually a sample rate in OFDM terminology. The basic principle discussed here also applies to LTE and WiFi.

timing correction input are different in such a choice. Fortunately, the conversion is easy. In this section, we will be general about such choice and ignore the conversion issue.

7.3.1 Feedforward (Open Loop) Timing Correction

Feedforward, or open loop, timing correction refers to the method of correcting sample timing after the samples are digitized. Timing errors are estimated from the same data to be corrected, that is, before timing correction. Figure 7.3 shows the block diagram for such a scheme. Both Carrier timing correction and Symbol timing correction blocks process digital samples after Mixer/Sampler, in baseband. The order of these two blocks is not important.

    The method of adjusting timing in digital baseband is useful for receiver implementations, as discussed in this chapter, as well as in expressing timing errors in a simulation system.

7.3.1.1 Carrier Timing Correction

Carrier timing concerns the local carrier clock, which is used to down-convert the passband signal to baseband. The process is described in Chapter 3. Here, we summarize the formulation in the context of carrier timing correction.

    Let f_c be the carrier frequency used by the transmitter. As discussed in Chapter 3, the transmitted passband signal, in complex form, is

$$s(t) = s_b(t)e^{j2\pi f_c t}, \tag{7.11}$$

where $s_b(t)$ is the baseband signal at the transmitter. At the receiver, a different carrier frequency, f_{cr}, is used to down-convert the signal to baseband. Thus, the received baseband signal $s_{br}(t)$ is

$$s_{br}(t) = s(t)e^{-j2\pi f_{cr}t} = s_b(t)e^{j2\pi(f_c - f_{cr})t} = s_b(t)e^{j2\pi\Delta f_c t}, \tag{7.12}$$

where

$$\Delta f_c \stackrel{\text{def}}{=} f_c - f_{cr} \tag{7.13}$$

is the carrier frequency error at the receiver (i.e., CFO). After sampling, we have

$$S_{brn} \stackrel{\text{def}}{=} s_{br}(nT_s) = S_{bn}e^{j2\pi\Delta f_c nT_s}, \tag{7.14}$$

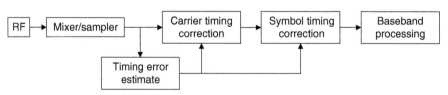

Figure 7.3   Feedforward timing correction architecture.

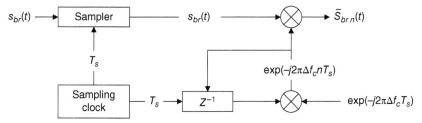

Figure 7.4   Carrier timing correction configuration.

where S_{brn} is the sample value with index n at the receiver, $S_{bn} \overset{\text{def}}{=} s_b(nT_s)$ is what the sample value should be without CFO, and T_s is the sample period. Here, we use samples instead of symbols to allow for over-sampling of the baseband signal. Obviously, the effect of CFO is a phase rotation, and it can be corrected by the corresponding phase rotation:

$$\bar{S}_{brn} \overset{\text{def}}{=} S_{brn} e^{-j2\pi\Delta f_c nT_s} = S_{bn}, \tag{7.15}$$

where $\bar{S}_{brn}$ is the sample with index n after carrier timing correction. Figure 7.4 shows an implementation of this method. CFO correction is achieved by the multiplier in the top row, which performs phase rotation expressed in (7.15). The multiplier at the bottom row advances the phase by T_s for each sample period, as stipulated by (7.15). The sampling clock drives the advancement of the sampler and the one-tap delay Z^{-1}.

7.3.1.2   Symbol Timing Correction

Conceptually, symbol timing correction involves resampling the baseband samples. Namely, we need to reconstruct the time-continuous signal $\bar{s}_{br}(t)$ from the received samples $\{S_{brn}\}$, such that[4]

$$\bar{s}_{br}(nT_{sr}) = S_{brn}, \tag{7.16}$$

where T_{sr} is the sample period at the receiver. This conceptual signal $\bar{s}_{br}(t)$ is then resampled at the correct time to form timing-corrected samples $\{\bar{S}_{brm}\}$:

$$\bar{S}_{brm} \overset{\text{def}}{=} \bar{s}_{br}(mT_s - d), \tag{7.17}$$

where T_s is the sample period at the transmitter (i.e., the correct one). m is the sample index. d is the initial timing offset when $m = 0$. According to Nyquist sampling theory (see Chapter 3), $\bar{s}_{br}(t)$ can be constructed by filtering the sample sequence[5]:

$$\bar{s}_{br}(t) = \sum_n h\left(\frac{t - nT_{sr}}{T_{sr}}\right) S_{brn}, \tag{7.18}$$

---

[4] Note that we use sample period instead of symbol period in this section, allowing for a more general case where the baseband signal is an over-sampled version of the symbols.

[5] In this section, limits of the summations depend on filter lengths, which are chosen in particular designs. For simplicity, summation limits are not expressed in the equations.

where h is a sinc function:

$$h(x) \overset{\text{def}}{=} \text{sinc}(x) = \frac{\sin(\pi x)}{\pi x}. \tag{7.19}$$

The sinc function has a relatively long impulse response and thus requires many taps to implement. In practice, alternative filters (unknown as interpolation filters) with similar frequency characteristics but shorter impulse responses may be used for h. Design methods for such filters are beyond the scope of this book. Furthermore, in some systems, the interpolation filter can be combined with the matched filter that is required in a digital receiver (see Chapter 4).

So far we have shown that the sample sequence with the correct timing $\{\bar{S}_{br\,m}\}$ can be conceptually obtained by first constructing a time-continuous function $\bar{s}_{br}(t)$ by filtering, then sampling such function at the time $\{mT_s\}$. However, $\bar{s}_{br}(t)$ needs to be constructed with fine enough time resolution to provide the samples at exactly the instants $\{mT_s\}$. Such requirement leads to high complexity in implementation.

Now let us go back to (7.17) and examine two practical implementations (the polyphase filter and the Farrow structure interpolator) in Sections 7.3.1.3 and 7.3.1.4.

7.3.1.3 Polyphase Filter

Let us first go through the formulation of the polyphase filter timing correction. A conceptual explanation will follow.

Combining (7.17) and (7.18), the time-corrected sample with index m can be expressed as the output of a discrete-time filter operating on the input samples:

$$\bar{S}_{br\,m} = \bar{s}_{br}(mT_s - d) = \sum_n h\left(\frac{mT_s - nT_{sr} - d}{T_{sr}}\right) S_{br\,n} = \sum_k H_{mk} S_{br\,m-k};$$

$$H_{mk} \overset{\text{def}}{=} h\left(\frac{mT_s - (m-k)T_{sr} - d}{T_{sr}}\right) = h\left(k + \frac{m\Delta T - d}{T_{sr}}\right); \tag{7.20}$$

$$\Delta T \overset{\text{def}}{=} T_s - T_{sr}.$$

We can view $\{H_{mk}\}$ as a filter whose taps are indexed with k. However, this filter changes with m; namely, for each new output sample, a new filter is required. As m progresses, new filter tap values H_{mk} can be computed from (7.20).

To avoid real-time computation of filter tap values, we can use a finite set of filters as an approximation [8]. This method is referred to as the polyphase filter bank method.

Suppose the desired timing alignment accuracy is $T_s/2L$. Construct a bank of filters $\left\{g_k^{[l]}, l=0,1,\ldots,L-1\right\}$, with the following tap values:

$$g_k^{[l]} \overset{\text{def}}{=} h\left(k+\frac{l}{L}\right).$$

(7.21)

$g_k^{[l]}$, with fixed l, is a sampled version of $h(x)$ introduced in (7.18) with a constant sampling period of 1. However, there is an additional phase offset in sampling, given by l/L. Equation (7.20) can be written as

$$\bar{S}_{br\,m} = \sum_k h\left(k+\frac{m\Delta T-d}{T_{sr}}\right) S_{br\,m-k} \approx \sum_k h\left(k+i_m+\frac{l_m}{L}\right) S_{br\,m-k} = \sum_j g_j^{[l_m]} S_{br\,m+i_m-j},$$

(7.22)

where

$$i_m \overset{\text{def}}{=} \left\lfloor \frac{m\Delta T-d}{T_{sr}} \right\rfloor$$

$$l_m \overset{\text{def}}{=} round\left[L\left(\frac{m\Delta T-d}{T_{sr}-i_m}\right)\right].$$

(7.23)

$\lfloor a \rfloor$ denotes the largest integer that is no larger than a. *round* is a function that yields the closet integer of the variable. In the above expressions, $i_m + (l/L)$ is an approximation of $(m\Delta T - d)/T_{sr}$, with an error not exceeding $1/(2L)$. Equation (7.22) is a regular filtering operation with filter $\left\{g_j^{[l_m]}\right\}$ given by (7.21). i_m represents an additional offset in the delay line.

As m progresses, $(m\Delta T - d)/T_{sr}$ increases or decreases, depending on the sign of ΔT. Usually, $\Delta T/T_{sr}$ is a small number (on the order of 10 or 100 ppm). Therefore, in most cases, i_m remains constant, and l_m changes with m. In this case, the delay line advances with m just as in a normal filter, whereas the filter tap set is selected by l_m. Occasionally, i_m would increase of decrease by 1. This means the delay line would make an extra advance (skipping) or omit an advance (stuffing).

This process is illustrated in Figure 7.5. Input samples $\{S_{br}\}$ are fed into the delay line for filtering. Advancement of this delay line is controlled by the shift clock, which normally advances in step with the output clock. However, when i_m changes value, the shift clock can produce an additional advancement (skipping) or omit an advancement (stuffing). The filter tap set is selected by the Selector based on l_m. i_m and l_m are computed by the compute block, according to (7.23). The samples in the delay line are multiplied to the corresponding taps of the selected filter, and the

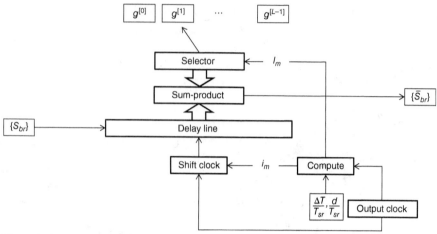

Figure 7.5 Polyphase filter for interpolation.

products are summed to form the output sample $\{\bar{S}_{br}\}$. This computation is performed in the sum-product block.

Conceptually, this timing correction process can be viewed as up-sampling the input samples (with a sample period T_{sr}) by a factor of L, and then down-sampling it to pick the correct sampling time [8]. This process includes three steps. First, up-sampling is achieved by inserting $L - 1$ zeros after each input sample. Next, the resulting sequence is filtered by h (with adjustment to the sampling rate) to reject aliasing. Lastly, output samples that represent the best timing match are chosen.

For the second step (filtering), because most of the input samples are zero, we only have to use the subset of the filter taps that are aligned with the nonzero samples. Furthermore, not all filter outputs are necessary; only the desired ones need to be computed. Therefore, the actual filter used, $\{g_j^{[l_m]}\}$, is a subset (i.e., a "phase") of the entire filter, which is a sampled version of h given by (7.19). l_m picks the "phase" to be used for the current output sample. Therefore, such a filtering structure is referred to as a polyphase filter bank.

A polyphase filter bank is efficient in computation because it does not compute filter taps in real time. However, it needs a large amount of memory to store the multiple sets of filter tap values and a mechanism to switch them on a sample-by-sample basis. Alternatively, one can compute filter tap values in real time through some simplification. One such technique is the polynomial interpolator with a Farrow structure [9] shown in Section 7.3.1.4.

7.3.1.4 Farrow Structure Interpolator
Equation (7.20) can be written as

$$\bar{S}_{br\,m} = \sum_k h\left(k + \frac{m\Delta T - d}{T_{sr}}\right) S_{br\,m-k} = \sum_k h(k + i_m + \mu) S_{br\,m-k} = \sum_j h(j + \mu) S_{br\,m+i_m-j},$$

$$(7.24)$$

where

$$i_m \overset{\text{def}}{\equiv} \left\lfloor \frac{m\Delta T - d}{T_{sr}} \right\rfloor .$$

$$\mu \overset{\text{def}}{\equiv} \frac{m\Delta T - d}{T_{sr}} - i_m$$

(7.25)

$h(t)$ is given in, for example, (7.19). Filter tap values $\{h(j+\mu)\}$ can be approximated as polynomials of μ:

$$h(j+\mu) \overset{\text{def}}{\equiv} p_j(\mu) = \sum_l c_{jl}\mu^l.$$

(7.26)

Coefficients c_{kl} can be optimized in various ways. An example was discussed in [10, sec. 9.1]; more analyses can be found in [11].

Equations (7.24) and (7.26) can be combined as

$$\bar{S}_{br\,m} = \sum_j p_j(\mu) S_{br\,m+i_m-j} = \sum_j \sum_l c_{jl}\mu^l S_{br\,m+i_m-j} = \sum_l \mu^l \left(\sum_j c_{jl} S_{br\,m+i_m-j} \right).$$

(7.27)

Namely, the time-corrected samples can be expressed as a polynomial of μ, whose coefficients are the result of the input samples being filtered by a bank of filters $\{c_{kl}, l = 0, 1, \ldots\}$. The computation of the polynomial can be further simplified by nested evaluation:

$$\sum_l a_l\mu^l = a_0 + \mu[a_1 + \mu(a_2 + \ldots)].$$

(7.28)

Namely, successive multiplications of μ are used to build up the term with μ^l. Such derivation leads to the Farrow structure of interpolator, as shown in Figure 7.6.

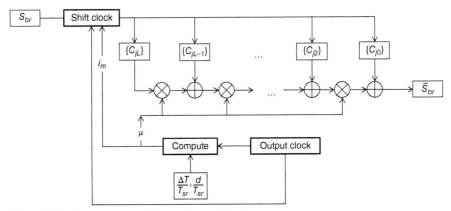

Figure 7.6 Farrow structure interpolator architecture.

As shown in the figure, the input samples, under the control of the shift clock, are fed to a bank of $L + 1$ filters, where L is the order of the polynomial. The shift clock controls the normal advance, skipping, or stuffing of the filter delay line, depending on i_m. The filter output data are summed with multiplication with μ, as given in (7.28). The compute block calculates the values of μ and i_m from (7.25), whereas the output clock block controls the advancement of index m.

In summary, feedforward symbol timing correction is an interpolation problem, which can be formulated as time-varying filtering and be implemented in various ways. The two structure presented here, the polyphase filter and the Farrow structure, exemplifies the tradeoff between of computation complexity and memory demand.

7.3.2   Feedback (Closed Loop) Timing Recovery

Feedback, or closed loop, timing recovery refers to the method of adjusting sample timing based on *previous* timing error estimation. Figure 7.7 shows two possible architectures. In architecture A, timing adjustment is performed through the clock that drives the mixer (down-converter) and sampling circuitry. In architecture B, timing adjustment is made in the baseband, in the same ways as described in Section 7.3.1. In both cases, the timing error estimate block operates on the samples whose timing has already been corrected. Therefore, the resulting estimate reflects residual timing error, which drives the timing correction for *future* samples.

The Sections 7.3.2.1 and 7.3.2.2 focus on clock timing adjustment techniques (panel A of Figure 7.7) and the loop filter issue. Architecture B uses techniques similar to those discussed in Section 7.3.1.

7.3.2.1   Clock Timing Adjustment
The most common clock timing adjustment method is the voltage-controlled oscillator (VCO) [12]. The oscillation frequency of a VCO can be adjusted by an external voltage. There are several types of VCO, the most common one for our application being the voltage–controlled crystal oscillators (VCXOs). A typical VCXO has a

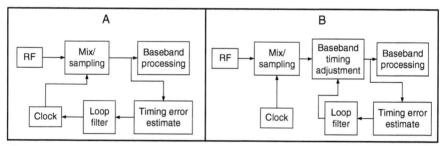

Figure 7.7   Closed loop timing recovery architectures. A: timing correction through sampling clock; B: timing correction through postprocessing in baseband.

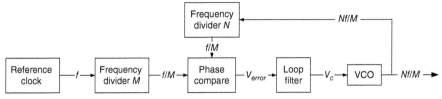

Figure 7.8   Phase-locked loop frequency synthesizer.

frequency adjustment range of 100 ppm and a responding speed of 1 ms.[6] These characteristics are suitable for timing recovery applications.

VCO can also be used together with a reference clock to form a more stable and digitally controllable clock. Figure 7.8 shows an example of the commonly used phase-locked loop (PLL) synthesizer [13].

In this system, the reference clock serves as a time base. It produces a stable signal at frequency f. However, the output frequency can be different by a rational factor, which is controlled by the two frequency dividers. If the desired frequency is $(N/M)f$, the two frequency dividers produce $1/M$ of the reference frequency (f) and $1/N$ of the output frequency, as shown in the figure. Frequency dividers can easily be implemented, for example, by digital counters. The phases of the two outputs from the two frequency dividers are compared by the phase compare block. A phase error generates a proportional voltage. This voltage is low-pass filtered by the loop filter block to remove short-term fluctuations. The output of the filter is a control voltage fed into the VCO. Therefore, in equilibrium, the outputs of the two frequency dividers are driven to maintain a constant phase difference and thus the same frequency. This means the output frequency of the VCO is kept at $(N/M)f$, and the signal has a predictable phase relationship with the reference clock (thus the term "phase lock"). In other words, the VCO output is slaved to the reference clock, although they have different frequencies. The output frequency can be adjusted through the values of N and M, which are be digitally controlled.

Another way for timing adjustment is known as direct digital synthesis (DDS) [13]. Figure 7.9 shows the structure. For every clock tick, the phase accumulator advances the current phase value (a value between 0 and 2π) with a step size controlled by the control word input. The waveform read-only memory (ROM) stores a waveform lookup table, which converts the phase value into the corresponding amplitude value. The output of the waveform ROM is converted to analog by the digital-to-analog (D/A) converter, whose output is further filtered to remove spurious components of the signal. The final output is a high-quality sine wave, whose frequency is determined by the input clock frequency and the phase advance step size as controlled by the control word.

A DDS can have precise frequency control and can smoothly change its frequency without causing phase disruption. However, its frequency range is limited because the filter at the final stage is fixed. On the other hand, PLL produces

---

[6] For example, model VX-805 made by Vectron has parameters similar to such values.

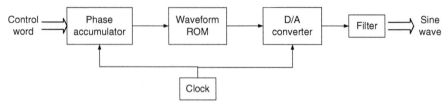

Figure 7.9 DDS block diagram.

high-quality clock output over a large frequency range. However, the frequency setting is coarse because the numbers N and M for the frequency divider cannot be too large. A combination of both, where a DDS serves as the clock input of a PLL, combines the advantages of both systems.

7.3.2.2 Loop Filters and the Vector Tracking Filter (VTF)

The output of the timing error estimate block in Figure 7.7 usually contains strong random components (considered as noise), as is discussed in more detail in Section 7.4. To achieve stable time control, a loop filter is added at the output of the timing error estimate, as shown in Figure 7.7. For feedforward schemes, the loop filter is simply a low-pass filter. For feedback schemes, however, the loop filter is a part of the dynamic control loop. Therefore, its design and tuning are more complicated.

This section does not discuss the theoretical treatment of loop filters, which is a topic of control theory and dynamic systems. Interested readers can consult books and papers on PLL and other control loops. Here, we present some practical considerations and present the vector tracking filter (VTF) as an example of practical innovation.

The simplest loop filter is a low-pass filter. It rejects the random changes while preserving the slow-varying timing error component. Because its bandwidth is very narrow, such a filter is commonly implemented in an infinite impulse response (IIR) form. The simplest form is a first-order IIR, also referred to as a leaky integrator, as shown in Figure 7.10. Its operation is expressed as:

$$y_n = (1-\beta)y_{n-1} + \beta x_n, \tag{7.29}$$

where $\{x_n\}$ is the input data. $\{y_n\}$ is the output data. n is the sample index, and $0 < \beta < 1$ is a parameter controlling bandwidth. A smaller β puts more weight on the memory y_{n-1},

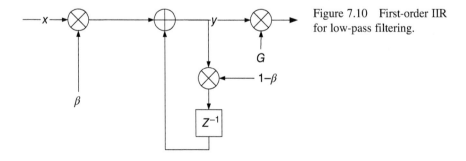

Figure 7.10 First-order IIR for low-pass filtering.

resulting in a narrower bandwidth. In our application, y is proportional to the frequency error to be corrected. The actual correction is Gy, where G is the loop gain.

More elaborate filters can be used to control the response better. However, in most cases the first-order IIR is satisfactory.

A narrowband filter has a long delay time. Namely, it has a long "memory," and its output does not respond immediately to the input. Such a delay time causes two problems.

1. For initial convergence, the filter needs time to build up the proper output to correct timing error. A larger gain G helps to correct the error quickly because it delivers more correction for the same registered error.

2. For steady state, overcompensation may be a problem. During timing adjustment, the new timing may already be correct. However, the filter, because of its long memory, continues to output the old correction "command," causing the clock or interpolator to over-correct, resulting in a new timing error in the opposite direction. This timing error will keep growing until it is finally reflected at the loop filter output. Then the process happens again in the opposite direction, resulting in an oscillation in timing error. Therefore, we would want the timing correction to be done in small steps to wait for the loop filter to update properly; namely, a smaller G is desired.

One way to resolve such conflicting requirements on loop gain G is "gear shifting." Namely, G is changed from a large value to a smaller one, possibly in several steps, as the system progresses from the initial acquisition stage to steady state. This helps to achieve fast initial convergence and a stable symbol time afterward. The gear shifting schedule is usually determined by experiment. Sometimes the filter bandwidth, controlled by β in (7.29), is also changed in the gear shifting process.

Furthermore, for a proportionally controlled feedback loop, the residual error is not zero. For example, a VCO requires a certain control voltage to produce the correct frequency. However, a certain frequency error is needed to produce such voltage in the feedback loop. Namely, at equilibrium, a finite residual error still remains. To reduce such residual error, one needs to have a large filter gain, so that a small frequency error can produce a large correction voltage. The extreme case is an integrator

$$y_n = y_{n-1} + x_n. \tag{7.30}$$

An integrator can be viewed as a low-pass filter (7.29) with $\beta = 0$, and x_n is multiplied by an infinitely large gain before entering the filter. As previously discussed, such large gains may cause overcompensation leading to oscillation.

The vector tracking filter (VTF) [14] is a technique to improve filter response. Figure 7.11 shows an example. The detected timing error passes through a standard one-tap IIR low-pass filter, whose output feeds into the correction decision block. The correction decision block then generates the correction signal and sends it to the clock. This is a standard timing recovery loop filter. The unique feature of VTF is the feedback link from correction decision to the low-pass filter. The correction decision block predicts the change in timing error that the current correction signal will cause, and

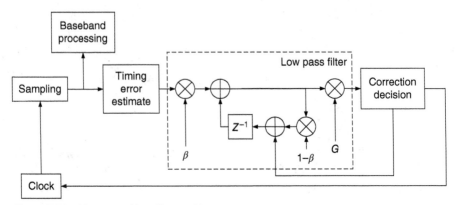

Figure 7.11 Vector tracking filter architecture.

injects it to the state memory of the IIR. This way, the IIR is "updated" with the current clock change, instead of letting such change build up over time through the low-pass filter. The VTF is very effective for improving filter response time, thus allowing for narrower filter bandwidth, resulting in both fast converging and stable steady-state operation.

7.3.3 Comparison and Combination of Feedforward and Feedback Methods

Sections 7.3.1 and 7.3.2 describe two schemes for timing correction, both with pros and cons. The feedforward method performs timing error estimation and timing error correction on the same set of signal samples. It is especially suitable for short bursts of signals (e.g., a frame or a packet). Such a collection of signals can be stored after sampling. Timing error is then estimated and corrected on the stored samples. On the other hand, the feedback method is suitable for continued updating of a stable link, where a relatively long initial acquisition period is allowed. In this case, the timing error is estimated after clock correction, and the estimation result is used to correct the timing of *future* samples.

Furthermore, the feedforward method requires accurate timing error estimate, as corrections are solely based on such estimate. The feedback method is iterative. Therefore, the timing error estimate does not need to be very accurate. In fact, as long as the estimated error has a monotonic relationship with the actual error and has the correct sign, feedback timing error correction will eventually drive the timing error toward zero. On the other hand, a more accurate error estimate is beneficial to timing correction performances such as convergence speed and stability.

Because of the relative strength and weakness of the two methods, sometimes it is advisable to use the combination, known as hybrid timing correction. In this scheme, timing error estimate results are used for both clock adjustment (feedback) and postsampling timing correction (feedforward).

7.4   TIMING ERROR ESTIMATE

This section addresses the timing error estimate block in Figures 7.3 and 7.7. The goal is estimating the timing error in terms of frequency offset or phase difference. The accuracy requirement of timing error estimate depends on the accuracy requirement of synchronization, as discussed in Section 7.2.3. Such requirement also depends on whether feedforward or feedback scheme is used, as discussed in Section 7.3.3.

In general, timing error estimation methods can be divided into three categories, depending on receiver knowledge of the signal [15]. If the exact transmitted data are known, it can be compared with the actual received signal for timing error estimate. This method is referred to as data-aided (DA). DA is typically used for initial acquisition phase, where known symbols (also referred to as preambles or training symbols) are transmitted (see Section 7.5). DA is also applicable when dedicated timing reference signals are transmitted along with data (e.g., in some orthogonal frequency division multiplexing [OFDM] systems, see Chapter 10). Another class is decision directed (DD). In this case, the receiver makes a decision about the transmitted data based on the received signal and uses its decision to help with timing recovery. DD is often used in data mode continuous adaptation, where demodulation decisions are mostly correct. If no knowledge of data is used, the receiver relies only on general knowledge of transmitted signal (e.g., nominal timing, pulse shaping function, and modulation). Such "blind" estimation method is referred to as non-data-aided (NDA). NDA is useful for initial acquisition when known symbols are not available or when DD does not have a significant advantage (such as in symbol timing estimation).

Section 7.4.1 lays down the foundation by discussing the optimal timing error estimation method. Different practical methods are used for carrier timing and symbol timing, as described in Sections 7.4.2 and 7.4.3.

7.4.1   Optimal Timing Estimate

This section discusses the optimum timing estimate based on the maximum a posteriori (MAP) principle (see Chapter 4 for more discussions on MAP). The basic idea is to search through all possible timing parameter values to find a set that best matches the received signal. Although this searching process is not always practical for complexity reasons, some of the practical timing estimate methods can be linked to such a concept. Therefore, we discuss the optimal timing estimate method as a starting point.

The problem of timing estimate is a parameter estimation problem [15]. The received signal is modeled as

$$r(t) = s(t; \theta) + n(t), \tag{7.31}$$

where $s(t; \theta)$ is the expected signal at time t, with timing parameters (carrier and symbol timing errors) collected in a vector parameter θ; $n(t)$ is the additive noise at time t. $r(t)$ is the received signal at t.

The optimum estimation of timing error parameters is the parameter values that make the received signal most likely to happen (maximum a posteriori estimation, or MAP). Namely,

$$\hat{\theta} = arg\,max_\theta P[\theta \,|\, r(t)], \tag{7.32}$$

where $P[\theta \,|\, r(t)]$ is the probability of timing parameter being θ, given that the received signal $r(t)$.[7]

From Bayes' theorem,

$$P[\theta \,|\, r(t)] = \frac{P[r(t) \,|\, \theta]P(\theta)}{P[r(t)]}, \tag{7.33}$$

where $P[r(t) \,|\, \theta]$ is the probability of receiving $r(t)$ given timing error parameter θ. $P(\theta)$ and $P[r(t)]$ are a priori probabilities for θ and $r(t)$, respectively.

Because we have no knowledge of $P(\theta)$, we assume it is a uniform distribution over some range. The denominator in (7.33) is independent of θ, and thus has no effect on the argument of maxima operation. Therefore, the MAP estimation becomes the maximum likelihood (ML) estimation:

$$\hat{\theta} = arg\,max_\theta P[r(t) \,|\, \theta]. \tag{7.34}$$

In the common case where $n(t)$ is a white Gaussian distribution, the probability distribution of noise, $p_n[n(t)]$ is given by (see Chapters 4 and 6 for noise characterization):

$$p_n[n(t)] = Ae^{-\int_0^T \frac{|n(t)|^2}{N_0}}. \tag{7.35}$$

Here, observation time is from $t = 0$ to $t = T$. N_0 is the noise power spectral density. A is a normalization constant. From (7.31), we have

$$p[r(t) \,|\, \theta] = p_n[r(t) - s(t;\theta)] = Ae^{-\int_0^T \frac{|r(t) - s(t;\theta)|^2}{N_0}}. \tag{7.36}$$

Furthermore, in the exponent,

$$|r(t) - s(t;\theta)|^2 = |r(t)|^2 - 2Re[r(t)s(t;\theta)] + |s(t;\theta)|^2. \tag{7.37}$$

Timing errors are usually small; thus, we can assume $\int_0^T |s(t;\theta)|^2 dt$ is independent of θ. Therefore, for the purpose of computing argument of maxima, only the middle term in (7.37) is important. Furthermore, because the exponential function is monotonic, the argument of maxima can be applied to the exponent. Namely,

$$\hat{\theta} = arg\,max_\theta \int_0^T Re[r(t)s(t;\theta)]dt. \tag{7.38}$$

---

[7] Here $r(t)$ is a function of time, instead of a value at a given time.

In other words, timing error estimating is made by choosing the timing assumption that maximizes the cross-correlation between the assumed received signal and the actual one. Therefore, under certain commonly applicable assumptions, the optimal MAP estimation (7.32) becomes ML estimation (7.34), which leads to the maximization of a cross-correlation (7.38). This argument is very similar to the derivation of the matched filter in Chapter 4.

ML estimation assumes the receiver knows the expected baseband signal $s(t; \theta)$. This is true in the case of DA and initial acquisition (Section 7.5). More often than not, however, the transmit symbols are unknown to the receiver at the timing recovery stage (i.e., the NDA case). In such cases, one can derive the likelihood function similar to (7.36), although $s(t; \theta)$ is unknown. Such a function can then be averaged over all possible transmitted data to form a likelihood function that depends on θ only. For constant-power signals such as phase shift keying (PSK) modulation, the result is a detection method similar to the baud tone technique described in Section 7.4.3.3 [16]. However, other techniques are often used in such cases, as well.

Furthermore, timing parameter θ contains both CFO and symbol timing error. According to the MAP or ML method, they should be estimated jointly based on (7.32) or (7.34). However, in practice, as discussed in Section 7.2.4, CFO and symbol timing error can be estimated separately.

7.4.2   Carrier Timing Estimation

The goal of carrier timing estimation is obtaining the CFO or, equivalently, the instantaneous phase difference between the expected signal and the actual one. As discussed in Section 7.3.1.1, CFO is reflected in the baseband signal as a phase rotation. Such phase rotation provides the basis for CFO estimation in baseband.

7.4.2.1   ML Method

The ML estimation method described in Section 7.4.1 can be used for CFO estimation. Equation (7.38) can be written as

$$\Delta \hat{f}_c = arg \max_{\Delta f_c} \Gamma(\Delta f_c)$$
$$\Gamma(\Delta f_c) \overset{\text{def}}{=} \left| \int_0^T r(t) \tilde{s}_{br}^*(t, \Delta f_c) dt \right| , \tag{7.39}$$

where $\Delta \hat{f}_c$ is the estimated CFO. The estimation spans over time T, which usually contains multiple symbols. $r(t)$ is the received baseband signal, and $\tilde{s}_{br}(t, \nu)$ is the expected received signal assuming the CFO is ν.

Based on the model given by (7.12), the expected signal can be written as

$$\tilde{s}_{br}(t, \Delta f_c) = s_b(t) e^{j2\pi\Delta f_c t} = e^{j2\pi\Delta f_c t} \sum_k c_k g(t - k T_{Sym}), \tag{7.40}$$

where $s_b(t)$ is the transmitted baseband signal, $g(t)$ is the pulse shaping function. T_{Sym} is the symbol period, and $\{c_k\}$ is the transmitted symbol sequence (see Chapter 3). Therefore, (7.39) becomes

$$\Gamma(\Delta f_c) = \left| \sum_k c_k^* e^{-j(2\pi k\Delta f_c T_{Sym})} \int_{-kT_S}^{T-kT_S} r(\tau + kT_{Sym}) e^{-j(2\pi \Delta f_c \tau)} g^*(\tau) d\tau \right|. \tag{7.41}$$

As a result of (7.9), we can make the following approximation within the support of $g(t)$, whose length is on the order of T_{sym}:

$$e^{-j2\pi \Delta f_c \tau} \approx 1. \tag{7.42}$$

Therefore,

$$\Gamma(\Delta f_c) \approx \left| \sum_k e^{-j2\pi \Delta f_c kT_{Sym}} y_k c_k^* \right|. \tag{7.43}$$

$$y_k \overset{\text{def}}{=} \int r(t + kT_{Sym}) g^*(t) dt$$

The integration limits are naturally determined by the support of $g(t)$. y_k is actually the output of the matched filter at time kT_{Sym} (see Chapter 3). Therefore, Γ in (7.43) is a symbol-by-symbol product between the transmit symbol c_k and the output of the matched filter y_k, summed with expected phase correction. From these equations, we can search for the Δf_c that maximizes Γ, thus estimating CFO.

To prepare for later discussions, let us examine more properties of $\{y_k\}$. Based on the same model as in (7.40), we have

$$r(t + kT_{Sym}) = e^{j2\pi \Delta \bar{f}_c (t + kT_{Sym})} \sum_l c_{l+k} g[t - lT_{Sym}] + n(t + kT_{Sym}), \tag{7.44}$$

where $n(t)$ is the added noise and $\Delta \bar{f}_c$ is the true CFO (as opposed to the assumed CFO Δf_c used previously). Furthermore, typically the pulse shaping function satisfies the Nyquist criterion (Chapter 3). Namely,

$$\int g(t) g^*(t - kT_{Sym}) dt = \delta_{k,0} \forall k \tag{7.45}$$

Therefore, with the same approximation (7.42), we have

$$y_k = c_k e^{j(2\pi \Delta \bar{f}_c kT_{Sym})} + n_k, \tag{7.46}$$

where $n_k \overset{\text{def}}{=} \int n(t + kT_S) g^*(t) dt$ is added noise for the digital samples. Equation (7.46) shows that the matched filter output $\{y_k\}$ is the transmitted symbol with phase rotation caused by CFO, plus additive noise. The correlation function (7.43) becomes

$$\Gamma(\Delta f_c) = \left| \sum_k e^{-j2\pi(\Delta f_c - \Delta \bar{f}_c)kT_{Sym}} |c_k|^2 + \sum_k n_k c_k^* \right|. \tag{7.47}$$

Ignoring the noise, we see that $\Gamma(\Delta f_c)$ reaches maximum when[8]

---

[8] This maximization condition is obvious because we would like to have all the terms in the summation to have the same phase. It can also be proven using the Cauchy–Schwarz inequality.

$$\Delta f_c = \Delta \bar{f}_c. \tag{7.48}$$

In other words, the ML estimation searches for the best correction of such phase rotation, to yield the maximum cross-correlation between received and expected symbol sequences. Sections 7.4.2.2 – 7.4.2.4 discuss practical methods of CFO estimate, depending on the receiver's knowledge of the signal [7].

7.4.2.2 Data-Aided (DA)

In this case, the receiver knows the transmitted signal. The ML method described in Section 7.4.2.1 can be easily applied. However, searching for Δf_c to maximize Γ is not necessarily the most efficient operation. Described next are two other practical methods.

This section assumes transmitted symbols c_k have unit power, as in the case of PSK modulations:

$$|c_k| = 1. \tag{7.49}$$

This assumption is not critical to the methods described here. However, it is an assumption made in the literature to simplify analyses. Otherwise, symbol power can be normalized at the receiver processing. Such normalization causes noise power to be unequal among the samples and introduces additional complication in optimization and performance analyses.

Let

$$z_k \stackrel{\text{def}}{=} y_k c_k^* = e^{j\left(2\pi\Delta\bar{f}_c kT_{Sym}\right)} + n_k', \tag{7.50}$$

where n_k' is the added noise:

$$n_k' \stackrel{\text{def}}{=} n_k c_k^*. \tag{7.51}$$

Except for the noise, z_k carries the phase rotation between the received symbol y_k and the expected symbol c_k. Based on such formulation, two methods are described below.

Method 1: averaging the successive phase rotation. The estimated CFO $\Delta\hat{f}_c$ is given by

$$\Delta\hat{f}_c = \frac{1}{2\pi T_{Sym}} \sum_{k=1}^{L_0-1} \gamma_k \, arg\left(z_k z_{k-1}^*\right). \tag{7.52}$$

Expression $arg(\cdot)$ computes the phase angle (in radians) of the variable. L_0 is the number of symbols used in estimation, and $\{\gamma_k\}$ is an averaging filter. For constant-power symbols satisfying (7.49), it is shown that the optimal filter (in ML sense) is [7, sec. 3.2.3]:

$$\gamma_k = \frac{3}{2} \frac{L_0}{L_0^2 - 1} \left[1 - \left(\frac{2k - L_0}{L_0}\right)^2\right], k = 1, 2, \dots, L_0 - 1. \tag{7.53}$$

In this method, $arg(z_k z_{k-1}^*)$ is the phase rotation between the two consecutive symbols. According to (7.50), it should be $2\pi T_{Sym}\Delta\bar{f}_c$. This quantity is filtered in (7.52) to reduce noise and is divided by $2\pi T_{Sym}$ to recover the CFO.

Method 2: averaging over the samples, instead of their phases. Let

$$R_m \stackrel{\text{def}}{=} \frac{1}{L_0 - m} \sum_{k=m}^{L_0-1} z_k z_{k-m}^*. \tag{7.54}$$

From (7.50), we have

$$R_m = e^{j2\pi m\Delta\bar{f}_c T_{Sym}} + n_m, \tag{7.55}$$

where n_m is the noise component, which is reduced from that of z_k by averaging. Therefore, ignoring the noise term, we have

$$\sum_{m=1}^{N} R_m = \sum_{m=1}^{N} e^{j2\pi m\Delta\bar{f}_c T_{Sym}} = \frac{\sin\left(\pi N\Delta f_c T_{Sym}\right)}{\sin\left(\pi\Delta f_c T_{Sym}\right)} e^{j\pi(N+1)\Delta\bar{f}_c T_{Sym}}. \tag{7.56}$$

The first factor is a positive real number as long as

$$N\Delta\bar{f}_c T_{Sym} \leq 1. \tag{7.57}$$

The sum limit N is chosen so that (7.57) is satisfied. Under such condition,

$$\Delta\hat{f}_c = \frac{1}{\pi(N+1)T_{Sym}} arg\left(\sum_{m=1}^{N} R_m\right). \tag{7.58}$$

Comparing the two methods, method 2 involves another layer of summation and therefore may have higher computational complexity. However, method 2 behaves better under low SNR, when the computation of $arg(z_k z_{k-1}^*)$ may be off by multiples of 2π (the wrapping problem). At high SNR, the two methods yield near-optimal performance [7, sec. 3.2.7]. More comparison between the two methods is provided in Section "Homework".

7.4.2.3   Decision Directed (DD)

DD CFO estimation is used in cases where the transmitted data $\{c_k\}$ are unknown to the receiver, but can be derived with reasonable confidence from the received signal. Assuming all data are recovered correctly, the DA methods described in Section 7.4.2.2 can be used. However, a few practical issues need to be considered.

In the presence of CFO, data decisions are likely to be incorrect over a long block because of the accumulated phase rotation. Therefore, estimation should rely only on relative phase rotations among neighboring symbols. For example, small values for N should be used in (7.58). This degrades performance when all symbols are derived correctly (see homework). However, it reduces the chances that wrong decisions introduce large estimation errors.

On the other hand, DD CFO estimation is typically conducted in a feedback timing recovery architecture (Section 7.3). In this case, estimation accuracy is not critical. Furthermore, DD estimation is typically performed continuously during data receiving, instead of one-shot. Therefore, complexity is more important. These concerns may drive different engineering choices for DD CFO estimation.

During data receiving, especially in the case of coherent demodulation, we are usually more concerned about locking the carrier phase than the frequency. Therefore, estimating phase error is often more relevant. Phase error can also be estimated based on data decision:

$$\Delta \hat{\phi}_{ck} = arg(z_k), \tag{7.59}$$

where $\Delta \hat{\phi}_{ck}$ is the estimated phase error for the kth symbol; z_k is given in (7.50), where c_k is the symbol *decision* from the receiver. $\Delta \hat{\phi}_{ck}$ can be fed to the feedback PLL, which adjusts the timing correction module accordingly to ideally keep such phase error at zero.

DD timing estimation is vulnerable to decision errors. Therefore, a low-pass filter is needed to combat these (hopefully) occasional errors. Furthermore, one can limit the range of the phase error [17]. When the absolute value of phase error is larger than a chosen threshold, the current phase error estimate is discarded, and the previous one is used. Since erroneous decisions are often accompanied with large phase errors, such limit can help prevent decision errors from corrupting CFO estimation.

7.4.2.4   Non-data-Aided (NDA)

NDA CFO estimation can be formulated based on the ML method, where the likelihood function is averaged over all possible transmitted signals (see [18, sec. 5.2-5]). However, in practice, some other methods may be more attractive.

An obvious approach for NSA CFO estimation is converting the problem into DA (i.e., with known symbols) by taking a power of the received signal. This method works for PSK-type modulation. In a M-ary phase shift keying (M-PSK) modulation, the possible phases for the transmitted signal c_k are $2\pi k/M$, where k is an integer determined by the underlying modulating data. Therefore, $\bar{c}_k \stackrel{\text{def}}{=} [c_k]^M$ has a constant phase of 0. Based on (7.46), we can consider the received signal with CFO (ignoring noise for simplicity)

$$\bar{y}_k \stackrel{\text{def}}{=} [y_k]^M = [c_k]^M e^{j(2\pi M \Delta \bar{f}_c kT_s)} = \bar{c}_k e^{j(2\pi M \Delta \bar{f}_c kT_s)}. \tag{7.60}$$

Therefore, the same methods in Section 7.4.2.2 can be used, resulting in the estimation of $M\Delta \bar{f}_c$. Similar operations can be applied to z_k as well, as detailed in [7, sec. 3.4.3].

For quadrature amplitude modulation (QAM) signals, the phase of each symbol has many possible values and thus cannot be resolved in the same way. However, a more predictable phase estimation can be achieved [7, sec. 5.7.5].

7.4.3   Symbol Timing Estimate

There is no easy way to estimate symbol timing for the feedforward timing correction architecture, where the timing error needs to be quantitatively correct. The most common method is DA using a bank of matched filters, to be described in Section 7.5.

This section discusses a few techniques for symbol timing estimation to be used with the feedback timing correction architecture. As discussed in Section 7.3.3, this architecture does not require the estimation to be very accurate, because timing errors are measured and corrected iteratively.

7.4.3.1 ML Estimation (DA Method)

When the transmitted signal is known, Section 7.4.1 gives the ML detection Equation (7.38). In the case of symbol timing estimate, we have

$$\hat{\tau} = argmax[L(\tau)]$$

$$L(\tau) \stackrel{\text{def}}{=} \int_0^T Re\,[r(t)s(t;\tau)]dt \qquad (7.61)$$

The integral limit T is chosen so that the integral covers the duration of the known signal. $r(t)$ is the received signal and $s(t; \tau)$ is the expected signal, assuming a symbol timing offset τ, which is assumed to be constant during the integration:

$$s(t;\tau) = \sum c_k g\left(t - kT_{Sym} - \tau\right), \qquad (7.62)$$

where c_k is the transmitted signal at kT_{Sym}, and T_{Sym} is symbol period. Consider the derivative of $L(\tau)$:

$$L'(\tau) = \int_0^T Re\left[r(t)\frac{ds(t;\tau)}{d\tau}\right]dt = \sum c_k \int_0^T r(t)g'\left(t - kT_{Sym} - \tau\right)dt. \qquad (7.63)$$

The integral in the last expression is a filter $g'(t)$ (time derivative of g) applied to the received signal, sampled at times $kT_{Sym} + \tau$. Therefore, $L'(\tau)$ can be computed through a filter structure with $g'(t)$ pre-computed as tap values. At the correct timing $\hat{\tau}$, $L'(\tau)$ is zero. Therefore, $\hat{\tau}$ can be found by searching for the zero of $L'(\tau)$ or the maximum of $L(\tau)$. This can be done by filter banks, similar to that described in Section 7.5.

In a feedback architecture, only the sign of the timing error is important. Because $L'(\tau)$ is typically a monotonic function of τ around $\hat{\tau}$, $L't(\tau)$ or its sign can be used as an error signal for the symbol timing adjustment loop. Figure 7.12 shows a block diagram of a timing adjustment loop based on this method. Such feedback loop drives $L'(\tau)$ toward zero. See [7, sec. 7.4.1], [18, sec. 5.3-1] for more details.

Note that this ML estimator monitors symbol time offset τ, instead of symbol frequency. It is essentially a phase lock loop.

Figure 7.12   DD ML estimator.

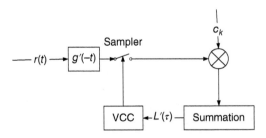

7.4.3.2 Early–Late Gate (NDA Method)

Early–late gate [18, sec. 5.3-2] is an NDA symbol timing estimation method. It is based on the knowledge that a typical pulse shaping function is symmetric and approximately bell-shaped near the center. Therefore, if the correct timing is $\hat{\tau}$, then, for the output of the matched filter $y(t)$, $y\left(kT_{Sym} + \hat{\tau} + \delta\right)$ and $y\left(kT_{Sym} + \hat{\tau} - \delta\right)$ should be the same for all $\delta \ll T_{Sym}$. Otherwise, the sampling point with larger amplitude is closer to the correct symbol time. Obviously, this method does not accurately estimate the timing error. It only indicates the sign of the timing error, which can be used to drive a timing adjustment loop in a feedback architecture.

Figure 7.13 shows a structure for early–late gate symbol timing estimate [18]. The symbol template produces the pulse shaping function, which, after delay and advance, forms two matched filters with different timing offsets. The input signal is filtered by the two matched filters, and the outputs are compared in amplitude. The comparison results are fed into a loop filter, which adjusts the symbol timing clock, driving the sampling time toward a symmetric point. In other words, the difference between the two matched filter outputs is driven toward zero by timing adjustment. Similar to the ML estimator described in Section 7.4.3.1, the early–late gate estimator also monitors symbol time offsets, instead of symbol frequency.

7.4.3.3 Baud Tone (NDA Method)

The baud tone technique is another NDA symbol timing estimate method that applies to a large class of modulations that use pulse shaping functions (see Chapter 3). This method requires no information about the signal other than the nominal symbol period and is immune to common channel distortions. This method uses a tone at the symbol rate (also called baud rate), which is generated after some signal processing (described later in this section).

The baud tone technique started with a simple squaring of digital baseband signal [19] and was refined by adding band edge filters to the scheme [20, 21]. It has broad

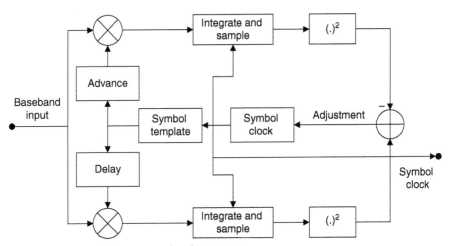

Figure 7.13 Early–late gate synchronizer.

applications in communications systems such as the Digital Subscriber Line (DSL) and the Ethernet [22]. This section derives the method and discusses some practical aspects.

7.4.3.3.1 Cyclostationarity and Modulated Signals

We consider a general class of modulated signal, including PSK, QAM, and so on. Such modulated baseband signals can be expressed as (see Chapter 3):

$$s(t) = \sum_n a_n g(t - nT_{Sym}),$$ (7.64)

where $s(t)$ is the modulated signal. g is the pulse shaping function. n is the symbol index. a_n, which is unknown to the receiver, is a complex number carrying the modulating data. T_{Sym} is the symbol period.

The modulating data are assumed to be uncorrelated:

$$E(a_n a_m^*) = P\delta_{n,m},$$ (7.65)

where P is the transmit power. We can see that the autocorrelation

$$R_S(t, \tau) \stackrel{def}{=} E(s(t)s^*(t - \tau)) = P\sum_n g(t - nT_{Sym})g^*(t - \tau - nT_{Sym})$$ (7.66)

depends on t. Therefore, it is not a wide-sense stationary process (Chapter 4). However, $R(t, \tau)$ has a period of T_{Sym} with regard to t. Such a process is said to be cyclostationary.

The autocorrelation in the frequency domain is

$$R_F(\omega, \Omega) \stackrel{def}{=} E[\tilde{s}(\omega)\tilde{s}^*(\omega - \Omega)],$$ (7.67)

where $\tilde{s}(\omega)$ is the Fourier transform of $s(t)$ (Chapter 3):

$$\tilde{s}(\omega) = \int_{-\infty}^{\infty} s(t)e^{-j\omega t}\, dt.$$ (7.68)

Therefore,

$$\begin{aligned} R_F(\omega, \Omega) &= E\left[\int s(t)s^*(t - \tau)e^{-j\omega t}e^{j(\omega - \Omega)(t - \tau)}\, dt d\tau\right] \\ &= \int R_S(t, \tau)e^{-j\Omega t}e^{j(\omega - \Omega)\tau}\, dt d\tau \end{aligned}.$$ (7.69)

As discussed in Chapter 3, the fact that $R_S(t, \tau)$ has a period of T_{Sym} means that the integration $\int R_S(t, \tau)e^{-j\Omega t}dt$ has peaks at $\Omega = 2\pi k/T_{Sym}$ for any integer k. These are known as the baud tones in our context.

7.4.3.3.2 Baud Tone

Section 7.4.3.3.1 pointed out that a cyclostationary signal has baud tones in its spectral correlation function. In this section, we study the baud tone in more details using the signal form given by (7.64). The perspective used here is different from, yet equivalent to, Section 7.4.3.3.1. The Fourier transform of the signal $s(t)$ is

$$\tilde{s}(\omega) = \tilde{a}(\omega)\tilde{g}(\omega),$$ (7.70)

where $\tilde{s}$, $\tilde{a}$, and $\tilde{g}$ are Fourier transforms of s, a, and g, respectively:

$$\tilde{s}(\omega) \overset{\text{def}}{=} \int s(t)e^{j\omega t}dt$$

$$\tilde{a}(\omega) \overset{\text{def}}{=} \sum_n a_n e^{j\omega n T_{Sym}}. \tag{7.71}$$

$$\tilde{g}(\omega) \overset{\text{def}}{=} \int g(t)e^{j\omega t}dt$$

Note that $\tilde{a}(\omega)$ is periodic:

$$\tilde{a}\left(\omega + \frac{2\pi}{T_{Sym}}\right) = \tilde{a}(\omega). \tag{7.72}$$

Furthermore, because $g(t)$ has a limited support in the time domain, $g(\omega)$ is usually a smooth function in the frequency domain.

Figure 7.14 illustrates the typical behavior of $\tilde{g}(\omega)$ and $\tilde{a}(\omega)$.[9] The thick line shows $\tilde{g}$, which remains relatively flat at low frequency and rolls off at the Nyquist frequencies $\pm\pi/T_{Sym}$. The part of nonzero $\tilde{g}$ that extends beyond the Nyquist frequency range is referred to as the "excess bandwidth" and is critical to the baud tone concept. $\tilde{a}(\omega)$ is plotted in three parts. The central part (solid thin line) extends between the plus and minus Nyquist frequencies. The other parts (dashed line on the left side and dotted line on the right side) are the same as the center part, translated by $\pm 2\pi/T_S$, following (7.72). The signal spectrum $\tilde{s}$ is the product of $\tilde{g}$ and $\tilde{a}$.

Next, consider the autocorrelation of $\tilde{s}$ in the frequency domain[10]:

$$\tilde{S}(\Omega) \overset{\text{def}}{=} \int \tilde{s}(\omega)\tilde{s}^*(\omega + \Omega)d\omega = \int \tilde{g}(\omega)\tilde{g}^*(\omega + \Omega)\tilde{a}(\omega)\tilde{a}^*(\omega + \Omega)d\omega. \tag{7.73}$$

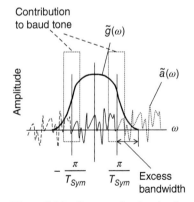

Figure 7.14   Spectra of pulse shaping function and data.

[9] Both $\tilde{g}(\omega)$ and $\tilde{a}(\omega)$ are complex numbers; their real and imaginary parts behave as depicted in Figure 7.14.
[10] Unlike in Section 7.4.3.3.1, here the autocorrelation is defined with an integral over ω instead of the expectation value.

The result would usually be close to zero because of the randomness of $\tilde{a}$. However, when the relative frequency offset is $\Omega = 2\pi/T_{Sym}$, we have $\tilde{a}(\omega) = \tilde{a}(\omega + \Omega)$. Therefore, a strong peak, known as the baud tone, would result at $\Omega = 2\pi/T_{Sym}$, indicating the symbol period T_{Sym}.

Baud tone is contributed by the parts of the spectrum in the dashed boxes in Figure 7.14, where both $\tilde{g}(\omega)$ and $\tilde{g}(\omega + \Omega)$ have significant nonzero values. Because $\tilde{g}(\omega)$ is a smooth function, the factor $\tilde{g}(\omega)\tilde{g}^*(\omega + \Omega)$ is not qualitatively important to the formation of the baud tone.

7.4.3.3.3 Symbol Timing Estimation Based on Baud Tone

Section 7.4.3.3.2 described the basic mechanism behind the baud tone. To summarize, the baud tone is generated by $\tilde{S}$, which is the autocorrelation of $\tilde{s}$ in the frequency domain. When the lag equals to $2\pi/T_{Sym}$, a strong peak (baud tone) appears in $\tilde{S}$. Its position indicates the symbol rate (i.e., the baud rate). To implement baud tone as a method for symbol timing estimate, a few practical problems need to be addressed.

First, we would like to implement the technique in the time domain to avoid Fourier transform computations. From the property of the Fourier transform, autocorrelation in the frequency domain is equivalent to a square operation in the time domain:

$$\tilde{S}(\Omega) = \int \tilde{s}(\omega)\tilde{s}^*(\omega + \Omega)d\omega = \int s(t)s^*(t)e^{j\Omega t}dt. \tag{7.74}$$

Therefore, a peak for $\tilde{S}(\Omega)$ means a tone in the time domain signal

$$S(t) \stackrel{\text{def}}{=} |s(t)|^2. \tag{7.75}$$

Second, note that only signals at the band edge (i.e., near the Nyquist frequency, as indicated by the dashed boxes in Figure 7.14) contribute to the baud tone. All other signals form a random background that obscures the baud tone. Therefore, to improve SNR for the baud one, we should perform filtering before the square operation.

Third, a filter at the output of the square operation is necessary to reject time domain noises at other frequencies.

Lastly, the baud tone frequency can be compared with the current symbol timing clock. The difference can be used as a timing error to drive clock adjustment.

Figure 7.15 shows an implementation block diagram. Note that in order to realize the spectrum beyond the Nyquist frequencies, as shown in Figure 7.14, the Sampler must perform over-sampling (l samples per symbol, with $l > 1$). The input signal

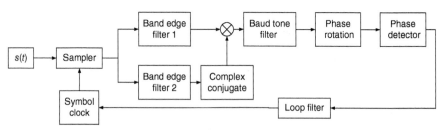

Figure 7.15   Baud tone symbol timing recovery.

$s(t)$, given by (7.64), is fed into the two band edge filters, tuned at π/T_{Sym} and $-\pi/T_{Sym}$, respectively. These filters can be implemented as one-tap IIR with

$$y_n = x_n + \alpha e^{(j\phi)} y_{n-1}, \tag{7.76}$$

where $0 < \alpha < 1$ determines the bandwidth. Note that these filter have a gain larger than 1 (determined by α). ϕ is the phase rotation for each clock advance:

$$\phi = \pm \frac{\pi}{l}. \tag{7.77}$$

Plus and minus signs in (7.77) are assigned to band edge filters 1 and 2, respectively. The complex conjugate block and the multiplier in Figure 7.15 implement (7.75). The output is filtered again at baud rate by the baud tone filter block. It can also be a one-tap IIR given by (7.76), except at double the frequency:

$$\phi = \frac{2\pi}{l}. \tag{7.78}$$

The output of the baud tone filter contains a sharp peak (the baud tone) at angular frequency $2\pi/T_{Sym}$. T_{Sym} is the true symbol period. This means the phases of the samples are

$$\varphi_n = \frac{\bar{T}_{Sym}}{T_{Sym}} \frac{2\pi}{l} n, \tag{7.79}$$

where $\bar{T}_{Sym}$ is the symbol period according to the local sampling clock. The phase rotation block imposes a progressive phase rotation of the samples:

$$z_n = y_n e^{-j\frac{2\pi}{l}n}, \tag{7.80}$$

where y_n and z_n are input and output of the phase rotation block. The resulting phase θ_n for the nth output sample z_n is

$$\theta_n \stackrel{\text{def}}{=} \varphi_n - \frac{2\pi}{l} n = \frac{\Delta T}{T_{Sym}} \frac{2\pi}{l} n, \tag{7.81}$$

$$\Delta T \stackrel{\text{def}}{=} \bar{T}_{Sym} - T_{Sym}$$

where ΔT is the symbol period error. Therefore, symbol timing error causes a phase rotation in successive samples. This is similar to the effect of CFO discussed in Section 7.4.2. From this result, ΔT can be estimated using the same methods outlined in Section 7.4.2. However, in this case, there may be relatively high noise level caused by other frequency components that pass the filters. Such noise depends on the filter designs and the signal pulse shaping function. Therefore, expected performance is best determined by simulation.

In practice, the feedback architecture is often used for baud tone symbol timing recovery. In this case, as shown in Figure 7.15, the phase detector block detects the phase change given by (7.81). A phase deviation from zero drives symbol clock frequency change. Of course, a loop filter block (described in Section 7.3.2.2) is needed to achieve a smooth and stable adjustment. Because the phase is controlled to be close to zero, the phase detector block can simply use the imaginary part of the signal to approximate its phase (with a constant scaling factor, which can be absorbed into the loop gain).

There are four filters involved in the baud tone timing recovery scheme. The bandwidth of the two band edge filters depends on the excess bandwidth of the pulse shaping filter. As shown in Figure 7.14, these band edge filters should center at the Nyquist frequency and admit the frequency components that have significant signal energy. The bandwidth of the baud tone filter, which centers at twice the Nyquist frequency, should be determined by the expected frequency error. Lastly, the loop filter design is as discussed in Section 7.3.2.2.

Figure 7.16 shows sample output spectra of a baud tone filter shown in Figure 7.15. The left panel shows the normal baud tone result. The sharp spike at

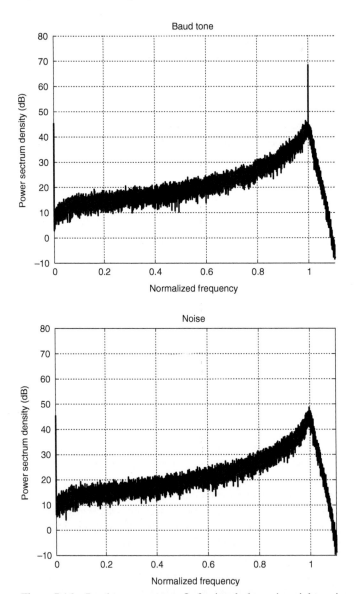

Figure 7.16   Baud tone spectrum. Left: signal plus noise; right: noise only.

normalized frequency 1 (i.e., the symbol rate) is the baud tone. The right panel shows the result of feeding noise (instead of modulated symbols) into the sampler. The baud tone is absent in this case. Such comparison shows that the sharp baud tone is indeed caused by the cyclostationarity property of the signal, instead of the band edge filters and the baud tone filter.

7.5   INITIAL ACQUISITION

The synchronization techniques discussed in Sections 7.3 and 7.4 all assume that the timing errors are small, which is the case if synchronization works as desired. However, when a receiver begins to acquire a signal, the clocks from the two sides can be far off. At this acquisition phase, different strategies are needed to achieve fast synchronization.

In many communications systems, transmissions are not continuous but come in bursts. For example, for time-division multiple access (TDMA) systems, signals are organized into timeslots or frames. Successive frames may be transmitted by different transmitters and thus have different timing errors. A receiver must achieve synchronization with a frame of signal with little prior knowledge about its timing error. Furthermore, such synchronization must be achieved quickly (i.e., without too many initial symbols out of synchronization) so that a larger portion of the frame can be received correctly.

On the other hand, if a frame is relatively short, accumulated timing drift during the frame time is not large. In this case, initial synchronization would be enough to put the whole frame in reasonably good synchronization, and no further timing adjustment is necessary, as discussed in more details in Section 7.2.3.

Suffice it to say that initial acquisition plays a critical role in communications systems. Sections 7.5.1 – 7.5.3 describe practical methods for initial synchronization. The methods rely on known symbols at the beginning of a frame (i.e., DA method). These known symbols are referred to as preambles or training symbols. There are other initial acquisition techniques that do not depend on the known symbols. They are referred to as blind synchronization. However, they are not widely used in practice, and therefore are not covered in this chapter.

7.5.1   Correlator Bank

For initial acquisition, the receiver needs to determine two parameters: the starting time of the frame and the CFO.[11] This two-dimensional search problem can be addressed by a correlator bank, based on the known transmit symbols (preambles).

---

[11] Another timing parameter is symbol frequency (baud rate). However, in most systems it is not as important. Symbol frequency is tied to carrier frequency in many systems (Section 7.2) and thus is taken care of when CFO is corrected. In some other systems, the clock tolerance is good enough for symbol synchronization without corrections (Section 7.2.3). In case symbol frequency does need to be corrected, the correlator bank method discussed here can be easily extended for such task.

The correlator bank is a collection of correlators, whose tap values are the expected received signal, with different assumptions of carrier frequency offset. In baseband, carrier frequency offset causes phase rotation of the samples, as described in Section 7.3.1.1.

When the preamble samples are $\{p_n\}$, and the assumed CFO is Δf_c, construct a sequence $\{c_n(\Delta f_c)\}$:

$$c_n(\Delta f_c) \overset{\text{def}}{=} p_n e^{j2\pi \Delta f_c T_s n}. \qquad (7.82)$$

T_s is the sample period, and n is the sample index. Larger index means later in time. A correlator is a filter whose tap values are $\{c_n^*(\Delta f_c)\}$. Multiple correlators are constructed with various values of Δf_c. Such collection of correlators is known as a correlator bank.

If the received sample sequence is $\{r_n\}$, the correlator output is

$$y_n(\Delta f_c) = \sum_k r_{n+k} c_k^*(\Delta f_c). \qquad (7.83)$$

The received signal is fed to all filters in the correlator bank. Outputs of the correlators are analyzed. Position (in time) of the peak indicates the time mark, at which a received preamble is aligned with the data pattern in the correlator [23]. This time mark is a reference point for symbol timing. The correlator that yields the maximum peak height indicates the correct CFO. The process is shown in Figure 7.17.

In addition to position and amplitude of the peaks, the phase of the peak also provides a phase reference of the received signal. This phase reference can be used as the starting point of equalizer training (Chapter 9). Therefore, a correlator bank can determine all timing parameters: frame starting time, CFO, and initial phase.

Obviously, this method follows the ML principle outlined in Section 7.4.1. Performance assessment is discussed in Section 7.5.3.

7.5.2   Coarse and Fine Acquisitions

As described in Section 7.5.1, CFO is determined by the identity of the correlator that yields the largest peak amplitude. The accuracy of such determination is limited by the

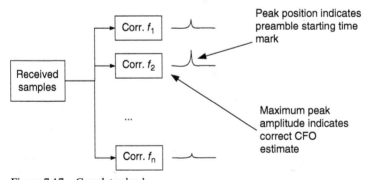

Figure 7.17   Correlator bank.

granularity of correlators, that is, the difference between the successive assumed CFO values among the correlators. On the other hand, the collection of all assumed CFO must cover the range of clock tolerance. When such a range is large and the requirement for accuracy is high, the number of correlators can be too large to be practical. In this case, a two-stage synchronization strategy can be used.

In this strategy, the first (coarse) stage uses large granularity in CFO estimation; thus, fewer correlators are required to cover the whole range. The resulted estimate may have large errors. At the second (fine) stage, a correlator bank is chosen to cover only CFO values around the coarse estimate, with finer granularity. Because the covering range is smaller, the number of correlators is still small. Therefore, this two-stage strategy can reduce the total computation load. If a bank of N filters is used for each stage, a two-stage acquisition is equivalent to a one-stage acquisition with N^2 filters. More stages can be used to achieve even finer resolution, until the desired accuracy of acquisition is achieved.

7.5.3   Preamble Design

Preambles are the known symbols, usually placed at the beginning of a frame, to help with initial acquisition. An ideal preamble sequence $\{p_n\}$ should have the following autocorrelation function:

$$E(p_n p_m) = a\delta_{n,m}. \tag{7.84}$$

This way, the correlation peak can be detected with no self-noise. Factor a is determined by the transmission power.

A frequent choice is the maximum length sequence (MLS), also referred to as the m-sequence [24]. It is a family of binary sequences (with values ± 1), with a length of $2^m - 1$, where m is an integer. The details of the MLSs can be found easily in the literature. The MLSs have the following properties. First, its autocorrelation property is near ideal:

$$\sum_{n=1}^{N} p_n p_{n+k} = \begin{cases} N & k=0 \\ -1 & \text{otherwise} \end{cases}, \tag{7.85}$$

where N is the length of the sequence. The summation is circular. Namely, p_n is extended in a periodic way:

$$p_{n+lN} = p_n, \forall l = \ldots -1, 0, 1, 2, \ldots, n = 0, 1, \ldots, N-1. \tag{7.86}$$

Second, an MLS has a small average value (1 or -1). That is, there are approximately equal numbers of 1s and -1s in the sequence.

Third, an MLS can be implemented easily. It can be generated by a set of shift registers. Therefore, one does not need to store a long sequence in memory.

However, (7.85) shows that the MLS must be long enough to exhibit good autocorrelation property; otherwise, the residual correlation of -1 becomes important. If the preamble is short, better sequences can be found through a computer search.

The length of the sequence depends on the desired performance. The ability of a receiver to correctly acquire preamble is measured by two parameters: the probability of detection P_a and the probability of false alarm P_f. The former is the probability for the receiver to declare acquisition at the correct preamble timing. The latter is the probability for the receiver to declare acquisition at any other time. Naturally, we desired that P_f is low and P_a is high. As described in Section 7.5.1, the receiver determines acquisition based on correlation peaks. The receiver sets a threshold. Acquisition is declared only when the correlation peaks exceed such threshold. The role of the threshold is to prevent false detection in the absence of preamble, when noise may accidentally produce some peaks at the correlator output. When the threshold is higher, both P_a and P_f become lower. Therefore, the choice of threshold represents a compromise between P_a and P_f. On the other hand, for a given threshold, P_f increases when SNR is lower because the noise has more chance to create false correlation peaks. For a given P_f and SNR, longer preamble produces higher P_a because the correlation peak is higher (see homework for an example). Therefore, the length of the preamble depends on the expected SNR at the receiver and the desired P_a and P_f.

7.6   SUMMARY

This chapter covered some issues one needs to consider when designing a successful synchronization scheme for a communications system. These issues are recapped from the designer's perspective.

First, we need to assess the conditions and constraints, which lead to the three questions considered in Section 7.2:

1. What are the accuracy and stability of the clocks in the system?
2. What are the required accuracies in carrier and symbol timing synchronization?
3. Are carrier and symbol clocks locked together?

Other constraints include complexity, available memory, and protocol overhead (the portion of the transmitted signal intended for synchronization assistance). It is helpful to separate these constraints into the initial acquisition phase and the regular operation phase. For initial acquisition, the synchronization task is more challenging because we start with a relatively larger timing error. However, because most of the operations for demodulation and decoding are not conducted at that time, more resources are available for synchronization.

Next, we should consider the choice between feedforward and feedback architectures (Section 7.3), which depends on the signal structure (burst or continuous) as well as the hardware design of the system (whether the clock frequency can be adjusted). Such architecture choice affects timing correction strategies as well as the choice of timing error estimation methods, as outlined in Section 7.3.3.

The choice of timing error estimation algorithm is arguably the most impactful and challenging design decision about synchronization. Sections 7.4 and 7.5 provide some options.

Synchronization is a broad topic including design and choice of clock circuitry, loop filter designs, and various timing error estimation algorithms. This chapter covers only a portion of the relevant subjects, selecting the knowledge deemed most useful in practice. For more detailed and complete treatment, the users are referred to specialized books in this area, such as [7, 10, 25]. On the other hand, for many engineering design cases, the general theories can only serve as a guide. Simulations with reasonable fidelity are needed to evaluate various architectural and algorithmic choices and to determine parameter choices.

REFERENCES

1. ETSI, *Interactive Satellite System (DVB-RCS2); Part 2: Lower Layers for Satellite Standard*. ETSI EN 301 545-2 V1.1.1, 2012.
2. Philips Semiconductors, "Clock Regenerator with Crystal-Controlled," Philips Application Notes AN182, 1991.
3. Silicon Labs Product Datasheet, "Crystal Oscillator (XO) 100 kHz to 250 MHz Si510/511," 2017.
4. Magic Xtal Ltd., "MXO37/R Ultra-High Stability Low Power Consumption Series OCXO" [Online]. Available http://magicxtal.com/products/?S=18&C=32&I=73 [Accessed March 5, 2018].
5. IEEE, *802.11-2016—IEEE Standard for Information Technology—Telecommunications and Information Exchange Between Systems Local and Metropolitan Area Networks—Specific Requirements—Part 11: Wireless LAN Medium Access Control (MAC) and Physical Layer (PHY)*. IEEE, 2016.
6. K. Bucket and M. Moeneclaey, "Effect of Random Carrier Phase and Timing Errors on the Detection of Narrowband M-PSK and Bandlimited DS/SS M-PSK Signals," *IEEE Trans. Commun.*, vol. 43, no. 2/3/4, pp. 1260–1263, 1995.
7. U. Mengali and A. N. D'Andrea, *Synchronization Techniques for Digital Receivers*. New York: Plenum Press, 1997.
8. F. J. Harris and M. Rice, "Multirate Digital Filters for Symbol Timing Synchronization in Software Defined Radios," *IEEE J. Sel. Areas Commun.*, vol. 19, no. 12, pp. 2346–2357, 2001.
9. L. Erup, F. M. Gardner, and R. A. Harris, "Interpolation in Digital Modems Part II: Implementation and Performance," *IEEE Trans. Commun.*, vol. 41, no. 6, pp. 998–1008, 1993.
10. H. Meyr, M. Moeneclaey, and S. A. Fechtel, *Digital Communication Receivers: Synchronization, Channel Estimation, and Signal Processing*. New York: Wiley, 1998.
11. J. Vesma, M. Renfors, and J. Rinne, "Comparison of Efficient Interpolation Techniques for Symbol Timing Recovery," in *Proceedings of GLOBECOM'96. 1996 IEEE Global Telecommunications Conference*, 18–28 November, London, UK, 1996, pp. 953–957.
12. G.-C. Hsieh and J. C. Hung, "Phase-Locked Loop Techniques—A Survey," *IEEE Trans. Ind. Electron.*, vol. 43, no. 6, pp. 609–615, 1996.
13. C. W. Sayre, *Complete Wireless Design*, 2nd ed. New York: McGraw-Hill, 2008.
14. C. W. Farrow, "Vector Tracking Filter," US Patent No. 5,963,594, 1999.
15. C. Herzet, H. Wymeersch, M. Moeneclaey, and L. Vandendorpe, "On Maximum-Likelihood Timing Synchronization," *IEEE Trans. Commun.*, vol. 55, no. 6, pp. 1116–1119, 2007.
16. M. Morelli, A. N. D'Andrea, and U. Mengali, "Feedforward ML-Based Timing Estimation with PSK Signals," *IEEE Commun. Lett.*, vol. 1, no. 3, pp. 80–82, 1997.
17. H. Sari and S. Moridi, "New Phase and Frequency Detectors for Carrier Recovery in PSK and QAM Systems," *IEEE Trans. Commun.*, vol. 36, no. 9, pp. 1035–1043, 1988.
18. J. Proakis and M. Salehi, *Digital Communications*, 5th ed. McGraw-Hill Education, 2007.
19. M. Oerder and H. Meyr, "Digital Filter and Square Timing Recovery," *IEEE Trans. Commun.*, vol. 36, no. 5, pp. 605–612, 1988.
20. W. A. Gardner, "The Role of Spectral Correlation in Design and Performance Analysis of Synchronizers," *IEEE Trans. Commun.*, vol. 34, no. 11, pp. 1089–1095, 1986.

21. C. W. Farrow, D. J. Udovic, V. Marandi, M. S. Mobin, K. Mondal, and K.-C. Wong, "Fixed Clock Based Arbitrary Symbol Rate Timing Recovery Loop," US Patent 6,295,325 B1, 2001.

22. K. Kim, Y. Song, B. Kim, and B. Kim, "Symbol Timing Recovery Using Digital Spectral Line Method for 16-CAP VDSL System" in *IEEE GLOBECOM 1998*, 8–12 November, Sydney, Australia, 1998, pp. 3467–3472.

23. J. L. Massey, "Optimum Frame Synchronization," *IEEE Trans. Commun.*, vol. 20, no. 2, pp. 115–119, 1972.

24. D. V. Sarwate and M. B. Pursley, "Crosscorrelation Properties of Pseudorandom and Related Sequences," *Proc. IEEE*, vol. 68, no. 5, pp. 593–619, 1980.

25. J. A. López-Salcedo, J. A. D. Peral-Rosado, and G. Seco-Granados, "Survey on Robust Carrier Tracking Techniques," *IEEE Commun. Surv. Tutorials*, vol. 16, no. 2, pp. 670–688, 2014.

HOMEWORK

7.1   To achieve fast convergence and stability, one can use two parallel loop filters, one with wide bandwidth (for fast reaction) and the other with narrow bandwidth (for noise rejection). For the ease of analysis, let us take the extreme case, where the fast filter is merely a direct feed, and the slow filter is an integrator. Namely, the sum of the output is

$$y_n = \alpha x_n + \beta \sum_{k=0}^{n} x_k.$$

Consider the case where the timing error is represented by phase, and the filter output y drives the frequency change of the clock. Namely, the timing error changes by

$$x_n - x_{n-1} = -y_{n-1} = -\alpha x_{n-1} - \beta \sum_{k=0}^{n-1} x_k$$

Or, in continuous-time approximation (when quantities change very slowly in each time step),

$$\frac{dx}{dt} = -\alpha' x - \beta' \int_0^t x(t')dt',$$

where α' and β' are, respectively, α and β multiplied by a constant, which represents the timescale. Let $z(t) \overset{\text{def}}{=} \int_0^t x(t')dt'$ and take derivative over t on both sides, we get

$$\frac{d^2 z}{dt^2} = -\frac{\alpha' dz}{dt} - \beta' z$$

a. Show that $z(t) = Ae^{ct}$ is a solution to the differential equation. Here, c is a complex number depending on α' and β' and A is a constant depending on the initial condition.

b. Find the condition of α' and β' that determines whether the system has oscillation. Look up "critical damping" to understand how the choices of filter gains (α', β') affect the dynamics of clock correction.

c. Write a simple simulation program for the original loop (i.e., without the continuous-time approximation) to show that a discrete system also displays the critical damping behavior. Experiment with the parameters and report any interesting findings.

7.2   Compare the performance of the two DA CFO estimation methods described in Section 7.4.2.2. For notations in this problem, see Section 7.4.2.2.

a. Construct a simple model by ignoring the pulse shaping function (i.e., use delta-function as g). Use a random sequence of ± 1 as symbols, and consider one sample per symbol. Let $\Delta f_c/T_s = 0.01$. Use (7.46) to construct the input signal $\{y_k\}$.

b. Perform CFO estimation using the two equations in Section 7.4.2.2: (7.52) and (7.58). Use the following parameters: $L_0 = \{128, 1024\}$; $N = \{1, 32, 64\}$. Perform estimations at various SNR values from -20 dB to 30 dB at 0.5-dB intervals.

c. For each parameter setting (L_0, N and SNR), perform CFO estimation 1000 times under different realizations of signal and noise. Observe the performance by recording the mean-squared error and mean error among the 1000 trials.

d. Plot the mean-squared error and mean error versus SNR.

e. For the first method, there appear to be three regions for mean-squared error performance. At low SNR, the estimation error stays at a high level. As SNR increases, the estimation error experiences a fast drop and then drops almost linearly at a higher SNR region. Cross-reference such behavior with the mean error plot and explain. In the first two regions, why does the second method seem advantageous?

f. Try $\Delta f_c/T_s = 0.02$. Explain the different behavior observed.

g. Experiment with other parameters (L_0, N) and report any interesting observations.

7.3 The length of the pseudorandom noise (PN) sequence depends on the requirement of the system. Consider a receiver with input $y_k = x_k + n_k$, where x_k is a PN sequence and n_k is the noise, obeying a zero-mean Gaussian distribution.

a. We wish to detect the presence of the PN sequence, which indicates the beginning of a frame, against the alternative hypothesis of nothing being transmitted. The relevant metrics are SNR, false alarm rate (the probability of claiming detection while nothing is transmitted) P_f, and the detection rate (the probability of claiming detection while the signal is transmitted) P_d. Given P_f, intuitively describe how the required length of the PN sequence change with SNR and P_d.

b. Try to develop a quantitative relationship between these quantities. Hint: you need to know the output SNR of a correlator, given the input SNR and the length of the PN sequence. You also need to know about the Q functions, which describe the probability of a Gaussian variable exceeding some threshold value. See Chapter 4 for more details.

c. Write a simple simulation to verify your formulas. You can use a sequence of uniformly distributed random numbers of ± 1 to approximate a PN sequence for the simulation. Be careful on how you compute the SNR (see Chapters 4 and 6).

7.4 An alternative to the filter bank method for CFO estimation is to let $s(t)$ be the expected received signal without any frequency offset (i.e., no phase rotation). Multiply each received sample with the complex conjugate of the corresponding sample in $s(t)$. Collect the phase. Perform a linear fit of the phase, and determine the frequency offset from its slope.

a. Simulate both methods. Use a sequence of uniformly distributed random numbers as the input sequence.

b. Compare the performance under various SNRs for frequency estimation accuracy. (Of course, the granularity of the filters in the filter bank must be small enough to ensure accuracy.)

7.5   Write a MATLAB code to reproduce the baud tune detection spectrum (Figure 7.16). Use a raised cosine filter for pulse shaping, then experiment with the simulation to gain more understanding.

   a. Add noise to the modulated signal and see how it changes the final spectra.

   b. Under different noise levels, see how the spectra respond to filter bandwidth.

   c. Does the optimal value of α depend on the excess bandwidth of the pulse shaping function?

7.6   How does the Doppler effect affect symbol period? Derive an equation to connect relative motion speed with the symbol period change.

CHAPTER 8

ADAPTIVE FILTER

8.1 INTRODUCTION

The adaptive filter is a filter whose coefficients (tap values) are updated during operation, to achieve some optimal results. Such updating can be one-time, for example, at the beginning of a communication session, or continuous, for example, as new input samples are collected.

Figure 8.1 shows the high-level physical layer architecture, discussed in Chapter 1. This chapter addresses a part of the demodulation module, the adaptive filter.

Figure 8.2 shows two adaptive filter block diagrams. Panel A is a general case. Input signal x is filtered by the adaptive filter W, producing output d. d is also fed into a decision block D, which produces an adjustment signal to update the tap values of W. Panel B shows a more specific example. In this case, the filter is optimized so that its output d matches another sequence $\hat{d}$. Therefore, the updating is driven by the error signal $e = d - \hat{d}$, and the goal is minimizing $|e|$. This example is detailed in Section 8.2 and is the focus of this chapter.

Adaptive filters can be used in communication systems in several ways. The two major applications are equalizers (Chapter 9) and self-interference canceller (Section 8.8). There are other applications in special cases.

Adaptive filter is a big subject in signal processing with many specialized books and research works. This chapter covers only the knowledge that a typical communications engineer would need to know. Section 8.9 will briefly cover some other topics and resources for further study.

In keeping with the tradition in most literature, this chapter deals with real-valued signals and filter taps. Although baseband signals are complex-valued, they can be viewed as two real-valued signals with cross-links (Chapter 3). Therefore, the methods discussed here are applicable to baseband signal processing. Chapter 9 deals with complex-valued adaptive filters. Results in this chapter can be extended to complex-valued cases, with some precaution [1, ch. 14].

Digital Communication for Practicing Engineers, First Edition. Feng Ouyang.
© 2020 by The Institute of Electrical and Electronics Engineers, Inc.
Published 2020 by John Wiley & Sons, Inc.

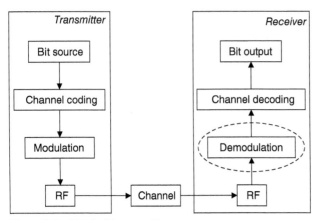

Figure 8.1   Physical layer architecture.

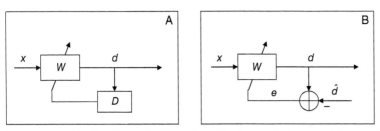

Figure 8.2   Adaptive filter. A: General architecture and B: a special case where D is a comparator.

Many variables and parameters used in this chapter are vectors and matrixes. Notation conventions outlined in Chapter 1 are used here. Especially, vectors are expressed in bold-face lowercase letters, while matrices are in uppercase letters.

For the rest of the chapter, Section 8.2 formulates the adaptive filter as an optimization problem, whose optimal solution is provided in Section 8.3. Section 8.4 describes an iterative solution that avoids matrix inversion operation. Such a method also leads to the sample-by-sample adaptation method, which does not need a priori knowledge of signal statistics. This broadly used algorithm is known as the least mean squares (LMS) and is described in Section 8.5. Section 8.6 introduces a block-based adaptation algorithm known as least squares (LS). The matrix inversion involved in the LS algorithm can be built up iteratively, leading to the recursive least squares (RLS) algorithm described in Section 8.7. After explaining the various algorithms, Section 8.8 outlines a sample application, the full-duplex radio. This example shows practical considerations, challenges, and power of adaptive filters. A summary is given in Section 8.9.

8.2   ADAPTIVE FILTER OVERVIEW

8.2.1   Problem Formulation

This chapter considers only the finite impulse response (FIR) filters. The filter operation can be formulated as vector inner product:

$$d_n = w^H x(n) = \sum_{l=1}^{L} w_l x_{n-l+1}. \tag{8.1}$$

Here, d_n is the output sample from the filter at time step n. w is a vector of dimension L, denoting the filter tap values $\{w_l, l = 1, 2, \ldots, L\}$, L being the filter length. $x(n)$ is a vector of dimension L, representing the input data residing in filter delay line at time step n.

$$x(n) \overset{\text{def}}{\equiv} \begin{pmatrix} x_n \\ x_{n-1} \\ \vdots \\ x_{n-L} \end{pmatrix}. \tag{8.2}$$

x_n is the input data at step n. $\{x_n\}$ are the samples of a wide-sense stationary (WSS) stochastic variable.[1] Note that elements of x are arranged backward: the most recent sample is the first element.

In general, an adaptive filter has the optimal tap values satisfying the following condition:

$$w_o = arg \min_w c(\{d_n, n = 1, 2, \ldots\}). \tag{8.3}$$

Here, the optimization target c is known as the cost function.[2] To summarize, an adaptive filter consists of three parts: the filter (8.1), the cost function, and the adaptation algorithm that computes w_o from (8.3).

In this chapter, we consider a special case, where we desire that the output sequence $\{d_n\}$ matches a given WSS sequence $\{\hat{d}_n\}$. Define the matching error as

$$e_n \overset{\text{def}}{\equiv} d_n - \hat{d}_n. \tag{8.4}$$

---

[1] More discussions on WSS stochastic variables can be found in Chapter 4.
[2] Another equivalent formulation is $w_o = arg \max_w u(\{d_n, n = 1, 2, \ldots\})$, where u is known as the utility function.

We choose to use the following cost function, known as the mean-squared error (MSE), defined as

$$c(\{d_n\}) \stackrel{\text{def}}{=} E(e_n^2). \tag{8.5}$$

Since $\{x_n\}$ and $\{\hat{d}_n\}$ are WSS, so are $\{d_n\}$ and $\{e_n\}$. Therefore, the right-hand side in (8.5) is independent of n, as to be expected from a cost function. Since $\{d_n\}$ is a function of w, so is c. Equations (8.1) and (8.3)–(8.5) formulate the problem to be addressed in the rest of this chapter.

8.2.2   Application Examples

Figure 8.3 shows three examples of adaptive filter application. Panel A shows an adaptive filter w trying to match a system filter w_s, so that their outputs are close to each other. This case is further discussed in Section 8.3.3. An example of such configuration is the self-interference canceller discussed in Section 8.8. Panel B shows an adaptive filter trying to "reverse" the effect of a system filter, so that its output matches

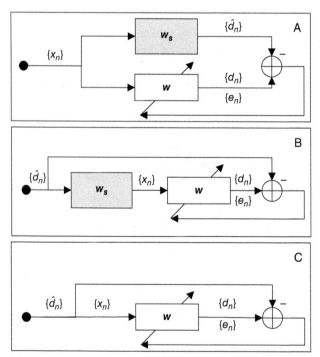

Figure 8.3   Adaptive filter application examples. A: matching w with w_s; B: using w to cancel w_s; and C: using w to predict input samples.

original data. The linear channel equalizer, discussed in Chapter 9, is an example. Panel C shows an adaptive filter trying to "predict" the input signal, so that its output cancels the *next* input sample. An example of Panel C is the noise predictive filter, as discussed in Chapter 9.

8.3   OPTIMAL SOLUTION

In this section, we study the solution to the optimization problem formulated in Section 8.2.1. All input signals $\{x(n)\}$ and $\{\hat{d}_n\}$ are WSS, with known statistical properties. As stated before, in this chapter we focus on the MSE optimization (8.5).

8.3.1   Wiener–Hopf Equation

Given $\{\hat{d}_n\}$ and $\{x_n\}$, c is a function of w. Therefore, the minimum point in (8.3) is given by

$$\nabla_w c(w) = 0. \tag{8.6}$$

In the special case of MSE optimization, Equations (8.1), (8.4), and (8.5) lead to

$$\nabla_w c(w) = 2E(e_n \nabla_w e_n) = 2E[e_n x(n)]. \tag{8.7}$$

Thus, the optimum condition (8.6) implies

$$E[e_n x(n)] = 0. \tag{8.8}$$

With (8.1) and (8.4), (8.8) can be written as

$$E\{x(n)[x^T(n)w - \hat{d}_n]\} = 0 \tag{8.9}$$

or

$$R_{xx}w = r_{xd}$$
$$R_{xx} \overset{\text{def}}{=} E[x(n)x^T(n)]. \tag{8.10}$$
$$r_{xd} \overset{\text{def}}{=} E[x(n)\hat{d}_n]$$

Because $\{x(n)\}$ and $\{\hat{d}_n\}$ are WSS, their autocorrelation and correlation, R_{xx} and r_{xd}, are independent of n. Equation (8.10) is known as the Wiener–Hopf equation. The optimal filter is thus[3]

---

[3] Here, R_{xx} is assumed to be full rank. Otherwise, the basic scheme still works but the mathematics is more involved.

$$w_o = R_{xx}^{-1} r_{xd} \tag{8.11}$$

Equation (8.11) shows that the optimal solution depends only on the second-order statistics of input data, that is, R_{xx} and r_{xd}. Under solution (8.11), the cost function from (8.4) and (8.5) is

$$
\begin{aligned}
c_o \stackrel{\text{def}}{=} E\left(e_n^2\right)\big|_{w=w_o} &= E\left\{\left[w_o^T x(n) - \hat{d}_n\right]^2\right\} \\
&= w_o^T E\left[x(n) x^T(n)\right] w_o - 2 w_o^T E\left[x(n)\hat{d}_n\right] + \hat{d}_n^2 \\
&= r_{xd}^T R_{xx}^{-T} R_{xx} R_{xx}^{-1} r_{xd} - 2 r_{xd}^T R_{xx}^{-T} r_{xd} + \hat{d}_n^2 \\
&= \hat{d}_n^2 - r_{xd}^T R_{xx}^{-T} r_{xd}
\end{aligned} \tag{8.12}
$$

This result can also be written as

$$c_o = E\left(e_n \hat{d}_n\right)\big|_{w=w_o}. \tag{8.13}$$

8.3.2 Principle of Orthogonality

Equation (8.8) can be written more explicitly:

$$E(e_n x_{n-l+1}) = 0, \forall l = 1, 2, \ldots, L. \tag{8.14}$$

This property is known as the principle of orthogonality [2, sec. 2.2]. Equations (8.8) and (8.14) hold only when optimization (8.6) is achieved.

The principle of orthogonality means the error is uncorrelated with the input data within the "horizon" of the filter. This property can be understood with the following intuition: if the correlation between e_n and $x(n)$ is nonzero, it means we can predict a part of e_n from $x(n)$. Since $x(n)$ is known, we can further reduce e_n based on such information (see homework).

Equation (8.14) also implies

$$E(e_n d_n) = 0. \tag{8.15}$$

8.3.3   Unbiased Estimation

Now consider the case where $\{\hat{d}_n\}$ is actually produced by a filter:

$$\hat{d}_n = \bar{w}^T x(n) + z_n = x^T(n)\bar{w} + z_n, \tag{8.16}$$

where $\bar{w}$ is the "true" filter. z_n is the noise at step n, which is uncorrelated with $x(n)$. The questions are how w_o given by (8.11) is related to $\bar{w}$. Namely, how does the optimal filter match the "true" filter in a linear system. From (8.16), we have

$$r_{xd} = E\left[x(n)\hat{d}_n\right] = E\left[x(n)x^T(n)\bar{w}\right]. \tag{8.17}$$

Therefore, (8.11) leads to

$$w_o = \bar{w}. \tag{8.18}$$

In other words, even in the presence of noise, the adaptive filter can still reproduce the "true" filter of a linear system. Such property is independent of the autocorrelations of input data $\{x(n)\}$ and that of the noise $\{z_n\}$, as long as $\{x(n)\}$ and $\{z_n\}$ are uncorrelated. From (8.18), the optimal filter w_o is said to be an "unbiased" estimate of the true filter $\bar{w}$.

In this case, the remaining error is just the added noise:

$$e_n = d_n - \hat{d}_n = z_n. \tag{8.19}$$

Note that the adaptive filter works even if (8.16) does not hold. For example, $\{\hat{d}_n\}$ may have a nonlinear relationship with $\{x_n\}$. Or, even if they are linearly dependent, the true filter $\bar{w}$ may be longer than the adaptive filter length L. In such cases, w_o *approximates* the true system response.

8.4   ITERATIVE SOLUTION: SPEEDIEST DESCENT (SD)

Section 8.2 formulates the adaptive filter problem with (8.3) and (8.5). The optimal solution (8.11) is derived in Section 8.3. In practice, however, computation of matrix inversion can be expensive in resource demand. In this section, we study an alternative approach: the iterative solution. In addition to reducing complexity, the iterative method can also respond to system changes, as shown in Section 8.5.

8.4.1   The Steepest Descent (SD) Method

Consider an optimization problem:

$$w_o = arg\,\min_w c(w). \tag{8.20}$$

Here, w is a vector variable. c is the cost function to be minimized, and w_o is the optimal point we are looking for. The iterative method starts with some value of w and updates the value in multiple steps, in the hope of approaching the optimal point. The updating process subtracts a displacement δ from the previous value of w:

$$w^{(n)} = w^{(n-1)} - \delta^{(n)}. \tag{8.21}$$

Here, $w^{(n)}$ and $\delta^{(n)}$ are w and the displacement at step n. With the steepest descent (SD) method, the displacement $\delta^{(n)}$ is chosen so that c decreases the fastest along this direction. Namely,

$$\delta^{(n)} = \nu \nabla_w c(w)|_{w=w^{(n-1)}}, \tag{8.22}$$

where ν is a positive scalar representing step size. The gradient operator ∇ is defined as

$$\nabla_w c(w) \overset{\text{def}}{=} \begin{pmatrix} \dfrac{\partial c}{\partial w_1} \\[6pt] \dfrac{\partial c}{\partial w_2} \\[6pt] \dfrac{\partial c}{\partial w_3} \\[6pt] \vdots \\[6pt] \dfrac{\partial c}{\partial w_L} \end{pmatrix} \tag{8.23}$$

From calculus, when ν is small enough, (8.21) leads to

$$c\left(w^{(n)}\right) = c\left(w^{(n-1)}\right) - \nabla_w^T c\left(w^{(n-1)}\right) \delta_n. \tag{8.24}$$

From (8.22),

$$c\left(w^{(n)}\right) = c\left(w^{(n-1)}\right) - \nu \left|\nabla_w c\left(w^{(n-1)}\right)\right|^2 \tag{8.25}$$

Equation (8.25) shows that when ν is small enough, the cost function c always decreases through iteration, unless a minimum point is reached so that $\nabla_w c(w)$ is zero. Therefore, with small enough ν, SD can always reach a minimum point. However, it could be a local one instead of the global one.

8.4.2   Application to the Adaptive Filter Problem

For the adaptive filter problem (8.5),

$$\nabla_w c(w) = 2R_{xx}w - 2r_{xd}. \tag{8.26}$$

Therefore, SD algorithm (8.21) and (8.22) can be written as

$$w^{(n)} = w^{(n-1)} - \mu\left(R_{xx}w^{(n-1)} - r_{xd}\right). \tag{8.27}$$

Here, $\mu = 2\nu$ compared to (8.22). Since the cost function (8.5) has only one minimum point, SD always converges correctly when μ is small enough.

Equation (8.27) can also be written in terms of the optimal solution (8.11):

$$w^{(n)} = w^{(n-1)} + \mu R_{xx}\left(w_o - w^{(n-1)}\right). \tag{8.28}$$

8.4.3   Convergence Speed and Step Size

So far, we have seen that the SD method updates the filter taps w, which converges to the optimal value if the step size μ is small enough. In this section, we study how to choose μ so that an optimal solution can be obtained with fewer steps of iteration.

Let us define $u^{(n)}$ as the difference between a filter vector and the optimal one at the nth step:

$$u^{(n)} \stackrel{\text{def}}{=} w_o - w^{(n)}. \tag{8.29}$$

Equation (8.28) leads to

$$u^{(n)} = \left(1 - \mu R_{xx}\right)u^{(n-1)}. \tag{8.30}$$

Obviously, $\{u^{(n)}\}$ converges to zero. Let us study its convergence behavior in terms of the eigenvalues and eigenvectors of $(1 - \mu R_{xx})$. Since R_{xx} is symmetric, so is $(1 - \mu R_{xx})$. Therefore, its eigenvectors form a complete orthogonal basis, upon which $\{u^{(n)}\}$ can be projected.

The eigendecomposition of R_{xx} is

$$R_{xx} = U\Lambda U^T. \tag{8.31}$$

Here, U is an orthogonal matrix, whose columns are the eigenvectors of R_{xx}. Λ is a diagonal matrix, whose diagonal values are the eigenvalues of R_{xx}. It is easy to see that

$$(1 - \mu R_{xx}) = U(I - \mu\Lambda)U^T. \tag{8.32}$$

Let $v^{(n)}$ be the result of $u^{(n)}$ projected onto the eigenvectors in U:

$$v^{(n)} \stackrel{\text{def}}{=} U^T u^{(n)}. \tag{8.33}$$

$\{\mathbf{v}^{(n)}\}$ also converges to zero. Because U is orthogonal, we have

$$\mathbf{u}^{(n)} = U\mathbf{v}^{(n)}. \tag{8.34}$$

Equations (8.30), (8.32), and (8.33) lead to

$$\mathbf{v}^{(n)} = (I - \mu\Lambda)\mathbf{v}^{(n-1)}. \tag{8.35}$$

Since matrix $(I - \mu\Lambda)$ is diagonal, we can write the iteration separately for each component:

$$v_l^{(n)} = (1 - \mu\lambda_l)v_l^{(n-1)} \forall l = 1, 2, \ldots, L, \tag{8.36}$$

where λ_l is the lth eigenvalue of R_{xx} (i.e., the lth diagonal element in Λ). Define the maximum and minimum eigenvalues λ_{max} and λ_{min}:

$$\begin{aligned} \lambda_{max} &\overset{\text{def}}{=} \max\{\lambda_l\} \\ \lambda_{min} &\overset{\text{def}}{=} \min\{\lambda_l\} \end{aligned}. \tag{8.37}$$

Because R_{xx} is positive definite,[4] all of its eigenvalues $\{\lambda_l\}$ are positive. Equation (8.36) leads to

$$v_l^{(n)} = (1 - \mu\lambda_l)^n v_l^{(0)} \forall l = 1, 2, \ldots, L. \tag{8.38}$$

Here, $v_l^{(0)}$ is the initial value of v_l. The iteration converges if and only if

$$|1 - \mu\lambda_l| < 1 \forall l = 1, 2, \ldots, L, \tag{8.39}$$

which leads to

$$0 < \mu < \frac{2}{\lambda_{max}}. \tag{8.40}$$

Even within this range, μ can be optimized for faster convergence. For the lth component, v_l converges faster when $|1 - \mu\lambda_l|$ is smaller. However, the overall convergence speed is limited by the slowest converging component, which is associated with either λ_{max} or λ_{min}. Therefore, optimum convergence behavior can be obtained when

$$|1 - \mu\lambda_{max}| = |1 - \mu\lambda_{min}|. \tag{8.41}$$

---

[4] R_{xx} is assumed full rank.

This relationship leads to the optimal step size:

$$\mu_o = \frac{2}{\lambda_{max} + \lambda_{min}}. \tag{8.42}$$

Of course, λ_{max} and λ_{min} can be expensive to compute in real time. Usually, the step size is determined during the design, based on expected system characteristics.

The optimal convergence speed depends on the difference between λ_{max} and λ_{min} Such difference can be measured by the *condition number* of R_{xx}, noted as κ:

$$\kappa(R_{xx}) \overset{\text{def}}{=} \left| \frac{\lambda_{max}}{\lambda_{min}} \right|. \tag{8.43}$$

Convergence speed is faster when κ is close to 1, and slower for larger κ values.

If $\{x_n\}$ are independent samples, then R_{xx} is an identical matrix multiplied by a constant scalar λ. This is the most advantageous case in SD convergence. In fact, when μ is chosen to be $1/\lambda$, the iteration can converge in just one step. Of course, this effectively returns to the one-step solution (8.11).

Let us look at an example with $L = 2$. The two eigenvalues of R_{xx} are $\lambda_1 = 1.0$ (the horizontal dimension) and $\lambda_2 = 0.2$ (the vertical dimension). The optimal step size given by (8.42) is approximately 1.67. The maximum step size from (8.40) is 2.

Figure 8.4 shows the convergence trace of $v^{(n)}$ given by (8.38) with different step sizes, for the first 10 steps. The dashed line ($\mu = 1.90$) has a step size larger than optimal. It converges well at the vertical dimension (i.e., approaching 0 quickly). However, at the horizontal dimension, it oscillates with a slowly decreasing amplitude (note that $1 - \mu\lambda_1 < 0$). The dotted line ($\mu = 0.83$) has a step size smaller than optimal. It converges well at the horizontal dimension. However, it moves slowly at the vertical dimension. The solid line ($\mu = 1.67$) is the optimal choice, where converging speeds at both dimensions are balanced.

8.5 SAMPLE-BY-SAMPLE ADAPTATION: LEAST MEAN SQUARES (LMS) ALGORITHM

Sections 8.3 and 8.4 present an optimal solution to the adaptive filter problem, based on signal statistics, as shown in (8.11) and (8.27). However, in practice, these statistics are not always known or constant over time. In this section, we discuss an iterative optimization method, which builds up signal statistics during iteration. This method is known as the least mean squares (LMS). It is a stochastic gradient adaptive algorithm.

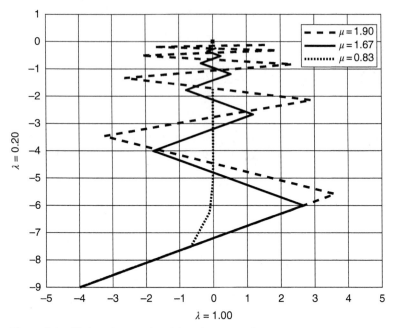

Figure 8.4   SD iteration results with various step sizes.

8.5.1 LMS Algorithm

The SD algorithm (8.27) requires the knowledge of R_{xx} and r_{xd}, as defined in (8.10). When they are not available, we can approximate them with current sample estimates $x(n)x^T(n)$ and $x(n)\hat{d}_n$. Therefore, (8.27) becomes

$$w^{(n)} = w^{(n-1)} - \mu \left[x(n)x^T(n)w^{(n-1)} - x(n)\hat{d}_n \right]. \tag{8.44}$$

Recalling (8.1) and (8.4), (8.44) can be written as

$$w^{(n)} = w^{(n-1)} - \mu x(n)e'_n. \tag{8.45}$$

Note that e'_n is the error given by (8.4) based on the filter estimate at the *last step* $w^{(n-1)}$[5]:

$$e'_n \overset{\text{def}}{=} x^T(n)w^{(n-1)} - \hat{d}_n. \tag{8.46}$$

---

[5] Be careful when comparing with literatures; the definition of error e_n may have a different sign in other publications.

Equation (8.45) is the LMS algorithm.

To understand the effect of filter update, let us compare the cost functions when $w^{(n)}$ or $w^{(n-1)}$ are applied to $x(n)$.

$$\Delta c \overset{\text{def}}{=} \left[x^T(n)w^{(n)} - \hat{d}_n \right]^2 - \left[x^T(n)w^{(n-1)} - \hat{d}_n \right]^2 \tag{8.47}$$

Recalling (8.45) and (8.46), (8.47) leads to

$$\Delta c = \left[e_n' - \mu x(n)^T x(n) e_n' \right] - e_n'^2 = -2\mu x(n)^T x(n) e_n'^2 + \mu^2 \left[x(n)^T x(n) \right]^2 e_n'^2 \tag{8.48}$$

When μ is small enough, Δc is always smaller than 0. This means the cost function decreases after each filter updating. It should be pointed out, however, that in reality, the filter does not always apply to the same input signal. With a different instance of the input signal, the cost function could increase.

The SD algorithm in Section 8.4 iterates based on deterministic data R_{xx} and r_{xd}, while LMS is based on stochastic data $x(n)$ and $\hat{d}_n$. Intuitively, these stochastic data fluctuate around their expectation values, which are used by SD. Over the course of iteration, such fluctuation largely cancels out, allowing LMS to reach the optimal result.

Another feature of LMS is responding to the change of statistics. Since input samples $\{x(n)\}$ and $\{\hat{d}_n\}$ are used in the updating process (8.45), change of their statistics affects future updating. Therefore, the resulting w "follows" the change of input signals.

In an environment where fast adaptation is desired, the LMS algorithm can be modified, so that old training results are "forgotten" to some extent. Such thought leads to the *leaky LMS*, where the updating is

$$w^{(n)} = (1 - \mu\beta)w^{(n-1)} - \mu x(n)e_n. \tag{8.49}$$

The newly introduced coefficient β causes "forgetfulness" of the previous filter tap values, putting more weight on the new signal. Leaky LMS increases adaption speed, but they introduce an additional "pressure" to reduce the filter tap values. On the other hand, such treatment also introduces a bias to the adaptation result, increasing the cost function. Therefore, the value of β should be kept small.

Even with a stationary input, leaky LMS may still be desirable. As discussed in Section 8.4.3, convergence speed is different among the eigenvectors of R_{xx}, and can be quite slow if $1 - \mu\lambda_l$ is close to ± 1 for some l, where λ_l is the corresponding eigenvalue of R_{xx}. Therefore, if a random fluctuation in the input "kicks off" an error proportional to that eigenvector, it will take a long time for the error to dissipate. Leaky LMS helps to shorten such recovering process.

8.5.2   Choice of Step Size

As with the SD algorithm, step size in LMS is important for performance. Detailed analyses can be found in the literature [2, sec. 4.5]. On the other hand, in most practical cases, the step size is best chosen through simulation. The following considerations may be helpful in guiding the search for optimal step sizes.

1. As with SD, if the step size is too big, the LMS adaption will be unstable. Since instability may also be caused by implementation errors, it would be a good idea to start testing with a small step size to simplify diagnostic.

2. Even when adaptation is stable and converged, a large step size leads to higher "residual error." With large step sizes, the adaptive filter is more sensitive to "kicks" from noises and fluctuations, thus is more likely to deviate from its optimal solution.

3. If the step size is too small, adaptation converges slowly in initial training. The filter is also slow to respond to signal statistic changes.

Because of these conflicting requirements, the practice of "gear shifting" is often used. The filter operation is divided into a training phase and a steady phase. A large step size is used in the training phase to induce fast convergence, while a smaller step size in the steady phase reduces residual error. More stages of gear shifting can be designed to achieve a more optimal convergence schedule. Typically, such a strategy and schedule are predetermined through simulation. Gear shifting can also be adaptive, based on current error level.

Another factor affecting convergence is the filter length L. Longer filters usually require smaller step sizes for stability, and they converge slower. Therefore, it sometimes pays to use a shorter filter, even when it leads to a higher residual error.

Another issue for stochastic training is the "colored input" [3]. As discussed in Section 8.4.3, convergence speed depends on the condition number of R_{xx}. If the input signal has independent samples (said to be "white"), the condition number is 1, which is the most favorable situation. If we have some knowledge about R_{xx} during the design process, it may be beneficial to use a prefilter, in order to produce more "white" input to the adaptive filter. Such filter is known as a prewhitening filter.

8.5.3   Variances of LMS Algorithm

The LMS algorithm can be varied for several purposes.

The sign error algorithm (SEA) is a variation to reduce computational load. The iteration Equation (8.45) is modified to

$$w^{(n)} = w^{(n-1)} - \mu x(n) \, sign\left(e'_n\right) \qquad (8.50)$$

By replacing e'_n with $sign\left(e'_n\right)$, multiplication operation is avoided.[6] This also has the benefit of clipping the large errors, making the adaptation process more stable.

---

[6] Multiplying by μ can be implemented as a binary shift operation when μ is in the form of 2^{-k} and k is an integer.

Another way to reduce computational load is replacing $x(n)$ with its sign, known as the sign data algorithm (SDA).

$$w^{(n)} = w^{(n-1)} - \mu sign[x(n)]e'_n. \tag{8.51}$$

Both SEA and SDA have poorer convergence performances and higher levels of residue error [2, sec. 4.7].

Another variation is the normalized LMS (NLMS). Its iteration equation is

$$w^{(n)} = w^{(n-1)} - \frac{\mu}{|x(n)|^2 + \delta} x(n)e'_n, \tag{8.52}$$

where δ is a small number for avoiding divergence when $|x(n)|^2$ is zero. Comparing with LMS, NLMS provides larger step size under weak signal, improving convergence speed. As shown in (8.42), the optimal step size is inversely proportional to the eigenvalues of R_{xx}. Such property applies to LMS, as well. When x is scaled, the eigenvalues of R_{xx} scales with $|x(n)|^2$. Therefore, NLMS is able to maintain optimal step size in the cases that the average power of $\{x(n)\}$ is not stationary (such as in speech processing).

NLMS has other mathematical justifications and implications, as detailed in the literature [2, sec. 4.2], [4].

8.6 BLOCK-BASED ADAPTATION: LEAST SQUARES (LS) ALGORITHM

The sample-by-sample adaptation discussed in Section 8.5 builds up signal statistics while performing adaptation. It has low computational complexity and continuously responds to changes in signal statistics. However, it is in general slow in convergence. In this section, we discuss an alternative method: block-based adaptation. As in Section 8.5, we do not rely on the prior knowledge about signal statistics. However, adaptation is based not on the current sample, but a block of collected samples.

8.6.1 Problem Formulation

We consider a "block," which collects data from M filter operations. These operations follow (8.1), with n running from 1 to M. Let X be a matrix collecting input signal $\{x(n)\}$. Namely,

$$X \stackrel{\text{def}}{=} \begin{pmatrix} x^T(1) \\ x^T(2) \\ \vdots \\ x^T(M) \end{pmatrix} \tag{8.53}$$

X is of size $M \times L$. L is the length of the filter. Similarly, we define vectors d and $\hat{d}$ to collect filter output and data to be matched over the M time steps.

$$d \stackrel{\text{def}}{=} \begin{pmatrix} d_1 \\ d_2 \\ \vdots \\ d_M \end{pmatrix}$$

$$\hat{d} \stackrel{\text{def}}{=} \begin{pmatrix} \hat{d}_1 \\ \hat{d}_2 \\ \vdots \\ \hat{d}_M \end{pmatrix}$$

(8.54)

Similar to (8.1), the filter output can be expressed as

$$d = Xw. \tag{8.55}$$

Similar to (8.4) and (8.5), the cost function is

$$c \stackrel{\text{def}}{=} \left| d - \hat{d} \right|^2. \tag{8.56}$$

Note that unlike (8.5), there is no expectation value in (8.56). In other words, the optimization scope is limited to the current block.

8.6.2   The Least Squares (LS) Algorithm

We can apply the same method as in Section 8.3 to get the optimal solution. Equations (8.55) and (8.56) can be written as

$$c = w^T X^T X w - w^T X^T \hat{d} - \hat{d}^T X w + \hat{d}^T \hat{d}. \tag{8.57}$$

The optimal condition is

$$c \nabla_{w^T} c \big|_{w = w_o} = 0, \tag{8.58}$$

where w_o is the optimal filter. This leads to[7]

---

[7] Here, we treat w and w^T as independent variables. This is a common practice when dealing with quadratic forms such as (8.58). The validity of such practice can be proved rigorously, in Homework of this chapter.

$$X^T X w_o = X^T \hat{d}. \tag{8.59}$$

Equation (8.59) is known as the normal equation, which can be used to compute the optimal filter w_o. Note that because of the size of X, its rank is no more than $\min(M, L)$. On the other hand, the size of $X^T X$ is $L \times L$. Therefore, when $M < L$, $X^T X$ is not full rank. However, (8.59) is always solvable regardless of the rank of X, as to be shown below.

Perform singular value decomposition (SVD) to X:

$$X = U \Sigma V^H. \tag{8.60}$$

Here, U and V are the unitary matrices of sizes $M \times M$ and $L \times L$, respectively. Σ is a diagonal matrix of size $M \times L$, whose diagonal elements are the singular values of X. If K of the singular values are nonzero, the rank of X is K. Without losing generality, we assume the first K diagonal values of Σ are nonzero. Taking advantage of the properties of unitary matrices, (8.59) can be written as

$$\begin{aligned} \Lambda V^H w_o &= \Sigma U^H \hat{d} \\ \Lambda &\stackrel{\text{def}}{=} \Sigma^H \Sigma \end{aligned}. \tag{8.61}$$

Λ is a diagonal matrix of size $L \times L$, with the first K diagonal values being nonzero. Equation (8.61) represents K linear equations for $\{ w_l, l = 1, 2, \dots, L \}$. Since $K \le L$, there are always nontrivial solutions. When $K < L$, this is an underdetermined problem. Namely, when $M < L$, there are multiple filters yielding the same minimum cost function value.

In parallel with Section 8.3.3, let us see how LS estimates a linear system. In this case, assume

$$\hat{d} = X \bar{w} + z, \tag{8.62}$$

where $\bar{w}$ is the actual filter in the system. z is a vector representing noise. Equation (8.59) yields (assuming $X^T X$ is full rank for simplicity)

$$w_o = \bar{w} + \left(X^T X \right)^{-1} z. \tag{8.63}$$

Due to noise, w_o is not exactly the same as $\bar{w}$. However, we still have (when the noise has zero mean)

$$E(w_o) = \bar{w}. \tag{8.64}$$

Therefore, LS is still an unbiased estimate of the true filter. In this case, LS is actually the best linear estimator in the sense that the estimation error

$$g \stackrel{\text{def}}{=} E \left[|w_o - \bar{w}|^2 \right] \tag{8.65}$$

is the lowest among all w_o generated by various estimation methods [2, sec. 5.4].

8.7 BLOCK-BASED ITERATION: RECURSIVE LEAST SQUARES (RLS) ALGORITHM

While LS provides an optimal filter based on a block of data, it is a "one-shot" operation: the next update requires another complete computation. LS algorithm also involves matrix inversion, which has high computational complexity. In this section, we introduce a more practical algorithm: recursive least squares (RLS). There are many versions of RLS, with different pros and cons. We will look at the exponentially weighted RLS (EWRLS).

The derivation of the RLS algorithm is somewhat cumbersome. However, the resulted algorithm is rather straightforward.

8.7.1 Problem Formulation

RLS is an iterative algorithm, with a solution that evolves with time. We start with rewriting the problem formulation in Section 8.2.1, with a slightly different notation for clarity. Equation (8.1) is rewritten as

$$d^{(n)} = x^T(n)w^{(n)}. \tag{8.66}$$

Here, $d^{(n)}$ is the filter output at step n. $x(n)$ represents the input signal in the filter delay line at step n. $w^{(n)}$ is the filter computed at step n. The output is to match with the expected output $\hat{d}^{(n)}$, and the error at step n is defined as

$$e^{(n)} \overset{\text{def}}{=} d^{(n)} - \hat{d}^{(n)}. \tag{8.67}$$

Now, similar to (8.53) and (8.54), we accumulate all of these data from step 1 to n and define the following vectors and matrices:

$$
\begin{aligned}
d(n) &\overset{\text{def}}{=} \left(d^{(1)}, d^{(2)}, \ldots, d^{(n)}\right)^T \\
\hat{d}(n) &\overset{\text{def}}{=} \left(\hat{d}^{(1)}, \hat{d}^{(2)}, \ldots, \hat{d}^{(n)}\right)^T \\
X(n) &\overset{\text{def}}{=} (x(1), x(2), \ldots, x(n))^T \\
e(n) &\overset{\text{def}}{=} \left(e^{(1)}, e^{(2)}, \ldots, e^{(n)}\right)^T
\end{aligned}
\tag{8.68}
$$

Here, $d(n)$, $\hat{d}(n)$, and $e(n)$ are vectors of size n. X is a matrix of size $n \times L$. These variables grow in size as n increases. However, the quantities we will eventually use in RLS have constant sizes.

Equation (8.67) can thus be written in vector form:

$$e(n) = d(n) - \hat{d}(n) = X(n)w^{(n)} - \hat{d}(n) \tag{8.69}$$

The cost function is defined as

$$c^{(n)} \overset{\text{def}}{=} e^T(n)\Lambda(n)e(n) \tag{8.70}$$

Comparing with the cost function for LS (8.56), an additional matrix $\Lambda(n)$ is introduced. $\Lambda(n)$ is an $n \times n$ diagonal matrix, with its diagonal elements determining the weights of the errors at each step. Equation (8.70) can be written as

$$c^{(n)} = \sum_{i=1}^{n} \lambda_i^{(n)} \left(e^{(i)} \right)^2,$$

(8.71)

where $\lambda_i^{(n)}$ is the ith diagonal element of $\Lambda(n)$. For EWRLS, the weights can be chosen as

$$\lambda_i^{(n)} = \alpha^{n-i},$$

(8.72)

where $0 < \alpha \leq 1$ is a positive constant controlling the memory length. Such expression allows the adaptive filter to put more weights on recent samples, and thus to better respond to the system changes. When α is 1, Λ is an identity matrix. This is the case of the classical RLS.

Similar to Section 8.6.2, the equation for the optimal filter is

$$X^T(n)\Lambda(n)X(n)w_o^{(n)} - X^T(n)\Lambda(n)\hat{d}(n) = 0$$

(8.73)

The solution is (assuming all matrices are full rank)

$$w_o^{(n)} = R(n)^{-1}s(n)$$

$$R(n) \overset{\text{def}}{=} X^T(n)\Lambda(n)X(n).$$

$$s(n) \overset{\text{def}}{=} X^T(n)\Lambda(n)\hat{d}(n)$$

(8.74)

$R(n)$ is of size $L \times L$. $s(n)$ is of size L. They all have fixed sizes, although the sizes for variables in (8.68) grow with n.

8.7.2 Derivation of RLS Algorithm

The key of the RLS algorithm is avoiding the matrix inversion computation in (8.74). Instead, the solution at the nth step is built from that at the $n-1$ step. In this section, we derive these iterative relationships.

From (8.74), (8.72), and (8.68),

$$R(n) = \sum_{i=1}^{n} \alpha^{n-i} x(i) x^T(i)$$

$$= \alpha \sum_{i=1}^{n-1} \alpha^{n-1-i} x(i) x^T(i) + x(n) x^T(n) = \alpha R(n-1) + x(n) x^T(n)$$

(8.75)

In addition, from the same equations,

$$s(n) = \sum_{i=1}^{n} \alpha^{n-i} x(i) \hat{a}^{(i)}$$

$$= \alpha \sum_{i=1}^{n-1} \alpha^{n-i-1} x(i) \hat{a}^{(i)} + x(n) \hat{a}^{(n)} = \alpha s(n-1) + x(n) \hat{a}^{(n)} \qquad (8.76)$$

To get the iteration for inversion, we start from a well-known equation (see homework):

$$(A + BDC)^{-1} = A^{-1} - A^{-1} B (D^{-1} + CA^{-1}B)^{-1} CA^{-1}. \qquad (8.77)$$

This equation is mapped to our problem as follows.

$$\begin{aligned} A &\rightarrow \alpha R(n-1) \\ B &\rightarrow x(n) \\ C &\rightarrow x^T(n) \\ D &\rightarrow 1 \end{aligned} \qquad (8.78)$$

Equation (8.75) leads to

$$R(n)^{-1} = \left[\alpha R(n-1) + x(n) x^T(n) \right]^{-1}. \qquad (8.79)$$

Applying (8.77) to the right-hand side of (8.79):

$$\left[\underbrace{(\alpha R(n-1))}_{A} + \underbrace{x(n)}_{B} \underbrace{x^T(n)}_{C} \right]^{-1} = \underbrace{\alpha^{-1} R(n-1)^{-1}}_{A^{-1}}$$

$$\underbrace{- \alpha^{-1} R(n-1)^{-1}}_{A^{-1}} \underbrace{x(n)}_{B} \left[\underbrace{1}_{D^{-1}} + \underbrace{x^T(n)}_{C} \underbrace{\alpha^{-1} R(n-1)^{-1} x(n)}_{A^{-1}} \underbrace{x(n)}_{B} \right]^{-1} \underbrace{x^T(n)}_{C} \underbrace{\alpha^{-1} R(n-1)^{-1}}_{A^{-1}} \qquad (8.80)$$

In (8.80), mappings in (8.78) are indicated by the under-braces. This seemingly cumbersome expression actually can be simplified as shown below. Note that $[1 + \alpha^{-1} x^T(n) R(n-1)^{-1} x(n)]$ is a scalar. Therefore, its inversion is just a division. Equations (8.79) and (8.80) lead to

$$R(n)^{-1} = \alpha^{-1} R(n-1)^{-1} - \frac{\alpha^{-2} R(n-1)^{-1} x(n) x^T(n) R(n-1)^{-1}}{1 + \alpha^{-1} x^T(n) R(n-1)^{-1} x(n)}. \qquad (8.81)$$

Equation (8.81) expresses $R(n)^{-1}$ in terms of $x(n)$ and $R(n-1)^{-1}$. Both quantities are known at step n. This equation involves matrix and vector multiplications and addition, but not matrix inversion.

Now return to (8.74) for $w_o^{(n)}$. From (8.81) and (8.76), we have

$$
w_o^{(n)} = R(n)^{-1} s(n)
$$
$$
= \left[\alpha^{-1} R(n-1)^{-1} - \frac{\alpha^{-2} R(n-1)^{-1} x(n) x^T(n) R(n-1)^{-1}}{1 + \alpha^{-1} x^T(n) R(n-1)^{-1} x(n)} \right] \left[\alpha s(n-1) + x(n) \hat{d}^{(n)} \right].
$$
(8.82)

Using (8.74) for step $n-1$, the two terms on the right-hand side of (8.82) can be simplified:

$$
\left[\alpha^{-1} R(n-1)^{-1} - \frac{\alpha^{-2} R(n-1)^{-1} x(n) x^T(n) R(n-1)^{-1}}{1 + \alpha^{-1} x^T(n) R(n-1)^{-1} x(n)} \right] \alpha s(n-1)
$$
$$
= w_o^{(n-1)} - \frac{\alpha^{-1} R(n-1)^{-1} x(n)}{1 + \alpha^{-1} x^T(n) R(n-1)^{-1} x(n)} x^T(n) w_o^{(n-1)}
$$
(8.83)

$$
\left[\alpha^{-1} R(n-1)^{-1} - \frac{\alpha^{-2} R(n-1)^{-1} x(n) x^T(n) R(n-1)^{-1}}{1 + \alpha^{-1} x^T(n) R(n-1)^{-1} x(n)} \right] x(n) \hat{d}^{(n)}
$$
$$
= \frac{\alpha^{-1} R(n-1)^{-1} x(n) + \alpha^{-2} x^T(n) R(n-1)^{-1} x(n) R(n-1)^{-1} x(n) - \alpha^{-2} R(n-1)^{-1} x(n) x^T(n) R(n-1)^{-1} x(n)}{1 + \alpha^{-1} x^T(n) R(n-1)^{-1} x(n)} \hat{d}^{(n)}
$$
(8.84)

In (8.84), note that $x^T(n) R(n-1)^{-1} x(n)$ in the numerator is a scalar, which can be moved within a product in the order. Therefore, the second and third terms in the numerator cancel out. Namely,

$$
\left[\alpha^{-1} R(n-1)^{-1} - \frac{\alpha^{-2} R(n-1)^{-1} x(n) x^T(n) R(n-1)^{-1}}{1 + \alpha^{-1} x^T(n) R(n-1)^{-1} x(n)} \right] x(n) \hat{d}^{(n)}
$$
$$
= \frac{\alpha^{-1} R(n-1)^{-1} x(n)}{1 + \alpha^{-1} x^T(n) R(n-1)^{-1} x(n)} \hat{d}^{(n)}
$$
(8.85)

Note that the first factor in (8.85) is the same as that in the second term of (8.83). Combining (8.82), (8.83), and (8.85),

$$
w_o^{(n)} = w_o^{(n-1)} + \frac{\alpha^{-1} R(n-1)^{-1} x(n)}{1 + \alpha^{-1} x^T(n) R(n-1)^{-1} x(n)} \left[\hat{d}^{(n)} - x^T(n) w_o^{(n-1)} \right].
$$
(8.86)

Define the vector

$$
k^{(n)} \overset{\text{def}}{=} \frac{\alpha^{-1} R(n-1)^{-1} x(n)}{1 + \alpha^{-1} x^T(n) R(n-1)^{-1} x(n)}
$$
(8.87)

and the scalar

$$\xi^{(n)} \stackrel{\text{def}}{=} \hat{d}^{(n)} - x^T(n) w_o^{(n-1)}. \tag{8.88}$$

Equation (8.86) is thus written as

$$w^{(n)} = w^{(n-1)} + k^{(n)} \xi^{(n)}. \tag{8.89}$$

$k^{(n)}$ is known as the Kalman gain. $\xi^{(n)}$ is known as a priori estimation error, which is the error when the previous filter is used on the current data. With such definition, (8.81) can be written as

$$R(n)^{-1} = \alpha^{-1} R(n-1)^{-1} - \alpha^{-1} k^{(n)} x^T(n) R(n-1)^{-1}. \tag{8.90}$$

8.7.3   RLS Summary and Discussion

As derived in Section 8.7.2, the RLS algorithm is as follows:

$$
\begin{aligned}
& w^{(n)} = w^{(n-1)} + k^{(n)} \xi^{(n)} \\
& R(n)^{-1} = \alpha^{-1} R(n-1)^{-1} - \alpha^{-1} k^{(n)} x^T(n) R(n-1)^{-1} \\
& k^{(n)} \stackrel{\text{def}}{=} \frac{\alpha^{-1} R(n-1)^{-1} x(n)}{1 + \alpha^{-1} x^T(n) R(n-1)^{-1} x(n)} \\
& \xi^{(n)} \stackrel{\text{def}}{=} \hat{d}^{(n)} - x^T(n) w_o^{(n-1)}
\end{aligned}
\tag{8.91}
$$

At step n, we need to compute $k^{(n)}$ and $\xi^{(n)}$ from the current data and the previous matrix $R(n-1)^{-1}$. Then we update $w^{(n)}$ and $R(n)^{-1}$.

$R(1)$ can be initialized as identity matrix times a small factor:

$$R(1) = \delta I, \tag{8.92}$$

where $0 < \delta \ll 1$. This ensures that $R(n)$ is always full-rank during the iterations.

Properties of RLS have been the subject of many research works, which usually involve sophisticated mathematical analyses. Here, we list some interesting results [2, sec. 5.5].

- The computational load for each update is on the order of L^2. In comparison, the LS computation requires operations on the order of L^3. The number of operations for an LMS update is on the order of L.

- The initialization factor δ should be large enough, so that $R(n)$ continues to be full-rank at the initial steps. Note that from (8.74), $R(n)$ is always rank-deficient when $n < L$. In this case, the initial setting "cheats the system" and forces the iteration result to be full rank. As iteration goes on, the effect of the initial value choice will be "forgotten" in EWRLS.

- For a linear system (8.62), RLS results in an unbiased estimate of the true filter $\bar{w}$, if the noise is zero-mean.

- The convergence speed of RLS is an order of magnitude (or more) faster than the LMS.

- The convergence of RLS is insensitive to the condition number of the input correlation matrix (see Section 8.5.2 for LMS behavior). Effectively, in the context of the discussions in Section 8.5.2, the use of Kalman gain applies different step sizes along different eigenvectors. Therefore, we no longer need to compromise between the larger and smaller eigenvalues.

Overall, both RLS and LMS are iterative adaptation algorithms that are capable of responding to system changes. RLS converges faster, yet with higher complexity. A common practice is using RLS in initial training, to allow the filter to adapt to the system state quickly. LMS is then used in steady-state mode, to follow any possible change and to fine-tune the filter tap values. Often, more computing resources are available during initial training because some receiver modules are not yet operational. Such an arrangement, therefore, reaches a better compromise between resource demand and performance.

Besides practical advantages, RLS is also a fascinating theoretical subject. It is closely related to the Kalman filter, which has profound theoretical significance in several related fields. There are also numerous variances of RLS for complexity reduction, convergence improvement, and robust enhancement reported in the literature.

8.8   CASE STUDY: FULL-DUPLEX RADIO AND SELF-INTERFERENCE CANCELLATION

This section briefly describes an application of adaptive filters: self-interference cancellation in full-duplex radio (FDR). Only an online in the context of the adaptive filter is provided here.

Presently for most of the radios, transmission and receiving operations are separated either by time or by frequency. These two separation methods are referred to as time division duplex (TDD) and frequency division duplex (FDD), respectively. For example, the wireless local area network (WLAN) uses time-division multiple access (TDMA), where each radio transmits in turn. TDMA is a generalized version of TDD. On the other hand, most fourth generation (4G) cellular systems use FDD, where downlink (base station to mobile devices) and uplink use different frequency bands.[8] Instead, a full-duplex radio (FDR) transmits and receives at the same time and frequency.

FDR is attractive because it provides a higher data rate. By using the same time and frequency resource for both transmission and receiving, an FDR doubles its

---

[8] The prevailing 4G standard Long-Term Evolution (LTE) supports both FDD and TDD operations. The TDD version of LTE is used only in a few countries.

overall spectral efficiency comparing to TDD or FDD. In comparison, doubling spectral efficiency within the TDD or FDD scheme would require doubling SNR in dB scale. As an example, consider a modern digital communications system using 256-QAM (such as in the latest WLAN and LTE-advanced standards). With a coding rate of 1/2, the spectral efficiency is 4 bits per second per Hertz.[9] The Shannon bound for SNR (Chapter 2) is

$$SNR = 2^R - 1, \tag{8.93}$$

where R is the spectral efficiency. In our case, SNR is 15, or 11.8 dB. To double the spectral efficiency and assuming the noise level remains the same, we would need to increase transmit power by 15 times. This usually comes with a significant penalty in size, weight, power, and cost of the device, not to mention spectrum management complications (see Chapter 6 for more discussions). Furthermore, FDD and TDD systems typically require guard bands or guard time between transmission and receiving operations. An FDR system does not need such overhead.

In addition, FDR enables a transceiver to obtain channel state information (CSI) from the receiving path and to use it for transmission optimization, based on the reciprocal principle. For the conventional systems, obtaining transmitter-side CSI is more difficult. For FDD, the transmitter relies on the receiver to report back its channel measurement results. CSI can be obtained through reciprocity at the transmitter for TDD, but significant latency is involved.

Another advantage of FDR is that each transceiver "announces its presence" in the operational band, helping cognitive radios (Chapter 13) to detect other users for interference avoidance.

Unfortunately, FDR also encounters a fundamental difficulty: self-interference (SI) caused by the transmitted signal leaking into the receiving path. To get a feeling of the magnitude of the problem, let us look at an example [5]. Consider a WLAN transceiver with a transmit power of 20 dBm and a bandwidth of 80 MHz. The thermal noise power is −95 dBm. With a noise figure of 5 dB, the noise at the receiver is −90 dBm.[10] Therefore, in order to reduce SI to the same level as receiver noise (thus raising the overall noise level by only 3 dB), we need 110 dB of SI cancellation. Such an estimate is actually optimistic. In reality, receiver noise level may be lower if the noise figure is lower, and the bandwidth is smaller (e.g., if the noise figure is 1 dB and the bandwidth is 20 MHz, the noise power level would be −100 dBm). The transmit power can also be higher (e.g., LTE mobile equipment has a maximum transmit power of 23 dBm, while the maximum transmitted power for WLAN devices in 2.5 GHz band is 30 dBm). Therefore, the required SI reduction ratio is very large.

The adaptive filter is a natural choice for SI cancellation. The notional SI canceller has a configuration shown in panel A of Figure 8.3, where the input $\{x_n\}$ is the transmitted signal, and $\left\{\hat{d}_n\right\}$ is the leaked transmitted signal in the receive chain

---

[9] This assumes Nyquist symbol rate, without consideration of guard bands. Namely, the symbol rate is the same as passband bandwidth.

[10] See Chapter 6 for more details about noise power calculation.

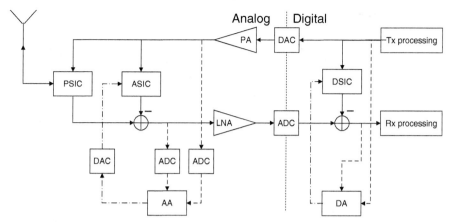

Figure 8.5   SI canceller architecture.

(the real received signal serves as noise in $\{\hat{d}_n\}$). However, achieving 110 dB (i.e., a factor of 10^{11}) or more cancellation is no mean feat. Therefore, the SI cancellation is more involved than a standard adaptive filter.

Because of the difficulty in SI cancellation, FDR is not widely used in wireless communications today. However, FDR has been gaining traction in the past decade and is identified as one of the enabling technologies for 5G cellular systems [6–8].[11] FDR is currently an active research field [5, 9–11]. Full-duplex systems have been used in voiceband modem and Ethernet for several decades, where the channels are much more predictable than in wireless communications. For these systems, SI cancellers are known as echo cancellers.

Figure 8.5 shows the SI canceller architecture [12]. The system is divided into analog and digital domains by the vertical dotted line across the analog-to-digital converter (ADC) and digital-to-analog converter (DAC) blocks in the middle. At the transmitting side, baseband digital signals are converted to RF analog by the DAC block, which, for our discussion, includes digital-to-analog conversion, up-converting, amplification, and other operations.[12] The resulting signal is then sent to the power amplifier (PA). At the receiving side, the RF analog signal output from the low noise amplifier (LNA) is converted to baseband digital samples by the ADC block, which includes amplification, down-converting, analog-to-digital conversion, and other operations. In both domains, the solid arrowed lines in Figure 8.5 indicate the flow of signals. Dashed lines show data used for adaptation. Dash-dotted lines show the path of control signals for the adaptive filters. As will be explained later, SI cancellation is performed in both analog and digital domains.

In the analog domain (left side of Figure 8.5), the transmitted signal from the power amplifier (PA) and the received signal going into the low noise amplifier (LNA)

---

[11] At present, the most mature 5G technology proposal, the New Radio (NR) by 3GPP, appears not to have included FDR.

[12] See Chapter 3 for a detailed transceiver architecture description.

share the same antenna, through a passive SI canceller (PSIC). The PSIC reduces the leakage from the transmit side to the receive side. The remaining SI is further cancelled by the analog SI canceller (ASIC), which takes the transmitted signal and subtracts it from the received signal after filtering. The ASIC is controlled by the analog adaptation (AA). The AA module works in baseband digital domain. The transmitted and received signals pass through the dedicated ADC blocks, which include down-converting and analog-to-digital conversion, to form baseband digital samples. These samples drive the adaptation process in AA, whose output is sent through the digital-to-analog converter (DAC) to control the ASIC filter.

In the digital domain (right side of Figure 8.5), a digital SI canceller (DSIC) feeds filtered transmitted signal to the receiving path to cancel any remaining SI. This filter is controlled by the digital adaptation (DA), which takes input from the transmitted signal and the received signal (after SI canceller).

The PSIC is the first line of defense. It is typically a circulator or a hybrid circuit. Ideally, it ensures that the transmitted signal flows *into* the antenna, without getting into the receiving path. However, usually, due to antenna impedance mismatch and reflection from surrounding objects, the rejection of PSIC is limited to about 20 dB [12].

Although signal processing in the digital domain is usually more straightforward, the ASIC is necessary for two reasons. First, as described above, the SI is much higher than the received signal. To send both SI and the received signal to the digital domain without distortion, a very large dynamic range is required for the entire receiving path. Such a dynamic range, even when possible, dramatically increases the cost of the receiver. Second, the analog transmitting path usually includes nonlinearity, especially in low-cost devices. Such nonlinearity cannot be captured in the digital domain. Instead, one must extract the transmitted signal for SI cancellation near the antenna port (i.e., at the output of the PA), so that the rest of the SI path (the PSIC and antenna) can be cancelled effectively with a linear SI canceller. Such need leads to an ASIC. ASCI can be built in many ways, but it is basically an analog filter, whose tap values are controlled by the AA module. Typically, an ASIC can achieve 40–70 dB of SI cancellation [10].

ASIC is indispensable, yet not sufficient enough. Because the analog filter cannot be accurately controlled, there is still residual SI when the received signal enters digital domain. A digital SI canceller (DSIC) is thus required to remove SI further. A DSIC can be a conventional LMS filter described in Section 8.5. However, in some cases residue nonlinearity still needs to be addressed in the digital domain. A nonlinear SI canceller can be built by using $\{x_n\}$, $\{|x_n|^2 x_n\}$, and so on as inputs. A DSIC can add an additional 30–60 dB of SI rejection.

As an adaptive filter, the DSIC faces another difficulty. In the context of the system model (8.16), the "noise" $\{z_n\}$ (in this case the actual received signal) can be a few tens of dBs higher than the "signal" $\{\bar{w}^T x(n)\}$ (in this case the remaining SI). This means the adaptation step sizes must be small, to average out the "noise" (see Section 8.5.2). Furthermore, the adaptation only works well when the signal and noise are uncorrelated. This assumption is generally true for data communication.

However, during initial training, both transmitted and received signals may be the same training symbols (preambles), causing correlation. Some systems (such as DSL and voiceband modem) have a half-duplex phase (i.e., only one side transmits) during startup, to allow for SI canceller training. A half-duplex training phase allows for a larger step size and thus a faster initial convergence. The correlation issue is also avoided by half-duplex training.

With the combination of these three SI cancellers, the goal of more than 100 dB total SI rejection can be achieved, making FDR feasible.

8.9 SUMMARY

This chapter outlines several strategies for adaptive filtering. In an ideal case, if the signal and error statistics are known, the optimization problem can be solved analytically, as described in Section 8.3. Alternatively, a block of data can be used to derive the optimal solution as in Section 8.6. Matrix inversions are used in both methods. These approaches are OK if the solution is to be computed once. However, the computational load may be too much for continuous updating. LMS (Section 8.5) and RLS (Section 8.7), on the other hand, are practical methods for iterative optimization. The LMS operation is based on the current sample, while the RLS keeps a running record of the past block of samples. RLS is more complex in both formulation and computation. However, it converges much faster and is insensitive to the condition number of the autocorrelation matrix of the input signal.

Section 8.8 presents an application example of adaptive filters: the FDR. Besides being an important candidate technology for the 5G cellular systems, FDR also demonstrates the power of adaptive filtering: being able to achieve more than 100 dB of self-interference cancellation. Another sample of adaptive filters, the channel equalizer, is discussed in Chapter 9.

The adaptive filter has a long history of research and innovation, and it remains an active research field in signal processing. There are many applications beyond digital communications. For example, acoustic echo cancellers and active acoustic noise cancellers are vital parts of telephone and video equipment [13, ch. 6]. Many automatic control systems and robotic technologies also rely on adaptive filters. Interested users can consult many specialized books on the subject [1, 2, 13].

This chapter covers only the types of adaptive filters commonly used in digital communication systems. Topics not covered include infinite impose response (IIR) adaptive filters [1, ch. 10], which face unique challenges on stability and convergence speed. Another notable exclusion is the Kalman filter [1, ch. 17]. Kalman filter is used for system estimation when the input is state-driven and nonstationary. Kalman filter is a powerful tool that can be understood with several frameworks. Its application in communications systems is actively researched, but not widespread in practice yet.

MATLAB provides support for LMS and RLS filters. For more information, search for "overview of adaptive filters and applications" in MATLAB documentation.

REFERENCES

1. P. S. R. Diniz, *Adaptive Filtering*. Springer, 2013.
2. L. R. Vega and H. Rey, *A Rapid Introduction to Adaptive Filtering*. Berlin, Heidelberg: Springer, 2013.
3. D. R. Morgan, "Slow Asymptotic Convergence of LMS Acoustic Echo Cancelers," *IEEE Trans. Speech Audio Process.*, vol. 3, no. 2, pp. 126–136, 1995.
4. D. T. M. Slock, "On the Convergence Behavior of the LMS and the Normalized LMS Algorithms," *IEEE Trans. Signal Process.*, vol. 41, no. 9, pp. 2811–2825, 1993.
5. Z. Zhang, K. Long, A. V. Vasilakos, and L. Hanzo, "Full-Duplex Wireless Communications: Challenges, Solutions, and Future Research Directions," *Proc. IEEE*, vol. 104, no. 7, pp. 1369–1409, 2016.
6. F. Evaluation, "Next Generation 5G Wireless Networks: A Comprehensive Survey," *IEEE Commun. Surv. Tutorials*, vol. 18, no. 3, pp. 2005–2008, 2016.
7. S. Hong et al., "Applications of Self-interference Cancellation in 5G and Beyond," *IEEE Commun. Mag.*, vol. 52, no. 2, pp. 114–121, 2014.
8. Z. Zhang, X. Chai, K. Long, A. V. Vasilakos, and L. Hanzo, "Full Duplex Techniques for 5G Networks: Self-interference Cancellation, Protocol Design, and Relay Selection," *IEEE Commun. Mag.*, vol. 53, no. 5, pp. 128–137, 2015.
9. G. Y. Li, M. Bennis, and G. Yu, "Full Duplex Communications [Guest Editorial]," *IEEE Commun. Mag.*, vol. 53, no. 5, pp. 90–90, 2015.
10. D. Kim, H. Lee, and D. Hong, "A Survey of In-Band Full-Duplex Transmission: From the Perspective of PHY and MAC Layers," *IEEE Commun. Surv. Tutorials*, vol. 17, no. 4, pp. 2017–2046, 2015.
11. A. Sabharwal, P. Schniter, D. Guo, D. W. Bliss, S. Rangarajan, and R. Wichman, "In-Band Full-Duplex Wireless: Challenges and Opportunities," *IEEE J. Sel. Areas Commun.*, vol. 32, no. 9, pp. 1637–1652, 2014.
12. D. Korpi et al., "Full-Duplex Mobile Device: Pushing the Limits," *IEEE Commun. Mag.*, vol. 54, no. 9, pp. 80–87, 2016.
13. B. Kovačević, Z. Banjac, and M. Milosavljević, *Adaptive Digital Filters*. Berlin, Heidelberg: Springer, 2013.

HOMEWORK

8.1 In the context of Section 8.3.2, show the following. If $r_{ex} \stackrel{\text{def}}{=} E[x(n)e_n)]$ is not zero, then there is a filter v such that $E\{[e_n - v^T x(n)]\} < E(e_n^2)$. Namely, the residual error e_n can be reduced by using a new filter $w' = w - v$, where w is the filter that produced e_n. Hint: use the same approach as Section 8.3 to find the optimal v.

8.2 In the context of Section 8.3.3, consider the case where $\bar{w}$ in (8.16) is longer than the w to be optimized.

    a. What would be the optimal solution w_o?

    b. What would be the remaining error e_n?

8.3 Prove (8.26) based on (8.23). Note that R_{xx} is symmetric.

8.4 Express the cost function (8.5) in terms of u and v, defined by (8.29) and (8.33).

8.5 In the context of Section 8.4.3, the convergence of v can be expressed as $|v^{(n)}| \sim \alpha^n |v^{(0)}|$, where α is determined by the slowest converging component(s) of v. When optimal step size (8.42) is used, express α in terms of the condition number κ in (8.43).

8.6 Write a program to reproduce Figure 8.4. Play with other parameters (e.g., the condition number) to see how they affect convergence.

8.7 Write (8.57) in component form. Minimize the cost function in terms of partial derivatives on $\{w_l, l = 1, 2, \ldots, L\}$. Show that such operations lead to the normal equation (8.59).

8.8 Prove (by explicit multiplication) the matrix inversion lemma used in RLS algorithm (8.77): $[A^{-1} - A^{-1}B(D^{-1} + CA^{-1}B)^{-1}CA^{-1}](A + BDC) = I$.

CHANNEL EQUALIZATION

9.1 INTRODUCTION

As described in Chapter 6, the channel may be time-dispersive in a scattering environment. Namely, there are multiple copies of the transmitted signal arriving at the receiver with different time delays. As a result, the signal from one symbol is mixed with that from the neighboring ones, causing what is known as the inter-symbol interference (ISI). For the typical demodulators discussed in Chapter 4, ISI causes receiver performance degradation. Channel equalization is a technique used to reduce or remove ISI.

Figure 9.1 shows the high-level physical layer architecture, discussed in Chapter 1. This chapter addresses a part of the demodulation module, the channel equalizer.

Channel equalizers were developed in the mid-twentieth century to combat time dispersion and associated ISI in transmission lines (mainly telephone lines) [1]. It was used for wireless communications in the 1990s, for some of the second generation of cellular systems (using standards IS-136 in the United States and GSM in the rest of the world, see Chapter 12). While powerful in reducing ISI, a channel equalizer requires relatively long training time. Therefore, it is difficult to use equalizers with rapid-changing wireless channels. In the third generation cellular systems, code-division multiple access (CDMA) was used, which deals with ISI differently.[1] In the fourth generation cellular systems and other modern wireless systems such as the Worldwide Interoperability for Microwave Access (WiMAX) and most flavors of the Wireless Local Area Network (WLAN), the ISI issue is addressed with a different technology, the orthogonal frequency division modulation (OFDM), which is discussed in Chapter 10.

Channel equalizers are a very useful tool in today's digital communication systems. For wireless systems, although OFDM is currently the prevailing solution, equalizers may still be suitable for some applications. For example, the American digital TV standard Advanced Television Systems Committee (ATSC) benefits from equalizers [2, 3].

---

[1] CDMA was also used in some second generation cellular systems in North America (standard IS-95). See Chapter 12 for more details.

Digital Communication for Practicing Engineers, First Edition. Feng Ouyang.
© 2020 by The Institute of Electrical and Electronics Engineers, Inc.
Published 2020 by John Wiley & Sons, Inc.

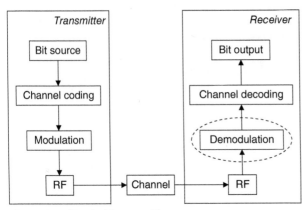

Figure 9.1 Physical layer architecture.

Channel equalizers are widely used in wired systems (such as the Ethernet) and are being adopted for optical communication systems. Furthermore, equalization represents an important milestone in the development of digital communication. Concepts utilized by channel equalization, such as adaptation, decision feedback, noise whitening, and minimum mean-squared error, are applicable to other areas of digital communications, especially multiple-in multiple-out (MIMO) technologies (Chapter 11). Therefore, equalizer remains an important part of the digital communications toolkit.

In the rest of this chapter, Section 9.2 starts from the channel model described in Chapter 6 to establish a discrete channel model, whose output is a signal sequence with optimal sampling and white noise. Such output becomes the input to the various equalization algorithms discussed in this chapter. Section 9.3 discusses the optimal detection method with dispersive channels, the maximum likelihood sequence estimation (MLSE). Sections 9.4 and 9.5 describe the linear equalizer and the decision feedback equalizer, which are the focus of this chapter. Section 9.6 introduces the Tomlinson–Harashima precoding (THP), which can be viewed as an alternative form of the decision feedback equalizer. Section 9.7 presents the concept of the fractionally spaced equalizer (FSE), which combines the equalizer with other filters in the receiver chain. Section 9.8 provides summary and closing remarks. There are some mathematics and derivation details included in the appendices (Sections 9.9–9.13).

Equalizers are special forms of adaptive filters, which are discussed in more details in Chapter 8. It is helpful to refresh the content in Chapter 8 before reading the current one. Especially, adaptation and convergence behaviors discussed therein are not repeated in this chapter.

Discussions in this chapter are in baseband, using complex-valued signals. x^* means the complex conjugate of x. There are some subtleties about maximizing or minimizing functions with complex-valued variables, as discussed in an appendix (Section 9.10).

Many discussions in this chapter are in the form of the z-transform of sequences. Readers not familiar with the z-transform can consult an appendix (Section 9.9). In this chapter, a sequence of samples is expressed as $\{x_n\}$, where n is the index for sequence terms. In some cases, the range for n is given explicitly, such as $\{x_n, n = 1, 2, \ldots, N\}$. The z-transform of $\{x_n\}$ is noted as $\bar{x}(z)$. The correspondence of a sequence and its z-transform is implied by using the same letter for both.

9.2 CHANNEL DISPERSION FORMULATION

In this section, we start from the channel model in Chapter 6 and the digital modulation model in Chapter 3 to arrive at a baseband channel model connecting transmitted symbols and received signals. Furthermore, the maximum likelihood detection is applied, resulting in a discrete-time channel model.

9.2.1 Time-Continuous Channel Model

Let us first review the time-continuous baseband channel model presented in In Chapter 6.

$$y(t) = \int g(t,\tau)d(t-\tau)d\tau + v(t), \tag{9.1}$$

where $d(t)$ and $y(t)$ are the transmitted and received signals. $v(t)$ is the additive noise at the receiver. $g(t, \tau)$ is the dispersive channel gain. For simplicity, we consider only a constant channel $g(\tau)$.

As described in Chapter 3, the baseband transmitted signal is

$$d(t) = \sum_m d_m s(t - mT_{Sym}), \tag{9.2}$$

where $\{d_m\}$ represents the transmitted symbols. $s(t)$ is the pulse shaping function. Combining $s(t)$ and $g(\tau)$ into a response function $q(t)$, (9.1) and (9.2) yields

$$y(t) = \sum_m d_m \int g(\tau)s(t - mT_{Sym} - \tau)d\tau + v(t) = \sum_m d_m q(t - mT_{Sym}) + v(t)$$
$$q(t) \stackrel{\text{def}}{=} \int g(\tau)s(t-\tau)d\tau \tag{9.3}$$

We further single out the contribution from the desired symbol d_m.

$$y^{(m)}(t) = d_m q(t - mT_{Sym}) + \sum_{m' \neq m} d_{m'} q(t - m'T_{Sym}) + v(t). \tag{9.4}$$

The second term in (9.4) is the contribution to $y^{(m)}$ from other transmitted symbols $\{d_{m'}, m' \neq m\}$, that is, the inter-symbol interference (ISI). The third term is the natural noise.

At the receiver, we can consider the ISI as a part of the noise. This way, (9.4) is the same as the channel model used in Chapter 4, except that the noise level is increased by the ISI. Furthermore, the ISI is proportional to signal power. Therefore, when ISI dominates the noise level, the resulting SNR cannot be improved by increasing signal power. In other words, with the channel model of (9.3), demodulation methods in Chapter 4 can still be used, albeit with degraded performance. In this chapter, we consider alternative detection methods to deal with the ISI more effectively.

9.2.2 Maximum Likelihood Detection

Having established the time-continuous baseband channel model (9.3), we apply the maximum likelihood (ML) detection (discussed in Chapter 4) to derive a discrete-time channel model. As pointed out in Chapter 4, ML detection is optimal when the transmitted symbols are equiprobable.

The ML detection finds the transmitted symbol that maximizes the probability of received signal:

$$\{d_m\}_{ML} = \arg\max_{\{d_m\}} p(y \mid \{d_m\}). \tag{9.5}$$

Here, $\{d_m\}_{ML}$ is the transmitted symbol sequence as determined by the ML receiver. Note that with the channel model here, the received signal depends not on a single symbol, but on the sequence of transmitted symbols. Therefore, the ML detection needs to consider the whole sequence. From (9.3), we have

$$p(y(t) \mid \{d_m\}) = p_v\left[y(t) - \sum_m d_m q(t - mT_{Sym})\right], \tag{9.6}$$

where p_v is the probability density function (pdf) of the noise $v(t)$. If we consider signal over time, the conditional probability is a product of the above probability over t. Therefore, if the noise has a Gaussian distribution, we have

$$p(y \mid \{d_m\}) \propto \exp\left[-\frac{\int \left|y(t) - \sum_m d_m q(t - mT_{Sym})\right|^2 dt}{2\sigma_v^2}\right], \tag{9.7}$$

where σ_v^2 is the variance of the noise. Here, $p(y \mid \{d_m\})$ denotes the probability of receiving the signal $y(t)$ over a given time period, given that the transmitted symbol sequence is $\{d_m\}$. In (9.7), we ignore the issue of probability normalization, as it does not affect our results.

With the same approach as in Chapter 4, Equations (9.7) and (9.5) lead to

$$\{d_m\}_{ML} = \arg\max_{\{d_m\}} \left\{ \int 2Re\left[y(t) \sum_m d_m^* q^*(t-mT_{Sym}) \right] dt \right. $$
$$\left. - \int \sum_{m,m'} d_m d_m'^* q(t-mT_{Sym}) q^*(t-m'T_{Sym}) dt \right\} \qquad (9.8)$$

Equation (9.8) can be further written as

$$\{d_m\}_{ML} = \arg\max_{\{d_m\}} \left[2Re \sum_m y_m' d_m^* - \sum_{m,m'} d_m d_{m'}^* x_{m'-m} \right]$$
$$y_m' \overset{\text{def}}{=} \int y(\tau) q^*(\tau-mT_{Sym}) d\tau \qquad (9.9)$$
$$x_n \overset{\text{def}}{=} \int q(\tau) q^*(\tau-nT_{Sym}) d\tau$$

Equation (9.9) indicates that ML detection can be performed based on discrete-time samples $\{y_m'\}$. All other variables in (9.9) are independent of the received signal. Therefore, in the scope of ML, $\{y_m'\}$ contains all of the necessary information about the received signal $y(t)$. Equations (9.3) and (9.9) lead to:

$$y_m' = \int \left[\sum_{m'} d_{m'} q(t-m'T_{Sym}) + v(t) \right] q^*(t-mT_{Sym}) dt = \sum_{m'} x_{m-m'} d_{m'} + v_m'$$
$$v_m' \overset{\text{def}}{=} \int v(t) q^*(t-mT_{Sym}) dt \qquad (9.10)$$

Equation (9.10) is a discrete channel model, where v_m' is the discrete noise. Similar to the case in Chapter 4, y_m' is obtained by passing the received signal $y(t)$ through the matched filter $q^*(-t)$, and sampling the output at a rate of $1/T_{Sym}$.

9.2.3 Discrete Channel Model with White Noise

In Equation (9.10), the noise samples $\{v_m'\}$ are correlated, even when $v(t)$ is white. In fact, if

$$E[v(t)v^*(t_1)] = N_0\delta(t-t_1), \qquad (9.11)$$

then

$$E\left(v_m' v_{m'}'^* \right) = N_0 x_{m-m'}. \qquad (9.12)$$

In analyses, it is more convenient to deal with white (i.e., uncorrelated) noise. To achieve that, we can add another whitening filter $\{f_n\}$, as shown in an appendix (Section 9.9.5). The noise samples at the output of the filter become uncorrelated:

$$v_m \stackrel{\text{def}}{=} \sum_k f_k v'_{m-k} ,$$
$$E\left(v_m v^*_{m'}\right) = N_0 \delta_{m,m'} \tag{9.13}$$

Of course, the same filter applies to the received signal:

$$y_m = \sum_k f_k y'_{m-k}. \tag{9.14}$$

Filter $\{f_k\}$ can be combined with the discrete channel response $\{x_k\}$ used in (9.10) to form the total filter

$$h_m \stackrel{\text{def}}{=} \sum_k x_k f_{m-k}. \tag{9.15}$$

With that, we have the discrete channel model

$$y_m = \sum_k h_k d_{m-k} + v_m. \tag{9.16}$$

Here, $\{h_k\}$ is the equivalent channel; they are assumed known to the receiver, except when we discuss adaptation. $\{v_m\}$ is white Gaussian noise. The combination of the matched filter $q^*(-t)$ and the whitening filter $\{f_k\}$ is known as the whitened matched filter (WMF).

For a channel $\{h_k\}$ with length L, a natural formulation is let k range between 0 and $L-1$. Namely, (9.16) can be written explicitly as

$$y_m = \sum_{k=0}^{L-1} h_k d_{m-k} + v_m. \tag{9.17}$$

Now redefine the received symbol with a time shift M:

$$y''_m \stackrel{\text{def}}{=} y_{m+M}$$
$$v''_m \stackrel{\text{def}}{=} v_{m+M} . \tag{9.18}$$

Equation (9.17) can thus be written as

$$y''_m = \sum_{k=0}^{L-1} h_k d_{m+M-k} + v''_m \underset{j \stackrel{\text{def}}{=} k-M}{=} \sum_{j=-M}^{L-M-1} h''_j d_{m-j} + v''_m . \tag{9.19}$$
$$h''_j \stackrel{\text{def}}{=} h_{j+M}$$

Namely, by shifting the indices of the received symbol, we effectively get a new channel $\{h''_n\}$, which is a shifted version of the original channel. In other words, we have the freedom to select the alignment between the transmitted and received signals, as long as the channel taps are defined accordingly. We can define such alignment by specifying the position of h_0.

9.2.4   The Total Channel and the Physical Channel

Sections 9.2.1–9.2.3 construct the discrete channel model for a dispersive baseband channel. The process is summarized in Figure 9.2. Panel A shows the actual channel, described by (9.2), (9.3), (9.10), and (9.14). $q(t)$ is referred to as the physical channel. Panel B shows the time-discrete model in (9.10). Panel C shows the final time-discrete channel model including noise whitening, expressed in (9.16). $\{h_n\}$ is referred to as the total channel. The rest of the chapter is based on the discrete channel model (9.16).

The receiver components of the total channel $\{h_n\}$, that is, the matched filter and the whitening filter, all depend on the physical channel $q(t)$. Therefore, the total channel can be expressed in terms of $q(t)$. In fact, as shown by (9.212) in an appendix (Section 9.9.6),

$$
\bar{h}(z)\bar{h}^*\left(z^{*-1}\right)\Big|_{z=\exp(j\omega T_{Sym})} = \left|\bar{h}\left[\exp\left(j\omega T_{Sym}\right)\right]\right|^2
$$

$$
= \bar{x}\left[\exp\left(j\omega T_{Sym}\right)\right] = \frac{1}{T_{Sym}}\sum_m \left|\tilde{q}\left(\omega - \frac{2\pi m}{T_{Sym}}\right)\right|^2. \quad (9.20)
$$

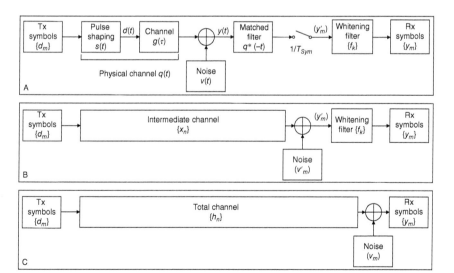

Figure 9.2   Development of the discrete channel model. The definition of channel extends to include more parts in the signal chain. Definition of the added noise changes accordingly.

Here, $\bar{h}(z)$ and $\bar{x}(z)$ are the z-transform of $\{h_n\}$ and $\{x_n\}$, respectively, as defined in an appendix (Section 9.9.1). $\tilde{q}(\omega)$ is the Fourier transform of the physical channel $q(t)$. T_{Sym} is the symbol period.

In practice, the channel response $g(\tau)$ is not known at design time. Therefore, the matched filter and the whitening filter must be set at runtime based on a separate channel estimation process or through adaptation (not covered in this chapter). In most practical systems, however, the matched filter and the whitening filter in panel A are not separate entities but a part of the equalizers discussed in Sections 9.4 and 9.5. Section 9.7 provides more details on this issue.

9.3 MAXIMUM LIKELIHOOD SEQUENCE ESTIMATION (MLSE)

In Section 9.2, it is shown that the ML detection with a dispersive channel (9.1) depends only on the time-discrete samples $\{y_m\}$, which can be modeled with a time-discrete channel (9.16). In this section, we present the optimal detection algorithm based on $\{y_m\}$. This algorithm is known as the maximum likelihood sequence estimation (MLSE), which is based on the Viterbi algorithm (VA) described in Chapter 5.

Let L be the length of the channel. The nonzero taps range from index 0 to index $L-1$. Namely,

$$h_k = 0 \forall k \geq L \text{ or } k < 0. \tag{9.21}$$

With such setup, y_m is the first received symbol that d_m influences. The channel model is expressed by (9.17).

With the same approach as in Section 9.2.2, we can continue to perform ML detection based on (9.16). As discussed in Section 9.2.2, under the ML criterion and with Gaussian noise, we look for the transmitted sequence $\{d_m\}$ so that the received symbol sequence $\{y_m\}$ is the closest to the expectation. Namely,[2]

$$\{d_m\}_{ML} = \arg\min_{\{d_m\}}(c_m)$$

$$c_m \overset{\text{def}}{=} \sum_{n=0}^{m} \left| y_n - \sum_{k=0}^{L-1} h_k d_{n-k} \right|^2. \tag{9.22}$$

The cost function $\{c_m\}$ is additive at each step:

$$c_m = \left| y_m - \sum_{k=0}^{L-1} h_k d_{m-k} \right|^2 + c_{m-1}. \tag{9.23}$$

The additional term at step m depends on y_m, as well as the past L symbols in the hypothetical sequence $\{d_{m-k}, k = 0, 1, \ldots, L-1\}$. The previous function c_{m-1} does

---

[2] For simplicity, we assume that d_n exists when n is negative.

not depend on y_m. Therefore, such detection can be performed with the Viterbi algorithm (VA) described in Chapter 5. To apply VA to MLSE, we specify a state with the L most recent symbols. For an M-ary modulation, there are M^L possible states. In this sense, the discrete channel (9.17) can be viewed as a convolutional code with coding rate 1 and constraint length L, but the "code bits" are not binary.

MLSE with VA is an elegant concept and can be used in practice if L is not too long. However, for larger L and M values, the complexity of VA becomes prohibitive. In those cases, the linear equalizer and the decision feedback equalizer discussed in Sections 9.4 and 9.5 are more attractive.

Above formulation depends on the fact that the noise is white. Otherwise, ML detection does not lead to (9.22). There is a more general treatment of the MLSE problem [4]. In that paper, the author first derived an optimal MF under correlated noise. At the output of such MF, the noise samples are still correlated. Therefore, (9.22) needs to be modified to account for such correlation. The result is a little more cumbersome, yet still tractable.

MLSE can also be adaptive to channel conditions and the timing offset between the transmitter and receiver. One such approach was proposed by [4]. In this method, the transmitted sequence is assumed known, either by using predetermined training symbols or by using the previous demodulation decisions (i.e., decision-directed), as described in Sections 9.4–9.6. Adaptation is performed to both the MF coefficients and the coefficients used in VA to account for noise correlation, similarly as the LMS method described in Chapter 8. One can also use a separate channel estimator, which is essentially an adaptive filter modeling the actual channel response, as described in Chapter 8. A more detailed description can be found in [5, sec. 10.1-7].

9.4 LINEAR EQUALIZER (LE)

Section 9.3 shows the optimal sequence estimation, the MLSE algorithm, whose complexity is too high for many practical cases. For the rest of the chapter, we examine the equalizer as an alternative method of dealing with ISI. An equalizer attempts to remove or reduce ISI through filtering. With an equalizer, the total channel can be approximated as additive white Gaussian noise (AWGN), that is, without ISI.[3] Therefore, data recovery can be achieved through demodulation (Chapter 4) and decoding (Chapter 5). Typically, an equalizer is adaptive, without a separate channel estimation module.

This section discusses the simpler form of equalizers, the linear equalizer (LE). The more sophisticated decision feedback equalizer (DFE) is discussed in Section 9.5. While a DFE is superior in performance, it is difficult to integrate with modern channel coding schemes. DFE adaptation convergence is also slower. Therefore, in addition to being the foundation of DFE, LEs are still broadly used in practice.

---

[3] As will be shown later in this chapter, some types of equalizers have residual ISI at the output. With interleaving, such residual ISI is treated as white noise by the subsequent processing modules (demodulator or channel decoder).

9.4.1 General Formulation

We start with the discrete channel model (9.16). An LE is an adaptive filter applied to the channel output $\{y_m\}$, as shown in Figure 9.3. The left part, up to $\{y_m\}$, depicts (9.16). $\{y_m\}$ is fed through the equalizer, whose tap values are $\{a_n\}$. The output of the equalizer $\{r_m\}$ is sent to the demodulator, which is described in Chapter 4. $\{\hat{d}_m\}$ is the demodulator's estimate of transmitted symbol $\{d_m\}$. $\{\hat{d}_m\}$ is sent to further processing such as channel decoding.[4] The adaptation of the equalizer is based on the error sequence $\{e^{(m)}\}$, which is the difference between $\{r_m\}$ and $\{d'_m\}$. $\{e_m\}$ includes both noise and residual ISI. The goal of the equalizer, generally speaking, is reducing $\{e_m\}$, and thus reducing the error probability of the demodulator that follows it.

The adaptation of the equalizer, to be detailed in Sections 9.4.3 and 9.4.4, is based on $\{e^{(m)}\}$. At the receiver, $\{e^{(m)}\}$ is given by[5]

$$e^{(m)} \overset{\text{def}}{=} d'_m - r_m. \tag{9.24}$$

$\{d'_m\}$ is obtained in two ways. During the training phase (usually at the beginning of a communication session), a known sequence of $\{d_m\}$ is transmitted. Therefore, $\{d_m\}$ is known to the receiver and can be used as $\{d'_m\}$ for the computation of $\{e_m\}$ as shown by the dashed line in Figure 9.3. During data transmission phase (the normal communication mode, also known as "decision-directed" phase), the receiver does not know $\{d_m\}$, but can approximate it with the demodulator output $\{\hat{d}_m\}$. In this case, $\{d'_m\}$ is taken from the demodulator output $\{\hat{d}_m\}$. It is different from d_m only

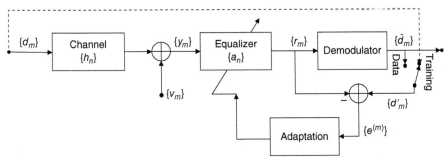

Figure 9.3   Linear equalizer architecture.

---

[4] If soft-input decoding is used, $\{r_m\}$ is directly sent to the decoder. However, typically a demodulator is still used, to support equalizer adaptation.

[5] Note that in some literature, $e^{(m)}$ is defined as $r_m - \hat{d}_m$. Such difference in sign is carried into the formulation of adaptation algorithm.

when the demodulator makes an error, which is a rare event if the system works properly. The two modes of obtaining $\{e_m\}$ are indicated by the double throw switch in Figure 9.3. In our analyses, it is assumed that the demodulator always makes correct decisions. Therefore, $\{d_m\}$ and $\{d'_m\}$ are considered the same.

The adaptation block also takes other quantities as input, as to be detailed in Sections 9.4.3.2 and 9.4.4.3. Such data paths are not shown in Figure 9.3.

The output of the equalizer can be modeled with the equivalent channel Q and equivalent noise $\{u_m\}$:

$$r_m = \sum_k a_k y_{m-k} = \sum_n Q_n d_{m-n} + u_m$$

$$Q_n \stackrel{\text{def}}{=} \sum_k a_k h_{n-k}$$

$$u_m \stackrel{\text{def}}{=} \sum_n a_n v_{m-n}$$

$$(9.25)$$

Figure 9.3 and (9.25) provide the general formulation of the LEs. Furthermore, LEs can be classified according to their optimization criteria. The two most common types are the zero forcing (ZF) and the minimum mean-squared error (MMSE) equalizers, which are described in Sections 9.4.3 and 9.4.4.

9.4.2   Notes on Tap Alignments

As discussed in Section 9.2.3, we are free to choose the position of tap h_0 among the L taps. In Section 9.3, h_0 is chosen to be the first tap of the channel filter. In Section 9.4, h_0 is chosen to be the tap with maximum modulus. Namely,

$$|h_0| = \max_n |h_n|. \tag{9.26}$$

We also choose a tap in the equalizer to be a_0, known as the center tap. For the total filter Q, (9.25) implies

$$Q_0 = \sum_k h_k a_{-k}. \tag{9.27}$$

Intuitively, when the output of h containing $d_m h_0$ is aligned with tap a_0, the output of the equalizer corresponds to r_m. In other words, r_m in (9.25) contains the term $h_0 a_0 d_m$.

Figure 9.4 illustrates the alignment between tap and data. The left side shows the channel filter. The top row is the taps (only three taps around h_0 are shown). The next three rows show the delay line contents at three consecutive sample times. The outputs are, respectively, y_{m-1}, y_m, and y_{m+1} according to (9.16). These outputs are fed into the delay line of the equalizer, shown on the right side. The left column is the delay line, while the right column is the tap values (only three taps around a_0 are shown).

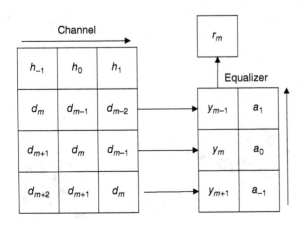

Figure 9.4   Data and tap alignment for linear equalizer.

The output of this particular state is r_m, according to (9.25). Data in the channel delay line move to the right, while those in the equalizer move upward.

In the above scheme, the position of h_0 depends on the channel characteristics. However, the position of a_0 is chosen during receiver design. When the equalizer is long, the choice of a_0 position is not very critical. However, for short equalizers, the a_0 position is an important design parameter.

In Sections 9.4.3 and 9.4.4, we consider two optimization criteria for the LE: the zero forcing (ZF) and the minimum mean-squared error (MMSE).

9.4.3   Zero Forcing (ZF) Solution

9.4.3.1   ZF LE Optimization Criterion

One natural optimization criterion is minimizing the maximum possible ISI. The maximum ISI happens when all symbols take the maximum amplitude and the appropriate phase so that contributions from all symbols add up constructively. In this case, the ISI is

$$ISI_{\max} = A \sum_n |Q_n|, \tag{9.28}$$

where A is the maximum amplitude of transmitted symbols. The zero forcing (ZF) equalizer minimizes ISI_{max} normalized by the signal amplitude. In other words, the ZF equalizer minimizes the following cost function:

$$c_{ZF} \overset{\text{def}}{=} \frac{ISI_{\max}}{Ah_0} = \frac{1}{h_0} \sum_n |Q_n|. \tag{9.29}$$

Here, h_0 is the tap of channel filter $\{h_n\}$ with the maximum modulus (see Section 9.4.2). Such cost function is known as the "maximum distortion criterion"

[5, sec. 9.4-1]. It is shown in an appendix (Section 9.11) that if the channel is "weakly dispersive," that is,

$$|h_0| > \sum_{n \neq 0} |h_n|, \tag{9.30}$$

then such criterion leads to the following optimization result: all taps in $\{Q_n\}$ that are contributed by h_0 are zero, except for Q_0. To be more specific, if the equalizer is given by $\{a_n, n = -N_1, -N_1 + 1, \ldots, 0, \ldots, N_2\}$ where N_1 and N_2 are positive integers, then optimization that minimizes the cost function given by (9.29) leads to[6]

$$Q_n = 0, \forall -N_1 \leq n \leq N_2, n \neq 0. \tag{9.31}$$

Therefore, such an equalizer is named "zero forcing": it drives ISI to zero within certain tap positions.

In addition, we can choose a normalization condition for the equalizer, to remove ambiguity about equalizer gain:

$$Q_0 = 1. \tag{9.32}$$

Equations (9.31) and (9.32) contain a total of $N_2 + N_1 + 1$ equations. They can be used to solve the $N_2 + N_1 + 1$ equalizer tap values $\{a_n\}$.

9.4.3.2   ZF LE Adaptation

The optimal solution of ZF equalizer can be obtained through adaptation [5, sec. 10.1-1], [6, sec. 7.3.1.1]. The tap values $\{a_n\}$ are iterated as follows:

$$a_n^{(m+1)} = a_n^{(m)} + \mu e^{(m)} d_{m-n}^{'*}. \tag{9.33}$$

Here, in reference to Figure 9.3 in Section 9.4.1, $a_n^{(m)}$ is the nth equalizer tap at the mth iteration step. r_m is the equalizer output at the mth step, given by (9.25). $e^{(m)}$ is the equalizer error at the mth step, given by (9.24). $d_m^{'*}$ is the mth transmitted symbol as determined by the receiver as indicated in Figure 9.3. μ is a positive step size chosen for adaptation.

At step m, the maximum response from symbol d_{m-n}, namely, the product of d_{m-n} and the channel center tap h_0, resides in the equalizer delay line position corresponding to a_n. This can serve as an intuitive guide for signal alignment in implementation.

Note that depending on how the center tap of the equalizer is selected (see Section 9.4.2), n can be negative. A negative n refers to a symbol that the modulator

---

[6] Note that filter $\{Q_n\}$ is longer than this range. In other words, the "edge taps" of filter $\{Q_n\}$ are not constrained by (9.31).

has not yet determined in (9.33). In this case, proper delay and storage of $\{d_n\}$ and $\{e^{(n)}\}$ are necessary to perform the adaptation after all data are available.

From (9.33), we can intuitively see that such adaptation has a stochastic fixed point, where

$$E\left(e^{(m)}\hat{d}_{m-n}^{*}\right) = 0. \tag{9.34}$$

Namely, when the optimal state is reached (at the fixed point), the residual error is uncorrelated with the transmitted symbols. Equivalently, ISI is removed within the range of n, which is to be expected from (9.31).

The step sizes can be chosen following the principles discussed in Chapter 8. It should be noted, however, that this adaptation method is not guaranteed to converge. In general, if the channel is "weakly dispersed" as stated in (9.30), then a ZF equalizer is likely to work well.

9.4.3.3   ZF LE Performance

Intuitively, the function of a ZF equalizer is reversing the channel before it, so that all ISIs are removed. Since the noise is white at the equalizer input, the output noise PSD is the square of the equalizer transfer function.

As shown in (9.20), the spectrum of the total channel $\{h_m\}$ is proportional to the folded spectrum of the physical channel $q(t)$. Therefore, the effect of the ZF equalizer is inverting the folded version of the physical channel $q(t)$. In the following, we use z-transform to study the output SNR of ZF equalizer with infinite-length.[7] With z-transform, the ZF equalizer response is

$$\bar{a}(z) = \bar{h}(z)^{-1}. \tag{9.35}$$

Remember that the noise at the equalizer input $\{v_m\}$ is white (9.11):

$$E\left(v_m v_n^{*}\right) = N_0 \delta_{m,n}. \tag{9.36}$$

The noise power at the output of the equalizer is thus

$$E\left(\left|e^{(m)}\right|^2\right) = E\left(\sum_{n,k} a_n a_k^* v_{m-n} v_{m-k}^*\right) = N_0 \sum_{n} |a_n|^2, \tag{9.37}$$

Define the following sequence:

$$A_n \overset{\text{def}}{=} \sum_{k} a_k a_{k-n}^*. \tag{9.38}$$

---

[7] See Section 9.9 for definitions and properties of z-transform.

Obviously, $\sum_n |a_n|^2$ is the same as A_0. On the other hand, $\{A_n\}$ is $\{a_n\}$ convoluted with its time-reversal and complex conjugate. Therefore, as shown in an appendix (Section 9.9.1), the z-transform of $\{A_n\}$ is

$$\bar{A}(z) = \bar{a}(z)\bar{a}^*\left(z^{*-1}\right). \tag{9.39}$$

From (9.35),

$$\bar{A}(z) = \left[\bar{h}(z)\bar{h}^*\left(z^{*-1}\right)\right]^{-1}. \tag{9.40}$$

From (9.195) in an appendix (Section 9.9.3.4) and (9.40), we have

$$\begin{aligned}
A_0 &= \frac{T_{Sym}}{2\pi} \int_{-\pi/T_{Sym}}^{\pi/T_{Sym}} \bar{A}\left[\exp\left(j\omega T_{Sym}\right)\right] d\omega \\
&= \frac{T_{Sym}}{2\pi} \int_{-\pi/T_{Sym}}^{\pi/T_{Sym}} \left|\bar{h}\left[\exp\left(j\omega T_{Sym}\right)\right]\right|^{-2} d\omega
\end{aligned} \tag{9.41}$$

From (9.20) in Section 9.2.4, we can replace $\bar{h}\left[\exp\left(j\omega T_{Sym}\right)\right]$ with the physical channel response $\tilde{q}(\omega)$, leading to

$$A_0 = \frac{T_{Sym}}{2\pi} \int_{-\pi/T_{Sym}}^{\pi/T_{Sym}} \left[\frac{1}{T_{Sym}} \sum_m \left|\tilde{q}\left(\omega - \frac{2\pi m}{T_{Sym}}\right)\right|^2\right]^{-1} d\omega. \tag{9.42}$$

To compute the SNR, one should remember that the signal power at the equalizer output matches the input signal power, which is σ_x^2. The noise power is given by (9.37) and (9.42). Therefore, the SNR at the ZF equalizer output is [6, sec. 7.3.1.1]

$$SNR_{ZF} = \frac{\sigma_x^2}{N_0} \left\{\frac{T_{Sym}}{2\pi} \int_{-\pi/T_{Sym}}^{\pi/T_{Sym}} \left[\sum_n \frac{1}{T_{Sym}} \left|\tilde{q}\left(\omega - \frac{2\pi n}{T_{Sym}}\right)\right|^2\right]^{-1} d\omega\right\}^{-1}. \tag{9.43}$$

We can change integration parameter from ω to frequency $f = \omega/2\pi$:

$$SNR_{ZF} = \frac{\sigma_x^2}{N_0} \left\{T_{Sym} \int_{-1/2T_{Sym}}^{1/2T_{Sym}} \left[\sum_n \frac{1}{T_{Sym}} \left|\tilde{q}_f\left(f - \frac{n}{T_{Sym}}\right)\right|^2\right]^{-1} df\right\}^{-1}. \tag{9.44}$$

Here, $\tilde{q}_f$ is the Fourier transform of $q(t)$ in terms of frequency.

The ZF equalizer attempts to remove all ISI within its coverage. While this sounds like a good idea, such an optimization process does not consider the noise. As a result, while ISI is suppressed, noise may be amplified. Consider an extreme case where the physical channel has nulls at some frequencies. To produce a flat channel (i.e., remove ISI), the equalizer provides large gains at these null frequencies. This behavior results in high amplification of noise, reducing the resulting SNR, as can be seen from the above derivation. If the modulus of $\sum_n (1/T_{Sym}) |\tilde{q}[\omega - (2\pi n/T_{Sym})]|^2$ is very small (or even zero) at some frequency, the integrals in (9.43) and (9.44) would become very large, and thus SNR_{ZF} would be very small. In other words, ZF equalizer does not work well if there are nulls in channel frequency response.

Above analysis assumes that the equalizer is infinitely long. Otherwise, it cannot exactly reverse the effect of the channel. However, the problem with channel nulls and noise amplification still exists with finite-length ZF equalizers.

9.4.3.4 ZF LE Example

In this section, we look at an example of the ZF filter. The channel has six taps with values $(0.2, -0.3, 1, 0.2, -0.3, 0.1)$, with the third one being the center tap. The input signal is a sequence of independent complex Gaussian symbols. The ZF filter adaptation step size is 0.001. The results are shown in Figure 9.5. Except for plot C, no noise is added to the simulation.

In Figure 9.5, for plots A–C, the equalizer has 50 taps, with its center tap at 20. Plot A shows the change of the relative error $E(e^{(m)2})/E(|d_m|^2)$ with sample index. The errors are low-pass filtered to reduce sample-by-sample fluctuation. This plot shows adaptation reduces the equalization error over time. Note that the adaptation step size is not optimized in this case. Therefore, this plot does not represent the maximum convergence speed. Plot B shows the final tap values of the total response $\{Q_n\}$ with equalizer length of 50. As expected, all taps have very low values except for the center one, consistent with the "zero forcing" concept. In plot C, white Gaussian noise is added with various SNR values. The output SNR values are evaluated after the equalizer adaptation. In this case, the output SNR matches input SNR within approximately 1 dB. Note that this result cannot be compared with (9.43) or (9.44) because the input SNR is measured before the equalizer; it is not necessarily σ_x^2/N_0. For plot D, the equalizer length varies between 10 and 80, with center taps located in the middle. The result shows the output SNR steadily increasing with the filter length, even when the filter length (up to 80) is much larger than the channel length (6).

9.4.4 Minimum Mean-Squared Error (MMSE) Solution

9.4.4.1 MMSE Optimization Criterion

Section 9.4.3 describes an LE under the maximum distortion, or zero forcing (ZF), criterion. While removing ISI, a ZF equalizer suffers from noise amplification, because its optimization criterion does not consider noise. In this section, we discuss an alternative optimization criterion: the minimum mean-squared error (MMSE). An MMSE equalizer simply minimizes the total squared equalization error:

$$c_{MMSE} \overset{\text{def}}{=} E\left(\left|e^{(m)}\right|^2\right). \tag{9.45}$$

where $e^{(m)}$ is the equalization error at time m, defined by (9.24).

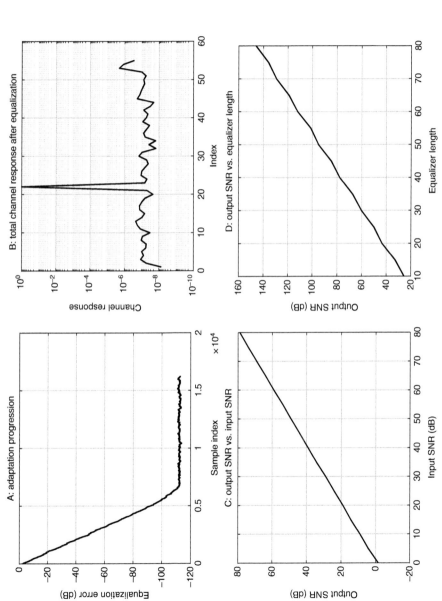

Figure 9.5 ZF equalizer example. A: initial convergence: error decreases with the sample index. B: output channel response in time domain, resembling a delta function (i.e., no dispersion). C: output SNR vs. input SNR. D: output SNR versus equalizer length. A, B, and C uses an equalizer with 50 taps.

We start our discussion with the formulation of the MMSE equalizer. From the equalizer model (9.24) and (9.25), we have

$$e^{(m)} = d_m - \sum_k a_k y_{m-k} = d_m - \sum_{k,n} a_k h_{n-k} d_{m-n} - \sum_k a_k v_{m-k}. \quad (9.46)$$

Here, $\{d_m\}$ is the transmitted symbols. $\{a_k\}$ is the equalizer tap values. $\{h_n\}$ is the total channel before the equalizer. $\{v_m\}$ is the (white) noise at the equalizer input. Since we do not consider demodulator errors, we consider $\{d'_m\}$ and $\{d_m\}$ to be the same in this section. With proper scaling of the whitening filter, we can assume the following statistics:

$$E(v_n v_m^*) = N_0 \delta_{n,m}$$
$$E(d_n d_m^*) = \sigma_x^2 \delta_{n,m} . \quad (9.47)$$
$$E(d_n v_m^*) = 0$$

Therefore, the cost function can be written as

$$C_{MMSE} = \sigma_x^2 \left(1 - \sum_k a_k h_{-k} - \sum_k a_k^* h_{-k}^* + \sum_{k,k',n} a_k a_{k'}^* h_{n-k} h_{n-k'}^* \right) + N_0 \sum_k a_k a_k^* \quad (9.48)$$

As discussed in an appendix (Section 9.10), the optimal solution is given by the following equation:

$$\frac{\partial C_{MMSE}}{\partial a_k^*} = 0 \forall k. \quad (9.49)$$

Equation (9.49) leads to

$$\sigma_x^2 \left(-h_{-k}^* + \sum_{k',n} a_{k'} h_{n-k'} h_{n-k}^* \right) + N_0 a_k = 0 \forall k. \quad (9.50)$$

This equation determines the equalizer tap values $\{a_n\}$.

Now, to evaluate the residual error level c_{MMSE} with optimized tap values, we look at the optimization from a different angle. Since $e^{(m)}$ depends on $\{a_k\}$ but not $\{a_k^*\}$, (9.45) and (9.49) lead to

$$E\left(e_o^{(m)} \frac{\partial}{\partial a_k^*} e_o^{(m)^*} \right) = 0. \quad (9.51)$$

From (9.46),

$$\frac{\partial}{\partial a_k^*} e^{(m)^*} = -y_{m-k}^*. \quad (9.52)$$

Therefore, at the optimum point,

$$E\left(e_o^{(m)}y_k^*\right) = 0 \forall k. \tag{9.53}$$

Note that this result is in parallel with (9.34) for the ZF equalizer. With this result, (9.45) and (9.46) lead to

$$c_{MMSE} = E\left[e^{(m)}\left(d_m^* - \sum_k a_k^* y_{m-k}^*\right)\right] = E\left(e^{(m)}d_m^*\right). \tag{9.54}$$

With (9.46), (9.54), and (9.47), we have

$$c_{MMSE} = E\left[\left(d_m - \sum_{k,n} a_k h_{n-k}d_{m-n} - \sum_k a_k u_{m-k}\right)d_m^*\right]$$
$$= \sigma_x^2\left(1 - \sum_k a_k h_{-k}\right) = \sigma_x^2(1 - Q_0) \tag{9.55}$$

Here, $\{Q_n\}$ is the combination of channel and equalizer, given in (9.25). Equation (9.55) is valid only when the equalizer taps $\{a_k\}$ are optimized.

9.4.4.2   MMSE LE Solution

In Section 9.4.4.1, we derived the expressions for the MMSE cost function and its minimization condition. The tap values can be solved by (9.50). In this section, we continue to look at the expression and properties of the optimal solution. Let us first study the case of infinite equalizer length, where z-transform can be used conveniently. Taking z-transform of (9.50), we obtain[8]

$$\sigma_x^2\left[\bar{a}(z)\bar{h}(z)\bar{h}^*\left(z^{*-1}\right) - \bar{h}^*\left(z^{*-1}\right)\right] + N_0\bar{a}(z) = 0. \tag{9.56}$$

Therefore, the equalizer tap values are given by the z-transform:

$$\bar{a}(z) = \frac{\bar{h}^*\left(z^{*-1}\right)}{\bar{h}(z)\bar{h}^*\left(z^{*-1}\right) + N_0/\sigma_x^2}. \tag{9.57}$$

---

[8] See Section 9.9 in appendix for the definition and properties of z-transform.

In contrast to the ZF equalizer, MMSE equalizer gain will not go to infinity, even when channel response $\bar{h}(z)$ has nulls.

We can also take the z-transform of (9.46), to get the residual error $\bar{e}_o(z)$ with optimized tap values:

$$\bar{e}_o(z) = \bar{d}(z) - \bar{a}(z)\bar{h}(z)\bar{d}(z) - \bar{a}(z)\bar{v}(z)$$

$$= \frac{N_0/\sigma_x^2}{\bar{h}(z)\bar{h}^*(z^{*-1}) + N_0/\sigma_x^2}\bar{d}(z) - \frac{\bar{h}^*(z^{*-1})}{\bar{h}(z)\bar{h}^*(z^{*-1}) + N_0/\sigma_x^2}\bar{v}(z). \tag{9.58}$$

The first term is the residual ISI. The second term is the noise. In contrast to the ZF equalizer, an MMSE equalizer does not eliminate ISI, even with infinite equalizer length. Instead, it balances residual ISI and noise to achieve minimum mean-squared error. On the other hand, the intensity of ISI is proportional to N_0/σ_x^2. It approaches zero under high SNR. In fact, from (9.57) and (9.25), we have the z-transform of the total filter Q:

$$\bar{Q}(z) = \bar{a}(z)\bar{h}(z) = \frac{\bar{h}(z)\bar{h}^*(z^{*-1})}{\bar{h}(z)\bar{h}^*(z^{*-1}) + N_0/\sigma_x^2}. \tag{9.59}$$

With very high SNR, we can ignore the term N_0/σ_x^2, leading to

$$\bar{Q}(z) = 1, \tag{9.60}$$

which is the same as the ZF solution (9.35). In other words, at high SNR, the difference between the ZF and the MMSE solutions vanish.

Note that the above solution (9.57) is for infinitely long equalizers. For finite-length MMSE equalizers, the tap values can be obtained by solving the system of linear Equation (9.50).

9.4.4.3 MMSE LE Adaptation

Adaptation of an MMSE equalizer is very similar to that of the LMS adaptive filter discussed in Chapter 8. From (9.45) and (9.52), we have

$$\frac{\partial C_{MMSE}}{\partial a_k} = -E\left(e^{(m)^*}y_{m-k}\right) \tag{9.61}$$

With the speediest descent (SD) method discussed in Chapter 8 extended to complex numbers, we have the iteration

$$a_k^{(m+1)} = a_k^{(m)} - \mu\left(\frac{\partial C_{MMSE}}{\partial a_k}\right)^* = a_k^{(m)} + \mu E\left(e^{(m)}y_{m-k}^*\right), \tag{9.62}$$

where $a_k^{(m)}$ is the tap value a_k at the mth iteration. μ is a positive step size. Following the LMS approach, we replace the expectation value with the current value, and obtain

$$a_k^{(m+1)} = a_k^{(m)} + \mu e^{(m)} y_{m-k}^*. \qquad (9.63)$$

We can compare the values of c_{MMSE} with the two versions of the equalizer tap values, while keeping the content of the delay line unchanged. Let

$$e' \overset{\text{def}}{=} e^{(m)} \Big|_{\{a_k^{(m)}\}}$$
$$e'' \overset{\text{def}}{=} e^{(m)} \Big|_{\{a_k^{(m+1)}\}} \qquad (9.64)$$

be the error values using tap values $\{a_k^{(m)}\}$ and $\{a_k^{(m+1)}\}$, respectively. From (9.46) and (9.63) we have

$$e'' = e' - \sum_k \mu e' |y_{m-k}|^2 = e'\left(1 - \sum_k \mu e' |y_{m-k}|^2\right). \qquad (9.65)$$

Therefore, the new cost function is (again, replacing the expectation with current value)

$$c_{MMSE}^{(m)} \Big|_{\{a_k^{(m+1)}\}} = |e''|^2 = |e'|^2 \left|1 - \sum_k \mu e' |y_{m-k}|^2\right|^2$$
$$= c_{MMSE}^{(m)} \Big|_{\{a_k^{(m)}\}} \left|1 - \sum_k \mu e' |y_{m-k}|^2\right|^2. \qquad (9.66)$$

When μ is small enough, the factor $1 - \sum_k \mu e' |y_{m-k}|^2$ is a positive number smaller than 1. Therefore, c_{MMSE} always decreases after iteration, until minimum is reached.

Comparing with the adaptation method for ZF equalizer (9.33), the only difference in MMSE adaptation (9.63) is that y_{m-k}^*, instead of d_{m-k}^*, is used to multiply with $e^{(m)}$. For the updating of a_k, the content in delay line position k is used. The alignment between $\{y_m\}$ and $\{d_m\}$ is specified by the center tap position, as explained in Section 9.4.2. Such alignment does not change during adaptation. For finite-length MMSE equalizer, the position of the center tap is important to performance, as shown in the examples in Section 9.4.4.4.

9.4.4.4 MMSE LE Performance

Let us continue to look at the expected performance of MMSE in terms output SNR. Here we only consider the case with infinite length. Recall (9.55) in Section 9.4.4.2. With the optimal solution, the mean squared error is

$$c_{MMSEo} = \sigma_x^2(1 - Q_0). \qquad (9.67)$$

Q_0 can be computed according to (9.195) in an appendix (Section 9.9.3.4).

$$Q_0 = \frac{T_{Sym}}{2\pi} \int_{-\pi/T_{Sym}}^{\pi/T_{Sym}} \bar{Q}\left[\exp\left(j\omega T_{Sym}\right)\right] d\omega. \tag{9.68}$$

At the optimal point, $\bar{Q}(z)$ is given by (9.59) in Section 9.4.4.2. Therefore,

$$Q_0 = \frac{T_{Sym}}{2\pi} \int_{-\pi/T_{Sym}}^{\pi/T_{Sym}} \frac{\left|\bar{h}\left[\exp\left(j\omega T_{Sym}\right)\right]\right|^2}{\left|\bar{h}\left[\exp\left(j\omega T_{Sym}\right)\right]\right|^2 + N_0/\sigma_x^2} d\omega. \tag{9.69}$$

We thus get c_{MMSEo} from (9.67):

$$c_{MMSEo} = \sigma_x^2(1 - Q_0) = N_0 \frac{T_{Sym}}{2\pi} \int_{-\pi/T_{Sym}}^{\pi/T_{Sym}} \left\{\left|\bar{h}\left[\exp\left(j\omega T_{Sym}\right)\right]\right|^2 + N_0/\sigma_x^2\right\}^{-1} d\omega. \tag{9.70}$$

From (9.20) in Section 9.2.4, we can replace $\bar{h}\left[\exp\left(j\omega T_{Sym}\right)\right]$ with the physical channel response $\tilde{q}(\omega)$, leading to

$$c_{MMSEo} = N_0 \frac{T_{Sym}}{2\pi} \int_{-\pi/T_{Sym}}^{\pi/T_{Sym}} \left[\frac{1}{T_{Sym}} \sum_m \left|\tilde{q}\left(\omega - \frac{2\pi m}{T_{Sym}}\right)\right|^2 + N_0/\sigma_x^2\right]^{-1} d\omega. \tag{9.71}$$

Or, in terms of frequency $f = \omega/2\pi$,

$$c_{MMSEo} = N_0 T_{Sym} \int_{-1/2T_{Sym}}^{1/2T_{Sym}} \left[\frac{1}{T_{Sym}} \sum_m \left|\tilde{q}_f\left(f - \frac{m}{T_{Sym}}\right)\right|^2 + N_0/\sigma_x^2\right]^{-1} df, \tag{9.72}$$

where $\tilde{q}_f(f)$ is the Fourier transform of $q(t)$ in terms of f. The SNR at the output of the equalizer is

$$\begin{aligned}
SNR_{MMSE} &= \frac{\sigma_x^2}{c_{MMSEO}} - 1 \\
&= \frac{\sigma_x^2}{N_0} \left\{ \frac{T_{Sym}}{2\pi} \int_{-\pi/T_{Sym}}^{\pi/T_{Sym}} \left[\frac{1}{T_{Sym}} \sum_m \left|\tilde{q}\left(\omega - \frac{2\pi m}{T_{Sym}}\right)\right|^2 + N_0/\sigma_x^2\right]^{-1} d\omega \right\}^{-1} - 1 \\
&= \frac{\sigma_x^2}{N_0} \left\{ T_{Sym} \int_{-1/2T_{Sym}}^{1/2T_{Sym}} \left[\frac{1}{T_{Sym}} \sum_m \left|\tilde{q}_f\left(f - \frac{m}{T_{Sym}}\right)\right|^2 + N_0/\sigma_x^2\right]^{-1} df \right\}^{-1} - 1
\end{aligned} \tag{9.73}$$

Comparing with the output SNR of ZF equalizer (9.44) in Section 9.4.3.3, there is an additional term N_0/σ_x^2 here. Therefore, even when $\tilde{q}(f)$ is zero, the integral would not diverge. In other words, noise amplification is limited and bounded for an MMSE equalizer.

The term -1 in (9.73) is due to a subtlety of MMSE detectors [6, sec. 7.3.1.2]. Simply put, the overall gain of an MMSE detector is slightly smaller than one. Therefore, the signal component in the output is slightly less than σ_x^2, as discussed in [7–9] and explained in more details in an appendix (Section 9.12). Nevertheless, SNR is usually much larger than one. Therefore, the difference here is typically inconsequential.

9.4.4.5 MMSE LE Example

MATLAB provides a function lineareq to support LEs. It supports LMS adaptation methods, which is the same as the MMSE adaptation in Section 9.4.4.3. Figure 9.6 shows some sample results. The channel used has six taps: $(0.73 + 0.53j, -0.06 + 0.09j, 0.31 + 0.95j, 0.90, 0.03 + 0.10j, 0.50j)$. The third tap is considered as the center tap of the channel. White Gaussian noise is injected after the channel. Adaption step size is 10^{-4}. Adaptation is considered stable after 6.7 million adaptation steps.

In Figure 9.6, plots A and B show the results of an MMSE equalizer, which is 2048 taps long. Its center tap is located at 45, which is the optimal position as shown by plot D below. Plot A shows the frequency response of the equalizer at various SNR levels as indicated in the legend. The frequency response of the channel (the thick line) is also plotted for comparison. We can see that under high SNR (the dashed line), the equalizer response is the inverse of the channel. Such response tends to completely remove ISI, as to be expected from a ZF equalizer. However, at lower SNR, the equalizer responses do not show sharp peaks at channel null, indicating that an MMSE equalizer balances the needs of removing ISI (i.e., inverting the channel) and avoiding noise amplification. Plot B shows the output SNR of this equalizer under various input SNR. Note that we cannot compare this performance with the ZF equalizer performance as shown in plot C of Figure 9.5, because the channels are different. In fact, when using the current channel, a ZF equalizer cannot even converge, as the condition in (9.30) is not satisfied. Furthermore, this performance cannot be compared with (9.73) in Section 9.4.4.4, because the "input SNR" here is the SNR after the channel, not σ_x^2/N_0.

Plot C shows the output SNR values from equalizers with various lengths. The center taps are at 45. The input SNR is 120 dB. We see that performance improves with equalizer length, even when its length is far larger than the channel length (six taps).

In plot D, we examine the output SNR as a function of the center tap position, with an equalizer of length 2048 and no added noise. The result shows that the optimal tap position is at 45, far away from the center. Such a result depends on the channel and is not universal. Nevertheless, we can see how sensitive the performance is to the center tap position, even for a very long equalizer (2048 taps comparing to the channel length of six taps).

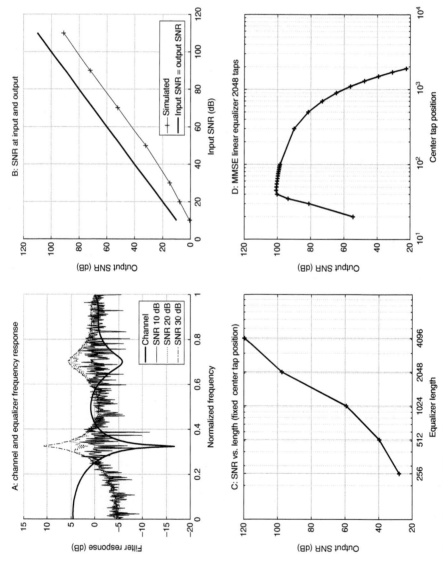

Figure 9.6 MMSE equalizer examples. See text for more explanations.

9.4.5   Summary of LE

The concept of linear equalization is straightforward. With an additional filter (the equalizer), we hope to reverse the dispersive effect of the channel, thus mitigating ISI. However, such a scheme does not work perfectly in practice. The main reason is that a finite-length filter cannot completely reverse channel dispersive effects, even when the channel is very short. Furthermore, LEs produce colored noise, which is suboptimal for the demodulator.

This section introduces two types of LEs: the ZF equalizer and the MMSE equalizer. They are optimized under different criteria, yet follow the same architecture shown in Figure 9.3. The adaptation schemes of the two equalizers are also similar. A ZF equalizer focuses on removing ISI completely, while an MMSE equalizer compromises between the needs of reducing ISI and limiting noise amplification. At a high SNR, optimal solutions of the two equalizers become the same.

LEs are a form of adaptive filter discussed in Chapter 8. Therefore, all adaptation methods discussed there can be applied. While the adaptation methods discussed in this section are widely used, some systems do use other adaptations for lower complexity or faster convergence at startup. For example, recursive least squares (RLS) introduced in Chapter 8 can be applied to linear equalizers, as a block-based fast adaptation technique [10, 11]. These techniques are typically application-specific.

In Section 9.5, we discuss another form of equalizer: the decision feedback equalizer (DFE).

9.5   DECISION FEEDBACK EQUALIZER (DFE)

In Section 9.4, we discussed two forms of the linear equalizers. A common shortcoming of linear equalizers is that the output noise is not white, providing further rooms to improve SNR. Such an opportunity is leveraged by the decision feedback equalizer (DFE).

In this section, we start in Section 9.5.1 with a noise predictor (NP) that improves SNR by taking advantage of the correlation between noise samples. In Section 9.5.2, we show that the NP is equivalent to a DFE. The DFE solution and its adaptation are discussed in Sections 9.5.3–9.5.5. Sections 9.5.6–9.5.8 discuss a few additional topics relating to the DFE. Section 9.5.9 provides some examples.

While DFE significantly improves performance comparing to the linear equalizers, its operation depends on the correctness of demodulator decisions, which feed the feedback portion of the equalizer. This issue is known as the error propagation problem, discussed in Section 9.5.6. Therefore, the DFE is somewhat limited in its application. Section 9.6 presents the Tomlinson–Harashima precoding, which is based on the same principles as DFE, yet avoiding the error propagation problem.

As in the case of the linear equalizers, a DFE can be optimized under various criteria, the most common ones including MMSE and ZF. In this section, we consider only the MMSE criterion, as it is most commonly used and has better performance. ZF DFE formulations can be found in the literature and textbooks [12, ch. 3], [13].

There are many ways to formulate DFE in the literature [5, 7–9, 14]. This section follows the formulations of [7, 9].

9.5.1 Colored Noise and Noise Predictor

As stated in Section 9.4, linear equalizers produce correlated (colored) noise samples. In this section, we will show that the noise power can be further reduced by removing such correlation.

Let us first consider an idealized case, where we know the noise of the *previous* samples. Consider a wide-sense stationary (WSS) noise sequence $\{u_k\}$ with known autocorrelation[9]:

$$E\left(u_k u_l^*\right) = r_{k-l}. \tag{9.74}$$

Suppose we know all the previous noise samples $\{u_l, l = 0, 1, \dots, k-1\}$. We wish to reduce the power of noise c_{NP} by subtracting from it our prediction, which is a filtered version of the previous noise samples:

$$u_k' = u_k - \sum_{l=1}^{L} f_l u_{k-l}$$

$$c_{NP} \overset{\text{def}}{=} E\left(\left|u_k'\right|^2\right) \tag{9.75}$$

Here, $\{f_l\}$ are the NP tap values. L is the length of the NP filter. C_{NP} is the cost function that we wish to minimize. The optimization condition is

$$\frac{\partial c_{NP}}{\partial f_l^*} = E\left(u_k' u_{k-l}^*\right) = 0 \forall l = 1, 2, \dots, L. \tag{9.76}$$

Or more explicitly,

$$r_l - \sum_{j=1}^{L} f_j r_{l-j} = 0 \forall l = 1, 2, \dots, L. \tag{9.77}$$

Equation (9.77) is a system of L linear equations, from which $\{f_l\}$ can be solved. For the purpose of later use, we can write (9.77) in vector form:

$$Rf = r. \tag{9.78}$$

Here, R is a matrix of size $L \times L$. f and r are vectors of size L. They are defined as follows:

$$f \overset{\text{def}}{=} (f_1, f_2 \dots, f_L)^T$$

$$r \overset{\text{def}}{=} (r_1, r_2, \dots, r_L)^T \tag{9.79}$$

$$R_{i,j} \overset{\text{def}}{=} r_{i-j}.$$

[9] The concept of wide sense stationary (WSS) is discussed in Chapter 4.

We can see that R is the autocorrelation matrix of $\{u_k\}$. Therefore, it is positive-definite.

Equation (9.77) can also be expressed in z-transform for infinitely long filters, as discussed in Section 9.4. We will not go into details.

Let us return to (9.75). Because of (9.76), we have at the optimum point

$$c_{NPo} = E\left(u'_k u'_k{}^*\right) = E\left(u'_k u_k^*\right) = E\left(u_k u_k^*\right) - \sum_l f_l E\left(u_{k-l} u_k^*\right) = r_0 - \sum_l f_l r_l^*. \tag{9.80}$$

From (9.78) and (9.79), (9.80) can be expressed in vector form:

$$c_{NPo} = r_0 - r^H f = r_0 - r^H R^{-1} r. \tag{9.81}$$

Since R is positive definite, so is R^{-1}. Therefore, the second term in (9.81) is always negative:

$$c_{NPo} \leq r_0. \tag{9.82}$$

But r_0 is none other than the original noise power $E(|u_k|^2)$. Therefore,

$$E\left(\left|u'_k\right|^2\right) \leq E\left(\left|u_k\right|^2\right). \tag{9.83}$$

Namely, NP reduces noise power. In fact, (9.81) shows that $c_{NPo} = E(|u_k|^2)$ is minimum if and only if r is zero, meaning $\{u_k\}$ is uncorrelated with offsets between 1 and L.

The next task is getting previous noise samples $\{u_l, l = k-1, k-2, \ldots, k-L\}$, where k is the current sample count. Previous noise samples can be extracted by comparing the input and output of the demodulator, which is assumed to make correct decisions. Namely,

$$u_l = w'_l - d_l, \tag{9.84}$$

where w'_k is the input to the demodulator (before the injection from NP). d_l is either the demodulator decision or the predetermined training symbol. The architecture is shown in Figure 9.7.

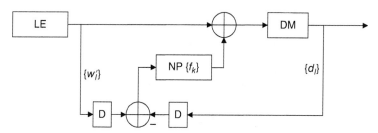

Figure 9.7 Noise predictor architecture. LE, linear equalizer; NP, noise equalizer; DM, demodulator (e.g., slicer). D, one-symbol delay.

In Figure 9.7, LE is a linear equalizer. DM is the demodulator. D means one symbol delay. NP is the noise predictive filter. NP filter can also take slightly different architectures, for example, in [5, sec. 9.5-3]. The form in Figure 9.7 is chosen for the purpose of Section 9.5.2.

9.5.2   Decision Feedback Equalizer (DFE)

Section 9.5.1 shows that an NP after a linear equalizer can help improving performance by reducing the noise power. In the following, we will show that such an NP architecture is equivalent to the decision feedback equalizer (DFE).

Figure 9.8 shows the equivalency between the NP and DFE architectures. Panel A shows the NP architecture, the same as in Figure 9.7. Panel B shows an equivalent architecture where the NP is split into two. Instead of having an NP applied to the difference between $\{w'_k\}$ and $\{d_k\}$ in (9.84), we have two copies of the NP acting on w'_k and d_k separately. The two blocks marked by the dashed lines are equivalent

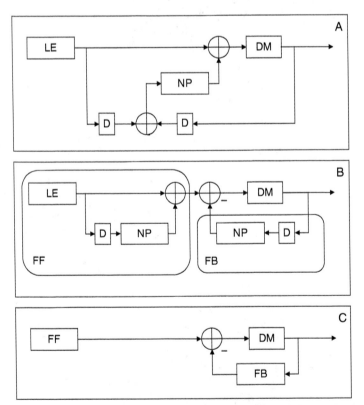

Figure 9.8   NP and DFE equivalency. A: linear equalizer (LE) combined with noise predictor (NP). B: NP is split into feedforward and feedback paths. C: feedforward NP and LE are combined into the feedforward filter of DFE.

filters, known as the feedforward (FF) filter and the feedback (FB) filter. Panel C shows the DFE architecture, where the FF and FB blocks in panel B are represented as single filters.

Furthermore, in theoretical analyses of DFE, we usually assume all demodulator decisions are correct (more on this topic in Section 9.5.6). Therefore, we can ignore the demodulator and replace its decisions with the transmitted signal. Similar to Figure 9.3, the model for DFE can be shown in Figure 9.9. In doing so, we need to be careful about causality. Namely, only past transmitted symbols are available to FB and adaptation.

In Figure 9.9, the transmitted symbols $\{d_m\}$ pass the channel $\{h_n\}$ with noise $\{v_m\}$ injected, to form $\{y_m\}$ according to (9.16). The signal then passes the feedforward (FF) filter, which is a part of the DFE. Separately, $\{d_m\}$ (mimicking the demodulator output) is also sent to the feedback (FB) filter, whose output is subtracted from the FF output to form $\{r_m\}$, which is the DFE output. $\{d_m\}$ is also used to compute the error $\{e^{(m)}\}$ for adaptation. As will be shown in Section 9.5.5, adaptation also takes other inputs ($\{y_m\}$ and $\{d_m\}$ for MMSE adaptation) that are not shown in Figure 9.9. Based on this architecture, we establish the model for DFE as follows.

The input signal to the DFE is $\{y_m\}$, given by (9.16):

$$y_m = \sum_k h_k d_{m-k} + v_m. \tag{9.85}$$

Here, $\{h_k\}$ is the total channel, including the pulse shaping filter, the propagation channel, the matched filter, and the noise whitening filter. $\{d_m\}$ is the transmitted symbols sequence and $\{v_m\}$ is the injected noise. They have the statistics

$$E\left(d_m d_n^*\right) = \sigma_x^2 \delta_{m,n}$$

$$E\left(v_m v_n^*\right) = N_0 \delta_{m,n} . \tag{9.86}$$

$$E\left(d_m v_n^*\right) = 0$$

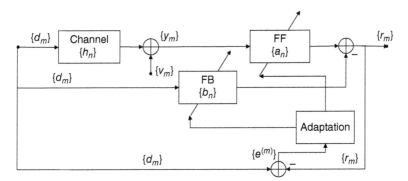

Figure 9.9   Equivalent DFE without decision errors.

From Figure 9.9, we can see that

$$r^{(m)} = \sum_k a_k y_{m-k} - \sum_{k=1} b_k d_{m-k}$$

$$e^{(m)} = d_m - r_m = d_m - \left(\sum_k a_k y_{m-k} - \sum_{k=1} b_k d_{m-k} \right). \tag{9.87}$$

Note that in (9.87), for the summation $\sum_k a_k y_{m-k}$, index k can run from negative to positive values, depending on the length of the FF filter and the position of the center tap. For the summation $\sum_{k=1} b_k d_{m-k}$, regardless of the FB filter length, index k always starts from 1 due to causality.

Since we will be focusing on $e^{(m)}$, it is convenient to introduce an effective FB coefficient $\{b'_n\}$:

$$b'_0 \overset{\text{def}}{=} 1$$
$$b'_n \overset{\text{def}}{=} b_n \forall n \geq 1. \tag{9.88}$$

The z-transform of $\{b'_n\}$ is $\bar{b}'(z)$. It has the following properties: there are no negative-order terms, and the zeroth order coefficient must be 1. Functions with such property are known as *monic*. With the introduction of $\{b'_n\}$, (9.87) leads to

$$e^{(m)} = \sum_{k=0} b'_k d_{m-k} - \sum_k a_k y_{m-k}. \tag{9.89}$$

Equation (9.89) provides the DFE model for our discussion. Next, we examine the formulation, optimal solutions, and properties of the MMSE DFE, in the cases of infinite-length filters (Section 9.5.3) and finite-length filters (Section 9.5.4).

9.5.3 Minimum Mean Squared Error DFE (Infinite-Length Filters)

9.5.3.1 Formulation in Z-Transform
In the case with of infinitely long equalizers, z-transform is a useful tool (see an appendix Section 9.9). We write (9.89) and (9.85) in z-transform:

$$\bar{e}(z) = \bar{b}'(z)\bar{d}(z) - \bar{a}(z)\bar{y}(z) = \bar{b}'(z)\bar{d}(z) - \bar{a}(z)[\bar{h}(z)\bar{d}(z) + \bar{v}(z)]. \tag{9.90}$$

Here, $\bar{e}(z)$, $\bar{b}'(z)$, $\bar{a}(z)$, $\bar{d}(z)$, and $\bar{v}(z)$ are z-transforms of $\{e^{(m)}\}$, $\{b'_n\}$, $\{a_n\}$, $\{d_m\}$, and $\{v_m\}$, respectively. Remember that $\bar{b}'(z)$ has to be monic, due to the constraint in (9.88). The cost function for the MMSE optimization is

$$c_{MMSE} \overset{\text{def}}{=} E\left(\left|e^{(m)}\right|^2\right). \tag{9.91}$$

We introduce the autocorrelation of the error sequence $\{e^{(m)}, m = \ldots, -1, 0, 1, \ldots\}$, based on (9.181) in Section 9.9.2[10]:

$$R_{een} \overset{\text{def}}{=} E\left(e^{(m)}e^{(m-n)*}\right)$$
$$\bar{R}_{ee}(z) = \frac{1}{M}E\left[\bar{e}(z)e^*\left(\bar{z}^{-1*}\right)\right]. \tag{9.92}$$

Here, M is the length of the sequence $\{e^{(m)}\}$. $\bar{R}_{ee}(z)$ is the z-transform of $\{R_{een}\}$. Therefore,

$$c_{MMSE} = R_{ee0}, \tag{9.93}$$

Given such formulation, we go on to look for the optimal solutions for $\bar{a}(z)$ and $\bar{b}'(z)$. We start by optimizing $\bar{a}(z)$ for any given $\bar{b}'(z)$ (Section 9.5.3.2). This standard minimization problem can be solved by letting the corresponding derivative be zero. As a result, c_{MMSE} can be expressed in terms of $\bar{b}'(z)$. c_{MMSE} is further optimized with regard to $\bar{b}'(z)$ in Section 9.5.3.3. Because $\bar{b}'(z)$ must be monic, a different approach is used. Afterward, Section 9.5.3.4 provides some discussions on the error and cost function at the optimum point. Section 9.5.3.5 provides more discussions on DFE performance in terms of output SNR.

9.5.3.2 Optimization with Regard to $\bar{a}(z)$
Let us try to minimize the whole $\bar{R}_{ee}(z)$ with regard to $\bar{a}(z)$, while holding $\bar{b}'(z)$ constant. From (9.92) and (9.89), the optimal condition is, according to (9.221) in Section 9.10,

$$\frac{\partial \bar{R}_{ee}(z)}{\partial \bar{a}^*\left(z^{-1*}\right)} = \frac{1}{M}E\left[\bar{e}(z)y^*\left(z^{-1*}\right)\right] = \frac{1}{M}E\left\{\bar{e}(z)\left[\bar{h}^*\left(z^{-1*}\right)\bar{d}^*\left(z^{-1*}\right) + \bar{v}^*\left(z^{-1}\right)\right]\right\} = 0. \tag{9.94}$$

---

[10] Since both transmitted symbols and noise samples are WWS, so is $\{e^{(n)}\}$.

Namely,

$$E\left\{\left[\bar{b}'(z)\bar{d}(z)-\bar{a}(z)\bar{h}(z)\bar{d}(z)-\bar{a}(z)\bar{v}(z)\right]\left[\bar{h}^*\left(z^{-1*}\right)\bar{d}^*\left(z^{-1*}\right)+\bar{v}^*\left(z^{-1}\right)\right]\right\}=0. \quad (9.95)$$

Remembering that signal and noise are uncorrelated, (9.95) leads to

$$
\begin{aligned}
\bar{a}(z) &= \frac{\bar{b}'(z)\bar{R}_{dd}(z)\bar{h}^*\left(z^{-1*}\right)}{\bar{h}(z)\bar{h}^*\left(z^{-1*}\right)\bar{R}_{dd}(z)+\bar{R}_{vv}(z)} \\
\bar{R}_{dd}(z) &\stackrel{\text{def}}{=} \frac{1}{M}E\left[\bar{d}(z)\bar{d}^*\left(z^{-1*}\right)\right] \\
\bar{R}_{vv}(z) &\stackrel{\text{def}}{=} \frac{1}{M}E\left[\bar{v}(z)\bar{v}^*\left(z^{-1*}\right)\right]
\end{aligned}
\quad (9.96)
$$

Note that $a(z)$ is a rational function, that is, a fraction whose denominator and numerator are polynomials. This form implies that $\{a_n\}$ is a filter with infinite length. Substituting (9.96) into (9.90), we have

$$\bar{e}(z)=\bar{b}'(z)\left\{\bar{d}(z)-\frac{\bar{R}_{dd}(z)\bar{h}^*\left(z^{-1*}\right)}{\bar{h}(z)\bar{h}^*\left(z^{-1*}\right)\bar{R}_{dd}(z)+\bar{R}_{vv}(z)}\left[\bar{h}(z)\bar{d}(z)+\bar{v}(z)\right]\right\} \quad (9.97)$$

Because of (9.94), the optimal $\bar{R}_{ee}(z)$ is

$$
\begin{aligned}
\bar{R}_{ee}(z) &= \frac{1}{M}E\left[\bar{e}(z)\bar{b}'^*\left(z^{-1*}\right)\bar{d}^*\left(z^{-1*}\right)\right] \\
&= \frac{1}{M}\bar{b}'(z)\bar{b}'^*\left(z^{-1*}\right)M\bar{R}_{dd}(z)\left\{1-\frac{\bar{R}_{dd}(z)\bar{h}(z)\bar{h}^*\left(z^{-1*}\right)}{\bar{h}(z)\bar{h}^*\left(z^{-1*}\right)\bar{R}_{dd}(z)+\bar{R}_{vv}(z)}\right\} \\
&= \bar{b}'(z)\bar{b}'^*\left(z^{-1*}\right)\bar{R}_{dd}(z)\frac{\bar{R}_{vv}(z)}{\bar{h}(z)\bar{h}^*\left(z^{-1*}\right)\bar{R}_{dd}(z)+\bar{R}_{vv}(z)} = \bar{b}'(z)\bar{b}'^*\left(z^{*-1}\right)\bar{R}_{d/y}(z)
\end{aligned}
$$

$$\bar{R}_{d/y}(z)\stackrel{\text{def}}{=}\frac{\bar{R}_{dd}(z)\bar{R}_{vv}(z)}{\bar{h}(z)\bar{h}^*\left(z^{-1*}\right)\bar{R}_{dd}(z)+\bar{R}_{vv}(z)}$$

$$(9.98)$$

In (9.98), $\bar{R}_{ee}(z)$ is expressed in terms of $\bar{b}'(z)$, assuming $\bar{a}(z)$ is optimized according to (9.96).

The definition of $\bar{R}_{dd}(z)$ and $\bar{R}_{vv}(z)$ depends on the sequence length M, and so is $\bar{R}_{d/y}(z)$. M is a rather arbitrary number. As Section 9.5.3.5 will show, $\bar{R}_{dd}$ and $\bar{R}_{vv}$, and thus all the results we are interested in, are independent of M.

9.5.3.3   Optimization Regarding $\bar{b}'(z)$

Using the same treatment as for $a(z)$ in Section 9.5.3.2, we get the "optimal" $\bar{b}'(z)$ to be zero. This solution makes sense in light of (9.98) because a zero $\bar{b}'(z)$ leads to zero $\bar{R}_{ee}$. However, it is obviously not correct. The reason is that according to (9.88), $\bar{b}'(z)$ has a constraint that it must be monic (i.e., $\{b'_n\}$ is a causal sequence with 1 as the first term). Fortunately, there is an easy way to deal with such a constraint.

Because of (9.96), the factor $\bar{R}_{d/y}(z)$ has the "complex reversal symmetry"[11]:

$$\bar{R}_{d/y}(z) = \bar{R}^*_{d/y}\left(z^{-1*}\right) \tag{9.99}$$

According to (9.199) in an appendix (Section 9.9.4), we can factorize $\bar{R}_{d/y}(z)$ into:

$$\bar{R}_{d/y}(z) = S_0 \bar{G}(z) \bar{G}^*\left(z^{-1*}\right), \tag{9.100}$$

where $\bar{G}(z)$ is monic. S_0 is a constant that will be examined in Section 9.5.3.5. With such factorization, (9.98) can be written as

$$\bar{R}_{ee}(z) = S_0 \bar{s}(z)\bar{s}^*\left(z^{-1*}\right) \\ \bar{s}(z) \overset{\text{def}}{=} \bar{b}'(z)\bar{G}(z) \tag{9.101}$$

Since both $\bar{b}'(z)$ and $\bar{G}(z)$ are monic, so is $\bar{s}(z)$. With (9.101), the zeroth order coefficient of $\bar{R}_{ee}$ is

$$R_{ee0} = S_0\left(1 + \sum_{k=1}^{\infty} |s_k|^2\right), \tag{9.102}$$

where $\{s_k, k = 1, 2, \ldots\}$ represents the coefficients in $\bar{s}(z)$. Therefore, the optimum is reached when

$$s_k = 0 \forall k = 1, 2, \ldots. \tag{9.103}$$

Namely, the optimal FB filter is[12]

$$\bar{b}'(z) = \bar{G}(z)^{-1}. \tag{9.104}$$

With such a solution, the optimal cost function C_{MMSEo} is, from (9.93) and (9.101),

$$C_{MMSEo} = R_{ee0} = S_0. \tag{9.105}$$

---

[11] This symmetry is to be expected because $\bar{R}_{ee}(z)$ has such symmetry by definition (9.92).
[12] It can be shown that the inverse of a monic filter is also monic.

Although $\bar{G}(z)$, and thus $\bar{b}'(z)$, cannot be expressed explicitly beyond (9.100), this result is very useful in analyzing the DFE, as will be shown in Section 9.5.3.4.

9.5.3.4 Discussions on Optimal Solution

So far we obtained the optimal solution for $\bar{a}(z)$ and $\bar{b}'(z)$. Let us continue to examine some properties of the infinite-length DFE.

Substituting solution (9.104)–(9.101), we have

$$\bar{R}_{ee}(z) = S_0. \tag{9.106}$$

The fact that $\bar{R}_{ee}(z)$ is a constant implies that

$$E\left(e^{(m)} e^{(n)^*}\right) \propto \delta_{m,n}. \tag{9.107}$$

Namely, the residual error (noise plus ISI) is white. This characteristic is important for the infinite-length DFE.

We can actually write out $\bar{e}(z)$. From (9.90) and (9.96):

$$
\begin{aligned}
e(z) &= \bar{b}'(z)\bar{d}(z) - \bar{a}(z)[\bar{h}(z)\bar{d}(z) + \bar{v}(z)] \\
&= \bar{b}'(z)\bar{d}(z) - \bar{b}'(z)\frac{\bar{R}_{dd}(z)\bar{h}(z)\bar{h}^*(z^{-1^*})}{\bar{h}(z)\bar{h}^*(z^{-1^*})\bar{R}_{dd}(z) + \bar{R}_{vv}(z)}\bar{d}(z) + \bar{b}'(z)\frac{\bar{R}_{dd}(z)\bar{h}^*(z^{-1^*})}{\bar{h}(z)\bar{h}^*(z^{-1^*})\bar{R}_{dd}(z) + \bar{R}_{vv}(z)}\bar{v}(z) \\
&= \bar{b}'(z)\frac{\bar{R}_{vv}(z)}{\bar{h}(z)\bar{h}^*(z^{-1^*})\bar{R}_{dd}(z) + \bar{R}_{vv}(z)}\bar{d}(z) + \bar{b}'(z)\frac{\bar{R}_{dd}(z)\bar{h}^*(z^{-1^*})}{\bar{h}(z)\bar{h}^*(z^{-1^*})\bar{R}_{dd}(z) + \bar{R}_{vv}(z)}\bar{v}(z).
\end{aligned}
\tag{9.108}
$$

Equation (9.108) shows that in addition to the filtered noise, the output error contains residual ISI (the first term in the last line), which is proportional to noise power $\bar{R}_{vv}(z)$. Equations (9.107) and (9.108) show that the FF filter balances ISI and noise to result in a white residual error.

Let us further examine the residual ISI term in (9.108). From (9.98), we can write

$$\frac{\bar{R}_{vv}(z)}{\bar{h}(z)\bar{h}^*(z^{-1^*})\bar{R}_{dd}(z) + \bar{R}_{vv}(z)} = \bar{R}_{dd}^{-1}(z)R_{d/y}(z) \tag{9.109}$$

We consider the case where the transmitted symbols are white. In this case (see Section 9.5.3.5),

$$\bar{R}_{dd}(z) = \sigma_x^2. \tag{9.110}$$

With (9.110), (9.109), (9.100), and (9.104), the residual ISI term in (9.108) can be written as

$$\bar{b}'(z)\frac{\bar{R}_{vv}(z)}{\bar{h}(z)\bar{h}^*(z^{-1*})\bar{R}_{dd}(z) + \bar{R}_{vv}(z)}\bar{d}(z) = \sigma_x^{-2}\bar{b}'(z)R_{d/y}(z)\bar{d}(z)$$

$$= \sigma_x^{-2}\left(\bar{G}(z)^{-1}\right)S_0\bar{G}(z)\bar{G}(z^{-1*})\bar{d}(z). \quad (9.111)$$

$$= \sigma_x^{-2}S_0\bar{G}(z^{-1*})\bar{d}(z)$$

Since $\bar{G}(z)$ is a monic filter, $\bar{G}(z^{-1*})$ is an anticausal filter. Namely, the residual ISI in (9.111) contains only contributions from "future" transmitted symbols, that is, $\{d_n, n = m + 1, \ldots\}$, where m denotes the "current symbol" used to compute error in (9.87). In other words, ISI caused by the "previous" transmitted symbols are completely removed by the FB filter. This observation makes sense because this part of the ISI is covered by the FB filter. Unlike the FF filter, which needs to balance between ISI cancellation and noise whitening, the FB filter does not need to care about noise. Therefore, the optimal solution for the FB filter is to cancel all ISI within its coverage.

A "future" transmitted symbol contributes to the current error with the part of its waveform that precedes the "main peak." Therefore, ISI from "future" symbols are known as *precursor* ISI. Correspondingly, ISI from "previous" symbols are known as *postcursor* ISI. In an MMSE DFE, postcursor ISI is canceled by the FB filter output.

The above analyses show that, as in the case of an MMSE linear equalizer, an MMSE DFE seeks a compromise between reducing both ISI and noise contribution to the mean squared error. However, an infinite-length MMSE DFE reaches the optimal uncorrelated noise at the output. The FF filter has the freedom of reducing *both* noise and precursor ISI at the expense of increased postcursor ISI, which is canceled by the FB filter. Therefore, DFE can achieve better performance than a linear equalizer.

9.5.3.5 *Performance of MMSE DFE*

Section 9.5.3.4 shows that for infinite-length DFE, the total output noise is white. Such noise includes residual precursor ISI, whose intensity is proportional to the input noise power. The postcursor ISI is completely canceled out by the FB filter. In this section, we continue to look at the values of C_{MMSE} and output SNR at the optimum.

From (9.199) in an appendix (Section 9.9.4), S_0, as introduced in (9.100), can be expressed as

$$S_0 = \exp\left(\frac{1}{2\pi}\int_{-\pi}^{\pi} \ln\left|R_{d/y}\left(e^{j\theta}\right)\right| d\theta\right). \quad (9.112)$$

Let us consider the case where both signal and noise are white. Namely,

$$E(d_n d_{n'}^*) = \sigma_x^2 \delta_{n,n'}$$
$$E(v_n v_{n'}^*) = N_0 \delta_{n,n'} \tag{9.113}$$

Thus from (9.96),

$$\bar{R}_{dd}(z) = \sigma_x^2$$
$$\bar{R}_{vv}(z) = N_0. \tag{9.114}$$

Note that $\bar{R}_{dd}(z)$ and $\bar{R}_{vv}(z)$ are constants and are independent of M as in (9.96). With these expressions, (9.98) and (9.112) lead to

$$S_0 = \exp\left\{ -\frac{1}{2\pi} \int_{-\pi}^{\pi} \ln\left[\sigma_x^2 \left(1 + \frac{\sigma_x^2 |\bar{h}(\exp(j\theta))|^2}{N_0} \right) \right] d\theta \right\}$$

$$\underset{\omega \overset{\text{def}}{=} \theta/T_{Sym}}{=} \sigma_x^2 \exp\left[-\frac{T_{Sym}}{2\pi} \int_{-\pi/T_{Sym}}^{\pi/T_{Sym}} \ln\left(1 + \frac{\sigma_x^2 |\bar{h}[\exp(j\omega T_{Sym})]|^2}{N_0} \right) d\omega \right]. \tag{9.115}$$

From (9.20) in Section 9.2.4, we can replace $\bar{h}[\exp(j\omega T_{Sym})]$ with the physical channel response $\tilde{q}(\omega)$, leading to

$$S_0 = \sigma_x^2 \exp\left[-\frac{T_{Sym}}{2\pi} \int_{-\pi/T_{Sym}}^{\pi/T_{Sym}} \ln\left(1 + \frac{\sigma_x^2}{N_0 T_{Sym}} \sum_m \left| \tilde{q}\left(\omega - \frac{2\pi m}{T_{Sym}} \right) \right|^2 \right) d\omega \right]. \tag{9.116}$$

The DFE input SNR at angular frequency ω is

$$SNR_{in}(\omega) = \frac{\sigma_x^2}{N_0 T_{Sym}} \sum_m \left| \tilde{q}\left(\omega - \frac{2\pi m}{T_{Sym}} \right) \right|^2. \tag{9.117}$$

Therefore, (9.116) leads to

$$S_0 = \sigma_x^2 \exp\left\{ -\frac{T_{Sym}}{2\pi} \int_{-\pi/T_{Sym}}^{\pi/T_{Sym}} \ln[1 + SNR_{in}(\omega)]\, d\omega \right\}. \tag{9.118}$$

The "nominal SNR" at the output of the DFE is, from (9.105):

$$SNR_{DFE} \overset{\text{def}}{=} \frac{\sigma_x^2}{C_{MMSEo}} = \frac{\sigma_x^2}{S_0} = \exp\left\{ \frac{T_{Sym}}{2\pi} \int_{-\pi/T_{Sym}}^{\pi/T_{Sym}} \ln[1 + SNR_{in}(\omega)] d\omega \right\}. \tag{9.119}$$

As discussed in an appendix (Section 9.12), MMSE detectors such as the MMSE DFE produce a biased output. Such bias increases the resulting nominal SNR, but the bit error rate is actually increased. The optimal receiver (optimal in the sense of minimizing bit error rate) is an unbiased detector, that is, with a gain of 1.

Let us calculate the gain of the DFE, which is noted as g. The contribution of d_m to the error e is

$$e_d = d_m - gd_m = (1-g)d_m. \tag{9.120}$$

Since $e(z)$ is known, we can figure out g by looking at its zeroth coefficient. Leveraging the results (9.108)–(9.111) in Section 9.5.3.4, ignoring the noise part, and remembering that $\bar{G}(z^{-1})$ is a monic filter, we have

$$e_d = \sigma_x^{-2} S_0 d_m. \tag{9.121}$$

Therefore, the gain of an MMSE equalizer is

$$g = 1 - \frac{S_0}{\sigma_x^2}. \tag{9.122}$$

To achieve the unbiased DFE, we insert a gain g^{-1} between the DFE output and the demodulator [9]. The SNR for this unbiased detector is [9]

$$SNR_{DFEU} = SNR_{DFE} - 1. \tag{9.123}$$

Equation (9.119) thus leads to (with a change of logarithm base on both sides)

$$Log_2(1 + SNR_{DFEU}) = \frac{T_{Sym}}{2\pi} \int_{-\pi/T_{Sym}}^{\pi/T_{Sym}} \log_2[1 + SNR_{in}(\omega)]\, d\omega$$

$$\underset{f \triangleq \omega/2\pi}{=} T_{Sym} \int_{-1/T_{Sym}}^{1/T_{Sym}} \log_2[1 + SNR_{in}(f)]\, df. \tag{9.124}$$

Recall the expression of channel capacity for complex additive white Gaussian noise (AWGN) channels (Chapter 2):

$$C = \log_2(1 + SNR). \tag{9.125}$$

More detailed analyses show the following results [8]:

1. Equation (9.125) gives the capacity with DFE if SNR_{DFEU} is used as SNR. Therefore, (9.124) suggests that the channel capacity at the output of DFE is the average of input channel capacity over frequency. In other words, DFE does not cause degradation in channel capacity. This conclusion is based on the assumption that no detection error occurs, as implied in (9.87).

2. Furthermore, channel capacity can be achieved with the same coding techniques as in AWGN channels [8].

3. Therefore, DFE achieves essentially the same performance as the optimal MLSE receiver (Section 9.3), except that DFE is vulnerable to decision errors, as will be discussed in Section 9.5.6.

The additional gain at the DFE output can be understood as changing the decision rule (i.e., the scales in the slicer, see Chapter 4) in the demodulator to achieve lower bit error rate. Since our analyses assume perfect decisions (see Figure 9.9), the optimal solutions presented in this section are still optimal with such additional gain. In practice, SNR is usually much larger than 1. Therefore, the difference between SNR_{DFE} and SNR_{DFEU} is negligible. However, the introduction of SNR_{DFEU} helps in demonstrating the nice relationship (9.124). The introduction of the unbiased demodulator makes a difference in some cases, especially when channel coding allows for lower SNR [15].

9.5.4   Minimum Mean Squared Error DFE (Finite-Length Filters)

Section 9.5.3 formulates and analyzes DFE with infinite filter lengths. To achieve the predicted performance, equalizer filters, especially the FF filter, may need to be very long, even when the channel dispersion is short (see Section 9.4.4.5 for examples with the linear equalizer). Therefore, in reality, we often need to consider the limitation on filter lengths.

It is not easy to impose length constraint when using z-transform. Therefore, DFE with finite-length filters can be analyzed using vectors and matrices. An appendix (Section 9.13) provides detailed derivations on finite-length DFE. This section only summarizes some important results.

9.5.4.1   Formulation

Finite-length DFE can be formulated by (9.87) in Section 9.5.2. However, since the filters are relatively short, alignment among them becomes important and thus needs to be explicitly specified.

The channel model is given by (9.16). The relationship between received symbol $\{y_m\}$, transmitted symbols $\{d_m\}$, and noise $\{v_m\}$ is given by

$$y_m = \sum_{k=0}^{N_H} h_k d_{m-k} + v_m. \tag{9.126}$$

Here, $\{h_m\}$ is the tap sequence of the total channel (from transmitter symbol to the input of the DFE), with a length of $N_H + 1$. The choice of summation limits implies that d_m is the latest transmit symbol that contributes to the received symbol y_m. Such alignment (shown in Figure 9.10) can be achieved by properly defining $\{h_m\}$.

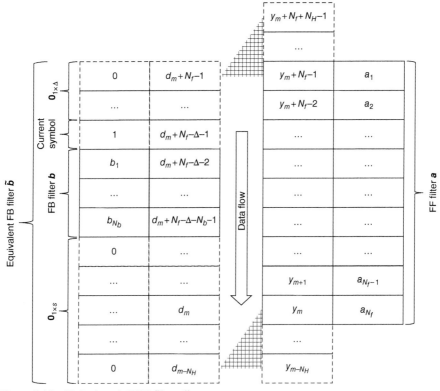

Figure 9.10   Finite-length DFE filter alignments.

This model can be expressed in vector form[13]:

$$y = Hd + v. \tag{9.127}$$

The vectors can be defined as follows, to be consistent with (9.126).

$$(y)_i = y_{m+N_f-i}, i = 1, 2, \ldots, N_f$$

$$(d)_i = d_{m+N_f-i}, i = 1, 2, \ldots, N_f + N_H. \tag{9.128}$$

$$(v)_i = v_{m+N_f-i}, i = 1, 2, \ldots, N_f$$

---

[13] In this chapter, bold-faced lowercase letters indicate vectors.

Here, $(y)_i$ is the ith component of vector y, while y_k is the kth symbol in sequence $\{y_m\}$. The same convention applies to vectors d and v. N_f is the length of the feedforward (FF) filter. H in (9.127) is a Toeplitz matrix of size $N_f \times (N_f + N_H)$ representing the channel. Comparing (9.126) and (9.127), components of H, $\{H_{i,j}\}$, are given by

$$H_{i,j} = \begin{cases} h_{j-i} & 0 \leq j - i \leq N_H \\ 0 & \text{otherwise} \end{cases}. \tag{9.129}$$

Equations (9.127) and (9.128) describe a process where $N_f + N_H$ transmitted symbols in d gererate N_f received symbols in y.

Now let us move from the channel model to the DFE model. A DFE is depicted in Figure 9.9 in Section 9.5.2. As discussed therein, we introduce an additional tap with value 1 in the feedback (FB) filter to represent the true current symbol. This addition makes expressing error easier. The detailed alignments between the current symbol, the FF filter, and the FB filter in our formulation are described below.

The tap values of the FF and FB filters are represented as vectors

$$a \overset{\text{def}}{=} (a_1, a_2, \ldots, a_{N_f})^T$$
$$b \overset{\text{def}}{=} (b_1, b_2, \ldots, b_{N_b})^T \tag{9.130}$$

N_b is the length of the FB filter. For our formulation, the FB filter b is imbedded in an equivalent filter $\tilde{b}$ of length $N_f + N_H > N_b$ with an offset Δ:

$$\tilde{b} \overset{\text{def}}{=} (0_{1 \times \Delta}, 1, b^T, 0_{1 \times s})^T, \tag{9.131}$$

$0_{1 \times \Delta}$ and $0_{1 \times s}$ are row zero-vectors of lengths Δ and s, respectively.

$$s = N_f + N_H - \Delta - N_b - 1. \tag{9.132}$$

The 1 in $\tilde{b}$ represents the direct path for error computation, as discussed above. It is assumed that N_f, N_H, and Δ are properly chosen so that $s \geq 0$.

At time m, the FF and FB filter delay lines are filled with vectors y and d, respectively. Therefore, the output of the DFE, as shown in (9.87) in Section 9.5.2, can be written as

$$r^{(m)} = \sum_{n=1}^{N_f} a_n y_{m+N_f-n} - \sum_{n=1}^{N_b} b_n d_{m+N_f-\Delta-1-n}$$

$$e^{(m)} = d_{m+N_f-\Delta-1} - r^{(m)} = d_{m+N_f-\Delta-1} - \sum_{n=1}^{N_f} a_n y_{m+N_f-n} + \sum_{n=1}^{N_b} b_n d_{m+N_f-\Delta-1-n}$$

. (9.133)

These equations can be written in vector form, using definitions in (9.128), (9.130), and (9.131):

$$r^{(m)} = a^H y - \tilde{b}^H d + d_{m+N_f-\Delta-1}$$

$$e^{(m)} = \tilde{b}^H d - a^H y$$

. (9.134)

Figure 9.10 illustrates the alignments between data and filter taps for DFE at time m. The center two columns show data d and y in the delay lines, as given by (9.128). Data points flow from top to bottom in the delay lines. The shaded triangles show the received symbols that a transmitted symbol influences, as described by (9.126). For example, d_{m-N_H} contributes to y_m to y_{m-N_H}. The right column shows the taps of the FF filter a, as defined by (9.130). The left column shows the actual and equivalent FB filters b and $\tilde{b}$, as defined by (9.130) and (9.131), respectively. Boxes with dashed lines are for logical illustration only; they are not included in the actual implementation.

We further introduce the following correlation matrices:

$$R_{dd} \overset{\text{def}}{=} E\left(dd^H\right)$$

$$R_{dy} \overset{\text{def}}{=} E\left(dy^H\right)$$

$$R_{yy} \overset{\text{def}}{=} E\left(yy^H\right)$$

$$R_{vv} \overset{\text{def}}{=} E\left(vv^H\right)$$

. (9.135)

v is uncorrelated with d. The autocorrelation matrices R_{dd}, R_{yy}, and R_{vv} are assumed to be positive-definite, and therefore nonsingular. Furthermore, the channel model (9.127) leads to

$$R_{dy} = R_{dd} H^H$$

$$R_{yy} = H R_{dd} H^H + R_{vv}$$

. (9.136)

The cost function is

$$c_{MMSE} \overset{\text{def}}{=} E\left(\left|e^{(m)}\right|^2\right). \tag{9.137}$$

With above definitions and the model (9.134), the cost function can be written as

$$c_{MMSE} = \tilde{\boldsymbol{b}}^H R_{dd}\tilde{\boldsymbol{b}} + \boldsymbol{a}^H R_{yy}\boldsymbol{a} - \tilde{\boldsymbol{b}}^H R_{dy}\boldsymbol{a} - \boldsymbol{a}^H R_{dy}^H \tilde{\boldsymbol{b}}. \tag{9.138}$$

9.5.4.2 Optimal Solution

As detailed in an appendix (Section 9.13), DFE filters $\boldsymbol{a}$ and $\tilde{\boldsymbol{b}}$ can be optimized similarly to the infinite-length case. To present the results, we need to first introduce a few more definitions. Let

$$R_{d/y} \overset{\text{def}}{=} R_{dd} - R_{dy}R_{yy}^{-1}R_{dy}^H. \tag{9.139}$$

It is a Hermitian matrix of size $(N_f + N_H) \times (N_f + N_H)$. The inverse of $R_{d/y}$, also Hermitian, can be factorized into

$$R_{d/y}^{-1} = L\Lambda L^H. \tag{9.140}$$

Here, L is a lower-triangular matrix, with all diagonal elements equal to 1:

$$\begin{aligned} L_{i,j} &= 0 \forall i < j \\ L_{i,i} &= 1 \forall i \end{aligned}. \tag{9.141}$$

It is easy to see that L^{-1} has these properties, as well. Λ is a diagonal matrix, whose diagonal elements are noted as $\{\lambda_i, i = 1, 2, \ldots, N_f + L_H\}$:

$$\Lambda_{i,j} = \begin{cases} \lambda_i & i = j \\ 0 & \text{otherwise} \end{cases}. \tag{9.142}$$

The ith column of L are noted as vector $\boldsymbol{l}^{(i)}$. It has $i - 1$ leading zeros, followed by 1. The rest of its components can have any value.

Further define vector $\boldsymbol{u}^{(i)}$ of size $N_f + N_H$, whose elements are all zero except for the ith, which is 1:

$$u_n^{(i)} = \delta_{n,i}. \tag{9.143}$$

With these definitions, the optimal solution is (Section 9.13)

$$\begin{aligned} \boldsymbol{a}_o &= R_{vv^{-1}}HL^{-H}\lambda_{\Delta_o+1}^{-1}\boldsymbol{u}^{(\Delta_o+1)} \\ \tilde{\boldsymbol{b}}_o &= \boldsymbol{l}^{(\Delta_o+1)} \\ c_{MMSEo} &= \lambda_{\Delta_o+1}^{-1} \\ \Delta_o &\overset{\text{def}}{=} \arg\max_\Delta \lambda_{\Delta+1} \end{aligned} \tag{9.144}$$

Here, Δ_o indicates the optimal alignment between the samples in FF delay line and the current symbol, as illustrated in Figure 9.10.

9.5.4.3 Properties

As derived or explained in an appendix (Section 9.13), the finite-length DFE has the following properties:

1. When $N_b = N_H$ and with white input and noise, it can be shown that $\Delta_o > N_f - 2$. Multiple simulations also show that $\Delta_o = N_f - 1$ holds for most practical channel and noise scenarios and reasonably long feedforward filters [7]. This means $s = 0$ as given by (9.132). In such alignment, the FB filter covers all transmitted symbols that contribute to the data in the FF filter delay line. Therefore, any longer FB filters are unnecessary. Other works use different Δ values, for example, [16, sec. II-B].

2. As in the case of infinitely long DFE, the residual ISI at the output of the FF filter can be divided into precursor (i.e., caused by the "future" symbols) and postcursor (i.e., caused by the "past" symbols). The latter is completely removed by the FB filter, if it is long enough.

3. As in the case of infinitely long DFE, the residual precursor ISI approaches zero at high SNR. In general, the FF filter makes a compromise between residual ISI and noise.

4. Unlike an infinite-length DFE, the total noise at a finite-length DFE output is not exactly white.

9.5.5 Adaptation of DFE

Similar to the LEs (Section 9.4.4.3), the FF and FB filters of a DFE can be adapted to near-optimal values. Following the same approach as in Section 9.4.4.3, the following iteration can be derived under MMSE criteria [6, sec. 7.3.2]:

$$
\begin{aligned}
a_k^{(m+1)} &= a_k^{(m)} + \mu_a e^{(m)} y_{m+N_f-k}^* \\
b_k^{(m+1)} &= b_k^{(m)} - \mu_b e^{(m)} d_{m+N_f-\Delta-1-k}^*
\end{aligned}
\tag{9.145}
$$

Here, $a_k^{(m)}$ and $b_k^{(m)}$ are tap values of a_k and b_k at time m, respectively. μ_a and μ_b are positive step sizes for FF and FB filter adaptation, respectively. To verify such algorithm, we can look at the updated error $e^{(m)'}$, which is the error using updated tap values but the same delay line content. From (9.133):

$$
\begin{aligned}
e'^{(m)} &= e^{(m)} - \mu_a e^{(m)} \sum_k \left| y_{m+N_f-k} \right|^2 - \mu_b e^{(m)} \sum_k \left| d_{m+N_f-\Delta-1-k} \right|^2 \\
&= e^m \left(1 - \mu_a \sum_k \left| y_{m+N_f-k} \right|^2 - \mu_b \sum_k \left| d_{m+N_f-\Delta-1-k} \right|^2 \right)
\end{aligned}
\tag{9.146}
$$

The mean squared errors before and after updating are

$$
c_{MMSE}^{(m)} \stackrel{\text{def}}{=} \left| e^{(m)} \right|^2
$$

$$
c_{MMSE}'^{(m)} \stackrel{\text{def}}{=} \left| e'^{(m)} \right|^2
$$

$$
c_{MMSE}'^{(m)} = \left(1 - \mu_a \sum_k \left| y_{m+N_f-k} \right|^2 - \mu_b \sum_k \left| d_{m+N_f-\Delta-1-k} \right|^2 \right)^2 c_{MMSE}^{(m)}
$$

(9.147)

When the step sizes are small enough, the first factor in (9.147) is a positive number smaller than 1. Therefore, (9.147) shows that the mean squared error decreases after updating.

In general, different step sizes are used for FF and FB filters, as filter lengths and data characteristics (such as the average power of $\{y_n\}$ and $\{d_n\}$) may be different in the two filters.

As in the case of LE, the adaptation of DFE involves training and data phases.[14] In the training phase, the transmitted symbols are predetermined and known to the receiver. Therefore, decision errors do not exist. After a reasonably good performance is reached through training, the DFE enters the data phase (also known as the decision-directed phase), where demodulator decisions are used for the FB filter and for further adaptation. On the other hand, comparing to the LE, a DFE requires a lower rate of decision errors in order to work properly, due to the issue of error propagation (Section 9.5.6). Therefore, for the same filter lengths, a DFE usually needs a longer training phase comparing to an LE.

In many applications, the training phase causes significant overhead in operation. Therefore, it is important to shorten the required training period. There are many tricks for such purpose; some are situation-specific. The most general ones include:

- Use gear-shifting, as discussed in Chapter 8
- Pass the saved training data multiple times through the DFE, to refine training. Note that since each pass experiences the same noise, such repeated training may "mislead" the DFE. Therefore, the number of repeats should be carefully tuned.
- Train the FF filter first, while keeping the FB filter taps to zero. Unfreeze and update the FB taps only when the decision error rate is low enough. This approach may require a longer FF filter, in order to achieve good enough performance at the first step.
- Prepopulate the filters with "likely" tap values (e.g., from the last communication session if the channel changes slowly) as a starting point for adaptation.

---

[14] It is possible to adapt DFE without training phase, known as blind equalization [5, sec. 10.5]. However, it is not covered in this book.

There are few more points to be noted in studying DFE training.

- The position of the center tap (controlled by Δ in our formulation) is predetermined in our adaptation scheme. As stated in Section 9.5.4.3, in most channel conditions, the optimal Δ value is $N_f - 1$.
- Data for the FF and FB filters ($\{y_n\}$ and $\{d_m\}$) are strongly correlated. Therefore, there may be strong interaction or "competition" between the adaptations of these two filters.[15] One can avoid such competition by designing a gear-shifting schedule such that one filter may "dominate" adaptation at a given time.

Adaptation strategy depends on many factors such as channel characteristics, SNR, length of the training phase, speed of channel changes, the length of the filters, and so on. Although there are theoretical studies in the literature, usually the best adaptation strategy is found through simulations.

Besides the MMSE/LMS adaptation described above, there are many other adaptation methods. Most notably, the recursive least squares (RLS) block-based adaptation method described in Chapter 8 can be applied to DFE adaptation [1, 5]. RLS provides significantly faster adaptation, at the expense of higher complexity. RLS and related fast-training techniques make a DFE practical even for fast-changing channels [17].

9.5.6 Error Propagation

The DFE analyses presented in Sections 9.5.3 and 9.5.4 assume all demodulator decisions are correct. Under such an assumption, a DFE is equivalent to the architecture shown in Figure 9.9 in Section 9.5.2. Since it is a linear system as expressed in (9.87), optimization and performance evaluation are relatively straightforward.

In reality, however, the demodulator (DM in Figure 9.8 in Section 9.5.2) makes decision errors from time to time. When erroneous decisions are fed into the FB filter, the output of the DFE is degraded, possibly causing more decision errors before the system recovers (i.e., all erroneous decisions are purged from the FB delay line). Worse, the system may never recover to the relatively error-free state. Such a problem is referred to as error propagation.[16] Theoretical analyses in the presence of propagation error are complicated [18, 19].

For most communications systems, low symbol error rates (on the order of 10^{-5}–10^{-7}) are desired. For uncoded systems, the demodulator output has the same symbol error rates. Such low error rates usually do not produce significant error propagation problem. Therefore, analyses under the perfect decision assumption can be useful approximations. On the other hand, if a strong error correction code is used (Chapter 5), the symbol error rate at the demodulator output can be much higher

---

[15] We can understand the correlation issue in another way, by conceptually combining the FF and FB filters into a long filter. Correlation between data in the two filters means the autocorrelation matrix of the data in this long filter is near-singular. Namely, its eigenvalues have a large spread. Such large spread causes slow convergence of LMS algorithm, as discussed in Chapter 8.

[16] Decision error also has an impact on the adaptation in both LE and DFE cases. However, such effect lasts only for one symbol, and is not as damaging as the FB filter error propagation discussed here.

(e.g., on the order of 10^{-3}). In such systems, error propagation can be a serious problem.

There are two possible solutions. One is using the Tomlinson–Harashima precoding, which effectively moves the feedback filter to the transmitter side (see Section 9.6). This would eliminate the error propagation problem, as the transmitter knows the true transmitted symbols. However, if channel condition changes quickly, it would be difficult for the transmitter to update the precoding in time.

The other method is treating the dispersive channel and the (typically) convolutional encoder as two concatenated coders, very similar to the turbo code discussed in Chapter 5. Therefore, an iterative "decoder" can be used to address both ISI and convolutional coding. Such a technique is known as "turbo equalization" [5, 20, 21], can further improve performance by using the soft decoding methods inherent to the turbo decoders. However, the complexity of this technique is much higher than an ordinary DFE.

An alternative method of alleviating ISI is multicarrier modulation, particularly the OFDM modulation, to be discussed in Chapter 10. This method does not suffer from the error propagation problem and is thus attractive for systems with powerful error correction codes.

A DFE may never be able to recover from an error propagation event. Therefore, if error propagation is to be tolerated, the DFE should be frequently restarted (i.e., trained with known symbols, see Section 9.5.5) [17].

9.5.7 Transmitter Optimization

As discussed in Chapter 6, a dispersive channel has a nonflat frequency response. As pointed out in Chapter 2, to maximize capacity under the constraint of total transmission power, the transmitted spectrum should follow the water filling principle, instead of being flat. For DFE, as stated in Section 9.5.3.5, the achievable channel capacity is the same as that of the original channel, as shown in (9.124) therein. Therefore, to maximize total capacity with a DFE receiver, the transmitted spectrum should be optimized according to the water filling principle [14]. Since (9.124) deals with the folded spectrum given by (9.117), achievable capacity also depends on the symbol period T_{Sym}.

The transmitter optimization problem was further studied in [8]. It was concluded that optimization of the transmission spectrum is not important in general. While T_{Sym} can be optimized (depending on the SNR), resulting capacity does not change much when T_{Sym} varies around the optimal value.

On the other hand, in order to optimize itself, a transmitter must have channel state information. It may cost complexity and bandwidth to acquire such information. Changing T_{Sym} involves changes in many parts of the transmitter and the receiver. Therefore, while a theoretical possibility, optimizing transmitter is in general not worth the cost.

9.5.8   Hybrid DFE

It was shown in Section 9.5.2 that a noise predictor (NP) is equivalent to a DFE. However, as stated in Sections 9.5.3.4 and 9.5.4.3, DFE output noise (additive noise plus residual ISI) is white for infinite-length DFE but not necessarily white for finite-length DFE. Namely, for finite-length DFE, there is an opportunity to further improve performance using an NP filter. Such consideration motivates a hybrid DFE as shown in Figure 9.11, where an additional NP filter is added to the standard DFE [22–24]. Overall, depending on the channel condition, a hybrid DFE may achieve better performance than a finite-length DFE with the same total number of taps.

Adaptation of the NP filter is similar to (9.145):

$$c_k^{(m+1)} = c_k^{(m)} + \mu_c e^{(m)} u_{m-k}^* \qquad (9.148)$$

Here, $c_k^{(m)}$ is the kth tap value of NP filter at time m. μ_c is the step size. u_{m-k} is the estimated noise at the kth position of the NP filter tap line.

However, it should be noted that $\{u_k\}$ is strongly correlated with $\{y_k\}$ in the FF filter. Therefore, there may be strong competition between NP filter adaptation and the other filters. To avoid the issue, for example, one can freeze the NP adaptation until the FF and FB reach a steady state. After that, NP uses a much smaller step size, to "follow" the other two filters in adaptation and to pick up additional improvements. Detailed analyses are difficult. Simulation studies may be more helpful.

9.5.9   DFE Examples

DFE performances in various applications can be found in the literature, for example, [5, 17, 25, 26].

MATLAB supports DFE simulation with function dfe. It supports several adaptation methods including the LMS, same as the MMSE adaptation discussed in Section 9.5.5. However, it uses the same step size for FF and FB filters. Therefore, the convergence speed may not be optimal. The following plots are generated with MATLAB to show some DFE properties. The channel used is the same as that for the linear equalizer simulation (Section 9.4.4.5), with six tap values $(0.73 + 0.53j, -0.06 + 0.09j, 0.31 + 0.95j, 0.90, 0.03 + 0.10j, 0.50j)$. No noise was added in these simulations.

Figure 9.12 shows the simulation results. Plot A shows the output SNR with various FF filter lengths N_f. The FB filter length N_b is fixed at 6 (the same as the

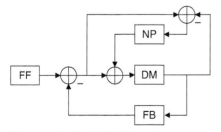

Figure 9.11   Hybrid DFE architecture.

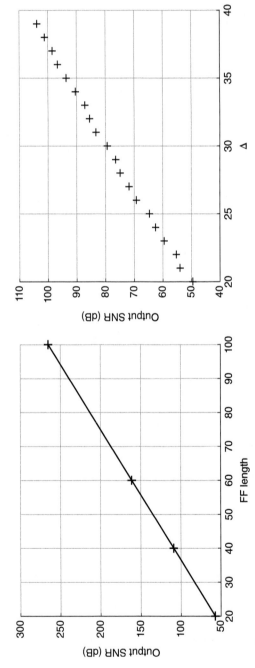

Figure 9.12   DFE examples. A: Output SNR vs. FF length. B: SNR vs. Δ, FF length 40.

channel length). Δ is $N_f - 1$. These are optimal settings according to Section 9.5.4.3. As expected, SNR increases as N_f increases. The performances are excellent (remember that for most common modulations, the targeted SNR is far below 50 dB). Such performances are in contrast with the results of the MMSE LE (Section 9.4.4.5), where in order to reach an SNR of 40 dB, the filter length needs to be 512 (see Plot C of Figure 9.6). Therefore, although a DFE uses one more filter, the total number of taps for a DFE is much smaller than an LE for the same performance.

Plot B shows the output SNR at different Δ values, ranging from $N_f/2$ to $N_f - 1$. N_f is 40. The FB filter length N_b is 20 for this simulation. As shown, performance steadily improves as Δ approaches $N_f - 1$, as expected from Section 9.5.4.3. Furthermore, at the optimal point, the output SNR is 105 dB, the same as the SNR for $N_f = 40$ in Plot A, where $N_b = 6$. Therefore, the additional FB filter taps do not help DFE performance. In other words, $N_b = N_H$, as assumed in Section 9.5.4.3, is indeed the best choice.

9.6 TOMLINSON–HARASHIMA PRECODING (THP)

As discussed in Section 9.5.6, error propagation may be a significant problem for DFE performance in some cases. Instead of canceling ISI using demodulator decisions (and thus suffering from the error propagation), it would be nice to use known symbols at the transmitter side, instead. This idea motivates the Tomlinson–Harashima precoding, also known as TH precoding or THP [27, 28].

The concept of "presubtraction" of ISI at the transmitter sounds intuitive and straightforward. However, such an operation can potentially change the peak-to-average ratio (PAR), power, and power spectrum density of the transmitted signal, making the transmission less optimal. Fortunately, THP avoids such drawbacks through an ingenious precoding algorithm, to be detailed in Section 9.6.1.

Now that the DFE is effectively split between the transmitter and the receiver, its training and updating become an issue. Section 9.6.2 describes some adaptation strategies.

THP solves the problem of error propagation in DFE. Therefore, it can be used with strong error correction coding schemes. The THP concept can be extended beyond the problem of ISI mitigation. However, THP requires channel information at the transmitter. This and some other implementation issues limit THP applications. Section 9.6.3 provides discussions on these topics.

9.6.1 Precoding Formulation

This section discusses in detail the THP formulation. We describe the THP transmitter and receiver separately, before showing how they work together. For simplicity, we consider real-valued signals. The results can be easily extended to complex-valued signals [29, 30]. The following discussions are all in baseband.

9.6.1.1   THP Transmitter

Figure 9.13 shows the THP transmitter. Sequence $\{d_n\}$ is the transmitted symbols. The modules inside the dashed-line box are the THP. Sequence $\{d'_n\}$ is the output of the THP, which undergoes the normal pulse shaping and other processes. The modulo-2M module in THP is a nonlinear device, which "folds" the data into a specified range:

$$d'_n = u_n + 2Mj_n$$
$$-M \le d'_n < M$$

(9.149)

Integer j_n is chosen by the modulo-2M module to satisfy the inequality for output $\{d'_n\}$. M is a constant, not necessarily integer. j_n can be explicitly expressed as

$$j_n = \left\{-\frac{u_n}{2M}\right\},$$

(9.150)

where $\{a\}$ denotes the operation of taking the nearest integer to a. The "Z^{-1}" block in Figure 9.13 indicates a delay of one symbol. The "P" block is the precoding filter, with tap values $\{P_n, n = 0, 1, 2, \dots\}$. We will show later that $\{P_n\}$ is related to the FB filter $\{b_n\}$ in DFE. Note that since P is a feedback filter, it must be causal (i.e., the tap value starts from $n = 0$) to be realizable.

As discussed in an appendix (Section 9.9), the input and output relationship for THP can be expressed in z-transform:

$$\bar{d}'(z) = \frac{\bar{d}(z) + 2M\bar{j}(z)}{1 - z^{-1}\bar{P}(z)}.$$

(9.151)

Here, $\bar{d}(z), \bar{j}(z), \bar{P}(z)$, and $\bar{d}'(z)$ are z-transforms of $\{d_n\}$, $\{j_n\}$, $\{P_n\}$, and $\{d'_n\}$, respectively.

At first glance, the THP precoder is an IIR filter. As such, the peak output value and power spectrum density are different from the input. Because of the modulo-2M operation, however, THP does not change the output characteristics in a transmitter, as shown below.

In reference to Figure 9.13, let us assume for the moment that the input to THP $\{d_n\}$ has a uniform distribution between $[-M, M)$. For a given q_n as the output of the

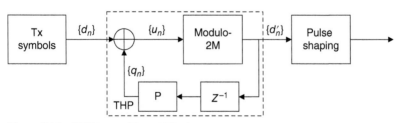

Figure 9.13   THP transmitter architecture.

precoding filter P, the summation u_n has a uniform distribution in $[-M + q_n, M + q_n)$. The modulo-2M operation "folds" this distribution back into a uniform distribution in $[-M, M)$. Therefore, the output of THP $\{d_n'\}$ obeys uniform distribution in $[-M, M)$, independent of the value of q_n. It follows that $\{d_n'\}$ has the same probability distribution as $\{d_n\}$, as long as the latter is uniform. Therefore, THP does not change the statistics and power level of the transmitted signal.

Furthermore, since each output symbol of THP obeys the same uniform distribution independent of the feedback q_n, there is no correlation among the symbols. Namely, the probability distribution of one symbol does not depend on the value of another symbol. Therefore, the output power spectral density is flat (i.e., with a white spectrum). More importantly, no additional ISI is introduced by the THP.

With above discussion, we can see that the THP does not change ISI, amplitude statistics, power and PSD of the transmitted signal. This statement hinges on the assumption that the input sequence $\{d_n\}$ has a uniform distribution and a white spectrum. In reality, $\{d_n\}$ is drawn from a finite alphabet, whose members appear in equal probability. With a reasonably high-order modulation and a proper selection of M, uniform distribution is a good approximation.

9.6.1.2   THP Receiver

Turning to the receiver, let us first review the standard DFE operation. As discussed in Section 9.5.3.4, the FF filter output contains remaining ISI, which can be broken down to the precursor and postcursor parts, the latter being completely canceled by the FB filter output. Thus, the output at the FF filter can be written in z-transform as

$$\bar{w}_{FF}(z) = \bar{d}(z)[1 + \bar{H}_1(z) + \bar{H}_2(z)] + \bar{u}(z). \qquad (9.152)$$

Here, $\bar{H}_1(z)$ and $\bar{H}_2(z)$ represents the postcursor and precursor ISI, respectively. $\bar{u}(z)$ is the noise component at this point. $\bar{d}(z)$ is the transmitted symbol sequence. The optimal FB filter is

$$\bar{b}(z) = \bar{H}_1(z). \qquad (9.153)$$

Combining the two filters and assuming input to the FB filter is also $\bar{d}(z)$ (i.e., no decision error), the final output of the DFE is (see (9.87) in Section 9.5.2).

$$\bar{w}_{DFE}(z) = \bar{w}_{FF}(z) - \bar{b}(z)\bar{d}(z) = \bar{d}(z) + \bar{d}(z)\bar{H}_2(z) + \bar{u}(z). \qquad (9.154)$$

In the result above, the first term is the desired signal. The second and third terms are the residual precursor ISI and noise, respectively.

A DFE can use various optimization criteria, the most common ones being zero forcing (ZF) and minimum mean-squared error (MMSE). This chapter focuses on the MMSE type. However, Equations (9.152)–(9.154) apply to all DFE types.

With a THP, the receiver structure is shown in Figure 9.14. Comparing to the standard DFE structure in Figure 9.8, the FB filter is omitted. A modulo-2M module,

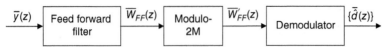

Figure 9.14 THP receiver.

similar to the one at the transmitter, is inserted at the output of the FF filter. The FF filter is trained by a DFE structure. Therefore, Equation (9.152) is valid for the THP receiver.

The output of the modulo-2M module is

$$\bar{w}'_{FF}(z) = \bar{w}_{FF}(z) + 2M\bar{j}_r(z)$$
$$j_m = \left\{ -\frac{\bar{w}_{FFn}}{2M} \right\}. \tag{9.155}$$

Every sample of the sequence $\{w'_{FFn}\}$ falls within $[-M, M)$.

9.6.1.3 THP Transmitter and Receiver Working Together

Now let us consider the entire signal train. To include the effect of the THP precoder, we replace $\bar{d}(z)$ in (9.152) with $\bar{d}'(z)$, given by (9.151). Furthermore, the precoding coefficients are chosen to be the same as the FB filter:

$$\bar{P}(z) = -z\bar{b}(z) = -z\bar{H}_1(z). \tag{9.156}$$

The last equation above is based on (9.153). The extra factor z reflects the fact that $\{b_n\}$ starts with $n = 1$, while $\{P_n\}$ starts with $n = 0$.

Combining (9.151), (9.152), and (9.156), we obtain

$$\bar{w}'_{FF}(z) = \frac{\bar{d}(z) + 2M\bar{j}(z)}{1 + \bar{H}_1(z)}[1 + \bar{H}_1(z) + \bar{H}_2(z)] + \bar{u}(z) + 2M\bar{j}_r(z). \tag{9.157}$$

The first two terms in the square bracket cancels out the filter at the denominator, yielding

$$\bar{w}'_{FF}(z) = \bar{d}(z) + \bar{d}'(z)\bar{H}_2(z) + \bar{u}(z) + 2M[\bar{j}_r(z) + \bar{j}(z)], \tag{9.158}$$

where $\bar{d}'(z)$ is expressed in (9.151). As shown in Section 9.6.1.1, $\{d'_n\}$ has the same statistics as $\{d_n\}$. Therefore, term $\bar{d}'(z)\bar{H}_2(z)$ has the same statistics as the precursor ISI $\bar{d}(z)\bar{H}_2(z)$ in (9.154). The last term $2M[\bar{j}_r(z) + \bar{j}(z)]$ ensures that $\{w'_{FFn}\}$ falls between $[-M, M)$. If M is chosen so that $\{d_n\}$ always fall within $[-M, M)$ and when the noise and ISI terms are small, we have

$$\bar{j}_r(z) + \bar{j}(z) = 0. \tag{9.159}$$

In this case, (9.158) becomes

$$\bar{w}'_{FF}(z) = \bar{d}(z) + \bar{d}'(z)\bar{H}_2(z) + \bar{u}(z). \tag{9.160}$$

The residual ISI $\bar{d}'(z)\bar{H}_2(z)$ and the noise $\bar{u}(z)$ in $\bar{w}'_{FF}(z)$ have the same statistics as the corresponding terms in (9.154). Therefore, a THP system yields the same performance as a DFE system, except that it is not vulnerable to error propagation.

The THP operation contains two phases. In phase 1 (training), a standard DFE is used at the receiver, and the THP module is bypassed at the transmitter. Typically, the DFE training is conducted with known symbols to ensure fast and correct convergence. At the end of the training, the FB filter tap values are sent to the transmitter where they are used to populate the recoding filter P according to (9.156). Phase 2 is then commenced, in which the transmitter and receiver as shown in Figures 9.13 and 9.14 are used.

It should be noted that if the residual ISI and noise in (9.158) are strong enough, (9.159) may not be true. In this case, there would be a big difference (on the order of $2M$) between w'_{FF} and d. Such an event is known as "data flipping." We should recognize that such strong noise probably already causes demodulation error even in DFE. Therefore, data flipping does not cause significant degradation in performance. Furthermore, such an effect can be alleviated by using a larger M at the receiver and expanding the constellation space at the demodulator, as detailed in [31].

9.6.2   THP Updating

As described in Section 9.6.1, a THP operates in two phases. In phase 1, a DFE is trained. Its FF filter remains at the receiver, while its FB filter is transferred to the transmitter as the precoder. In phase 2, the transmitter and receiver operate with the THP architecture. In this section, we consider how THP filters can be updated in phase 2 as the channel changes with time.

In a standard THP structure, the precoding filter P is fixed. The FF filter at the receiver side can be adaptive, to partially compensate for channel changes. To enhance adaptation capability, the receiver can even use the standard DFE structure (with the additional Modulo-2M operation), with the FB filter canceling additional postcursor ISI caused by channel change. Because of the precoding, the FB filter tap values are likely to be close to zero. Therefore, the effect of error propagation is likely to be small. However, the FB input should be the "true symbols," that is, symbols without the modulo-2M operation at the receiver. Therefore, we need to remove the modolo-2M effect in the feedback loop. The architecture shown in Figure 9.15 can be used [29].

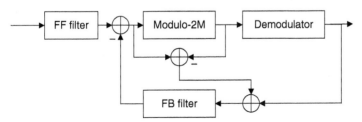

Figure 9.15   DFE architecture under THP.

The strategy above may work well when the channel fluctuates around some averaged state. However, if the channel continues to drift in a particular direction (e.g., when a mobile user moves from one scattering environment to another), the accumulative adaptation at the receiver may grow large, diminishing the benefit of THP. In this case, it becomes necessary to update the precoder itself. Let us first consider a straightforward approach [29].

As shown in (9.152), the channel H after the FF filter can be expressed as

$$H = 1 + H_1 + H_2, \tag{9.161}$$

where H_1 and H_2 represent postcursor and precursor ISI, respectively. As the result of channel change, A_1 and A_2 are added to H_1 and H_2, respectively. The new channel H' is thus[17]

$$H' = 1 + H_1 + A_1 + H_2 + A_2. \tag{9.162}$$

With a THP optimized for H, the received signal under the new channel is, according to (9.157):

$$
\begin{aligned}
\bar{w}'_{FF}(z) &= \frac{\bar{d}(z) + 2M\bar{j}(z)}{1 + \bar{H}_1(z)}[1 + \bar{H}_1(z) + \bar{A}_1(z) + \bar{H}_2(z) + \bar{A}_2(z)] + \bar{u}(z) + 2M\bar{j}_r(z) \\
&= \bar{d}(z) + \bar{d}(z)\frac{\bar{A}_1(z)}{1 + \bar{H}_1(z)} + \bar{d}(z)\frac{\bar{H}_2(z) + \bar{A}_2(z)}{1 + \bar{H}_1(z)} + \bar{u}(z) + 2M[\bar{j}_r(z) + \bar{j}(z)]
\end{aligned}
\tag{9.163}
$$

The third term is the new precursor ISI, which is not examined here. The factor in the second term

$$\bar{C}(z) \stackrel{\text{def}}{=} \frac{\bar{A}_1(z)}{1 + \bar{H}_1(z)} \tag{9.164}$$

is the new postcursor ISI, which is estimated by the receiver using a DFE structure shown in Figure 9.15. According to (9.156), the new precoder should be

$$\bar{P}'(z) = -z[\bar{H}_1(z) + \bar{A}_1(z)] = \bar{P}(z) - z[1 + \bar{H}_1(z)]\bar{C}(z). \tag{9.165}$$

Therefore, the precoder update can be performed in the following steps.

1. The receiver estimates the channel to obtain $\bar{C}(z)$. It then computes $[1 + \bar{H}_1(z)]\bar{C}(z)$.

2. The result is sent to the transmitter.

3. The transmitter switches to the new precoder given by (9.165). *At the same time*, the receiver sets the FB filter taps to zero.

4. The receiver continues to estimate the channel in preparation for the next update.

---

[17] Note that this new channel already includes the effect of the FF filter optimization in response to physical channel change. Therefore, we focus only on the resulting postcursor ISI change.

There are two difficulties in this scheme. The first is rather apparent: at step 3 above, the transmitter and receiver need to be synchronized in updating. Such synchronization requires some protocol support, which may introduce overhead or cause disruption in data transmission. Such consideration limits the frequency of updates. The second difficulty is subtler. Usually, the postcursor H_1 and A_1 are rightfully assumed to have limited length. The FB filter should have the same length as the postcursor, as discussed in Section 9.5.4. However, C, as defined by (9.164), is the result of A_1 filtered by an IIR. Therefore, it can be very long. This means that the receiver needs a very long FB filter in order to correctly estimate C, leading to high complexity and low adaptation speed. Furthermore, the precoder update $[1 + \bar{H}_1(z)]\bar{C}(z)$ may be longer than A_1 due to estimation errors. Therefore, the resulting precoder would keep growing in length. It can be truncated; however, it is not clear that the associated error is bounded during continuous updating.

Concerning the first difficulty, the transmitter can conduct "gradual" updates, where the required updates are added to the existing tap values in multiple small steps. This way, the receiver can automatically adapt to the new precoder, without disrupting data transmission. Tight synchronization in the updating process is thus not required. The price to pay is a longer updating process, which may be problematic if the channel changes fast.

To address the second difficulty, we can use an alternative method and have the FB filter estimate A_1 directly. For this purpose, we want to use d' given by (9.151), instead of d, as input to the FB filter. Namely, the architecture shown in Figure 9.16 is used. With such structure, the result in FB filter is simply A_1, which can be directly used for precoding update at the transmitter without concerns about filter length.

Above discussions are based on an uncoded system. When coding is combined with THP, there is a delay in the decision, which causes more complication in adaptation. However, it can still be done by using a "shadow DFE" operating on delayed symbols [32]. There are also other updating schemes reported in the literature, for example, [33, 34].

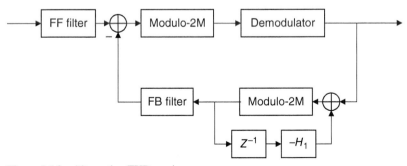

Figure 9.16   Alternative THP receiver.

9.6.3 General Remarks on THP

The Tomlinson–Harashima precoding was first invented in the early 1970s to address the ISI problem [27, 28]. The original work was for real-valued signals for phase-amplitude modulation (PAM) and was later extended to complex-valued signals for quadrature amplitude modulation (QAM) [35]. It was further combined with error correction coding [15]. Later, this precoding technique is used in the spatial dimension to cancel multiuser interference (so-called dirty-paper coding [36]) and mutual interference among antennas for MIMO system [37] (see Chapter 11).

Since THP achieves the same performance as DFE without error propagation, it seems that under MMSE optimization, a THP system can reach channel capacity with proper error correction coding (see Section 9.5.3.5). However, there are a few subtleties as discussed in [38].

- At the limit of high SNR, the difference between the THP capacity and the capacity of an AWGN channel with ISI (with the same average signal and noise powers) is exactly the shaping gain $(1/2)\log_2(\pi e/6)$ discussed in Chapter 5. Simply put, as discussed in Section 9.6.1.1, THP requires a rectangular constellation in signal space, because its input obeys uniform distribution in each dimension. This distribution is not optimal, resulting in the loss of shaping gain.

- At very low SNR levels, the TH precoding system also has lower capacity than AWGN channel, although the probability distribution of transmit signal is not important. This capacity difference is due to the non-Gaussian distribution of the noise, caused by the modulo-2M operation.

To address the loss of shaping gain, THP is further combined with other coding techniques to form signals beyond the square constellation map of QAM. Such "signal shaping" techniques have been used in the voice band modem standards [39, 40].

In some cases, for example, to keep the receiver complexity low, one can further move the FF filter to the transmitter side. However, the transmitted PSD is altered by the (transmitter side) equalization [16, 30].

The THP technique is most suitable for channels that do not change rapidly over time because the transmitter needs to know the channel state. Therefore, it is widely used in wired line communications such as voice-band modem, symmetric high-speed digital subscriber line (HDSL2 and SHDSL), and Ethernet. However, THP can also be applied to relatively stable wireless channels such as indoor channels [16].

Despite its beauty, there are a few complications in THP implementations.

- The receiver peak-to-average ratio (PAR) is likely to increase. At the receiver, the resulting signal (before the modulo-2M operation) contains the desired signal and the addition of $\{2j_nM\}$ caused by precoding. Therefore, the signal can be very large in value. Receiver design must accommodate larger PAR than a DFE system.

- If the postcursor ISI is not perfectly canceled due to imperfect training, limit of precision, or channel drifting, the residual error affects the tap values of the precoder, which is effectively an IIR. Therefore, the resulting error at the receiver is not necessarily bounded (although its second order statistics is small).

- Adaptation and updating are cumbersome and need transmitter/receiver exchange. Therefore, THP is difficult to apply to rapidly changing channels.

9.7   FRACTIONALLY SPACED EQUALIZERS

So far, the discrete model presented in Section 9.2.3 is used in our discussions. As shown in Figure 9.2, the received signal is filtered by a matched filter before being sampled at the rate of $1/T_{Sym}$. The sampled data then passes a noise whitening filter before arriving at the equalizer input. At this point, the equalizer experiences a composite channel, which is a combination of the pulse shaping filter at the transmitter, the propagation channel, the matched filter, and the whitening filter.

While such a model is convenient for analyses, it is difficult to implement. The main issue is that the matched filter and the noise whitening filter, which are considered as fixed and optimized in our analyses, actually depend on the physical channel. Therefore, they need to be optimized in real time. It may seem that, at least in some cases, an approximation can be made by assuming the propagation channel to be flat, that is, nondispersive. Under such assumption, the matched filter and noise whitening filter can be designed based on the transmitter pulse shaping filter.

However, there is the issue of timing synchronization between the transmitter and the receiver [1]. As pointed out in Chapter 7, it is easier to get the correct clock *frequency*, that is, to get the receiver and transmitter clocks running with a constant yet unknown time offset. Removing this time offset, however, is more difficult. With such timing offset, the receiver experiences an effective pulse shaping function that is a time-shifted version of the one used at the transmitter, as shown by (9.19) in Section 9.2.3. Therefore, even with a flat channel, the approximation used above may be invalid. On the other hand, the solution presented below provides good performance over a broad range of timing offsets [41, 42].

The obvious solution is including the matched filter and the noise whitening filter into the adaptation process. Namely, these filters should be treated as a part of the equalizer. However, as shown in Section 9.2, the matched filter is supposed to be placed before the sampling process. According to the Nyquist sampling theorem (Chapter 3), the matched filter can operate in the digital domain, but the sampling rate needs to be high enough to preserve all information:

$$\frac{1}{T_s} > 2f_M \tag{9.166}$$

T_s is the sample period. f_M is the baseband signal bandwidth, which is typically larger than the Nyquist frequency dictated by symbol rate (Chapter 3):

$$f_M = (1+\alpha)\frac{1}{2T_{Sym}}. \tag{9.167}$$
$$\alpha > 0$$

α is known as the excess bandwidth factor. Above two equations lead to

$$T_s < \frac{T_{Sym}}{1+\alpha}. \tag{9.168}$$

Therefore, to include the matched filter in the equalizer, the sampling period needs to be a fraction of the symbol period. Such a filter architecture is known as the fractionally spaced equalizer (FSE).

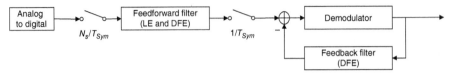

Figure 9.17   Fractionally spaced equalizer architecture.

Figure 9.17 shows the architecture of an FSE. The feedforward filter operates at a multiple of the symbol rate. This filter is common to both LE and DFE. The feedback filter is specific to DFE and operates at the symbol rate. Typically, α in (9.167) is less than 1. Therefore, the rate multiplier N_s in Figure 9.17 can be chosen as 2. However, N_s can be a fraction, as well [42]. Using fractional N_s may save some taps (as N_s can be smaller than 2), at the expense of more complicated controlling logics.[18]

As to adaptation, the LMS adaptation methods discussed in Sections 9.4.4.3 and 9.5.5 can be directly extended to FSE. However, the convergence behavior may be different. The Nyquist bandwidth of an FSE is usually larger than the signal bandwidth. Therefore, signal PSD is zero in a part of the operation band. If the noise there is also very small, the frequency response of the equalizer in this part of the band does not significantly affect the mean squared error. In other words, there are "degenerated" solutions for equalizer: multiple solutions all yielding optimal performance. This situation is equivalent to the singularity of autocorrelation matrix discussed in Chapter 8, resulting slow convergence in adaptation.[19] Special techniques to deal with this issue are discussed in [1, 41, 43].

FSE is a common architecture in today's equalizer designs. In fact, DFE analyses are often based on FSE structure [7, 14]. Simulations show that FSE performance is better than symbol-spaced DFE when the FF filter covers the same amount of symbols [41], which means more filter taps for FSE. However, since no separate matched filter is required for FSE, the overall resource requirement may be lower. It is also possible to further reduce the tap numbers by splitting the FF filter into fractionally spaced and symbol spaced parts, each trained separately [44].

9.8   SUMMARY

This section discusses ways to mitigate signal degradation due to ISI. The optimal method is the maximum likelihood sequence estimation (MLSE, Section 9.3). However, equalizers are commonly used because of their lower complexity and separable operation from error correction coding.

There are two types of equalizers: the linear equalizer (LE, Section 9.4) and the decision feedback equalizer (DFE, Section 9.5). Each type can be optimized based on a number of criteria, the most common ones being zero forcing (ZF) and minimum

---

[18] See Chapter 7 for more discussions on fractional rate-conversion filters.

[19] In other words, because of the high sampling rate, the consecutive samples are correlated.

mean-squared error (MMSE). LE is a filter in the signal path. While conceptually simple, the filter may need to be much longer than the channel dispersion for good performance. The output from LE has nonwhite noise, indicating more room for SNR improvement. DFE has an additional filter, which takes the decisions from the demodulator and feeds them back to the demodulator input. If the feedforward filter is long enough, DFE can produce white noise. In fact, infinite-length DFE provides the same channel capacity as the raw channel, when the effect of error propagation is ignored. Even with limited filter lengths, DFE typically achieves better performance than LE of similar complexity, if error propagation is not severe.

The Tomlinson–Harashima precoding (THP, Section 9.6) builds upon the DFE concept and moves the feedback filter to the transmitter side. Its ingenious design allows for removing of postcursor ISI without altering transmitted signal statistics. THP avoids the error propagation problem and thus allows for powerful error correction coding. However, the standard THP cannot reach channel capacity, because it requires a uniformly distributed signal, thus forgoing the shaping gain. Later techniques combine THP with signal shaping, recovering most of the shaping gain. Updating precoder during communication to cope with channel changes is another challenge.

To analyze the ISI problem, one typically divides the receiving signal processing into the matched filter, noise whitening, and MLSE or equalizer (Section 9.2). In practice, on the other hand, it makes sense to implement and adapt all these filters as one piece. A fractionally spaced equalizer (or fractionally spaced MLSE) can achieve such goal, as discussed in Section 9.7. One main advantage of such architecture is that constant timing offset can be compensated for, making synchronization design easier.

In equalizer designs, we need to know the optimal performance and adaptation behavior. Optimal performances can be obtained either by theoretical results presented in this chapter (for infinitely long equalizers) or by solving the optimal tap values (for finitely long equalizers). Adaptation is usually studied through simulation (with static or time-varying channels), with the help of guidelines provided in this chapter and Chapter 8.

While much attention has been paid (especially in Sections 9.4 and 9.5) to getting high equalizer performance in terms of mitigating ISI, one should recognize that equalizer output SNR is often limited by its input SNR. In such cases, residual ISI may not be critical in overall performance if it is already below the noise level. Therefore, when designing an equalizer, especially when making the tradeoff between performance and complexity, it is essential to keep the context of operating SNR in mind. Furthermore, equalizer performances are sensitive to channel condition (the delay spread). Therefore, one should exercise caution when comparing reported performances from the literature, as they may be related to different channel conditions. It is important to establish a realistic channel model for the particular application before selecting equalizer parameters through simulation.

When surveying the literature, attention should also be paid to details in the formulation as authors may follow different conventions. For example, in Sections 9.4 and 9.5, the error $\{e^{(m)}\}$ is defined as the true symbol minus the equalizer output. In some works, the error is defined with an opposite sign: the equalizer output minus the true symbol, leading to a difference in adaptation formulation.

While this chapter covers the basic concepts and formulation of MLSE and equalization, there are a number of topics left to further exploration. They include:

1. Details of adaptation behavior. As pointed out in Chapter 8, adaptive filtering is a broad subject, on which many books and papers have been published. Especially, DFE and FSE adaptations have some subtle issues due to the correlation among the samples. For practicing engineers in communications, one is more likely to understand the system by simulation. For more theoretical guidance beyond Section 9.5.5, consult other textbooks such as [5, ch. 10].

2. Block-based training. This chapter focuses on LMS adaptation for LE and DFE. As mentioned in Section 9.5.5, block-based training such as the recursive least squares (RLS) algorithm is very useful in improving convergence speed, at the expense of increased complexity. Many flavors of the RLS algorithm have been proposed in the literature with different tradeoffs between complexity and performance. The reader may use Chapter 8 as a starting point and search in the literature for the most suitable solution to a particular application.

3. Blind equalization. Blind equalization refers to the initial training of equalizers without known symbols. Such techniques have been researched extensively, but not broadly used at present.

4. Turbo equalization. As mentioned in Section 9.5.6, joint equalization and decoding can be performed to avoid the error propagation problem under strong channel coding. This technique is further extended to address ISI in MIMO systems, known as generalized DFE, or GDFE (see Chapter 11). More information about turbo equalization solutions can be found in the literature, for example, [5, 20, 21].

As mentioned in Section 9.1, the problem of ISI can also be addressed with multicarrier modulations, especially the orthogonal frequency division multiplexing (OFDM). OFDM is the topic of Chapter 10.

9.9 APPENDIX: Z-TRANSFORM AND RELATED RESULTS

9.9.1 Z-Transform

As summarized in Chapter 3, Fourier transform is a powerful tool for representing and analyzing time-continuous waveforms. Especially, a time domain convolution operation (e.g., filtering) can be represented as a product in the frequency domain. Z-transform is a similar tool for discrete signal sequences [45, ch. 6]. In this section, z-transform and its properties relevant to the equalizer discussions are summarized for reference.

Any complex-valued sequence $\{x_n, n = 1, 2, \dots\}$ can be expressed by its z-transform:

$$\bar{x}(z) \overset{\text{def}}{=} \sum_n x_n z^{-n}. \tag{9.169}$$

Here, z is a complex-valued variable. $\bar{x}$ are known as the transfer function of $\{x_n\}$. When $|z| = 1$, the transfer function becomes the discrete Fourier transform, as will be shown later. Conversely, $\{x_n\}$ can be obtained from $\bar{x}(z)$ with inverse z-transform, which is formulated in Section 9.9.3.

Some interesting properties of z-transform are listed below:

$$y_n = x_{n-d} \Leftrightarrow \bar{y}(z) = z^{-d}\bar{x}(z)$$

$$y_n = \begin{cases} x_r & n = rk \\ 0 & \text{otherwise} \end{cases} \Leftrightarrow \bar{y}(z) = \bar{x}\left(z^k\right)$$

$$y_n = a^n x_n \Leftrightarrow \bar{y}(z) = \bar{x}\left(a^{-1}z\right)$$

$$y_n = x_{-n} \Leftrightarrow \bar{y}(z) = \bar{x}\left(z^{-1}\right)$$

$$y_n = x_n^* \Leftrightarrow \bar{y}(z) = \bar{x}^*\left(z^*\right)$$

(9.170)

The last two lines above show the property with time reversal and complex conjugate. These are most relevant to our discussions in this chapter.

The most interesting property of z-transform is that filtering can be easily expressed. It can be shown that given

$$y_n = \sum_{k=K_0}^{K_1} h_k x_{n-k}, \qquad (9.171)$$

then

$$\bar{y}(z) = \bar{h}(z)\bar{x}(z), \qquad (9.172)$$

where $\bar{y}$, $\bar{h}$, and $\bar{x}$ are z-transforms of y, h, and x, respectively.

The filtering expressed in (9.171) is known as the finite impulse response (FIR) filter, because the filter taps $\{h_k\}$ span a finite range. Now let us look at an infinite impulse response (IIR) filter shown in Figure 9.18. It has a feedforward filter b

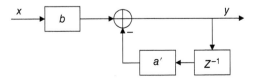

Figure 9.18   IIR filter.

and a feedback filter a'. The z^{-1} block reflects a delay of one clock cycle. The time domain expression is

$$y_n = \sum_k b_k x_{n-k} - \sum_{k=0} a'_k y_{n-k-1} \tag{9.173}$$

For $\{a'_k\}$, the index starts from 0 in order to be implementable. Applying (9.172), we get the z-transform version

$$\bar{y}(z) = \bar{b}(z)\bar{x}(z) - z^{-1}\bar{a}'(z)\bar{y}(z), \tag{9.174}$$

which leads to

$$\bar{y}(z) = \frac{\bar{b}(z)}{1 + z^{-1}\bar{a}'(z)}\bar{x}(z) = \frac{\bar{b}(z)}{\bar{a}(z)}\bar{x}(z)$$

$$\bar{a}(z) \stackrel{\text{def}}{=} 1 + z^{-1}\bar{a}'(z) \tag{9.175}$$

The factor $\bar{b}/\bar{a}$ in (9.175) can be connected to the definition (9.169) in terms of Tylor expansion, resulting in an infinitely long sequence. This infinity is to be expected: the IIR structure in Figure 9.18 has an infinitely long response in the time domain.

As an important special case, the inverse filter of $\bar{h}(z)$ is simply $1/\bar{h}(z)$, also noted as $\bar{h}^{-1}(z)$. $\bar{h}^{-1}$ can be realized as IIR filter, as long as the first tap value of h is 1. This value can be achieved by scaling. Therefore, even if a dispersive channel can be modeled as a short filter, its inverse can be infinitely long. Namely, the ISI it causes cannot be completely removed by a finite-length linear equalizer.

From (9.175), a filter can be expressed in general as

$$\bar{f}(z) = \frac{\bar{b}(z)}{\bar{a}(z)}, \tag{9.176}$$

where $\bar{a}$ and $\bar{b}$ are polynomials of z^{-1}, and the constant term in $\bar{a}$ is 1. The zeros of $\bar{b}(z)$ are the zeros of $\bar{f}(z)$. The zeros of $\bar{a}(z)$ are the divergent points of $\bar{f}(z)$, known as the poles. Recall from the theory of polynomials that a polynomial is defined, up to a constant factor, by its zeros. More specifically, if $\{z_k, k = 1, 2, \dots, K\}$ is a collection of zeros of polynomial $p(z)$, then

$$p(z) = P \prod_{k=1}^{K} (z - z_k), \tag{9.177}$$

where P is an arbitrary constant. Therefore, a filter is determined by its zeros and poles, up to a constant factor. Many filter properties can be extracted from the distribution of zeros and poles. Here, we only mention two of them.

A causal filter does not rely on input samples in the infinite future. In other words, K_0 in (9.171) is a finite number (although it can be negative). Similarly, an anticausal filter has K_1 in (9.171) as a finite number. Obviously, FIR filters are both causal and anticausal. In terms of z-transfers, the transfer function of a causal filter is well-defined when $|z| > 1$. Namely, all poles are located inside or on the unit circle.

The transfer function of an anticausal filter is well-defined when $0 < |z_0| < |z| < 1$, where z_0 is some constant.

Furthermore, a causal IIR is stable (i.e., the output is always finite when the input is finite) if and only if all of its poles are inside the unit circle.

9.9.2 Z-Transform of Autocorrelation

A special case of filtering (9.172) is the autocorrelation function. The autocorrelation of WSS sequence $\{x_n\}$ is defined as

$$R_n \stackrel{\text{def}}{=} E\left(x_m x^*_{m-n}\right). \tag{9.178}$$

Because $\{x_n\}$ is WSS, R_n defined above is independent of m. Therefore, we can take a summation over m:

$$R_n = \frac{1}{M} E\left(\sum_{m=1}^{M} x_m x^*_{m-n}\right) \tag{9.179}$$

The summation limit M can be any number.
We can write R_n in the form of filtering, using (9.171):

$$R_n = \frac{1}{M} E\left(\sum_{m=1}^{M} x_m y_{n-m}\right). \tag{9.180}$$

$$y_n \stackrel{\text{def}}{=} x^*_{-n}$$

From the properties in (9.172) and (9.170),

$$\bar{R}(z) = \frac{1}{M} E[\bar{x}(z)\bar{y}(z)] = \frac{1}{M} E\left[\bar{x}(z)\bar{x}^*\left(z^{*-1}\right)\right]. \tag{9.181}$$

Here, the summations for z-transform in (9.169) runs for M samples. M should be much larger than any n of interest in (9.181), so that the edge cases are unimportant. $\bar{R}(z)$ has the "complex reversal symmetry":

$$R_n = R^*_{-n}$$
$$\bar{R}(z) = \bar{R}^*\left(z^{*-1}\right). \tag{9.182}$$

This symmetry will be discussed in more details in Section 9.9.4.

9.9.3 Z-Transform and Fourier Transform

Assume sequence $\{s_n\}$ is sampled from a time-continues function $s(t)$. In this section, we establish the connection between the spectrum of $s(t)$ and the z-transform of $\{s_n\}$. Furthermore, we show how to obtain $\{s_n\}$ from its z-transform (inverse z-transform). For clarity, our derivation is broken down into several small steps as shown below.

9.9.3.1 Review of Fourier Transform Formulation for Time-Continuous Functions

As given in Chapter 3, the Fourier transform of function $s(t)$ is $\tilde{s}(\omega)$:

$$\tilde{s}(\omega) \stackrel{\text{def}}{=} \int_{-\infty}^{\infty} s(t)\exp(-j\omega t)dt, \tag{9.183}$$

with the inverse Fourier transform:

$$s(t) = \frac{1}{2\pi} \int_{-\infty}^{\infty} \tilde{s}(\omega)\exp(j\omega t)d\omega. \tag{9.184}$$

9.9.3.2 A Sequence as Function Samples

Define an alternative function $s'(t)$:

$$s'(t) \stackrel{\text{def}}{=} s(t) \sum_n \delta(t - nT) = \sum_n s(nT_s)\delta(t - nT_s), \tag{9.185}$$

where T_s is the sampling period. $s'(t)$ represents a collection of impulse functions modulated by $\{s(nT_s)\}$. Therefore, it connects the time-continuous function $s(t)$ with a sequence $\{s_n\}$:

$$s_n \stackrel{\text{def}}{=} s(nT_s)$$
$$s'(t) = \sum_n s_n \delta(t - nT_s) \cdot \tag{9.186}$$

9.9.3.3 Fourier Transform of $s'(t)$

Now we study the Fourier transform of $s'(t)$ and its inversion. The results are linked with the Fourier transform of the original function $s(t)$. According to (9.183), the Fourier transform of $s'(t)$ is

$$\tilde{s}'(\omega) = \int s'(t)\exp(-j\omega t)dt = \sum_n s_n\exp(-jn\omega T_s). \tag{9.187}$$

On the other hand, according to the Nyquist sampling theory presented in Chapter 3, $\tilde{s}'(\omega)$ is the folded version of $\tilde{s}(\omega)$[20]:

$$\tilde{s}'(\omega) = \frac{1}{T_s} \sum_m \tilde{s}\left(\omega - \frac{2\pi m}{T_s}\right). \tag{9.188}$$

Given $\tilde{s}'(\omega)$, we can obtain $\{s_n\}$ from inverse Fourier transform. From (9.184) and (9.187),

$$s'(t) = \frac{1}{2\pi}\int_{-\infty}^{\infty} \tilde{s}'(\omega)\exp(j\omega t)d\omega = \frac{1}{2\pi}\int \sum_n s_n \exp[-j\omega(nT_s - t)]d\omega. \tag{9.189}$$

From (9.187), we can see that $\tilde{s}'(\omega)$ has a period of $2\pi/T_s$. Therefore, (9.189) can be written as

$$s'(t) = \frac{1}{2\pi}\int_{-\pi/T_s}^{\pi/T_s} \tilde{s}'(\omega)\exp(j\omega t)d\omega \sum_m \exp\left(\frac{j2\pi m}{T_s}t\right). \tag{9.190}$$

The last summation is [46, sec. 1.17][21]

$$\sum_m \exp\left(\frac{j2\pi m}{T_s}t\right) = T_s \sum_k \delta(t - kT_s). \tag{9.191}$$

Therefore,

$$s'(t) = \frac{T_s}{2\pi}\sum_k \int_{-\pi/T_s}^{\pi/T_s} \tilde{s}'(\omega)\exp(j\omega kT_s)d\omega \delta(t - kT_s). \tag{9.192}$$

Comparing (9.192) with (9.186) and using (9.188), we have

$$\begin{aligned} s_n &= \frac{T_s}{2\pi}\int_{-\pi/T_s}^{\pi/T_s} \tilde{s}'(\omega)\exp(j\omega nT_s)d\omega \\ &= \frac{1}{2\pi}\int_{-\pi/T_s}^{\pi/T_s}\sum_m \tilde{s}\left(\omega - \frac{2\pi m}{T_s}\right)\exp(j\omega nT_s)d\omega \end{aligned} \tag{9.193}$$

$\tilde{s}'(\omega)$ is often referred to as the Fourier transform of sequence $\{s_n\}$. Equations (9.187) and (9.193) thus give the Fourier transform and inverse Fourier transform between $\{s_n\}$ and $\tilde{s}'(\omega)$.

[20] Equation (9.188) can be briefly proved as follows: $\tilde{s}(\omega) = \sum_n s(nT_s)\exp(-jn\omega T_s) = \sum_n \left[\frac{1}{2\pi}\int d\omega' \tilde{s}(\omega')\right.$ $\exp(jnT_s\omega')]\exp(-jn\omega T_s) = \frac{1}{2\pi}\int d\omega' \tilde{s}(\omega)\sum_n \exp[jnT_s(\omega - \omega')] = \frac{1}{2\pi}\int d\omega' \tilde{s}(\omega')\frac{2\pi}{T_s}\sum_m \delta\left(\omega' - \omega + \frac{2\pi m}{T_s}\right) = \frac{1}{T_s}$ $\sum_m \tilde{s}\left(\omega - \frac{2\pi m}{T_s}\right)$. The second equality uses (9.184); the fourth equality uses (9.191).
[21] See appendix in Chapter 3 for mathematical details.

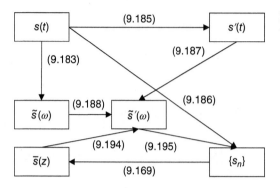

Figure 9.19 Connection among the various quantities.

9.9.3.4 Connection Between Z-Transform and Fourier Transform

Equation (9.193) connects $\{s_n\}$, which is a sequence of samples of function $s(t)$, with the spectrum of $s(t)$. In this section, we further obtain s_n from its z-transform. From (9.187) and the definition of z-transform (9.169), we have

$$\tilde{s}'(\omega) = \bar{s}[\exp(j\omega T_s)]. \tag{9.194}$$

Here, $\bar{s}(z)$ is the z-transform of $\{s_n\}$ given by (9.186). Equation (9.193) thus becomes[22]

$$s_n = \frac{T_s}{2\pi}\int_{-\pi/T_s}^{\pi/T_s} \bar{s}[\exp(j\omega T_s)]\exp(j\omega n T_s)\,d\omega \underset{\theta \overset{\text{def}}{=} \omega T_s}{=} \frac{1}{2\pi}\int_{-\pi}^{\pi} \bar{s}[\exp(j\theta)]\exp(jn\theta)\,d\theta. \tag{9.195}$$

9.9.3.5 Summary

Figure 9.19 summarizes the relationships established by the various equations in Section 9.9.3.

9.9.4 Z-Transform Factorization

In this section, we consider a sequence $\{R_n\}$ with the "complex reversal symmetry":

$$R_n = R^*_{-n}. \tag{9.196}$$

---

[22] There is a more general formulation of the "inverse z-transform," that is, obtaining $\{s_n\}$ from $\bar{s}(z)$. However, for the purpose of this chapter, (9.195) is enough.

From (9.170), its z-transform has the following symmetry:

$$\bar{R}(z) = \bar{R}^* \left(z^{*-1}\right).$$ (9.197)

Equation (9.197) implies that if $\bar{R}(z)$ has a zero or pole at z_k, it also has a zero or pole at z_k^{*-1}. In this pair of zeros or poles, one is within the unit circle, and the other is outside of it.[23] If we collect one zero or pole from each pair to construct a function $\bar{g}(z)$ (see Section 9.9.1), then the rest of the zero and pole would contribute to function $\bar{g}^* \left(z^{*-1}\right)$. Since all zeros and poles of $\bar{R}$ are collected into either $\bar{g}(z)$ or $\bar{g}^* \left(z^{*-1}\right)$, with a proper choice of the constant factor in (9.177), we have

$$\bar{R}(z) = \bar{g}(z)\bar{g}^* \left(z^{*-1}\right).$$ (9.198)

A stronger conclusion was stated as follows (lemma 1 in [9, sec. II-A]). If $\left|\bar{R}(e^{j\theta})\right|$ and $\ln[|R(e^{j\theta})|]$ are both integrable over $-\pi < \theta \le \pi$, then $\bar{R}(z)$ can be factorized as[24]

$$\bar{R}(z) = R_0 g(z)\bar{g}^* \left(z^{*-1}\right)$$

$$R_0 \stackrel{\text{def}}{=} \exp\left(\frac{1}{2\pi}\int_{-\pi}^{\pi} \ln|\bar{R}[\exp(j\theta)]|d\theta\right).$$ (9.199)

Here, $g(z)$ is monic, that is, the corresponding time-domain sequence $\{g_n\}$ has the following property[25]:

$$g_n = 0 \forall n < 0$$
$$g_0 = 1$$ (9.200)

Namely, $\{g_n\}$ is a causal sequence that starts with 1. An informal justification of the value of R_0 in (9.199) is provided in [14].

An important example of the complex reversal symmetric sequence is the auto-correlation sequence given by (9.178) and (9.179). Section 9.9.5 demonstrates an application of such factorization. Another important application of the factorization can be found in the MMSE DFE discussion in Section 9.5.3.3.

---

[23] We ignore the case where $\bar{R}(z)$ has zeros or poles at the unit circle. However, the arguments here can be easily extended to include such case.

[24] Note that (9.195) is not applicable here, since R_0 is not a term in $\{R_n\}$.

[25] $\{g_n\}$ is also minimum-phase, that is, all poles are outside the unit circle. We are not concerned with such property in this chapter.

9.9.5   Noise Whitening Filter

In this section, we apply the factorization of Section 9.9.4 to derive the noise whitening filter. Given a WSS sequence $\{v'_n\}$ with the following autocorrelation:

$$x'_{n-n'} \stackrel{\text{def}}{=} E\left(v'_n v'^*_{n'}\right), \tag{9.201}$$

we wish to find a filter so that the output sequence $\{v_m\}$ is uncorrelated. Namely, we want to find a filter $\{f_k\}$, so that

$$v_n = \sum_k f_k v'_{n-k}, \tag{9.202}$$

and[26]

$$x_{n-n'} \stackrel{\text{def}}{=} E\left(v_n v^*_{n'}\right) = \delta_{n,n'}. \tag{9.203}$$

Let us consider the problem with z-transform. From (9.172),

$$\bar{v}(z) = \bar{f}(z)\bar{v}'(z). \tag{9.204}$$

We can then compute the autocorrelation based on (9.181):

$$\begin{aligned}\bar{x}(z) &= \frac{1}{M}E\left[\bar{v}(z)\bar{v}^*\left(z^{*-1}\right)\right] = \frac{1}{M}\bar{f}(z)E\left[\bar{v}'(z)\bar{v}'^*\left(z^{*-1}\right)\right]\bar{f}^*\left(z^{*-1}\right) \\ &= \bar{f}(z)\bar{f}^*\left(z^{*-1}\right)\bar{x}'(z) \end{aligned} \tag{9.205}$$

Here, M is the sequence length. Since $\{v_n\}$ is supposed to be white, (9.203) implies

$$\bar{x}(z) = \bar{f}(z)\bar{f}^*\left(z^{*-1}\right)\bar{x}'(z) = 1. \tag{9.206}$$

$\bar{x}'(z)$ is a correlation function, with the complex reversal symmetry discussed in Section 9.9.4. It can thus be factorized according to (9.198):

$$\bar{x}'(z) = \bar{g}(z)\bar{g}^*\left(z^{*-1}\right). \tag{9.207}$$

One obvious choice of $\bar{f}(z)$ is thus

$$\bar{f}(z) = \bar{g}^{*-1}\left(z^{*-1}\right) \tag{9.208}$$

Note that such factorization is not unique. If there are N pairs of the zeros and poles for $\bar{x}'(z)$, then there are 2^N ways to partition them between $\bar{g}(z)$ and $\bar{g}^*\left(z^{*-1}\right)$. Equation (9.198) is still valid for these alternative $\bar{g}$ and $\bar{f}$ functions. However, there is one construction that makes $\bar{f}(z)$ a stable filter [5, sec. 9.3-2], [6, sec. 7.2.1].

---

[26] For analyses in this chapter, we typically use the noise correlation $E\left(v_n v^{*'}_n\right) = N_0\delta_{n,n'}$. Its difference from (9.203) can be easily reconciled by scaling the filter $\{f_k\}$.

9.9.6 Total Filter

Let us apply the whitening filter derived in Section 9.9.5 to the channel model in Section 9.2.3, and derive the total channel $\{h_n\}$ in (9.16).

As shown in (9.12) in Section 9.2.3, the noise correlation $\{x_n\}$ in (9.201) is given by (9.9):

$$x_n = \int q(t)q^*\left(t - nT_{Sym}\right)dt. \tag{9.209}$$

$q(t)$ is the channel response including the pulse shaping filter and the propagation channel. T_{Sym} is the symbol period. From the filter cascade (9.15) and the whitening filter results (9.207) and (9.208), we have

$$\bar{h}(z) = \bar{x}(z)\bar{f}(z) = \bar{g}(z)\bar{g}^*\left(z^{*-1}\right)\bar{g}^{*-1}\left(z^{*-1}\right) = \bar{g}(z). \tag{9.210}$$

A commonly used quantity is $\bar{h}(z)\bar{h}^*\left(z^{*-1}\right)$. From (9.210),

$$\bar{h}(z)\bar{h}^*\left(z^{*-1}\right) = \bar{g}(z)\bar{g}^*\left(z^{*-1}\right) = \bar{x}(z). \tag{9.211}$$

From its definition in (9.9), x_n is the sampling of the squared channel $\int q(\tau)q^*$ $(\tau - t)d\tau$ at nT_{Sym}. Therefore, according to (9.194) and (9.188),

$$\bar{h}(z)\bar{h}^*\left(z^{*-1}\right)\Big|_{z=\exp\left(j\omega T_{Sym}\right)} = \left|\bar{h}\left[\exp\left(j\omega T_{Sym}\right)\right]\right|^2$$

$$= \bar{x}\left[\exp\left(j\omega T_{Sym}\right)\right] = \frac{1}{T_{Sym}}\sum_m \left|\tilde{q}\left(\omega - \frac{2\pi m}{T_{Sym}}\right)\right|^2. \tag{9.212}$$

9.10 APPENDIX: OPTIMIZATION OF FUNCTIONS WITH COMPLEX VARIABLES

In Chapter 8, we study optimization of functions $c(w)$, where both the scalar c and the vector w are real-valued. The optimization involves taking derivative of c with regard to w, or $\nabla_w c$. In this chapter, we extend the operations to the cases where w may be complex-valued. For this purpose, we need to mathematically define and justify derivative operations with regard to a complex vector w. In the following, we extend the derivatives with regard to complex variables [47, ch. 14].

First, consider the case where w is a scalar. Namely, consider function $c(w)$, and

$$c = u + jv$$
$$w = x + jy. \tag{9.213}$$

u and v are the real and imaginary parts of c. x and y are the real and imaginary parts of w. A straightforward extension of the derivative definition is

$$\frac{dc}{dw} \overset{\text{def}}{=} \lim_{\Delta w \to 0} \frac{c(w + \Delta w) - c(w)}{\Delta w} \tag{9.214}$$

This definition is meaningful only when the limit does not depend on the phase of Δw as its modulus approaches zero. Functions with such property are known as analytical functions. A function c is analytical if and only if the Cauchy–Riemann equation holds:

$$\begin{cases} \dfrac{\partial u}{\partial x} = \dfrac{\partial v}{\partial y} \\[2mm] \dfrac{\partial u}{\partial y} = -\dfrac{\partial v}{\partial x} \end{cases} \tag{9.215}$$

In our problems, the cost functions to optimize are real-valued. Namely, $v \equiv 0$. Equation (9.215) thus dictates that u must be a constant. In other words, a real-valued and nonconstant function cannot be analytical. Therefore, we must find an alternative definition of the derivative.

Let us return to the basics and express the change in terms of $\partial c/\partial x$ and $\partial c/\partial y$. The first-order change in c due to the change of w by $\Delta x + j\Delta y$ is:

$$\Delta c = \frac{\partial c}{\partial x} \Delta x + \frac{\partial c}{\partial y} \Delta y. \tag{9.216}$$

Now introduce partial derivative $\partial c/\partial w$ while w^* is kept constant, and similarly define $\partial c/\partial w^*$. They are known as the Wirtinger derivatives. From the standard calculus rule and (9.213), we get

$$\frac{\partial c}{\partial x} = \frac{\partial c}{\partial w}\frac{\partial w}{\partial x} + \frac{\partial c}{\partial w^*}\frac{\partial w^*}{\partial x} = \frac{\partial c}{\partial w} + \frac{\partial c}{\partial w^*}$$
$$\frac{\partial c}{\partial y} = \frac{\partial c}{\partial w}\frac{\partial w}{\partial y} + \frac{\partial c}{\partial w^*}\frac{\partial w^*}{\partial y} = j\left(\frac{\partial c}{\partial w} - \frac{\partial c}{\partial w^*}\right) \tag{9.217}$$

Or,

$$\frac{\partial c}{\partial w} = \frac{1}{2}\left(\frac{\partial c}{\partial x} - j\frac{\partial c}{\partial y}\right)$$
$$\frac{\partial c}{\partial w^*} = \frac{1}{2}\left(\frac{\partial c}{\partial x} + j\frac{\partial c}{\partial y}\right) \tag{9.218}$$

The first-order change in (9.216) can thus be written as

$$
\begin{aligned}
\Delta c &= \frac{\partial c}{\partial x}\Delta x + \frac{\partial c}{\partial y}\Delta y \\
&= \left(\frac{\partial c}{\partial w} + \frac{\partial c}{\partial w^*}\right)\Delta x + j\left(\frac{\partial c}{\partial w} - \frac{\partial c}{\partial w^*}\right)\Delta y \\
&= \frac{\partial c}{\partial w}(\Delta x + j\Delta y) + \frac{\partial c}{\partial w^*}(\Delta x - j\Delta y) \\
&= \frac{\partial c}{\partial w}\Delta w + \frac{\partial c}{\partial w^*}\Delta w^*
\end{aligned}
\tag{9.219}
$$

Equation (9.219) shows that when computing the first-order change of c, we can use the normal calculus rules while considering w and w^* as independent variables. Furthermore, when $c(w, w^*)$ is real, both $\partial c/\partial x$ and $\partial c/\partial y$ are real. Therefore, (9.218) implies

$$
\frac{\partial c}{\partial w} = \left(\frac{\partial c}{\partial w^*}\right)^*.
\tag{9.220}
$$

According to standard calculus principles, if $(w_s, c(w_s))$ is an extreme point of c (w), then

$$
\begin{aligned}
\frac{\partial c}{\partial w}\bigg|_{w = w_s} &= 0 \\
\frac{\partial c}{\partial w^*}\bigg|_{w^* = w_s^*} &= 0
\end{aligned}
\tag{9.221}
$$

On the other hand, when $c(w)$ is real, (9.220) implies that the two equations in (9.221) are equivalent. Therefore, to find the extreme points, only one of them is needed.

These results can be extended to the cases where the variable is a vector w of dimension N. The first-order change is thus

$$
\Delta c = \left[\nabla_w^T c(w, w^*)\right]\Delta w + \left[\nabla_{w^*}^T c(w, w^*)\right]\Delta w^*.
\tag{9.222}
$$

Here

$$
\nabla_w c(w, w^*) \stackrel{\text{def}}{=}
\begin{pmatrix}
\frac{\partial c}{\partial w_1}\big|_{w^*} \\
\frac{\partial c}{\partial w_2}\big|_{w^*} \\
\frac{\partial c}{\partial w_3}\big|_{w^*} \\
\vdots \\
\frac{\partial c}{\partial w_N}\big|_{w^*}
\end{pmatrix},
\tag{9.223}
$$

and $\nabla_{w^*}c(w,w^*)$ is similarly defined. An extreme point $(w_s, c(w_s))$ has the following properties:

$$\nabla_w c(w,w^*)\big|_{w=w_s} = 0$$
$$\nabla_{w^H} c(w,w^*)\big|_{w=w_s} = 0 \qquad (9.224)$$

The two equations in (9.224) are equivalent when $c(w, w^*)$ is real-valued. Therefore, only one of them is needed.

9.11 APPENDIX: OPTIMAL SOLUTION OF ZERO FORCING LINEAR EQUALIZER

This section proves that the zero forcing (ZF) linear equalizer has the optimal solution that sets ISI within a certain range to zero. The proof approach was adopted from the literature [48]. For the convenience of cross-referencing, we use the same notation as in the paper [48]. Such notations may be different from the rest of this chapter.

9.11.1 System Model

In the treatment of ZF equalizers, we focus on ISI and do not consider noise. Let the original time-discrete channel be $\{x_n\}$. The linear equalizer taps are $\{c_n\}$. The total channel (including the equalizer) is given by (9.25):

$$h_n = \sum_j c_j x_{n-j}. \qquad (9.225)$$

Let $\{a_n\}$ and $\{y_n\}$ be the transmitted and received signals, respectively. We have from (9.25):

$$y_n = \sum_j h_j a_{n-j} \qquad (9.226)$$

Note that indices for h, x and c can be any integer, including negative ones.

9.11.2 Zero Forcing Optimization

By definition (Section 9.4.3.1), a zero forcing equalizer minimizes the maximum distortion

$$D \overset{\text{def}}{=} \frac{1}{h_0} \sum_{n \neq 0} |h_n|, \qquad (9.227)$$

where $n = 0$ is the main tap of the total channel. Further, assume (by scaling) that the center tap for the channel before the equalizer

$$x_0 = 1. \tag{9.228}$$

We also define D_0 as the distortion without equalizing:

$$D_0 \stackrel{\text{def}}{=} \sum_{n \neq 0} |x_n|. \tag{9.229}$$

The equalizer is further normalized so that the center tap value is 1:

$$h_0 = 1. \tag{9.230}$$

With such normalization, (9.227) leads to

$$D = \sum_{n \neq 0} |h_n|. \tag{9.231}$$

In the following, we express this constraint by the dependency of c_0 to other tap values. The remaining tap values are thus allowed to be optimized without worrying about the constraint.

From (9.225):

$$h_0 = \sum_{k} c_j x_{-j}. \tag{9.232}$$

Considering (9.228), normalization (9.230) becomes

$$c_0 = 1 - \sum_{j \neq 0} c_j x_{-j}. \tag{9.233}$$

From (9.233) and (9.225), we get

$$
\begin{aligned}
h_n &= c_0 x_n + \sum_{j \neq 0} c_j x_{n-j} = \left(1 - \sum_{j \neq 0} c_j x_{-j}\right) x_n + \sum_{j \neq 0} c_j x_{n-j} \\
&= x_n + \sum_{j \neq 0} c_j \left(x_{n-j} - x_n x_{-j}\right)
\end{aligned}
\tag{9.234}
$$

Note that (9.234) and (9.228) automatically implies (9.230). Therefore, there is no more constraint on $\{c_n, n \neq 0\}$. From (9.234) and (9.230), the optimization target (9.231) can be written as

$$
\begin{aligned}
D &= \sum_{n \neq 0} h_n \, sgn(h_n) \\
&= \left[\sum_{n \neq 0} x_n \, sgn(h_n) + \sum_{j \neq 0} c_j \sum_{n \neq 0} \left(x_{n-j} - x_n x_{-j}\right) sgn(h_n)\right].
\end{aligned}
\tag{9.235}
$$

Here, $sgn(x)$ is 1 or -1, depending on the sign of x. A ZF equalizer chooses tap values $\{c_n\}$ to minimize D. Since h_n depends on $\{c_n\}$, the minimization is not as easy as it appears.

9.11.3 Minimization Condition

In this section, we prove that if D_0, as defined by (9.229), meets the following condition

$$D_0 < 1 \tag{9.236}$$

(i.e., with weak channel dispersion), then the optimal solution for $\{c_n\}$ would set all $\{h_n, n \neq 0\}$ taps influenced by x_0 to zero. There are N of such $\{h_n\}$ taps in addition to h_0, if the length of the equalizer is $N + 1$.

The idea of the proof is that if $h_{k \neq 0}$ is nonzero, we can always modify the equalizer tap value c_k to achieve a lower distortion. The proof goes as follows.

Assume there is a set of equalizer tap values $\{c_n\}$, such that

$$h_k \neq 0 \tag{9.237}$$

for some $k \neq 0$, and the corresponding distortion function (9.231) is D. We select a set of new tap values $\{c_n^*\}$ such that[27]

$$\begin{aligned}
c_j^* &= c_j \forall j \neq k \\
c_k^* &= c_k - \Delta \, sgn(h_k)
\end{aligned} \tag{9.238}$$

where Δ is a small step size:

$$0 < \Delta < \frac{1}{2}|h_k|. \tag{9.239}$$

Note that in this derivation, c^* does not mean the complex conjugate of c.

Modification (9.239) also implicitly changes the value of c_0 because of the constraint (9.230). We use (9.233) in the following derivations to include such an effect.

The choice of modification (9.238) can be intuitively understood as follows. The contribution of c_k to h_k is $c_k x_0 = c_k$. Therefore, modification (9.238) would reduce the value of $|h_k|$, thus reducing D. Such modification also affects the contribution of c_k to other taps of $\{h_n\}$. However, because of the weak dispersion condition (9.236), the sum of such contribution, even in the worst case, is smaller than the reduction of h_k. Therefore, the net effect is still reducing D. Another complication is that c_0 is also changed by such modification due to the normalization constraint. Therefore, we need to evaluate the total effect carefully, as shown below.

According to (9.231), (9.234), and (9.238), we have

$$\begin{aligned}
D^* &= \sum_{n \neq 0} \left| x_n + \sum_{j \neq 0} c_j^* \left(x_{n-j} - x_n x_{-j} \right) \right| \\
&= \sum_{n \neq 0} \left| x_n + \sum_{j \neq 0} c_j \left(x_{n-j} - x_n x_{-j} \right) - \Delta \, sgn(h_k) \left(x_{n-k} - x_n x_{-k} \right) \right|
\end{aligned} \tag{9.240}$$

---

[27] Note that tap value modification implies that the range of k is within the equalizer coverage. Namely, c_k is not always zero.

The first part in the summation is no other than h_n according to (9.234). Therefore, we get

$$D^* = \sum_{n \neq 0} |h_n - \Delta\, sgn(h_k)(x_{n-k} - x_n x_{-k})|. \tag{9.241}$$

In the second term, $\Delta\, sgn\ (h_k)x_{n-k}$ represents the change in the contribution from c_k. $\Delta\, sgn\ (h_k)x_n x_{-k}$ represents the change in the contribution from c_0. Let us first single out the term $n = k$ and remember $x_0 = 1$. Remember that $h_k = |h_k|\, sgn\ (h_k)$.

$$\begin{aligned}
D_k &\stackrel{\text{def}}{=} |h_k - \Delta\, sgn(h_k)(x_0 - x_k x_{-k})| = |h_k - \Delta\, sgn(h_k)(1 - x_k x_{-k})| \\
&= |[|h_k| - \Delta(1 - x_k x_{-k})]\, sgn(h_k)| \\
&= ||h_k| - \Delta(1 - x_k x_{-k})|
\end{aligned} \tag{9.242}$$

Based on the weak channel dispersion condition (9.236), we have

$$-1 < x_k x_{-k} < 1. \tag{9.243}$$

Considering (9.239) and (9.243), the variable inside the absolute value operation of (9.242) is guaranteed to be positive. Therefore,

$$D_k = |h_k| - \Delta(1 - x_k x_{-k}). \tag{9.244}$$

Next, we consider other terms in (9.241) with $n \neq k$.

$$\begin{aligned}
D_n &\stackrel{\text{def}}{=} |h_n - \Delta\, sgn(h_k)(x_{n-k} - x_n x_{-k})| \leq |h_n| + |\Delta\, sgn(h_k)(x_{n-k} - x_n x_{-k})| \\
&= |h_n| + |\Delta(x_{n-k} - x_n x_{-k})|
\end{aligned} \tag{9.245}$$

Therefore, from (9.241), (9.244), and (9.245),

$$\begin{aligned}
D^* &= \sum_{n \neq 0} D_n \\
&\leq \sum_{n \neq 0} |h_n| + \sum_{n \neq k, n \neq 0} |(x_{n-k} - x_n x_{-k})| - \Delta(1 - x_k x_{-k})
\end{aligned} \tag{9.246}$$

The first term in the last line is the original distortion function D. Therefore, considering (9.227),

$$\begin{aligned}
D^* &= D + D_R \\
D_R &\stackrel{\text{def}}{=} \Delta \sum_{n \neq k, n \neq 0} |(x_{n-k} - x_n x_{-k})| - \Delta(1 - x_k x_{-k})
\end{aligned} \tag{9.247}$$

From the property of absolute values, we have the following inequalities:

$$\begin{aligned}
|x_{n-k} - x_n x_{-k}| &\leq (|x_{n-k}| + |x_n||x_{-k}|) \\
x_k x_{-k} &\leq |x_k||x_{-k}|
\end{aligned} \tag{9.248}$$

Next, we will establish an upper bound of D_R and show such bound is less than zero. Let us study the summation terms. Using the first inequality in (9.248),

$$\sum_{n\neq k, n\neq 0} |(x_{n-k}-x_n x_{-k})| \leq \sum_{n\neq k, n\neq 0} |x_{n-k}| + |x_{-k}| \sum_{n\neq k, n\neq 0} |x_n|. \tag{9.249}$$

For the first summation above, we substitute the summing variable, and remove the exclusion $n \neq 0$ with a correction term:

$$\sum_{n\neq k, n\neq 0} |x_{n-k}| = \sum_{n\neq k} |x_{n-k}| - |x_{-k}| \underset{m \overset{\text{def}}{=} n-k}{=} \sum_{m\neq 0} |x_m| - |x_{-k}|. \tag{9.250}$$

For the second summation in (9.249), we remove (with correction) the exclusion $n \neq k$:

$$|x_{-k}| \sum_{n\neq k, n\neq 0} |x_n| = |x_{-k}| \sum_{n\neq 0} |x_n| - |x_{-k}||x_k|. \tag{9.251}$$

Adding (9.250) and (9.251) together, (9.249) becomes

$$\sum_{n\neq k, n\neq 0} |(x_{n-k}-x_n x_{-k})| \leq \sum_{n\neq k, n\neq 0} |x_{n-k}| + |x_{-k}| \sum_{n\neq k, n\neq 0} |x_n|$$

$$= (1 + |x_{-k}|) \sum_{n\neq 0} |x_n| - |x_{-k}| - |x_{-k}||x_k|. \tag{9.252}$$

$$= (1 + |x_{-k}|)D_0 - |x_{-k}| - |x_{-k}||x_k|$$

D_0 is defined by (9.229). Going back to D_R given by (9.247) and considering the second inequality in (9.248):

$$D_R \leq \Delta[(1 + |x_{-k}|)D_0 - |x_{-k}| - |x_{-k}||x_k|] - \Delta(1 - x_k x_{-k})$$
$$\leq \Delta[(1 + |x_{-k}|)D_0 - |x_{-k}| - |x_{-k}||x_k| - 1 + |x_k||x_{-k}|] \tag{9.253}$$
$$= \Delta[(1 + |x_{-k}|)D_0 - |x_{-k}| - 1] = \Delta(1 + |x_{-k}|)(D_0 - 1)$$

Because of (9.236), (9.253) leads to

$$D_R < 0. \tag{9.254}$$

Therefore, (9.247) implies

$$D^* < D. \tag{9.255}$$

In other words, if $|h_k| > 0$, the corresponding equalizer cannot be optimal. Therefore, an optimal equalizer must satisfy

$$h_k = 0 \,\forall k \neq 0. \tag{9.256}$$

In general, an equalizer satisfying (9.256) can be found because (9.256) leads to N equations for the N taps of the equalizer $\{c_n, n \neq 0\}$.

9.12   APPENDIX: GAIN OF AN MMSE EQUALIZER

In this section, we show that an equalizer optimized under the MMSE criterion has a gain smaller than 1. In other words, an MMSE equalizer is biased, in the sense that its output error has a nonzero mean for a given input.

We look at a very simple MMSE equalizer, with only one tap. Consider a single-symbol channel model

$$y = x + n. \tag{9.257}$$

Here, x is the transmitted symbol, with a variance of σ_x^2. n is the noise, with a variance of N_0. y is the received symbol.

At the receiver, the equalizer would have one tap f. The equalizer output signal r is

$$r = fy = f(x + n) \tag{9.258}$$

With MMSE criterion, we want to choose f to minimize the cost function

$$c \overset{\text{def}}{=} E\left(|r - x|^2\right) = E\left(|f(x + n) - x|^2\right) = (f - 1)^2 \sigma_x^2 + f^2 N_0. \tag{9.259}$$

Therefore, the optimal f is

$$f_o = \frac{\sigma_x^2}{\sigma_x^2 + N_0}. \tag{9.260}$$

The corresponding cost function is

$$c_o = (f_o - 1)^2 \sigma_x^2 + f_o^2 N_0 = \frac{\sigma_x^2 N_0}{\sigma_x^2 + N_0}. \tag{9.261}$$

The output SNR is thus

$$SNR_{MMSE} = \frac{\sigma_x^2}{c_0} = \frac{\sigma_x^2}{N_0} + 1. \tag{9.262}$$

This result is counterintuitive. First, we would expect the optimal gain f_o to be 1, so that r is exactly x plus noise. Second, it is surprising that the output SNR is larger than the input SNR, which is σ_x^2 / N_0.

The reason for such a result is that an MMSE receiver aims at reducing the mean squared error. By using a gain smaller than 1, the variance of noise is reduced, at the expense of additional error caused by $E(r) \neq x$. An MMSE receiver balances these two effects to arrive at the optimal gain as given by (9.260). Note that with this solution, the mean of the error is not zero given x. Therefore, the analysis of bit error rates in Chapter 4 is no longer valid. In fact, the bit error rate under MMSE optimization is higher than the choice of $f = 1$. However, since SNR is usually much larger than 1, f_0 is very close to 1. MMSE is still an optimal solution in practice.

The SNR given by (9.262) is not the "true SNR" in the sense that it includes the "systematic" error between $E(r)$ and x. The true SNR, defined as the ratio between the powers of the signal and noise components in r, is obviously still σ_x^2/N_0. Therefore, the "true SNR" is

$$SNR = SNR_{MMSE} - 1. \tag{9.263}$$

Again, because SNR is usually much larger than 1, the difference here is not practically significant. Equation (9.263) is very general [9, sec. III-B]. This section provides a simple example to illustrate the concept.

By the way, (9.260) and (9.261) lead to

$$f_o = 1 - \frac{c_o}{\sigma_x^2}. \tag{9.264}$$

This relationship agrees with those in MMSE DFE analyses, (9.122) in Section 9.5.3.5 and (9.308) in Section 9.13.4.

9.13   APPENDIX: DETAILED DERIVATION OF FINITE-LENGTH DFE

In this section, we restate a part of the derivations given in [7], in support of the results summarized in Section 9.5.4. For simplicity, we only consider symbol-spaced sampling. Namely, the sampling rate of the feedforward (FF) filter is the same as the symbol rate.

Other than the obvious exceptions, boldface lowercase letters represent vectors. Uppercase letters represent matrixes.

9.13.1   The Channel and Equalizer Models

The channel and equalizer models are described in Section 9.5.4.1. Only major results are summarized here.

The channel model can be expressed in vector form:

$$y = Hd + v. \tag{9.265}$$

The vectors at time m can be defined as follows:

$$(y)_i = y_{m+N_f-i}, i = 1, 2, \ldots, N_f$$

$$(d)_i = d_{m+N_f-i}, i = 1, 2, \ldots, N_f + N_H. \tag{9.266}$$

$$(v)_i = v_{m+N_f-i}, i = 1, 2, \ldots, N_f$$

Here, $(y)_i$ is the ith component of vector y, while y_k is the kth symbol in sequence $\{y_n\}$. The same convention applies to vectors d (for transmitted symbols) and v (for noise). $\{d_n\}$, $\{y_n\}$, and $\{v_n\}$ are sequences representing the transmitted symbols, received symbols at the input of the DFE, and the noise at the input of the DFE, respectively. N_f is the length of the feedforward (FF) filter. H in (9.265) is a Toeplitz matrix representing the channel, whose components $\{H_{i,\,j}\}$ are defined as follows:

$$H_{i,j} = \begin{cases} h_{j-i} & 0 \leq j-i \leq N_H \\ 0 & \text{otherwise} \end{cases}. \tag{9.267}$$

At time m, vectors y and d fill the delay lines for the FF and the feedback (FB) filters, respectively. The tap values of the FF and FB filters are represented as vectors

$$\begin{aligned} a &\stackrel{\text{def}}{=} (a_1, a_2, \ldots, a_{N_f})^T \\ b &\stackrel{\text{def}}{=} (b_1, b_2, \ldots, b_{N_b})^T \end{aligned}. \tag{9.268}$$

For our formulation, the FB filter b (of length N_b) is imbedded in an equivalent filter $\tilde{b}$ of length $N_f + N_H$, with an offset Δ:

$$\tilde{b} \stackrel{\text{def}}{=} (0_{1\times\Delta}, 1, b^T, 0_{1\times s})^T, \tag{9.269}$$

$0_{1\times\Delta}$ and $0_{1\times s}$ are row zero-vectors of length Δ and s, respectively.

$$s = N_f + N_H - \Delta - N_b - 1. \tag{9.270}$$

The 1 in $\tilde{b}$ represents the direct path for error computation, as discussed above.

The output sample of the DFE $r^{(m)}$ and error $e^{(m)}$ can thus be written in vector form, using definitions in (9.266) and (9.268):

$$\begin{aligned} r^{(m)} &= a^H y - \tilde{b}^H d + d_{m+N_f-\Delta-1} \\ e^{(m)} &= \tilde{b}^H d - a^H y \end{aligned}. \tag{9.271}$$

We further introduce the following correlation matrices:

$$\begin{aligned} R_{dd} &\stackrel{\text{def}}{=} E\left(dd^H\right) \\ R_{dy} &\stackrel{\text{def}}{=} E\left(dy^H\right) \\ R_{yy} &\stackrel{\text{def}}{=} E\left(yy^H\right) \\ R_{vv} &\stackrel{\text{def}}{=} E\left(vv^H\right) \end{aligned}. \tag{9.272}$$

v is uncorrelated with d. The autocorrelation matrices R_{dd}, R_{yy}, and R_{vv} are assumed to be positive-definite, and therefore nonsingular. Of course, these autocorrelation matrices are all Hermitian. For now, we keep R_{dd} and R_{vv} in general forms.

The special case of white input and white noise (9.86) is to be discussed later. Furthermore, the channel model (9.265) leads to

$$R_{dy} = R_{dd}H^H$$
$$R_{yy} = HR_{dd}H^H + R_{vv}$$

(9.273)

The cost function is

$$c_{MMSE} \stackrel{\text{def}}{=} E\left(\left|e^{(m)}\right|^2\right).$$

(9.274)

With above definitions and the model (9.271), the cost function can be written as

$$c_{MMSE} = \tilde{b}^H R_{dd}\tilde{b} + a^H R_{yy}a - \tilde{b}^H R_{dy}a - a^H R_{dy}^H \tilde{b}.$$

(9.275)

In the following, we examine the optimal filter tap values a and $\tilde{b}$ for DFE. As in the case of infinite-length DFE (Section 9.5.3), we first optimize the FF filter a given the FB filter $\tilde{b}$ (Section 9.13.2), leading to an expression of c_{MMSE} in terms of $\tilde{b}$. Next, we optimize $\tilde{b}$ (Section 9.13.3) and have some discussions on the resulting properties (Section 9.13.4). Some mathematical details are provided separately in Sections 9.13.5 and 9.13.6.

9.13.2 Optimization of the FF Filter

We minimize C_{MMSE} in terms of the FF filter a, by requiring the derivative with regard to a^H to be zero (see (9.224) in Section 9.10):

$$\nabla_{a^H} c_{MMSE} = R_{yy}a - R_{dy}^H \tilde{b} = 0.$$

(9.276)

This equation leads to the optimal FF filter (in terms of $\tilde{b}$)

$$a_o = R_{yy}^{-1} R_{dy}^H \tilde{b}.$$

(9.277)

From (9.271) and (9.274), the same optimization also leads to the orthogonality principle at the optimal point

$$E\left(e^{(k)}y\right) = 0.$$

(9.278)

Substituting a_o in (9.277)–(9.275), we get

$$c_{MMSE} = \tilde{b}^H R_{d/y}\tilde{b}$$
$$R_{d/y} \stackrel{\text{def}}{=} R_{dd} - R_{dy}R_{yy}^{-1} R_{dy}^H.$$

(9.279)

With (9.273), we have

$$R_{d/y} = R_{dd} - R_{dd}H^H \left(HR_{dd}H^H + R_{vv} \right)^{-1} HR_{dd} = \left(R_{dd}^{-1} + H^H R_{vv}^{-1} H \right)^{-1}. \quad (9.280)$$

The last equality can be proved by matrix multiplication, as shown in Section 9.13.5.

9.13.3 Optimization of the FB Filter

The FB filter is constrained to the form of (9.269). Therefore, we cannot optimize it with the derivative. On the other hand, note that $R_{d/y}$ is Hermitian. Therefore, based on (9.280), it can be factorized as

$$R_{d/y}^{-1} = \left(R_{dd}^{-1} + H^H R_{vv}^{-1} H \right) = L\Lambda L^H. \quad (9.281)$$

Here, L is a lower triangular matrix, with all diagonal elements equal to 1:

$$L_{i,j} = 0 \, \forall \, i < j$$
$$L_{i,i} = 1 \, \forall i \quad (9.282)$$

It is easy to see that L^{-1} has these properties, as well. Λ is a diagonal matrix, whose diagonal elements are noted as $\{\lambda_i, \, i = 1, 2, \dots, N_f + L_H\}$:

$$\Lambda_{i,j} = \begin{cases} \lambda_i & i = j \\ 0 & \text{otherwise} \end{cases}. \quad (9.283)$$

Equation (9.281) is a standard factorization in numerical linear algebra, known as the LDL decomposition or LDLT decomposition [49]. It is closely related to the commonly used Cholesky factorization.

With (9.281), (9.279) can be written as

$$c_{MMSE} = \tilde{b}^H L^{-H} \Lambda^{-1} L^{-1} \tilde{b}. \quad (9.284)$$

The columns of L form $N_f + N_H$ linearly independent vectors $\{l^{(i)}, \, i = 1, 2, \dots, N_f + N_H\}$ in an $N_f + N_H$ dimensional space. $l^{(i)}$ has $i - 1$ leading zeros, followed by 1. The rest of the components can take any value:

$$l_n^{(i)} = 0 \, \forall \, n < i$$
$$l_i^{(i)} = 1 \quad (9.285)$$

We can project $\tilde{b}$ onto such a basis:

$$\tilde{b} = \sum_{i=1}^{N_f + L_H} \beta_i l^{(i)}. \quad (9.286)$$

Because

$$L^{-1}L = I_{N_f + N_H},\tag{9.287}$$

we have

$$L^{-1}l^{(i)} = u^{(i)},\tag{9.288}$$

where $u^{(i)}$ is a vector of size $N_f + N_H$, whose elements are all zero except for the ith, which is 1:

$$u_n^{(i)} = \delta_{n,i}.\tag{9.289}$$

Recall that $\tilde{b}$ must be in the form of (9.269). Namely, its elements must have Δ leading zeros, followed by 1.[28] Therefore, (9.286) is limited to the form[29]

$$\tilde{b} = l^{(\Delta + 1)} + \sum_{k = \Delta + 2}^{N_f + N_H} \beta_k l^{(k)}.\tag{9.290}$$

Equations (9.284), (9.290), and (9.288) lead to

$$c_{MMSE} = \lambda_{\Delta + 1}^{-1} + \sum_{k = \Delta + 2}^{N_f + N_H} \lambda_k^{-1} |\beta_k|^2 \geq \lambda_{\Delta + 1}^{-1}.\tag{9.291}$$

The equality holds (i.e., c_{MMSE} is minimized) when all β_k are zero. Namely,

$$\tilde{b} = l^{(\Delta + 1)}.\tag{9.292}$$

Equation (9.292) expresses the optimal $\tilde{b}$ for given Δ. Furthermore, Δ can be optimized to minimize c_{MMSE}. The final optimal solution can be expressed as

$$\begin{aligned} \tilde{b}_o &= l^{(\Delta_o + 1)} \\ \Delta_o &\overset{\text{def}}{=} arg\max_\Delta \lambda_{\Delta + 1} \end{aligned}.\tag{9.293}$$

The optimal cost function is

$$c_{MMSEo} = \lambda_{\Delta_o + 1}^{-1}.\tag{9.294}$$

---

[28] We consider only the case where s in (9.269) is 0. As will be shown later, such choice is optimal. Therefore, we do not consider the constraint that the last s elements of $\tilde{b}$ must be zero.

[29] Note that we count all indices from 1, in contrast to [7], which counts from zero. Therefore, all Δ in [7] are replaced by $\Delta + 1$ here.

Unfortunately, it is difficult to go further and connect $\lambda_{\Delta_o+1}^{-1}$ with channel conditions. Most performance evaluations of finite-length DFE are performed through simulation.

Reference [7] states that

- When $N_b = N_H$ and with white input and noise, $\Delta_o > N_f - 2$.
- Multiple simulations show that $\Delta_o = N_f - 1$ holds for most practical channel and noise scenarios and reasonably long feedforward filters.

These two statements imply $s = 0$ as given by (9.270).

Now return to the optimal FF filter solution. From (9.277) and (9.273),

$$
\begin{aligned}
a_o &= R_{yy}^{-1} R_{dy}^H \tilde{b} = \left(HR_{dd}H^H + R_{nn}\right)^{-1} HR_{dd}\tilde{b} \\
&= R_{vv}^{-1} H \left(R_{dd}^{-1} + H^H R_{vv^{-1}H}\right)^{-1} \tilde{b}.
\end{aligned}
\tag{9.295}
$$

The last equation is proven in Section 9.13.6. With (9.293), (9.288), and (9.281), the above expression (9.295) can be further simplified to

$$
a_o = R_{vv}^{-1} HL^{-H} \Lambda^{-1} L^{-1} l^{(\Delta_o+1)} = R_{vv}^{-1} HL^{-H} \lambda_{\Delta_o+1}^{-1} u^{(\Delta_o+1)},
\tag{9.296}
$$

where $u^{(\Delta_o+1)}$ is defined in (9.289). λ_{Δ_o+1} is defined in (9.283). Equation (9.296) shows that a_o is the $(\Delta_o + 1)$th column of matrix $R_{vv}^{-1} HL^{-H}$, multiplied by the scaler $\lambda_{\Delta_o+1}^{-1}$.

9.13.4 Discussion About the Residual ISI

In this section, we show that as in the case of infinitely long DFE, the output of the FF filter contains the desired signal, noise, postcursor ISI, and precursor ISI. The postcursor ISI is completely canceled by the output from the FB filter.

We consider the special case of white input and white noise. Namely, the correlation matrices defined in (9.272) are

$$
\begin{aligned}
R_{dd;} &= \sigma_x^2 I_{N_f+N_H} \\
R_{vv} &= N_0 I_{N_f}
\end{aligned}
\tag{9.297}
$$

σ_x^2 and N_0 are energy per symbol for signal and noise at the DFE input. Equations (9.280) and (9.281) thus become

$$
R_{d/y} = \left(\sigma_x^{-2} I_{N_f+v} + N_0^{-1} H^H H\right)^{-1} = \left(L\Lambda L^H\right)^{-1}.
\tag{9.298}
$$

Let us define the input SNR:

$$
SNR_{in} \stackrel{\text{def}}{=} \frac{\sigma_x^2}{N_0}.
\tag{9.299}
$$

Equation (9.298) can thus be rewritten as

$$H^H H = N_0 L \Lambda L^H - \frac{1}{SNR_{in}} I_{N_f + \nu}. \tag{9.300}$$

From (9.265) and (9.296), let us look at the scalar output of the FF filter at time m:

$$w^{(m)} \stackrel{\text{def}}{=} a_o^H y = u^{(\Delta_0 + 1)H} \lambda_{\Delta_o}^{-1} L^{-1} H^H R_{vv}^{-1} (Hd + v), \tag{9.301}$$

where $u^{(i)}$ is defined in (9.289). Equation (9.297) leads to

$$w^{(m)} = \frac{1}{N_0} u^{(\Delta_0 + 1)H} \lambda_{\Delta_o + 1}^{-1} L^{-1} H^H Hd + \frac{1}{N_0} u^{(\Delta_0 + 1)H} \lambda_{\Delta_o + 1}^{-1} L^{-1} H^H v. \tag{9.302}$$

The signal component can be rewritten using (9.300):

$$\begin{aligned} w^{(m)} &= u^{(\Delta_0 + 1)H} \lambda_{\Delta_o + 1}^{-1} \Lambda L^H d - \frac{1}{N_0} u^{(\Delta_0 + 1)H} \lambda_{\Delta_o + 1}^{-1} L^{-1} \frac{1}{SNR_{in}} d \\ &\quad + \frac{1}{N_0} u^{(\Delta_0 + 1)H} \lambda_{\Delta_o + 1}^{-1} L^{-1} H^H v \end{aligned} \tag{9.303}$$

For the first term in (9.303), remembering the definition of $u^{(i)}$ in (9.289) and that of Λ in (9.283), we have

$$\begin{aligned} u^{(\Delta_0 + 1)H} \lambda_{\Delta_o + 1}^{-1} \Lambda L^H d &= \lambda_{\Delta_o + 1}^{-1} \left(u^{(\Delta_0 + 1)H} \Lambda L^H \right) d \\ &= \lambda_{\Delta_o + 1}^{-1} \left(\lambda_{\Delta_o + 1} l^{(\Delta_o + 1)H} \right) d = l^{(\Delta_o + 1)H} d \end{aligned}, \tag{9.304}$$

where $l^{(\Delta_o + 1)}$ is the $(\Delta_o + 1)$th column of L, with the property shown in (9.285). Remembering the definition of d in (9.266), $l^{(\Delta_o + 1)H} d$ captures the contribution from transmitted symbols from $d_{m + N_f - \Delta_0 - 1}$ back to $d_{m - N_H}$, which are the current symbol and all of the symbols arriving *before* it:

$$l^{(\Delta_o + 1)H} d = d_{m + N_f - \Delta_0 - 1} + \sum_{n = \Delta_o + 2}^{N_f + N_H} L_{n, \Delta_o + 1}^* d_{m + N_f - n}, \tag{9.305}$$

where $L_{i, j}$ denotes elements in L. The first term is the desired symbol. The second is the postcursor ISI discussed in Section 9.5.3. According to the equalizer model (9.271) and the FB filter solution (9.293), the postcursor ISI is canceled out by the FB filter.

For the second term in (9.303), recognize that $u^{(\Delta_0 + 1)H} L^{-1}$ is the $(\Delta_o + 1)$th row of L^{-1}. L^{-1} is also a lower-triangular matrix with diagonal elements being 1, as depicted by (9.282). Therefore, for the $(\Delta_o + 1)$th *row* of L^{-1}, only the first $(\Delta_o + 1)$ elements are nonzero, while the $(\Delta_o + 1)$th element is 1. Therefore, this term can be written more explicitly as

$$\frac{1}{N_0} u^{(\Delta_0+1)H} \lambda_{\Delta_o+1}^{-1} L^{-1} \frac{1}{SNR_{in}} d$$

$$= \frac{1}{N_0 SNR_{in} \lambda_{\Delta_o+1}} \left(\sum_{n=1}^{\Delta_o} L_{\Delta_o+1,n}^{-1} d_{m+N_f-n} + d_{m+N_f-\Delta_o-1} \right). \tag{9.306}$$

Here, $L_{i,j}^{-1}$ denodes elements in L^{-1}. Equation (9.306) shows that only the current symbol $d_{k+N_f-\Delta_o-1}$ and symbols *after* it (i.e., the precursor ISI) contribute to this term.

The precursor ISI given by (9.306) approaches zero when SNR_{in} approaches infinity. This result is to be expected because, at high SNR, an MMSE equalizer approaches a ZF equalizer, where all ISIs are removed.

From (9.299) and (9.294), the first factor in (9.306) can be simplified, yielding the expression

$$\frac{1}{N_0} u^{(\Delta_0+1)H} \lambda_{\Delta_o+1}^{-1} L^{-1} \frac{1}{SNR_{in}} d = \frac{c_{MMSEo}}{\sigma_x^2} \left(\sum_{n=1}^{\Delta_o} L_{\Delta_o,n}^{-1} d_{m+N_f-n} + d_{m+N_f-\Delta_o+1} \right). \tag{9.307}$$

It is interesting to look at the contribution from the current symbol in $w^{(m)}$. From (9.303), (9.305), and (9.307), such contribution is

$$d_{m+N_f-\Delta_o-1} - \frac{c_{MMSEo}}{\sigma_x^2} d_{m+N_f-\Delta_o-1} = \left(1 - \frac{c_{MMSEo}}{\sigma_x^2} \right) d_{m+N_f-\Delta_o-1}. \tag{9.308}$$

Since the current symbol does not contribute to the feedback path, its total contribution to the DFE output is given by (9.308). Therefore, an MMSE DFE has a gain slightly less than 1, similar to the case of infinite-length DFE as shown in (9.122) in Section 9.5.3.5. This gain is explained in an appendix (Section 9.12).

The third term in (9.303) is the contribution of input noise, which is not discussed in detail here. It is worth pointing out, though, that even if H is singular, $L^{-1}H^H$ is always bounded. Therefore, the DFE does not have excessive noise amplification.

The interesting characteristics of the finite-length DFE can be summarized as below.

1. The output of the FF filter contains residual ISI in addition to the contribution from input noise. See (9.303).

2. The postcursor ISI (contributed by symbols preceding the current one) is canceled by the FB filter. See (9.304).

3. The precursor ISI (contributed by symbols after the current one) remains as part of the output error. It approaches zero when the input SNR approaches to infinity. See (9.307).

4. The MMSE DFE has a gain that depends on SNR and is usually slightly smaller than 1. See (9.308).

Furthermore, the following properties of finite-length DFE have been shown in [7]. They are briefly stated here without derivation.

1. As in the case of infinite-length DFE, an FF filter can be understood as a cascade of a noise whitening filter, a matched filter, and a symbol-spaced ISI canceller/reducer.[30] This relationship is hinted by (9.296) but is shown more clearly when the FF filter sampling rate is a multiple of the symbol rate, as studied in [7].

2. Unlike infinite-length DFE, the total noise at DFE output (i.e., the summation of residual ISI and filtered noise) is not exactly white.

There are some more subtleties discussed in [7]. However, the above two points have the most practical value.

9.13.5   Appendix: Proof of (9.280)

In this section, we prove (9.280), namely,

$$R_{dd} - R_{dd}H^H \left(HR_{dd}H^H + R_{vv} \right)^{-1} HR_{dd} = \left(R_{dd}^{-1} + H^H R_{vv}^{-1} H \right)^{-1}. \tag{9.309}$$

We start with the equation

$$HR_{dd}H^H R_{vv}^{-1} H + H = H + HR_{dd}H^H R_{vv}^{-1} H. \tag{9.310}$$

Manipulating on both sides:

$$(HR_{dd}H^H + R_{vv})R_{vv}^{-1} H = HR_{dd} \left(R_{dd}^{-1} + H^H R_{vv}^{-1} H \right)$$

$$R_{vv}^{-1} H = \left(HR_{dd}H^H + R_{vv} \right)^{-1} HR_{dd} \left(R_{dd}^{-1} + H^H R_{vv}^{-1} H \right) \tag{9.311}$$

$$R_{dd}H^H R_{vv}^{-1} H = R_{dd}H^H \left(HR_{dd}H^H + R_{vv} \right)^{-1} HR_{dd} \left(R_{dd}^{-1} + H^H R_{vv}^{-1} H \right)$$

Moving everything to the left-hand side, and adding an identity matrix to both sides:

$$I + R_{dd}H^H R_{vv}^{-1} H - R_{dd}H^H \left(HR_{dd}H^H + R_{vv} \right)^{-1} HR_{dd} \left(R_{dd}^{-1} + H^H R_{vv}^{-1} H \right) = I$$

$$R_{dd} \left(R_{dd}^{-1} + H^H R_{vv}^{-1} H \right) - R_{dd}H^H \left(HR_{dd}H^H + R_{vv} \right)^{-1} HR_{dd} \left(R_{dd}^{-1} + H^H R_{vv}^{-1} H \right) = I.$$

$$\left[R_{dd} - R_{dd}H^H \left(HR_{dd}H^H + R_{vv} \right)^{-1} HR_{dd} \right] \left(R_{dd}^{-1} + H^H R_{vv}^{-1} H \right) = I$$

$$\tag{9.312}$$

---

[30] In our discussion, noise whitening filter and matched filter are not a part of the DFE. Especially, while the matched filter is at higher sampling rate, the DFE is symbol spaced. However, in [7] the FF is fractional symbol spaced. As will be outlined in Section 9.7, in this case we can combine these filters into the FF filter, as [7] does.

Right-multiplying $\left(R_{xx}^{-1}+H^{H}R_{vv}^{-1}H\right)^{-1}$ to both sides, we get

$$R_{dd}-R_{dd}H^{H}\left(HR_{dd}H^{H}+R_{vv}\right)^{-1}HR_{dd}=\left(R_{dd}^{-1}+H^{H}R_{vv}^{-1}H\right)^{-1}. \tag{9.313}$$

QED.

9.13.6 Appendix: Proof of (9.295)

In this section, we prove the equation used in (9.295):

$$\left(HR_{dd}H^{H}+R_{vv}\right)^{-1}HR_{dd}=R_{nn}^{-1}H\left(R_{dd}^{-1}+H^{H}R_{vv}^{-1}H\right)^{-1}. \tag{9.314}$$

Start with the equality

$$H+HR_{dd}H^{H}R_{vv}^{-1}H=H+HR_{dd}H^{H}R_{vv}^{-1}H \tag{9.315}$$

Introduce some pairs of inverse matrices:

$$HR_{dd}R_{dd}^{-1}+HR_{dd}H^{H}R_{vv}^{-1}H=R_{vv}R_{vv}^{-1}H+HR_{dd}H^{H}R_{vv}^{-1}H \tag{9.316}$$

Regroup both sides:

$$HR_{dd}\left(R_{dd}^{-1}+H^{H}R_{vv}^{-1}H\right)=\left(R_{vv}+HR_{dd}H^{H}\right)R_{vv}^{-1}H. \tag{9.317}$$

Left multiply $(R_{vv}+HR_{dd}H^{H})^{-1}$ and right multiply $\left(R_{dd}^{-1}+H^{H}R_{vv}^{-1}H\right)^{-1}$ to both sides:

$$\left(R_{vv}+HR_{dd}H^{H}\right)^{-1}HR_{dd}=R_{vv}^{-1}H\left(R_{dd}^{-1}+H^{H}R_{vv}^{-1}H\right)^{-1}. \tag{9.318}$$

QED.

REFERENCES

1. S. U. H. Qureshi, "Adaptive Equalization," *Proc. IEEE*, vol. 73, no. 9, pp. 1349–1387, 1985.
2. M. S. Richer, G. Reitmeier, T. O. M. Gurley, G. A. Jones, J. Whitaker, and R. Rast, "The ATSC Digital Television System," *Proc. IEEE*, vol. 94, no. 1, pp. 37–42, 2006.
3. Y. Wu, X. Wang, R. Citta, B. Ledoux, S. Laflèche, and B. Caron, "An ATSC DTV Receiver with Improved Robustness to Multipath and Distributed Transmission Environments," *IEEE Trans. Broadcast.*, vol. 50, no. 1, pp. 32–41, 2004.
4. G. Ungerboeck, "Adaptive Maximum-Likelihood Receiver for Carrier-Modulated Data-Transmission Systems," *IEEE Trans. Commun.*, vol. 22, no. 5, pp. 624–636, 1974.
5. J. Proakis and M. Salehi, *Digital Communications*, 5th ed. McGraw-Hill Education, 2007.
6. G. L. Stuber, *Principles of Mobile Communication*. New York: Springer, 2011.
7. N. Al-Dhahir and J. M. Cioffi, "MMSE Decision-Feedback Equalizers: Finite-Length Results," *IEEE Trans. Inf. Theory*, vol. 41, no. 4, pp. 961–975, 1995.
8. J. M. Cioffi, G. P. Dudevoir, M. V. Eyuboglu, and G. D. Forney, "MMSE Decision-Feedback Equalizers and Coding. II. Coding Results," *IEEE Trans. Commun.*, vol. 43, no. 10, pp. 2595–2604, 1995.
9. J. M. Cioffi, G. P. Dudevoir, M. Vedat Eyuboglu, and G. D. Forney, "MMSE Decision-Feedback Equalizers and Coding. I. Equalization Results," *IEEE Trans. Commun.*, vol. 43, no. 10, pp. 2582–2594, 1995.

10. M. F. Mosleh and A. H. AL-Nakkash, "Combination of LMS and RLS Adaptive Equalizer for Selective Fading Channel," *Eur. J. Sci. Res.*, vol. 43, no. 1, pp. 127–137, 2010.

11. J. G. Proakis, "Adaptive Equalization for TDMA Digital Mobile Radio," *Veh. Technol. IEEE Trans.*, vol. 40, no. 2, pp. 333–341, 1991.

12. G. E. Bottomley, *Channel Equalization for Wireless Communications*. Hoboken, NJ: Wiley, 2011.

13. D. G. Messerschmitt, "A Geometric Theory of Intersymbol Interference," *Bell Syst. Tech. J.*, vol. 52, no. 9, pp. 1483–1519, 1973.

14. J. Salz, "Optimum Mean-Square Decision Feedback Equalization," *Bell Syst. Tech. J.*, vol. 52, no. 8, pp. 1341–1373, 1973.

15. M. V. Eyuboglu and G. D. Forney, "Trellis Precoding: Combined Coding, Precoding and Shaping for Intersymbol Interference Channels," *IEEE Trans. Inf. Theory*, vol. 38, no. 2, pp. 301–314, 1992.

16. M. R. Gibbard and A. B. Sesay, "Asymmetric Signal Processing for Indoor Wireless LANs," *IEEE Trans. Veh. Technol.*, vol. 48, no. 6, pp. 2053–2064, 1999.

17. E. Eleftheriou and D. Falconer, "Adaptive Equalization Techniques for HF Channels," *IEEE J. Sel. Areas Commun.*, vol. 5, no. 2, pp. 238–247, 1987.

18. D. Duttweiler, J. Mazo, and D. Messerschmitt, "An Upper Bound on the Error Probability in Decision-Feedback Equalization," *IEEE Trans. Inf. Theory*, vol. 20, no. 4, pp. 490–497, 1974.

19. N. C. Beaulieu, "Bounds on Recovery Times of Decision Feedback Equalizers," *IEEE Trans. Commun.*, vol. 42, no. 10, pp. 2786–2794, 1994.

20. C. Douillard et al., "Iterative Correction of Intersymbol Interference: Turbo-Equalization," *Eur. Trans. Telecommun.*, vol. 6, no. 5, pp. 507–511, 1995.

21. M. Tuchler, R. Koetter, and A. C. Singer, "Turbo Equalization: Principles and New Results," *IEEE Trans. Commun.*, vol. 50, no. 5, pp. 754–767, 2002.

22. B. Daneshrad and H. Samueli, "1.6 Mbps Digital-QAM System for DSL Transmission," *IEEE J. Sel. Areas Commun.*, vol. 13, no. 9, pp. 1600–1610, 1995.

23. Jin-Der Wang, "Hybrid Equalizer Arrangement for Use in Data Communications Equipment," US Patent No. 5,604,769A, 1997.

24. B. R. Saltzberg, "Comparison of Single-Carrier and Multitone Digital Modulation for ADSL Applications," *IEEE Commun. Mag.*, vol. 36, no. 11, pp. 114–121, 1998.

25. P. Monsen, "Theoretical and Measured Performance of a DFE Modem on a Fading Multipath Channel," *IEEE Trans. Commun.*, vol. 25, no. 10, pp. 1144–1153, 1977.

26. P. Bisaglia, R. Castle, and S. H. Baynham, "Channel Modeling and System Performance for HomePNA 2.0," *IEEE J. Sel. Areas Commun.*, vol. 20, no. 5, pp. 913–922, 2002.

27. M. Tomlinson, "New Automatic Equaliser Employing Modulo Arithmetic," *Electron. Lett.*, vol. 7, no. 5–6, p. 138, 1971.

28. H. Harashima and H. Miyakawa, "Matched-Transmission Technique for Channels with Intersymbol Interference," *IEEE Trans. Commun.*, vol. 20, no. 4, pp. 774–780, 1972.

29. J. E. Smee and S. C. Schwartz, "Adaptive Compensation Techniques for Communications Systems with Tomlinson-Harashima Precoding," *IEEE Trans. Commun.*, vol. 51, no. 6, pp. 865–869, 2003.

30. R. F. H. Fischer and J. B. Huber, "Comparison of Precoding Schemes for Digital Subscriber Lines," *IEEE Trans. Commun.*, vol. 45, no. 3, pp. 334–343, 1997.

31. A. K. Aman, R. L. Cupo, and N. A. Zervos, "Combined Trellis Coding and DFE Through Tomlinson Precoding," *IEEE J. Sel. Areas Commun.*, vol. 9, no. 6, pp. 876–884, 1991.

32. M. Glavin and E. Jones, "Equalization of Digital Subscriber Lines Under Dynamic Channel Conditions," *Signal Process.*, vol. 84, no. 5, pp. 853–864, 2004.

33. J. P. Meehan and A. D. Fagan, "Precoding Over a Dynamic Continuous Wave Digital Mobile Channel," in *Personal, Indoor and Mobile Radio Communications, 1997 (PIMRC '97)*, 1–4 September, Helsinki, Finland, 1997, pp. 1140–1144.

34. H. Laamanen, "Method and Apparatus for Implementing Adaptive Tomlinson-Harashima Precoding in a Digital Data Link," US Patent No. 7,409,003 B2, 2008.

35. J. Mazo and J. Salz, "On the Transmitted Power in Generalized Partial Response," *IEEE Trans. Commun.*, vol. 24, no. 3, pp. 348–352, 1976.

36. U. Erez and S. ten Brink, "A Close-to-Capacity Dirty Paper Coding Scheme," *IEEE Trans. Inf. Theory*, vol. 51, no. 10, pp. 3417–3432, 2005.

37. R. F. H. Fischer, C. Windpassinger, A. Lampe, and J. B. Huber, "Space-Time Transmission Using Tomlinson-Harashima Precoding," in *ITG FACHBERICHT*, 28–30 January, Berlin, Germany, 2002, pp. 139–148.

38. R. D. Wesel and J. M. Cioffi, "Achievable Rates for Tomlinson-Harashima Precoding," *IEEE Trans. Inf. Theory*, vol. 44, no. 2, pp. 824–831, 1998.

39. R. F. H. Fischer, *Precoding and Signal Shaping for Digital Transmission*. New York: Wiley-Interscience, 2002.

40. R. Laroia, "Coding for Intersymbol Interference Channels-Combined Coding and Precoding," *IEEE Trans. Inf. Theory*, vol. 42, no. 4, pp. 1053–1061, 1996.

41. R. D. Gitlin and S. B. Weinstein, "Fractionally-Spaced Equalization: An Improved Digital Transversal Equalizer," *Bell Syst. Tech. J.*, vol. 60, no. 2, pp. 275–296, 1981.

42. G. Ungerboeck, "Fractional Tap-Spacing Equalizer and Consequences for Clock Recovery in Data Modems," *IEEE Trans. Commun.*, vol. 24, no. 8, pp. 856–864, 1976.

43. G. Long, F. Ling, and J. G. Proakis, "Fractionally-Spaced Equalizers Based on Singular Value Decomposition," in *ICASSP-88, International Conference on Acoustics, Speech, and Signal Processing*, 11–14 April, New York, 1988, pp. 1514–1517.

44. F. Ouyang, "Combined Feedforward Filter for a Decision Feedback Equalizer," US Patent No. 7,031,414 B2, 2006.

45. P. Prandoni and M. Vetterli, *Signal Processing for Communications*. CRC Press, 2008.

46. DLMF, "NIST Digital Library of Mathematical Functions" [Online]. Available at http://dlmf.nist.gov/ [Accessed December 13, 2016].

47. P. S. R. Diniz, *Adaptive Filtering*. Springer, 2013.

48. R. W. Lucky, "Automatic Equalization for Digital Communication," *Bell Syst. Tech. J.*, vol. 44, no. 4, pp. 547–588, 1965.

49. S. S. Niu, L. Ljung, and A. Bjorck, "Decomposition Methods for Solving Least-Squares Parameter Estimation," *IEEE Trans. Signal Process.*, vol. 44, no. 11, pp. 2847–2852, 1996.

HOMEWORK

9.1 Derive (9.9) from (9.8) in Section 9.2.2.

9.2 Read [4] and provide the following details in derivations therein. Equation numbers below refer to those in [4]

a. Follow the system model description and complete the details in derivation up to Equation 24. Compare it with the discussions on MF in Chapter 4.

b. Complete the details from equation (57) to equations (58) to (61). Compare it with the LMS method discussed in Chapter 8.

9.3 In the context of Section 9.4.4, provide a detailed derivation of (9.56), based on (9.50).

9.4 Compare (9.53) in Section 9.4.4.1 with the principle of orthogonality for the MSE adaptive filter introduced in Chapter 8. Are they equivalent? If not, can one of them be derived from another?

9.5 With (9.73) (Section 9.4.4.4) and (9.193) (Section 9.9.3.3), show that when the channel $q(t)$ satisfies the Nyquist criteria given in Chapter 4 (i.e., when there is no ISI from the channel), SNR_{MMSE} is $\left(\sigma_x^2/N_0\right)$, as expected.

9.6 In the context of Section 9.5.3.3, show that:

a. The product of two monic functions (in z-transform) is also a monic function;

b. The inverse of a monic function is a monic function.

9.7 Show that (9.116) in Section 9.5.3.5 is consistent with equation 7.131 in [6, sec. 7.3]. Note the change in integration variable: $f = \omega/2\pi$.

9.8 In Figure 9.15 (Section 9.6.2), what should be the error signal for adaptation (see Section 9.5.5), in the presence of the modulo-2M operation?

9.9 Study MATLAB support of equalization (starting with "doc equalization" and choosing the function in the communications system toolbox). Try to reproduce the simulation results in Sections 9.4.4.4 and 9.5.9. Play with various parameter settings and report any interesting results. Especially, study the effect of decision errors on adaptation and feedback path. Optional: compare simulated performance (output SNR) of linear equalizers and DFEs with the theoretical values given in this chapter.

9.10 In Chapter 3, it is stated that if the symbols are transmitted faster than the Nyquist rate, then the ISI cannot be completely recovered. Provide a support to this statement based on our knowledge of the equalizers and ISI reduction.

ORTHOGONAL FREQUENCY DIVISION MULTIPLEXING (OFDM)

10.1 INTRODUCTION

Chapter 9 discusses a way to mitigate the dispersive channel effect, that is, the inter-symbol interference (ISI). This chapter introduces an alternative technique, the orthogonal frequency division multiplexing (OFDM) modulation.

Figure 10.1 shows the high-level physical layer architecture, discussed in Chapter 1. This chapter addresses the modulation and demodulation blocks at the receiver.

OFDM is a multicarrier modulation technique. Its signal is divided into subcarriers, each modulated and demodulated with the methods discussed in Chapters 3 and 4. This chapter focuses on how to combine the subcarriers at the transmitter and how to separate them at the receiver.

The concept of dividing signals into a collection of subcarriers separated by their center frequencies has a long history. It was initially used to divide up spectrum resource to multiple users. Therefore, such a technique is known as the frequency division multiplexing (FDM). In the late 1960s and early 1970s, it was realized that if the multiple subcarriers are used by the same transceiver (i.e., with close coordination among the subcarrier signals), interference between symbols separated by time and subcarriers can be zero, even with a dispersive channel [1, 2]. Furthermore, such signal can be generated and received with relatively low complexity, using the Fourier transform [3]. The current form of OFDM was proposed in 1980 [4]. For a more comprehensive historical review of OFDM, see [5].

The first commercial application of OFDM was in the asymmetric digital subscriber line (ADSL) technology [6]. ADSL is a subclass of the digital subscriber line (DSL) technologies, which provide data services through the telephone lines, using the spectrum beyond the 0–4 kHz bandwidth designated for voice services. OFDM is known as the discrete multitone (DMT) modulation in the ADSL framework [7, 8]. There was a fierce competition between the DMT technology and the traditional modulation and equalization techniques (known as single-carrier

Digital Communication for Practicing Engineers, First Edition. Feng Ouyang.
© 2020 by The Institute of Electrical and Electronics Engineers, Inc.
Published 2020 by John Wiley & Sons, Inc.

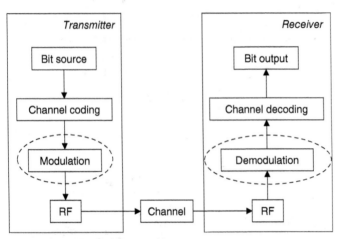

Figure 10.1   Physical layer architecture.

modulations) described in other chapters of this book [9]. Eventually, DMT was selected for ADSL, while another flavor of DSL used a single-carrier modulation. Later on, ADSL became the most prevalent DSL technology. Its successor very high bit–rate digital subscriber line (VDSL) also uses DMT modulation.

In the late 1990s, some companies started developing OFDM technologies for cellular applications [10], while the wireless local area network (WLAN) standard IEEE 802.11a, which uses OFDM, was ratified in 1999 [11]. OFDM later became the modulation technology of the long-term evolution (LTE), which is the prevailing fourth generation (4G) cellular standard [12]. The main competitor to LTE for 4G cellular technology was the Worldwide Interoperability for Microwave Access (WiMAX) standard, which also uses OFDM. Furthermore, OFDM is the modulation chosen by the European digital TV broadcasting technology (standards DVB-T/DVB-T2 for terrestrial transmission and DVB-H/DVB-NGH for handheld devices) [13, 14]. (The digital TV standard in the United States is ATSC, which uses single career modulation.) The upcoming cellular standard 5G also uses OFDM modulation (Chapter 12).

In the following, we introduce the formulation for OFDM in Section 10.2. Section 10.3 discusses time domain equalization, which is an addition to the basic OFDM receiver to deal with excessive channel dispersion. Section 10.4 compares OFDM with the traditional single-carrier modulation in performance, enhancement, and complexity. Practical issues with OFDM are discussed in Section 10.5 (channel estimation and synchronization). Section 10.6 presents two OFDM implementation challenges, the peak-to-average ratio and the sidelobes. Improvement alternatives are considered in that section. Section 10.7 extends OFDM to multiuser cases and discusses the Orthogonal Frequency Division Multiple Access (OFDMA) scheme. OFDMA has the additional challenge of synchronization among multiple transmitters. Such challenge motivates a generalization of OFDM, the filter bank multicarrier (FBMC) modulation, which is the topic of Section 10.8. Finally, Section 10.9 provides a summary.

10.2 OFDM FORMULATION

10.2.1 Channel Dispersion Revisited

Chapter 9 points out that a dispersive channel causes inter-symbol interference (ISI). Namely, the signal contribution from one symbol overlaps that from the next one, confusing the receiver. An equalizer can be used to address the ISI problem.

An alternative way to deal with ISI is adding guard time between the symbols. With guard time, the overlap between the symbols can be removed, even with a dispersive channel. Figure 10.2 shows this concept.

In Panel A of Figure 10.2, the top picture shows the traditional transmitted signal containing two symbols, each with a period of T_{Sym}. The middle picture shows the channel response in the time domain, the delay spread being T_H. The bottom picture shows the received signal. Because of the channel dispersion, the signal contribution from the first symbol overlaps with that from the second symbol, causing ISI. In Panel B, a guard time of T_G is introduced in between the two consecutive transmitted symbols. As shown in the bottom picture, because of the guard time, the channel delay spread does not cause ISI as long as

$$T_G > T_H. \tag{10.1}$$

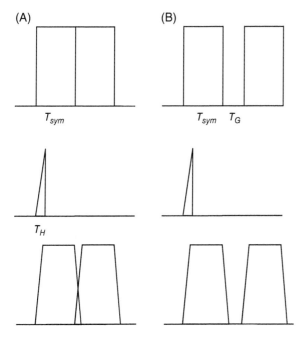

(A)   (B)

T_{sym} T_{sym} T_G

T_H

Figure 10.2 Guard time. A: consecutive symbols causing ISI. B: guard time between the symbols avoids ISI.

The problem with such a scheme is that the guard time lowers the symbol rate, which is proportional to the data rate. To reduce such overhead effect, we should choose

$$T_{Sym} \gg T_G > T_H. \tag{10.2}$$

Therefore, given T_H from the channel, T_{Sym} can be very large. From the Nyquist theorem discussed in Chapter 3, a large T_{Sym} means a small signal bandwidth with the typical modulation design. If such a bandwidth is smaller than the allocated one, we waste spectrum resources and get suboptimal performance.

To fully use the bandwidth, we can transmit multiple streams of signals in parallel, each up-converted to different center frequencies (see Chapter 3 for the concept of up-conversion). These signal streams are referred to as subcarriers. With multiple subcarriers, a higher total data rate can be achieved by better utilization of the allocated bandwidth.

However, as discussed in Chapter 3, modulated subcarriers necessarily have excess bandwidths. Guard bands between the subcarriers are thus needed to avoid mutual interference among them (known as inter-carrier interference, or ICI). As the guard time, Guard bands are also an overhead, reducing the data rate.

The OFDM modulation solves such a problem. With OFDM, the subcarriers are "orthogonal" and thus can be placed next to each other without mutual interference.

Figure 10.3 shows the spectra of the various schemes discussed above. The vertical dashed lines show allocated spectrum range. Panel A shows the traditional (single carrier) signal. The signal, including its excess bandwidth, occupies the entire allocated bandwidth. Panel B shows the case with very a large T_{Sym}, resulting in a narrow signal spectrum. Panel C shows the multicarrier scheme, where the subcarriers are

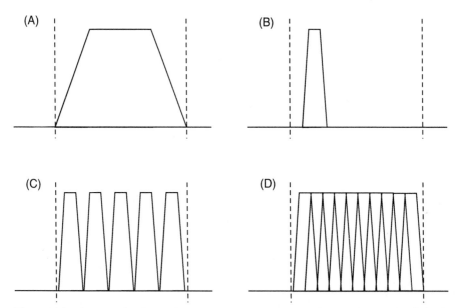

Figure 10.3 Spectra of various schemes. A: Full-band single carrier. B: Narrow-band single carrier. C: multiple subcarriers spaced to avoid mutual interference. D: OFDM, where the subcarriers are more densely placed.

spaced with guard bands to avoid ICI. Panel D shows the OFDM scheme, where the excess bandwidth of a subcarrier overlap with the next subcarrier, increasing the number of subcarriers allowed. The rest of Section 10.2 details how OFDM avoids ICI.

The idea of using a large T_{Sym} in a dispersive channel can also be understood in the frequency domain. As discussed in Chapter 6, a dispersive channel has an uneven frequency response. However, a narrow-band signal experiences a flat channel (i.e., no channel dispersion) as long as its bandwidth is less than the channel coherent bandwidth. Therefore, the ISI problem can be alleviated by dividing the transmitted signal into narrow-band subcarriers. OFDM is a scheme for removing the ICI among these subcarriers by introducing a small overhead in the time domain (equivalent to the guard time).

For the discussions in this chapter, we stay in the time domain.

10.2.2 Discrete Fourier Transform (DFT)

Before presenting OFDM formulation, let us first get familiar with the discrete Fourier transform.

10.2.2.1 Discrete Fourier Transform and Its Inverse

For a block of N samples $\{a_n, n = 0, 1, \dots, N-1\}$, define the discrete Fourier transform (DFT) as

$$\tilde{a}_k \overset{\text{def}}{=} \sum_{n=0}^{N-1} a_n \exp\left(-j\frac{2\pi kn}{N}\right), k = -N/2, -N/2+1, \dots, N/2-1. \tag{10.3}$$

Here, $j \overset{\text{def}}{=} \sqrt{-1}$ is the unit imaginary number. N is an even number. The inverse discrete Fourier transform (IDFT) is

$$\bar{a}_n \overset{\text{def}}{=} \frac{1}{N} \sum_{k=-N/2}^{N/2-1} \tilde{a}_k \exp\left(j\frac{2\pi kn}{N}\right), n = 0, 1, \dots, N-1 \tag{10.4}$$

With the following property

$$\sum_{n=0}^{N-1} \exp\left(j\frac{2\pi nk}{N}\right) = N \sum_{m=-\infty}^{\infty} \delta_{k,mN}, \tag{10.5}$$

it is easy to verify that in (10.3) and (10.4)

$$\bar{a}_n = a_n. \tag{10.6}$$

Namely, DFT and IDFT are indeed a pair of inverse transforms.

Note that in the definitions of DFT and IDFT, the factor $1/N$ in (10.4) can be distributed differently. For example, we can have a factor of $1/\sqrt{N}$ in both DFT (10.3) and IDFT (10.4), or have a factor of $1/N$ in (10.3) and a factor of 1 in (10.4). The choice made here is consistent with MATLAB functions fft and ifft for DFT and IDFT, respectively. Note that in MATLAB documentation, the summation over k in (10.4) is from 0 to $N-1$. In this case, $\tilde{a}_k$ for $k \geq N/2$ can be obtained by extending the range of k in (10.3), or by the periodicity

$$\tilde{a}_{k+N} = \tilde{a}_k. \tag{10.7}$$

10.2.2.2 Connection with the Fourier Transform

We now connect DFT with the continuous-time Fourier transform formulated in Chapter 3. For that purpose, consider a function $a(t)$, constructed from the samples $\{a_n\}$:

$$a(t) = \sum_{n=0}^{N-1} a_n \delta(t - nT_s), \tag{10.8}$$

where T_s is the sampling period. Its Fourier transform, according to Chapter 3, is

$$\tilde{a}(\omega) \overset{\text{def}}{=} \int a(t)\exp(-j\omega t)dt = \sum_{n=0}^{N-1} a_n \exp(-j\omega n T_s) \tag{10.9}$$

Comparing (10.9) with (10.3), we see that

$$\tilde{a}_k = \tilde{a}(k\omega_s)$$
$$\omega_s \overset{\text{def}}{=} \frac{2\pi}{NT_s}. \tag{10.10}$$

Therefore, $\{\tilde{a}_k\}$ is a sample sequence from the spectrum of $a(t)$. The sampling space ω_s is inversely proportional to NT_s, which is the time span of $a(t)$. The frequency range of interest $N\omega_s$ is inversely proportional to the time domain sampling period T_s.

10.2.2.3 DFT of Convolution

There are many interesting properties of DFT. The following one is the most relevant to our discussion on OFDM.

Consider a sequence in time domain $\{a_n\}$, where n extends from $-\infty$ to ∞. Further, assume that $\{a_n\}$ has a period of N:

$$a_{n+N} = a_n. \tag{10.11}$$

Now consider a short sequence $\{h_n, n = 0, 1, \ldots, N_H - 1\}$ where N_H is smaller than N. A segment of the convolution of $\{a_n\}$ and $\{h_n\}$ is $\{b_n, n = 0, 1, \ldots, N-1\}$:

$$b_n \overset{\text{def}}{=} \sum_{m=0}^{N_H-1} h_m a_{n-m}. \tag{10.12}$$

The DFT of $\{b_n\}$ is, according to (10.3):

$$
\begin{aligned}
\tilde{b}_k &= \sum_{n=0}^{N-1}\sum_{m=0}^{N_H-1} h_m a_{n-m} \exp\left(-j\frac{2\pi nk}{N}\right) \\
&= \sum_{m=0}^{N_H-1} h_m \exp\left(-j\frac{2\pi mk}{N}\right) \sum_{n=0}^{N-1} a_{n-m} \exp\left(-j\frac{2\pi(n-m)}{N}\right). \\
&\overset{l\overset{\text{def}}{=}n-m}{=} \sum_{m=0}^{N_H-1} h_m \exp\left(-j\frac{2\pi mk}{N}\right) \sum_{l=-m}^{N-m-1} a_l \exp\left(-j\frac{2\pi kl}{N}\right) \\
&= \tilde{h}_k \tilde{a}_k
\end{aligned}
\tag{10.13}
$$

Here, $\tilde{h}_k$ and $\tilde{a}_k$ are samples in the DTF of $\{h_n\}$ and $\{a_n\}$, respectively, according to (10.3). The last step in (10.13) considered the fact that both $\{a_n\}$ and $\{\exp(-j(2\pi n/N))\}$ have a period of N. Therefore, a summation from $-m$ to $N - m - 1$ is the same as the one from 0 to $N - 1$. Equation (10.13) shows a convolution in the time domain can be expressed as a product in the frequency domain through DFT. The same property exists for continuous-time Fourier transform as discussed in Chapter 3. However, (10.13) is valid only if $\{a_n\}$ has the same period as the DFT block. The convolution theorem for the continuous-time Fourier transform, on the other hand, is valid for general functions.[1]

10.2.3 OFDM in Discrete-Time: Transmit and Receive

In this section, we describe how to generate and demodulate OFDM signal in the baseband. Section 10.2.4 will examine the behavior of the OFDM signal in the case of dispersive channels.

Let N be the number of subcarriers.[2] For each OFDM symbol, each subcarrier is modulated by some data bits, in the same way as described in Chapter 3. The modulation result is a block of N subcarrier symbols $\{\tilde{a}_k, k = -N/2, -N/2+1, ..., N/2-1\}$. IDFT is performed to this block according to (10.4), obtaining the time domain samples $\{a_n, n = 0, 1, ..., N - 1\}$.

In the next step, we insert what is known as the cyclic prefix (CP), the reason for which will become clear in Section 10.2.4. CP is added in front of the time domain sample block by repeating the last samples in the block. Namely, we form a new

---

[1] Of course Fourier transform itself is only defined for integrable functions. We will not get into the details in this book.

[2] In OFDM jargon, subcarriers are also known as bins or tones.

sample block $\{a'_n, n = 0, 1, \ldots, N + N_c - 1\}$. N_c is a parameter specifying the length of the CP.

$$a'_n \overset{\text{def}}{=} \begin{cases} a_{N-N_c+n} & n = 0, 1, \ldots, N_c - 1 \\ a_{n-N_c} & n = N_c, N_c + 1, \ldots, N + N_c - 1 \end{cases}. \tag{10.14}$$

As will be shown in Section 10.2.4, CP enables an easy correction of channel dispersion.

Figure 10.4 shows the construction relationships expressed in (10.14). The last N_c samples in $\{a_n\}$ (shaded in the figure) is replicated as the beginning par to $\{a'_n\}$, forming the CP. Due to the periodicity of $\exp(j(2\pi kn/N))$, construction (10.14) can also be written as

$$d'_n = \frac{1}{N} \sum_{k=-N/2}^{N/2-1} \tilde{a}_k \exp\left(j\frac{2\pi(n-N_c)k}{N}\right), n = 1, 2, \ldots, N + N_c - 1. \tag{10.15}$$

Equation (10.15) is the same as (10.3), except for the range extension and an offset of N_c introduced to index n.

The sequence $\{a'_n\}$ represents signal samples in the time domain. Section 10.2.6 will show how the continuous-time signal is generated based on this sequence. For now, we will stay with discrete-time samples. In OFDM terms, $\{\tilde{a}_k\}$ is known as the *frequency domain data*, each k representing a subcarrier. The whole

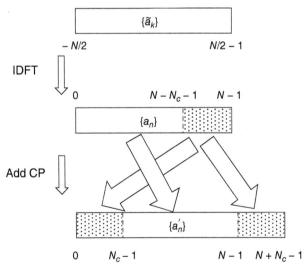

Figure 10.4   OFDM symbol construction process.

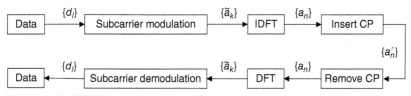

Figure 10.5   OFDM transmitter and receiver.

block of $\{\tilde{a}_k\}$ or $\{a'_n\}$ is an *OFDM symbol*. Each a'_n represents an *OFDM sample*. There are $N + N_c$ samples in an OFDM symbol. In the time domain, the OFDM samples play the same role as the symbols in single-carrier modulation discussed in Chapters 3 and 4.

At the receiver, suppose we can recover the time domain samples $\{a'_n\}$. The CP is discarded. $\{\tilde{a}_k\}$ can be recovered by performing DFT to the sequence $\{a'_n, n = N_c, \ldots, N + N_c - 1\}$, which is the same as $\{a_n\}$. Therefore, the scheme described above can send the frequency domain data $\{\tilde{a}_k\}$ from the transmitter to the receiver. $\{\tilde{a}_k\}$ at the receiver can be further demodulated in the same way as described in Chapter 4 to recover the underlying data. The process is shown in Figure 10.5. The upper row shows the transmitter, while the bottom one shows the receiver.

Note that in practice, the direct current (DC) component ($k = 0$) is often not loaded (i.e., kept at zero) [15, sec. 6.12]. This is based on analog circuit designs that have difficulty dealing with very low–frequency signals. Alternatively, it can be modulated at the transmitter, while the receiver may choose to ignore the DC subcarrier and rely on error correction coding to recover the data therein [16, sec. 5.3, 8.1.4]. In another approach, the whole spectrum is shifted by half the subcarrier bandwidth, so that no subcarrier is centered at DC [15, sec. 5.6].

10.2.4   OFDM and Channel Dispersion

Section 10.2.3 described OFDM transmission and reception without considering the channel. In this section, we add a dispersive channel into the process.

As shown in Chapter 9, for an optimal receiver, the received continuous-time signal passes a matched filter before being sampling at the symbol rate. After sampling, the signal can be expressed as

$$b_n = \sum_{k=0}^{N_H - 1} h_k a'_{n-k+N_c} + z_n. \tag{10.16}$$

Here, $\{b_n\}$ and $\{a'_n\}$ are the received and transmitted symbols, respectively. In our context, they would be OFDM samples. z_n is the added noise at the receiver. The offset N_c in the index of $\{a'_n\}$ in the summation is due to the removal of CP at the receiver as described in Section 10.2.3. $\{h_k\}$ represents the taps of the dispersive channel, which includes the transmitter pulse shaping, the propagation channel, and the matched filter.[3] N_H is the channel length.[4] It is assumed that

$$N_H \le N_c + 1. \tag{10.17}$$

With the condition (10.17) and considering the construction of $\{a'_n\}$ given by (10.14), (10.16) can be written as

$$b_n = \sum_{k=0}^{N_H-1} h_k a_{n-k} + z_n, \tag{10.18}$$

with the periodic range extension for $\{a_n\}$:

$$a_n \overset{\text{def}}{=} a_{n+N} \forall -N_c \le n < 0. \tag{10.19}$$

Let us ignore the noise term, which will be studied in Section 10.2.5. Equation (10.18) is thus the same as equation (10.12) in Section 10.2.2.3. However, unlike the case therein, for an OFDM symbol, $\{a'_n\}$ does not extend infinitely. Fortunately, given the limit on channel length (10.17) and the range extension (10.19), (10.16) and (10.18) are valid. Therefore, (10.13) is valid:

$$\tilde{b}_k = \tilde{h}_k \tilde{a}_k. \tag{10.20}$$

Here, $\{\tilde{b}_k\}$, $\{\tilde{a}_k\}$, and $\{\tilde{h}_k\}$ are DFT of $\{b_n\}$, $\{a_n\}$, and $\{h_n\}$, respectively. $\{\tilde{h}_k\}$ is assumed to be known to the receiver. Therefore, the frequency domain data $\{\tilde{a}_k\}$ can be recovered by a simple gain compensation

$$\tilde{a}_k = \frac{\tilde{b}_k}{\tilde{h}_k}. \tag{10.21}$$

Above discussion shows that as long as the channel length does not exceed that of the CP, the orthogonality among subcarriers is maintained even with a dispersive channel. Namely, $\tilde{b}_k$ depends only on $\tilde{a}_k$ and not on other transmitted symbols. Therefore, the channel effect can be compensated by subcarrier-specific gain values at the receiver. The processing in (10.21) is also known as "frequency domain equalization," where $\{\tilde{h}_k^{-1}\}$ can be considered as one-tap equalizers for the subcarriers.

---

[3] In Chapter 9, we also included a conceptual noise whitening filter. Here, however, we do not care if the noise is white. Therefore, we do not worry about noise whitening.

[4] Some authors ignore the first channel tap when defining N_H. In that case, the summation in (10.16) should be from 0 to N_H.

This result can also be understood in the frequency domain. As pointed out in Chapter 6, in the frequency domain, the received signal can be written as (ignoring noise)

$$\tilde{b}(\omega) = \tilde{h}(\omega)\tilde{a}(\omega).$$
(10.22)

Here, $\tilde{b}(\omega)$ and $\tilde{a}(\omega)$ are the spectra of the received and transmitted signals. $\tilde{h}(\omega)$ is the channel frequency response. This is also a multiplicative relationship as in (10.20). Therefore, one possible way of removing ISI caused be dispersive channels is processing received signal in the frequency domain:

$$\tilde{a}(\omega) = \frac{\tilde{b}(\omega)}{\tilde{h}(\omega)}.$$
(10.23)

However, this processing requires taking the entire signal sequences in the time domain and perform Fourier transform. In practice, signal processing can only be processed in blocks. Therefore, there still is ISI at the boundaries of the blocks. The beauty of OFDM is by introducing CP, there is no interference among the consecutive blocks (i.e., OFDM symbols), as long as the channel is shorter than the CP period.

Therefore, CP is effectively a guard period, which adds length to an OFDM symbol. The ISI from the previous OFDM symbol is absorbed in the CP, which is then discarded for the processing of the current OFDM symbol.

By using the cyclic extension instead of, for example, a silent period, subcarrier orthogonality is preserved even when the starting time of a symbol is estimated with a small offset. When $N_H + 1$ is strictly less than N_c, there is some leeway in the alignment between the transmitted and received signals (10.16) (see homework 1). Therefore, the synchronization accuracy requirement can be relaxed (Chapter 7).

In all discussions in this chapter, we assume that the channel does not change during an OFDM symbol. Such an assumption is supported by our parameter choices of the modulation (see Section 10.2.10). If the channel does change within an OFDM symbol, ICI will occur. Of course, if the channel (as a function of time) is known, such ICI can be predicted and corrected [17, sec. 6.5.2]. However, it is often impractical to track channel changes within an OFDM symbol.

10.2.5   OFDM Noise Characteristics

Now let us consider the noise. When discussing equalizers in Chapter 9, we paid much attention to noise correlation. A correlated noise implies further opportunities to improve performance. Therefore, let us examine the correlation between noises in the various subcarriers.

After DFT of the received samples, the noise for the kth subcarrier is

$$\tilde{z}_k = \sum_{n=0}^{N-1} z_n \exp\left(-j\frac{2\pi kn}{N}\right),$$
(10.24)

where $\{z_n\}$ is the noise in time domain samples. From (10.24),

$$E\left(\tilde{z}_k \tilde{z}_l^*\right) = \sum_{n=0}^{N-1}\sum_{m=0}^{N-1} E\left(z_n z_m^*\right)\exp\left(-j\frac{2\pi(kn-lm)}{N}\right)$$

$$= \sum_{i=n-m}^{N-1}\sum_{n=0}^{n}\sum_{i=n-N+1}^{n} E\left(z_n z_{n-i}^*\right)\exp\left(-j\frac{2\pi[n(k-l)+il]}{N}\right) \tag{10.25}$$

Assuming that the time domain noise $\{z_n\}$ is wide-sense stationary (WSS) (see Chapter 4), we have

$$E\left(z_n z_{n-i}^*\right) = r_i, \tag{10.26}$$

which is independent of n. Equation (10.25) can thus be written as

$$E\left(\tilde{z}_k \tilde{z}_l^*\right) = \sum_{n=0}^{N-1}\exp\left(-j\frac{2\pi n(k-l)}{N}\right)\sum_{i=n-N+1}^{n} r_i\exp\left(-j\frac{2\pi il}{N}\right). \tag{10.27}$$

The summation over i depends on n only in its summing limits. We hope to show that this summation is in fact independent of n. Therefore, the summation over n yields a delta function.

Assume the span of r_i is small, namely,

$$r_i = 0 \forall |i| > L_r$$
$$L_r \ll N \tag{10.28}$$

With (10.28), the summation over i is

$$\sum_{i=n-N+1}^{n} r_i\exp\left(-j\frac{2\pi il}{N}\right) = \sum_{i=-L_r+1}^{L_r-1} r_i\exp\left(-j\frac{2\pi il}{N}\right)$$

$$- \sum_{i=-L_r+1}^{n-N} r_i\exp\left(-j\frac{2\pi il}{N}\right) - \sum_{i=n+1}^{L_r-1} r_i\exp\left(-j\frac{2\pi il}{N}\right) \tag{10.29}$$

The two terms in the second line of (10.29) are the correction terms in the cases when the new summation range $[-L_r+1, L_r-1]$ exceeds the original one $[n-N+1, n]$. These terms are not zero only when

$$N > n \geq N-L_r+1 \text{ or } 0 \leq n < L_r-1. \tag{10.30}$$

Furthermore, recognize that (10.26) holds only when both n and $n-i$ are between 0 and $N-1$, while $|i| \leq L_r$. In other words, (10.26) does not hold for the following n values:

$$N > n \geq N-L_r \text{ or } 0 \leq n < L_r. \tag{10.31}$$

Namely, at the edges of the block (10.30) and (10.31), there some complications. When $L_r \ll N$, only a small portion of the n values satisfy (10.30) and (10.31). Therefore, we can approximately disregard the correction terms in (10.29) and consider (10.26) to be true.

With such approximations, (10.27) becomes

$$E\left(\tilde{z}_k \tilde{z}_l^*\right) = \sum_{n=0}^{N-1} \exp\left(-j\frac{2\pi n(k-l)}{N}\right) \sum_{i=-l_r+1}^{L_r-1} r_i \exp\left(-j\frac{2\pi i l}{N}\right) \qquad (10.32)$$

Now that the second sum is independent of n, the two summations can be evaluated independently:

$$E\left(\tilde{z}_k \tilde{z}_l^*\right) = N\tilde{r}_l \delta_{k,l}. \qquad (10.33)$$

$\{\tilde{r}_l\}$ is the DFT of $\{r_i\}$. Equation (10.33) shows that noise from different subcarriers is uncorrelated.

Therefore, we conclude that frequency domain noise $\{\tilde{z}_k\}$ are mutually independent, given that the time domain noise $\{z_n\}$ is WSS with correlation span much less than N. Such conditions are commonly met in communication systems.[5] The limited time range of correlation (10.28) also implies that the noise correlation across multiple OFDM symbols can be ignored.

10.2.6 OFDM in Continuous-Time

In this section, we study in detail how OFDM symbols are constructed in continuous-time while staying in the baseband. We first present two approaches in Section 10.2.6.1 and show they are mutually equivalent within a single OFDM symbol. To study the waveform across multiple symbols and prepare for the study of OFDM properties in later sections, we follow a multistep approach. Starting from an infinite repeating of a symbol (Section 10.2.6.2), we study the effect of restrictions in the frequency domain (Section 10.2.6.3) and the time domain (Section 10.2.6.4). Section 10.2.6.5 summarizes the results and shows that the two construction approaches are approximately the same for multisymbol waveforms.

10.2.6.1 *Pulse Shaping and Tonal Formulations*

In Sections 10.2.3 to 10.2.5, we show that the OFDM modulation has the following properties. The subcarrier data are orthogonal; they can be demodulated independently from each other. Furthermore, such property holds even with dispersive channels. Those analyses are based on discrete-time OFDM samples. In other words, we assume that the OFDM samples $\{a'_n\}$ are transmitted and received (with a dispersive channel and additive noise). In this section, we consider how continues-time OFDM signal is constructed.

Time domain operation of OFDM is very similar to that of the single-carrier modulation described in Chapters 3 and 4, where the OFDM samples play the role

---

[5] In some cases, the correlation span limitation (10.28) is not true. For example, for powerline communications, some noise components can be modeled as single tones, which imply long-range correlation in the time domain [18]. In these cases, further noise reduction may be possible.

of the symbols in the single-carrier modulation [19, sec. 3.5], [20]. Following the single-carrier modulation formulation, the baseband time continuous signal is

$$a(t) = \sum_{m} \sum_{n=0}^{N+N_c-1} a_n'^{(m)} s(t-nT_s-mT_O-N_cT_s) \tag{10.34}$$

Here, index m counts the OFDM symbols, while n counts the OFDM samples within an OFDM symbol. T_s is the sample period. T_O is the time length of an OFDM symbol and has the value $(N+N_c)T_s$. $s(t)$ is the baseband pulse shaping function. $\{a_n'^{(m)}, n=0,1,...,N+N_c-1\}$ is the OFDM samples of the mth OFDM symbol.

However, the continuous-time OFDM signal is typically formulated in another way (e.g., see [15, sec. 6.12] for LTE and [21] for WLAN). The formulation in [15, sec. 6.12] is (with simplification)

$$x'(t) = \frac{1}{N} \sum_{k=-N/2}^{N/2-1} \exp[j2\pi k\Delta f(t-N_cT_s)] \forall t \in [0, (N_c+N)T_s). \tag{10.35}$$

Here, Δf is the subcarrier spacing, which is the inverse of NT_s:

$$\Delta f \overset{\text{def}}{=} \frac{1}{NT_s}. \tag{10.36}$$

Equation (10.35) gives the time domain signal $x'(t)$ within one OFDM symbol time. Signals from successive OFDM symbols are transmitted consecutively. In such formulation, each frequency domain data $\tilde{a}_k$ modulates a tone at frequency $k\Delta f$. This picture is consistent with the multichannel frequency division multiplexing (FDM) view, mentioned in Section 10.1 and illustrated in Panel C of Figure 10.3.

Recalling (10.3) and (10.14), (10.35) can be written as

$$
\begin{aligned}
x'(t) &= \frac{1}{N} \sum_{k=-N/2}^{N/2-1} \left[\sum_{n=0}^{N-1} a'_{n+N_c} \exp\left(-j\frac{2\pi kn}{N}\right) \right] \exp[j2\pi k\Delta f(t-N_cT_s)] \\
&\underset{l \overset{\text{def}}{=} n+N_c}{=} \frac{1}{N} \sum_{l=N_c}^{N+N_c-1} \sum_{k=-N/2}^{N/2-1} d'_l \exp\left(-j\frac{2\pi k(l-N_c)}{N}\right) \exp[j2\pi k\Delta f(t-N_cT_s)]
\end{aligned}
\tag{10.37}
$$

Note that the quantity to be summed has a period of N regarding l. Therefore, the summation limits can be changed to 0 to $N-1$.

$$
\begin{aligned}
x'(t) &= \frac{1}{N} \sum_{l=0}^{N-1} \sum_{k=-N/2}^{N/2-1} d'_l \exp\left(-j\frac{2\pi k(l-N_c)}{N}\right) \exp[j2\pi k\Delta f(t-N_cT_s)] \\
&= \sum_{l=0}^{N-1} d'_l \left\{ \sum_{k=-N/2}^{N/2-1} \exp\left[\frac{j2\pi k}{NT_s}(\Delta f NT_s t - lT_s)\right] \frac{1}{N} \exp\left[\frac{j2\pi k}{N} N_c(NT_s\Delta f - 1)\right] \right\}
\end{aligned}
\tag{10.38}
$$

Further, consider (10.36). Above result can be written as

$$
\begin{aligned}
x'(t) &= \sum_{l=0}^{N-1} a'_l r(t - lT_s - N_c T_s) \\
r(t) &\overset{\text{def}}{=} \frac{1}{N} \sum_{k=-N/2}^{N/2-1} \exp\left[\frac{j2\pi k}{NT_s}(t + N_c T_s)\right]
\end{aligned}
\tag{10.39}
$$

Equation (10.39) is derived from (10.35). It has the same form as (10.34) for the first symbol (i.e., $m = 0$). Therefore, the two constructions (10.35) and (10.34) are equivalent within an OFDM symbol. However, as to be shown below, there are some differences between the two when we cross symbol boundaries. In the subsequent sections, the formulation (10.35) is built in several steps.

10.2.6.2 Spectrum of OFDM Symbol with Periodic Extension

In this section, we show that an OFDM symbol with periodic extension in the time domain has a tonal spectrum.

Consider a periodic extension of the time domain samples $\{a_n, n = \ldots, -1, 0, 1, \ldots\}$ based on IDFT (10.4):

$$
a_n = \frac{1}{N} \sum_{k=-N/2}^{N/2} \tilde{a}_k \exp\left(j\frac{2\pi kn}{N}\right), n = \ldots, -1, 0, 1, \ldots.
\tag{10.40}
$$

Consider the time domain signal with a sequence of impulse functions modulated by $\{a_n\}$:

$$
u(t) \overset{\text{def}}{=} \sum_{n=-\infty}^{\infty} a_n \delta(t - nT_s - N_c T_s),
\tag{10.41}
$$

where T_s is the sample period. The term $N_c T_s$ in the delta function accounts for the time shift between $\{a_n\}$ and $\{a'_n\}$ due to CP insertion as formulated by (10.14). The spectrum of $u(t)$ is, from the Fourier transform formulated in Chapter 3 and restated in (10.9),

$$
\tilde{u}(\omega) = \int u(t)\exp(-j\omega t)dt = \exp(-j\omega N_c T_s) \sum_{n=-\infty}^{\infty} a_n \exp(-j\omega n T_s).
\tag{10.42}
$$

Recall that a_n has a period of N. By expressing n as $mN + l$ where l is between 0 and $N - 1$, the summation in (10.42) can be written as

$$
\sum_{n=-\infty}^{\infty} a_n \exp(-j\omega n T_s) = \sum_{m=-\infty}^{\infty} \exp(-j\omega mN T_s) \sum_{l=0}^{N-1} a_l \exp(-j\omega l T_s).
\tag{10.43}
$$

The two summations can be performed separately. From [22, sec. 1.17]:

$$
\frac{1}{2\pi} \sum_{n=-\infty}^{\infty} e^{jnx} = \sum_{k=-\infty}^{\infty} \delta(x - 2k\pi), \forall x \in (-\infty, \infty).
\tag{10.44}
$$

Apply this to the summation over m in (10.43):

$$\sum_{m=-\infty}^{\infty} \exp(-j\omega m N T_s) = 2\pi \sum_{k=-\infty}^{\infty} \delta(\omega N T_s - 2k\pi) = \frac{2\pi}{N T_s} \sum_{k=-\infty}^{\infty} \delta\left(\omega - \frac{2\pi k}{N T_s}\right). \quad (10.45)$$

Equation (10.42) can thus be written as

$$\tilde{u}(\omega) = \frac{2\pi}{N T_s} \exp(-j\omega N_c T_s) \sum_{k=-\infty}^{\infty} \delta\left(\omega - \frac{2\pi k}{N T_s}\right) \sum_{l=0}^{N-1} a_l \exp\left(-j2\pi\frac{kl}{N}\right)$$

$$= \frac{2\pi}{N T_s} \exp(-j\omega N_c T_s) \sum_{k=-\infty}^{\infty} \tilde{a}'_k \delta\left(\omega - \frac{2\pi k}{N T_s}\right) \qquad . \quad (10.46)$$

$$\tilde{a}'_k \overset{\text{def}}{=} \sum_{l=0}^{N-1} a_l \exp\left(-j2\pi\frac{kl}{N}\right)$$

Since $\{a_l, l = 0, 1, \ldots, N-1\}$ is the IDFT of subcarrier symbols $\{\tilde{a}_k\}$ according to (10.40), $\{\tilde{a}'_k\}$ is the same as $\{\tilde{a}_k\}$ based on the DFT (10.3) with periodic extension:

$$\tilde{a}'_k = \tilde{a}_{k-N\lfloor k/N + 1/2\rfloor}, k = \ldots, -1, 0, 1, \ldots. \quad (10.47)$$

$\lfloor x \rfloor$ means the largest integer that is smaller than x. Equation (10.46) shows that each subcarrier data $\tilde{a}'_k$ drives one tone at frequency

$$\omega_k \overset{\text{def}}{=} \frac{2\pi k}{N T_s} = 2\pi k \Delta f, \quad (10.48)$$

where Δf is defined in (10.36). Therefore, we show that when a time domain signal is constructed by (10.41), its spectrum contains a sequence of tones that are driven by $\{\tilde{a}_k\}$, as suggested by the OFDM construction (10.35). In other words, the FT of the continuous-time signal $u(t)$ is connected with the DFT of the time domain samples $\{a_n\}$.

10.2.6.3   Limiting the Spectrum Span

We recognize that the spectrum $\tilde{u}(\omega)$ given by (10.46) contains a set of tones, whose frequency range extends from $-\infty$ to ∞. As discussed in Chapter 3, we want to limit the transmitted frequency ω within the Nyquist frequency range for the sampling period T_s, or $[-\pi/T_s, \pi/T_s)$. In other words, we want to limit the range of k within $-N/2$ to $N/2 - 1$. This can be achieved by applying a filter $s(t)$, whose frequency response is

$$\tilde{s}(\omega) = \begin{cases} 1 & \omega \in [-\pi/T_s, \pi/T_s) \\ 0 & \text{otherwise} \end{cases}. \quad (10.49)$$

Such filtering in effects limits the range of k for the summation in (10.46) to $\{-N/2, -N/2+1, \ldots, N/2-1\}$. Within this range of k, $\{\tilde{a}'_k\}$ and $\{\tilde{a}_k\}$ are the same. Therefore, (10.46) can be written as

$$\tilde{x}(\omega) \overset{\text{def}}{=} \tilde{u}(\omega)\tilde{s}(\omega) = \frac{2\pi}{N T_s} \exp(-j\omega N_c T_s) \sum_{k=N/2-1}^{N/2} \tilde{a}_k \delta(\omega - 2\pi k \Delta f), \quad (10.50)$$

where Δf is defined in (10.48). Taking an inverse Fourier transform of $\tilde{x}(\omega)$, we get

$$x(t) = \frac{1}{2\pi} \int \tilde{x}(\omega) \exp(j\omega t) dt = \frac{1}{NT_s} \sum_{k=-N/2}^{N/2} \tilde{a}_k \exp[j2\pi k \Delta f(t-N_c T_s)]. \qquad (10.51)$$

As shown in Chapter 3, $s(t)$ is a sinc function

$$s(t) = \frac{1}{T_s} \text{sinc}\left(\frac{t}{T_s}\right)$$

$$\text{sinc}(v) \overset{\text{def}}{=} \frac{\sin(\pi v)}{\pi v} \qquad (10.52)$$

The transmit signal $x(t)$ is $u(t)$ in (10.41) convoluted with $s(t)$:

$$x(t) = \int u(t-\tau)s(\tau)d\tau = \sum_n a_n s(t - nT_s - N_c T_s). \qquad (10.53)$$

Equation (10.53) is the same as (10.34) in form. In other words, the tonal construction of $x(t)$ given by (10.51) can be expressed in the form of pulse shaping (10.34), where the pulse shaping function is a sinc function.

10.2.6.4   Time Domain Window Function

So far we have shown that a pulse shaping construction of OFDM signal (10.53) yields the same result as the tonal construction (10.51). However, (10.51) is still different from the "standard" OFDM symbol construction given in (10.35). $u(t)$ in (10.41), and thus $x(t)$ in (10.51), extends to the entire time range, based on the periodic extension of $\{a_n\}$ (10.40). On the other hand, the signal for one OFDM symbol in (10.35) is limited within $t \in [0, T_s(N+N_s))$. Therefore, we need to multiply $x(t)$ with a window function $w(t)$ in the time domain to match the $x'(t)$ in (10.35).

$$x'(t) = x(t)w(t)$$

$$w(t) \overset{\text{def}}{=} \begin{cases} 1 & t \in [0, T_s(N+N_s)) \\ 0 & \text{otherwise} \end{cases}. \qquad (10.54)$$

Figure 10.6 summarizes the two construction methods based on (10.35) (panel A) and (10.54) (panel B). Constant factors in $x'(t)$ are ignored in this figure. The "PE" block in panel B denotes periodic extension (10.40).

(A)

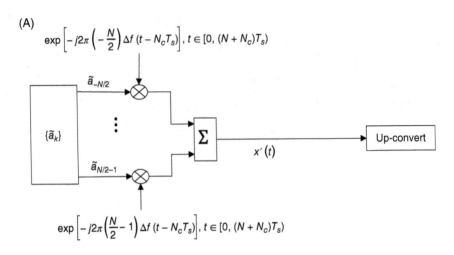

(B)

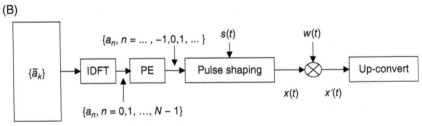

Figure 10.6   OFDM signal construction process.

10.2.6.5   Summary

As shown in the previous sections, the time domain signal of an OFDM sample can be generated in two ways, using either (10.54) or (10.35). These two methods yield the same results: a collection of tones within the duration of the OFDM symbol. The tones are driven by the frequency domain symbols $\{\tilde{a}_k\}$.

Two new parameters, the subcarrier spacing Δf and the sample period T_s, are introduced in the conversion from discrete-time OFDM symbols to the continuous-time ones. They are linked by (10.36).

Two points of difference should be noted between the construction (10.35) and the single-carrier modulation (10.34). The first point is the time domain windowing (10.54). The second point is that in (10.35), the edge samples of an OFDM symbol experience contributions (through convolution of the pulse shaping filter) from its periodic repetitions in (10.53), instead of from its neighboring symbols as in the case of single-carrier modulations (10.34).

Despite such difference, the receiver can still use $s^*(-t)$ as a matched filter before sampling [20]. Therefore, most of the performance analyses can be based on the discreet-time model presented in Sections 10.2.3–10.2.5.

10.2.7 Spectrum of OFDM Modulation

Let us look at the spectrum of an OFDM modulation, based on the constructions discussed in Section 10.2.6. As shown in Section 10.2.6, $x(t)$ given by (10.51) and (10.53) has a spectrum that contains N discrete tones within the transmission bandwidth, separated by Δf. The final transmit signal $x'(t)$ given by (10.54) is the product of $x(t)$ and the window function $w(t)$. Therefore, the final spectrum is a convolution of the multiple tones and the spectrum of $w(t)$, which is a sinc function:

$$\tilde{w}(\omega) = [T_s(N+N_c)]\text{sinc}\left(\frac{\omega T_s(N+N_c)}{2\pi}\right). \tag{10.55}$$

Therefore, an OFDM modulation always has sidelobes governed by (10.55), even when the pulse shaping function is the perfect sinc function (10.49) [23, sec. 5.1.1.1].

10.2.8 OFDM Example

Figure 10.7 shows an example of OFDM modulation. For this particular case, $N = 256, N_c = 8$. The time domain signal is over-sampled by a factor of 16 (i.e., 16 samples for every OFDM sample). Three time domain signals are generated: x_1 by (10.34), x_2 by (10.54), and x_3 by (10.35). Simulation results show that x_2 and x_3 are the same, as expected. x_1 and x_2 have some differences, especially at the edges of the symbols. This is also to be expected.

Plot A in Figure 10.7 shows the time domain signal level of x_1 over an OFDM symbol period, which extends from 0 to 1 on the x axis. To reduce random fluctuation, the data are filtered by a low pass filter:

$$y_n = 0.1x_n + 0.99y_{n-1} \tag{10.56}$$

where $\{x_n\}$ and $\{y_n\}$ are input and output data of the filter, respectively.

In Plot A, the center OFDM symbol is set to zero. Therefore, the signal in this symbol period represents ISI from the previous and following symbols. The signal level beyond this period indicates the "normal" signal level. The plot shows that the ISI level can be about 30 dB below the normal signal level at the symbol edges. Such ISI level may be too high for some modulation schemes. Therefore, signal construction (10.34), where the pulse shaping filtering run across OFDM symbols, has a disadvantage of introducing residual ISI, despite the CP. Such ISI is eliminated in the "time-windowed" construction of (10.34) and (10.54).

Plot B in Figure 10.7 shows the power spectrum density (PSD) of the OFDM signal x_2, constructed by (10.54). The PSD was generated by MATLAB function pwelch, with a moving Fourier transform window of size 2^{14} samples. As mentioned above, x_2 is the same as x_3. The normalized frequency in the plot is

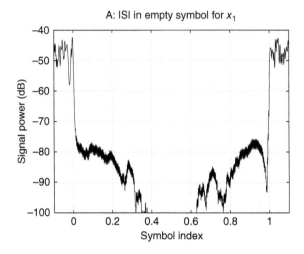

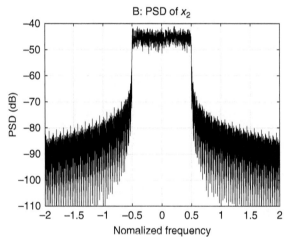

Figure 10.7   OFDM examples in time domain and frequency domain. A: Signal power in time domain of x_1 (10.34), with the center symbol set to zero. B: Signal power in frequency domain of x_2 (10.54).

$$f_n \overset{\text{def}}{=} \frac{\omega T_s}{2\pi}. \tag{10.57}$$

The main part of the spectrum is in the range $|f_n| < 0.5$. This is equivalent to

$$-\frac{\pi}{T_s} \leq \omega \leq \frac{\pi}{T_s} \tag{10.58}$$

Such a range is consistent with the tone range in (10.35), considering (10.36). The sidelobes are caused by the windowing function (10.54). The sidelobes will be discussed more in Section 10.6.2.

Reference [19, ch. 6] provides more information on the simulation of OFDM systems.

10.2.9   Capacity of OFDM Modulation

We have seen OFDM provides an effective and straightforward way to deal with a dispersive channel. In this section, we will see that OFDM achieves such task without significant loss of performance. To show such optimality, we compare the capacity of the channel and that with the OFDM modulation.

Consider an OFDM modulation with N subcarriers, N_c CP samples, and a sampling period of T_s. As discussed above, the system can be viewed as having N independent subcarriers. Let SNR_k be the SNR for subcarrier k. According to the Shannon theorem in Chapter 2, we can transmit up to M_k bits per channel use (i.e., per OFDM symbol):

$$M_k = \log_2(1 + SNR_k). \tag{10.59}$$

The OFDM symbol rate is

$$R_s = \frac{1}{(N + N_c)T_s}. \tag{10.60}$$

Therefore, the maximum total data rate is

$$R_{OFDM} = R_s \sum_{k=-N/2}^{N/2} M_k = \frac{1}{(N + N_c)T_s} \sum_{k=-N/2}^{N/2} (1 + SNR_k)$$

$$= \frac{N}{N + N_c} \Delta f \sum_{k=-N/2}^{N/2} (1 + SNR_k) \tag{10.61}$$

where Δf is given by (10.36).

On the other hand, we can compute the total capacity of the channel. From (10.49), the OFDM frequency range (in linear frequency) is $[-1/2T_s, 1/2T_s)$, or $[-\Delta fN/2, \Delta fN/2)$. From the Shannon theorem (Chapter 2), the total capacity is

$$C = \int_{-\Delta fN/2}^{\Delta fN/2} \log_2[1 + SNR(f)] df. \tag{10.62}$$

Here, $SNR(f)$ is the SNR at frequency f. To compare the two capacities R_{OFDM} and C, note that when $N \to \infty$ while the total bandwidth ΔfN is held constant, we have

$$\Delta f \sum_{k=-N/2}^{N/2} \log_2(1 + SNR_k) = \int_{-\Delta fN/2}^{\Delta fN/2} \log_2[1 + SNR(f)] df. \tag{10.63}$$

Therefore, under such limitations, we have

$$R_{OFDM} = \frac{N}{N + N_c} C. \tag{10.64}$$

Namely, an OFDM system can achieve channel capacity, except for the overhead introduced by N_c. Note that such result is based on (10.59), which assumes there is a coding method to reach the Shannon bound in a subcarrier.

Therefore, if $N_c \ll N$, an OFDM modulation does not bring loss to channel capacity. In other words, OFDM is an optimal modulation method. Later in this chapter, we will discuss the advantages and disadvantages of the OFDM modulation, comparing to the traditional single-carrier modulation.

To understand the role of N_c in (10.64) intuitively, recall (Chapter 2) that Shannon theorem

$$C = W\log_2(1 + SNR) \tag{10.65}$$

gives the capacity of a channel with a bandwidth W. This relationship assumes that the symbols are transmitted at the Nyquist rate (Chapter 3) of W. In the case of OFDM, the symbol rate is $1/(N + N_c)T_s$, or $[N/(N + N_c)]\Delta f$, while the bandwidth for a subcarrier is Δf. Therefore, the symbol rate is below the Nyquist frequency, leading to a capacity loss. As stated above, such loss is negligible when $N_c \ll N$.

Although (10.63) is true only when $N \to \infty$, the value of N is not necessarily very large for (10.63) to serve as a good approximation. In fact, (10.63) holds as long as $SNR(f)$ is constant within the subcarrier $f \in [(k - 1/2)\Delta f, (k + 1/2)\Delta f]$. Such condition is met when the channel response in the subcarrier is flat, and the noise does not change dramatically with frequency (e.g., no narrow-band noises). As discussed in Chapter 6, the requirement that all subcarriers are flat implies that channel dispersion is much shorter than the OFDM symbol duration. Such a condition is usually met with a proper choice of OFDM parameters, as will be discussed in Section 10.2.10. Therefore, the conclusion (10.64) is valid in general.

10.2.10   Selection of Parameters

As a summary of the OFDM formulation, let us review the various parameters of an OFDM modulation. Table 10.1 shows the parameters commonly used to define and describe an OFDM modulation. As shown in the table, these parameters are not independent. In fact, of the six parameters listed here, only three can be independently selected. Here, the first three parameters, N, N_c, and T_s are shown as independent ones, and the rest can be derived from them. However, such a choice is not unique. For example, one can replace N or N_c with T_O, or set Δf or W instead of T_s.

TABLE 10.1   OFDM parameters.

Symbol	Description	Relationship with others
N	Number of subcarriers	
N_c	Number of CP samples	
T_s	Sample period	
T_O	OFDM symbol length	$T_O = (N + N_c)T_s$
Δf	Subcarrier spacing	$\Delta f = (NT_s)^{-1}$
W	Total bandwidth	$W = N\Delta f = T_s^{-1}$

The choice of modulation parameters depends on various considerations. The following are the primary considerations:

- The cyclic prefix (CP) length $N_c T_s$ should be larger than the channel delay spread T_H, as indicated by (10.1), to remove ISI between consecutive OFDM symbols.

- The total bandwidth W is usually limited by regulation (more details in Chapter 6).

- The OFDM symbol length T_O should be less than the channel coherence time (as defined in Chapter 6). In other words, the channel should not have significant change within an OFDM symbol.

There are some secondary concerns, as well.

- The CP period should be much less than the main part of an OFDM symbol to reduce the overhead caused by the CP as indicated in (10.64). Namely, $N_c \ll N$ as implied in (10.2). Inequalities in (10.2) also imply that the subcarrier spacing Δf is smaller than channel coherence frequency. Namely, the subcarriers experience flat channels.

- The complexity of OFDM is proportional to $W\log_2 N$. This relationship puts a limit to the choice of N.

- The subcarrier spacing Δf should be much larger than potential carrier frequency errors, caused either by hardware limitations or by the Doppler effect. See Section 10.5.2 for more discussions.

The above parameters specify the OFDM modulation structure. In addition, each subcarrier is modulated (possibly with channel coding) according to the traditional single-carrier modulation described in Chapter 3. Such subcarrier modulation and coding scheme are governed by another set of parameters, which can be different among the subcarriers.

10.3   TIME DOMAIN EQUALIZATION

Section 10.2 shows that with OFDM modulation, channel dispersion effect (delay spread) can be addressed in each subcarrier with a one-tap equalizer. To achieve such simplification, (10.1) must hold. Namely, the guard time (populated by the CP) must be longer than the channel delay spread. On the other hand, the CP introduces overhead and reduces the data rate, as shown in (10.64). Therefore, in some systems, it is necessary to choose a CP length smaller than the channel delay spread, resulting in residual ISI between OFDM symbols. Such ISI can be further reduced or removed by an additional time domain equalizer (TDE), which is the subject of this section.

The simplest form of TDE is shown in Figure 10.8 [24]. This TDE is optimized under the minimum mean-squared error (MMSE) criterion discussed in detail in Chapter 9.

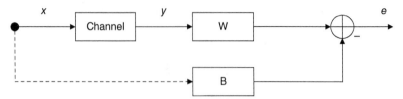

Figure 10.8 MMSE time domain equalizer architecture.

In Figure 10.8, two filters W and B constitute the TDE. The TDE operations are in two modes. In the training mode, a signal known to the receiver is transmitted. The TDE adjusts its filter taps to optimize performance based on the received signal. Figure 10.8 shows the training mode connection. In data mode, the data-carrying signal is transmitted and demodulated by the receiver. Filter W applies to the signal stream to reduce the residual ISI, while filter B is disconnected after the training is completed.

During the training mode, the transmitted time domain training signal x, which is known to the receiver, passes through the channel to become received signal y, which passes the TDE W. The input to filter B is a copy of the training signal provided by the receiver. We can equivalently consider such input as coming from the transmitter, as indicated by the dashed arrow in Figure 10.8.

The goal of an optimum TDE is minimizing the ISI outside a time window T_{CP}, which is covered by the CP. ISI inside the T_{CP} window is of no concern, as it can be dealt with by the cyclic prefix (CP), as described in Section 10.2.4. To facilitate such an optimization goal, the helper filter B is introduced, which takes the same x as input. B spans the same time range as the CP and is adapted along with filter W, and thus cancels the ISI within the window T_{CP} at the output. Therefore, any ISI within that window would not contribute to the error signal e, and thus would not affect the optimization result for W. One particular tap in filter B is fixed as 1 to mimic the correct signal, so that the output of the comparator is indeed the equalization error. After the training (i.e., in the data mode), the filter B is removed, while the output from W is used for further processing. The adaptation algorithm is very similar to those in Chapter 9, and is not repeated here. Note, however, that data fed to W and B are highly correlated. Therefore, proper adaptation strategies, especially the step sizes, should be chosen carefully.

Although the MMSE equalizers in Chapter 9 are optimal for single-carrier modulation, MMSE is not strictly optimal for TDE. One reason is that the remaining ISI from such TDE is not stationary. Instead, ISI is stronger at the boundaries of an OFDM symbol and weaker in the middle. Another reason is that that the data rate loss due to ISI depends on the SNR in a nonlinear way (Chapter 4). Therefore, the overall performance depends on not only the average residual ISI level but also how ISI is distributed among the subcarriers. A minimum mean-squared error does not necessarily translate to the maximum achievable data rate. The strictly optimal solution of TED is computationally prohibitive. Several alternative optimization techniques have been proposed, for example, in [25, 26]. Furthermore, the decision feedback equalizer (DFE) structure has also been suggested to address the residual ISI problem [27, 28].

As shown above, a TDE usually requires a separate training phase and does not adapt during normal operations. Therefore, it is not suitable for rapidly changing channels. As a result, although the TDE is widely used in wired systems such as the ADSL, it is not typically used for wireless systems in the form described here.

10.4   OFDM ADVANTAGES AND ENHANCEMENTS

Section 10.2 describes the basic OFDM operations, while Section 10.3 discusses the additional module of a time domain equalizer. Overall, we see that OFDM can address channel dispersion in a simple way, and can reach near-optimum data rate. In this section, we provide a more detailed comparison between OFDM and the single-carrier modulation with the decision feedback equalizer (DFE) discussed in Chapter 9 [9].

10.4.1   Advantages and Disadvantages of OFDM

As we have seen so far, both OFDM and DFE can reach near-optimum performance. In other words, with proper modulation and channel coding, their data rates can be close to the Shannon limit. However, as pointed out in Chapter 9, DFE has the problem of error propagation, which is more severe when a powerful channel coding is used. OFDM, on the other hand, suffers from the overhead of CP, as pointed out in Section 10.2.9. In most cases, the performance of the two schemes is reasonably close to each other [29–31].

As shown in Section 10.4.2, OFDM has lower computational complexity at the receiver. Furthermore, OFDM offers more flexibility in optimizing the transmission scheme according to the channel condition, as to be outlined in Section 10.4.3. OFDM also provides the basis of a multiaccess scheme: the orthogonal frequency division multiple access (OFDMA) (Section 10.7).

On the other hand, OFDM has some implementation challenges, the most important ones being sensitivity to carrier frequency offset (CFO), high peak-to-RMS ratio, and spectral sidelobes. These issues will be discussed in Section 10.6.

Channel dynamics also play a role in the comparison. As discussed in Section 10.2.10, OFDM symbol period should be much longer than the channel delay spread to reduce the overhead from CP. On the other hand, the OFDM symbol period should be shorter than the channel coherence time, so that the channel is constant within an OFDM symbol. Therefore, channel coherence time must be much longer than channel delay spread to allow for OFDM modulation. In other words, the channel cannot change too quickly. DFE does not have such a constraint. However, DFE adaptation is usually slow. For fast-changing channels, one must reduce filter length, sacrificing performance for faster adaptation. It is difficult to draw general conclusions in such comparison; simulation with specific channel models may be the best way to investigate.

It is possible to combine the ideas of OFDM and single-carrier modulation in achieving better solutions. For example, one can introduce guard time and CP in single-carrier modulation to form "block-based" transmission. Frequency domain equalization can thus be performed to such signals. Such a scheme may combine the simplicity and flexibility of OFDM without the high peak-to-RMS ratio problem [32, 33]. The fourth generation (4G) cellular systems, known as the long-term evolution (LTE), use such a modulation scheme. In LTE, the uplink (i.e., from a mobile device to a base station) modulation is known as "single-carrier frequency division multiple access (SC-FDMA)" [34–36]. In SC-FDMA, a single-carrier modulated signal (time domain) is broken into blocks. Each block undergoes a DFT to form a frequency domain symbol (similar to the subcarrier data in OFDM modulation). These symbols can be reallocated among the subcarriers for multiple access (see OFDMA in Section 10.7). The result is converted back to the time domain via IDFT and appended with CP, similar to the case of OFDM. SC-FDMA combines the low peak-to-RMS ratio of single-carrier modulation with the flexibility and simplicity of OFDM.

Furthermore, with multiple-in-multiple-out (MIMO) technologies (Chapter 11), it is very difficult to deal with ISI in single-carrier modulation. With OFDM, on the other hand, each subcarrier can be processed separately as a flat-channel transmission. Therefore, OFDM is very attractive when MIMO technologies are used. Besides the complexity issue, single-carrier and OFDM schemes have the same optimal performance in point-to-point MIMO communications (ignoring the overhead caused by CP). However, in multipoint-to-point and point-to-multipoint scenarios, OFDM is theoretically better in capacity than a single-carrier modulation [37, sec. 3.6].

10.4.2   Complexity

In addition to the conceptual simplicity in addressing channel delay spread, OFDM also results in lower receiver complexity, compared with single-carrier modulation with a DFE receiver. The key advantage for OFDM in this aspect is that the DFT and IDFT operations can be implemented with the fast Fourier transform (FFT) techniques. For a block of N data points, the number of multiplications required for DFT or IDFT computation is $N\log_2(N)$ with the FFT algorithm.

Let us focus on the data mode operations and ignore adaptation. For an OFDM modulation with N subcarriers, the receiver needs to perform DFT, which takes approximately $N\log_2(N)$ multiplications.[6] Each subcarrier needs a one-tap equalizer, resulting in another N multiplications. Therefore, the total number of multiplications is approximately $N + N\log_2(N)$. For DFE, assuming the total filter length is L, we need NL multiplications to pass the N samples through the filters. The length L depends on performance requirements and channel delay spread length. A rule of thumb estimate is that $L = 4N_H$, N_H being the channel delay spread measured in symbol periods.

Let us take the LTE as an example and compare the number of multiplications needed to process the signal during one OFDM symbol period. For downlink at 20 MHz bandwidth, the OFDM modulation has $N = 2048$ with 15 kHz subcarrier spacing. There are 1200 useable subcarriers spanning 18 MHz. The shortest CP duration is 144 OFDM samples [15, sec. 6.12]. This duration is assumed to be the channel

---

[6] There may be a factor for such estimate, depending on the actual algorithm and whether N is an integer power of 2. However, for our purposes, we assume such factor is 1.

dispersion length that LTE is designed for. Therefore, for OFDM, the number of multiplications for each OFDM symbol is

$$M_{OFDM} = N + N\log_2(N) = 2048 + 2048\log_2(2048) = 24576. \qquad (10.66)$$

In the case of DFE, since there are only 1200 useful subcarriers (spanning to 18 MHz), we can use a slower symbol rate, with a total of 1200 symbols for an OFDM symbol period. The length of the channel dispersion in symbol counts is thus[7]

$$N_H = 144\frac{1200}{2048} \approx 84.4. \qquad (10.67)$$

Assuming a DFE receiver using symbol-spaced filtering, with a total length of $4N_H$. The total number of multiplication for processing such block of data is

$$M_{DFE} = 4NN_H = 4 \times 1200 \times 84.4 = 405120. \qquad (10.68)$$

In this case, DFE takes approximately 16 times of the processing power compared to OFDM. Another comparison can be made pertaining the 802.11 standard at 24 Mbps. In that case, DFE requires 960 million multiplications per second, while OFDM requires 96 million multiplications per second [38, ch. 2.7]. A more detailed comparison in a different case can be found in [39]. Of course, for more accurate estimations we need to include the adaptation. However, the conclusion remains that OFDM has a significantly lower complexity comparing to DFE.

10.4.3   OFDM Enhancements

In addition to lower complexity, OFDM also provides extra flexibility in various optimizations. Since an OFDM symbol is organized in subcarriers, the power of each subcarrier can be adjusted independently. Therefore, the transmit spectrum of the OFDM signal can be modified easily, by merely changing the transmitting and receiving gains of the various subcarriers, a process known as power allocation. In comparison, for a single-carrier modulation system to change the transmit spectrum, the pulse shaping function needs to be replaced. Since pulse shaping functions are designed with various considerations (Chapter 4), they are difficult to change during operation. The length of a pulse shaping function is usually limited by complexity constraint. Therefore, complicated transmit spectrum (such as a notch in the band) is challenging to implement.

One application of the power allocation is maximizing data rate under the total power constraint [40, 41]. As pointed out in Chapter 2, under total power constraint, maximum channel capacity is achieved when the transmitted power is allocated across the operating frequency band according to the water-filling principle. Detailed discussions are provided in Chapter 2. Simply put, the power allocated to subcarrier k should be

$$S_k = \max\left(0, C - \frac{N_{0k}}{|h_k|^2}\right), \qquad (10.69)$$

---

[7] Here, we ignore the change in OFDM symbol length due to the addition of CP.

where S_k is the power at subcarrier k, N_{0k} and h_k are the noise power and channel gain at subcarrier k. C is a constant determined by the total power constraint:

$$\sum_k S_k = P_T, \tag{10.70}$$

where P_T is the constant total power. With such a water-filling technique, more power is allocated to the subcarriers with better channel conditions. With OFDM, it is easy to adjust power allocation among the subcarriers to satisfy (10.69).

In OFDM, as with a single-carrier modulation, modulation and coding schemes (MCS) for the subcarriers need to match the available SNR, in order to achieve the highest data rate with the target probability of error, as discussed in Chapter 3. However, MCS matching has a certain granularity. For example, for QAM modulation with a fixed channel coding, the required SNR changes by 3 dB when the order of constellation changes by 1 bit (Chapter 4). Therefore, it is possible to "waste" up to 3 dB of SNR in such matching result. For OFDM, however, we can fine-tune power allocation to match the required margin levels while observing the total power constraint. This way, small excess power from several subcarriers can be pooled together to support one additional bit in modulation for one of the subcarriers. Such a scheme is used in ADSL.

Bit allocation among the subcarriers requires negotiation between the transmitter and receiver, and some significant computation. Therefore, it does not happen too often. However, when the channel or noise level changes slightly, power allocation can be fine-tuned to maintain designed SNR level for each subcarrier. Power fine-tuning can be initiated by the transmitter based on channel quality report from the receiver. The receiver can adapt to small power changes without the need for handshaking or other negotiation.

In addition to maximizing data rate, spectrum manipulation can also be used to control mutual interference among the users. For example, various spectral profiles can be selected for ADSL services over long distances. Some of them have notches to comply with interference limitations [42]. The LTE standard also has provisions allowing for "notching," that is, turning off some subcarriers, to deal with mutual interference issues. Such agility in spectrum choice is especially useful when an OFDM system needs to share the spectrum with other systems in a scheme known as dynamic spectrum access (DSA) [43, ch. 6], [44]. DSA is further discussed in Chapter 13.

On the other hand, as shown in Section 10.2.7, OFDM modulation generates spectral sidelobes as a result of the time domain windowing. Such sidelobe puts a limit on the "sharpness" of the notches. Section 10.6.2 will discuss more on this issue.

10.5   RECEIVER TRAINING AND ADAPTATION

10.5.1   Channel Estimation

As in the case of single-carrier modulation (Chapter 9), channel estimation for OFDM has two phases. When a link is established (training mode), the receiver needs to obtain a channel estimate that is close enough to recover most of the data correctly. Such estimation is typically done through known symbols (training symbols or pre-ambles). In the next phase (data mode), the channel estimate is continuously updated. In the following, we first discuss a simpler approach: decision-directed channel estimation, which is used in data mode. We will then consider channel estimation with known symbols.

10.5.1.1 Decision-Directed Channel Estimation

Decision-directed channel estimation is similar to the ways described in Chapter 9. Such adaptation is performed independently in the various subcarriers. Each subcarrier has only one parameter to adapt: the complex gain, that is, the tap value of the one-tap equalizer. Therefore, implementation and analyses of adaptation are a simple special case of the equalizers discussed in Chapter 9.

The most straightforward channel estimate is

$$\hat{\tilde{h}}(k,l) = \frac{\tilde{b}(k,l)}{\tilde{a}(k,l)}. \tag{10.71}$$

Here, $\tilde{a}(k,l)$ and $\tilde{b}(k,l)$ are transmitted (obtained from demodulator output) and received symbols for subcarrier k and OFDM symbol l, respectively. $\hat{\tilde{h}}(k,l)$ is the estimate of the channel response for the same subcarrier and OFDM symbol.

Such estimation is simple but not optimal because it does not use our a priori knowledge about the channel. For example, we know that channel responses in neighboring subcarriers and successive OFDM symbols can be strongly correlated, as to be discussed in Section 10.5.1.2. Therefore, a two-dimensional filter can be applied to such estimate, to get a smoother response and reduce the effect of noise. The method for constructing such filter is the same as discussed in Section 10.5.1.3. In fact, such filter can be used to "predict" future channel responses from past experiences. For more details, see [45, ch. 15].

10.5.1.2 Placement of Known Signals

In addition to the decision-directed channel estimation discussed in Section 10.5.1.1, OFDM systems typically provide known signals not only at the beginning of the connection (i.e., training mode) but also throughout the data transmission. Such known signals are referred to as reference signals (RS), to borrow the LTE terminology. In this section, we provide some more discussion on RS and their utilization. For notations used in this section, see Section 10.2.

OFDM systems typically embed RS with data transmission to assist channel estimation. To study such embedding, we can divide the OFDM signal into resource elements (borrowing the term from LTE terminology). A resource element (RE) refers to a particular subcarrier of a particular OFDM symbol. It is a unit of time and frequency in OFDM. REs are shown as rectangular cells in Figures 10.9 and 10.10. An RE can be allocated for either data or RS transmission. The goal of channel estimation is determining channel responses for all REs.

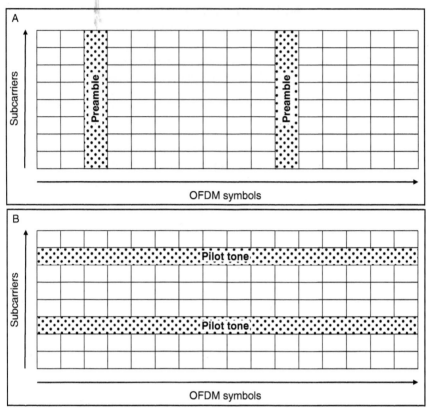

Figure 10.9   Training symbols or preambles (A) and pilot tones (B).

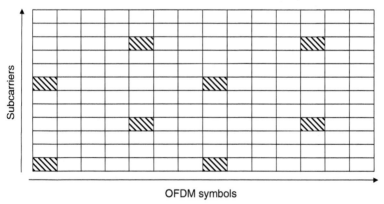

Figure 10.10   Scattered reference REs for LTE.

For our discussion, REs used to transmit RS are referred to as "reference REs." To reduce overhead, we would like to minimize the number reference REs, while maintaining adequate channel estimation functionality. Since the channel change has a finite rate in both time and frequency, channel responses in neighboring REs are correlated. Therefore, reference REs can be placed sparsely. Some smoothing interpolation can be used to obtain channel response for the rest of the REs.

A straightforward approach is placing the reference REs across all subcarriers in particular symbols or across all symbols at some specific subcarriers, as shown in Figure 10.9, where the shaded blocks indicate the reference REs. Panels A and B show the schemes with training symbols or preambles (i.e., an entire OFDM symbol is known) and pilot tones (i.e., a subcarrier across all OFDM symbols is known), respectively.

Such structures are used in the ADSL standard [6, ch. 7.11.1.2] and WLAN (802.11) [21].[8] Since ADSL transmits through telephone lines, where the channel is relatively stable, full-fledged channel estimation with elaborate training symbols is conducted at the link initiation (training mode). In data mode, one pilot tone is inserted in the downstream signal, mainly to ensure frequency synchronization. WLAN, on the other hand, expects frequent channel variations. Therefore, training symbols are placed at the beginning of each time slot. Also, there are four pilot tones over the 20 MHz bandwidth, spaced by 14 subcarriers. These pilot subcarriers transmit pseudo-random bit sequences known to the receiver.

We can further reduce the resources reference RE by allowing discontinuity in *both* frequency and time. Figure 10.10 shows an example, where shaded blocks indicate the reference REs. This particular example is for the LTE downlink, cell-specific reference signal [15, fig. 6.10.1.2-1]. This picture shows a group of 12 subcarriers and 14 symbols. Such a structure is repeated over the entire bandwidth and time.

As pointed out in Chapter 6, the rate of channel change in the time domain is characterized by the Doppler spread F_D, while that in the frequency domain is characterized by the delay spread length T_H. According to the Nyquist theory (Chapter 3), the sampling interval should be smaller than the inverse of the maximum frequency, in order to capture the entire waveform. Therefore, a good rule of thumb for the density of the reference REs is [45, sec. 14.5]:

$$\Delta s_f = \frac{1}{T_H \Delta F}$$
$$\Delta s_t = \frac{1}{F_D (N + N_c) T_s}$$

(10.72)

Here, Δs_f is the maximum spacing of the reference REs in the frequency domain, in terms of subcarriers. Δs_t is the maximum spacing of the reference REs in the time domain, in terms of OFDM symbols. T_H is the channel dispersion length. Δf is the subcarrier spacing. F_D is the Doppler spread (see Chapter 6). T_s is the OFDM sample period. N and N_c are the numbers of subcarriers and CP samples, respectively.

---

[8] 802.11 has multiple operational modes using OFDM, specified in various sections of the standard. The references given here is only for the basic OFDM operation (formally known as 802.11g).

According to the Nyquist theorem, channel response for the rest of REs can be derived from estimates based on these reference REs.

Equation (10.72) assumes that the channel responses at the reference REs can be determined precisely. However, in reality, channel measurements are contaminated by noise. Therefore, it is helpful to increase the density of the reference REs to provide some redundancy for noise immunity.

We can compare the above parameter selection rules with practical systems. For WLAN OFDM mode [21, sec. 17.3], the subcarrier spacing Δf is 0.3125 MHz. The maximum channel dispersion length, based on indoor measurements, can be taken as $T_H = 25$ ns [46]. Therefore, according to (10.72), we have

$$\Delta s_f = \frac{1}{25 \times 10^{-9} \times 0.3125 \times 10^6} = 128. \tag{10.73}$$

Therefore, the spacing of 14, described above, provides plenty of redundancy.

For LTE, the subcarrier spacing Δf is 15 kHz [15, sec. 6.2.3]. The average OFDM symbol length is $(N + N_c)T_s = 71.4$ μs. The mean value of the delay spread in the worst case (urban macrocell) is $10^{-6.44}$ seconds, or 363 ns [47, sec. A1–1.3.2.1].[9] The maximum speed is 350 km/h, or $v = 97.2$ m/s [47, sec. A1–1.2.4]. At a carrier frequency of $f_c = 2$ GHz [47, sec. 8.4], these values lead to a Doppler spread of

$$F_D = 2\frac{v}{c}f_c = 1.3\,\text{kHz}. \tag{10.74}$$

From these numbers, (10.72) yields

$$\begin{aligned} \Delta s_f &= \frac{1}{363 \times 10^{-9} \times 15 \times 10^3} = 184 \\ \Delta s_t &= \frac{1}{1.3 \times 10^3 \times 71.4 \times 10^{-6}} = 11 \end{aligned}. \tag{10.75}$$

From Figure 10.10, we see that the reference REs in LTE are six subcarriers and seven OFDM symbols apart. Therefore, there is still much redundancy in the frequency dimension.[10]

Besides their placement, the design of training signals is also a research topic [49, 50].

10.5.1.3   MMSE Channel Estimate
As described in Section 10.5.1.2, the RS embedded in OFDM transmission allows us to get channel response at those particular REs. With these measurements, channel response at all REs can be computed with various interpolation methods.[11] This section provides some more details on the interpolation process.

---

[9] Another estimation of delay spread is based on street microcell measurements [48]. At 2.44 GHz and distance of 140 m, RMS delay spread is given as 200 ns.

[10] Note that the channel spread quoted above is mean length. The maximum length can be five times higher.

[11] For simplicity, in our discussions we assume N is an even number. This is true for almost all practical systems.

The most straightforward way is estimating the time domain channel $\{h_n, n = 0, 1, \ldots, N_H - 1\}$. N_H is the time domain channel length in OFDM symbols. By doing it in the time domain, the constraint of channel length can be applied directly. Van de Beek et al. proposed several estimation methods in this aspect in 1995 [51]. Here, we only present the minimum mean-squared error (MMSE) method, which has the best performance. This method minimizes the mean squared error between the estimate and true channel responses

$$J \stackrel{\text{def}}{=} E\left(\sum_{n=0}^{N_H - 1} |\hat{h}_n - h_n|^2 \right). \tag{10.76}$$

In the following, we express all quantities in the forms of vectors and matrices. Matrices are expressed by capital letters, while vectors are expressed by bold-face lower case letters. Let

$$\boldsymbol{h} \stackrel{\text{def}}{=} \begin{pmatrix} h_0 \\ h_1 \\ \vdots \\ h_{N_H - 1} \end{pmatrix} \tag{10.77}$$

be the channel vector in the time domain and

$$\tilde{\boldsymbol{y}} \stackrel{\text{def}}{=} \begin{pmatrix} \tilde{y}_{-\frac{N}{2}} \\ \tilde{y}_{-\frac{N}{2}+1} \\ \vdots \\ \tilde{y}_{\frac{N}{2}-1} \end{pmatrix} \tag{10.78}$$

be the received signal at the various subcarriers. To perform DFT on $\boldsymbol{h}$, define the DFT matrix F of size $N \times N_H$, whose elements are[12]

$$F_{n,m} = \exp\left(-j\frac{2\pi(n-1)(m-1)}{N} \right). \tag{10.79}$$

Further, consider the transmission of training symbols (panel A of Figure 10.9). Let $\{x_k, k = -N/2, -N/2 + 1, \ldots, N/2 - 1\}$ be the RS in the frequency domain. Define a diagonal matrix X of size $N \times N$ with the following elements:

$$X_{l,k} = \delta_{l,k} x_{k+N/2+1}. \tag{10.80}$$

---

[12] Matrix F expresses circular DFT. Namely, $\{h_n\}$ is assumed to be extended periodically similar to (10.11) in Section 10.2.2.3.

As shown in Section 10.2.4, the channel effect is multiplicative in the frequency domain. Equation (10.20) therein can be expressed in our notation:

$$\tilde{y} = XFh + \tilde{n}, \tag{10.81}$$

where $\tilde{n}$ is the noise vector in the frequency domain:

$$\tilde{n} \stackrel{\text{def}}{=} \begin{pmatrix} \tilde{n}_{-\frac{N}{2}} \\ \tilde{n}_{-\frac{N}{2}+1} \\ \vdots \\ \tilde{n}_{\frac{N}{2}-1} \end{pmatrix}. \tag{10.82}$$

$\{\tilde{n}_k\}$ is the noise elements in the subcarriers, whose variance is σ_n^2. With such notation, the MMSE optimal estimate of $\hat{h}$ is [51]:

$$\begin{aligned} \hat{h} &= R_{hy}R_{yy}{}^{-1}\tilde{y} \\ R_{hy} &\stackrel{\text{def}}{=} E(hy^H) = R_{hh}F^H X^H \\ R_{yy} &\stackrel{\text{def}}{=} E(yy^H) = XFR_{hh}F^H X^H + \sigma_n^2 I \\ R_{hh} &\stackrel{\text{def}}{=} E(hh^H) \end{aligned} \tag{10.83}$$

Here, I is the identity matrix of size $N \times N$. R_{hh} is a statistical property of the channel and is assumed to be known to the receiver. Once $\hat{h}$ is found, the frequency domain channel responses $\left\{ \hat{\tilde{h}}_k \right\}$ can be obtained by DFT.

When the reference REs do not occupy a whole OFDM symbol but are spread across frequency and time as shown in Figure 10.10, we would like to get estimates $\hat{h}$ at a specific time and frequency (i.e., at a specific RE) by two-dimensional filtering:

$$\hat{\tilde{h}}(k,l) = \sum_{k',l'} w^*(k,l,k',l')r(k',l'). \tag{10.84}$$

Here, $\hat{h}(k,l)$ is the channel estimate for the RE at the kth subcarrier and lth OFDM symbol. $r(k', l')$ is the measured channel response at reference RE (k', l'). The summation is over all reference REs relevant to the estimation of $\hat{h}(k,l)$. $w(k, l, k', l')$ is the tap values of a two-dimensional filter, to be determined by the MMSE principle.

Let us collect all measured samples $r(k', l')$ into a vector r. Correspondingly, the filter can also be written as a vector $w(k, l)$. Equation (10.84) becomes

$$\hat{\tilde{h}}(k,l) = w^H(k,l)^T r. \tag{10.85}$$

The mean squared error for channel estimation at (k, l) is

$$J(k,l) = E\left(\left| \tilde{h}(k,l) - \hat{\tilde{h}}(k,l) \right|^2 \right) = E\left\{ \left[\tilde{h}(k,l) - w^H(k,l)^T r \right] \left[\tilde{h}^*(k,l) - r^H w(k,l) \right] \right\}, \tag{10.86}$$

where $\tilde{h}(k,l)$ is the true value of channel response at RE (k, l). The MMSE criterion leads to (see Chapter 9 for details)

$$\nabla_{w^H} J(k,l) = 0. \tag{10.87}$$

Namely,

$$EE\left\{ r\left[\tilde{h}^*(k,l) - r^H w(k,l)\right] \right\} = 0. \tag{10.88}$$

The solution is [45, sec. 14.6.1][13]:

$$w(k,l) = \Phi_{rr}^{-1}\phi(k,l)_{rh}$$

$$\Phi_{rr} \overset{\text{def}}{=} E(rr^H) \tag{10.89}$$

$$\phi(k,l)_{rh} \overset{\text{def}}{=} E\left(\tilde{h}^*(k,l)r\right)$$

With proper processing of the received signal, we can model r as the true value $\tilde{h}'$ at these reference REs with an additive noise n, with covariance of σ_n^2 [14]:

$$r = \tilde{h}' + n$$

$$E(nn^H) = \sigma_n^2 I \tag{10.90}$$

In this case, we have

$$\Phi_{rr} = E\left(\tilde{h}'\tilde{h}'H\right) + \sigma_n^2 I$$

$$\phi(k,l)_{rh} = E\left(\tilde{h}^*(k,l)\tilde{h}'\right) \tag{10.91}$$

The quantities on the right sides of the above equations are statistical properties of the channel, which is assumed to be known to the receiver.

10.5.1.4 Other Estimation Techniques

OFDM channel estimation is still an active research area. While the MMSE estimation presented above serves as a starting point, there are many later innovations in this field. For example, to reduce complexity, the two-dimensional filtering method above can be substituted by two one-dimensional filtering operations with minor performance degradation [53]. Other variations and improvements can be found in literature, for example [45, ch. 14]. Reference [54] compares several channel estimation methods in the context of LTE downlink implementation.

In addition to the maximum channel delay spread and the statistical properties, other prior knowledge of the channel can be used in the estimation process. For example, for sparse channels (i.e., a relatively large delay spread with a small number of

---

[13] This result was probably first derived by [52], although it assumed that the channel response was WSS in both time and frequency domain, and arrived at a slightly different result.
[14] Note that $\tilde{h}$ is a vector covering the channel response at the reference REs. It is difference from $\tilde{h}(j,k)$ that we are evaluating.

scattering paths), the technique of compressive sensing can be applied [55]. Channel estimation and data decoding can also be performed jointly, in an iterative manner [37, sec. 3.1.3], [56]. Blind channel estimation is also investigated by many authors, for example, [57].

Reference [45, sec. 15.1] provides an excellent history recount of OFDM channel estimation research. For a more recent literature review and survey, [58] is a good starting point.

10.5.2 Synchronization

As shown in Chapter 7, the task of receiver synchronization has two parts. One is symbol time synchronization, meaning that the matched filter output is sampled at the correct instances. The other part is carrier frequency synchronization, meaning that the down-conversion uses the correct frequency. Such general scheme applies to OFDM modulation, as well. This section discusses some issues specific to OFDM.

Synchronization for OFDM is an active research area. There are many innovative techniques for better performance and lower complexity. However, for standards-based communications systems, conventional methods discussed in this section and Chapter 7 should be enough. For more details and references, see [19, ch. 4], [23, ch. 5].

10.5.2.1 Symbol (Sample) Timing Synchronization

In the OFDM context, symbol timing discussed in Chapter 7 refers to the timing of OFDM samples. There are two tasks here. The first is ensuring each OFDM sample is sampled at the correct time instance. The Second is recognizing the beginning of an OFDM symbol.

Concerning the first task, a constant timing offset in sampling is equivalent to a change of channel dispersion response. An OFDM receiver is able to deal with dispersive channels (Section 10.2.4). Therefore, timing offset within an OFDM sample is unimportant, as long as it is constant across all OFDM samples. From a timing point of view, an OFDM sample is similar to a symbol of single-carrier modulation. Therefore, all techniques and considerations discussed in Chapter 7 apply here.

Concerning the second task, OFDM symbol alignment is typically done with some known symbols. In this scheme, the transmitter sends some known symbols (i.e., training symbols or preambles), either during link initiation phase (training mode) or periodically (such as in every frame or time slot). By detecting these known symbols, a receiver can identify the timing.

Alternatively, an OFDM receiver can look at the autocorrelation of time domain samples. Let

$$C_n \overset{\text{def}}{=} \sum_{k=1}^{N_c} a'_n a'_{n+N}{}^*. \tag{10.92}$$

Here, N and N_c are the number of subcarriers and the number of samples in CP, respectively. $\{a'_n\}$ is the time domain OFDM samples, formulated by (10.14) in Section 10.2.3. Because the first and the last N_c samples are the same in an OFDM symbol, C_n reaches a maximum when $n = 0$. Therefore, a peak in $\{C_n\}$ corresponds to the beginning of an OFDM symbol.

10.5.2.2 Carrier Frequency Synchronization

As in the case of single-carrier modulation discussed in Chapter 7, the objective for carrier frequency synchronization is to have the correct down-conversion frequency. The remaining frequency error is known as carrier frequency offset (CFO). A nonzero CFO causes a baseband symbol (or OFDM samples) to have a phase rotation relative to the previous one.

For OFDM, the effect of CFO can also be viewed as introducing inter-carrier interference (ICI). Namely, the signal from one subcarrier is mixed with that from its neighboring subcarriers, causing degradation of the demodulation performance. Such ICI can be quantitatively estimated, even corrected [19, sec. 4.1].

The degradation caused by CFO is determined by the ratio between the CFO and the subcarrier spacing Δf. Therefore, an OFDM system with more subcarriers (and thus smaller Δf) is more vulnerable to degradation caused by CFO. Figure 10.11 shows a simulation plot for a QPSK OFDM system in an AWGN channel [23, fig. 5.5 (a)]. The legend shows the various CFO values, relative to the sub-carrier spacing Δf. Even a CFO of $0.1\Delta f$ results in a degradation of 3 dB in SNR at the bit error rate (BER) of 10^{-5}. The degradation increases at lower BER levels and for higher order modulations (not shown in this figure). Eventually, there is a BER "floor" caused by CFO, meaning that BER does not improve as SNR increases. More examples on the impact of synchronization errors can be found in [59, sec. III].

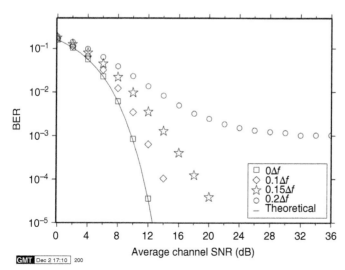

Figure 10.11 CFO and additional BER [23, fig. 5.5(a)]. Reproduced with permission. The legend shows the various CFO values, relative to the subcarrier spacing Δf.

Fortunately, CFO can be estimated and corrected to a level well below $0.1\Delta f$ for OFDM systems. CFO is more of an issue with OFDMA (Section 10.7), where signals of different subcarriers may be transmitted by different users, and thus carries different CFO.

As discussed in Chapter 7, CFO, once known, can be corrected in several ways. One can directly adjust the down-conversion frequency to eliminate such error (feedback) or add a compensational phase rotation to the baseband symbols (feedforward). For OFDM systems, there are several ways to estimate CFO.

First, CFO can be estimated using known signals (reference signals or RS). As discussed in Section 10.5.1, a typical OFDM system inserts RS in the transmission to assist channel estimation. Such signals can be used for CFO estimation, as well. Especially, RS are important in the coarse estimation of CFO, whose goal is limiting the residual error to less than Δf. Under the assumption of random noise and uniform distribution of CFO, maximum likelihood (ML) estimation can be performed. In an ML estimation, one searches of the CFO value that maximizes the probability for the received RS to happen. There are other simplified methods. Details depend on the design of the RS.

Second, once the receiver is working properly, CFO estimation can be conducted continuously to track any clock frequency drifts. Such tracking can be done by decision-directed estimation. As discussed in Chapter 7, CFO introduces symbol phase rotation. For OFDM, such phase rotation happens to every subcarrier, from one OFDM symbol to the next. With decision-directed estimation, phase rotation caused by CFO can be tracked by computing the phase difference between the input and output of the demodulator (i.e., the phase in the noise). Although such phase difference is mostly driven by additive noise, the effect of CFO can be extracted by averaging over subcarriers and OFDM symbols. This method depends on the assumption that the demodulator makes correct decisions. Therefore, it only works if the CFO is not too large.

Incidentally, symbol time offset (Section 10.5.2.1) also introduces a phase rotation in the various subcarriers. Unlike CFO, though, such phase rotation does not change from one OFDM symbol to the next. Instead, the phase rotation caused by symbol time offset increases linearly with the subcarrier index (see homework). Therefore, symbol time offset can also be tracked through decision-directed estimation. As stated in Section 10.5.2.1, a constant time offset is not a problem. However, if such offset changes with time, it should be corrected either by adjusting the sampling clock (feedback) or performing interpolation after sampling (feedforward). More details are presented in Chapter 7.

Third, the autocorrelation method in Section 10.5.2.1 can also be used for CFO estimation. As given by (10.92), The phase of C_n at its peak reflects the phase rotation over N OFDM samples. It can thus be used to estimate the CFO. However, note that the phase shift is $2\pi\delta f N T_s$, where δf is the CFO. If δf is larger than Δf given by (10.36) in Section 10.2.6, the phase shift would be larger than 2π, and thus cannot be estimated this way. Therefore, the first method is still needed to limit the range of the remaining CFO.

For more detailed discussions on the various synchronization challenges and solutions, see [59].

10.5.2.3 Phase Noise

An issue related to synchronization is phase noise, which refers to the random phase changes in the received signal, due to transmitter or receiver clock instability. Phase noise causes instantaneous and random frequency drifts. Correction of such effects is possible but difficult. Fortunately, for the frequency range currently used (up to 5 GHz), phase noise is not an important problem. For higher frequencies (such as 80 GHz) and lower subcarrier spacing Δf, phase noise may degrade OFDM performance. For more information, see [45, sec. 3.6], [48].

10.6 IMPLEMENTATION ISSUES

So far we have seen how OFDM works and the various benefits of such modulation scheme. In this section, we discuss two significant practical challenges with OFDM: peak-to-average ratio (PAR) and sidelobes.

10.6.1 Peak-to-Average Ratio (PAR)

Let us study the relative peak values of a signal $s(t)$:

$$
\begin{aligned}
R_P &\stackrel{\text{def}}{=} \frac{\max\left(|s(t)|^2\right)}{E\left(|s(t)|^2\right)} \\
R_A &\stackrel{\text{def}}{=} \frac{\max(|s(t)|)}{\sqrt{E\left(|s(t)|^2\right)}}
\end{aligned}
\tag{10.93}
$$

The first definition R_P is the ratio between the peak and average power values, known as the peak-to-average ratio (PAR) or peak-to-average power ratio (PAPR). The second definition R_A expresses the ratio between the peak and root mean square amplitude values, known as the peak-to-root mean square ratio or the crest factor. Both ratios are the same when expressed in dB. In this book, we use the term PAR.

PAR is important in communication systems because it affects the dynamic range of the signal paths. More importantly, it affects the requirement of the power amplifier (PA) at the transmitter. Ideally, the PA should remain a linear response up to the output level of $\max(|s(t)|)$. Therefore, given the output power level $E(|s(t)|^2)$, a larger PAR means the PA must have a larger linear range. This usually leads to a higher cost, a higher power consumption, and a larger size. Insufficient PA linear range leads to distortion of the output signal, resulting in higher sidelobe (i.e., out-of-band emission) and degraded performance at the receiver.

OFDM inherently has higher PAR than a single-carrier modulation, because signals from the multiple subcarriers have a chance to add together in-phase, resulting in large peaks.

We often express PAR in terms of the complementary cumulative distribution (CCDF) of signal peak values. Define the one-symbol PAR for an OFDM modulation with N subcarriers:

$$P_N \stackrel{\text{def}}{=} \frac{\max_{nT_O < t \le (n+1)T_O} |s(t)|^2}{E\left(|s(t)|^2\right)}. \tag{10.94}$$

T_O is the time period for one OFDM symbol. P_N is the PAR R_P given in (10.93) while limited to one OFDM symbol.

It has been shown that when N is large, the CCDF of P_N can be well approximated by the following [60]:

$$p(P_N > y) \approx 1 - \exp\left[-\exp(-y)N\sqrt{\frac{\pi}{3}\ln(N)}\right]. \tag{10.95}$$

Equation (10.95) shows that PAR approximately increases with the logarithm of N when N approaches infinity. In other words, $p(P_N > y)$ is approximately a function of $y/\ln(N)$.

Figure 10.12 shows the CCDF of the PAR for various modulations. The two left traces are single-carrier modulations QPSK and 256QAM, pulse-shaped with a root–raised cosine pulse filter with a roll-off factor of 0.3 and a filter span of 20 symbols (see Chapter 3 for root–raised cosine filters). The two right traces are OFDM modulations with 64 and 2048 subcarriers, respectively. 256-QAM is used for subcarrier

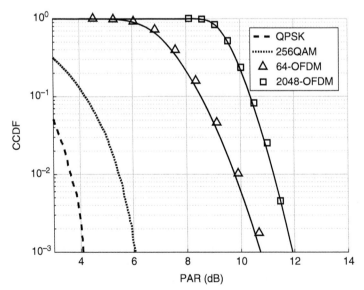

Figure 10.12   PAR of various modulations.

modulations, although theory and simulation show PAR of OFDM is insensitive to subcarrier modulation schemes. The markers show simulation results, while the lines show theoretical results from (10.95). We can see that the PARs for OFDM are significantly higher than those for single-carrier modulations. For more PAR comparisons among the various modulation schemes, see [36, sec. 7.3].

Reducing PAR for OFDM modulations is an active research area [45, 61–64]. Most of the techniques belong to one of three categories as briefly described below.

1. Coding. When OFDM modulation is combined with channel coding (Chapter 5), we can exclude from codewords those bit combinations causing high PAR. In other words, the codeword set selection is optimized for reducing PAR. The price to pay is that such codeword set may be suboptimal regarding noise immunity.

2. Alternatives in modulation. The transmitter chooses one of the several alternatives in modulation that yields the minimum PAR. Such alternatives can include additional phase shifts to the subcarriers, different ordering of bits, and so on. The selection of the alternative needs to be communicated to the receiver through a side channel. Therefore, "useful" data rate is reduced as a price.

3. Redundancy. Some of the modulation resources are reserved for PAR reduction. For example, one can expand the constellation set to have a set of multiple constellation points representing the same data. Constellation points within the set can be chosen for PAR minimization, without the need to inform the receiver of such choice. Such a scheme is similar to the Tomlinson–Harashima precoding described in Channel 9. Another way is reserving some of the subcarriers to carry "dummy symbols," whose sole purpose is reducing PAR. Of course, total capacity is reduced due to redundancy.

At first glance, each of the above methods involves optimization or searching among a potentially large set of alternatives. However, keep in mind that we do not need to minimize the PAR of every OFDM symbol. Instead, we only need to avoid the worst situations. Therefore, the complexity is not prohibitive.

Also, one can preemptively "clip" the signal so that PAR is kept within a limit. Such clipping may introduce wideband noise, which increases out-of-band transmission. Additional filtering can be applied to suppress such transmission, at the risk of "re-growing" the PAR.

Various PAR reduction techniques result in various levels of gain, typically 1–2 dBs. These techniques are used in some practical systems today. The digital TV broadcast standard DVT-2 supports two PAR reduction techniques [13, sec. 9.6]. The LTE cellular system uses single-carrier FDMA (SC-FDMA) for uplink to reduce PAR [35, 36].

10.6.2   Sidelobes

As pointed out in Section 10.2.7, OFDM modulation typically has significant sidelobes, because each symbol is cut off sharply in the time domain. The sidelobe shape is given by (10.55) in Section 10.2.7. Such sidelobe causes interference to other systems in the neighboring frequency band. Sidelobes present severe challenges when frequency notches are used for coexistence with other systems as described in Section 10.4.3.

Note that such sidelobes do not introduce inter-carrier interference. As shown in Section 10.2, signals from difference subcarriers are mutually orthogonal, although their spectra overlap due to the sidelobes. This orthogonality holds even when the subcarriers are transmitted by different transmitters as in the case of uplink OFDMA (Section 10.7, as long as the correct timing synchronization among the subcarriers is maintained. On the other hand, sidelobes cause mutual interference among uncoordinated systems, even if they all use OFDM.

A common practice for addressing sidelobes is leaving a guard space between the frequency bands used by different systems. Of course, such a method reduces system capacity because subcarriers in the guard space cannot be used for data transmission.

For example, in one of the LTE downlink normal cyclic prefix modes, subcarrier spacing is $\Delta f = 15$ kHz. Each resource block (RB) contains $N_{SC}^{RB} = 12$ subcarriers [15, sec. 6.2.3]. For 20 MHz bandwidth, a total of $N_{RB} = 100$ RBs are transmitted [65, sec. 5.6]. Therefore, the total transmission bandwidth (including the nontransmitting DC subcarrier) is

$$W_T = \Delta f \left(N_{SC}^{RB} N_{RB} + 1 \right) = 18.015 \, \text{MHz}. \qquad (10.96)$$

The remaining approximately 2 MHz is the guard band. For WLAN (IEEE 802.11) OFDM mode (20 MHz channel spacing), the subcarrier spacing is $\Delta f = 0.3125$ MHz. There are a total of $N_{ST} = 52$ transmitting subcarriers, including $N_{SD} = 48$ data-carrying subcarriers and $N_{SP} = 4$ pilot subcarriers [21]. Therefore, the total transmission bandwidth (including the nontransmitting DC subcarrier) is

$$W_T = \Delta f (N_{ST} + 1) = 16.5625 \, \text{MHz}. \qquad (10.97)$$

The remaining approximately 3.5 MHz is the guard band.

Suppressing OFDM sidelobes without adversely reducing data capacity is an active research area. A widely studied technique is active cancellation [66–68]. In this method, some subcarriers are reserved for sidelobe suppression. Their transmission is determined by the content in other subcarriers, optimized to reduce sidelobe. Alternatively, data-carrying subcarriers can be assigned to different power levels (e.g., edge subcarriers get lower power) to reduce sidelobe [69]). Such technique causes a slight increase in bit error rate as a cost of reducing sidelobes by about 10 dB. One can also reserve "sidelobe suppression resources" in the time domain. In this scheme, extra time domain samples are inserted between the OFDM symbols [70]. These time domain samples are optimized based on the two neighboring OFDM symbols to minimize sidelobes. The time domain cancellation signal can overlap with the data samples [71]. This way, no additional guard time is required. However, the cancellation signal introduces (controllable) interference to the data. Another way is performing an optimizing linear transform of the transmitted subcarrier signals. Such a method does not add any transmission overhead, but the signal is slightly distorted [72]. Nevertheless, a substantial reduction in sidelobes can be achieved without significant BER penalty.

Yet another method is using extended windowing. The OFDM symbols described in Section 10.2.6 has a time domain window (10.54), which causes the sidelobe. Instead, we can use a different time domain window with a gentler roll-off at the edges [38, sec. 2.4], [73, 74]. Windowing reduces the sidelobes but causes intersymbol interference (ISI) as the OFDM symbols overlap with each other. Also, windowing may also cause inter-carrier interference (ICI), because the effective channel gain changes during an OFDM symbol. Windowing can even be applied to different subcarriers differently, putting more emphasis on the edge subcarriers [75]. Windowing technique is used in the digital video broadcast return channel (DVB-RCT) standard [76, sec. 6.9]. The 5G standard envisions sidelobe reduction techniques (filtering or windowing) implemented at the transmitter side [16, sec. 6.1.3]. These schemes are considered as transparent to the receiver. Namely, the receiver does not try to remove the resulting ISI.

Unique word OFDM (UW-OFDM) is a more sophisticated OFDM scheme with several advantages, including lower sidelobes [77].

10.7 ORTHOGONAL FREQUENCY DIVISION MULTIPLE ACCESS (OFDMA)

Orthogonal frequency division multiple access (OFDMA) can be viewed as a natural extension of OFDM. In OFDMA, multiple users share the bandwidth occupied by an OFDM modulation signal. Typically, OFDMA is used in a star-spoke network, where a star node (a server, a base station, or an access point) communicates with multiple spoke nodes (clients, user equipment devices, or subscriber stations). The OFDM resources (subcarriers and symbols) are dynamically allocated to the various links, coordinated by the center (star) node. Borrowing terms from LTE, we call a time-frequency unit (one subcarrier in one OFDM symbol length) a resource element (RE). Transmission from the star node is referred to as a downlink (DL), and those from the spoke nodes are uplink (UL).

For DL, the star node essentially transmits an OFDM signal, except that different REs are meant for different receivers. A spoke node receives the entire OFDM signal and performs DFT. After that, it can choose to demodulate only the REs addressed to it. For UL, a spoke node transmits the OFDM signal, but only modulates the REs assigned for its use. All other REs are set to zero in its transmission. The star node receives the superposition of all transmissions and demodulates it as one OFDM signal.

OFDMA has several advantages, comparing to other multiple access technologies such as time-division multiple access (TDMA) and frequency division multiple access (FDMA).

First, OFDMA is very flexible in allocating the capacity resource. Allocation of REs among the users can be changed every frame. The users do not need to modify any hardware-related operations. Therefore, the resource allocation process can be highly dynamic and optimal for the current traffic load.

Second, OFDMA does not require any overhead (such as guard band for FDMA and guard time for TDMA) when dividing up the resource. Therefore, resource partition can be done at very fine levels, that is, more users can share the channel.

Third, the resource allocation scheme can be optimized in various ways. For example, in a selective fading environment, a particular subcarrier may provide a good channel condition to one user but a bad one to another. Subcarrier allocation can be made according to these channel conditions [78]. Or, we can take interference into account. If a subcarrier is more prone to interference to or from the neighboring cell for a particular user, we can avoid such interference through resource allocation.

The concept of OFMDA sounds simple enough. However, there are complications for OFDMA, in addition to the practical issues for OFDM (Section 10.6). Especially in UL, since different REs are transmitted by different users, time and frequency synchronization become tricky. At the symbol level, the relative timing offsets among the subcarriers, together with the channel delay spread, need to fit into the CP period. Therefore, we need tight synchronization to ensure that OFDM symbols transmitted by the various spoke notes arrive at the star node simultaneously. Symbol level synchronization can be achieved through closed-loop control. Each spoke node aligns its transmission timing with that of the received OFDM symbols. The relative timing offset between transmission and receiving is derived from the timing advance command sent by the star node. Such synchronization mechanism is used by LTE.[15]

A more severe issue is carrier frequency offset (CFO) at the UL. Although the spoke nodes can lock its transmission frequency with the receiving frequency (reflecting clock frequency at the star node) thus eliminating any CFO due to clock frequency difference, Doppler shifts cause transmitter-specific CFO. As a result, the OFDM symbol received by the star node carries CFO values that are different among subcarriers. For example, in the LTE system with a speed of $v = 500$km/h at the user terminal and a carrier frequency of $f_c = 2$ GHz, the CFO caused by Doppler effect is

$$f_D = \frac{v}{c}f_c \approx 1\,\text{kHz}. \tag{10.98}$$

This Doppler effect, compared to the subcarrier spacing of 15 kHz, would cause moderate performance degradation, as discussed in Section 10.5.2.2.

It is possible to estimate and correct the multiple CFO in an OFDMA system [79–81]. CFO causes inter-carrier interference (ICI) in the frequency domain. Such interference can be modeled as the original signal multiplied by a matrix. CFO correction thus involves multiplying the received signal with the inverse matrix. ICI caused by CFO can be corrected in several other ways, as well [59, sec. VII]. However, due to its complexity, CFO correction is not widespread in practical systems.

---

[15] LTE uplink does not use OFDMA, but a similar modulation. The synchronization principle remains the same.

10.8 FILTER BANK MULTICARRIER (FBMC) MODULATION

So far we have seen that OFDM is an attractive modulation scheme, because of its simplicity in dealing with channel dispersion effects and its flexibility in optimizing transmission and accommodating multiple users. However, OFDM has a significant disadvantage: it has relatively high sidelobes, as discussed in Sections 10.2.7. Such sidelobes not only cause interferences to other systems in the neighboring frequency band but also, as mentioned in Section 10.7, cause interference among the multiple users in the same system (i.e., in OFDMA) unless their transmissions are highly synchronized. Such synchronization may be challenging, especially in highly mobile situations. The filter bank multicarrier (FBMC) modulation is a generalization of OFDM for addressing the sidelobe issue.

The FBMC modulation can be formulated in general terms [74, 82], [83, ch. 3]. The modulated signal can be written as

$$x(t) = \sum_{m=-\infty}^{\infty} \sum_{k=0}^{N-1} X_{m,k} g_{m,k}(t). \tag{10.99}$$

Here, $X_{m,k}$ are the transmitted symbols, modulated in the same way as the single-carrier modulation (Chapter 3). $g_{m,k}(t)$ is known as the synthesis functions, which maps the symbols $\{X_{m,k}\}$ to a waveform (i.e., a function of time). For FBMC, the synthesis functions take a special form:

$$g_{m,k}(t) = p_{tx}(t - mT_{Sym}) \exp(j2\pi k \Delta ft). \tag{10.100}$$

$p_{tx}(t)$ is known as the transmitter prototype filter, or pulse shape. T_{Sym} and Δf are parameters chosen for the modulation. We can see that $g_{m,k}(t)$ is the prototype filter translated on a time-frequency grid, with the time interval of T_{Sym} and frequency interval of Δf. Same wise, we can view m and k as indices in time and frequency dimensions. A family of $\{g_{m,k}\}$ in the form of (10.100) is known as a Gabor system.

At the receiver, the symbols are recovered with a bank of matched filters.

$$\bar{X}_{n,l} = \int y(t) \gamma_{n,l}^*(t) dt. \tag{10.101}$$

Here, $y(t)$ is the received signal, including channel effects and noise. $\gamma_{n,l}(t)$ are known as the analysis functions, in the form of

$$r_{n,l}(t) = p_{rx}(t - nT_{Sym}) \exp(j2\pi l \Delta ft). \tag{10.102}$$

$p_{rx}(t)$ is known as the receiver prototype filter.

As a special case, the OFDM modulation follows this general form. The prototype filters for OFDM are:

$$p_{tx}(t) = \begin{cases} 1 & 0 < t \le (N + N_c)T_s \\ 0 & \text{otherwise} \end{cases}$$

$$p_{rx}(t) = \begin{cases} 1 & N_c T_s < t \le (N + N_c)T_s \\ 0 & \text{otherwise} \end{cases} \tag{10.103}$$

An FBMC modulation is said to be biorthogonal if

$$\int g_{m,k}(t)\gamma_{n,l}^*(t)dt = \delta_{m,k}\delta_{n,l}. \tag{10.104}$$

In this case, in the absence of channel effects, recovered symbol $\bar{X}_{n,l}$ is the same as the transmitted symbol $X_{n,l}$. Especially, a biorthogonal FBMC is said to be orthogonal if, in addition,

$$p_{tx}(t) = p_{rx}(t). \tag{10.105}$$

Obviously, OFDM is a biorthogonal FBMC with an immaterial time shift.

It is desirable that the family of $\{g_{m,k}\}$ form an orthonormal basis for the whole functional space. The reason for orthogonality was explained above. Normality is just a scaling for convenience. The fact that $\{g_{m,k}\}$ forms a basis means for any function $x(t)$, there is a unique set of $\{X_{m,k}\}$ so that (10.99) holds. The existence of such a set means all possible waveforms can be used to transmit some data symbol. The uniqueness means such data symbol can be recovered from the waveform.

Also, we would like the prototype functions to be local in both time domain and frequency domain. Namely, we wish that

$$\begin{aligned} \int t^2 p_{tx}(t)dt < \infty \\ \int f^2 \tilde{p}_{tx}(f)df < \infty \end{aligned}. \tag{10.106}$$

Here, $\tilde{p}_{tx}(f)$ is the Fourier transform of $p_{tx}(t)$ using linear frequency. The Balian–Low theorem says that if (10.106) holds and $\{g_{m,n}\}$ forms an orthonormal basis, then [84, ch. 15]

$$T_{Sym}\Delta f \geq 1. \tag{10.107}$$

As an example, in OFDM, we have[16]

$$T_{Sym} = (N + N_s)T_s. \tag{10.108}$$

From (10.36) in Section 10.2.6,

$$\Delta f = \frac{1}{NT_s}. \tag{10.109}$$

Therefore, (10.107) holds. However, OFDM is not localized in the frequency domain, because of its sidelobes discussed in Section 10.2.7.

Searching for prototype functions with desirable properties is an active research area. Ideally, we would like the prototype function to be limited to $[0, T_{Sym}]$ in the time domain, and its spectrum be limited to $[-\Delta f/2, \Delta f/2]$ in the frequency domain. This way, we have neither inter-symbol interference nor inter-carrier interference. However, in reality, compromises are necessary. For example, the prototype function for OFDM (10.103) is exactly limited in the time domain. However, in the frequency domain, as discussed in Section 10.2.7, its spectrum is extended.

---

[16] For the meanings of the various symbols, see Section 10.2.

When the prototype function is selected so that its spectrum is limited to $[-\Delta f/2, \Delta f/2]$ (or slightly exceeding it), different subcarriers (i.e., signal contributions in (10.99) with different k values) can be demodulated separately as in the case of OFDM. However, unlike OFDM, strict synchronization among the subcarriers is not necessary. This enables a multiaccess system where different subcarriers are transmitted by different users without closed-loop timing control. Such property makes the FBMC attractive. In addition, FBMC provides additional degrees of freedom in optimizing waveform properties. For example, one can design and FBMC waveform with low PAR (Section 10.6.1) [85].

On the other hand, FBMC is disadvantageous in complexity. Since it cannot use the highly efficient fast Fourier transform (FFT) algorithm, transmitter and receiver complexities are increased comparing with OFDM. With dispersive channels, more sophisticated equalizers are needed, further increasing receiver complexity. There are additional complications with multiple antenna transmitting and receiving schemes. For more details, see [74, 82, 86].

FBMC has been proposed as a candidate waveform for the 5G cellular systems [83, ch. 3–4], [87, 88].

10.9   SUMMARY

This chapter introduces OFDM, a modern modulation technique widely used in today's commercial communications systems (Section 10.2). OFDM loads data into the multiple subcarriers. Waveform contributions from the subcarriers are mutually orthogonal, even with a dispersive channel. Therefore, a receiver can easily separate the subcarriers from each other in demodulation. The major advantages of OFDM include lower complexity in dealing with channel dispersion thanks to the discrete Fourier transform (DFT) and the accompanying fast Fourier transform (FFT) algorithm. In addition to lower complexity, OFDM also provides more flexibility in optimizing the transmitted spectrum, either for capacity improvement or for interference avoidance (Section 10.4).

Although simple and elegant in theory, OFDM does have some challenges in implementation. OFDM performs channel estimation in the frequency domain and thus uses different reference signal forms. It also has tight synchronization requirements (Section 10.5). When the channel delay spread exceeds the length of the CP, time domain equalizer can be used to improve performance (Section 10.3). Furthermore, OFDM has the drawbacks of high peak-to-average ratio and high sidelobes. Various techniques have been devised to mitigate these issues (Section 10.6).

By partitioning transmitted signal into OFDM symbols (in the time dimension) and subcarriers (in the frequency dimension), OFDM offers a flexible way for multiple users to share the media, resulting in the OFDMA technique (Section 10.7). While advantageous in several ways, OFDMA presents the challenge of tight synchronization requirement, which may be difficult to satisfy for mobile users. A generalized modulation scheme, the FBMC, is being developed to address this issue (Section 10.8).

FBMC allows for the partition of subcarriers among multiple users without synchronization requirement. However, it has higher complexity than OFDMA.

OFDM, especially its implementation, is an active research area. This chapter only provides the basic concept and starting points on the various subtopics. For further exploration, the users are referred to the various references cited in this chapter, especially [19, 23, 37, 45, 89, 90]. Reference [48] provides some interesting discussion on the application of OFDM to the 5G cellular networks.

REFERENCES

1. R. W. Chang, "Synthesis of Band-Limited Orthogonal Signals for Multichannel Data Transmission," *Bell Syst. Tech. J.*, vol. 45, no. 10, pp. 1775–1796, 1966.
2. R. Chang and R. Gibby, "A Theoretical Study of Performance of an Orthogonal Multiplexing Data Transmission Scheme," *IEEE Trans. Commun. Technol.*, vol. 16, no. 4, pp. 529–540, 1968.
3. S. B. Weinstein and P. M. Ebert, "Data Transmission by Frequency-Division Multiplexing Using the Discrete Fourier Transform," *IEEE Trans. Commun. Technol.*, vol. 19, no. 5, pp. 628–634, 1971.
4. A. Peled and A. Ruiz, "Frequency Domain Data Transmission Using Reduced Computational Complexity Algorithms," in *IEEE International Conference on Acoustics, Speech, and Signal Processing*, 9–11 April, Denver, CO, 1980, pp. 964–967.
5. S. B. Weinstein, "The History of Orthogonal Frequency-Division Multiplexing," *IEEE Commun. Mag.*, vol. 47, no. 11, pp. 26–35, 2009.
6. ITU-T, *Asymmetric Digital Subscriber Line (ADSL) Transceivers*. ITU-T Recommendation G.992.1, 1999.
7. J. Bingham, "Multicarrier Modulation for Data Transmission: An Idea Whose Time Has Come," *IEEE Commun. Mag.*, vol. 28, no. 5, pp. 5–14, 1990.
8. K. Sistanizadeh, P. Chow, and J. Cioffi, "Multi-tone Transmission for Asymmetric Digital Subscriber Lines (ADSL)," in *Proceedings of ICC '93—IEEE International Conference on Communications*, 23–26 May, Geneva, Switzerland, 1993, vol. 2.
9. B. R. Saltzberg, "Comparison of Single-Carrier and Multitone Digital Modulation for ADSL Applications," *IEEE Commun. Mag.*, vol. 36, no. 11, pp. 114–121, 1998.
10. S. J. Vaughan-Nichols, "OFDM: Back to the Wireless Future," *IEEE Comput.*, vol. 35, no. 12, pp. 19–21, 2002.
11. U. Varshney, "The Status and Future of 802.11-Based WLANs," *IEEE Comput.*, vol. 36, no. 6, pp. 102–105, 2003.
12. D. Astély, E. Dahlman, A. Furuskär, Y. Jading, M. Lindström, and S. Parkvall, "LTE: The Evolution of Mobile Broadband," *IEEE Commun. Mag.*, vol. 47, no. 4, pp. 44–51, 2009.
13. European Telecommunications Standards Institute (ETSI), *Digital Video Broadcasting (DVB) Frame Structure Channel Coding and Modulation for a Second Generation Digital Terrestrial Television Broadcasting System (DVB-T2)*. ETSI EN 302 755 v1.4.1 (2015-07), 2015.
14. ETSI, *Digital Video Broadcasting (DVB); Framing Structure, Channel Coding and Modulation for Satellite Services to Handheld Devices (SH) Below 3 GHz*. V1.2.1. ETSI EN 302 583, 2011.
15. 3GPP, *3rd Generation Partnership Project; Technical Specification Group Radio Access Network; Evolved Universal Terrestrial Radio Access (E-UTRA); Physical Channels and Modulation*. TS 36.211 v13.0.0 (Release 13), 2015.
16. 3GPP, *3rd Generation Partnership Project; Technical Specification Group Radio Access Network; Study on New Radio Access Technology Physical Layer Aspects*. 3GPP Technical Report TR 38.802 V14.2.0 (Release 14), 2017.
17. T.-D. Chiueh and P.-Y. Tsai, *OFDM Baseband Receiver Design for Wireless Communications*. Wiley, 2007.

18. N. Andreadou and F.-N. Pavlidou, "Modeling the Noise on the OFDM Power-Line Communications System," *IEEE Trans. Power Deliv.*, vol. 25, no. 1, pp. 150–157, 2010.

19. A. B. Narasimhamurthy, M. K. Banavar, and C. Tepedelenlioğlu, *OFDM Systems for Wireless Communications*. Morgan & Claypool Publishers, 2010.

20. B. Saltzberg, "Performance of an Efficient Parallel Data Transmission System," *IEEE Trans. Commun. Technol.*, vol. 15, no. 6, pp. 805–811, 1967.

21. IEEE, *802.11-2016—IEEE Standard for Information Technology—Telecommunications and Information Exchange Between Systems Local and Metropolitan Area Networks—Specific Requirements—Part 11: Wireless LAN Medium Access Control (MAC) and Physical Layer (PHY)*. IEEE, 2016.

22. DLMF, "NIST Digital Library of Mathematical Functions" [Online]. Available at http://dlmf.nist.gov/ [Accessed December 13, 2016].

23. L. Hanzo and T. Keller, *OFDM and MC-CDMA A Primer*. Wiley, 2006.

24. J. S. Chow and J. M. Cioffi, "A Cost-Effective Maximum Likelihood Receiver for Multicarrier Systems," in *IEEE International Conference on Communications (ICC), Chicago, IL*, 1992, pp. 948–952.

25. J. S. Chow, J. M. Cioffi, and J. A. C. Bingham, "Equalizer Training Algorithms for Multicarrier Modulation Systems," in *IEEE International Conference on Communications (ICC), Geneva*, 1993, pp. 761–765.

26. M. Milosevic, L. F. C. Pessoa, B. L. Evans, and R. Baldick, "DMT Bit Rate Maximization with Optimal Time Domain Equalizer Filter Bank Architecture," in *Conference Record of the Thirty-Sixth Asilomar Conference on Signals, Systems and Computers, 2002*, 3–6 November, Pacific Grove, CA, 2002, pp. 377–382.

27. G. R. Parsace, A. Yarali, and H. Ebrahimzad, "MMSE-DFE Equalizer Design for OFDM Systems with Insufficient Cyclic Prefix," in *IEEE 60th Vehicular Technology Conference*, 26–29 September, Los Angeles, CA, 2004, pp. 3828–3832.

28. Z. Jie, S. Wee, and A. Nehorai, "Channel Equalization for DMT with Insufficient Cyclic Prefix," *Conf. Rec. Thirty-Fourth Asilomar Conf. Signals, Syst. Comput. (Cat. No.00CH37154), Pacific Grove, CA, USA*, vol. 2, pp. 951–955, 2000.

29. Y. Wu, E. Pliszka, B. Caron, P. Bouchard, and G. Chouinard, "Comparison of Terrestrial DTV Transmission Systems: The ATSC 8-VSB, the DVB-T COFDM, and the ISDB-T BST-OFDM," *IEEE Trans. Broadcast.*, vol. 46, no. 2, pp. 101–113, 2000.

30. J. M. Cioffi, G. P. Dudevoir, M. V. Eyuboglu, and G. D. Forney, "MMSE Decision-Feedback Equalizers and Coding. II. Coding Results," *IEEE Trans. Commun.*, vol. 43, no. 10, pp. 2595–2604, 1995.

31. N. Zervos and I. Kalet, "Optimized Decision Feedback Equalization Versus Optimized Orthogonal Frequency Division Multiplexing for High-Speed Data Transmission Over the Local Cable Network," in *IEEE International Conference on Communications (ICC)*, 11–14 June, Boston, MA, 1989, pp. 1080–1085.

32. J. Louveaux, L. Vandendorpe, and T. Sartenaer, "Cyclic Prefixed Single Carrier and Multicarrier Transmission: Bit Rate Comparison," *IEEE Commun. Lett.*, vol. 7, no. 4, pp. 180–182, 2003.

33. N. Benvenuto, R. Dinis, D. Falconer, and S. Tomasin, "Single Carrier Modulation with Nonlinear Frequency Domain Equalization: An Idea Whose Time Has Come—Again," *Proc. IEEE*, vol. 98, no. 1, pp. 69–96, 2010.

34. H. G. Myung, J. Lim, and D. J. Goodman, "Single Carrier FDMA for Uplink Wireless Transmission," *IEEE Veh. Technol. Mag.*, vol. 1, no. 3, pp. 30–38, 2006.

35. G. Berardinelli, L. A. M. R. de Temino, S. Frattasi, M. I. Rahman, and P. Mogensen, "OFDMA vs. SC-FDMA: Performance Comparison in Local Area imt-a Scenarios," *IEEE Wirel. Commun.*, vol. 15, no. 5, pp. 64–72, 2008.

36. H. G. Myung and D. J. Goodman, *Single Carrier FDMA: A New Air Interface for Long Term Evolution*. Wiley, 2008.

37. H. Rohling, *OFDM: Concepts for Future Communication Systems*, 1st ed. Heidelberg, Dordrecht: Springer, 2011.

38. R. van. Nee and R. Prasad, *OFDM for Wireless Multimedia Communications*. Boston: Artech House, 2000.

39. J. T. E. McDonnell and T. A. Wilkinson, "Comparison of Computational Complexity of Adaptive Equalization and OFDM for Indoor Wireless Networks," in *Proceedings of PIMRC '96—7th International Symposium on Personal, Indoor, and Mobile Communications*, 18 October, Taipei, Taiwan, 1996, vol. 3, pp. 1088–1091.

40. A. Leke and J. M. Cioffi, "A Maximum Rate Loading Algorithm for Discrete Multitone Modulation Systems," in *IEEE Global Telecommunications Conference. Conference Record*, 3–8 November, Phoenix, AZ, 1997, vol. 3, pp. 1514–1518.

41. J. Campello, "Practical Bit Loading for DMT," in *IEEE International Conference on Communications*, 6–10 June, Vancouver, BC, Canada, 1999, vol. 2, pp. 801–805.

42. F. Ouyang, P. Duvaut, O. Moreno, and L. Pierrugues, "The First Step of Long-Reach ADSL: Smart DSL Technology, READSL," *IEEE Commun. Mag.*, vol. 41, no. 9, pp. 124–131, 2003.

43. E. Hossain and V. K. Bhargava, *Cognitive Wireless Communication Networks*. Springer, 2007.

44. M. Zivkovic, D. Auras, and R. Mathar, "OFDM-Based Dynamic Spectrum Access," in *2010 IEEE Symposium on New Frontiers in Dynamic Spectrum (DySPAN)*, 6–9 April, Singapore, 2010, pp. 1–2.

45. L. Hanzo, M. Münster, B. J. Choi, and T. Keller, *OFDM and MC-CDMA for Broadband Multi-User Communications, WLANs and Broadcasting*. Chichester, UK: Wiley, 2003.

46. G. J. M. Janssen, P. A. Stigter, and R. Prasad, "Wideband Indoor Channel Measurements and BER Analysis of Frequency Selective Multipath Channels at 2.4, 4.75, and 11.5 GHz," *IEEE Trans. Commun.*, vol. 44, no. 10, pp. 1272–1288, 1996.

47. ITU-R, *Guidelines for Evaluation of Radio Interface Technologies for IMT-Advanced*. ITU-R Report M.2135-1, 2009.

48. A. A. Zaidi et al., "Waveform and Numerology to Support 5G Services and Requirements," *IEEE Commun. Mag.*, vol. 54, no. 11, pp. 90–98, 2016.

49. W. C. Chen and C. D. Chung, "Spectrally Efficient OFDM Pilot Waveform for Channel Estimation," *IEEE Trans. Commun.*, vol. 65, no. 1, pp. 387–402, 2017.

50. D. Hu, L. He, and X. Wang, "An Efficient Pilot Design Method for OFDM-Based Cognitive Radio Systems," *Wirel. Commun. IEEE Trans.*, vol. PP, no. 99, pp. 1–8, 2011.

51. J.-J. van de Beek, O. Edfors, M. Sandell, S. K. Wilson, and P. O. Borjesson, "On Channel Estimation in OFDM Systems," in *1995 IEEE 45th Vehicular Technology Conference: Countdown to the Wireless Twenty-First Century*, 25–28 July, Chicago, IL, 1995 vol. 2, pp. 815–819.

52. P. Hoeher, S. Kaiser, and P. Robertson, "Two-Dimensional Pilot-Symbol-Aided Channel Estimation by Wiener Filtering," in *1997 IEEE International Conference on Acoustics, Speech, and Signal Processing (ICASSP)*, 21–24 April, Munich, Germany, 1997, vol. 3, pp. 1845–1848.

53. P. Hoeher, S. Kaiser, and P. Robertson, "Pilot-Symbol-Aided Channel Estimation in Time and Frequency," in *Multi-Carrier Spread-Spectrum*, K. Fazel and G. P. Fettweis, Eds. Boston, MA: Springer US, 1997, pp. 169–178.

54. M. Simko, D. Wu, C. Mehlfuhrer, J. Eilert, and D. Liu, "Implementation Aspects of Channel Estimation for 3GPP LTE Terminals," in *17th European Wireless Conference*, 2011, pp. 1–5.

55. C. Peng et al., "Channel Estimation for OFDM Systems over Doubly Selective Channels: A Distributed Compressive Sensing Based Approach," *Commun. IEEE Trans.*, vol. 2, no. 10, pp. 4173–4185, 2003.

56. F. Sanzi, S. Jelting, and J. Speidel, "A Comparative Study of Iterative Channel Estimators for Mobile OFDM Systems," *IEEE Trans. Wirel. Commun.*, vol. 2, no. 5, pp. 849–859, 2003.

57. M. C. Necker and G. L. Stüber, "Totally Blind Channel Estimation for OFDM on Fast Varying Mobile Radio Channels," *IEEE Trans. Wirel. Commun.*, vol. 3, no. 5, pp. 1514–1525, 2004.

58. Y. Liu, Z. Tan, H. Hu, L. J. Cimini, and G. Y. Li, "Channel Estimation for OFDM," *IEEE Commun. Surv. Tutorials*, vol. 16, no. 4, pp. 1891–1908, 2014.

59. M. Morelli, C.-C. J. Kuo, and M.-O. Pun, "Synchronization Techniques for Orthogonal Frequency Division Multiple Access (OFDMA): A Tutorial Review," *Proc. IEEE*, vol. 95, no. 7, pp. 1394–1427, 2007.

60. S. Wei, D. L. Goeckel, and P. E. Kelly, "A Modern Extreme Value Theory Approach to Calculating the Distribution of the Peak-to-Average Power Ratio in OFDM Systems," *2002 IEEE Int. Conf. Commun. (ICC)*., vol. 3, pp. 1686–1690, 2002.

61. S. Han and J. Lee, "An Overview of Peak-to-Average Power Ratio Reduction Techniques for Multi-carrier Transmission," *Wirel. Commun. IEEE*, vol. 12, no. 2, pp. 56–65, 2005.

62. T. Jiang and Y. Wu, "An Overview: Peak-to-Average Power Ratio Reduction Techniques for {OFDM} Signals," *IEEE Trans. Broadcast.*, vol. 54, no. 2, pp. 257–268, 2008.

63. F. Sandoval, "Hybrid Peak-to-Average Power Ratio Reduction Techniques: Review and Performance Comparison," *IEEE Access*, vol. 5, pp. 27145–27161, 2017.

64. J. Proakis and M. Salehi, *Digital Communications*, 5th ed. McGraw-Hill Education, 2007.

65. 3GPP, *3rd Generation Partnership Project; Technical Specification Group Radio Access Network; Evolved Universal Terrestrial Radio Access (E-UTRA); Base Station (BS) Radio Transmission and Reception.* 3GPP Technical Specifications 36.104 V13.10.0, 2017.

66. S. Brandes, I. Cosovic, and M. Schnell, "Sidelobe Suppression in OFDM Systems by Insertion of Cancellation Carriers," in *VTC-2005-Fall. 2005 IEEE 62nd Vehicular Technology Conference*, 28 September, Dallas, TX, 2005, vol. 1, pp. 152–156.

67. A. Selim, I. Macaluso, and L. Doyle, "Efficient Sidelobe Suppression for OFDM Systems Using Advanced Cancellation Carriers," in *IEEE International Conference on Communications (ICC)*, 9–13 June, Budapest, Hungary, 2013, pp. 4687–4692.

68. J. F. Schmidt, D. Romero, and R. López-Valcarce, "Low-power active interference cancellation for OFDM spectrum sculpting," *IEEE Commun. Lett.*, vol. 18, no. 9, pp. 1543–1546, 2014.

69. I. Cosovic, S. Brandes, and M. Schnell, "Subcarrier Weighting: A Method for Sidelobe Suppression in OFDM Systems," *IEEE Commun. Lett.*, vol. 10, no. 6, pp. 444–446, 2006.

70. H. A. Mahmoud and H. Arslan, "Sidelobe Suppression in OFDM-Based Spectrum Sharing Systems Using Adaptive Symbol Transition," *IEEE Commun. Lett.*, vol. 12, no. 2, pp. 133–134, 2008.

71. D. Qu, Z. Wang, and T. Jiang, "Extended Active Interference Cancellation for Sidelobe Suppression in Cognitive Radio OFDM Systems with Cyclic Prefix," *IEEE Trans. Veh. Technol.*, vol. 59, no. 4, pp. 1689–1695, 2010.

72. J. Van De Beek, "OFDM Spectral Precoding with Protected Subcarriers," *IEEE Commun. Lett.*, vol. 17, no. 12, pp. 2209–2212, 2013.

73. M. Gudmundson and P.-O. Anderson, "Adjacent Channel Interference in an OFDM System," in *Proceedings of Vehicular Technology Conference—VTC*, vol. 2, pp. 918–922, 1996.

74. B. Farhang-Boroujeny, "OFDM Versus Filter Bank Multicarrier," *IEEE Signal Process. Mag.*, vol. 28, no. 3, pp. 92–112, 2011.

75. A. Sahin and H. Arslan, "Edge Windowing for OFDM Based Systems," *IEEE Commun. Lett.*, vol. 15, no. 11, pp. 1208–1211, 2011.

76. European Telecommunications Standards Institute (ETSI), *Digital Video Broadcating (DVB); Interaction Channel for Digital Terrestrial Television (RCT) Incorporating Multiple Access OFDM*, vol. 1.1.1. ETSI EN 201 958 v1.1.1 (2002-03), 2002.

77. M. Rajabzadeh and H. Steendam, "Power Spectral Analysis of UW-OFDM Systems," *IEEE Trans. Commun.*, vol. 66, no. 6, pp. 2685–2695, 2017.

78. D. Kivanc and H. Liu, "Subcarrier Allocation and Power Control for OFDMA," in *Conference Record of the Thirty-Fourth Asilomar Conference on Signals, Systems and Computers (Cat. No.00CH37154)*, Pacific Grove, CA, USA, 2000, pp. 147–151.

79. K. Lee, S. R. Lee, S. H. Moon, and I. Lee, "MMSE-Based CFO Compensation for Uplink OFDMA Systems with Conjugate Gradient," *IEEE Trans. Wirel. Commun.*, vol. 11, no. 8, pp. 2767–2775, 2012.

80. D. Huang and K. Ben Letaief, "An Interference-Cancellation Scheme for Carrier Frequency Offsets Correction in OFDMA Systems," *IEEE Trans. Commun.*, vol. 53, no. 7, pp. 1155–1165, 2005.

81. A. Farhang, N. Marchetti, and L. E. Doyle, "Low Complexity LS and MMSE Based CFO Compensation Techniques for the Uplink of OFDMA Systems," *2013 IEEE International Conference on Communications (ICC)*, 9–13 June, Budapest, Hungary, 2013, pp. 5748–5753.

82. A. Sahin, I. Guvenc, and H. Arslan, "A Survey on Multicarrier Communications: Prototype Filters, Lattice Structures, and Implementation Aspects," *IEEE Commun. Surv. Tutorials*, vol. 16, no. 3, pp. 1312–1338, 2014.

83. F.-L. Luo and C. Zhang, *Signal Processing for 5G: Algorithms and Implementations*. Wiley, 2016.

84. S. J. S. Eckhard Hitzer, *Quaternion and Clifford Fourier Transforms and Wavelets*. Basel: Springer, 2013.

85. D. Na and K. Choi, "Low PAPR FBMC," *IEEE Trans. Wirel. Commun.*, vol. 17, no. 1, pp. 182–193, 2018.

86. B. Farhang-Boroujeny and B. Farhang-Boroujeny, "Filter Bank Multicarrier Modulation: A Waveform Candidate for 5G and Beyond," *Adv. Electr. Eng.*, vol. 2014, p. e482805, 2014.

87. M. Schellmann et al., "FBMC-Based Air Interface for 5G Mobile: Challenges and Proposed Solutions," *Proc. 9th Int. Conf. Cogn. Radio Oriented Wirel. Networks*, vol. i, pp. 102–107, 2014.

88. C. Kim, Y. H. Yun, K. Kim, and J. Y. Seol, "Introduction to QAM-FBMC: From Waveform Optimization and to System Design," *IEEE Commun. Mag.*, vol. 54, no. 11, pp. 66–73, 2016.

89. B. Ahmed and M. A. Matin, *Coding for MIMO- OFDM in Future Wireless Systems*. Switzerland: Springer International Publishing AG, 2015.

90. Taewon Hwang, Chenyang Yang, Gang Wu, Shaoqian Li, and G. Ye Li, "OFDM and Its Wireless Applications: A Survey," *IEEE Trans. Veh. Technol.*, vol. 58, no. 4, pp. 1673–1694, 2009.

HOMEWORK

10.1 In the context of Section 10.2.4, let $\Delta j \overset{\text{def}}{=} N_c - N_H - 1$. Show that if $\Delta > 0$, (10.20) is still valid if the DFT is performed to the sample block $\{b_n, n = -\delta, -\delta - 1, \ldots, N - \delta - 1\}$ instead of the block $\{b_n, n = 0, 1, \ldots, N - 1\}$ as used in Section 10.2.4, as long as $\Delta \geq \delta \geq 0$. The only difference is that the effective channel $\{\tilde{h}_k\}$ would contain a phase shift, which depends on k and δ. This means there is a tolerance in alignment between the transmitted and received signals.

10.2 Study the OFDM modulation parameters (discussed in Section 10.2.10) for the LTE downlink FDD signal. Compare these parameters with the relevant channel characteristics. Related information can be found in the literature.

10.3 Write a program to reproduce the plots in Figure 10.7.

10.4 In the context of Section 10.5.2.1, show that a sample timing offset in the time domain produces a phase rotation in the frequency domain data $\{\tilde{b}_k\}$ given by (10.13). You can leverage the results from Problem 1.

10.5 (Optional) Consider the optimal filtering given in (10.89) of Section 10.5.1.3. How does the knowledge of channel delay spread T_H, which is explicitly used in the previous method (10.83), affect Φ_{rr} and $\phi(k, l)_{rh}$?

10.6 In the context of Section 10.6.2, check whether the guard band reserved in the LTE and 802.11 OFDM specification would make the OFDM spectrum satisfy the emission limits. Emission limits for LTE and 802.11 can be found in [65, sec. 6.6.2.1] and [21], respectively. You can do it either by analysis based on (10.55) or by simulation.

CHAPTER 11

MULTIPLE-INPUT MULTIPLE-OUTPUT (MIMO) TECHNOLOGY

11.1 INTRODUCTION

This chapter introduces the multiple-input multiple-output (MIMO) technologies. The goal is to prepare the reader for further study of the subject by explaining the most common MIMO schemes and providing a general theoretical framework. Provided first is a brief overview of the MIMO technology and its history (Section 11.2). Next, MIMO technologies are introduced through three examples: the array antennas (Section 11.3), the Bell Laboratories Layered Space-Time (BLAST) algorithm (Section 11.4), and the Alamouti code (Section 11.5). Section 11.6 describes the general theoretical framework. Section 11.7 describes a few other MIMO forms not covered by the previous examples. Section 11.8 briefly discusses some other MIMO topics. Section 11.9 provides some MIMO application examples, and Section 11.10 provides a summary. More details in some derivations are collected in the appendices (Sections 11.11 and 11.12).

MIMO is a family of multiple technologies, each in turn with numerous discoveries and innovations. Many books on MIMO provide various levels of details [1]. This chapter is only an introduction to the subject. In this chapter, readers will learn enough about MIMO basic concepts to decide whether MIMO is suitable for their specific applications. The basic concepts introduced will be helpful as the readers continue to study more specialized MIMO textbooks and research papers.

On the other hand, MIMO are also an interesting extension to the conventional communications theories. Some mathematical details in MIMO provide deeper insights that apply to other branches of communications as well. To that end, Sections 11.11 and 11.12 serve as appendices and provide some detailed mathematical derivations that may be of interest to some readers.

Discussions in this chapter are based on linear algebra. Chapter 1 gives the mathematical notations used. We use N_T and N_R for the number of antennas at the transmitter and receiver, respectively.

Our discussions are in the baseband. Therefore, all signals, noises, and channel coefficients are complex numbers.

Digital Communication for Practicing Engineers, First Edition. Feng Ouyang.
© 2020 by The Institute of Electrical and Electronics Engineers, Inc.
Published 2020 by John Wiley & Sons, Inc.

11.2   MIMO OVERVIEW

In the context of wireless communications, MIMO denotes a system where the transmitter and receiver have multiple antennas. The transmitter antennas transmit different, but coordinated, signals. The receiver processes signals from the multiple antennas jointly. In the context of MIMO, the conventional systems with a single-antenna transmitter and receiver are referred to as single-input, single-output (SISO). Similarly, systems with a single-antenna transmitter and a multi-antenna receiver are referred to as single-input, multiple-output (SIMO), and those with multi-antenna transmitters and single-antenna receivers are referred to as multiple-input, single-output (MISO). They can all be considered as special forms of MIMO.

The main advantage of a MIMO system is the great gains in channel capacity. Without increasing bandwidth and power, the capacity of a MIMO system may increase linearly with the number of antennas under certain channel conditions, as will be shown in Section 11.6. Consider a simple case of a wireless system using 16-quadrature amplitude modulation (16-QAM). To double the capacity (data rate), one can use a MIMO system with two antennas on each side. Alternatively, one can increase the transmission power and use higher-order modulation (256-QAM, or double the number of bits per symbol). In this option, to achieve the same bit error rate (BER), the transmit power needs to be increased by 12 dB, or more than 16 times (see Chapter 4)!

The history of MIMO can be traced back to the development of antenna arrays, primarily to form directed beams for radar applications. Radar systems with antenna arrays were an active research area during and after World War II. In the 1970s, the introduction of digital signal processors enabled more sophisticated adaptive processing. These techniques were vigorously developed for military applications.

In the 1990s, people started to realize that a MIMO system achieves huge gains in capacity in wireless communication. The MIMO capacity was studied by several AT&T Bell Labs researchers. Winters was among the first who proposed the concept of increased capacity with multiple antennas [2]. Salz and Wyner pointed out the capacity gain from a MIMO system [3]. These studies are more or less tied to specific schemes and receiver structures. Later in 1995, Telatar [4, 5] and Foschini and Gans [6, 7] provided more general formulation of channel capacity for Rayleigh channels with and without fading correlations.[1] In their studies, the channels were assumed to be stochastic, and the capacities were expressed in statistical senses. Their results are considered today as standard expressions of the MIMO capacities. Probably independently, Raleigh and Cioffi also studied MIMO capacity in 1996 [8]. They addressed the issue of delay-spread in time domain and proposed a solution based on orthogonal frequency division multiplexing (OFDM), which is discussed in Chapter 10.

One MIMO approach is reaping the capacity gain using parallel communication channels to increase data rate (see Section 11.4). In 1985, Salz of AT&T proposed a minimum mean-squared error (MMSE) receiver for such a MIMO system (he called it

---

[1] Channel fading is discussed in later parts of this chapter and Chapter 6.

cross-coupled linear channels) [9]. In 1996, the Bell Labs researchers devised a practical scheme, known as the Bell Laboratories Layered Space-Time (BLAST) [10]. A prototype BLAST system and associated field tests were reported in 2003 [11].

Another MIMO approach is using the added spatial dimension to gain diversity against channel fading (see Section 11.5). A typical technique is space-time coding. The most famous code is the Alamouti code, invented in 1997 [12, 13]. Subsequently, a general theory about the design and performance of space-time codes, as well as a different family of code construction, was published by Tarokh et al. of AT&T in 1998 [14].

The first commercial product using MIMO technology was introduced in 2002. Since then, MIMO has been applied in various commercial wireless standards and other products. Section 11.9 provides more information on MIMO applications.

More coverage of the emerging MIMO-related technologies can be found in Chapter 12.

11.3   A SIMPLE CASE OF MIMO: MULTIBEAM TRANSMISSION

This section presents a naïve idea of using multibeam transmission to increase system capacity. Such an idea introduces spatial multiplexing, one of the key MIMO concepts.

11.3.1   Phased-Array Antennas

It is well-known that with an antenna array and proper phase relationship among the antennas, we can steer the emission beam to one particular direction. Figure 11.1 shows the setup. The dots represent the antennas. The N_T antennas form a linear array, with inter-antenna spacing d. Transmission wavelength is λ.

When observation distance is much larger than antenna size, the amplitude distribution is[2]

$$A(\theta) = \sum_{k=1}^{N_T} a_k e^{-j2\pi k\epsilon \sin\theta},$$

(11.1)

$$\epsilon \stackrel{\text{def}}{=} d/\lambda$$

where $A(\theta)$ is the field strength (including amplitude and phase) at angle θ from the normal direction and $\{a_k\}$ is the complex amplitude of the signal applied to antenna element k. To focus the energy to one direction, we can form a special phase relationship among the transmitted signals $\{a_k\}$. For example, we may have

$$a_k = a e^{j2\pi k\epsilon \sin\gamma},$$

(11.2)

---

[2] For simplicity, we do not consider propagation amplitude attenuation in this analysis. Equivalently, we can say that a_k is scaled by a factor that reflects propagation attenuation. At a far distance, the propagation attenuation is approximately the same for all antenna elements and can thus be ignored in this discussion.

where γ is the pointing angle for the intended beam. From (11.1) and (11.2), the resulting field strength is

$$A(\theta) = ae^{j\pi(N_T+1)\epsilon(\sin\gamma-\sin\theta)} \frac{\sin[\pi N_T\epsilon(\sin\gamma-\sin\theta)]}{\sin[\pi\epsilon(\sin\gamma-\sin\theta)]}. \tag{11.3}$$

As indicated by the last factor, this is a "focused beam" facing angle γ. The beam width is determined by $N_T\epsilon$, which is the total width of the array measured in wavelength. To ensure the beam has a single peak (i.e., the denominator in (11.3) has only one zero point), we need to have

$$\epsilon < 0.5. \tag{11.4}$$

All of the following discussions assume (11.4) is true.

Now let us take a step further. An antenna array can also form multiple, independently controllable beams. For example, we may have multiple signals aiming at different directions:

$$a_k = \sum_l c_l e^{j2\pi k\frac{l+0.5}{N_T}}, \tag{11.5}$$

where index l denotes various data streams, which are modulated by $\{c_l\}$. The total signal becomes

$$A(\theta) = \sum_l c_l e^{j\pi(N_T+1)\epsilon\left(\frac{l+0.5}{\epsilon N_T}-\sin\theta\right)} \frac{\sin\left[\pi N_T\epsilon\left(\frac{l+0.5}{\epsilon N_T}-\sin\theta\right)\right]}{\sin\left[\pi\epsilon\left(\frac{l+0.5}{\epsilon N_T}-\sin\theta\right)\right]}. \tag{11.6}$$

We can see that c_l modulate a beam that centers at θ_l:

$$\sin\theta_l \stackrel{\text{def}}{=} \frac{l+0.5}{\epsilon N_T}. \tag{11.7}$$

This equation also implies the range for l:

$$\left|\frac{l+0.5}{\epsilon N_T}\right| \leq 1. \tag{11.8}$$

We can see in (11.6) that at θ_l, contributions from all other beams $l' \neq l$ are zero. This means there is no interference among the beams. If a receiver is put at the direction θ_l, it will only receive the beam carrying c_l. Furthermore, from (11.8) and (11.4), the maximum number of beams (i.e., the number of allowable values of l) is N_T, which is the number of antennas in the array.

The example shows that, with N_T antennas at the transmitter and N_T antennas at the receiver, we can simultaneously transmit N_T data streams.[3] In other words, with

---

[3] One might ask whether it is possible to transmit more than N_T data streams with N_T antennas on each side. The simple answer is that we cannot avoid interferences in that case (i.e., no perfect separation is possible). As shown later, further increase of the number of data streams does not increase overall capacity.

the same frequency allocation, the capacity is increased by N_T times compared to a single-antenna system.[4]

11.3.2   Mutual Interference and Zero-Forcing Receiver

11.3.2.1   Space-Limited Receiver

Section 11.3.1 introduces a simple MIMO system based on beamforming, where channel capacity can be increased as the number of antennas (and thus the number of beams) increases.[5] Capacity gain is achieved by constructing N_T spatially separated channels without mutual interference.

However, there is a practical problem. From (11.7), the N_T receive antennas are distributed around the semicircle. In other words, the size of the receiver must be comparable with the distance between the transmitter and receiver.

In a more practical configuration, the receiver has an antenna array similar to that of the transmitter (see Figure 11.1). The spacing between the antennas (on the order of a wavelength) is much smaller than the distance between the transmitter and the receiver. In this case, we can still perform beamforming as in Section 11.3.1. For example, we can replace (11.5) with

$$a_k = \sum_l c_l e^{j2\pi k \frac{\beta l + 0.5}{N_T}}.$$

(11.9)

$$\beta < 1$$

This results in a modified version of (11.6) and (11.7):

$$A(\theta) = \sum_l c_l e^{j\pi(N_T+1)\epsilon\left(\frac{\beta l+0.5}{\epsilon N_T}-\sin\theta\right)} \frac{\sin\left[\pi N_T \epsilon\left(\frac{\beta l+0.5}{\epsilon N_T}-\sin\theta\right)\right]}{\sin\left[\pi\epsilon\left(\frac{\beta l+0.5}{\epsilon N_T}-\sin\theta\right)\right]}.$$

(11.10)

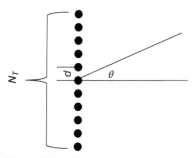

Figure 11.1   Phased array.

---

[4] To work out the details, we must also consider power normalizations when multiple antennas are introduced. However, for the purposes of illustration, the issue of normalization is not important.

[5] Here, we consider a system with the same number of antennas on both sides. However, this is not a general requirement. See Section 11.3.2.

The directions of the beams are given by θ_l:

$$\sin\theta_l = \frac{\beta l + 0.5}{\epsilon N_T}.$$ (11.11)

Equation (11.11) leads to a compressed angular spread, allowing for a smaller receiver. However, (11.10) shows that the beams will overlap with each other, causing mutual interference among the beams. This section analyzes the situation more carefully and attempts to devise a solution. We will show that a MIMO system can still work with a more sophisticated receiver. However, the performance gain is very limited.

11.3.2.2   Matrix Channel and System Equation

Before further analyses, let us switch to a more general formulation of a MIMO system. We use $H_{l,k}$ to represent the channel between the kth transmit element and the lth receive element. Let the transmission signal from the kth antenna be x_k and the received signal from the l antenna be y_l. We thus have

$$y_l = \sum_{k=1}^{N_T} H_{l,k} x_k + w_l, l = 1, 2, \ldots N_R,$$ (11.12)

where w_l is the noise at the lth receive antenna. We can rewrite (11.12) in matrix form:

$$y = Hx + w,$$ (11.13)

where matrix H (of size N_R by N_T) collects all $\{H_{l,k}\}$, vectors x (of size N_T) and y (of size N_R) collect $\{x_k\}$ and $\{y_l\}$, respectively. w (of size N_R) is a noise vector collecting $\{w_l\}$. N_T and N_R are the numbers of antennas at the transmitter and receiver, respectively. Equation (11.13) is the most basic equation for MIMO and is referred to as the "system equation" hereafter.

The most straightforward way to transmit data streams would be assigning each data stream to one component of x. However, this is not the only way. In general, we can write

$$x = Ss,$$ (11.14)

where s is a size N_D vector representing the independent data streams, N_D being the number of such data streams. S is an N_T by N_D matrix that projects the source signal s onto x. Such formulation covers the beamforming scheme described in (11.5) and (11.9), where $N_D = N_T$. Beamforming is discussed more in Section 11.7.3. The choice of S affects system performance, but at a less fundamental level. For now, consider S as an identity matrix. Therefore, we make no distinction between s and x. S is discussed in more details in Sections 11.3.3 and 11.6.2.

Consider a transmitter and receiver pair, both consisting of antenna arrays as shown in Figure 11.1 with N_T elements (i.e., $N_R = N_T$). The element spacing is $\epsilon\lambda$, where λ is the wavelength. The distance between the transmitter and the receiver is $\xi\lambda$. As in a typical system, we assume $\xi \gg \epsilon$. The following derivations in this subsection keep only the leading terms of ϵ/ξ.

The distance between the lth transmit and the kth receive antenna elements is

$$d_{l,k} = \lambda\sqrt{\xi^2 + \epsilon^2(l-k)^2} \approx \lambda\xi\left[1 + \frac{1}{2}\frac{\epsilon^2}{\xi^2}(l-k)^2\right]. \tag{11.15}$$

In the case of free-space propagation,[6] the channel causes a phase change proportional to the distance:

$$H_{l,k} = e^{-j\frac{2\pi}{\lambda}d_{l,k}}. \tag{11.16}$$

Using (11.15) and (11.16) and keeping only the leading term with regard to ϵ/ξ, we have

$$H_{l,k} \approx e^{-j2\pi\xi}e^{-j\pi\epsilon\frac{\epsilon^2}{\xi^2}(l-k)^2} \approx e^{-j2\pi\xi}\left[1 - j\pi\epsilon\frac{\epsilon^2}{\xi^2}(l-k)^2\right]. \tag{11.17}$$

Namely, $H_{l,k}$ equals to a constant $e^{-j2\pi\xi}$ plus a very small term.

11.3.2.3 Zero-Forcing Receiver

The simplest way to recover the transmitted signal x from received signal y is to apply the inverse of H:

$$\hat{x} \overset{\text{def}}{=} H^{-1}y = x + H^{-1}w. \tag{11.18}$$

The second equality comes from (11.13). This structure is known as a zero-forcing receiver.[7] It is optimal at the limit of high signal-to-noise ratio (SNR).

Therefore, it seems we can still transmit N_T data streams of communication even in the presence of mutual interference. However, (11.18) also transforms the noise. This transformation turns out to be critical to understanding the performance of such a system.

11.3.2.4 Noise Power

Let us consider the noise of the receiver as represented by (11.18). The total noise power after zero-forcing processing is

$$N_t = E\left(w^H H^{-H}H^{-1}w\right) = E\left[tr\left(H^{-1}ww^H H^{-H}\right)\right]. \tag{11.19}$$

---

[6] Here we only consider the phase. The amplitude is not important for our discussion. It does not change significantly among the elements of H.

[7] More generally, the numbers of transmit and receive antennas are not necessarily the same. Namely, H may not be a square matrix. In this case, the inverse is replaced by the pseudoinverse.

Consider the case where the channel noise w obeys independent and identically distributed (*iid*) zero mean Gaussian distribution with variance N_0. Namely,

$$E(ww^H) = N_0 I_{N_T}, \tag{11.20}$$

where I_{N_T} is an identity matrix of size $N_T \times N_T$. The total noise power is thus

$$N_t = N_0 tr(H^{-1} H^{-H}) = N_0 tr\left[(H^H H)^{-1}\right]. \tag{11.21}$$

Consider matrix $(H^H H)^{-1}$. Its trace is the sum of all of its eigenvalues (which are all nonnegative). On the other hand, as pointed out in (11.17), we can write

$$H = A + B. \tag{11.22}$$

A is a matrix with all entries equal to $e^{-j\,2\pi\xi}$. Its rank is 1. The amplitudes of all elements in matrix B are much smaller than 1 (on the order of $(\epsilon/\xi)^2$). Therefore, H is near rank 1. Likewise, $H^H H$ is also nearly rank 1. That is, all but one of its eigenvalues are very small. It follows that all but one of the eigenvalues for $(H^H H)^{-1}$ are very large. Therefore, $tr(H^{-1} H^{-H})$, which equals to the sum of these eigenvalues, is very large.

Figure 11.2 shows the eigenvalues for $(H^H H)^{-1}$, with marked parameters (both in terms of wavelengths for 2.5 GHz and in meters). The eigenvalues are computed from (11.16) directly. We can see that in all cases, there is one eigenvalue close to -18 dB (i.e., $10\log_{10}(1/64)$). All other eigenvalues are much, much larger, meaning there will be very high noise level associated with them. We further see that for the same distance (ξ), if we increase the antenna spacing (ϵ), we get better results (i.e., the eigenvalues are smaller). This suggests that antenna spacing is the key.

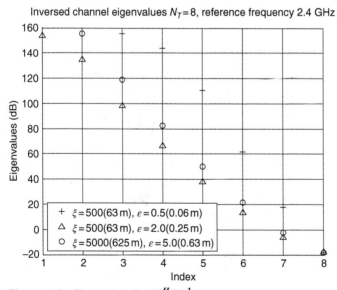

Figure 11.2   Eigenvalues for $(H^H H)^{-1}$. ξ is the distance between the transmitter and the receiver. ϵ is the antenna spacing. Both quantities are expressed in wavelengths and in meters.

Nevertheless, it is clear that for all practical purposes, our noise level will be too high. In other words, in the case of free-space propagation, the idea of using antenna array to increase channel capacity is impractical.

As mentioned in Section 11.3.2.2, here we consider only a specific way of transmitting the signal; the N_T data streams are applied directly to the individual antennas. In general, we can introduce a matrix S in "precoding" the transmitted signal, as shown in (11.14). Furthermore, we considered a particular form of receiver: the zero-forcing receiver. However, as will be shown later, the problem of H being near-singular is intrinsic to all MIMO schemes.

11.3.3 MIMO Channel Decomposition and Capacity

From the example shown in Section 11.3.2, we see that although theoretically we can transmit multiple data streams in a MIMO system, this scheme becomes impractical for a realistic channel (free space) because we encounter a much higher level of noise. Furthermore, we see that such difficulty is related to the near-singularity of the channel matrix.

The previous analysis shows another interesting point. In (11.21), the noise power for a zero-forcing receiver is related to the trace of a matrix, which is further related to the sum of its eigenvalues. In this process, is eigenvalue merely a mathematical tool to connect the trace to singularity, or does it play a more fundamental role in determining MIMO performance?

We also observed in Figure 11.2 that although other eigenvalues of $(H^H H)^{-1}$ are huge, there is one "normal" eigenvalue. We can actually achieve the SNR value as indicated by this eigenvalue by transmitting only one stream of data. In this scheme, we transmit the signal from only one of the transmit antennas (with eight times the previous per-antenna power, to keep the same total transmit power). The received signals from all receive antennas are summed together (after phase adjustment) to form the final received signal. Conversely, all transmit antennas can perform beamforming, toward one receive antenna. One data stream can thus be transmitted. Therefore, we see that although the overall MIMO channel is bad (i.e., yielding poor SNR), we have one "good" subchannel, which can be used to transmit one data stream. The question is: can we generalize such an observation to other eigenvalues? Does each eigenvalue represent a "separate" channel through which we can transmit one data stream?

The following Sections 11.3.3.1 and 11.3.2.2 present a general treatment of the problem. We will see that what was observed with a zero-forcing receiver actually reflects a fundamental property of the MIMO systems. That is, its capacity is determined by the eigenvalues of the channel matrix. Eigenvalues are indeed connected to subchannels for MIMO. Such analysis will also suggest a way to build an optimal MIMO receiver.

11.3.3.1 Singular Value Decomposition (SVD) of Channel Matrix
Let us repeat the system equation (11.13)

$$y = Hx + w, \tag{11.23}$$

where H is the channel matrix of size N_R by N_T; w is a random Gaussian vector of size N_R representing the noise; x and y are vectors of size N_T and N_R, respectively, representing the transmitted and received signals. Furthermore, assume the noise components have iid Gaussian distribution[8]:

$$E\left(ww^H\right) = N_0 I_{N_R},$$ (11.24)

where I_{N_R} is an identity matrix of size $N_R \times N_R$. Assume components in x are *iid* random numbers. Let E_s be the total signal power:

$$E\left(xx^H\right) = \frac{E_s}{N_T} I_{N_T}.$$ (11.25)

Perform SVD with channel matrix H:

$$H = UDV^H.$$ (11.26)

U and V are the unitary matrices of sizes N_R by N_R and N_T by N_T, respectively. Elements of D can be expressed as

$$D_{i,j} = \begin{cases} d_i \delta_{i,j} & \forall i = 1, 2, \ldots, r \\ 0 & \forall i > r \end{cases}.$$ (11.27)

The diagonal elements of D, $\{d_i\}$, are known as singular values of H. They are square roots of the eigenvalues of the matrix $H^H H$. r is the rank of matrix H. Naturally,

$$r \le \min(N_T, N_R).$$ (11.28)

Now, suppose we have an N_T-dimensional vector s to transmit. We can have the following transmit and receive method:

$$x = Vs$$
$$y = Hx + w.$$ (11.29)
$$\hat{s} = U^H y$$

The first line represents a precombining (known as precoding) at the transmitter. The second line represents the channel, same as in (11.23). The third line represents the receiver processing (known as combining), where $\hat{s}$ is the recovered data. Because V is unitary, (11.25) can be satisfied if and only if

$$E\left(ss^H\right) = \frac{E_s}{N_T} I_{N_T}.$$ (11.30)

---

[8] Such noise is described as "spatially white" and Gaussian. The theories discussed here can be extended to the cases where the noise components are correlated (or the noise is "spatially colored").

When these steps are combined with (11.26), we obtain

$$\hat{s} = Ds + z;$$
$$z \stackrel{\text{def}}{=} U^H w. \tag{11.31}$$

Also, because U is unitary and all components of w are iid, the output noise z has the same statistical property as w:

$$E\left(zz^H\right) = N_0 I_{N_R}. \tag{11.32}$$

Equation (11.31) can be written in component form:

$$\hat{s}_l = \begin{cases} d_l s_l + z_l & l = 1, 2, \ldots, r \\ z_l & l = r+1, r+2, \ldots, N_T \end{cases}. \tag{11.33}$$

Therefore, the MIMO scheme described by (11.29) is equivalent to a set of parallel subchannels without mutual interference. The lth component of s is transmitted in one of the subchannels, with channel gain as d_l and noise as z_l. Obviously, subchannel l can transmit data only if $d_l > 0$. Therefore, we have r "useful" subchannels, and can transmit r data streams simultaneously.[9]

To recover the transmit symbol s_l, we need to divide $\hat{s}_l$ by d_l. The noise presented to the demodulator thus becomes z_l/d_l. Therefore, small d_l values lead to poor SNR. This is the same noise amplification described in Section 11.3.2. In the free-space case, the channel is approximately rank 1. In other words, only one d_l is in nominal range; all others are very small. This is why there is one "good" subchannel. In general, the number of "good" subchannels is the number of singular values of H that yield acceptable SNR. At high SNR limit, this number equals to the rank of the channel.

To summarize, using SVD analysis, a MIMO channel is decomposed into r parallel subchannels, whose gain values are determined by the singular values of channel matrix H. We can potentially increase the total channel capacity by transmitting data simultaneously using these multiple subchannels.

11.3.3.2   MIMO Channel Capacity

Equation (11.33) shows that a MIMO channel can be decomposed into r parallel subchannels. With a gain of d_l and a noise of z_l, the SNR of subchannel l is, considering (11.30) and (11.32),

$$SNR_l = \frac{E_s}{N_T} |d_i|^2 \frac{1}{N_0}. \tag{11.34}$$

---

[9] In practice, there is no significant difference between a subchannel with zero gain and one with extremely small gain. Therefore, the number of usable subchannels actually depends on the SNR requirement for transmission. However, theoretical studies of MIMO systems often take the limit of high SNR. In this case, any subchannel with non-zero gain, no matter how small the gain would be, is usable in data transmission. In this sense, rank is an important measure of channel quality.

Remember from Chapter 2, the Shannon theory gives the channel capacity (in bits per channel use) as

$$C = \log_2(1 + SNR).$$ (11.35)

There is no factor 1/2, because we are considering complex-valued signals. The total capacity of the multiple parallel subchannels is the summation

$$C = \sum_{l=1}^{r} \log_2\left(1 + \frac{E_s}{N_T N_0}|d_i|^2\right).$$ (11.36)

C is the channel capacity in bits per channel use. This can be written in several equivalent forms [5, 6]:

$$C = \sum_{l=1}^{r} \log_2\left(1 + \frac{E_s}{N_T N_0}\lambda_l\right) = \log_2 det\left(I_{N_R} + \frac{E_s}{N_T N_0}DD^H\right)$$
$$= \log_2 det\left(I_{N_R} + \frac{E_s}{N_T N_0}HH^H\right)$$ (11.37)

D is the diagonal matrix from SVD of H specified in (11.26) and (11.27). $\{\lambda_l\}$ is given by

$$\lambda_l = |d_l|^2.$$ (11.38)

$\{\lambda_l\}$ are also the eigenvalues of matrices HH^H and DD^H. $det(A)$ denotes determinant of matrix A.

The derivation of (11.37) is based on a specific transmission scheme (11.29), which requires the transmit side to know the channel. However, the result is more general. In fact, because x and s differ by a linear transformation, the capacity of the channel between s and y, which is what we have derived, and the capacity of the channel between x and y are the same.[10] Therefore, the transmitter's knowledge of channel state information is not required for realizing such capacity, although the particular scheme (11.29) does require such knowledge. An appendix (Section 11.12) provides an alternative channel capacity derivation, which does not assume the transmitter's knowledge of channel state information.

On the other hand, if the transmitter does have channel state information, it can further optimize transmission by performing water-filling (i.e., power allocation) among the subchannels, violating (11.30) (see Section 11.6.2). Water-filling will

---

[10] Because there is a one-to-one mapping between s and x, we can easily see that the entropies concerning s and x are equal (Chapter 2). Namely, $H(x) = H(s)$, $H(x \mid y) = H(s \mid y)$. Here, $H(\cdot)$ represents the entropies discussed in Chapter 2.

further increase the capacity. This situation is analogous to the water-filling in frequency domain discussed in Chapter 2 and is discussed in more details in Section 11.6.2.1.

As shown in (11.37), when $SNR \gg 1$, the capacity is proportional to the logarithm of the SNR ($E_s/N_T N_0$), just as the conventional communications systems. However, the capacity is *linearly* proportional to the number of subchannels, r, assuming all nonzero eigenvalues are of the same order of magnitude. Therefore, the MIMO gain is impacted primarily by the channel rank r and the distribution of the channel singular values. This, in turn, is determined by the number of antennas and the scattering environment. As seen before, in free space, the approximate rank of the channel is always 1, no matter how many antennas are used. However, for a rich scattering environment, the rank of the channel is often limited by the number of antennas. Namely, the equality in (11.28) holds. In that case, MIMO channel capacity grows linearly with the number of antennas. More discussions on channel conditions can be found in Chapter 2 and Section 11.6.1.

In the MIMO literature, the following normalization is often used [1, ch. 3.1]. The channel and noise are normalized so that

$$\|\bar{H}\|_F^2 \overset{\text{def}}{=} \sum_{i,j} |\bar{H}_{i,j}|^2 = N_R N_T$$

$$E\left(|\bar{w}_i|^2\right) = 1 \qquad , \qquad (11.39)$$

$$E\left(|\bar{x}_i|^2\right) = 1$$

where $\bar{H}$, $\bar{w}$, and $\bar{x}$ are scaled versions of H, w, and x to satisfy (11.39). $\|\cdot\|_F$ denotes the Frobenius norm:

$$\|A\|_F \overset{\text{def}}{=} \sqrt{\sum_{n,m} |A_{n,m}|^2} = \sqrt{Tr(AA^H)}. \qquad (11.40)$$

Received signal y is also scaled accordingly into $\bar{y}$ so that (11.23) becomes

$$\bar{y} = \sqrt{\rho}\bar{H}\bar{x} + \bar{w}. \qquad (11.41)$$

ρ captures SNR. In this form, the channel capacity is expressed as

$$C = \log_2 det\left(I_{N_R} + \frac{\rho}{N_T}\bar{H}\bar{H}^H\right). \qquad (11.42)$$

11.3.4   Summary

Section 11.3 presents the basic concept of MIMO communications system. Starting from the naïve idea of multiple beamforming, we devise a MIMO system that allows transmission and reception of multiple data streams. We further developed the idea mathematically and showed that a MIMO channel can indeed be decomposed into parallel subchannels (or generalized "beams") without mutual interference. Following this approach, the channel capacity is derived. Through this process, we obtain the basic concept of how a MIMO system obtains its capacity gain, and how such gain depends on the system configuration and the properties of the channel. In addition to the SNR, the channel capacity depends on the rank and singular values of the channel.

In closing, it is important to point out that sending multiple data streams is just one of the ways to take advantage of a MIMO channel. Other benefits are discussed in later parts of this chapter.

11.4   SPATIAL MULTIPLEXING: BELL LABORATORIES LAYERED SPACE-TIME (BLAST)

Equation (11.29), in addition to showing channel capacity derivation, also suggests a transmitter–receiver structure (known as precoding) to realize the capacity. Unfortunately, the transmitter does not always know the channel state information, which is required when constructing the precoding matrix V. This is especially true for a mobile wireless link, whose gain can change rapidly. Section 11.3.2.3 presented an alternative, the zero-forcing receiver, which may amplify noise depending on channel characteristics.

This section builds on to the concept of decomposing a MIMO channel into subchannels and presents another practical solution where the transmitter does not know channel state information. A practical receiver is derived, known as the successive cancellation (SC). Based on such idea, we further describe a practical MIMO transceiver architecture, BLAST. In addition to its illustrational value, BLAST is also an important milestone in the development of MIMO technologies. It was first proposed in 1996 [10, 15].

Figure 11.3 shows a typical SC/BLAST MIMO architecture [15]. The transmitter is depicted on the left side. The input data go through a vector encoder to form separate data streams for each antenna. The simplest vector encoder is a demultiplexer that distributes input data among the antennas. The vector encoder can also perform channel encoding in various fashions, as described later in this section. The right-hand part is the receiver. Received signals from all antennas are processed jointly to recover the data. Note that the numbers of antennas on both sides are not necessarily equal. As in most communications systems, the receiver has the knowledge of channel state information.

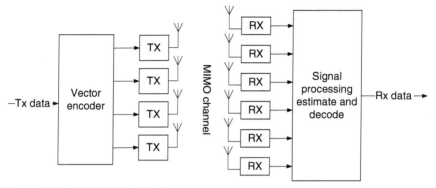

Figure 11.3   SC MIMO architecture.

11.4.1   Successive Cancellation (SC)

This section describes the successive cancellation (SC) technique, which is the basis for BLAST systems. As shown in Figure 11.3, the system contains N_T transmit antennas and N_R receive antennas. Each transmit antenna transmits an independent data stream (although the data may be coded across the antennas, as discussed in Section 11.4.3). We assume $N_T \leq N_R$ (otherwise, it is impossible to receive correctly the N_T data streams). SC is performed in steps, as illustrated in Figure 11.4.

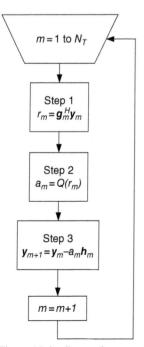

Figure 11.4   Successive cancellation processing flow.

Step 1: y_1 is the received vector. y_m is obtained iteratively in the process. y_m can be expressed as

$$y_m = \sum_{k=m}^{N_T} x_k h_k + w, \qquad (11.43)$$

where, referencing the system equation (11.13), x_k (a scalar) is the kth component of the transmitted signal vector x. h_k (a vector of size N_R) is the kth column of channel matrix H. w is the noise vector (size N_R). $x_k h_k$ is the contribution of transmitted signal x_k to y_m (size N_R). The summation in (11.43) starts with m. Contributions from $\{x_k, k < m\}$ are removed in other steps, as to be shown below. For the original signal, $m = 1$. Namely, all contributions are present.

The current task is to recover x_m by removing contributions (interferences) from other signal components $\{x_k, k = m + 1, m + 2, \ldots, N_T\}$. This is done by introducing a vector g_m of size N_R, so that

$$g_m^H h_k = \delta_{m,k}, \forall k = m, m + 1, \ldots, N_T. \qquad (11.44)$$

In other words, the orientation of g_m is chosen to be orthogonal to all h_k for $k > m$. The norm of g_m is normalized so that $g_m^H h_m = 1$. From (11.43) and (11.44), we can see that the result of step 1 is

$$r_m \overset{\text{def}}{=} g_m^H y_m = x_m + g_m^H w. \qquad (11.45)$$

In general, g_m that satisfies (11.44) is not unique. As shown in (11.45), we look for g_m with minimum norm to optimize SNR. The output of this step, r_m, is a scalar representing our estimate of x_m. Details on how to compute for g_m are not included here. An appendix (Section 11.11) provides an example in algorithm.[11]

Step 2: We reach a decision as to what is transmitted in x_m (by mapping the estimated signal r_m to the finite alphabet used for modulation at transmitter). This can be done with a slicer or a more elaborated decision device such as a decoder. The output a_m is our decision of the transmit symbol for component m.

Step 3: Once we have decided on the transmitted symbol a_m, its contribution to the received signal y is reconstructed and removed from y_m, yielding the received vector for the next iteration y_{m+1}.

$$y_{m+1} = y_m - a_m h_m. \qquad (11.46)$$

Equation (11.46) guarantees the validity of (11.43) through the iterative process, given that all decisions in step 2 are correct:

$$a_m = x_m. \qquad (11.47)$$

This brings us back to step 1 for $m + 1$. The loop continues until all N_T data streams are recovered.

---

[11] The algorithm provided in Section 11.11 does not exactly satisfy (11.44). Instead, it balances residual interference and noise to achieve MMSE. See derivations and discussions therein for more details.

To summarize, the receiver attempts to estimate and decode the transmitted data stream one by one. For each stream, the interferences from other streams are reduced in two ways. For the ones already demodulated or decoded, their interferences are estimated and subtracted (cancelled) in step 3. For the ones not yet decoded, their interferences are rejected (nulled) by projecting the received signal to a direction orthogonal to all interferers in step 1. Nulling and cancellation are key operations in SC.

The noise at iteration m is $\left|g_m^H w\right|^2$. Because w is a random vector, the expectation value of the noise is proportional to $|g_m|^2$. In general, the norm of g_m becomes smaller as the iteration progresses. This is because g_m is constrained by the set of vectors $\{h_k, k = m + 1, \ldots, N_T\}$ it must be orthogonal with. The number of such vectors decreases as the iteration progresses, affording g_m with more freedom to reduce its norm. In other words, the price of nulling is limiting the orientations that g_m can take, resulting in poorer SNR. Therefore, SNR is the worst at the beginning. Namely, decisions in step 2 are more likely to have errors at the beginning of the iteration process. Unfortunately, early errors are also more damaging because they invalidate (11.47), causing additional errors in subsequent iterations.[12] This dilemma can be called the "early error problem."

The early error problem can be addressed by ordering the data streams for SC or by error correction coding across the data streams. Sections 11.4.2 and 11.4.3 describe these two approaches.

11.4.2   Vertical BLAST (V-BLAST)

V-BLAST is the simpler form of BLAST [15, 16]. It was introduced for the construction of the first BLAST prototypes and remains a practical approach since then.

A V-BLAST system addresses the early error problem by reordering the data streams for SC. In system equation (11.13), the order of components in x can be chosen at will, as long as the order for the columns of H is adjusted accordingly. Because of channel variation, such ordering affects the SNR carried in r_m given in (11.45). Therefore, we can choose an optimal ordering, so that the SNR of r_m is higher for the smaller m values, namely in the earlier iterations. Such ordering helps to mitigate the early error problem.

In a V-BLAST receiver, at each step m, a search is conducted among the remaining data stream to be processed. Each of the remaining streams is assigned as data stream m and is processed as depicted by (11.43) and (11.45). The data stream with the highest SNR is the final choice for data stream m.[13]

---

[12] The damage caused by the early errors is similar to the "error propagation" issue with the decision feedback equalizer described in Chapter 9.

[13] SNR can be estimated using training symbols that are known to the receiver. Once the order is determined, they apply to subsequent symbols before the next channel update.

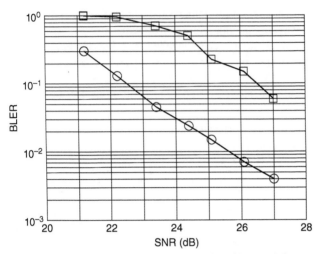

Squares = nulling alone; circles = nulling with cancellation

Figure 11.5   Block error rates for V-BLAST systems [15, fig. 2]. Reproduced with permission.

It turns out that such strategy yields a globally optimal ordering, although data stream selection is performed locally at each iteration.

Figure 11.5 [15] shows the block error rate (BLER) plots for the V-BLAST system. It was measured using a real implementation (as opposed to simulation) through an indoor channel. The system has 8 transmit antennas and 12 receive antennas transmitting eight data streams with uncoded 16-QAM. The total spectrum efficiency is 25.9 bit/s/Hz (4 bits per symbol per data stream, eight data streams, with a symbol rate of 24.3 ks/s occupying a 30-kHz bandwidth). The block length is 100 symbols. From the plot, we can see that at 27-dB average SNR, the BLER is about 4×10^{-3}, translating to approximately 10^{-6} in bit error rate (BER). In comparison, with an additive white Gaussian noise (AWGN) channel at the same SNR, a 16-QAM (spectrum efficiency of 3.24 bit/s/Hz) should achieve approximately 10^{-7} BER. Therefore, there is some degradation in BER performance compared with truly separated channels. However, the eightfold increase in spectrum efficiency is still very impressive. The plot also shows that the V-BLAST scheme (circles) results in an SNR gain of about 4 dB compared to a linear receiver, which is similar to the zero-forcing method described in Section 11.3.2.3.

V-BLAST uses symbol ordering to reduce the performance degradation caused by early errors. By processing data streams with high per-path SNR first, we reduce the probably of early errors. After cancelling interference from these data streams, the SNR of the remaining data streams can be improved, resulting in better overall performance. This scheme does not rely on channel coding. On the other hand, a cleverly designed channel coding scheme can help deal with the early error problem differently, as discussed next.

11.4.3   Diagonal BLAST (D-BLAST)

D-BLAST addresses the error propagation problem with an ingenious cross-stream coding scheme [17]. Figure 11.6 shows an example with four antennas.

The transmitted data stream is divided (demultiplexed) into N_T *layers* in coding. In the case shown in Figure 11.6, N_T is equal to 4. Each layer is coded independently in channel coding, as shown in the upper part of Figure 11.6. The coded and modulated symbols are sent to the Circulator, which applies the symbols to the antennas in a circular fashion.

The bottom half of Figure 11.6 shows how the Circulator works. In this part, each row represents an antenna (space). Each column represents a symbol slot (time). The letters at each space-time grid point indicates the identity of the layer being sent. As shown, the layers are rotated among the antennas as symbol time progresses. Each layer progresses diagonally in the space-time grid, thus the name diagonal-BLAST. This diagonal structure helps in addressing the early error problem as described below. There is a beginning period and an ending period for each frame, as shown on the left and right sides of the figure, where less than N_T symbols are transmitted for each symbol slot.

In contrast to V-BLAST, where detection is done symbol by symbol, the detection in D-BLAST is done layer by layer. At the beginning of the frame, a symbol in layer A (call it A_1) is transmitted alone. Without any interference, it achieves the highest possible SNR. For the next symbol slot, a symbol in layer A (call it A_2) is transmitted together with a symbol in layer D (call it D_1). The receiver can obtain an estimate of A_2, while the interference from D_1 is avoided by nulling (step 1 in Section 11.4.1). The SNR for A_2 suffers during the process. In the next symbol slot, three symbols (A_3, D_2, and C_1) are transmitted. Again, A_3 can be recovered by nulling against D_2 and C_1, with even more loss of SNR. The process continues until we receive N_T symbols in layer A. These N_T symbols typically have progressively degrading SNR. On the other hand, these symbols are protected by channel coding with

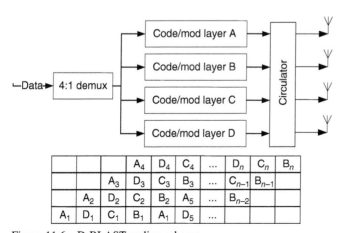

Figure 11.6   D-BLAST coding scheme.

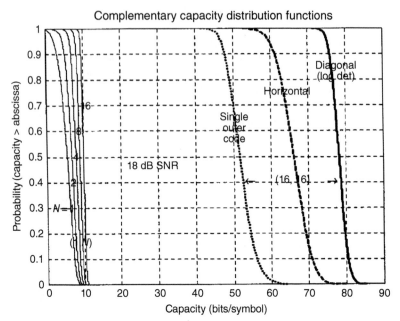

Figure 11.7   Outage capacity of MIMO systems [17, fig. 4]. Reproduced with permission.

interleaving and therefore can be decoded with the performance determined by the average SNR. In other words, the error rate for layer A is not dominated by the worst SNR. Once these symbols are determined by decoding, their interference to other symbols is cancelled (step 3 in Section 11.4.1). Now, layer D is in the same situation previously described for layer A. Namely, the first symbol D_1 is received without interference (after removal of the interference from A_2), and so on.

In addition to guarding against the SNR variation because of the detection order, such diagonal coding structure also guards against variation in gains of the subchannels. Through cross-antenna coding, a "good" subchannel can help a "bad" subchannel in achieving error-free transmission.[14] Therefore, in addition to increasing the data rate, D-BLAST also improves performance in the presence of channel fading. Such improvement is known as diversity, another important concept in MIMO that is discussed in Section 11.5.

Figure 11.7 [17] shows the performance of various MIMO systems. The channel is flat-fading Rayleigh, a stochastic channel whose gain is different in different realizations (Chapter 2 and Section 11.6.1). The average SNR is 18 dB. The x-axis indicates the desired capacity levels. The y-axis shows the probabilities of getting such capacity levels or better. This performance metric is known as "outage capacity" and is discussed in more details in Section 11.6.2. Obviously, a data point that is higher or to the right represents better performance.

---

[14] The subchannels here are different from those defined in Section 11.3.3. However, the concept remains the same.

The five SIMO curves on the left side are not relevant to this discussion. On the right side, the three curves show the BLAST systems (with 16 antennas on each side) with three different coding schemes. The "single outer code" is a straightforward SC system with single-stream channel coding. The "horizontal" is analogous to V-BLAST, with optimal ordering in detection and separate channel coding for each antenna. The "diagonal" is the D-BLAST. There is clearly a performance difference.

However, in this case there are a large number of antennas (16) on each side, whereas practical systems today use 2, 4, or 8 antennas. The performance advantage of D-BLAST is less pronounced with fewer antennas. In fact, D-BLAST is rarely used in today's commercial MIMO systems.

11.4.4   Generalized Decision Feedback Equalizer (GDFE)

The SC algorithm derived in Section 11.4.1 seems intuitive. Under the assumption of no decision errors, an optimal solution for the SC is derived in an appendix (Section 11.11) based on the MMSE criterion. The BLAST technologies provide additional ways to alleviate the problem of early errors or error propagation.

Mathematically, the SC algorithm is similar to the classic decision feedback equalizer (DFE) [18]. The nulling part corresponds to the feedforward filter, and the cancelling part corresponds to the feedback filter. The similarity is in concept as well as in the derivation of the optimal solutions. DFE is discussed in detail in Chapter 9.

Moreover, as in the case of DFE, one can move the cancellation part to the transmitter, if it has the channel state information [19]. This configuration eliminates the error propagation because the transmitter knows perfectly the interfering symbols. This concept is a generalization of the Tomlinson precoder in connection with DFE as described in Chapter 9. Unlike the linear precoding described in Section 11.3.3, Tomlinson precoding ensures that the peak signal levels at each transmit antenna are well controlled.

One can further combine the space (SC) and time (DFE), and build a generalized DFE (GDFE) to handle interferences in both dimensions [20, 21]. However, implementation is complex. There are usually better practical approaches (such is a MIMO-OFDM system). However, the concept is powerful and useful for some specific systems, such as the uplink of the long-term evolution (LTE)-advanced cellular system [22, 23].

11.5   SPATIAL DIVERSITY: SPACE-TIME CODING

This section examines the MIMO system from a different angle—diversity. We start with the basic concept of spatial diversity and then examine a simple yet practical embodiment—the Alamouti code.

11.5.1   Why Diversity?

In a typical mobile wireless communications application, the channel gain changes with time as a result of multipath, blockage, and other factors (Chapter 6).

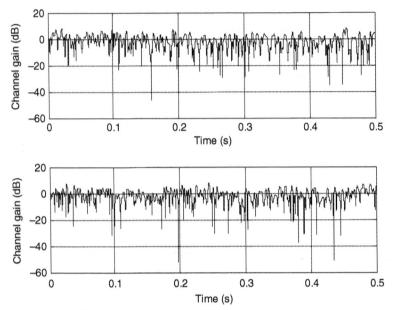

Figure 11.8 Channel gains for two mobile channels.

Figure 11.8 shows the plot of two mobile channels, based on the standard Rayleigh channel model. The carrier frequency is 2 GHz, and the vehicle speed is 100 km/h (approximately 60 mph). The channel gains are relative to the average gain. As shown, the channel gain fluctuates constantly. Although the channel gain is usually not very different from its average value, there are intervals when the channel gain is very low (referred to as deep fade).

Consider the case where the transmitter does not know the channel state. It must select a data rate that is lower than the channel capacity with proper modulation and coding scheme so as to achieve errorless transmission. However, the channel capacity changes with channel gain and can be very low at times. Therefore, to sustain uninterrupted communication, the data rate must accommodate the worst-case channel gain and be practically zero.

On the other hand, if we can tolerate some transmission failure (known as outage), a higher data rate can be chosen because the outage-causing "deep fade" is a rare event. Therefore, another metric, outage capacity, is introduced to measure performance under fading. Outage capacity is the capacity that can be sustained given a probability of outage. Obviously, outage capacity increases when a larger probability of outage is allowed. More discussions are presented in Section 11.6.2.2.

Now consider cases where we have two transmitter antennas and one receiver antenna available for the data stream. With this arrangement, there are two channels for transmission. Further assume these channels fade independently, as shown in the two plots in Figure 11.8. This assumption is true when the two transmitter antennas are reasonably separated in space (e.g., half wavelength. See Section 11.6.1.3). Because the two channels are not likely to get into deep fading at the same time, we may be able to use the "better channel" and thus avoid very low channel capacity. This consideration is the basic idea of diversity.

Figure 11.9 shows outage capacity (vertical axis) corresponding to the various outage probabilities (horizontal axis) under various schemes. A higher line represents better performance.

The two thin lines in Figure 11.9 show the outage capacities for the two Rayleigh channels in Figure 11.8 when they are used separately by a conventional SISO system. As shown, these capacities drop fast when low outage probability (i.e., higher reliability) is desired, due to deep fading.

To counter that problem, an ideal solution is picking the instantaneously better channel to transmit, leading to the performance shown by the thick solid line. We will still encounter deep fading if it happens to both channels at the same time. However, the probability of that happening would be much smaller. As expected, significant improvement is achieved. Unfortunately, such a scheme requires the transmitter to have instantaneous channel state information, which is impractical in many cases.

Without channel state information at the transmitter side, it may seem that the transmitter can do no better than sending the same signal through both channels. The effective channel in this case would be the sum of the two channels. The capacity for this scheme is shown in Figure 11.9 by the thick dashed line. It does not bring apparent improvement over the single-channel transmission, because the fading channels also have random phases. A simple sum would sometimes result in destructive interference.

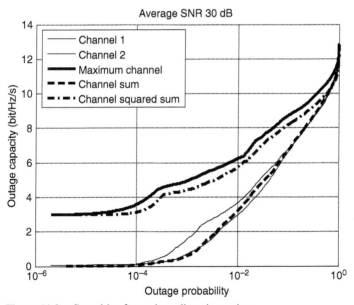

Figure 11.9   Capacities for various diversity options.

On the other hand, suppose we can have a channel that is the power sum of the two channels, as shown in (11.50) later, we would be able to avoid the random phase problem. The resulting outage capacity is shown in Figure 11.9 with the thick dash-dotted line. Its performance level is close to the ideal case (the thick solid line).

Such an effective channel could be achieved if we had two receiver antennas and used the maximum–ratio receiver combining (MRRC) technique (not discussed here). In the case of two transmitter antennas and one receiver antenna, however, MRRC cannot be done. However, a technique known as space-time coding can achieve the same result, as shown below.

11.5.2   Alamouti Code

Alamouti code is the first space-time code reported [13]. Starting as a clever design, it later turned out to be a member of a more general code class—the orthogonal space-time block code (STBC) (see Section 11.6.3.2). Alamouti originally proposed and analyzed several configurations, including multiple antennas on both the transmitter and receiver sides. Here we only describe a simple, yet widely used, case to illustrate the concept.

Consider the same configuration discussed in Section 11.5.1. It has a transmitter with two antennas and a receiver with one antenna. The transmitter does not have instantaneous channel state information, whereas the receiver has perfect channel state information. In the system configuration shown in Figure 11.10 [13], the two rows on top of the antennas show a symbol transmitted at the two symbol slots (symbol periods).

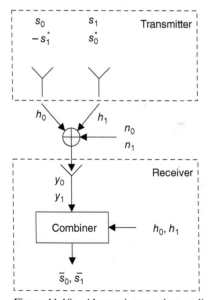

Figure 11.10   Alamouti space-time coding scheme [13].

In the first symbol slot, antenna 0 and antenna 1 transmit symbols s_0 and s_1, respectively. In the second symbol slot, the two antennas transmit $-s_1^*$ and s_0^*, respectively. The two channels from transmitter antennas to the receiver are h_0 and h_1, which remain constant over these two symbol slots. The additive noises that the receiver experiences in the two symbols slots are n_0 and n_1, which are assumed to be uncorrelated. Thus, the received signals in the two symbol slots y_0, y_1 are:

$$y_0 = h_0 s_0 + h_1 s_1 + n_0$$
$$y_1 = -h_0 s_1^* + h_1 s_0^* + n_1 \qquad (11.48)$$

At the receiver, these signals can be combined to form the final received symbols:

$$\tilde{s}_0 = h_0^* y_0 + h_1 y_1^*$$
$$\tilde{s}_1 = h_1^* y_0 - h_0 y_1^* \qquad (11.49)$$

Equations (11.48) and (11.49) lead to

$$\tilde{s}_0 = \left(|h_0|^2 + |h_1|^2 \right) s_0 + h_0^* n_0 + h_1 n_1^*$$
$$\tilde{s}_1 = \left(|h_0|^2 + |h_1|^2 \right) s_1 - h_0^* n_1^* + h_1 n_0 \qquad (11.50)$$

Therefore, we can recover the transmitted symbols with added noise. The effective channel is the power sum of the two channels h_0 and h_1, as described in Section 11.5.1 and plotted in Figure 11.9. This coding scheme provides the same total symbol rate as a SISO system (two symbols in two symbol slots) and achieves much better performance regarding to reliability.

Figure 11.11 [13] shows the simulated BER for binary phase shift keying (BPSK) modulation, under uncorrelated Rayleigh fading. The horizontal axis shows average SNR in decibels. The "MRRC" traces use the maximum–ratio receiver combining technique with multiple receive antennas (not discussed in this book). The "new scheme" is the Alamouti coding described in this section. We only discuss the two-transmitter antenna, one-receiver antenna (2 Tx, 1 Rx) case. As shown in the figure, space-time coding indeed replicates the MRRC performance, with a 3-dB loss, which results from power normalization considerations. Also, the BER versus SNR plots in Figure 11.11 shows three groups, each with different slopes. The slop is related to the measure of "diversity," as elaborated in Section 11.6.

Although Alamouti code was put forward in the context of MIMO systems, it is a more general diversity idea. For example, the space or time dimension can be replaced by the frequency dimension by using multiple subcarriers in an OFDM system (Chapter 10). Section 11.9.2 provides one such example.

So far, we have shown the concept of diversity and a particular way of achieving it (Alamouti code). The information discussed here is treated more generally in

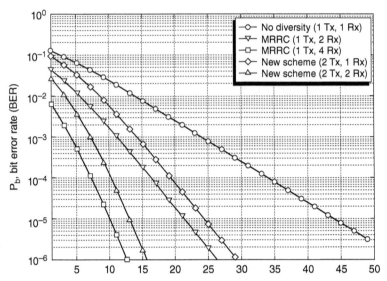

Figure 11.11   Simulated performance of Alamouti codes [13, fig. 4]. Reproduced with permission.

Section 11.6. It turns out that, in spite of its simplicity, Alamouti code is optimal in some sense. It is the most practical way of achieving space-time diversity.

11.6   THEORETICAL TREATMENTS OF MIMO TECHNIQUES

Sections 11.4 and 11.5 introduce the two most important concepts of MIMO (spatial multiplexing and space-time diversity) by describing two MIMO schemes (BLAST and Alamouti code). Although there are other forms of MIMO (as discussed in Section 11.7), most of the MIMO technologies are based on these two themes.

This section outlines the landscape of major theoretical topics in MIMO. Presented first in Section 11.6.1 are MIMO channel models. Similar to the conventional (SISO) wireless channels discussed in Chapter 6, MIMO channels are described by channel gains and their time-varying and statistical behaviors. However, MIMO channels have other characteristics, such as rank and fading correlation, which are not applicable to SISO channels. Section 11.6.2 discusses channel capacity. While capacity was derived for a fixed channel matrix in Section 11.3.3.2, here we move to stochastic channels, where channel capacity can be characterized by ergodic capacity and outage capacity. Section 11.6.3 describes the theoretical framework of space-time diversity and space-time coding. Lastly, Section 11.6.4 links the concepts of spatial multiplexing and space-time diversity and presents a trade-off bound.

11.6.1   Channel Models

Similar to the conventional wireless channel (a SISO channel), the key to understanding a MIMO channel is its stochastic nature or fading behavior [24]. This subsection reviews the fading properties of a conventional wireless channel (see Chapter 6) and extends the property to a MIMO channel. Also, the two additional key parameters for a MIMO channel (correlation and rank) are examined.

11.6.1.1   Matrix Channel
Recall that in the MIMO system equation (11.13), the channel is represented by a matrix of size N_R by N_T. An element $H_{i,j}$ represents the channel gain between transmit antenna j and receive antenna i. Usually, we use baseband representations, where all quantities involved are complex numbers.

11.6.1.2   Fading
Channel fading for single-antenna (SISO) systems are discussed in Chapter 6. In summary, one of the standard channel models is the Rician fading model. In this model, the channel gain $\{H_{i,j}\}$ has the probability distribution function (PDF):

$$f(\rho,\theta) = \frac{2(K+1)\rho}{\sigma^2}\exp\left(-K - \frac{(K+1)\rho^2}{\sigma^2}\right)I_0\left(2\sqrt{\frac{K(K+1)}{\sigma^2}}\rho\right), \qquad (11.51)$$

where ρ is the channel amplitude gain, and θ is the channel phase. Of the parameters, I_0 is the zeroth order–modified Bessel function of the first kind. K is the ratio between the power levels of the line-of-sight (LOS) and none line-of-sight (NLOS) components of the channel. σ^2 is the average channel power gain. The right-hand side of (11.51) is independent of θ; therefore, Rician fading has a uniform distribution regarding phase θ. A notable special case is when $K = 0$. Namely, there is no LOS channel. This is referred as Rayleigh fading, which is the most common model used in MIMO research.

It is customary to use normalized channels in MIMO research, as described in (11.39). For fading channels, the first equation in (11.39) is replaced by

$$E\left(\left|H_{i,j}\right|^2\right) = 1. \qquad (11.52)$$

In this case, σ^2 in (11.51) is set to 1. We need to explicitly add average channel gain, either in channel expression or in system equation such as (11.41).

Of course, all channel matrix elements do not necessarily have the same characteristics (such as average channel gain). They may also be correlated to each other. Section 11.6.1.3 discusses how to model such relationships.

11.6.1.3   Correlated Fading
A simple MIMO fading channel is expressed as a matrix whose elements undergo independent Rayleigh or Rician fading with the same parameters. Such a channel

is known as the independently fading channel. In many cases, however, the fading is not independent. Correlation in fading significantly impacts the channel capacity and diversity of MIMO systems.

At the level of second-order statistics, channel correlation can be described by the crosscorrelation between all elements. For the most general expression, we can use four indices to represent such correlation:

$$C_{(ij),(lk)} \overset{\text{def}}{=} E\left(H_{i,j}H_{lk}^*\right). \tag{11.53}$$

We can also stack all columns of H to form a vector of $N_T N_R$ elements. The correlation can thus be represented by a covariance matrix of size $N_T N_R \times N_T N_R$.[15] Such expression also allows the channel elements to have different average gains, as will be shown later.

A widely accepted simplification is known as the local scattering model, or Kronecker model [25]. This model assumes the scatterers that cause correlation are located either close to the transmitter or close to the receiver. The channel in between is a scattering-rich median that undergoes independent fading. The channel matrix under this model is

$$H = R_r^{1/2} G R_t^{1/2H}. \tag{11.54}$$

$R_r^{1/2}$ and $R_t^{1/2}$ are two deterministic matrices describing channel correlation. They are of sizes $N_R \times N_R$ and $N_T \times N_T$, respectively. G is an $N_R \times N_T$ matrix with each element undergoing independent fading:

$$E\left(G_{i,j}G_{n,m}^*\right) = \delta_{i,n}\delta_{j,m}. \tag{11.55}$$

The transmitter- and receiver-side correlation matrices are

$$R_r \overset{\text{def}}{=} R_r^{1/2} R_r^{1/2H}$$
$$R_t \overset{\text{def}}{=} R_t^{1/2} R_t^{1/2H}. \tag{11.56}$$

The resulting correlation is

$$E\left(H_{i,j}H_{n,m}^*\right) = R_{r\,i,n}R_{t\,j,m}^H. \tag{11.57}$$

Equation (11.57) can be verified by explicit expressions of (11.54) and (11.55). It can be interpreted as follows. R_r is a matrix specifying channel correlation between the various receiver antennas. Indeed, for a fixed pair of transmitter antennas, that is, fixed j and m in (11.57), correlation among the receiver antennas is specified by matrix R_r.

---

[15] Mathematically, there is a factor difference between covariance and correlation. However, because the variance of H elements is 1 after normalization, covariance and correlation are the same.

For different pairs of j and m, such correlation differs only by a scaling factor $R^H_{tj,m}$, which is common to all receiver antenna pairs. The underlying model is that R_r describes the scattering environment near the receiver. It thus influences the difference in correlations among the various receiver antenna pairs. R_t, on the other hand, is related to the scattering environment near the transmitter. These scatters are far away from the receiver; thus they affect only a common factor to all receiver antenna pairs. The same argument can be made regarding the transmitter antennas, with the roles of R_r and R_t reversed.

Note also that according to (11.57), the channel gain is

$$E\left(\left|H_{i,j}\right|^2\right) = R_{ri,i}R^H_{tj,j}. \tag{11.58}$$

Therefore, as in the general correlation model (11.53), the Kronecker model allows for different gains among channel elements. This model can be further extended to include LOS components. It can be shown that in the Kronecker model, correlation always reduces channel capacity [1, sec. 5.4].

Most studies of correlated MIMO channels use the Kronecker model. Theoretical studies and measurements have been taken to verify the validity of such a model for practical channels [25–28]. On the other hand, it should be recognized that the Kronecker model does limit the kind of channel correlations one can express. For example, it cannot describe the keyhole effect discussed in Section 11.6.1.4. Therefore, validity of Kronecker model depends on the particular situations. Other models have been proposed in the literature, for example, [29, 30].

11.6.1.4 Rank Deficiency

Rank deficiency is another important and unique characteristic of the MIMO channels. As shown in Section 11.3.3, the MIMO channel can be decomposed into parallel and independent channels, whose gains are given by the corresponding eigenvalues of the matrix HH^H. If the channel is full rank, the number of nonzero eigenvalues is $\min(N_T, N_R)$. Therefore, the capacity gain is determined by the number of antennas. On the other hand, if some of the eigenvalues are very small or zero, the corresponding subchannels are useless for data transmission. In these cases, the benefit of multiple antennas is not fully realized. These channels are known as rank-deficient channels.

The free-space case discussed in Section 11.3.2 is an example of rank-deficient channels, where the deficiency is caused by the lack of scatterers. Unless we have very large antenna spacing (compared to transmitter–receiver distance), there is little MIMO benefit in free space. Some other typical cases for channel deficiency are as follows:

1. *A few dominating scatterers near the transmitter or the receiver* [31]. This condition can be intuitively understood through an extreme case. If there is only one scatterer near the transmitter and all signal paths are bounced off by such scatterer, the rest of the channel would see all signals coming from the same location (where the strong scatterer is). Therefore, we effectively have only one transmit antenna instead of a MIMO system. Such an effect can be captured by R_t and R_r in the Kronecker channel model (11.54). If these matrices are not full rank, the entire channel will not be full rank.

2. *The keyhole effect.* If there is a spatially limiting point on the transmission path (a keyhole), the channel is also rank deficient [32]. The same phenomenon can appear in a dual case, where a dominating scatterer exists on the transmitting path. In this case, the channel cannot be modeled according to (11.54). More specifically, G will not contain independently fading elements.

3. *The total number of scatterers in the channel is limited (less than the number of antennas)* [33]. An extreme case is the free space discussed in Section 11.3. However, such a condition actually will happen to any channel if we continue to increase the number of antennas beyond the number of scatterers. Therefore, the statement that the capacity of a MIMO channel increases linearly with antenna numbers is true only to a certain extent.

Rank deficiency is another important factor to consider when evaluating the benefit of MIMO technology to a specific application.

11.6.1.5 Some Discussions on Correlation and Rank

Channel correlation and rank are often invoked together in MIMO discussions. They are sometimes even used interchangeably. However, correlation and rank are actually two separate concepts. Correlation is about how channel matrix coefficients change with time. It affects the diversity gain of a MIMO system. Rank can be applied to both static and time-varying channels. It is a key factor for spatial multiplexing gain. (The concepts of diversity gain and multiplexing gain will be explained in Sections 11.6.3 and 11.6.4.)

On the other hand, these two characteristics are connected as well. An uncorrelated channel likely will have full rank because rank deficiency requires channel coefficients to obey certain relationships, which is not maintained under independent fading. In fact, if we use the Kronecker model for channel fading (see Section 11.6.1.3), channel rank is usually determined by the ranks of the correlation matrices $R_R^{1/2}$ and $R_T^{1/2}$ in (11.54).

A commonly used channel model in MIMO research is the independent Rayleigh fading channel, where all channel coefficients are independent Rayleigh random variables with the same gain. This channel is both uncorrelated and full rank.

11.6.1.6 Frequency Selective Fading

So far, we consider MIMO channels with "flat fading." That is, the channel response is flat in frequency domain. In time domain, this corresponds to "no delay spread." That is, successively transmitted symbols are received without mutual interference. As the channel bandwidth increases, however, such assumption becomes unrealistic. Intersymbol interference (ISI) would occur. This situation is known as frequency selective fading, which is discussed for the SISO case in Chapter 6.

As discussed briefly in Section 11.4.4, one can extend the concept of equalization (Chapter 9) to MIMO systems to form the generalized equalizers, also known as space-time equalizers. However, another way (OFDM) is usually simpler in concept and implementation, which is described in more detail in Chapter 10. Simply put, OFDM divides the entire frequency band into noninterfering sub-bands, each being flat. Under this framework, frequency selectiveness is not an important issue. However, there are issues specific to MIMO-OFDM, primarily channel estimation and training.

Most MIMO literature uses flat fading channel models; however, in some cases, we do need to capture the delay spread of MIMO channel in channel expression. Several forms of formulation are used in literature, but they are not elaborated on here.

11.6.2 Channel Capacity

This section extends the results of MIMO channel capacity given in Section 11.3.3 and addresses the stochastic channel conditions.

11.6.2.1 Fixed Channel

First, we repeat the results in Section 11.3.3, where the channel is fixed. They were obtained by decomposing the MIMO channel into parallel subchannels without mutual interference. The resulting capacity (bits per channel use), as shown in (11.37), is

$$C = \log_2 det\left(I_{N_R} + \frac{E_s}{N_T N_0} H H^H\right) = \log_2 det\left(I_{N_R} + \frac{E_s}{N_T N_0} D D^H\right)$$
$$= \sum_{i=1}^{r} \log_2\left(1 + \frac{E_s}{N_T N_0}\lambda_i^2\right) \tag{11.59}$$

D is the diagonal matrix in SVD decomposition of H (see Section 11.3.3.1):

$$H = U D V^H. \tag{11.60}$$

U and V are unitary matrices, whereas D is a diagonal matrix. $\{\lambda_i\}$ are the eigenvalues of HH^H, or the square of the diagonal elements in D. r is the rank of matrix H. E_s is the total transmission power. N_T is the number of transmit antennas. N_0 is noise power spectral density, or noise power per symbol per antenna. See Section 11.3.3 for detailed definitions of E_s and N_0.

On the other hand, if the channel matrix is known to the transmitter, one can further optimize the transmission by performing "water filling" among the subchannels [5], similar to what was done in frequency domain in Chapter 2. In this case, we choose to transmit correlated symbols with covariant matrix R_{xx}:

$$R_{xx} \stackrel{\text{def}}{=} E\left(xx^H\right), \tag{11.61}$$

where x is the transmitted signal vector. The total transmission power is

$$E_s = tr(R_{xx}). \tag{11.62}$$

It can be shown that the channel capacity is[16]

$$C = \log_2 det\left(I_{N_R} + \frac{1}{N_0}HR_{xx}H^H\right).$$ (11.63)

We need to find the R_{xx} that maximizes capacity in (11.63) with the constraint (11.62). Recall the following properties of determinant. For square matrices A and B:

$$det(AB) = det(BA).$$ (11.64)

Furthermore, for nonnegative definite matrix A:

$$det(A) \le \prod_i A_{ii}.$$ (11.65)

Inequality (11.65) is a special case of the Hadamard's inequality. The equality holds when A is a diagonal matrix.

Referring back to (11.60) and using (11.64), (11.63) can be written as

$$C = \log_2 det\left(I + \frac{1}{N_0}DV^H R_{xx}VD^H\right).$$ (11.66)

According to (11.65),

$$C \le \log_2 \prod_i \left(1 + \frac{1}{N_0}|d_i|^2 Q_{ii}\right).$$ (11.67)

where $\{d_i\}$ are the diagonal elements of D, and $\{Q_{ii}\}$ are diagonal elements of Q:

$$Q \overset{\text{def}}{=} V^H R_{xx}V.$$ (11.68)

From (11.62) and the fact that V is unitary, we have

$$E_s = tr(R_{xx}) = tr(Q) = \sum_i Q_{ii}.$$ (11.69)

Therefore, to maximize capacity, we can choose Q to be a diagonal matrix, and choose its diagonal elements $\{Q_{ii}\}$ to maximize the following product under the constraint of (11.69):

$$C' \overset{\text{def}}{=} \prod_i \left(1 + \frac{1}{N_0}|d_i|^2 Q_{ii}\right).$$ (11.70)

---

[16] Equation (11.63) can be understood in the following way. Consider $x = Rs$, where s is the uncorrelated, data-carrying vector whose norm is 1, that is, $E(ss^H) = I_{N_T}$. The covariance of x is thus $R_{xx} = E(xx^H) = RR^H$. On the other hand, when x is transmitted through channel H, the system equation becomes $y = Hx + n = HRs + n$. Therefore, we can consider HR as the equivalent channel for s, leading to (11.63).

Such optimization can be done by the method of Lagrangian multipliers (see Chapter 2 for some more mathematical details). The solution is

$$Q_{ii} = \begin{cases} \mu - \dfrac{N_0}{|d_i|^2} & \text{if } \mu - \dfrac{N_0}{|d_i|^2} \geq 0 \\ 0 & \text{otherwise} \end{cases}. \tag{11.71}$$

where μ is determined by (11.69). Equation (11.71) shows that we should allocate more power into subchannels with larger gain (quantified by $|d_i|^2$). This formula is commonly known as water filling (see Chapter 2) because, for the nonzero elements, we have

$$Q_{ii} + \frac{N_0}{|d_i|^2} = \mu. \tag{11.72}$$

Namely, the values of Q_{ii} are chosen so that, when starting from term $\dfrac{N_0}{|d_i|^2}$, each "bin" is "filled" up to a constant level μ.

From (11.71), we can construct the diagonal matrix Q. Recalling that V is unitary, R_{xx} can be obtained from (11.68):

$$R_{xx} = VQV^H. \tag{11.73}$$

Recall the transmitter side precoding given in (11.29):

$$x = Vs. \tag{11.74}$$

where s is the data stream to be transmitted. Equation (11.73) can be satisfied if s has the following covariance:

$$E\left(ss^H\right) = Q. \tag{11.75}$$

In other words, s contains mutually independent components (since Q is diagonal) whose powers are allocated according to (11.71).

This result can also be understood based on the channel decomposition concept presented in Section 11.3.3 and the conventional concept of water filling among parallel subchannels discussed in Chapter 2.

Therefore, when the transmitter knows the channel, it can perform proper precoding to achieve capacity

$$C = \log_2 \prod_i \left(1 + \frac{1}{N_0}|d_i|^2 Q_{ii}\right), \tag{11.76}$$

where $\{Q_{ii}\}$ are given by (11.71). This is a higher capacity than what is given by (11.59), which is achievable without the channel state information at the transmitter.

11.6.2.2 Stochastic Channel: Ergodic Capacity and Outage Capacity
Section 11.6.2.1 provided the channel capacity expression for a fixed channel. In practice, a wireless channel is stochastic, as discussed in Section 11.6.1. There are two ways to characterize capacity for stochastic channels.

If the transmitter knows the channel, it can change its instantaneous data rate to match the instantaneous channel condition. We therefore experience the average of channel capacity over time, known as the ergodic capacity. Such capacity also applies to fast-fading channel conditions, when the transmitter does not know about the channel. A fast-fading channel experiences both good and bad conditions over a short time, compared to the time window for error correction mechanisms such as forward error correction (FEC). In this case, the burst errors caused by bad conditions can be compensated by the good symbols, through FEC. The system sees (approximately) the averaged channel capacity. Ergodic capacities of channels with specific statistics have been derived by various authors. For example, Telatar gave the ergodic capacity of a Rayleigh fading channel [5].

On the other hand, if the channel fades slowly and the modulation and coding scheme (and thus the data rate) is fixed, the errors caused by bad channel conditions cannot be corrected at the physical layer. If we want to ensure error-free transmission, the data rate must be 0 because the instantaneous channel capacity is possible to be 0. Alternatively, we define *outage capacity*, which is the data rate we can achieve, allowing *outage* at a given percentage of time. Outage capacity is detailed in Section 11.5.1.

Both ergodic capacity and outage capacity are widely used in MIMO channel studies. Evaluation of these capacities is based on the distributions of the channel matrix eigenvalues. For uncorrelated Rayleigh fading channels, the results were given by the early MIMO works. For Rician channels and correlated fading channels, analyses are more complex. Sometimes only bounds and asymptotic results are available.

11.6.3 Space-Time Diversity

Section 11.6.2 discussed the channel capacity for a MIMO system. To achieve channel capacity, we need to match the data rate to the channel statistics (if not the instantaneous channel states). By doing so, we can achieve error-free transmission (except for the outage periods). The capacity achieved is higher than that of a SISO (single antenna) system. Such gain in capacity is referred to as the *multiplexing gain*. Conceptually, multiplexing gain is achieved by creating parallel subchannels across the spatial dimension.

This section presents another way to leverage the spatial dimension. We use the additional subchannels not to increase the data rate, but to combat fading. This approach helps to reduce the error rate, which is inevitable with a fading channel. Such benefit of MIMO is known as *diversity gain*. Section 11.5 illustrated the issue and concept using an example system. In this section, we briefly introduce and describe the formal theory on diversity and show how to quantify the benefit.

11.6.3.1 Diversity Gain and Coding Gain

Following the arguments of Tarokh et al. [14], we will first find an expression of the error rate of a space-time block code (STBC). Consider an STBC that takes N_T transmit antennas and over L symbol slots (in time dimension) to transmit a codeword. In this case, a codeword can be expressed by an N_T by L matrix. Each column represents the symbols to be transmitted at a symbol slot using the N_T antennas. For example, the Alamouti code discussed in Section 11.5.2 has codewords in the following form:

$$C_A(s_0, s_1) = \begin{bmatrix} s_0 & -s_1^* \\ s_1 & s_0^* \end{bmatrix}. \tag{11.77}$$

Different values for s_0 and s_1 produce different codewords.

If there are M legitimate codewords, the data rate for such an STBC system is

$$R = \frac{\log_2 M}{L}. \tag{11.78}$$

Pairwise error rate $P(c, e)$ is defined as the probability of sending codeword c but receiving codeword e. Maximum pairwise error rate over all codeword pairs can be used as a good approximation of the actual error rate.

Obviously, $P(c, e)$ depends on the difference between these two codewords. Define the difference between the codewords as

$$X(c, e) \stackrel{\text{def}}{=} c - e, \tag{11.79}$$

where c and e are codeword matrices. Furthermore, define

$$A(c, e) \stackrel{\text{def}}{=} X(c, e) X^H(c, e). \tag{11.80}$$

Naturally, $A(c, e)$ is a semi-positive definite Hermitian matrix of size N_T by N_T. Let its eigenvalues be $\{\lambda_m(c, e), m = 1, 2, \dots N_T\}$.

It has been shown that for independent Rayleigh fading channels, the probability of error has an upper bound [14][17]

$$P(c, e) \leq \left[\prod_{m=1}^{N_T} \left(1 + \lambda_m(c, e) \frac{E_s}{4N_0} \right) \right]^{-N_R}. \tag{11.81}$$

To make significant contribution to the product, the corresponding eigenvalue must meet the following condition:

$$\lambda_m(c, e) \frac{E_s}{4N_0} \gg 1. \tag{11.82}$$

---

[17] This expression of error rate is different from the BER formulas in Chapter 4 because we consider Rayleigh fading channels here, while constant channels are used in Chapter 4.

At the limit of high SNR (i.e., $E_s/N_0 \to \infty$), (11.82) holds for all nonzero eigenvalues. There are q of such eigenvalues, q being the rank of $A(c, e)$. For these factors, we can ignore the leading 1 in (11.81). Thus,

$$P(c,e) \leq \left\{ \left[\prod_{m=1}^{q} \lambda_m(c,e) \right]^{\frac{1}{q}} \right\}^{-qN_R} \left(\frac{E_s}{4N_0} \right)^{-qN_R}. \tag{11.83}$$

Therefore, the pairwise error probability for independent Rayleigh fading channels, as approximated by its upper bound and at the limit of high SNR, has two factors. The first depends on the eigenvalues of matrix A, which is constructed from the codewords as shown in (11.79) and (11.80). The second factor depends only on q and the SNR. We call the first factor (i.e., the quantity within the curly bracket) the coding gain. It is the geometric mean of the nonzero eigenvalues. The second factor dictates how fast the error rate decreases as SNR increases. Its exponent qN_R is known as the *diversity gain*.

This result is only for errors between one pair of codewords. In a code with many codewords, the overall error rate is bounded by the sum of all pairwise error rates that involve one particular codeword (the union bound). At the limit of high SNR, overall error rate can also be approximated by the highest pairwise error rate (nearest neighbor approximation).[18] Therefore, we are interested in the minimum value of q among all codeword pairs.

To summarize, a good set of space-time code should have high coding gain and high diversity gain. Namely, the matrix A, constructed from each pair of the codewords, should have large eigenvalues and high rank.

Of course, this analysis is for a very special case—the independent Rayleigh fading channel. However, the method of analysis applies to other channels as well, although the results may be different.

When A is full rank for all code pairs (i.e., $q = N_T$), the diversity gain reaches the maximum value:

$$d_{\max} = N_T N_R. \tag{11.84}$$

Systems with $d = d_{max}$ are known as achieving full diversity. Its coding gain depends on the determinant of matrix A. Obviously, for A to be full rank, we must have $L \geq N_T$.

Coding difference matrix A defined by (11.80) is not to be confused with the channel matrix H. A is independent of the channel and is defined for each pair of the codewords. The rank of channel matrix H determines the maximum number of parallel subchannels, and it affects the spatial multiplexing gain. The rank of A determines diversity gain.

---

[18] The error rate is actually approximated by the highest pairwise error rate multiplied by the average number of such "nearest neighbors" of a codeword. Such multiplication is not important here, as we consider how error rate changes with SNR.

11.6.3.2   Orthogonal Space-Time Code

One of the simpler STBCs is the orthogonal space-time code [34]. The Alamouti code described in Section 11.5.2 is a special case of the orthogonal code. The following is an example of real-valued orthogonal code [34]

$$
\begin{bmatrix}
x_1 & x_2 & x_3 & x_4 \\
-x_2 & x_1 & -x_4 & x_3 \\
-x_3 & x_4 & x_1 & -x_2 \\
-x_4 & -x_3 & x_2 & x_1
\end{bmatrix},
\tag{11.85}
$$

where each column shows the signals applied to all transmit antennas (four of them in this case). The different columns are for different symbol slots. These four signals, x_1–x_4, can be modulated independently to carry data. Therefore, four independent symbols are transmitted during four symbol slots. The columns are mutually orthogonal, as are the rows.

For various codewords, the construction is the same, as defined by (11.85). However, signals x_1–x_4 can take various values. Therefore, such construction does not specify the number of legitimate codewords. The actual data rate is determined by the coding and modulation scheme, which defines the alphabet for the signals.

Obviously, because of the orthogonality, the code is full rank, or full diversity. The orthogonality also makes the decoding process simpler. However, there is no guarantee that an orthogonal code is also optimal with regard to coding gain.

On the other hand, the requirement of orthogonality restricts the choice of code. For this particular class (real-valued orthogonal code), designs only exist for antenna numbers of 2, 4, or 8. There are a number of generalizations and variations.

Thus far, we have focused on the STBC, but this is only one way to realize spatial diversity. There are other space-time coding families, such as the space-time trellis code [14]. However, STBC is more widely used.

11.6.4   Tradeoff between Spatial Multiplexing and Spatial Diversity

We have presented the two critical advantages of a MIMO system: greater capacity (spatial multiplexing) and greater reliability (spatial diversity). They appear to be two separate features, each requiring a different configuration of the system. In this section, we show that spatial multiplexing and spatial diversity are related by a tradeoff [35]. Let us start by quantitatively defining the diversity gain and multiplexing gain.

Section 11.6.3.1 defines the diversity gain based on how the error rate changes with SNR. The expression (11.83) can be simplified as

$$
P_e \approx G(SNR)^{-d},
\tag{11.86}
$$

where P_e is error rate, G is a constant containing coding gain and some other quantities, and d is the diversity gain, which is the same as qN_R in (11.83).

On the other hand, to achieve channel capacity (11.59) in the "pure" spatial multiplexing scheme (Section 11.4), one is expected to adjust modulation and coding rate with SNR to keep error rate P_e constant. Therefore, in this scheme the diversity gain is 0.

We can also define a multiplexing gain. Based on (11.59), consider an idealized channel where $\lambda_i = 1$ and $SNR \gg 1$. We have channel capacity C:

$$C = \sum_{i=1}^{r} \log_2 \left(1 + \frac{E_s}{N_T N_0} \lambda_i^2 \right) \approx r \log_2(SNR). \tag{11.87}$$

Motivated by this approximation form, multiplexing gain r is defined as:

$$r \stackrel{\text{def}}{=} \lim_{SNR \to \infty} \frac{C}{\log_2(SNR)}. \tag{11.88}$$

In the space-time coding scheme (Section 11.5), data rate is determined by (11.78) and is kept constant under various SNR. Therefore, spatial multiplexing gain is 0.

The spatial multiplexing and space-time coding schemes discussed in Sections 11.4 and 11.5 provide maximum multiplexing gain and diversity gain, respectively, while the other gain is 0. These two schemes can be viewed as two extremes. Theoretically, there is a continuum of tradeoff between spatial multiplexing and spatial diversity [35]. It has been shown that under independent Rayleigh fading and at the limit of high SNR, the optimal tradeoff is

$$d_u(r) = (N_T - r)(N_R - r), \tag{11.89}$$

where $d_u(r)$ is the upper bound of diversity gain when the multiplexing gain is r. For space-time coding, $r = 0$ and $d = N_T N_R$. For pure spatial multiplexing, $d = 0$ and $r = \min(N_T, N_R)$. Both cases reach the upper bound given by (11.89). This result was extended to the case of finite SNR and correlated fading channels [36].

We can envision a naïve scheme to achieve the tradeoff between multiplexing and diversity gains. Suppose we have N_T transmit antennas. We can divide them into groups of r antennas each. Within the group we use spatial multiplexing scheme to increase data rate. Among the groups we use space-time coding to obtain diversity gain. This scheme was proposed by Tarokh et al. in 1999 [37]. A similar scheme is actually used in LTE (see Section 11.9.2.1). Other, more optimized, space-time codes were also proposed to achieve the tradeoff between d and r [38, 39].

11.7   OTHER FORMS OF MIMO

Thus far, we have examined the two main forms of MIMO: spatial multiplexing and space-time coding. This section briefly describes several other MIMO forms. Although some of these forms are not widely used today, they demonstrate the versatility of MIMO concepts.

11.7.1   Antenna Combining and Selection

Antenna combining is a SIMO configuration, where the transmitter has only one antenna and the receiver has multiple antennas. The signals are combined by maximum-ratio combining (MRC) or other algorithms to achieve maximum SNR. MRC is effective in combating channel fading and interference [40].

A simpler form, antenna selection, can be done at either the transmitter or the receiver. One radio frequency (RF) chain is switched among multiple antennas, so that the most favorable channel at the time is used for communication. Although simple in concept, antenna selection requires channel state information, even for the antennas that are not connected. Some clever training processes can yield such information. Antenna selection can also be combined with other MIMO techniques by selecting a subset of antennas instead of a single antenna [41].

Antenna combining and selection occurred in early forms of multi-antenna systems. They are usually not considered as a part of the mainstream MIMO technology in research. However, they can be attractive in providing practical performance gains.

11.7.2   Space Shift Keying (SSK)

Space shift keying (SSK), also known as spatial modulation, is a simplified MIMO form [42, 43]. In this form, data are encoded in the transmission spatial pattern. In other words, different transmitter antennas are activated corresponding to different transmit data payload.

In general, the N_T antennas can form 2^{N_T} possible patterns, while each pattern can incorporate additional modulations as in conventional MIMO. However, most SSK configurations use one transmitter antenna at a time. Such choice simplifies the transmitter and receiver structure. The transmitter needs only one RF chain, which can be switched among the antennas. Therefore, there is no need to keep multiple coherent RF modules. For the receiver, there is no need to deal with mutual interference among the data streams because only one data stream is used. The data rate is higher than a conventional SISO system, although lower than spatial multiplexing. SSK is currently not widely used. However, there are continuing research activities on this topic.

11.7.3   Beamforming

In the spatial multiplexing scheme with precoding (Section 11.3.3), we use eigenvectors of the channel matrix as subchannels. Such a technique is also known as eigen-beamforming, or simply beamforming, in MIMO literature.

On the other hand, there is a more "pure" form of beamforming, where the receiver has fewer antennas (typically one or two) than the transmitter. In this case, the spatial multiplexing gain is limited by the receiver antenna number. However, the larger number of transmitter antennas enable the "focusing" of transmitted energy on to the receiver. The channel eigenvectors with the largest eigenvalues are used as subchannels. Conceptually, beamforming can be performed in the reverse link, where the receiver has more antennas than the transmitter. However, such configuration is more commonly referred to as antenna combining (Section 11.7.1).[19]

Beamforming is important in cellular communications systems, where the mobile devices have limited size, weight, and power, and thus limited number of antennas. Usually, the base station can have more antennas and can perform beamforming to improve performance. Beamforming can also be used to enable one base station to serve multiple users simultaneously, with limited mutual interference. Such multiuser MIMO configuration is discussed in Section 11.9.2.2. Furthermore, beamforming can be used when the channel rank is low because, typically, only one or two spatial channels are used in beamforming.

A critical issue with beamforming is that the transmitter must know channel state information. Such requirement adds complexity in system design (see Section 11.8.2). An extreme version of beamforming is massive MIMO, which is described briefly in Chapter 12.

The term beamforming seems to imply that a geometrical beam is formed, in which the transmitted energy is focused. This is indeed the case in free space (Section 11.3). However, as a special case of the MIMO precoding (or eigen-beamforming), beamforming is viable even in a scattering environment, where no geometrical beam is involved. In this case, the transmitter still maximizes the energy at the receiver by choosing the precoding vectors corresponding to the channel eigenvectors with the largest eigenvalues.

11.7.4 Multiuser MIMO (MU-MIMO)

Multiuser communication refers to situations where multiple users share the same media in wireless communications. For SISO systems, different pairs of user usually use mutually orthogonal subchannels (e.g., orthogonal frequency-division multiple access (OFDMA), time-division multiple access (TDMA), frequency division multiple access (FDMA), code-division multiple access (CDMA)) to avoid mutual interference.[20] MIMO spatial multiplexing techniques (discussed in Section 11.4) seem to be attractive for multiuser communications, since one can assign the different spatial data streams to different user pairs. Namely, we can consider the collection of all transmitters as one MIMO transmitter and that of all receivers as one MIMO receiver. However, the spatial multiplexing schemes in Section 11.4 require joint processing of signals received from the various antennas. Such joint processing is not possible when

---

[19] In the literature on massive MIMO (see Chapter 12), the term receiver beamforming is typically used.

[20] There are some other technologies such as underlying or overlaying, multiuser detection, and so on. They are not currently widely used in practice. See discussion of NOMA in Chapter 12.

the receiving antennas belong to different devices as in the case of multiuser communications. Therefore, special schemes and techniques are required for multiuser MIMO communications [44].

MU-MIMO focuses on a specific configuration where one party (the "access point" or the "base station") has simultaneous wireless links with multiple partners (the "users" or the "mobile terminals"). For uplink (users to access point), a joint detection with successive cancellation (SC) can be used just as described in Section 11.4. The downlink (access point to users) can use a dirty-paper coding scheme. Dirty-paper coding is basically presubtraction of potential interferences at the transmitter side similar to the Tomlinson–Harashima precoding discussed in Chapter 9 [45]. Such precoding method reaches channel capacity [46, 47]. However, it requires the transmitter to have channel state information. Dirty-paper coding and, more generally, multiuser broadcasting were once an active research topic. However, today's multiuser systems tend to use simpler solutions such as user-specific beamforming. Sections 11.9.1 and 11.9.2.2 provides some examples.

In literature, the uplink and downlink multiuser channels are often referred to as multiple-access channel and broadcast channel, respectively.

A critical issue with MU-MIMO is the optimization objective. The simplest optimization objective is the sum capacity (i.e., the sum of all user capacities). However, in a realistic application, there are other factors (e.g., fairness, throughput, instantaneous traffic demand, and priority) to be considered.

A typical multiuser system supports a total number of data streams (i.e., summation of data streams for all users) that is comparable with the number of antennas at the access point. These data streams are allocated to the various simultaneous users. A newer development in this area is the massive MIMO (introduced in Chapter 12), whose number of antennas at the access point far exceeds the number of data streams.

11.8   AREAS OF FURTHER EXPLORATION

This section briefly introduces some MIMO topics not covered thus far in this chapter. These topics are important to the design and implementation of a MIMO system. However, they are not critical for beginners in the MIMO field trying to grasp the basic concept.

11.8.1   Security

In wireless communications, security mainly refers to two aspects. One is covertness, which means an eavesdropper will have difficulty detecting the existence of the communication link (*detection*) and/or obtaining the content being communicated (*interception*).[21] The other aspect is robustness, meaning an adversary will have difficulty

---

[21] In the literature, terms such as low probability of detection (LPD), low probability of interception (LPI), and low probability of exploitation (LPE) are used for covertness, although the exact definitions may be inconsistent among the authors.

disrupting the communication link. In the context of this discussion, robustness primarily means anti-jamming. There are, of course, other security threats such as impersonation, penetration, identification, geolocation, and so on. However, this section focuses on covertness and robustness.

A common approach against interception is encryption, which is done in higher open systems interconnection (OSI) layers and involves secret key distribution. Namely, the transmitter and receiver share some knowledge (i.e., encryption key) that the eavesdropper does not possess.[22] Furthermore, spread-spectrum technologies can be used against detection and jamming. In this case, the transmitter and receiver share information about how transmission is constructed (spreading code or frequency hopping sequence), while such information is kept away from the adversary. With MIMO technology, another layer of security can be added, through spatial pattern designs. Such a scheme can be completely open to everyone. The eavesdropper is at a disadvantage, not because he/she is unaware of some secrete, but because he/she experiences a different spatial channel than the receiver. Such an approach is referred to as physical layer security [48].

According to the secrecy capacity theory, one can design a transmit coding to make it impossible for an eavesdropper to decode transmitted data, even when the eavesdropper has knowledge of the code [49–51]. With such "perfect secrecy," the achievable capacity (referred to as the "secret capacity") is [50]

$$R_S = max_{R_{xx}} \left[\log_2 det \left(I + H_M R_{xx} H_M^H \right) - \log_2 det \left(I + H_E R_{xx} H_E^H \right) \right], \tag{11.90}$$

where R_S is the maximum achievable data rate under perfect secrecy. Channel matrices H_M and H_E depict transmitter-to-receiver and transmitter-to-eavesdropper channels, respectively. R_{xx} is the covariance matrix for transmitted signal. The maximization is performed over the choices of R_{xx}, subjected to constraints such as total power level. I stands for the identity matrix of appropriate sizes. The system is normalized such that the noise covariance matrix (assumed to be spatially white and the same for the receiver and the eavesdropper) is an identity matrix. Therefore, noise power N_0 is not included in the equation.

For those MIMO systems where the transmitter knows the channel state for both H_M and H_E, the transmitter can optimize R_{xx} so that the difference between the two determinants in (11.90) is maximized. Namely, the transmitter would want to deliver more energy to the intended receiver and less to the eavesdropper, essentially conducting beamforming. Because of the second term in (11.90) and because R_{xx} is not optimized for maximum SNR at the intended receiver, the presence of an eavesdropper does reduce system data rate, even if perfect secrecy is maintained.

In addition to the transmitter's action to thwart eavesdropping, the eavesdropper also faces the practical challenge of acquiring channel state information without

---

[22] Some encryption schemes such as the public-key cryptography do not require secret key distribution. However, they still rely on the fact that the communication parties possess some secret that the eavesdropper does not have.

transmitter cooperation, making it difficult for the eavesdropper to decode the signal. For MIMO systems, blindly (i.e., without training signals) estimating channel state information is more difficult than in the SISO case. This difficulty is another advantage of MIMO regarding covertness.

Beamforming also helps with detection avoidance because the eavesdropper sees less transmitted energy and thus has more difficulty distinguishing transmission from noise. Furthermore, a MIMO transmitter can use lower transmitted power to achieve the same data rate, compared to a SISO system, resulting in another advantage against detection. However, the probability of detection cannot be reduced to zero.

Another tactic to enhance covertness is using artificial noise [52]. In this configuration, the transmitter performs beamforming for the intended receiver, while transmitting an artificial noise that interferes with the eavesdropper. The artificial noise is transmitted with a spatial vector that is orthogonal to the beamforming vector, and thus does not degrade performance at intended receiver.

Regarding robustness, MIMO has some advantages as well [53]. The jamming signal appears as spatially colored noise to the receiver. The MMSE optimization, detailed in an appendix (Section 11.11), automatically whitens such noise, reducing the total noise power. (See Chapter 9 for more discussions.) Therefore, an MMSE receiver is intrinsically resistant to jamming. With the same number of data streams, a receiver with more antennas does better in "tuning out" jamming.

On the other hand, a jammer with multiple antennas have advantages, as well. It can perform beamforming to focus jamming energy to the target. It can also create more "white" noise, which a MIMO receiver cannot tune out [54, 55].

As shown, MIMO has intrinsic advantages and unique strategies for security enhancements. There are many variations of the techniques discussed, depending on channel conditions, number and distribution of the adversaries, and especially the knowledge about channel and transmission signal that each party holds. This section provides only a brief overview. Further explorations can start at some general survey papers, for example, [56, 57] and special journal issues, for example, *IEEE Communications Magazine* June and December 2015 issues (vol. 53, nos. 6 and 12).

11.8.2 Partial Channel State Information at the Transmitter

Some MIMO techniques, for example, transmitter precoding (Section 11.3.3), require the transmitter to have channel state information. The adaptive modulation and coding scheme, which selects modulation and coding parameters based on receiver SNR, also needs channel statistics (such as received power and SNR) at the transmitter. However, in practical situations, there are limitations to the availability of channel state information. Such limitations can be classified into three categories:

1. *Inaccuracy due to channel estimation process.* Noise at the receiver affects channel estimation accuracy.
2. *Limited reverse channel.* There is not enough bandwidth to pass the channel state information data (estimated by the receiver) to the transmitter.

3. *Time delay*. For a time-varying channel, channel estimation may be outdated. Time delay is exacerbated when the channel state information obtained at the receiver needs to be conveyed to the transmitter.

Not much can be done against the first limitation, other than using proper channel estimation methods and training signals. Section 11.8.5 expands on this topic.

One way to overcome the second limitation is using "codebook," in effect sending a quantized version of the channel estimation [58, 59]. In the simplest version, a predetermined codebook can be chosen, defining a list of precoding matrices, noted as *V* in (11.29), each associated with an identification number. During operation, the receiver chooses the best precoding matrix from the codebook based on the current channel estimate. The corresponding identification number is then sent to the transmitter. The use of codebook greatly reduces the volume of data to be sent. The number of entries in the codebook presents a compromise between reverse channel traffic load and precoding accuracy.

The codebook method still carries some open questions, such as the optimal design of the codebook entries. However, it is already used in LTE MIMO operations (Section 11.9.2).

As for the third limitation, there are some studies on channel prediction techniques, but they are not very successful and not widely used.

11.8.3   Situation-Specific Channel Models

In addition to the general channel models discussed in Section 11.6.1, channel models that better describe specific situations (such as indoor and urban) and specific frequency bands are also valuable [30]. These channel models can be constructed by high-fidelity simulations such as ray tracing [24, 60] or from measurement data [61]. Specific technologies such as wireless local area network (WLAN) and LTE also establish situation-specific channel models during the course of standard development [62].

Some channel models further include details of the transmitter and receiver electronics [24, 63]. When antennas are placed in close proximity, emission from one antenna causes induced current in other antennas, resulting in "reflection" from the positions of other antennas. A similar effect happens at the receiver end. Such coupling increases correlation between the antennas and should be included into the channel model in some cases.

11.8.4   Receivers for Spatial Multiplexing

The optimal receiver for MIMO is the maximum likelihood (ML) detector [64], whose general principle is discussed in Chapter 4. For spatial multiplexing, recall that the transmitted vector x has a dimension of N_T, and each component is drawn from M possible values for M-ary modulation. Therefore, an ML detector needs to compare the received vector y to the M^{N_T} possible values. Its complexity thus grows exponentially with the number of antennas. In the early years of MIMO development, ML was considered to be too complex. SC methods such as BLAST (Section 11.4) were viewed as the primary option for implementation.

However, in the past two decades, widespread MIMO adoptions are limited to two or four antennas for each side, with eight-antenna systems emerging in recent years. For these configurations, the complexity of ML is not as formidable, and the advantage of SC over linear processing (i.e., only nulling, no cancellation) is limited. Therefore, various other receiver structures are studied.

There are also some simplifications to ML methods. For example, instead of searching for the best match over all possible signal points, one can search only ones that appear to be close to the "best guess." One such method is spherical decoding [38, 65]. Combining MIMO with channel coding is another implementation issue [66].

11.8.5   Channel Estimation

Channel estimation is an important topic for wireless communications. Because the channel changes with time, the channel estimate needs to be updated frequently, yet without consuming too much bandwidth and processing resources. For MIMO, channel estimation is even more challenging because there are $N_T N_R$ channel elements to be estimated and tracked [1, 67, 68].

Typically, channel estimation is assisted by known transmit symbols (often referred to as training symbols, reference symbols, or pilot symbols). Transmit symbols are interleaved with data-carrying symbols in frequency (for OFDM) or time, or in both. Furthermore, it is desirable that training symbols from different transmit antennas be mutually orthogonal (i.e., without mutual interference), so that the receiver can easily distinguish the response of different channel elements.

A common tradeoff is between the training overhead and the resulting performance. Furthermore, analytical evaluation of training methods is usually based on mean-squared error (MSE). However, such a metric is not easy to be translated to receiver performance. Therefore, studies in this area are usually based on simulation.

11.9   MIMO APPLICATIONS

From the beginning, the potential capacity gain from MIMO has been fascinating to the industry. Bell Labs, the early inventor of major MIMO schemes, boasted in the 1990s that their technology was the "next revolution in communications." However, there was skepticism about the practical value of MIMO, which was much more complex than its contemporaries. Also, the claims of benefits were based on channel characteristics, especially fading correlation and channel rank, which have not been field tested.

Two decades later, with technology and demand growing, MIMO technologies have entered the mainstream wireless communications systems. This section briefly reviews a few applications. These applications all use OFDM modulations described in Chapter 10. Some OFDM concepts and terminologies are used in this section.

11.9.1   Wireless Local Area Network (WLAN): 802.11n and 802.11ac

Wireless local area network (WLAN) is a wireless system for short-distance (primarily indoor) networks. The prevailing industrial standard is IEEE 802.11. Today, it has very wide applications, mostly serving as the last link from a core network (fiber or cable) to a portable or mobile device. WLAN, through the 802.11n specification, was the first large-scale commercial adopter of MIMO technology. The indoor environment for most WLAN devices has favorable channels conditions (semi-stationary, scattering-rich) to harness the benefits of MIMO.

The 802.11n standard is an enhancement to the 802.11 WLAN platform to enable higher throughput (up to 600 Mbps). Development of 802.11n started in 2003, about 5 years after the first seminal MIMO papers were published. This enhancement was finalized in 2009 as Amendment 5 to the 802.11-2007 standard and was fully incorporated in the next standard revision (802.11-2012) and subsequent versions [69, ch. 19]. However, for historical and marketing reasons, people still refer to the MIMO features in WLAN as 802.11n, or "n" for short. The major physical layer enhancement from 802.11n is the addition of high throughput (HT) mode. Compared to the legacy 802.11g operations, HT mode supports higher bandwidth (up to 40 MHz), the same order of modulation (up to 64-QAM), and MIMO operations (up to four antennas on each side). Additionally, 802.11n provides some enhancement to the medium access control (MAC) protocols.

The 802.11n standard provides a unified architecture for the various MIMO operations. Because 802.11n uses OFDM, all operations described next apply to each subcarrier. Figure 11.12 shows a simplified flow chart for the 802.11n physical layer [69, sec. 19.3.3]. The symbols on the connections indicate the number of data streams. There are one or two parallel FEC encoders, with convolutional coding or low-density parity check (LDPC) coding. After channel coding, in the Steam Parser block, the output bits are split into multiple bit streams in a round-robin fashion. These bits are mapped into constellation points, forming N_{SS} spatial streams. The spatial streams pass through the STBC block, where possible space-time coding is performed, yielding N_{STS} space-time streams. After cyclic shift diversity (CSD), the space-time streams are processed by the spatial mapping block into N_{TX} transmit chains, which are converted to time domain signals through inverse fast Fourier transform (IFFT) and sent to RF processing and transmit antennas. (These processing blocks are

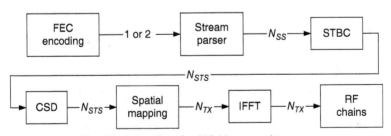

Figure 11.12   Simplified data flow for 802.11n transmitter.

described in more detail later.) The values for N_{SS}, N_{STS} and N_{TX} range from 1 to 4. Space-time coding, CSD, and spatial mapping are further described below.

Space-time coding is supported by block STBC. One space-time coding group takes one spatial stream and performs two-antenna Alamouti coding (Section 11.5) to generate two space-time streams. As explained in Section 11.5, a coding block spans over two consecutive symbols in time. A mixed operation (i.e., some data streams are space-time coded whereas others are not) is also allowed. Therefore, we have

$$N_{STS} = 2n_c + (N_{SS} - n_c), \tag{11.91}$$

where n_c is the number of spatial streams to be coded by STBC; n_c can be 0, 1, or 2.

When multiple antennas transmit the same signal (as for some packet header fields and in spatial expansion mode explained below), unintentional beamforming may occur, resulting in dead spots in certain receiving area, where signals from different antennas form destructive interference. To avoid such phenomenon, 802.11n inserts a CSD block, which adds a time delay between the transmit streams. In frequency domain, CSD effectively introduces frequency-dependent phase shift among the transmit streams. This way, destructive interference is unlikely to happen to all OFDM subcarriers for any location. For more details on the OFDM modulation and its properties, see Chapter 10.

The spatial mapping block at the transmitter performs the following operation:

$$x = Qs, \tag{11.92}$$

where x is a vector of size N_{TX}, representing the transmit stream to be sent to the antennas for transmission. s is a vector of size N_{STS}, representing the input space-time streams. s carries the modulated complex symbols; the modulation schemes can be different among the components of s. Q is a transformation matrix. In pure spatial multiplexing, $N_{STS} = N_{TX}$, and Q is an identity matrix. In so-called spatial expansion mode, some space-time streams are repeated at multiple transmit chains (CSD becomes important in this case). In this case, Q is populated with 1s and 0s with $N_{TX} > N_{STS}$. Alternatively, matrix Q can implement precoding MIMO (beamforming steering in 802.11 terminology) with $N_{TX} \geq N_{STS}$.

Beamforming and general precoding requires channel state information at the transmitter. In 802.11n, such information is obtained in one of the two ways: implicit feedback or explicit feedback. In implicit feedback, the transmitter measures the channel directly based on the reciprocity. Namely, the channel it experiences while receiving is related to the one it uses for transmission. The 802.11n protocol supports the required RF chain calibration for this purpose. In explicit feedback, the receiver sends the measured channel matrix back to the transmitter. Alternatively, the receiver can send the desired precoding matrix or a compressed (approximated) version of precoding matrix. The compression is performed to reduce control traffic load.

The 802.11ac standard provides further enhancements for even higher throughput (up to 2340 Mbps). It was approved in 2013, as Amendment 4 of 802.11-2012 [70] and a part of 802.11-2016 [69, ch. 21]. The 802.11ac standard adds a new mode to 802.11 operations—very high throughput (VHT). This mode uses higher bandwidth (up to two channels, each with 80-MHz bandwidth), higher modulation orders (up to

256-QAM), and more MIMO antennas (up to eight). VHT mode also made some simplifications compared to the HT mode defined in 802.11n.

Furthermore, 802.11ac also supports multiuser MIMO (MU-MIMO). In this case, the N_{STS} streams are divided among up to four users, each performing individual beamforming. This scheme is beneficial when the transmitter can support a larger N_{STS} while the receivers have fewer antennas (and thus smaller N_{STS} for each). The transmitter can also unilaterally modify receiver-provided precoding matrices to further reduce mutual interference among the concurrent users. The 802.11ac standard (VHT mode) supports only MU-MIMO transmission, not receiving.

11.9.2 Cellular Systems

The cellular systems undergo continuous development and evolution. Currently, the prevailing standard for cellular systems was developed by the Third-Generation Partnership Project (3GPP). MIMO technologies were introduced to the cellular systems since the third generation (3G) cellular standard. In the generation that followed, the long-term evolution (LTE), MIMO technologies became a basic component. Further enhancements were added in the follow-up standard, LTE-Advanced. MIMO, especially its multiuser form, is one of the technology focus areas for the development of the future fifth generation (5G) cellular technologies (Chapter 12).

The remainder of this section discusses in more detail the MIMO technologies supported by LTE-Advanced, based on 3GPP Release 13 (2016) specification [71]. The focus is on the shared channels that carry data traffic. Other control channels have some restrictions on the MIMO operations they support. For a general introduction to the cellular technologies, see Chapter 12.

LTE-Advanced supports several types of MIMO operations including single-user MIMO, multiuser MIMO (MU-MIMO), and coordinated multipoint (CoMP). An important part of the MIMO support in LTE-Advanced is the design of various training symbols (reference signals in LTE terminology). However, such issue is not covered in this chapter.

11.9.2.1 Single-User MIMO

LTE-Advanced supports up to four antennas on each side for uplink (from a mobile user to a base station) and eight antennas on each side for downlink (from base station to mobile user).

As with the WLAN (Section 11.9.1), spatial multiplexing with or without precoding is supported. Unlike the WLAN, however, receiver feedback for precoding matrix is codebook-based (Section 11.8.2). That is, the receiver sends an index identifying one of the predefined precoding matrixes. Such scheme reduces overhead at the expense of precoding optimality.

Spatial diversity is also supported in LTE-Advanced, using Alamouti code. Instead of using multiple symbol slots in time as described in Section 11.5, multiple subcarriers in frequency are used. In the two-antenna case, two symbols are coded with Alamouti code and transmitted through two antennas and two subcarriers in one symbol slot. A four-antenna version of spatial diversity is also supported. In this case, four antennas and four subcarriers are used to transmit four symbols in one symbol slot. There are two coding groups, each performing Alamouti coding to transmit two symbols with two antennas and two subcarriers. The antennas do not transmit anything in the subcarriers belonging to the other group.

While supporting beamforming as a special case for precoding (i.e., with one data stream and multiple antennas), LTE-Advanced also supports another form of beamforming for downlink, which is to a large extent transparent to the receiver. In the dedicated beamforming mode, training symbols are sent with the same beamforming (precoding) vector. Therefore, to the receiver, the beamforming processing appears to be a part of the channel. The receiver "sees" one effective transmit antenna, but with different training symbols from other modes.

The LTE-Advanced downlink uses OFDM, as in WLAN. However, the uplink uses a different modulation: single-carrier frequency division multiple access (SC-FDMA). Both OFDM and SC-FDMA feature a structure of subcarriers and are protected by cyclic prefix (see Chapter 10). Therefore, as far as MIMO is concerned, each subcarrier can be viewed as a flat-fading channel. For example, MIMO precoding can be subcarrier-dependent. However, in SC-FDMA, each subcarrier cannot be demodulated independently. Instead, a time-domain signal needs to be recovered from the combination of subcarriers and be demodulated afterward. Therefore, the MIMO receiver for LTE-Advanced uplink requires different design from the conventional MIMO-OFDM.

11.9.2.2 *Multiuser MIMO (MU-MIMO)*

LTE-Advanced supports MU-MIMO for uplink and downlink, enabling a base station to serve multiple mobile users at the same frequency and time [72]. LTE-Advanced enables simultaneous users to use distinct reference signals, so that their transmitting and receiving channels can be estimated independently from each other.

For the downlink, the base station can transmit signals for multiple users at the same time and frequency while relying on precoding (in this case functioning like beamforming) to keep mutual interference under control. This method is similar to the MU-MIMO in WLAN (Section 11.9.1). LTE-Advanced can support up to four mobile users with one data stream for each user, or two mobile users with two data streams for each user. For the base station to unilaterally optimize precoding in this case, LTE-Advanced supports another set of reference signals, which is transmitted with the same precoding as the data symbols. They enable a receiver to estimate the total channel including precoding, instead of relying on a priori knowledge about the precoding matrix.

For the uplink, LTE-Advanced can receive from multiple users at the same time while utilizing the different spatial signatures from different users to separate the signals. This task can be accomplished by receiver beamforming (similar to antenna combining in Section 11.7.1, while considering signals from other users as noise) or by multiuser detection. Such processing is transparent to the mobile users.

Inter-user interference in MU-MIMO, and thus achievable performance, depends on the characteristics of the various channels. Intuitively, the more different the channels, the easier it will be to separate them. Therefore, selecting the mobile users for a MU-MIMO group is very important to the performance. Such scheduling strategies are an active research area.

11.9.2.3   Coordinated Multipoint (CoMP)

In LTE-Advanced systems, neighboring cells use the same frequency (i.e., with a frequency reuse factor of 1 as explained in Chapter 12). While such arrangement allows efficient frequency utilization, it causes inter-cell interference (ICI). In particular, users located at the cell edge experience less signal power to and from the desired cell and stronger interference to and from the neighboring cell. Such ICI limits system throughput and degrades user experience. When the number of cellular users increases, smaller cells are used to increase frequency reuse and thus overall throughput. Under such conditions, ICI could surpass natural noise and become a limiting factor of system performance. Therefore, it is important to mitigate ICI.

Coordinated multipoint (CoMP) allows multiple cells to cooperate in combating ICI [73–76]. The neighboring base stations work together by exchanging information through the backbone network. Through such a connection, they can conceptually act as the multiple antennas in a MIMO system. CoMP, in the context of LTE-Advanced, includes multiple technologies, conceptually classified as follows:

1. *Scheduled interference avoidance.* Edge users in neighboring cells that can potentially interfere with each other are not scheduled for the same frequency and time.

2. *Joint beamforming.* A base station can perform beamforming to its own user while building a null toward the user in a neighboring cell. Such beamforming can be done in both transmission (downlink) and receiving (uplink).

3. *Joint processing.* The two base stations jointly transmit and receive from all users, in a similar manner as MU-MIMO (Section 11.9.2.2). In this case, both base stations act as a part of a joint MIMO transceiver.

Also, there is a technique of transmission point selection, where a signal to a user is transmitted by the most favorable base station. Such selection can adjust quickly in response to channel condition changes.

CoMP involves wireless interface designs as well as system-level designs and optimization. It is still under research and evolution.

11.9.3   Other MIMO Applications

The Defense Advanced Research Projects Agency (DARPA) has been closely monitoring MIMO technologies from the start. In the early years of MIMO development, several DARPA programs of various scales studied MIMO. The highest profile program was the Mobile Network MIMO (MNM) [77]. The program consists of two phases, which were led by Lucent Technologies and Silvus Technologies, respectively [78, 79]. MIMO has been an important component of subsequent DARPA programs in communications. For example, the Wireless Network after Next (WNaN) program produced low-cost, handheld radios that incorporate several advanced communications technologies, including MIMO [80]. DARPA's Mobile Ad-Hoc Interoperability Gateway (MAINGATE) program developed a vehicle-based mobile gateway that also supports MIMO in its physical layer [81].

Compared to commercial applications, military radios often operate in different environments (e.g., open fields and urban), with different radio profiles (e.g., both nodes are mobile with low antennas), and with different frequency bands (may be as low as 400 MHz). MIMO performance under such conditions has been assessed by experiments [82, 83]. With 2-by-2 MIMO systems, significant performance gains were observed for both spatial multiplexing and spatial diversity. Performance gains are higher in scattering-rich conditions (such as foliage) as expected, but also exist for approximately line-of-sight situations (such as wide road). Interestingly, multiplexing gain does not change significantly when antenna spacing changes from two wavelengths to one-quarter wavelength. Therefore, even for lower frequency applications, the system size can still be manageable. Of course, these results are from two-antenna systems. They cannot be easily extrapolated to cases with more antennas.

In addition to standardized wireless technologies (such as LTE and WiFi) and military tactical radios, MIMO technologies are used by some proprietary commercial mobile radios. For example, Silvus Technologies produces several mobile networking radios that support up to four antennas for various MIMO operations. Proxim Wireless also provides several MIMO radio series for high data rate microwave links.

As discussed in Section 11.6.1, the benefit of MIMO depends on the channel being high rank (for spatial multiplexing) and/or having uncorrelated fading (for spatial diversity). However, even with line-of-sight links, beamforming and receive antenna combining can still achieve some gain in SNR and reject directional interference. Furthermore, one can use dual-polarization to achieve a rank-2 channel with low cross-correlation in fading [84].

11.10 SUMMARY

MIMO is a family of techniques and algorithms based on multi-antenna configuration. The main benefits from MIMO technologies are higher data rate (spatial multiplexing) and higher reliability (spatial diversity).

In the 1990s, the communications industry was ready for another breakthrough. After turbo codes and low-density parity check (LDPC) codes approximately closed the performance gap from Shannon bound (although the block lengths required are impractically long for most applications), MIMO further multiplied the available throughput of wireless channels. Perhaps because of this, MIMO matured and was adopted quickly.

The MIMO technology family was built on previously established base technologies (Section 11.2) and the two early adoptions: BLAST (Section 11.4) and Alamouti code (Section 11.5). Aided by the more formal theoretical frameworks (Section 11.6), several MIMO technologies (schemes) were developed. These technologies led to large-scale commercial applications as well as tailored solutions (Section 11.9). Currently, most of the MIMO technologies are considered matured and the focus is on system design and implementation (Section 11.8). However, new developments are emerging in recent years, aiming at serving multiple users with high efficiency.

The aim of this chapter is getting a start in the field of MIMO. The chapter provides the knowledge framework needed to read specialized MIMO textbooks and research literature. However, it does not cover details of MIMO algorithms and implementations. To build a MIMO system, the reader needs to investigate more, especially on the tradeoff of performance and complexity and on RF issues.

11.11 APPENDIX: SUCCESSIVE CANCELLATION (SC) FORMULATION

As described in Section 11.4, successive cancellation (SC) is one of the major algorithm options for the realization of spatial multiplexing. Section 11.4.1 focuses on the conceptual construct of such algorithm. This appendix section presents detailed mathematical derivation based on the minimum mean-squared error (MMSE) optimization. Such details help to understand the concepts, and they reveal some interesting properties of this algorithm (Section 11.11.4). It is also interesting to draw parallels between SC and DFE, which is discussed and derived in Chapter 9.

This section discusses the pure SC, without the signal ordering discussed in Section 11.4. It is further assumed that all demodulation decisions are correct (i.e., no error propagation).

11.11.1   SC Problem Formularization

Figure 11.13 illustrates the MIMO system based on SC. The quantities within the parentheses indicate the dimensions of the variables.

The transmitted signal x is a vector with N_T components, transmitted by the N_T transmit antennas. After passing through the channel, this signal is received by N_R receive antennas and combined with noise n, to form vector y. The transfer function for the channel is represented by a matrix H. The system equation is (11.13):

$$y = Hx + n. \tag{11.93}$$

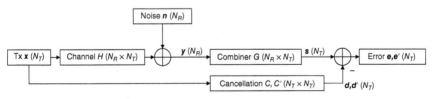

Figure 11.13   MIMO system with SC.

The receiver has a linear combiner matrix G (of size N_R by N_T) to produce signal s:

$$s \overset{\text{def}}{=} Gy. \tag{11.94}$$

This equation is the "nulling" part discussed in Section 11.4.1. C' is the cancellation matrix. Because we do not consider error propagation, we can consider the input to C' as coming from the transmitter. In reality, such input is generated by the receiver. Matrix C' takes the transmitted signal x and transforms it to d', imitating a part of the effect of the channel H:

$$d' \overset{\text{def}}{=} C'x. \tag{11.95}$$

The output of the SC processing is signal e', which goes to a slicer (or decoder) to recover the bit stream:

$$e' \overset{\text{def}}{=} s - d'. \tag{11.96}$$

C, d, and e are similar to C', d', and e', but for analysis purposes only, as shown next.

The SC works as follows. The first component of s is demodulated to form a decision. It is multiplied by the first column of C' to form a vector that is to be subtracted from all other components of s. This way, the interference from the first data stream is subtracted (cancelled) from all other streams (i.e., $s_2, s_3, \ldots, s_{N_T}$). Then the second component is demodulated. The decided symbol is multiplied by the second column of C', and the result is subtracted from s_3 to s_{N_T}. The process continues until all components are demodulated.

One can verify that, assuming the demodulated results are the same as transmitted signal x (i.e., no decision errors), this process is equivalent to (11.96). Furthermore, because subtraction based in x_i is performed only to $\{s_j, j > i\}$, C' must be a lower triangular matrix:

$$C'_{i,j} = 0 \, \forall i \leq j. \tag{11.97}$$

To analyze such receiver, we focus on the error instead of the output signal. The error is the difference between e' and x:

$$e \overset{\text{def}}{=} e' - x = s - (C' + I_{N_T})x. \tag{11.98}$$

For convenience, we define some new variables:

$$\begin{aligned} C &\overset{\text{def}}{=} C' + I_{N_T} \\ d &\overset{\text{def}}{=} (C' + I_{N_T})x = Cx \end{aligned} . \tag{11.99}$$

Equation (11.98) thus becomes

$$e = s - d = Gy - Cx. \tag{11.100}$$

Furthermore, because of (11.97) and (11.99), C must meet the following requirements:

$$
\begin{aligned}
C_{i,j} &= 0 \forall i < j \\
C_{i,i} &= 1 \forall i
\end{aligned}
\tag{11.101}
$$

The goal of optimization is to find G and C that minimize the mean-squared error (MSE):

$$
Z \overset{\text{def}}{=} E\left(|e|^2\right) = E\left[tr\left(ee^H\right)\right].
\tag{11.102}
$$

Such an optimization approach is known as the minimum mean-squared error (MMSE). It equivalently maximizes SNR at the SC output.

We assume the transmitted signal has a power of σ_x^2 per antenna and is uncorrelated:

$$
\Phi_{xx} \overset{\text{def}}{=} E\left(xx^H\right) = \sigma_x^2 I_{N_T}.
\tag{11.103}
$$

We also assume the noise is spatially white, that is, components of n are iid:

$$
\Phi_{nn} \overset{\text{def}}{=} E\left(nn^H\right) = \sigma_n^2 I_{N_R},
\tag{11.104}
$$

where σ_n^2 is the noise power for each antenna (same as N_0). Section 11.11.4 discusses the case where this assumption is removed. Furthermore, the noise is independent with the signal:

$$
\Phi_{nx} \overset{\text{def}}{=} E\left(xn^H\right) = 0.
\tag{11.105}
$$

The error correlation matrix is

$$
\Phi_{ee} \overset{\text{def}}{=} E\left(ee^H\right).
\tag{11.106}
$$

From (11.102) and (11.106),

$$Z = tr(\Phi_{ee}). \tag{11.107}$$

Therefore, our goal is minimizing means squared error Z, or the trace of Φ_{ee}, by choosing G and C with the constraint of (11.101).

11.11.2 Optimization of G

From (11.106) and (11.100), we can express the diagonal elements of the error covariance matrix as

$$(\Phi_{ee})_{i,i} = E\left(|e_i|^2\right) = E\left(\left|\sum_j G_{i,j}y_j - \sum_j C_{i,j}x_j\right|^2\right). \tag{11.108}$$

Note that $(\Phi_{ee})_{i,i}$ depends only on the ith row of G. Therefore, $(\Phi_{ee})_{i,i}$ can be minimized dependently from each other by selecting the values of the corresponding row in G. For such optimization, we require the following derivatives be zero[23]:

$$0 = \frac{\partial(\Phi_{ee})_{i,i}}{\partial G_{i,j}^*} = E\left[y_j^* \sum_k G_{i,k}y_k - y_j^* \sum_k C_{i,k}x_k\right] \forall i,j. \tag{11.109}$$

Equation (11.109) can be written in matrix form, where i denotes rows and j denotes columns:

$$G\Phi_{yy} - C\Phi_{xy} = 0$$
$$\Phi_{yy} \overset{\text{def}}{=} E(yy^H) \quad . \tag{11.110}$$
$$\Phi_{xy} \overset{\text{def}}{=} E(xy^H)$$

From (11.93) and (11.103) to (11.105), correlation functions introduced in (11.110) can be expressed as

$$\Phi_{yy} = H\Phi_{xx}H^H + \Phi_{nn} = \sigma_x^2 HH^H + \sigma_n^2 I_{N_R}, \tag{11.111}$$

$$\Phi_{xy} = \Phi_{xx}H^H = \sigma_x^2 H^H. \tag{11.112}$$

---

[23] Here, we treat $G_{i,j}$ and $G_{i,j}^*$ as independent variables for the purpose of minimization. This approach is further discussed in Chapter 9.

Therefore, (11.110) leads to

$$
\begin{aligned}
G &= C\Phi_{xy}\Phi_{yy^{-1}} \\
&= CH^H(HH^H + \xi I_{N_R})^{-1}. \\
\xi &\stackrel{\text{def}}{=} \sigma_n^2/\sigma_x^2
\end{aligned}
\tag{11.113}
$$

Equation (11.113) is the optimization condition for G, given C.

11.11.3 Optimization of C

This section considers the optimization of C, given that G is already optimized in terms of C in (11.113). Because of the constraints of (11.101), some algebraic manipulations are used for this task.

From (11.100), (11.106), (11.103), and (11.110) we have

$$
\Phi_{ee} = E\left[(Gy - Cx)(Gy - Cx)^H\right] = G\Phi_{yy}G^H - C\Phi_{xy}G^H - G\Phi_{xy}^H C^H + C\Phi_{xx}C^H.
\tag{11.114}
$$

The optimization result (11.110) leads to the cancellation of the first two terms. Therefore,

$$
\Phi_{ee} = \left(-G\Phi_{xy}^H + C\Phi_{xx}\right)C^H.
\tag{11.115}
$$

From (11.113),

$$
\Phi_{ee} = C\left[\Phi_{xx} - H^H(HH^H + \xi I_{N_R})^{-1}\Phi_{xy}\right]C^H.
\tag{11.116}
$$

For convenience, introduce $\Phi_{\tilde{e}\tilde{e}}$:

$$
\begin{aligned}
\Phi_{\tilde{e}\tilde{e}} &\stackrel{\text{def}}{=} \Phi_{xx} - H^H(HH^H + \xi I_{N_R})^{-1}\Phi_{xy}^H \\
\Phi_{ee} &= C\Phi_{\tilde{e}\tilde{e}}C^H
\end{aligned}
\tag{11.117}
$$

$\Phi_{\tilde{e}\tilde{e}}$ is independent of C.
From (11.111) and (11.112),

$$
\Phi_{\tilde{e}\tilde{e}} = \sigma_x^2\left[I_{N_T} - H^H(HH^H + \xi I_{N_R})^{-1}H\right].
\tag{11.118}
$$

This equation can be further simplified as

$$
\Phi_{\tilde{e}\tilde{e}} = \sigma_n^2(H^H H + \xi I_{N_T})^{-1}.
\tag{11.119}
$$

Equation (11.119) can be verified by multiplying the expression of $\Phi_{\tilde{e}\tilde{e}}$ in (11.118) by $(H^H H + \xi I_{N_T})$ and confirming that the result is $\sigma_n^2 I_{N_T}$.

Note that $\Phi_{\tilde{e}\tilde{e}}$ is Hermitian and positive-definite. Therefore, it can be factorized into

$$\Phi_{\tilde{e}\tilde{e}} = \sigma_n^2 S^{-1} V^{-1} S^{-H}, \tag{11.120}$$

where S is left-lower triangular matrix, with all diagonal elements being 1. This is the same constraint as C, described in (11.101). S^{-1} has the same property. V is a diagonal matrix with non-negative diagonal elements. Equation (11.120) is a standard factorization in numerical linear algebra, known as the LDL decomposition or the LDLT decomposition [85]. It is closely related to the commonly used LU decomposition.

The actual error correlation matrix is, from (11.117),

$$\Phi_{ee} = C\Phi_{\tilde{e}\tilde{e}}C^H = (CS^{-1})\sigma_n^2 V^{-1}(CS^{-1})^H = \sigma_n^2 T V^{-1} T^H. \tag{11.121}$$
$$T \overset{\text{def}}{=} CS^{-1}$$

Both C and S are lower triangular with unit diagonal elements. Thus, their product T satisfies the same constraint. Let us look at the diagonal elements of the error correlation.

$$[\Phi_{ee}]_{i,i} = \sigma_n^2 \sum_k T_{i,k} V_{k,k}^{-1} T_{i,k}^* = \sigma_n^2 \sum_k |T_{i,k}|^2 V_{k,k}^{-1} = \sigma_n^2 \left(\sum_{k \neq i} |T_{i,k}|^2 V_{k,k}^{-1} + V_{i,i}^{-1} \right). \tag{11.122}$$

The last equality is based on the fact that diagonal elements of T have value 1. Because all diagonal elements of V are non-negative, to minimize $[\Phi_{ee}]_{i,i}$, we should make

$$T_{i,k} = 0 \forall i \neq k. \tag{11.123}$$

That is, we should make T an identity matrix. Therefore, the optimal solution is

$$C = S. \tag{11.124}$$

The error correlation matrix is, from (11.121) and (11.124),

$$\Phi_{ee} = \sigma_n^2 V^{-1}. \tag{11.125}$$

Note that the error correlation matrix is diagonal, indicating that the residual noise at each component is mutually independent.

Going back to (11.113), we have the explicit solution for the combination matrix:

$$G = SH^H \left(HH^H + \xi I_{N_R} \right)^{-1} = S \left(H^H H + \xi I_{N_T} \right)^{-1} H^H. \tag{11.126}$$

The last equality is obtained by algebraic manipulation. One can left-multiply the identify matrix $\left(H^H H + \xi I_{N_T} \right)^{-1} \left(H^H H + \xi I_{N_T} \right)$ to $H^H \left(HH^H + \xi I_{N_R} \right)^{-1}$ and show that $\left(H^H H + \xi I_{N_T} \right) H^H = H^H \left(HH^H + \xi I_{N_R} \right)$.

11.11.4   Some Discussions

This section discusses the mathematical results obtained so far. It is interesting to compare these discussions with those about the decision feedback equalizer (DFE) in Chapter 9. Such exercise helps in understanding the concept of GDFE, outlined in Section 11.4.4.

11.11.4.1   Optimization of Matrix C in General

From (11.126) and the factorization of (11.120),

$$GH = S - \xi S \left(H^H H + \xi I_{N_T} \right)^{-1} = S - \xi V^{-1} S^{-H}. \tag{11.127}$$

In (11.127), the combination of the channel and the combiner GH is decomposed into two terms. The first is lower triangular and the second is upper triangular. The first term is the same as C.

$$GH = C + U$$
$$U \overset{\text{def}}{=} \xi V^{-1} S^{-H} \cdot \tag{11.128}$$

In the SC process, the feedback matrix C cancels out the lower triangular part. Namely, from (11.128), (11.93), and (11.100),

$$e = GHx + Gn - Cx = Ux + Gn. \tag{11.129}$$

In fact, this is true even when G does not take the optimal form. Recall (11.100)

$$e = Gy - Cx = (GH - C)x + Gn = (R - C)x + Gn$$
$$R \overset{\text{def}}{=} GH \tag{11.130}$$

Recalling correlations (11.103)–(11.105), the MSE is, from (11.107),

$$Z = E \left(\sum_i |e_i|^2 \right) = \sigma_x^2 \sum_{i,j} \left(R_{i,j} - C_{i,j} \right)^2 + \sigma_n^2 \sum_{i,j} |G_{i,j}|^2. \tag{11.131}$$

Therefore, to minimize Z, we need to choose C to set as many terms to zero as possible in the first summation. Because C is constrained to be lower triangular, it should cancel out the lower triangular part of R. This solution for C is independent of the optimization of G in (11.109). Such observation is important because (11.109) is not the only way to optimize G. An example of alternative is the so-called zero-forcing method (discussed next). Nevertheless, C is always optimized in the same way, given G and H.

11.11.4.2 MMSE and Zero-Forcing

As in the case of equalizers (Chapter 9), optimization of the successive canceller may be based on different criteria. Other than the MMSE method discussed here, a commonly used method is zero-forcing. In zero-forcing, mutual interferences, except for those cancelled by C, are completely removed by G, regardless of noise. The method used in the original BLAST work (Section 11.4.1) is zero-forcing, although it is easy to adopt MMSE into the BLAST technique. Zero-forcing does not yield the best SNR at the output because noise may be amplified during the process of eliminating mutual interference. However, as shown next, at high SNR, MMSE and zero-forcing methods merge to the same solution.

Consider the limit of high SNR. In this case, ξ is very small, and one can approximate the MMSE optimal G from (11.127) and (11.124):

$$G = S(H^H H)^{-1} H^H$$
$$GH = S = C = (I_{N_T} + C') \cdot \qquad (11.132)$$
$$e' = Gy - C'x = I_{N_T}x + Gn$$

This equation shows that G removes all mutual interference from H, except for those that can be removed by cancellation C'. Therefore, under high SNR, this MMSE solution becomes zero-forcing.

Interested readers can further think about the relationship between the first equation in (11.132) and the pseudoinverse of channel matrix H and how it is connected to the SC algorithm described in Section 11.4.1.

11.11.4.3 The Remaining Error

With the MMSE SC receiver, the remaining error matrix is given by (11.125):

$$\Phi_{ee} = \sigma_n^2 V^{-1}. \qquad (11.133)$$

In other words, Φ_{ee} is diagonal. This means residual noise, which is actually a combination of residual mutual interference plus noise, is spatially uncorrelated.

Furthermore,

$$E(ey^H) = E[(Gy - Cx)y^H] = G\Phi_{yy} - C\Phi_{xy} = 0. \qquad (11.134)$$

The last equality is based on (11.110). Equation (11.134) shows that the remaining error is uncorrelated with the received signal y. This is independent of the choice of C as long as G is optimized under MMSE.

Interested readers can further explore the similarity between SC and DFE in terms of remaining error characteristics.

11.11.4.4   The Case of Colored Noise

In the derivation so far, the noise is assumed to be spatially white, as in (11.104). However, results here can be generalized to colored noise. In this case, the noise covariance is a positive definite Hermitian matrix.

$$\Phi_{nn} \overset{\text{def}}{=} E\left(nn^H\right). \tag{11.135}$$

To extend the results to colored noise, a noise-whitening filter is conceptually introduced. Φ_{nn} can be factorized as

$$\Phi_{nn} = P^H P. \tag{11.136}$$

An example of such factorization is the Cholesky decomposition, where P is an upper-triangular matrix.

Recall the original system equation:

$$y = Hx + n. \tag{11.137}$$

Introduce noise-whitening filter P^{-H}. This filter is applied to y:

$$\tilde{y} \overset{\text{def}}{=} P^{-H}y = P^{-H}Hx + P^{-H}n. \tag{11.138}$$

This equation can be expressed in the form of a system equation with an effective channel $\tilde{H}$ and an effective noise $\tilde{n}$:

$$\tilde{y} = \tilde{H}x + \tilde{n}$$
$$\tilde{H} \overset{\text{def}}{=} P^{-H}H \ . \tag{11.139}$$
$$\tilde{n} \overset{\text{def}}{=} P^{-H}n$$

It can be verified that

$$\Phi_{\tilde{n}\tilde{n}} \overset{\text{def}}{=} E\left(\tilde{n}\tilde{n}^H\right) = I_{N_R}. \tag{11.140}$$

Therefore, (11.139) represents a MIMO system with white noise, which can be processed using the method described in this section. The MMSE optimization still yields spatially uncorrelated noise at the output of combination, as shown in (11.133).

11.12   APPENDIX: DERIVATION OF MIMO CHANNEL CAPACITY FOR FIXED CHANNEL

Section 11.3.3.2 provided a derivation for the MIMO channel capacity (11.37). That derivation is based on channel decomposition, showing that a MIMO channel is equivalent to a number of independent SISO channels whose capacities are readily expressed. This appendix section provides an alternative derivation of MIMO channel capacity, based on mutual information computation (see Chapter 2). This derivation was given by Telatar [4, 5] and independently by Foschini and Gans [6, 7].

11.12.1 Preliminaries

11.12.1.1 Capacity and Mutual Information
Recall the definition of mutual information (Chapter 2):

$$I(Y;X) = E\{\log_2[p(Y \mid X)] - \log_2[p(Y)]\}, \tag{11.141}$$

where Y is the received quantity, X is the transmitted quantity. p denotes probability distributions. The channel capacity C, in bits per channel use, is the mutual information, maximized over all possible probability distribution of X, $p_x(X)$.

$$C = \max_{p_x(X)} I(Y;X). \tag{11.142}$$

11.12.1.2 Gaussian Random Variables
Recall some properties of normal distribution (Gaussian distribution). Consider a random vector a of dimension K with the following distribution density function:

$$p(a) = (2\pi)^{-K/2} det(C_a)^{-1/2} \exp\left[-\frac{1}{2}(a-d_a)^H C_a^{-1}(a-d_a)\right], \tag{11.143}$$

where C_a is a semi positive definite Hermitian matrix of size $K \times K$, known as the covariance matrix. d_a is a vector of dimension K, representing the mean value of a. a is said to have a multivariate normal distribution with variance C_a and mean d_a, noted as

$$a \sim N(d_a, C_a). \tag{11.144}$$

Random vector a has the following properties:

$$\begin{aligned} E(a) &= d_a \\ E[(a-d_a)(a-d_a)^H] &= C_a \end{aligned} \tag{11.145}$$

Furthermore,[24]

$$E[(a-d_a)^H C_a^{-1}(a-d_a)] = K. \tag{11.146}$$

If $a \sim N(d_a, C_a)$ and $b \sim N(d_b, C_b)$ are random vectors of the same size and are mutually independent, then

$$a + b \sim N(d_a + d_b, C_a + C_b). \tag{11.147}$$

For complex variables, we consider only the circularly symmetric Gaussian random vectors [86–88].[25] A circularly symmetric Gaussian vector a always has zero mean. It is fully determined by its covariance matrix

$$C_a \overset{\text{def}}{=} E(aa^H). \tag{11.148}$$

---

[24] We can factorize C_a into $C_a = QQ^H$. Then let $y = Q^{-1}(a - d_a)$. We see that $y \sim N(0, I_K)$. Therefore, $E(yy^H) = I_K$. This leads to $E[(a-d_a)^H C_a^{-1}(a-d_a)] = E(y^H y) = tr[E(yy^H)] = K$.

[25] Circularly symmetric means if the vector is rotated by a constant phase, its distribution does not change.

The probability density distribution function of a is [86, 88]

$$p(a) = (\pi)^{-K} det(C_a)^{-1} \exp\left[-a^H C_a^{-1} a\right]. \tag{11.149}$$

Note that (11.149) is different from (11.143). However, properties (11.145)–(11.147) also apply to such complex Gaussian random variables. If a and b are circularly symmetric, then so is $a + b$. In MIMO studies, typically noise and channel elements are all circularly symmetric Gaussian.

11.12.2 MIMO Channel Capacity

11.12.2.1 Problem Setup
Recall the MIMO system equation (11.13):

$$y = Hx + n, \tag{11.150}$$

where x and y are vectors of size N_T and N_R, respectively, representing the transmitted and received signals. H is the channel matrix of size $N_R \times N_T$, and n is the noise vector of size N_R. It is assumed that the noise vector is circularly symmetric Gaussian with covariance of $\sigma_n^2 I_{N_R}$. Its probability density function p_n is

$$p_n(n) = (\pi)^{-N_R} \sigma_n^{-2N_R} \exp\left[-\frac{|n|^2}{\sigma_n^2}\right]. \tag{11.151}$$

It is also known (not proved here) that, as in the case of SISO, the optimal distribution for x to achieve capacity (i.e., maximum mutual information) is a circularly symmetric Gaussian distribution with covariance of $\sigma_x^2 I_{N_T}$, σ_x^2 being the transmit power per antenna.

11.12.2.2 Entropy Computations
Given such model, we have the conditional probability

$$p_{y|x}(y,x) = p_n(y - Hx) = (\pi)^{-N_R} \sigma_n^{-2N_R} \exp\left(-\frac{|y - Hx|^2}{\sigma_n^2}\right). \tag{11.152}$$

Without the conditioning on x, y is the sum of two independent random vectors Hx and n, both with circularly symmetric Gaussian distribution with covariance $\sigma_x^2 HH^H$ and $\sigma_n^2 I_{N_R}$, respectively. Therefore, y is also circularly symmetric Gaussian with covariance $\sigma_x^2 HH^H + \sigma_n^2 I_{N_R}$. The probability distribution is thus

$$p_y(y) = (\pi)^{-N_R} \left[det\left(\sigma_x^2 HH^H + \sigma_n^2 I_{N_R}\right)\right]^{-1} \exp\left[y^H \left(\sigma_x^2 HH^H + \sigma_n^2 I_{N_R}\right)^{-1} y\right]. \tag{11.153}$$

11.12.2.3 The Capacity
From (11.141) and (11.142), we have

$$C = E\left\{\log_2\left[p_{y|x}(y,x)\right] - \log_2\left[p_y(y)\right]\right\}. \tag{11.154}$$

From (11.152) and (11.153),

$$C = \left[-N_R \log_2 \pi - N_R \log_2 \sigma_n^2 + \frac{1}{\ln(2)} E\left(-\frac{|y-Hx|^2}{\sigma_n^2} \right) \right] -$$
$$\left\{ -N_R \log[\pi] - \log_2 det\left(\sigma_x^2 HH^H + \sigma_n^2 I_{N_R} \right) + \frac{1}{\ln(2)} E\left[y^H \left(\sigma_x^2 HH^H + \sigma_n^2 I_{N_R} \right)^{-1} y \right] \right\}$$

$$(11.155)$$

According to (11.146), the two expectation values in (11.155) have the same value and thus cancel out. Therefore,

$$C = \log_2 \left[det\left(I_{N_R} + \frac{\sigma_x^2}{\sigma_n^2} HH^H \right) \right]. \tag{11.156}$$

In communications, we usually use N_0, noise special density to express noise power. It is the same as σ_n^2 in the previous derivation. Furthermore, recall that σ_x^2 is the transmit power per antenna. The total transmit power is

$$E_s = N_T \sigma_x^2. \tag{11.157}$$

Therefore, (11.156) can also be written as

$$C = \log_2 \left[det\left(I_{N_R} + \frac{E_s}{N_T N_0} HH^H \right) \right]. \tag{11.158}$$

Equation (11.158) is the same as (11.37).

REFERENCES

1. J. R. Hampton, *Introduction to MIMO Communications*. Cambridge Universiity Press, 2014.
2. J. Winters, "On the Capacity of Radio Communication Systems with Diversity in a Rayleigh Fading Environment," *IEEE J. Sel. Areas Commun.*, vol. 5, no. 5, pp. 871–878, 1987.
3. J. Salz and A. D. Wyner, *On Data Transmission over Cross Coupled Multi-Input, Multi-Output Linear Channels with Applications to Mobile Radio*. AT&T Bell Labs. Tech. Memo, 1990.
4. E. Telatar, *Capacity of Multi-Antenna Gaussian Channels*. AT&T Bell Lab. Tech. Rep. BL0112170-950615-07TM, 1995.
5. E. Telatar, "Capacity of Multi-Antenna Gaussian Channels," *Eur. Trans. Telecommun.*, vol. 10, no. 6, pp. 585–595, 1999.
6. G. J. Foschini and M. J. Gans, "On Limits of Wireless Communications in a Fading Environment When Using Multiple Antennas," *Wirel. Pers. Commun.*, vol. 6, no. 3, pp. 311–335, 1998.
7. G. J. Foschini and M. J. Gans, *On Limits of Wireless Communications in a Fading Environment when Using Multiple Antennas*. AT&T Tech. Memo. 113490-950901-05-TM, 1995.

8. G. G. Raleigh and J. M. Cioffi, "Spatio-Temporal Coding for Wireless Communications," in *Proceedings of GLOBECOM'96. 1996 IEEE Global Telecommunications Conference*, 18–28 November, London, UK, 1996, vol. 3, pp. 1809–1814.

9. J. Salz, "Digital Transmission over Cross-Coupled Linear Channels," *AT&T Tech. J.*, vol. 64, no. 6, pp. 1147–1159, 1985.

10. G. J. Foschini, "Layered Space-Time Architecture for Wireless Communication in a Fading Environment When Using Multi-Element Antennas," *Bell Labs Tech. J.*, vol. 1, no. 2, pp. 41–59, 1996.

11. A. Adjoudani et al., "Prototype Experience for MIMO BLAST over Third-Generation Wireless System," *IEEE J. Sel. Areas Commun.*, vol. 21, no. 3, pp. 440–451, 2003.

12. S. Alamouti and V. Tarokh, "Transmitter Diversity Technique for Wireless Communications," US Patent No. 6,185,258 B1, 2001.

13. S. M. Alamouti, "A Simple Transmit Diversity Technique for Wireless Communications," *IEEE J. Sel. Areas Commun.*, vol. 16, no. 8, pp. 1451–1458, 1998.

14. V. Tarokh, N. Seshadri, and A. R. Calderbank, "Space-Time Codes for High Data Rate Wireless Communication: Performance Criterion and Code Construction," *IEEE Trans. Inf. Theory*, vol. 44, no. 2, pp. 744–765, 1998.

15. G. D. Golden, C. J. Foschini, R. A. Valenzuela, and P. W. Wolniansky, "Detection Algorithm and Initial Laboratory Results Using V-BLAST Space-Time Communication Architecture," *Electron. Lett.*, vol. 35, no. 1, p. 14, 1999.

16. C. J. Foschini, *Signal Detection Algorithm for Vertical BLAST*. Lucent Bell Lab. Technical Memo. BL01131K0-980825-02TM, 1998.

17. G. J. Foschini, D. Chizhik, M. J. Gans, C. Papadias, and R. A. Valenzuela, "Analysis and performance of some basic space-time architectures," *IEEE J. Sel. Areas Commun.*, vol. 21, no. 3, pp. 303–320, 2003.

18. G. Ginis and J. M. Cioffi, "On the Relation Between V-BLAST and the GDFE," *IEEE Commun. Lett.*, vol. 5, no. 9, pp. 364–366, 2001.

19. R. F. H. Fischer, C. Windpassinger, A. Lampe, and J. B. Huber, "Space-Time Transmission Using Tomlinson-Harashima Precoding," in *ITG FACHBERICHT*, 28–30 January, Berlin, Germany, 2002, pp. 139–148.

20. Y. C. Liang, S. Sun, and C. K. Ho, "Block-Iterative Generalized Decision Feedback Equalizers for Large Mimo Systems: Algorithm Design and Asymptotic Performance Analysis," *IEEE Trans. Signal Process.*, vol. 54, no. 6 I, pp. 2035–2048, 2006.

21. C. Tidestav, A. Ahlen, and M. Sternad, "Realizable MIMO Decision Feedback Equalizers: Structure and Design," *IEEE Trans. Signal Process.*, vol. 49, no. 1, pp. 121–131, 2001.

22. W. H. Gerstacker, P. Nickel, F. Obernosterer, U. L. Dang, P. Gunreben, and W. Koch, "Trellis-Based Receivers for SC-FDMA Transmission over MIMO ISI Channels," in *2008 IEEE International Conference on Communications*, 2008, pp. 4526–4531.

23. B. Dhivagar, K. Kuchi, and K. Giridhar, "An Iterative DFE Receiver for MIMO SC-FDMA Uplink," *IEEE Commun. Lett.*, vol. 18, no. 12, pp. 2141–2144, 2014.

24. M. A. Jensen and J. W. Wallace, "A Review of Antennas and Propagation for MIMO Wireless Communications," *IEEE Trans. Antennas Propag.*, vol. 52, no. 11, pp. 2810–2824, 2004.

25. J. P. Kermoal, L. Schumacher, K. I. Pedersen, P. E. Mogensen, and F. Frederiksen, "A Stochastic MIMO Radio Channel Model with Experimental Validation," *IEEE J. Sel. Areas Commun.*, vol. 20, no. 6, pp. 1211–1226, 2002.

26. C. Oestges, "Validity of the Kronecker Model for MIMO Correlated Channels," in *IEEE 63rd Vehicular Technology Conference*, 2006, pp. 2818–2822.

27. H. Tong and S. A. Zekavat, "On the Suitable Environments of the Kronecker Product Form in MIMO Channel Modeling," in *IEEE Wireless Communications and Networking Conference*, 31 March–3 April, Las Vegas, NV, 2008.

28. D. P. McNamara, M. A. Beach, and P. N. Fletcher, "Spatial Correlation in Indoor Mimo Channels," in *IEEE International Symposium on Personal, Indoor and Mobile Radio Communications, PIMRC*, 18 September, Pavilhao Altantico, Portugal, 2002, pp. 290–294.

29. A. K. Sadek, W. Su, and K. J. R. Liu, "Diversity Analysis for Frequency-Selective MIMO-OFDM Systems with General Spatial and Temporal Correlation Model," *IEEE Trans. Commun.*, vol. 54, no. 5, pp. 878–888, 2006.

30. P. Almers et al., "Survey of Channel and Radio Propagation Models for Wireless MIMO Systems," *EURASIP J. Wirel. Commun. Netw.*, vol. 2007, no. 1, p. 019070, 2007.

31. A. G. Burr, "Capacity Bounds and Estimates for the Finite Scatterers MIMO Wireless Channel," *IEEE J. Sel. Areas Commun.*, vol. 21, no. 5, pp. 812–818, 2003.

32. D. Chizhik, G. J. Foschini, M. J. Gans, and R. A. Valenzuela, "Keyholes, Correlations, and Capacities of Multielement Transmit and Receive Antennas," *IEEE Trans. Wirel. Commun.*, vol. 1, no. 2, pp. 361–368, 2002.

33. P. F. Driessen and G. J. Foschini, "On the Capacity Formula for Multiple Input-Multiple Output Wireless Channels: A Geometric Interpretation," *IEEE Trans. Commun.*, vol. 47, no. 2, pp. 173–176, 1999.

34. V. Tarokh, H. Jafarkhani, and A. R. Calderbank, "Space-Time Block Codes from Orthogonal designs," *IEEE Trans. Inf. Theory*, vol. 45, no. 5, pp. 1456–1467, 1999.

35. L. Zheng and D. N. C. Tse, "Diversity and Multiplexing: A Fundamental Tradeoff in Multiple-Antenna Channels," *IEEE Trans. Inf. Theory*, vol. 49, no. 5, pp. 1073–1096, 2003.

36. R. Narasimhan, "Finite-SNR Diversity–Multiplexing Tradeoff for Correlated Rayleigh and Rician MIMO Channels," *IEEE Trans. Inf. Theory*, vol. 52, no. 9, pp. 3965–3979, 2006.

37. V. Tarokh, A. Naguib, N. Seshadri, and A. R. Calderbank, "Combined Array Processing and Space-Time Coding," *IEEE Trans. Inf. Theory*, vol. 45, no. 4, pp. 1121–1128, 1999.

38. H. El Gamal, G. Caire, and M. O. Damen, "Lattice Coding and Decoding Achieve the Optimal Diversity–Multiplexing Tradeoff of MIMO Channels," *IEEE Trans. Inf. Theory*, vol. 50, no. 6, pp. 968–985, 2004.

39. R. Vaze and B. S. Rajan, "On Space–Time Trellis Codes Achieving Optimal Diversity Multiplexing Tradeoff," *IEEE Trans. Inf. Theory*, vol. 52, no. 11, pp. 5060–5067, 2006.

40. D. Lao and A. M. Haimovich, "Exact Closed-Form Performance Analysis of Optimum Combining with Multiple Cochannel Interferers and Rayleigh Fading," *IEEE Trans. Commun.*, vol. 51, no. 6, pp. 995–1003, 2003.

41. R. S. Blum and J. H. Winters, "On Optimum MIMO with Antenna Selection," in *2002 IEEE International Conference on Communications. Conference Proceedings. ICC 2002 (Cat. No.02CH37333)*, 2002, vol. 1, pp. 386–390.

42. R. Y. Mesleh, H. Haas, S. Sinanovic, C. W. Ahn, and S. Yun, "Spatial Modulation," *IEEE Trans. Veh. Technol.*, vol. 57, no. 4, pp. 2228–2241, 2008.

43. P. Yang et al., "Single-Carrier SM-MIMO: A Promising Design for Broadband Large-Scale Antenna Systems," *IEEE Commun. Surv. Tutorials*, vol. 18, no. 3, pp. 1687–1716, 2016.

44. D. Gesbert, M. Kountouris, R. Heath Jr., C. Chae, and T. Salzer, "Shifting the MIMO Paradigm," *IEEE Signal Process. Mag.*, vol. 24, no. 5, pp. 36–46, 2007.

45. W. Yu and J. M. Cioffi, "Trellis Precoding for the Broadcast Channel," in *GLOBECOM'01. IEEE Global Telecommunications Conference*, 2001, vol. 2, no. 4, pp. 1344–1348.

46. A. Goldsmith, S. A. Jafar, N. Jindal, and S. Vishwanath, "Capacity Limits of MIMO Channels," *IEEE J. Sel. Areas Commun.*, vol. 21, no. 5, pp. 684–702, 2003.

47. H. Weingarten, Y. Steinberg, and S. Shamai, "The Capacity Region of the Gaussian MIMO Broadcast Channel," *IEEE Trans. Inf. Theory*, vol. 52, no. 9, pp. 3936–3964, 2006.

48. M. Bloch and J. Barros, *Physical-Layer Security: From Information Theory to Security Engineering.* Cambridge University Press, 2011.

49. A. O. Hero, "Secure Space-Time Communication," *IEEE Trans. Inf. Theory*, vol. 49, no. 12, pp. 3235–3249, 2003.

50. F. Oggier and B. Hassibi, "The Secrecy Capacity of the MIMO Wiretap Channel," *IEEE Trans. Inf. Theory*, vol. 57, no. 8, pp. 4961–4972, 2011.

51. R. Liu, T. Liu, H. P. Vincent, and S. Shamai, "Multiple-Input Multiple-Output Gaussian Broadcast Channels with Confidential Messages," *IEEE Trans. Inf. Theory*, vol. 56, no. 9, pp. 4215–4227, 2010.

52. S. Goel and R. Negi, "Guaranteeing Secrecy Using Artificial Noise," *IEEE Trans. Wirel. Commun.*, vol. 7, no. 6, pp. 2180–2189, 2008.

53. E. Jorswieck and H. Boche, "Performance Analysis of Capacity of MIMO Systems Under Multiuser Interference Based on Worst-Case Noise Behavior," *EURASIP J. Wirel. Commun. Netw.*, vol. 2004, no. 2, p. 670321, 2004.

54. J. Gao, S. A. Vorobyov, H. Jiang, and H. V. Poor, "Worst-Case Jamming on MIMO Gaussian Channels," *IEEE Trans. Signal Process.*, vol. 63, no. 21, pp. 5821–5836, 2015.
55. Q. Liu, M. Li, X. Kong, and N. Zhao, "Disrupting MIMO Communications with Optimal Jamming Signal Design," *IEEE Trans. Wirel. Commun.*, vol. 14, no. 10, pp. 5313–5325, 2015.
56. Y. Liu, H.-H. Chen, and L. Wang, "Physical Layer Security for Next Generation Wireless Networks: Theories, Technologies, and Challenges," *IEEE Commun. Surv. Tutorials*, vol. 19, no. 1, pp. 347–376, 2017.
57. A. Mukherjee, S. A. Fakoorian, J. Huang, and A. L. Swindlehurst, "Principles of Physical Layer Security in Multiuser Wireless Networks: A Survey," *IEEE Commun. Surv. Tutorials*, vol. 16, no. 3, pp. 1550–1573, 2014.
58. V. K. N. Lau, Y. Liu, and T.-A. Chen, "Optimal Partial Feedback Design for MIMO Block Fading Channels with Feedback Capacity Constraint," in *IEEE International Symposium on Information Theory, 2003. Proceedings.*, 2003, p. 65.
59. A. Rosenzweig, Y. Steinberg, and S. Shamai, "On Channels with Partial Channel State Information at the Transmitter," *IEEE Trans. Inf. Theory*, vol. 51, no. 5, pp. 1817–1830, 2005.
60. T. Zwick, C. Fischer, and W. Wiesbeck, "A Stochastic Multipath Channel Model Including Path Directions for Indoor Environments," *IEEE J. Sel. Areas Commun.*, vol. 20, no. 6, pp. 1178–1192, 2002.
61. D. Chizhik, J. Ling, P. W. Wolniansky, R. A. Valenzuela, N. Costa, and K. Huber, "Multiple-Input—Multiple-Output Measurements and Modeling in Manhattan," *IEEE J. Sel. Areas Commun.*, vol. 21, no. 3, pp. 321–331, 2003.
62. C.-X. Wang, X. Hong, X. Ge, X. Cheng, G. Zhang, and J. Thompson, "Cooperative MIMO Channel Models: A Survey," *IEEE Commun. Mag.*, vol. 48, no. 2, pp. 80–87, 2010.
63. M. J. Gans, "Channel Capacity Between Antenna Arrays—Part I: Sky Noise Dominates," *IEEE Trans. Commun.*, vol. 54, no. 9, pp. 1586–1592, 2006.
64. X. Zhu and R. D. Murch, "Performance Analysis of Maximum Likelihood Detection in a MIMO Antenna System," *IEEE Trans. Commun.*, vol. 50, no. 2, pp. 187–191, 2002.
65. W. Zhao and G. B. Giannakis, "Sphere Decoding Algorithms with Improved Radius Search," *IEEE Trans. Commun.*, vol. 53, no. 7, pp. 1104–1109, 2005.
66. B. M. Hochwald and S. ten Brink, "Achieving Near-Capacity on a Multiple-Antenna Channel," *IEEE Trans. Commun.*, vol. 51, no. 3, pp. 389–399, 2003.
67. B. Hassibi and B. M. Hochwald, "How Much Training Is Needed in Multiple-Antenna Wireless Links?," *IEEE Trans. Inf. Theory*, vol. 49, no. 4, pp. 951–963, 2003.
68. D. Samardzija and N. Mandayam, "Pilot-Assisted Estimation of MIMO Fading Channel Response and Achievable Data Rates," *IEEE Trans. Signal Process.*, vol. 51, no. 11, pp. 2882–2890, 2003.
69. IEEE, *802.11-2016 - IEEE Standard for Information Technology—Telecommunications and Information Exchange Between Systems Local and Metropolitan Area Networks—Specific Requirements—Part 11: Wireless LAN Medium Access Control (MAC) and Physical Layer (PHY)*. IEEE, 2016.
70. IEEE, *802.11ac-2013—IEEE Standard for Information Technology—Telecommunications and Information Exchange Between Systems, Local and Metropolitan Area Networks—Specific Requirements—Part 11: Wireless LAN Medium Access Control (MAC) and Physical Layer (PHY)*. IEEE, 2013.
71. 3GPP, "Release 13" [Online]. Available at http://www.3gpp.org/release-13 [Accessed August 4, 2016].
72. C. Lim, T. Yoo, B. Clerckx, B. Lee, and B. Shim, "Recent Trend of Multiuser MIMO in LTE-Advanced," *IEEE Commun. Mag.*, vol. 51, no. 3, pp. 127–135, 2013.
73. M. Sawahashi, Y. Kishiyama, A. Morimoto, D. Nishikawa, and M. Tanno, "Coordinated Multipoint Transmission/Reception Techniques for LTE-Advanced [Coordinated and Distributed MIMO]," *IEEE Wirel. Commun.*, vol. 17, no. 3, pp. 26–34, 2010.
74. R. Irmer et al., "Coordinated Multipoint: Concepts, Performance, and Field Trial Results," *IEEE Commun. Mag.*, vol. 49, no. 2, pp. 102–111, 2011.
75. D. Lee et al., "Coordinated Multipoint Transmission and Reception in LTE-Advanced: Deployment Scenarios and Operational Challenges," *IEEE Commun. Mag.*, vol. 50, no. 2, pp. 148–155, 2012.
76. F. Qamar, K. Bin Dimyati, M. N. Hindia, K. A. Bin Noordin, and A. M. Al-Samman, "A Comprehensive Review on Coordinated Multi-Point Operation For LTE-A," *Comput. Netw.*, vol. 123, pp. 19–37, 2017.

77. R. A. Valenzuela, D. Chizhik, and J. Ling, "MIMO: A Disruptive Technology Enabling Very High Spectral Efficiency Beyond Conventional Limits," in *Proceedings. 2004 IEEE Radio and Wireless Conference (IEEE Cat. No.04TH8746)*, 2004, pp. 5–8.

78. Lucent Press Release, "Lucent Technologies Receives Advanced Mobile Network Contract from DARPA; Bell Labs' Innovations to Enhance the Capabilities of the U.S. Military's Communications," *Business Wire*, 2004 [Online]. Available at http://www.businesswire.com/news/home/20040209005505/en/Lucent-Technologies-Receives-Advanced-Mobile-Network-Contract [Accessed April 21, 2018].

79. Silvus Press Release, "Silvus Technologies Supports DARPA MNM (Mobile Networked MIMO) Program; Silvus Providing Multi-Antenna Radio Solution," *Market Wired*, 2008 [Online]. Available at http://www.marketwired.com/press-release/silvus-technologies-supports-darpa-mnm-mobile-net-worked-mimo-program-827666.htm. [Accessed March 13, 2018].

80. J. Redi and R. Ramanathan, "The DARPA WNaN Network Architecture," in *2011—MILCOM 2011 Military Communications Conference*, 7–10 November, Baltimore, MD, 2011, pp. 2258–2263.

81. H. S. Kenyon, "Wireless Gateway to Connect Warfighters," *Signal*, 2009 [Online]. Available at http://www.afcea.org/content/?q=wireless-gateway-connect-warfighters [Accessed February 15, 2018].

82. N. V. Saldanha et al., "Communications Performance Improvements of Mobile Networked MIMO in Army Operational Environments," in *MILCOM 2013—2013 IEEE Military Communications Conference*, 18–20 November, San Diego, CA, 2013, pp. 758–763.

83. H. Q. Lai et al., "Measurements of Multiple-Input Multiple-Output (MIMO) Performance Under Army Operational Conditions," in *Proceedings—IEEE Military Communications Conference MILCOM*, 31 October–3 November, San Jose, CA, 2010, pp. 2119–2124.

84. Y. Jiang, A. Tiwari, M. Rachid, and B. Daneshrad, "MIMO for Airborne Communications [Industry Perspectives]," *IEEE Wirel. Commun.*, vol. 21, no. 5, pp. 4–6, 2014.

85. S. S. Niu, L. Ljung, and A. Bjorck, "Decomposition Methods for Solving Least-Squares Parameter Estimation," *IEEE Trans. Signal Process.*, vol. 44, no. 11, pp. 2847–2852, 1996.

86. R. G. Gallager, "Circularly-Symmetric Gaussian Random Vectors," 2008. [Online]. Available at http://www.rle.mit.edu/rgallager/documents/CircSymGauss.pdf [Accessed January 1, 2016].

87. B. Picinbono, "On Circularity," *IEEE Trans. Signal Process.*, vol. 42, no. 12, pp. 3473–3482, 1994.

88. F. D. Neeser and J. L. Massey, "Proper Complex Random Processes with Applications to Information Theory," *IEEE Trans. Inf. Theory*, vol. 39, no. 4, pp. 1293–1302, 1993.

HOMEWORK

11.1 Use simulation to reproduce Figure 11.5. Note that you may get better performance because the figure was obtained from a real system instead of a simulation. Use independent fading, stationary Rayleigh channels, and average performance over many realizations.

a. Compare performance with and without error propagation.

b. Compare performance with and without reordering.

c. (optional) Compare with the analytical results in Section 11.11.

d. Play with parameters and report interesting findings.

11.2 In the context of Section 11.6.4, consider a single-antenna system, that is, $N_T = N_R = 1$. It can be considered as special cases of spatial multiplexing as well as space-time coding. From (11.87), we have $r = 1$. From (11.84), we get $d = 1$. However, such values of r and d violates (11.89). Explain such apparent contradiction.

11.3 (Optional) Extend the derivations in Section 11.12 to include correlated (colored) transmit signal and noise. You can either factorize the covariance matrices and include correlation to the equivalent channels (effectively adding a noise-whitening filter) or modify (11.151) and (11.152) to include correlation.

11.4 Consider the normalization described at the end of Section 11.3.3.2.

    a. Show that (11.39) did indeed lead to (11.42), based on (11.37).

    b. What is the relationship between ρ and E_s, N_0? What does ρ mean, in terms of the SNR at the receiver?

11.5 Provide detailed derivation from (11.54) to (11.57).

11.6 Provide detailed derivation of (11.66) and (11.67).

11.7 Show that the two determinants in (11.90) are related to the transmitted energy delivered to the receiver and the eavesdropper, respectively.

CHAPTER 12

5G CELLULAR SYSTEM RADIO INTERFACE TECHNOLOGY

12.1 INTRODUCTION

So far in this book, we have discussed various modules of the transceiver. The techniques discussed in those chapters are all mature, although some are still under active research. In this chapter, we provide a brief introduction to some emerging technologies in the context of the fifth generation (5G) cellular system. Other technologies not included in the 5G systems will be briefly discussed in Chapter 13.

The 5G cellular system is at an early development stage. Visions may vary among vendors and operators. This chapter is based on open literature available in March 2018.

For the rest of the chapter, we start in Section 12.2 with an overview of the cellular systems: their history, basic concepts, and technical challenges. Basic concepts introduced therein are helpful in understanding the rest of the chapter. We then shift focus to the 5G systems in Section 12.3 to discuss its standardization process and current status. Section 12.4 describes the currently prevailing 3GPP proposal. Section 12.5 discusses three of the new physical layer (i.e., radio interface) technologies included in the 3GPP proposal: massive multiple-input multiple-output (MIMO), millimeter wave (mmWave) communications, and nonorthogonal multiple access (NOMA). Finally, a summary is provided in Section 12.6.

12.2 CELLULAR SYSTEMS

Nowadays, cellular systems and mobile wireless communications are often used interchangeably. However, strictly speaking, the cellular system is a special form of mobile wireless communications, where the service areas are divided into "cells." Mobile users can move among the cells without interruption to their communication sessions. In this section, we provide an overview of the cellular technology in general.

Digital Communication for Practicing Engineers, First Edition. Feng Ouyang.
© 2020 by The Institute of Electrical and Electronics Engineers, Inc.
Published 2020 by John Wiley & Sons, Inc.

12.2.1   A Brief History of Cellular Systems

Let us start with a recount of the history of the cellular systems, especially the various "generations" of the cellular systems. Such recount provides the context for the 5G cellular system development. For more detailed descriptions of the prior cellular systems, see [1], [2, ch. 1], [3, 4].

The first large-scale commercial mobile service was provided by AT&T starting in 1946. It used a powerful radio tower to cover the entire city. As a result, only a few users can make the call at a given time due to the spectral limitation, resulting in long waiting queues for the service.

At about the same time, the Bell Labs conceived the concept of cellular service in 1947. Under this vision, mobile service coverage is divided into "cells." Neighboring cells may need to use different frequencies to avoid interference. However, interference at far-ways cells is weak enough so that the same frequency can be "reused." Such an architecture, to be detailed in Section 12.2.2, dramatically increases the number of concurrent users that can be supported. The first commercial cellular system was developed by the Nippon Telegraph and Telephone (NTT) in Japan in 1979. In 1983, AT&T launched its cellular service, known as the Advanced Mobile Phone Service (AMPS) [2, ch. 1].

AMPS, also known as the first generation (1G) cellular system, is an analog voice system, with each voice channel using a 30 kHz dedicated channel with frequency modulation (FM).[1] AMPS uses frequency division duplexing (FDD), that is, transmissions in both directions happen at the same time but with different frequencies. FDD has been used by most of the cellular services since then. At about the same period, other 1G systems were developed in Japan and Europe.

In the early 1990s, the second generation (2G) cellular systems began to emerge. There are several versions of 2G systems, but they share the following common traits.

- As the 1G systems, they are primarily circuit-switched voice services. Namely, each user is provided with a reserved channel, regardless of the actual traffic.

- They use digital modulations (Chapter 3) and forward error correction coding (Chapter 5).

- The voice is digitized by some voice encoder, which compresses the voice information, reducing the required bandwidth.

The first 2G system in the United States was introduced by AT&T in 1991, based on standard IS-54, which was later upgraded to IS-136. It is a time division multiple access (TDMA) system. Namely, multiple users share the same frequency channel but are separated into different time slots. This system uses the same 30 kHz channel bandwidth for backward compatibility with the AMPS. However, due to better modulation and compressive voice coding, each channel can accommodate three users.

---

[1] FM modulation is not covered in this book.

A competing 2G system in the United States was developed by the then-startup company Qualcomm, turning into standard IS-95. It uses code division multiple access (CDMA), where multiple users transmit at the same frequency and time [5, sec. 16.3]. The users are assigned different "codes," which enables the receiver to separate signals from different users. CDMA is not covered in this book, except for a brief description in Section 12.5.3.3.

In Europe, the unified 2G cellular standard is the Groupe Spécial Mobile (GSM), later renamed as Global System for Mobile Communications while keeping the same acronym. It was first launched in 1991 and is still widely used today [6]. A GSM channel is 200 kHz in bandwidth, shared by eight users through TDMA. The GSM standard is used in many parts of the world. It underwent subsequent upgrades and improvements to provide packet-based data services at the rates of tens and hundreds of kilobits per second.

The third-generation (3G) cellular systems evolved from the 2G systems following separate paths in Europe and North America. 3G systems are based on Internet Protocol (IP) network. Instead of circuit-switched connections as in 1G and 2G, an IP network is packet-switched. User data are organized into packets, each of them being sent independently through the network to their destinations. There are no capacity reservations; a user utilizes network capacity only when he has data to send.

In Europe, the Third Generation Partnership Project (3GPP) developed the widely adopted 3G cellular standard known as the wideband code division multiple access (WCDMA). The first version of the standard (known as 3GPP Release 99) was published in 2000. Several subsequent releases provided improvements and upgrades. In WCDMA, the users share channels of 5 MHz bandwidth through CDMA. The data rate is up to 384 kbps per user in both directions. Such speed was increased in later evolutions up to tens of Mbps. Multiple antenna techniques (multiple-input multiple-output [MIMO], described in Chapter 11) was also introduced in the later evolutions, further increasing data rates.

In North America, another 3G cellular standard was developed based on the 2G CDMA technology. The standard body was the Third Generation Partnership Project 2 (3GPP2), primarily sponsored by Qualcomm. The standard was known as CDMA2000-1X, and a later evolution is known as evolution-data only (EV-DO). In these systems, the users share channels of 1.25 MHz bandwidth through CDMA, just as in IS-95. EV-DO provides downlink rates up to 2.4 Mbps and uplink rates up to 153 kbps. A later enhancement, EV-DO Rev. A, increases the data rates to 3.07 and 1.8 Mbps for downlink and uplink, respectively.[2]

There is a third version of the 3G cellular system, known as the time division synchronous code division multiple access (TD-SCDMA), used by one of the Chinese cellular service providers. It is similar to WCDMA in many aspects, except that time division duplexing (TDD) is used between the uplink and downlink. In a TDD scheme, the uplink and downlink signals use the same frequency but different time

---

[2] There was a Rev. B for EV-DO, which further increases data rate. However, it was not widely deployed.

slots. TDD has several practical advantages. Notably, it allows for channel estimation based on reciprocity (to be discussed more in Section 12.5.1.4).

3G cellular systems were introduced around 2000. All systems underwent continuous enhancements and upgrades. There were several newcomers, most notably the Mobile Worldwide Interoperability for Microwave Access (mobile WiMAX), based on the IEEE standard 802.16e. At the same time, the advent of smartphones dramatically increased the demand for mobile data connections. Such demand was addressed by the 4G cellular systems.

Although there were several contenders for the 4G cellular systems, the Long-term evolution (LTE) standard developed by 3GPP has now become the dominating technology in this realm. Strictly speaking, the original LTE as defined by 3GPP specification Release 8 (2009) is not qualified as a 4G system. The subsequent enhanced version as defined by 3GPP specification Release 10 (2011), also known as LTE-Advanced or LTE-A, is a true 4G system. However, people often use the terms LTE and 4G interchangeably. In the following discussion, we do not distinguish between LTE and LTE-Advanced unless otherwise noted.

LTE uses a new modulation method known as orthogonal frequency division multiplexing (OFDM) (Chapter 10) and provides extensive support to the MIMO technologies (Chapter 11). Furthermore, LTE supports a variety of channel bandwidths: 1.4, 3, 5, 10, 15, and 20 MHz. Continuous or discontinuous channels can be aggregated to support a higher total data rate. These and other technological advances enable LTE-Advanced to provide up to 1 Gbps on the downlink and 500 Mbps on the uplink. These data rates are shared by all concurrent users and cannot be compared directly to the per-user data rates quoted for the 2G and 3G systems. However, it is evident that the 4G system represents a quantum leap in capacity.

12.2.2 Frequency Reuse and Handover

As stated in Section 12.2.1, cellular systems are based on the concept of "cell," first proposed in 1947 [7, 8]. It contains two fundamental concepts: frequency reuse and handover. These are briefly introduced in this section. More detailed introductions can be found in, for example, [4, ch. 3].

Frequency reuse means that multiple cells can use the same frequency, substantially increasing the number of users that can be served concurrently by the system.

The hexagons represent cells. Cells with a same marked number use the same frequency partition.

If a cellular system has a frequency reuse factor of C, the total frequency band is divided up to C nonoverlapping parts. Neighboring cells use different parts of such partition to avoid mutual interference. Figure 12.1 shows the frequency reuse patterns for $C = 1, 3, 7$. Other C values can also be used. In this figure, cells using frequency partition 1 are shaded. We can see that with a larger C, the distance between the cells using the same frequency is larger, reducing mutual interference. Of course, larger C also reduces system capacity, as each cell is allocated with a smaller portion of the total bandwidth.

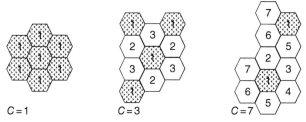

Figure 12.1   Frequency reuse patterns.

In cellular systems, the frequency reuse factor depends on propagation environments and cell designs. GSM typically uses $C = 7$. The 3G systems can use smaller frequency reuse factors such as 3 or 1, because the multiple users are separated by codes in the CDMA scheme. As long as no users in neighboring cells use the same code, there is no high interference. The 4G system (LTE-Advanced) uses a frequency reuse factor of 1. In fact, the LTE system may use "fractional frequency reuse," where mobile users at the center of a cell can use the same frequency allocation as those in a neighboring cell, while users near a cell boundary coordinate their frequency allocation to avoid overlapping with those on the other side of the boundary.

Handover techniques allow the mobile users to travel from one cell to another without losing connection. At a user crosses a cell boundary, his connection is switched from one base station to another in the handover process. Handover can be initiated by the base station based on its knowledge about the mobile user, initiated by the base station based on the mobile user report of signal qualities, or initiated by the mobile user. During a handover, a mobile user may break from his previous link and establish a new link (hard handover). Alternatively, a user may simultaneously maintain connectivity to both base stations during the process (soft handover).

12.2.3   Control and Data Traffic

For a cellular system to work properly, the mobile users and the base stations must exchange a set of control messages in addition to the user data traffic. Some of the control messages are exchanged only when establishing a connection, while others are sent and received at regular intervals. In some systems, dedicated channels are reserved for control messages. In others, control and data traffics are multiplexed within the same channel. Even in the latter case, control and data traffic flows may use different modulation and channel coding. In this chapter, we focus on data traffic only.

For example, in LTE, control and data traffics are separated into logical channels. At the physical layer, the downlink control messages use the physical downlink control channel (PDCCH), while the downlink data traffic uses the physical downlink shared channel (PDSCH). In the uplink, there are the corresponding physical uplink control channel (PUCCH) and physical uplink shared channel (PUSCH). There are several other control channels, but they are used less frequently.

In the downlink, certain subcarriers in certain symbols of the OFDMA modulation are reserved for control channels, while the rest is used for the shared channel (except for reference signals, see Chapter 10). In the uplink, a similar scheme is used.

12.2.4 Constraints to Mobile Devices

Another unique feature of the cellular system is the asymmetry between the two partners. A base station can afford to spend more power in transmission and processing, while a mobile device usually is very constrained in size and power consumption. Such feature drives some design considerations of the cellular technology.

12.3 THE 5G SYSTEM

This section provides a general overview of the 5G cellular system. The physical layer features will be discussed in more detail in Section 12.5.

12.3.1 Cellular System Standardization Framework

Starting from 3G, the international standardization of cellular systems is under the governance of the International Telecommunication Union Radio Communication Sector (ITU-R). ITU-R defines high-level performance requirements, spectrum allocations, and networking specifics to ensure that the various cellular systems can work together. These standards belong to the family of "international mobile telecommunication systems," or IMT. The ITU-R working party 5D (WP 5D) is responsible for developing the IMT standards. The standards for 3G, 4G, and 5G cellular systems are IMT 2000, IMT-Advanced, and IMT 2020, respectively. Each standard contains multiple ITU-R Recommendations.

For each generation of cellular systems, multiple technologies that meet IMT requirements can be proposed as candidates and be approved by ITU-R. As Section 12.2.1 described, there were several competing technologies for the 3G systems. For 4G, LTE-Advanced (developed by 3GPP) is the predominant technology. It is therefore expected that the dominant 5G technology will also come from 3GPP. This chapter focuses on the 5G proposal from 3GPP (Section 12.4). For more information on the IMT standard framework, see [9].

12.3.2 5G Requirements from ITU-R

The current 4G cellular system was developed to satisfy the rapidly growing need for mobile data services fueled by the proliferation of smartphones and other mobile devices. Subsequent enhancements of 4G also address needs of other mobile applications such as public safety, vehicle-to-vehicle (V2V) communications, and Internet of Things (IoT). The 5G vision is [10] "to expand and support diverse usage scenarios and applications that will continue beyond the current IMT." More explicitly, the envisioned 5G applications include three classes [11].

1. Enhanced mobile broadband (eMBB). eMBB is the traditional mobile data service, but at much higher data rates to support smart mobile devices, mobile computing and emerging applications such as virtual reality (VR).

2. Massive machine type communications (mMTC). This class of applications includes IoT and smart city. Comparing to eMBB, per-device data rate and latency requirements are relaxed for mMTC. However, the cellular network needs to support a vast number of users per area and to allow the user devices to operate with low power consumption.

3. Ultra-reliable and low latency communications (URLLC). Examples of such applications include medical devices, vehicle-to-vehicle communications, and industrial automation. The data rates are low to medium, while latency and reliability requirements are very stringent, even in highly mobile applications.

ITU-R specified a number of specific technical performance requirements [12]. They are summarized in Table 12.1. Data rates and spectral efficiencies are defined separately for downlink (DL) and uplink (UL).

To support 5G cellular system deployment, various countries have started the process of allocating spectrum bands for 5G, including the traditional cellular bands (below 6 GHz) and the millimeter wave bands (above 24 GHz) [13].

12.3.3   Current Standardization Status

According to the ITU-R plan, 5G requirement development at ITU-R was finished in mid-2017. Technology candidate proposal submission and evaluation are conducted in 2018 and 2019 [14].

12.4   HIGHLIGHTS OF 3GPP PROPOSAL

The current 4G cellular market is dominated by LTE and LTE-Advanced systems, whose standards were developed by 3GPP. It is expected that 3GPP will play a crucial role in developing the 5G standards and technologies. The only competitor to LTE in the 4G arena is the WiMAX based on the IEEE standard 802.16. However, IEEE, while actively engaged in 5G standard development with ITU-R and other partners, so far has not indicated its intention of submitting 802.16 as a competing technology.

In October 2017, ITU-R held a workshop on 5G to share information about the requirements and the proposal submission [15]. Its presenters included four potential proposers. Among them, 3GPP provided a relatively complete vision of the standard it planned to submit [16]. Other two presenters represented the Chinese IMT-2020 Promotion Group (a government, academia and industrial consortium), and the Korean Ministry of Science and ICT. Both of them expressed intentions to contribute to the 5G standard development while aligning with the 3GPP framework. The Korean presentation indicated their interest in the millimeter wave communications, which have already been demonstrated in Korea. The fourth presentation was on the DECT

TABLE 12.1 5G system key technical performance requirements [12].

Parameter	Value	Brief definition	Target application
Peak data rate (Gbps)	20 downlink (DL), 10 uplink (UL)	Data rate under ideal conditions, when one mobile user has all radio resources. Physical layer overhead is excluded	Enhanced mobile broadband (eMBB)
Peak spectral efficiency (b/s/Hz)	30 DL, 15 UL	Under ideal conditions, excluding physical layer overhead. Assuming MIMO configurations of 8 by 8 for DL and 4 by 4 for UL	eMBB
User data rate (Mbps)	100 DL, 50 UL	5% point of CDF of the user throughput at Layer 3 in the dense urban use case	eMBB
User spectral efficiency (b/s/Hz)		Based on 5% point of CDF of the user throughput at Layer 3	eMBB
	0.3 DL, 0.21 UL	Indoor hotspot	
	0.225 DL, 0.15 UL	Dense urban	
	0.12 DL, 0.045 UL	Rural	
Average spectral efficiency (b/s/Hz)		Average spectral efficiency of all users at Layer 3	eMBB
	9 DL, 6.75 UL	Indoor hotspot	
	7.8 DL, 5.4 UL	Dense urban	
	3.3 DL, 1.6 UL	Rural	
Area traffic capacity (Mbit/s/m^2)	10	Total traffic throughput per geographic area for indoor hotspot	eMBB
Latency (ms)	4 for eMBB, 1 for URLLC, 20 for control plane	One-way at the application layer for user latency; transition time from idle state to active state for the control plane	eMBB, Ultra-reliable and low latency communications (URLLC)
Connection density (user per km^2)	1,000,000	Total number of devices fulfilling a specific level of quality of service (QoS)	mMTC
Energy efficiency	Not specific yet	Features such as sleep mode, efficient transmission, and so on	eMBB
Success rate	$1 - 10^{-5}$	Success rate of Layer 2/3 packet delivery	URLLC
Mobility (km/h)		The speed ranges for the mobile station in mobility classes	eMBB
	0	Stationary	
	0–10	Pedestrian	
	10–120	Vehicular	
	120–500	High-speed vehicular (such as high-speed trains)	
		For the following mobility (km/h):	eMBB

TABLE 12.1   (**Continued**)

Parameter	Value	Brief definition	Target application
Mobility spectral efficiency (b/s/Hz)	1.5	10	
	1.12	30	
	0.8	120	
	0.45	500	
Minimum bandwidth (MHz)	100	Aggregated system bandwidth supported by single or multiple radio frequency carriers. At higher frequency bands (e.g., above 6 GHz), the system shall support higher bandwidths up to 1 GHz	All

standard, which is used for cordless telephone and some other applications. It was envisioned that DECT could be a part of the 5G standard for mMTC support. Therefore, as of October 2017, 3GPP appears to be the only organization with a plan for a complete proposal for the 5G system.

In January 2018, 3GPP formally submitted to ITU-R its initial proposal for 5G candidate technologies [17]. This proposal will lead to 3GPP's first 5G specification release (Release 15), to be completed by June 2018. At the same time, China and Korea submitted their proposals. The Chinese proposal is based on the 3GPP proposal with no significant modification [18]. The Korean proposal is also based on the 3GPP proposal but with some simplifications. It limits operational frequencies to 3.3–3.8 GHz and 24.25–29.5 GHz [19] (the 3GPP proposal lists many more supported bands below 6 GHz, as well as a band from 37 to 40 GHz [20]).

In 3GPP terminology, the 5G base station (corresponding to the eNodeB in LTE) is known as gNB. The mobile device (corresponding to the UE in LTE) is known as UE. The radio interface technology (i.e., air interface or PHY layer) is known as the New Radio (NR) [21, sec. 6].

The 3GPP proposal provides a roadmap of evolution from the current LTE system to the new 5G system. The gNB will first be connected to an LTE core network, with limited features. More features will be introduced at later stages (probably in Release 16) as the 5G core network is deployed. Since we focus on the radio technology NR, we will not get into the details of the core network evolution.

According to the 3GPP proposal, the following technologies are included in the NR [20]. The Item Numbers in the list bellow refer to those in [20]. We focus on the data channel (as opposed to the control channels) only. The item numbers in the list below refer to those in the ITU-R questionnaire as submitted by 3GPP. An overview of the 3GPP Release 15 can be found at [22].

- Both uplink (UL) and downlink (DL) data channels support the following subcarrier modulations: QPSK, 16QAM, 64QAM, and 256QAM, see Chapter 3 (Item Number 5.2.3.2.2.2.1).

- Both UL and DL data channels use low-density parity code (LDPC) channel coding of various coding rates, see Chapter 5. DL control channel uses Polar coding (Item Number 5.2.3.2.2.3.1). Note that turbo codes, instead of LDPC codes, are used in LTE.

- The NR provides multiple reference signals for both UL and DL assisting the various channel estimation tasks (Item Number 5.2.3.2.3).

- The NR supports flexible ways of utilizing spectrum. Consecutive or discontinuous frequency bands can be aggregated together for up to 6.4 GHz of total bandwidth. Multiple subcarrier spacing values can be used in a single band to support different services (Item Number 5.2.3.2.8.1) [23].

- Channel bandwidths can be 5–100 MHz for frequency range 450 MHz to 6 GHz, and up to 400 MHz for frequency range 24.25–52.6 GHz (Item Number 5.2.3.2.8.2).

- The NR uses the traditional OFDM (known as CP-OFDM in NR terminology) modulation for both UL and DL (see Chapter 10). Also, the SC-FDMA waveform used for LTE UL is supported for NR UL single stream transmissions and is known as DFT-spread OFDM or DFT-s-OFDM in NR terminology (Item Number 5.2.3.2.2.1). For more information, see [24, secs. 6.1.3 and 8.1.3].

- OFDM sidelobe suppressing techniques such as filtering and windowing (see Chapter 10) can be used at the transmitter as enhancements transparent to the receiver. Such techniques allow a reduction of the guard band comparing to LTE (Item Number 5.2.3.2.2.1). For more information, see [24, secs. 6.1.3 and 8.1.3].

- The NR supports various MIMO schemes including spatial multiplexing (open loop and closed loop), spatial transmit diversity, and hybrid beamforming (see Chapter 11) (Item Number 5.2.3.2.9.1). More information about MIMO schemes can be found in [24, secs. 6.1.6 and 8.1.6].

- The NR supports up to 32 antenna ports is DL and up to 4 antenna ports in UL. Antennas can be grouped into panels in NR operations. Therefore, the number of antenna elements may be larger than the number of antenna ports (Item Number 5.2.3.2.9.2).

- For spatial multiplexing, NR supports up to eight layers in DL and up to four layers in UL. Twelve orthogonal ports (with up to four per UE) are supported for multiuser MIMO. The NR further supports coordinated multipoint (CoMP) transmission and reception, see Chapter 11 (Item Number 5.2.3.2.9.4).

- The NR supports remote antennas and distributed antennas (Item Number 5.2.3.2.9.5).

- The NR provides various special modes and mechanisms to support low latency, low-cost and low-power operations required by the 5G URLLC and eMTC services (Item Number 5.2.3.2.23.2).

12.5   5G PHYSICAL LAYER TECHNOLOGIES

In the early development of 5G, many candidate technologies were proposed [25–27]. As described in Section 12.4, 3GPP has presented their plan for the 5G radio air interface, the NR. In this section, we discuss some of the new technologies included in the NR. They are examples of the emerging technologies in the wireless communications. Chapter 13 will briefly discuss some other emerging technologies not included in the NR.

Discussion of these technologies is on generic levels, reflecting the current literature. They are not specific to the schemes or options chosen by the NR. For a comprehensive overview of new physical layer features introduced in NR, see [28].

For the rest of this section, vectors are expressed by bold lowercase letters, while matrices are expressed by uppercase letters.

12.5.1   Massive MIMO

Chapter 11 described multiple-input multiple-output (MIMO) technologies, which use multiple antennas for a wireless link. MIMO technologies include spatial multiplexing, space-time diversity, and beamforming. Massive MIMO (also known as large-scale MIMO or large-scale antenna systems) is a hybrid of the spatial multiplexing and beamforming technologies.

12.5.1.1   System Configuration

Massive MIMO is typically used in a cellular setting, where the base station employs a large number of antenna elements. The base station supports multiple data streams with spatial multiplexing (Chapter 11). However, the number of antennas is much larger than that of the data streams. The data streams can be sent to one or more mobile users, while each mobile user is equipped with a small number of antennas.

Figure 12.2 shows massive MIMO configurations. The broken lines show cell boundaries. The triangles at the cell centers denote the base stations. The squares denote mobile users (or "users" for short). Panel A shows the downlink case, where the base station transmits to multiple users using multiple "beams." Panel B shows the uplink case, where the mobile users transmit with omnidirectional antennas, while the

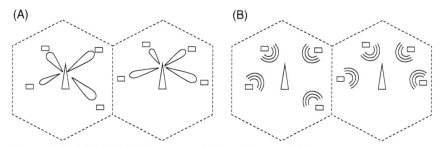

Figure 12.2   Massive MIMO systems. A: Downlink. B: Uplink.

base stations receive signals from multiple users. A typical configuration in the literature is a base station with M antennas, communicating with K mobile users, each with a single-antenna radio. Such configuration is shown in Figure 12.2 and is the focus of this chapter. Discussions and results are easily extended to cases with multi-antenna users.

Furthermore, in this book, we consider only the flat channels, that is, without inter-symbol interference (ISI) (Chapter 6). OFDM modulation (Chapter 10) can be used in channels with ISI, while our discussions apply to the OFDM subcarriers. There are other ways to address ISI [29], but they are not discussed in this book.

12.5.1.2 Single User

As described in Chapter 11, the equation describing a MIMO system with N_T transmitter antennas and N_R receiver antennas is

$$y = Hx + n. \tag{12.1}$$

Here, x and y are transmitted and received signal vectors of sizes N_T and N_R, respectively. n is the noise vector of size N_R. H is the channel matrix of size $N_R \times N_T$. Massive MIMO with a single user can be viewed as a special case of such a MIMO system. For the downlink, $N_R = 1$, $N_T = M$. Equation (12.1) can be written as

$$y = h^T x + n$$
$$x = ps \tag{12.2}$$

Here, y, x, and n are received symbol, transmitted symbol, and noise. y and n are scalars since $N_R = 1$. The channel matrix H in (12.1) has only one row, denoted as the row vector h^T in (12.2). In the second line in (12.3), the transmitted symbol vector x is formed based on a scalar data symbol s. p is known as the precoder or precoding vector. The signal transmitted by antenna element m is $p_m s$. Both vectors h and p are of size M.

For the uplink, $N_T = 1$, $N_R = M$. Equation (12.1) thus becomes

$$y = hx + n. \tag{12.3}$$

Here, y and n are the received signal and noise vectors of size M. Scalar x represents the transmitted signal. The channel matrix H in (12.1) has only one column, denoted by the vector h.

For the downlink, the precoder should be optimized to yield maximum signal power at the receiver. Namely,

$$p_o = \operatorname{argmax} \left| h^T p \right|^2$$
$$|p_o|^2 = 1 \tag{12.4}$$

The second equation above reflects the constraint on total power. From the Cauchy–Schwarz inequality, we have

$$\left| h^T p \right|^2 \le |h|^2 |p|^2. \tag{12.5}$$

The equality holds when p and h^* are in parallel. Therefore, the optimal precoder is

$$p_o = \frac{h^*}{|h|}. \tag{12.6}$$

This precoding result is known as the matched filter (MF) beamforming.[3]
For the uplink, a linear receiver is expressed as follows:

$$r = g^T y. \tag{12.7}$$

r is the scalar signal to be demodulated. g is the combination vector of size M. Equations (12.3) and (12.7) lead to

$$r = g^T h x + g^T n. \tag{12.8}$$

The first term on the right-hand side is the signal, while the second term is the noise. We would like select g to maximize the SNR:

$$g_o = argmax \frac{E\left(|g^T h x|^2\right)}{E\left(|g^T n|^2\right)}. \tag{12.9}$$

Assume the noise is spatially white, that is,

$$E\left(nn^H\right) = N_0 I_M, \tag{12.10}$$

where N_0 is the noise energy per receiving antenna. Further, assume that the transmitted power $E|x|^2$ is a constant. Equation (12.9) is then equivalent to

$$\begin{aligned} g_o &= argmax|g^T h|^2 \\ |g_o|^2 &= 1 \end{aligned}. \tag{12.11}$$

Equation (12.11) is the same as (12.4). Therefore, the optimal solution is

$$g_o = \frac{h^*}{|h|}. \tag{12.12}$$

Such a solution is known as the maximum ratio combining (MRC).

Recall that for a conventional MIMO system as discussed in Chapter 11, the channel capacity grows linearly with the minimum of the number of antennas at the transmitter and receiver. For the massive MIMO system discussed here, the

---

[3] MF in the time domain is discussed in Chapter 4.

minimum number of antennas is always 1. However, one can easily see that the SNR at the receiver is proportional to $|h|^2$ in both downlink and uplink. Consider that

$$|h|^2 = \sum_{m=1}^{M} |h_m|^2, \tag{12.13}$$

where $\{h_m\}$ is the channel between antenna m at the base station and the mobile user. M is the number of antennas at the base station. Typically, $E(|h_m|^2)$ is independent of m. Therefore, the "ergodic," that is, average SNR is proportional to $E(|h_m|^2)$. Namely, in the case of $SNR \gg 1$, the Shannon capacity of the channel (see Chapter 2) increase with $\log_2(M)$.

On the other hand, as discussed in Chapter 11, the capacity of a conventional MIMO system depends also on the rank of the channel. For low-rank channels (such as the line of sight, or free space channel), the full MIMO capacity cannot be realized. However, for massive MIMO, the capacity gain described above is always realizable, even with low-rank channels.

The capacity increase in massive MIMO is due to the fact that in both downlink and uplink, signals transmitted or received by the multiple antennas are added with the same phase (i.e., constructive interference) at the receiver. In other words, the base station transmitter and receiver "focus" their "beam" onto the mobile user. Such focusing occurs in both line of sight and scattering propagation environments.

Note that such SNR increase also exists in the conventional MIMO systems. After all, single-user massive MIMO is a special case of the conventional MIMO. However, SNR increase due to MF precoding or MRC processing is not considered as the primary avenue for performance improvement for the conventional MIMO, which focuses more on spatial multiplexing and spatial diversity.

In Section 12.5.1.3, we discuss massive MIMO for multiple users, where the massive MIMO deviates from the conventional MIMO.

12.5.1.3 Multiple Users

Other than increasing SNR as discussed in Section 12.5.1.2, massive MIMO further increases the system capacity by serving K users at the same time. In such case, massive MIMO is a special case of the multiuser MIMO (MU-MIMO) discussed in Chapter 11. With multiple users, the system equation remains the same as (12.1). Including more processing, the uplink system can be expressed as

$$r = Gy = GH_u x + n. \tag{12.14}$$

Here, y and n are vectors of size M, denoting received signals and noises at the M receiver antennas. r and x are vectors of size K, denoting the transmitted and received symbols from the K users.[4] G is the combination matrix of size $K \times M$. H_u is the uplink channel matrix of size $M \times K$. Column k of H_u is the channel vector for the kth mobile user, as used in (12.3). The downlink system can be expressed as

$$y = H_d P s + n. \tag{12.15}$$

---

[4] Remember that all users have single antenna, as stated in Section 12.5.1.1.

Here, y, s, and n are vectors of size K, denoting the received symbols, transmitted symbols, and noise at the K mobile user receivers, respectively. P is the precoding matrix of $M \times K$ operating at the base station. H_d is the downlink channel matrix of size $K \times M$. Row k of H_d and column k of P represent the channel vector and precoding vector for user k as used in (12.2). Note that different components of y are received by different users. Therefore, they are not processed jointly as in a conventional MIMO system.

As shown in Section 12.5.1.2, analyses of uplink and downlink are symmetric. Therefore, we focus on downlink below.

With multiple users, the transmission may cause mutual interference among the users. To avoid such interference, we would like to make $H_d P$ in (12.15) a diagonal matrix. Since H_d is nonsquare, P should be the pseudo inverse of H_d:

$$P = \alpha H_d^H \left(H_d H_d^H \right)^{-1}. \tag{12.16}$$

This precoding is known as the zero-forcing (ZF). α is a normalization constant chosen to satisfy the transmit power constraint. ZF precoding focuses on avoiding mutual interference among the users, without considering delivering maximum power to each user as in the MF precoding discussed in Section 12.5.1.2. A compromise between ZF and MF precoding schemes is known as the regularized zero-forcing (RZF) precoding [29–32]. An RZF precoder is given as

$$P = \alpha H_d^H \left(H_d H_d^H + \delta I_K \right)^{-1}. \tag{12.17}$$

Here, δ is an adjustable constant. α provides normalization for total power constraint. An RZF precoder becomes a ZF precoder or an MF precoder when δ goes to zero or infinity, respectively.

Similar results can be derived for uplink combiners. See homework. The concept of zero-forcing is discussed also in Chapters 9 and 11.

As mentioned in Chapter 11, multiuser MIMO requires sophisticated processing such as multiuser detection and dirty-paper coding in order to achieve full performance. Complexities of these techniques typically grow exponentially with the number of users. On the other hand, the massive MIMO schemes discussed above use linear processing (precoding and combining), which are in general suboptimal. The beauty of massive MIMO is that when the number of antennas M is very large (i.e., $M \gg K$), the channels for different users are nearly orthogonal [29]. In this case, the mutual interference is not important, and the MF recoding and MRC combining are actually optimal. Therefore, massive MIMO provides a scalable solution to multiuser MIMO. Namely, the complexity is under control when the number of users K increases, as long as $K \ll M$.

12.5.1.4 Channel Estimation and Time Division Duplexing

So far, we discuss massive MIMO under the assumption that the base station has perfect knowledge of channel state information. In this section, we discuss more on channel estimation.

For the uplink, the base station can measure the channel matrix when each mobile user sends known symbols (known as pilots), as discussed in Chapters 6 and 9. The kth column of H_u can be estimated by the M receiver antennas based on the pilot sent by user k. To resolve the entire channel matrix H_u, the base station needs to separate the pilot signals sent by the different users. One simple way to achieve the separation is for the K users to send their pilot signals at K different points in time. Another way is using K mutually orthogonal sequences. These sequences are necessarily no less than K symbols in length. Either way, the overhead from data transmission is K symbols for each channel estimation operation. The frequency of channel estimation depends on the channel coherence time, which in turn depends on user mobility, as measured by Doppler spread (Chapter 6). Therefore, depending on the application, the estimation overhead may be significant. In fact, it was claimed that in some cases, maximum uplink capacity is achieved when approximately a half of the available time is spent on sending pilot signals [33, secs. IV–C]. Such rule of thumb helps to determine the optimum number of users K for a system.

For the downlink, it seems logical that channel estimation is conducted by the mobile users. For this purpose, each of the M base station antennas needs to transmit a distinguishable pilot. Therefore, M symbols of overhead is required for each channel estimation operation. Since $M \gg K$, much more overhead is needed comparing to uplink channel estimation. Furthermore, channel estimation results need to be reported back to the base station for precoding, incurring more overhead. These two factors make such channel estimation approach almost impractical. Therefore, almost all massive MIMO systems currently studied or demonstrated leveraged the channel reciprocity for channel estimation.

When reciprocity is used, the uplink and downlink channels share the same frequency in time division duplexing (TDD) mode. Because of the reciprocity property of the channel, downlink channel matrix H_d can be derived from its uplink counter part H_u [34]. Therefore, only uplink channel estimation, which requires much less overhead, needs to be conducted by the base station. Note that reciprocity only applies to the physical channel. Characteristics of the amplifiers, antenna coupling, and filters distort the reciprocity relationship and need to be calibrated in operation [35].

While TDD mode provides a nice solution for massive MIMO channel estimation, it brings some restriction in other aspects of the system. Although the 4G standard LTE supports both FDD and TDD modes, the former dominates current deployment. Therefore, massive MIMO in frequency division duplexing (FDD) mode is a research area [36, sec. 2–C]. One approach is reducing the degree of freedom of the estimation problem and thus the required overhead. For example, when all potential mobile users are located in a limited area with common scatterers, the channel vectors are limited to a lower-dimensional subspace (see Section 12.5.2.2). With such prior knowledge, the number of pilot signals needed for channel estimation is reduced [37, 38]. One can also explore the intrinsic similarities between the uplink and downlink channels such as angle of arrival and scattering cluster geometry, to approximately derive H_d from the measured H_u at a different frequency band [39–41]. Furthermore, compressive sensing techniques can be used to exploit the sparsity of the scattering environment when channel frequency response is to be measured [42]. Massive MIMO with FDD mode has been demonstrated in field tests [43–46].

12.5.1.5 Pilot Contamination

In the above discussions, we extend the system scope of massive MIMO from single user to multiple users. Now let us consider the case of multiple cells all using massive MIMO.

When neighboring cells all engage in massive MIMO operations, channel estimation has a complication known as the pilot contamination [36]. In TDD systems, all cells tend to conduct channel estimation at the same time for system design reasons. Because the number of orthogonal pilot sequences is limited, such sequences are reused among the cells. Therefore, a base station receives pilot signals from its own users as well as those in the neighboring cells. Such inter-cell interference (ICI) causes "contamination" to the channel estimate. As a result, the beamforming solution based on such channel estimate has a portion of the power focused on the users in neighboring cells, increasing ICI in data transmission and receiving.

The problem of pilot contamination has been identified and analyzed from the beginning of massive MIMO research [33]. Pilot contamination can be reduced in various ways, leveraging the known differences between the channels in different cells [47–50]. It was pointed out that with proper mitigation measures, pilot contamination will not be a capacity-limiting factor for massive MIMO systems as the number of antennas increases [51]. On the other hand, intentionally created pilot contamination, that is, false pilot signals can be used to attack a massive MIMO system [52].

12.5.1.6 Analog Front End

Massive MIMO technology also has unique considerations regarding the analog front end. Massive MIMO uses a large number (up to hundreds) of antennas, each with its own RF chain. Such configuration significantly increases the total cost of analog components. On the other hand, with the same total transmission power, each antenna transmits a much smaller power. Therefore, the cost of power amplifiers can be much lower. To further reduce the power amplifier costs, the precoding algorithm can take into account the resulting peak-to-average ratio (PAR) among the antennas and for each antenna over time [29, sec. 4–C], [53].

To further save the cost of analog modules, hybrid beamforming can be used. In this scheme, a part of the precoding is done in the analog domain with adjustable analog amplifiers and phase shifters. The number of complete RF chains is thus reduced. Hybrid beamforming is used more often with millimeter wave communications and will be discussed further in Section 12.5.2.3.

The impact of hardware impairments to massive MIMO performance has been studied in detail [54]. It was found that hardware imperfections impose limits on channel estimation accuracy and system capacity. On the other hand, as the number of antennas grows, the analog requirements for each antenna can be relaxed because the errors tend to average out. These considerations play important roles in assessing practicality and cost of massive MIMO systems. Of particular interest is using low-resolution, or even 1-bit, analog-to-digital converters (ADCs) for a massive MIMO receiver [55–57]. Using low-resolution ADCs also implies a relaxation of other system requirements such as amplifier linearity and automatic gain control.

12.5.1.7 Summary

Massive MIMO is a new system configuration, where a base station uses a large number of transmitting and receiving antennas to serve multiple mobile users at the same time. At the single-user level, the large antenna array performs beamforming to effectively achieve large antenna gains, improving link capacity. Such improvement is realizable in both free space and scattering environments. For multiple users, because of the large number of antennas, mutual interference can be effectively reduced with linear algorithms, or simply ignored. In a multicell cellular setup, pilot contamination is caused by the reuse of channel probing signals among the cells. Pilot contamination degrades the performance of channel estimation and needs to be mitigated in various ways.

Channel estimation is an important issue for massive MIMO. Because of the large number of antennas, downlink channel estimation through known symbols may bring unacceptable overhead. A popular strategy is using TDD transmission, where the uplink channel response is measured by the base station. The downlink channel response is derived by reciprocity. On the other hand, FDD massive MIMO systems have also been developed and demonstrated.

Although not elaborated in this section, another advantage of massive MIMO is diversity under channel fading. Fading experienced by the various antennas tend to average out, resulting in a stable link quality that does not change significantly over frequency and time.[5] Stable link quality simplifies modulation and coding scheme (MCS) selection as well as multiuser resource allocation.

There are some excellent overview paper on massive MIMO [29, 30, 33, 58, 59]. There are also many collections of massive MIMO papers. For example, *IEEE Journal of Selected Topics in Signal Processing* had a special issue on massive MIMO in 2017 (Volume 8, No. 5). Collection of massive MIMO literature can be found from some websites, for example, [60, 61].

Besides cellular applications, the concept of massive MIMO can be used in some other communication settings. Especially, massive MIMO can be used in dynamic spectrum access (DSA) schemes, to be briefly described in Chapter 13 [62–65]. Another application of massive MIMO is in the "cell-free" setting, where multiple access point (AP) nodes are geographically distributed but connected with high-speed backhaul network [66, 67]. They serve multiple mobile users through joint transmission and receiving.

---

[5] By link quality, we mean the achievable SNR under optimal beamforming conditions. The actual beamforming still depends on the channel response, which changes with time and frequency and needs to be tracked through channel estimation.

The 4G cellular system LTE has adopted a limited version of massive MIMO, known as full-dimensional MIMO (FD-MIMO) [68–71]. FD-MIMO uses a two-dimensional antenna array to achieve beamforming in both horizontal and vertical directions. The 5G NR standard enhanced the support to channel measuring reference signals, enabling more advanced massive MIMO technologies.

12.5.2   Millimeter Wave (mmWave) Communications

12.5.2.1   Overview

Millimeter Wave (mmWave) communications use the mmWave band, which refers to the frequency range between 30 and 300 GHz (wavelength between 10 and 1 mm). However, in the context of 5G, any frequency above 24 GHz is considered as mmWave. Furthermore, some authors consider the frequency range from 3 to 300 GHz as mmWave, due to similar channel properties [72].[6]

Early applications of mmWave include radar and sensing. mmWave communications have been envisioned at least a 60 years ago [73]. Applications of mmWave to modern mobile broadband cellular systems were envisioned in the early 2010s [72]. IEEE has developed the WiFi standard 802.11ad (also known as WiGig) using the 60 GHz mmWave bands and is working on its enhancement, the 802.11ay standard, which provides data rates up to hundreds of Gb/s [74]. mmWave communications is now an active research and development area [75, 76].

The potential of mmWave communications was recognized in the early development of 5G technologies [77–81]. The 5G proposal from 3GPP supports frequencies up to 40 GHz [20]. Adoption of mmWave technologies in 5G is motivated by the need for higher data rates and system capacities. While the lower frequency bands (below 6 GHz) are getting crowded, the mmWave bands are currently available in most regions. Furthermore, higher frequencies allow for larger bandwidths, leading to higher data rates. The size of antenna arrays become smaller with higher frequency, making beamforming and massive MIMO (Section 12.5.1) feasible for small devices, further improving link quality and achievable data rate. The ubiquitous use of beamforming allows for frequency reuse within the same area, as long as the beams do not overlap. Furthermore, atmospheric absorption causes faster signal attenuation, allowing for frequency reuse even at the direction of the beams.

The mmWave channel has some disadvantages, as well. As to be detailed in Section 12.5.2.2, mmWave channels are vulnerable to blocking and weather-related attenuation. Therefore, it is difficult to guarantee coverage with mmWave communications. Usually, satisfactory signal levels can only be obtained through beamforming, which requires the knowledge of channel state at both the transmitter and the receiver. Such requirement poses challenges to initial acquisition and tracking in mobile situations. mmWave is often used for specific applications (such as fixed point-to-point links) or in complement with other frequency bands. In 5G cellular, mmWave can be used for fixed and mobile access links as well as for backhaul links.

---

[6] Traditionally, frequency range from 3 to 30 GHz and from 30 to 300 GHz are referred to as super high frequency (SHF) and extremely high frequency (EHF), respectively (see Chapter 6).

12.5.2.2 MMWAVE Channels

The unique feature of mmWave communications is its channel. mmWave channels have been widely studied and modeled in the literature. A reader can start with some comprehensive reviews, for example, [82, ch. 4], [83].

mmWave signals do not propagate well except in the line of sight (LOS) conditions. Therefore, the LOS (or free space) channel is very important to mmWave communications. As discussed in Chapter 6, the free space channel attenuation is proportional to the square of the frequency. Therefore, mmWave channels experience much higher free space attenuation comparing to lower frequency bands.

On the other hand, shorter wavelength results in larger antenna gains for a given antenna size, as discussed in Chapter 6. For mmWave communications, array antennas are often necessary to achieve any meaningful link range. With fixed antenna sizes at both the transmitter and the receiver, the overall link quality (the totality of the antenna gains and path loss) is actually better at higher frequencies. The price to pay, of course, is a higher count of antenna elements (since antenna elements are typically spaced at half the wavelength), and thus higher complexity and cost.

In addition to the free space loss, mmWave links also suffer from environmental effects such as gaseous absorption, weather-induced loss, and foliage loss. Gaseous losses are mainly caused by oxygen and water absorptions. They are usually small (below 1 dB/km) except for some peak frequency ranges and frequencies above 100 GHz [82, fig. 1]. Weather-induced attenuations can be more severe but are not always present. For practical guides, ITU-R publishes various recommendations concerning the excess attenuation effects of cloud and fog [84], vegetation [85], atmosphere [86], precipitation [87], and so on.

When the LOS channel is not available, propagation relies on diffraction, scattering, and reflection. Due to their short wavelengths, mmWave signals experience virtually no beneficial diffraction. mmWave scattering and reflection are also highly dependent on the geometric laydown and material of the environment. Furthermore, mmWave signals suffer very high losses when penetrating building walls. Therefore, channel models for wwWave can be very different from those for lower frequencies. 3GPP has provided a reference channel model [88]. Other reference models are discussed in [83]. Many field measurements have been conducted to ascertain real-life situations [80].

When designing mmWave communication systems, the following general channel properties should be considered [83].

- mmWave links usually have short ranges. A typical outdoor cellular link in mmWave is below 200 m in range.

- Array antennas with limited beam widths are typically used for mmWave communications. Such configuration limits channel scattering. For example, for LOS connections, ground reflection can often be ignored. Antenna patterns must be incorporated into the channel model.

- mmWave signals do not penetrate building walls very well, posing unique challenges for links between indoor and outdoor nodes.

- mmWave links are sensitive to blocking, weather, and other factors. Therefore, link outages may be more frequent.

- Fast fading or Doppler spread (see Chapter 6) may be more significant for mmWave channels due to their high frequency. On the other hand, the limited scattering in these channels may reduce the fading effect.

The short-range nature of mmWave links provides the advantage of frequency reuse. Because mmWave signals decay fast in the presence of gaseous attenuation, weather factors, and non-LOS channels, mutual interference is limited within small areas. Therefore, frequency reuse can be applied in higher densities. This factor not only increases overall system capacity but also allows for "micro-licensing," which enables uncoordinated systems to operate in adjacent areas such as buildings and sports arenas [89].

Another important feature of the mmWave channel is its low dimension or sparsity. In general, a channel can be expressed as a matrix of size $N_T \times N_R$, where N_T and N_R are the number of antennas at the transmitter and the receiver, respectively (see Chapter 11). However, for mmWave channels, the number of significant scatterers is usually much smaller than the number of antennas. Therefore, the channel matrix is constrained into a lower-dimensional space [90]. Such property is important in beamforming and channel estimation considerations, as will be shown later. Low dimension or sparsity is a common property for massive MIMO channels. However, since mmWave systems tend to use more focused beams and are more likely to operate in free space of near-free space environments, such property is more pronounced in mmWave situations. The channel low-dimensionality is the same as the "rank deficiency" discussed in Chapter 11 for MIMO systems. For a MIMO system, channel rank deficiency limits the extent of spatial multiplexing, resulting in low channel capacity. For the mmWave communications discussed here, such a disadvantage is unimportant because very few data streams are used to start with. Therefore, channel low-dimensionality primarily impacts the channel estimation process. Because of the sparsity, compressive sensing is a useful mathematical tool for channel estimation [91, 92].

12.5.2.3 Beamforming

As discussed above, a typical mmWave communication system uses an array antenna to improve link quality. A transmitter array antenna can focus the transmitted energy to the receiver, while a receiver one can combine received energy over the array area to form a strong signal. Such techniques are commonly referred to as beamforming. Link improvement can be gained in both LOS channels and scattering channels.

The principle of beamforming is the same as in massive MIMO (Section 12.5.1). In a massive MIMO configuration, the base station has a large number of antennas while the mobile user has one or a few antennas. On the other hand, mmWave links typically have antenna arrays on both sides, albeit the sizes may be different.

When discussing massive MIMO, we usually assume that the optimal precoding and combination, namely p in (12.2) and g in (12.7), can be realized precisely. For example, precoding can be performed in the baseband digital domain. Such "digital beamforming" approach, however, is difficult to realize for mmWave systems. The reason is that RF chains, especially digital-to-analog converters (DACs) and analog-to-digital converters (ADCs) are expensive for mmWave because of its frequency and bandwidth. Digital beamforming requires an RF chain for each antenna element, leading to an impractical system cost. Another extreme is performing beamforming in the analog domain [93–95]. In this case, one RF chain is connected to all antennas, each being equipped with an analog multiplier that can changed the amplitude and phase of the signal. Obviously, analog beamforming can support only one data stream. Furthermore, due to technology limitations, beamforming in the analog domain is not perfect. For example, the choices of gain and phase shift may be quantized and inaccurate. Many systems use phase shifting only while keeping the amplitude gain constant. These limitations degrade system performance [83, sec. IV-A].

A compromising solution is hybrid beamforming, where both digital and analog beamforming modules are used [96]. The number of the RF chain is larger than 1 but smaller than that of the antennas.

Figure 12.3 shows the three beamforming configurations. Since hybrid beamforming is most commonly used in mmWave communications and the other two configurations can be viewed as special cases, we will focus on hybrid beamforming in the following. Figure 12.3 shows only the transmitter side. The process is reversed at

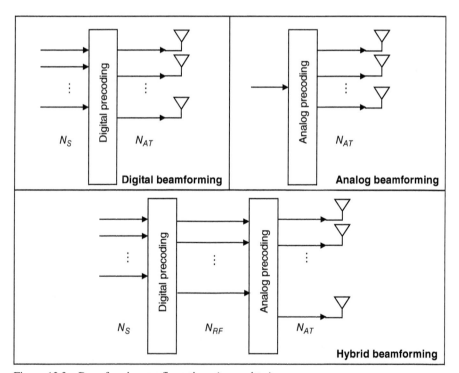

Figure 12.3   Beamforming configurations (transmitter).

the receiver. Note that beamforming configurations can be mixed-and-matched between the transmitter and receiver.

As shown in Figure 12.3, hybrid beamforming has two stages. At the transmitter, in stage 1, digital precoding is applied to the N_S data streams, to form N_{RF} RF streams. These streams are converted to analog and up-converted to mmWave frequency. Then analog precoding is applied to form N_{AT} signals, which are sent to the antennas for transmission. Typically,

$$N_S \leq N_{RF} \leq N_{AT}. \tag{12.18}$$

Digital and analog precodings can be expressed as matrix multiplications. Therefore, the output signal can be expressed as

$$x = WPs. \tag{12.19}$$

Here x is a vector of size N_{AT}, representing the output signals at the antennas. s is a vector of size N_S, representing the input data streams. P is a matrix of size $N_{RF} \times N_S$, representing digital precoding. W is a matrix of size $N_{AT} \times N_{RF}$, representing analog precoding. To further save hardware cost, the analog precoding may not be fully connected.

Figure 12.4 show three possible analog beamforming configurations on the transmitter side. In these configurations, the dots on the left-hand side represent outputs of the RF chain (logical output of the digital precoder). Each of such outputs is connected to multiple analog gain controllers, whose outputs are connected to the antennas. The analog gain controllers constitute the analog precoder. The left panel shows the fully connected configuration, where each RF chain is connected to all antennas through the analog gain controllers. This configuration is broadly studied in the literature. The middle panel shows a grouped configuration, where an RF chain is connected to a group of antennas, while each antenna is connected to only one RF chain [97]. The right panel shows a grouped configuration with overlap, where some antennas are connected to two or more RF chains, that is, shared by multiple groups [98].

Hybrid beamforming has been an active research area in mmWave [90, 97, 99–101]. As discussed in Chapter 11, the optimal precoding scheme is based on singular value decomposition (SVD) of the channel matrix. Namely, in the N_{AT} dimension channel space, we choose a subspace spanned by the N_S largest singular values to transmit the data. In principle, as long as N_{RF} is no less than N_S, a fully connected hybrid beamforming can produce such an optimal precoding. However, as discussed before, analog precoding has its inherent imperfection and constraints. Therefore, optimal solutions to hybrid precoding are different from the standard MIMO ones.

It turns out that the "phase-shifting only" constraint is not a big problem, as long as $N_{RF} \geq 2N_S$ [102]. Such lower bound can be further reduced considering that the channels are low-dimension (see the next paragraph). Alternatively, one can use two phase shifters for each RF chain to manipulate both amplitude and phase [103, 104]. This is easy to understand as the additional signal path provides the degree of freedom needed. When N_{RF} is not large enough or when the precoding is not fully connected, the beamforming problem becomes constrained optimization, which is

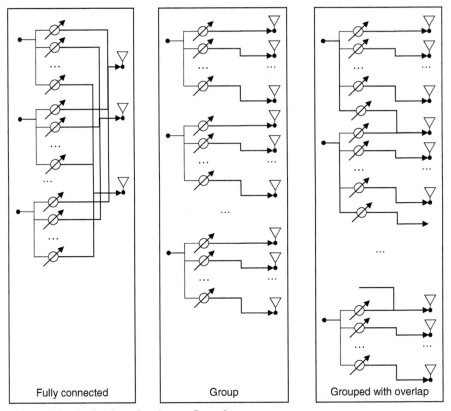

Figure 12.4   Analog beamforming configurations.

usually not mathematically tractable. Hence, only approximations to the real optimization problem can be solved. Therefore, suboptimal but low-complexity hybrid precoding/combining designs arc desirable [100].

On the other hand, the mmWave channels are often sparse, because the number of significant scatterers is usually small comparing to the number of antennas (Section 12.5.2.2). Under such conditions, hybrid beamforming can achieve performances close to those of ideal precoding [100]. Sparse channels are used for hybrid beamforming studies for single-user [105], multiuser [106], and frequency-selective fading channels [107].

In codebook-based beamforming, the precoding coefficients are chosen from a set of predetermined options (known as dictionary or codebook). The choice is informed by the receiver based either through computation (based on channel estimate) or through an exhaustive search. Codebook design has been studied in the literature, as well [108, 109]. Another option is directly solving for precoding coefficients (under the constraints discussed above) [110–114].

For wideband channels (Chapter 6), beamforming can be done in the frequency domain. For example, with OFDM (Chapter 10), precoding coefficients or phase shifts may be subcarrier-dependent. Beamforming can also be done in the time domain, where time delays, instead of phase shifts, are applied to the various antennas [115]. The latter approach is probably more straightforward for an analog domain implementation.

12.5.2.4 Front End

For mmWave communications, the analog front end circuitry is often expensive and imperfect. Therefore, special considerations are needed for the front end circuitry, including antennas.

One active research area is the design and performance with low-resolution analog-to-digital converters (ADCs) [100, 116–118]. Resolution and sampling speed are tradeoffs for many types of ADCs. Since mmWave systems typically have high sampling rates, using low-resolution ADC significantly reduces cost and power consumption.

The array antenna is another active topic for mmWave communications [119]. Because of the large number of antenna elements and the small spacing, wiring between the antenna elements and the rest of the circuits is difficult. A popular configuration is integrating the antenna elements and driving circuits on the same chip [120]. While the theory of antenna arrays is well-known and devices in mmWave frequency range have been used for decades, the applications in commercial communications pose unique challenges in cost reduction, size minimization, and other engineering issues. Analog front end prototypes for the 60 GHz band (for WiFi applications) and the 28 GHz band (for 5G applications) have been reported in the literature [121–123].

12.5.2.5 Spectrum Sharing

mmWave links typically use beamforming. In other words, the transmitted signal is "focused" onto the desired receiver. The receiver also accepts signal only at the direction of the transmitter. Such configuration significantly reduces interference among links using the same frequency, allowing multiple links to operate in the same area, using the same frequency and time resources. Such potential is leveraged in WiFi systems as well as 5G cellular systems [124–127], enabling spectrum sharing with or without coordination. Performance can be further improved when the beams are adaptive to interference situations, for example, using the zero-forcing algorithm discussed in Section 12.5.1.3 [128].

12.5.2.6 Channel Estimation and Initial Access

Because beamforming is used, channel estimation and initial access present unique challenges to mmWave communications.

Channel estimation is similar to those of massive MIMO, except that both sides have a large number of antenna elements. Therefore, using channel sparsity to reduce the amount of estimation is more important [100]. Furthermore, probing signals are received through the imperfect analog front end. Characterization and calibration of the front end thus become a part of the channel estimation task [129].

Initial access is even more challenging, as SNR at the receiver is usually too low before beamforming can be successfully established. Currently, several options are being explored [83, sec. VI-B], [130–132]:

- Brute-force beam searching. Test over a set of predetermined beams in the hope of getting a connection.
- Iterative beam searching. Start with a broad beam and progress to narrower ones when the signal is found.
- Two-stage access. Start with a omnidirectional broadcast at a lower frequency to exchange location information, which is then used for beamforming.

12.5.2.7 Summary

The primary advantage of mmWave communications is high capacity, resulting from the wide bandwidth and directional link (Sections 12.5.2.1 and 12.5.2.5). On the other hand, mmWave presents unique challenges. The mmWave channel is more suitable for shorter ranges (Section 12.5.2.2). Because of hardware constraints, mmWave transceivers typically use hybrid beamforming, whose optimization is still under research (Section 12.5.2.3). Currently, front end hardware is far from perfect due to power, size and cost constraints (Section 12.5.2.4).

Although mmWave communications is still an active research area, various product prototypes have been reported [123, 133, 134].

This section focuses on physical layer issues. On the other hand, mmWave communications also require tailored support from the higher layers [135, 136]. Since all signals use beamforming, control channels that usually use broadcasting need to be modified. In order to counter channel instability (blocking and fading), simultaneous maintenance of connections to multiple base stations is helpful for mobile users. The protocol design also needs to tolerate temporary channel outage.

There are still several open issues without satisfactory solutions [83]. They include:

- Indoor/outdoor connectivity. Because mmWave suffers high attenuation when penetrating building walls, connecting indoor users with outdoor base stations can be difficult.
- Integration with other radio access technologies. Since mmWave coverage is unreliable due to limited range and high blocking probabilities, it needs to work with other frequency bands to achieve seamless user experiences.
- Mobility support. Fast beam alignment updates are needed for mobile users.
- Handover. mmWave connectivity can be sporadic and unpredictable. When blocked from the current base station, a mobile user needs to conduct handover to a new base station quickly and without prior planning.

12.5.3 Nonorthogonal Multiple Access (NOMA)

12.5.3.1 Introduction

As described in Chapter 10, the 4G cellular system LTE uses OFDMA for multiple access, where different users are assigned different frequency and time resources blocks so that they can transmit without mutual interference. OFDMA is a special case of orthogonal multiple access (OMA), where the signals transmitted by different users are orthogonal (i.e., without interference). Such sharing of resources usually requires some central coordination in resource allocation. In the simplest forms of OMA, signals from different users are separated in frequency and/or time. A more sophisticated OMA method, the orthogonal code division multiple access (CDMA) will be briefly described in Section 12.5.3.3.

Nonorthogonal multiple access (NOMA) is a different approach [137–139]. With NOMA, signals from multiple users overlap in time and frequency, causing mutual interference. As long as such interference is below a certain threshold, communications can still be carried out. NOMA can achieve better overall capacity at least under some channel conditions.

In the context of 5G cellular, NOMA is particularly attractive because it allows for a large number of users to share the channel without preplanning and coordination. As described in Section 12.3.2, the mMTC application scenarios in 5G call for up to one million users per square kilometer. These users are primarily Internet of Things (IoT) devices, which typically transmit short bursts of data. NOMA provides quick and easy access to these users. 3GPP has been studying NOMA options in the past few years and plans to introduce NOMA in Release 16 [140–142].

In the following, we discuss two NOMA schemes: superposition and spreading. For the sake of simplicity, we examine single antenna cases only. These methods can be extended to MIMO and massive MIMO situations [70, sec. 7.5], [143, 144]. The IEEE Wireless Communications magazine had a special issue on NOMA for 5G in April 2018 (Volume 25, No. 2).

12.5.3.2 Superposition

In the superposition scheme, multiple users transmit at the same time and frequency. A receiver experiences the sum (superposition) of all signals. Successive interference cancellation (SIC) is used at the receiver to decode the signal. Such technique is also known as power domain NOMA. The first proposals of NOMA used superposition [145].

With SIC, the receiver first demodulates the strongest signal component, while treating all other signals as noise. After such demodulation, the strongest signal is reconstructed and subtracted from the received signal. The receiver then moves on to the second strongest signal, and so on. SIC is very similar to the successive cancellation (SC) method in the BLAST version of MIMO communication (Chapter 11).

To understand superposition NOMA, let us consider a case of two users [70, ch. 6]. User $\{n; n = 1, 2\}$ has transmit power P_n and channel gain h_n. We assume $|h_1| \leq |h_2|$.

The signals for user n are x_n. Let the transmit bandwidth be W, and the (white) noise power spectral density be N_0. Under these conditions, we compare the capacities of the superposition and the FDMA schemes.

Let us first consider downlink. With FDMA, the two users divide the bandwidth W. Assume the bandwidths assigned to users 1 and 2 are αW and $(1 - \alpha)W$, respectively. Transmitting power levels P_1 and P_2 can be chosen with the constraint

$$P_1 + P_2 \leq P_D, \tag{12.20}$$

where P_D is the total downlink power. The capacity of the two users are (see Chapter 2 for capacity computation):

$$R_{1,FDMA,DL} = \alpha W \log_2 \left(1 + \frac{|h_1|^2 P_1}{\alpha W N_0} \right)$$

$$R_{2,FDMA,DL} = (1 - \alpha) W \log_2 \left(1 + \frac{|h_2|^2 P_2}{(1 - \alpha) W N_0} \right) \tag{12.21}$$

With NOMA, the base station sends full-bandwidth signals x_1 and x_2 with power levels P_1 and P_2, also constrained by (12.20). The received signal for user n is

$$y_n = h_n \left(\sqrt{P_1} x_1 + \sqrt{P_2} x_2 \right) + W N_0. \tag{12.22}$$

For user 1, x_1 is demodulated while x_2 is considered as noise. Therefore, the capacity of user 1 is

$$R_{1,NOMA,DL} = W \log_2 \left(1 + \frac{|h_1|^2 P_1}{W N_0 + |h_1|^2 P_2} \right). \tag{12.23}$$

For user 2, x_1 is first demodulated and then removed from the received signal (12.22). Since $h_2 \geq h_1$, such demodulation is always possible. Therefore, x_2 can be demodulated without the interference from x_1. The capacity of user 2 is thus

$$R_{2,NOMA,DL} = W \log_2 \left(1 + \frac{|h_2|^2 P_2}{W N_0} \right). \tag{12.24}$$

For the uplink, FDMA works in a very similar way, except that total power is not constrained by (12.20). Instead, transmit power of each user is constrained to P_U.

$$P_1 \le P_U$$
$$P_2 \le P_U \tag{12.25}$$

Equation (12.21) applies to uplink as well:

$$R_{1,FDMA,UL} = \alpha W \log_2 \left(1 + \frac{|h_1|^2 P_1}{\alpha W N_0} \right)$$

$$R_{2,FDMA,UL} = (1-\alpha) W \log_2 \left(1 + \frac{|h_2|^2 P_2}{(1-\alpha) W N_0} \right). \tag{12.26}$$

With NOMA, the receiver (base station) can perform joint demodulation of x_1 and x_2, while the transmit power is constrained for the individual users (12.25). Therefore, the user capacities are limited by the joint capacity and individual capacities.

$$R_{1,NOMA,UL} \le W \log_2 \left(1 + \frac{|h_1|^2 P_1}{W N_0} \right)$$

$$R_{2,NOMA,UL} \le W \log_2 \left(1 + \frac{|h_2|^2 P_2}{W N_0} \right)$$

$$R_{1,NOMA,UL} + R_{1,NOMA,UL} \le W \log_2 \left(1 + \frac{|h_1|^2 P_1 + |h_2|^2 P_2}{W N_0} \right) \tag{12.27}$$

A common way to study multiple access system performance is capacity region. Capacity region is a region in the N-dimensional space, where N is the number of users. If a vector $(R_1, R_2, \dots, R_N)$ is within the capacity region, then it is possible to have a resource allocation so that user n has the capacity R_n. Therefore, capacity region reflects the tradeoff among user capacities. In the case of two-user FDMA, user capacity tradeoffs can be realized by adjusting P_1 and P_2 under the constraint of (12.20) for downlink and (12.25) for uplink. More importantly, the share of bandwidth for user 1 (α) can be adjusted between 0 and 1. For NOMA, user capacity trade off can also be realized by adjusting P_1 and P_2 under the same constraints (12.20) and (12.25) for downlink and uplink, respectively.

Figure 12.5 shows the capacity regions for two-user FDMA and NOMA, uplink (first row) and downlink (second row). The first column shows the cases with equal channel gains, while the second shows the ones where h_1 is 20 dBs lower than h_2. We can see that NOMA has larger capacity regions for both uplink and downlink. In other words, for the same R_1, NOMA allows for higher R_2. The advantage is more pronounced when the channel gains are different. Note that capacity regions are also affected by the SNR. Figure 12.5 shows the cases where the SNR is much larger than 1.

Figure 12.5 shows that the benefit of NOMA is higher for downlink. Furthermore, uplink NOMA requires the mobile users to coordinate their transmit power (i.e., power control), which leads to additional protocol overhead and latency. Therefore, superposition NOMA is typically used for downlink [137, 138], although there are proposals for uplink applications [146].

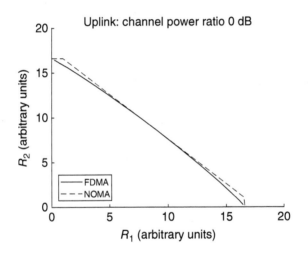

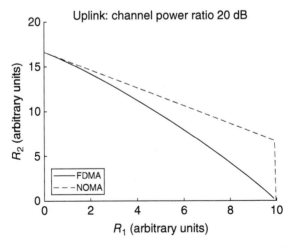

Figure 12.5   Capacity region comparison between FDMA and NOMA at different SNR levels.

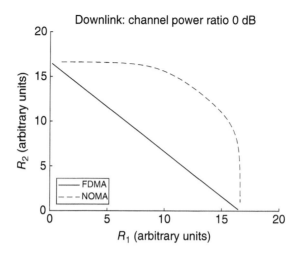

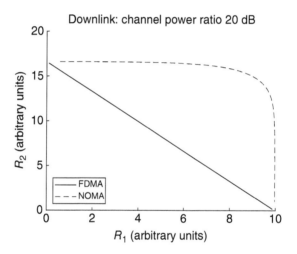

Figure 12.5 (Continued)

12.5.3.3 Spreading

To understand the spreading technique, we start with a brief description of the code division multiple access (CDMA) technique. As stated in Section 12.2.1, CDMA is a technology used in 2G and 3G cellular systems. It is not covered in detail in this book. For more information about CDMA techniques, see [147, ch. 12]

The basic concept of CDMA is multiplying the transmitted symbols with a "spreading code," which is a time domain sequence. Consider a particular symbol in time. For simplicity, we ignore the pulse shaping and consider the transmitted symbol to be a constant with a finite duration. Let s_n be the symbol sent by the nth user, lasting between time 0 and T_{Sym}. According to the Nyquist theorem (Chapter 3), the passband bandwidth needed for single-user transmission is

$$W_S = \frac{1}{T_{Sym}}. \qquad (12.28)$$

In CDMA, the transmitted signal is the product

$$u_n(t) \overset{\text{def}}{=} s_n c_n(t), \qquad (12.29)$$

where $c_n(t)$ is the spreading code assigned to user n. The received signal is (we do not consider channel dispersion for simplicity)

$$y(t) = \sum_{n=1}^{N} h_n u_n(t) + z(t). \qquad (12.30)$$

Here, N is the number of users. h_n is the channel gain for the nth user. $z(t)$ is the noise. To separate out the signal from the mth user, $y(t)$ is correlated with the complex conjugate of the corresponding spreading code:

$$y_m \overset{\text{def}}{=} \int_0^{T_{Sym}} y(t) c_m^*(t) dt. \qquad (12.31)$$

Given (12.29) and (12.30), (12.31) leads to

$$y_m = \sum_{n=1}^{N} h_n s_n d_{n,m} + z_m$$

$$d_{n,m} \overset{\text{def}}{=} \int_0^{T_{Sym}} c_n(t) c_m^*(t) dt. \qquad (12.32)$$

$$z_m \overset{\text{def}}{=} \int_0^{T_{Sym}} z(t) c_m^*(t) dt$$

If the interference-free transmission is desired, the spreading codes $\{c_n(t), n = 1, 2, \ldots, N\}$ can be chosen to be mutually orthogonal. Namely,[7]

$$d_{n,m} = \delta_{n,m}. \qquad (12.33)$$

In this case, (12.32) leads to

$$y_m = h_m s_m + z_m. \qquad (12.34)$$

Equation (12.34) shows that when the codes are orthogonal, the N users can transmit at the same time without mutual interference. Obviously, the maximum value of N depends on the number of orthogonal spreading codes available.

Assuming each spreading code $c_n(t)$ is constructed by N independent samples (known as chips in CDMA terminology) with pulse shaping (Chapter 3), the collection of the codes spans an N-dimensional linear space. Therefore, there are at most N

---

[7] To maintain (12.13), $c_n(t)$ and $c_m^*(t)$ need to be synchronized for all n. Such synchronization may be difficult in uplink cases, where different users have different propagation delays relative to the base station.

mutually orthogonal codes. On the other hand, the chip period is T_{Sym}/N. According to the Nyquist theorem (Chapter 3), the transmission passband bandwidth is

$$W_{CDMA} = \frac{N}{T_{Sym}}. \tag{12.35}$$

It turns out that the above analysis is true even if the chips are limited to binary values (1 and −1), for example, by using the Walsh code.

The above analysis shows that with such orthogonal CDMA, N users can share the channel without mutual interference using orthogonal spreading codes. However, the required bandwidth is increased by N times. Therefore, the overall capacity is the same as the case where N users equally share the channel with FDMA.

CDMA becomes more attractive when the orthogonality is relaxed. When the crosscorrelations among $\{d_{n,m}\}$ are not exactly delta functions as in (12.33), the number of spreading codes, thus the number of simultaneous users, can be increased significantly without increasing total bandwidth. The price to pay is the residual mutual interference. However, as long as the user numbers are not too high, mutual interference can be kept below a tolerable level. Such observation leads to the concept of "statistical multiplexing." With statistical multiplexing, a user is allowed to transmit at any time or frequency allocated to the system, with an assigned spreading code. However, the probability of a user to transmit at a given time and frequency is low. Therefore, the expected number of *active* users sharing the channel, and thus the mutual interference level, is controlled. Such control of mutual interference does not need any coordination among the users; it is a natural result from the statistical property of user transmission.

For 5G mMTC applications, the number of IoT users is huge, while each user typically transmits short bursts from time to time. Statistical multiplexing is thus sensible for such applications.

The concept of CDMA described above can be understood as "spreading" a symbol over multiple chip periods. The receiver "gathers" the spread transmission by correlation. Since crosscorrelation is typically low (if not zero, as in the case of orthogonal codes), mutual interference is suppressed in this practice (at the expense of expanded bandwidth). Such an idea can be extended. Instead of spreading over chip periods, one can spread the signal over OFDM subcarriers (Chapter 10) or spatial subchannels (Chapter 11). These extensions constitute the spreading techniques used in NOMA.

A receiver can actually do better than treating mutual interference as pure noise. Instead, multiuser detection (MUD) can be used. The SIC method discussed in Section 12.5.3.2 is an example of MUD, although there are more sophisticated ones. MUD, in general, provides better performance, although receiver complexity can be very high.

Another extension of NOMA over CDMA is the low-density codes. In a socalled low-density spreading (LDS) scheme, most of the chip values are zero [148]. Such choice reduces receiver complexity for MUD algorithms, similar to the complexity reduction of low-density parity code (LDPC) (Chapter 5).

A family of NOMA spreading techniques has been proposed for 5G [70, ch. 6], [137, 138]. Not all of them are well-developed and widely accepted. They include:

- LDS-CDMA and LDS-OFDM. LDS-CDMA is the same as CDMA discussed above, except that the spreading codes have low density. LDS-OFDM is similar, except that the spreading codes are across OFDM subcarriers, instead of being across the chip intervals.

- Sparse code multiple access (SCMA). This method combines spreading with channel code and constellation design [149, 150]. Such combination allows for codeword choices in a high-dimensional space, thus realizing some shaping gain (Chapter 5).

- Multiuser shared access (MUSA). MUSA uses multilevel (as opposed to binary) spreading codes for better performance [138]. The users can randomly choose codes from a pool and change the code from time to time. Therefore, no inter-user coordination is needed for code assignment. The spreading codes are designed for SIC receivers.

- Pattern division multiple access (PDMA) [138]. This method combines spatial patterning with superposition. It is used for downlink transmission. Spatial patterning means that in a multi-antenna transmitter, signals are transmitted by a user-specific subset of the antennas. By using multiple antennas, spatial diversity can be achieved (Chapter 11). SIC is envisioned as the receiver algorithm, although more sophisticated algorithms can be used, as well [70, ch. 6].

12.6 SUMMARY

This chapter provides an overview of 5G cellular standard development. Since its inception, cellular systems have undergone continuous evolution and enhancement. Its services have shifted from voice calls to data and networking (Section 12.2). The 5G system envisions an overarching mobile network covering not only the traditional data services provided by today's internet but also future applications such as IoT and critical ultra-reliable connections. The 5G standard development is led by the ITU-R and has world-wide participation (Section 12.3). Currently, the leading proposal comes from the 3GPP (Section 12.4). The 3GPP proposal highlights three new technical areas for the physical layer: massive MIMO, mmWave communications, and NOMA (Section 12.5). These technologies are all active research areas at present. Although they are discussed in the context of 5G systems, these technologies have applications in other fields, as well.

It should be pointed out that 5G also introduces many other technical innovations beyond the physical layer. They are not discussed in this book, which focuses on the physical layer. One technology worth mentioning, however, is the cloud radio access network (C-RAN) [151, 152].[8] C-RAN is an architecture for base stations,

---

[8] Simpler forms of C-RAN have been used in some 4G systems.

where radios and processors are not collocated. One or more centralized processing centers known as base band units (BBU) serve a set of radios known as remote radio heads (RRH), which are connected to the BBUs with high-speed optical fibers. C-RAN has some advantages such as reducing the size and cost of the base station units, pooling processing capabilities to save cost and power consumption, and achieving coordinated processing among multiple radio locations. The last feature is of special interest to physical layer designs. With central processing, multiple RRHs can participate in a coordinated service to one or multiple users similar to massive MIMO. The coordinated multipoint (CoMP) used in LTE is an example of such joint processing (Chapter 11). Currently, the 5G proposal from 3GPP supports remote antennas, which is a form of C-RAN (Section 12.4). However, to realize more benefit, unique support for C-RAN, especially channel estimation, may be added to the standard in the future.

Also, energy efficiency is an important performance goal for 5G, although both requirements and technical solutions are still open at present, as shown in Table 12.1 [12, 17]. The "energy" here includes both transmitted power and the power used for signal processing. Therefore, energy efficient radios (also known as "green communications" in the literature) cover innovation on many fronts [153, 154].

5G is still evolving at present. It attracts much research interest from the digital communication community. Monitoring 5G development is a good way to keep up with the cutting-edge technology development.

REFERENCES

1. M. Sauter, *From GSM to LTE-Advanced Pro and 5G*. Chichester, UK: John Wiley & Sons, Ltd, 2017.
2. A. Ghosh, *Fundamentals of LTE*. Upper Saddle River, NJ: Prentice Hall, 2011.
3. Qualcomm, "The Evolution of Mobile Technologies: 1G to 2G to 3G to 4G LTE," 2014 [Online]. Available at https://www.qualcomm.com/documents/evolution-mobile-technologies-1g-2g-3g-4g-lte [Accessed August 15, 2018].
4. T. S. Rappaport, *Wireless Communications*, 2nd ed. Prentice Hall PTR, 2002.
5. J. Proakis and M. Salehi, *Digital Communications*, 5th ed. McGraw-Hill Education, 2007.
6. P. Bell, "2G Is Fading Away, But It Might Outlive 3G in Europe," *TeleGeography*, 2017 [Online]. Available at https://blog.telegeography.com/2g-is-fading-away-but-it-might-outlive-3g-in-europe. [Accessed April 14, 2018].
7. V. H. MacDonald, "The Cellular Concept," *Bell Syst. Tech. J.*, vol. 58, no. 1, pp. 15–43, 1979.
8. A. C. Madrigal, "The 1947 Paper That First Described a Cell-Phone Network" [Online]. Available at https://www.theatlantic.com/technology/archive/2011/09/the-1947-paper-that-first-described-a-cell-phone-network/245222/ [Accessed April 17, 2018].
9. ITU-R, "ITU-R FAQ on International Mobile Telecommunications (IMT)," 2013 [Online]. Available at https://www.itu.int/en/ITU-R/Documents/ITU-R-FAQ-IMT.pdf [Accessed February 12, 2018].
10. ITU-R WP 5D, *IMT-2020 Background*, ITU-R Document IMT-2010/1-E, 2016.
11. ITU-R, *IMT Vision—Framework and Overall Objectives of the Future Development of IMT for 2020 and Beyond*. ITU-R M.2083-0. 2015.
12. ITU-R, *Minimum Requirements Related to Technical Performance for IMT-2020 Radio Interface(s)*. ITU-R M.2410-0, 2017.
13. J. Lee et al., "Spectrum for 5G: Global Status, Challenges, and Enabling Technologies," *IEEE Commun. Mag.*, vol. 56, no. 3, pp. 12–18, 2018.
14. ITU-R WP 5D, "ITU Towards 'IMT for 2020 and Beyond'" [Online]. Available at https://www.itu.int/en/ITU-R/study-groups/rsg5/rwp5d/imt-2020/Pages/default.aspx [Accessed February 25, 2018].

15. ITU-R WP 5D, "Workshop on IMT-2020 Terrestrial Radio Interfaces (Wednesday 4 October 2017, Munich, Germany)" [Online]. Available at https://www.itu.int/en/ITU-R/study-groups/rsg5/rwp5d/imt-2020/Pages/ws-20171004.aspx [Accessed February 28, 2018].

16. 3GPP, "Preparing the Ground for IMT-2020" [Online]. Available at http://www.3gpp.org/news-events/3gpp-news/1901-imt2020_news [Accessed February 28, 2018].

17. 3GPP, *Initial Description Template of 3GPP 5G Candidate for Inclusion in IMT-2020*. ITU-R Document 5D/817-E, ITU-R, 2018.

18. The People's Republic of China, *Initial Submission of Candidate Technology for IMT-2020 Radio Interface*. ITU-R Document 5D/838-E, ITU-R, 2018.

19. Republic of Korea, *Submission of a Candidate Technology of IMT-2020*. ITU-R Document 5D/819-E, ITU-R, 2018.

20. 3GPP, *Characteristics Template for NR RIT of '5G' (Release 15 and Beyond)*. ITU-R Document 5D/817 Attachment Part 1, 3GPP, 2018.

21. 3GPP, *3rd Generation Partnership Project; Technical Specification Group Radio Access Network; NG-RAN; Architecture Description*. 3GPP TS 38.401, 3GPP, 2018.

22. "3GPP Release 15 Overview," *IEEE Spectrum* [Online]. Available at https://spectrum.ieee.org/telecom/wireless/3gpp-release-15-overview [Accessed August 12, 2018].

23. A. A. Zaidi et al., "Waveform and Numerology to Support 5G Services and Requirements," *IEEE Commun. Mag.*, vol. 54, no. 11, pp. 90–98, 2016.

24. 3GPP, *3rd Generation Partnership Project; Technical Specification Group Radio Access Network; Study on New Radio Access Technology Physical Layer Aspects (Release 14)*. 3GPP Technical Report TR 38.802 V14.2.0, 3GPP, 2017.

25. C. Sexton, N. J. Kaminski, J. M. Marquez-Barja, N. Marchetti, and L. A. DaSilva, "5G: Adaptable Networks Enabled by Versatile Radio Access Technologies," *IEEE Commun. Surv. Tutorials*, vol. 19, no. 2, pp. 688–720, 2017.

26. Y. Cai, Z. Qin, F. Cui, G. Y. Li, and J. A. McCann, "Modulation and Multiple Access for 5G Networks," *IEEE Commun. Surv. Tutorials*, vol. 20, no. 1, pp. 629–646, 2018.

27. F. Boccardi, R. Heath, A. Lozano, T. L. Marzetta, and P. Popovski, "Five Disruptive Technology Directions for 5G," *IEEE Commun. Mag.*, vol. 52, no. 2, pp. 74–80, 2014.

28. A. Ghosh, "'5G New Radio (NR): Physical Layer Overview and Performance'," *IEEE Communication Theory Workshop 2018*, 2018 [Online]. Available at http://ctw2018.ieee-ctw.org/files/2018/05/5G-NR-CTW-final.pdf [Accessed December 13, 2018].

29. L. Lu, G. Y. Li, A. L. Swindlehurst, A. Ashikhmin, and R. Zhang, "An Overview of Massive MIMO: Benefits and Challenges," *IEEE J. Sel. Top. Signal Process.*, vol. 8, no. 5, pp. 742–758, 2014.

30. J. Hoydis, S. Ten Brink, and M. Debbah, "Massive MIMO in the UL/DL of Cellular Networks: How Many Antennas Do We Need?," *IEEE J. Sel. Areas Commun.*, vol. 31, no. 2, pp. 160–171, 2013.

31. L. Sanguinetti, E. Bjomson, M. Debbah, and A. L. Moustakas, "Optimal Linear Precoding in Multi-User MIMO Systems: A Large System Analysis," in *IEEE Global Communications Conference*, 8–12 December, Austin, TX, 2014, pp. 3922–3927.

32. H. Yang and T. L. Marzetta, "Performance of Conjugate and Zero-Forcing Beamforming in Large-Scale Antenna Systems," *IEEE J. Sel. Areas Commun.*, vol. 31, no. 2, pp. 172–179, 2013.

33. T. L. Marzetta, "Noncooperative Cellular Wireless with Unlimited Numbers of Base Station Antennas," *IEEE Trans. Wirel. Commun.*, vol. 9, no. 11, pp. 3590–3600, 2010.

34. J. Vieira, F. Rusek, O. Edfors, S. Malkowsky, L. Liu, and F. Tufvesson, "Reciprocity Calibration for Massive MIMO: Proposal, Modeling, and Validation," *IEEE Trans. Wirel. Commun.*, vol. 16, no. 5, pp. 3042–3056, 2017.

35. O. Raeesi, A. Gokceoglu, and M. Valkama, "Estimation and Mitigation of Channel Non-reciprocity in Massive MIMO," *IEEE Trans. Signal Process.*, vol. 66, no. 10, pp. 2711–2723, 2018.

36. O. Elijah, C. Y. Leow, T. A. Rahman, S. Nunoo, and S. Z. Iliya, "A Comprehensive Survey of Pilot Contamination in Massive MIMO-5G System," *IEEE Commun. Surv. Tutorials*, vol. 18, no. 2, pp. 905–923, 2016.

37. X. Xiong, X. Wang, X. Gao, and X. You, "Beam-Domain Channel Estimation for FDD Massive MIMO Systems with Optimal Thresholds," *IEEE Trans. Wirel. Commun.*, vol. 16, no. 7, pp. 4669–4682, 2017.

38. J. Fang, X. Li, H. Li, and F. Gao, "Low-Rank Covariance-Assisted Downlink Training and Channel Estimation for FDD Massive MIMO Systems," *IEEE Trans. Wirel. Commun.*, vol. 16, no. 3, pp. 1935–1947, 2017.

39. S. Imtiaz, G. S. Dahman, F. Rusek, and F. Tufvesson, "On the Directional Reciprocity of Uplink and Downlink Channels in Frequency Division Duplex Systems," in *IEEE International Symposium on Personal, Indoor and Mobile Radio Communications, PIMRC*, 2–5 September, Washington, DC, 2014, vol. 2014–June, pp. 172–176.

40. Y. Han, J. Ni, and G. Du, "The Potential Approaches to Achieve Channel Reciprocity in FDD System with Frequency Correction Algorithms," in *2010 5th International ICST Conference on Communications and Networking in China*, 2010, pp. 1–5.

41. J.-C. Shen, J. Zhang, E. Alsusa, and K. B. Letaief, "Compressed CSI Acquisition in FDD Massive MIMO: How Much Training Is Needed?," *IEEE Trans. Wirel. Commun.*, vol. 15, no. 6, pp. 4145–4156, 2016.

42. Z. Gao, L. Dai, W. Dai, B. Shim, and Z. Wang, "Structured Compressive Sensing Based Spatio-Temporal Joint Channel Estimation for FDD Massive MIMO," *IEEE Trans. Commun.*, vol. 64, no. 2, pp. 601–617, 2016.

43. "Telkomsel Started the Road to 5G by Demonstrating Indonesia's First FDD Massive MIMO Technology Together with Huawei," *Huawei*, 2017 [Online]. Available at http://www.huawei.com/en/press-events/news/2017/4/first-FDD-Massive-MIMO-technology [Accessed March 17, 2018].

44. "Telenet Belgium and ZTE Announce First FDD Massive MIMO Field Application in Europe," *ZTE*, 2017 [Online]. Available at http://www.zte.com.cn/global/about/press-center/news/201710ma/1019ma2 [Accessed March 17, 2018].

45. Vodafone Australia, "Vodafone Successfully Demonstrates FDD Massive MIMO," *Vodafone*, 2017 [Online]. Available at https://www.vodafone.com.au/media/vodafone-successfully-demonstrates-fdd-massive-mimo [Accessed March 17, 2018].

46. Verizon, "Verizon and Ericsson Team Up to Deploy Massive MIMO," *Cision PR Newswire*, 2017 [Online]. Available at https://www.prnewswire.com/news-releases/verizon-and-ericsson-team-up-to-deploy-massive-mimo-300544258.html [Accessed March 17, 2018].

47. E. Bjornson, J. Hoydis, and L. Sanguinetti, "Pilot Contamination Is Not a Fundamental Asymptotic Limitation in Massive MIMO," in *IEEE International Conference on Communications (ICC)*, 21–25 May, Paris, France, 2017.

48. K. Upadhya, S. A. Vorobyov, and M. Vehkapera, "Superimposed Pilots Are Superior for Mitigating Pilot Contamination in Massive MIMO," *IEEE Trans. Signal Process.*, vol. 65, no. 11, pp. 2917–2932, 2017.

49. C. Zhang and G. Zeng, "Pilot Contamination Reduction Scheme in Massive MIMO Multi-cell TDD Systems," *J. Comput. Inf. Syst.*, vol. 10, no. 15, pp. 6721–6729, 2014.

50. H. Yin, D. Gesbert, M. Filippou, and Y. Liu, "A Coordinated Approach to Channel Estimation in Large-Scale Multiple-Antenna Systems," *IEEE J. Sel. Areas Commun.*, vol. 31, no. 2, pp. 264–273, 2013.

51. E. Bjornson, J. Hoydis, and L. Sanguinetti, "Massive MIMO Has Unlimited Capacity," *IEEE Trans. Wirel. Commun.*, vol. 17, no. 1, pp. 574–590, 2018.

52. B. Akgun, M. Krunz, and O. O. Koyluoglu, "Pilot Contamination Attacks in Massive MIMO Systems," in *IEEE Conference on Communications and Network Security (CNS)*, 9–11 October, Las Vegas, NV, 9–11 October, Las Vegas, NV, 2017.

53. C. Mollén, E. G. Larsson, and T. Eriksson, "Waveforms for the Massive MIMO Downlink: Amplifier Efficiency, Distortion, and Performance," *IEEE Trans. Commun.*, vol. 64, no. 12, pp. 5050–5063, 2016.

54. E. Bjornson, J. Hoydis, M. Kountouris, and M. Debbah, "Massive MIMO Systems with Non-ideal Hardware: Energy Efficiency, Estimation, and Capacity Limits," *IEEE Trans. Inf. Theory*, vol. 60, no. 11, pp. 7112–7139, 2014.

55. Y. Li, C. Tao, G. Seco-Granados, A. Mezghani, A. L. Swindlehurst, and L. Liu, "Channel Estimation and Performance Analysis of One-Bit Massive MIMO Systems," *IEEE Trans. Signal Process.*, vol. 65, no. 15, pp. 4075–4089, 2017.

56. S. Jacobsson, G. Durisi, M. Coldrey, U. Gustavsson, and C. Studer, "Throughput Analysis of Massive MIMO Uplink with Low-Resolution ADCs," *IEEE Trans. Wirel. Commun.*, vol. 16, no. 6, pp. 4038–4051, 2017.

57. C. Mollen, J. Choi, E. G. Larsson, and R. W. Heath, "One-Bit ADCs in Wideband Massive MIMO Systems with OFDM Transmission," in *IEEE International Conference on Acoustics, Speech and Signal Processing (ICASSP)*, 20–25 March, Shanghai, China, 2016, vol. 2016–May, no. 1, pp. 3386–3390.

58. E. Björnson, E. G. Larsson, and T. L. Marzetta, "Massive MIMO: Ten Myths and One Critical Question," *IEEE Commun. Mag.*, vol. 54, no. 2, pp. 114–123, 2016.

59. T. L. Marzetta, "Massive MIMO: An Introduction," *Bell Labs Tech. J.*, vol. 20, pp. 11–22, 2015.

60. FP7 Project MAMMOET, "Massive MIMO Info Point" [Online]. Available at https://massivemimo.eu/ [Accessed March 20, 2018].

61. IEEE Communications Society, "Best Readings Topics on Massive MIMO" [Online]. Available at https://www.comsoc.org/best-readings/topics/massive-mimo [Accessed March 20, 2018].

62. L. Wang, H. Q. Ngo, M. Elkashlan, T. Q. Duong, and K.-K. Wong, "Massive MIMO in Spectrum Sharing Networks: Achievable Rate and Power Efficiency," *IEEE Syst. J.*, vol. 11, no. 1, pp. 20–31, 2017.

63. B. Kouassi, I. Ghauri, and L. Deneire, "Reciprocity-Based Cognitive Transmissions Using a MU Massive MIMO Approach," in *2013 IEEE International Conference on Communications (ICC)*, 9–13 June, Budapest, Hungary, 2013, pp. 2738–2742.

64. H. Xie, B. Wang, F. Gao, and S. Jin, "A Full-Space Spectrum-Sharing Strategy for Massive MIMO Cognitive Radio Systems," *IEEE J. Sel. Areas Commun.*, vol. 34, no. 10, pp. 2537–2549, 2016.

65. F. Ouyang, "Massive MIMO for Dynamic Spectrum Access," in *IEEE International Conference on Consumer Electronics (ICCE)*, 8–10 January, Las Vegas, NV, 2017.

66. H. Q. Ngo, A. Ashikhmin, H. Yang, E. G. Larsson, and T. L. Marzetta, "Cell-Free Massive MIMO versus Small Cells," *IEEE Trans. Wirel. Commun.*, vol. 16, no. 3, pp. 1834–1850, 2016.

67. E. Nayebi, A. Ashikhmin, T. L. Marzetta, and H. Yang, "Cell-Free Massive MIMO Systems," in *49th Asilomar Conference on Signals, Systems and Computers*, 8–11 November, Pacific Grove, CA, 2015.

68. G. Xu et al., "Full Dimension MIMO (FD-MIMO): Demonstrating Commercial Feasibility," *IEEE J. Sel. Areas Commun.*, vol. 35, no. 8, pp. 1876–1886, 2017.

69. H. Ji et al., "Overview of Full-Dimension MIMO in LTE-Advanced Pro," *IEEE Commun. Mag.*, vol. 55, no. 2, pp. 176–184, 2017.

70. F.-L. Luo and C. Zhang, *Signal Processing for 5G: Algorithms and Implementations*. Wiley, 2016.

71. Q. U. A. Nadeem, A. Kammoun, M. Debbah, and M. S. Alouini, "Design of 5G Full Dimension Massive MIMO Systems," *IEEE Trans. Commun.*, vol. 66, no. 2, pp. 726–740, 2018.

72. Z. Pi and F. Khan, "An Introduction to Millimeter-Wave Mobile Broadband Systems," *IEEE Commun. Mag.*, vol. 49, no. 6, pp. 101–107, 2011.

73. R. G. Fellers, "Millimeter Waves and Their Applications," *Electr. Eng.*, vol. 75, no. 10, pp. 914–917, 1956.

74. Y. Ghasempour, C. R. C. M. Silva, C. Cordeiro, and E. W. Knightly, "IEEE 802.11ay: 60 GHz Communication for 100 Gb/s Wi-Fi," *IEEE Commun. Mag.*, vol. 55, no. 12, pp. 186–192, 2017.

75. S. A. Busari, K. M. S. Huq, S. Mumtaz, L. Dai, and J. Rodriguez, "Millimeter-Wave Massive MIMO Communication for Future Wireless Systems: A Survey," *IEEE Commun. Surv. Tutorials*, vol. 20, no. 2, pp. 836–869, 2018.

76. I. A. Hemadeh, K. Satyanarayana, M. El-Hajjar, and L. Hanzo, "Millimeter-Wave Communications: Physical Channel Models, Design Considerations, Antenna Constructions and Link-Budget," *IEEE Commun. Surv. Tutorials*, vol. 20, no. 2, pp. 870–913, 2018.

77. F. Khan, Z. Pi, and S. Rajagopal, "Millimeter-Wave Mobile Broadband with Large Scale Spatial Processing for 5G Mobile Communication," in *2012 50th Annual Allerton Conference on Communication, Control, and Computing (Allerton)*, 2012, pp. 1517–1523.

78. M. Xiao et al., "Millimeter Wave Communications for Future Mobile Networks," *IEEE J. Sel. Areas Commun.*, vol. 35, no. 9, pp. 1909–1935, 2017.

79. Z. Gao, L. Dai, D. Mi, Z. Wang, M. A. Imran, and M. Z. Shakir, "MmWave Massive-MIMO-Based Wireless Backhaul for the 5G Ultra-Dense Network," *IEEE Wirel. Commun.*, vol. 22, no. 5, pp. 13–21, 2015.

80. S. Rangan, T. S. Rappaport, and E. Erkip, "Millimeter-Wave Cellular Wireless Networks: Potentials and Challenges," *Proc. IEEE*, vol. 102, no. 3, pp. 366–385, 2014.

81. Y. Niu, Y. Li, D. Jin, L. Su, and A. V. Vasilakos, "A Survey of Millimeter Wave Communications (mmWave) for 5G: Opportunities and Challenges," *Wirel. Networks*, vol. 21, no. 8, pp. 2657–2676, 2015.

82. ITU-R, *Technical Feasibility of IMT in the Bands Above 6 GHz*. ITU-R M.2376-0, 2015.

83. J. G. Andrews, T. Bai, M. Kulkarni, A. Alkhateeb, A. Gupta, and R. W. Heath, "Modeling and Analyzing Millimeter Wave Cellular Systems," *IEEE Trans. Commun.*, vol. 65, no. 1, pp. 403–430, 2017.

84. ITU-R, *Attenuation Due to Clouds and Fog*. Recommendation ITU-R P.840-6, 2013.

85. ITU-R, *Attenuation in Vegetation*. Recommendation ITU-R P.833-9, 2016.

86. ITU-R, *Attenuation by Atmospheric Gases*. Recommendation ITU-R P.676-11, 2016.

87. ITU-R, *Characteristics of Precipitation for Propagation Modelling*. ITU-R P.837-7, 2017.

88. 3GPP, *Study on Channel Model for Frequencies from 0.5 to 100 GHz (Release 14)*. 3GPP Technical Report TR 38.901 V14.3.0, 2017.

89. M. Matinmikko, M. Latva-aho, P. Ahokangas, and V. Seppänen, "On Regulations for 5G: Micro Licensing for Locally Operated Networks," *Telecomm. Policy*, no. August, pp. 1–14, 2017.

90. R. W. Heath Jr. et al., "An Overview of Signal Processing Techniques for Millimeter Wave MIMO Systems," *IEEE J. Sel. Top. Signal Process.*, vol. 10, no. 3, pp. 436–453, 2015.

91. A. Alkhateeb, O. El Ayach, G. Leus, and R. W. Heath, "Channel Estimation and Hybrid Precoding for Millimeter Wave Cellular Systems," *IEEE J. Sel. Top. Signal Process.*, vol. 8, no. 5, pp. 831–846, 2014.

92. X. Cheng, M. Wang, and S. Li, "Compressive Sensing-Based Beamforming for Millimeter-Wave OFDM Systems," *IEEE Trans. Commun.*, vol. 65, no. 1, pp. 371–386, 2017.

93. J. Wang et al., "Beam Codebook Based Beamforming Protocol for Multi-Gbps Millimeter-Wave WPAN Systems," *IEEE J. Sel. Areas Commun.*, vol. 27, no. 8, pp. 1390–1399, 2009.

94. X. Li, Y. Zhu, and P. Xia, "Enhanced Analog Beamforming for Single Carrier Millimeter Wave MIMO Systems," *IEEE Trans. Wirel. Commun.*, vol. 16, no. 7, pp. 4261–4274, 2017.

95. P. Xia, R. W. Heath, and N. Gonzalez-Prelcic, "Robust Analog Precoding Designs for Millimeter Wave MIMO Transceivers with Frequency and Time Division Duplexing," *IEEE Trans. Commun.*, vol. 64, no. 11, pp. 4622–4634, 2016.

96. T. Kim, J. Park, J. Y. Seol, S. Jeong, J. Cho, and W. Roh, "Tens of Gbps Support with mmWave Beamforming Systems for Next Generation Communications," in *GLOBECOM—IEEE Global Telecommunications Conference*, 9–13 December, Atlanta, GA, 2013, pp. 3685–3690.

97. S. Han, C. I, Z. Xu, and C. Rowell, "Large-Scale Antenna Systems with Hybrid Analog and Digital Beamforming for Millimeter Wave 5G," *IEEE Commun. Mag.*, vol. 53, no. 1, pp. 186–194, 2015.

98. N. Song, T. Yang, and H. Sun, "Overlapped Subarray Based Hybrid Beamforming for Millimeter Wave Multiuser Massive MIMO," *IEEE Signal Process. Lett.*, vol. 24, no. 5, pp. 550–554, 2017.

99. S. Kutty and D. Sen, "Beamforming for Millimeter Wave Communications: An Inclusive Survey," *IEEE Commun. Surv. Tutorials*, vol. 18, no. 2, pp. 949–973, 2016.

100. A. Alkhateeb, J. Mo, N. González-Prelcic, and R. W. Heath, "MIMO Precoding and Combining Solutions for Millimeter-Wave Systems," *IEEE Commun. Mag.*, vol. 52, no. 12, pp. 122–131, 2014.

101. A. F. Molisch et al., "Hybrid Beamforming for Massive MIMO: A Survey," *IEEE Commun. Mag.*, vol. 55, no. 9, pp. 134–141, 2017.

102. F. Sohrabi and W. Yu, "Hybrid Digital and Analog Beamforming Design for Large-Scale Antenna Arrays," *IEEE J. Sel. Top. Signal Process.*, vol. 4553, no. June 2015, pp. 1–1, 2016.

103. T. Bogale, L. Le, A. Haghighat, and L. Vandendorpe, "On the Number of RF Chains and Phase Shifters, and Scheduling Design with Hybrid Analog-Digital Beamforming," *IEEE Trans. Wirel. Commun.*, vol. 1276, no. c, pp. 1–1, 2016.

104. L. Zhao, D. W. K. Ng, and J. Yuan, "Multi-User Precoding and Channel Estimation for Hybrid Millimeter Wave Systems," *IEEE J. Sel. Areas Commun.*, vol. 35, no. 7, pp. 1576–1590, 2017.

105. O. El Ayach, S. Rajagopal, S. Abu-Surra, Z. Pi, and R. W. Heath, "Spatially Sparse Precoding in Millimeter Wave MIMO Systems," *IEEE Trans. Wirel. Commun.*, vol. 13, no. 3, pp. 1499–1513, 2014.

106. A. Alkhateeb, G. Leus, and R. W. Heath Jr., "Limited Feedback Hybrid Precoding," *IEEE Trans. Wirel. Commun.*, vol. 14, no. 11, pp. 6481–6494, 2015.

107. A. Alkhateeb and R. W. Heath, "Frequency Selective Hybrid Precoding for Limited Feedback Millimeter Wave Systems," *IEEE Trans. Commun.*, vol. 64, no. 5, pp. 1801–1818, 2016.

108. J. Song, J. Choi, and D. J. Love, "Codebook Design for Hybrid Beamforming in Millimeter Wave Systems," in *IEEE International Conference on Communications*, 8–12 June, London, UK, 2015, pp. 1298–1303.

109. Z. Xiao, P. Xia, and X. G. Xia, "Codebook Design for Millimeter-Wave Channel Estimation with Hybrid Precoding Structure," *IEEE Trans. Wirel. Commun.*, vol. 16, no. 1, pp. 141–153, 2017.

110. P. Xia, S. K. Yong, J. Oh, and C. Ngo, "Multi-Stage Iterative Antenna Training for Millimeter Wave Communications," in *GLOBECOM—IEEE Global Telecommunications Conference*, 30 November–4 December, New Orleans, LO, 2008.

111. W. Ni and X. Dong, "Hybrid Block Diagonalization for Massive Multiuser MIMO Systems," *IEEE Trans. Commun.*, vol. 64, no. 1, pp. 201–211, 2016.

112. Z. Wang, M. Li, X. Tian, and Q. Liu, "Iterative Hybrid Precoder and Combiner Design for mmWave Multiuser MIMO Systems," *IEEE Commun. Lett.*, vol. 21, no. 7, pp. 1581–1584, 2017.

113. S. Payami, M. Ghoraishi, and M. Dianati, "Hybrid Beamforming for Large Antenna Arrays With Phase Shifter Selection," *IEEE Trans. Wirel. Commun.*, vol. 15, no. 11, pp. 7258–7271, 2016.

114. X. Yu, J. C. Shen, J. Zhang, and K. B. Letaief, "Alternating Minimization Algorithms for Hybrid Precoding in Millimeter Wave MIMO Systems," *IEEE J. Sel. Top. Signal Process.*, vol. 10, no. 3, pp. 485–500, 2016.

115. Q. Xu, C. Jiang, Y. Han, B. Wang, and K. J. R. Liu, "Waveforming: An Overview With Beamforming," *IEEE Commun. Surv. Tutorials*, vol. 20, no. 1, pp. 132–149, 2018.

116. J. Mo, A. Alkhateeb, S. Abu-Surra, and R. W. Heath, "Hybrid Architectures with Few-Bit ADC Receivers: Achievable Rates and Energy-Rate Tradeoffs," *IEEE Trans. Wirel. Commun.*, vol. 16, no. 4, pp. 2274–2287, 2017.

117. K. Roth, J. A. Nossek, and L. Fellow, "Achievable Rate and Energy Efficiency of Hybrid and Digital Beamforming Receivers With Low Resolution ADC," *IEEE J. Sel. Areas Commun.*, vol. 35, no. 9, pp. 2056–2068, 2017.

118. L. F. Lin, W. H. Chung, H. J. Chen, and T. S. Lee, "Energy Efficient Hybrid Precoding for Multi-user Massive MIMO Systems Using Low-Resolution ADCs," *2016 IEEE International Workshop on Signal Processing Systems (SiPS)*, 26–28 October, Dallas, TX, 2016, pp. 115–120.

119. J. Zhang, X. Ge, Q. Li, M. Guizani, and Y. Zhang, "5G Millimeter-Wave Antenna Array: Design and Challenges," *IEEE Wirel. Commun.*, vol. 24, no. 2, pp. 106–112, 2017.

120. P. Smulders, "The Road to 100 Gb/s Wireless and Beyond: Basic Issues and Key Directions," *IEEE Commun. Mag.*, vol. 51, no. 12, pp. 86–91, 2013.

121. S. Emami et al., "A 60GHz CMOS Phased-Array Transceiver Pair for Multi-Gb/s Wireless Communications," in *IEEE International Solid-State Circuits Conference*, 20–24 February, San Francisco, CA, 2011, pp. 164–166.

122. M. Boers et al., "A 16TX/16RX 60 GHz 802.11ad Chipset with Single Coaxial Interface and Polarization Diversity," *IEEE J. Solid-State Circuits*, vol. 49, no. 12, pp. 3031–3045, 2014.

123. W. Hong, K. H. Baek, Y. Lee, Y. Kim, and S. T. Ko, "Study and Prototyping of Practically Large-Scale mmWave Antenna Systems for 5G Cellular Devices," *IEEE Commun. Mag.*, vol. 52, no. 9, pp. 63–69, 2014.

124. H. Shokri-Ghadikolaei, F. Boccardi, C. Fischione, G. Fodor, and M. Zorzi, "The Impact of Beamforming and Coordination on Spectrum Pooling in mmWave Cellular Networks," in *50th Asilomar Conference on Signals, Systems and Computers*, 2016, pp. 21–26.

125. H. Shokri-Ghadikolaei, F. Boccardi, C. Fischione, G. Fodor, and M. Zorzi, "Spectrum Sharing in mmWave Cellular Networks via Cell Association, Coordination, and Beamforming," *IEEE J. Sel. Areas Commun.*, vol. 34, no. 11, pp. 2902–2917, 2016.

126. A. K. Gupta, J. G. Andrews, and R. W. Heath, "On the Feasibility of Sharing Spectrum Licenses in mmWave Cellular Systems," *IEEE Trans. Commun.*, vol. 64, no. 9, pp. 3981–3995, 2016.

127. A. K. Gupta, A. Alkhateeb, J. G. Andrews, and R. W. Heath, "Gains of Restricted Secondary Licensing in Millimeter Wave Cellular Systems," *IEEE J. Sel. Areas Commun.*, vol. 34, no. 11, pp. 2935–2950, 2016.

128. A. H. Jafari, J. Park, and R. W. Heath, "Analysis of Interference Mitigation in mmWave Communications," in *IEEE International Conference on Communications*, 21–25 May, Paris, France, 2017, pp. 1–6.

129. X. Jiang and F. Kaltenberger, "Channel Reciprocity Calibration in TDD Hybrid Beamforming Massive MIMO Systems," *IEEE J. Sel. Top. Signal Process.*, vol. 12, no. 3, pp. 422–431, 2018.

130. M. Giordani, M. Mezzavilla, C. N. Barati, S. Rangan, and M. Zorzi, "Comparative Analysis of Initial Access Techniques in 5G mmWave Cellular Networks—2016," in *2016 Annual Conference on Information Science and Systems (CISS)*, 16–18 March, Princeton, NJ, 2016.

131. M. Giordani, M. Mezzavilla, and M. Zorzi, "Initial Access in 5G mmWave Cellular Networks," *IEEE Commun. Mag.*, vol. 54, no. 11, pp. 40–47, 2016.

132. C. N. Barati et al., "Initial Access in Millimeter Wave Cellular Systems," *IEEE Trans. Wirel. Commun.*, vol. 15, no. 12, pp. 7926–7940, 2016.

133. M. Cudak, T. Kovarik, T. A. Thomas, A. Ghosh, Y. Kishiyama, and T. Nakamura, "Experimental mm Wave 5G CELLULAR SYSTEM," *2014 IEEE Globecom Work. GC Wkshps* 2014, pp. 377–381, 2014.

134. W. Roh et al., "Millimeter-Wave Beamforming as an Enabling Technology for 5G Cellular Communications: Theoretical Feasibility and Prototype Results," *IEEE Commun. Mag.*, vol. 52, no. 2, pp. 106–113, 2014.

135. Y. Niu, Y. Li, D. Jin, L. Su, and D. Wu, "Blockage Robust and Efficient Scheduling for Directional mmWave WPANs," *IEEE Trans. Veh. Technol.*, vol. 64, no. 2, pp. 728–742, 2015.

136. H. Shokri-Ghadikolaei, C. Fischione, P. Popovski, and M. Zorzi, "Design Aspects of Short Range Millimeter Wave Networks: A MAC Layer Perspective," *IEEE Netw.*, vol. 30, no. 3, pp. 88–96, 2016.

137. Z. Ding, X. Lei, G. K. Karagiannidis, R. Schober, J. Yuan, and V. Bhargava, "A Survey on Non-orthogonal Multiple Access for 5G Networks: Research Challenges and Future Trends," *IEEE J. Sel. Areas Commun.*, vol. 35, no. 10, pp. 2181–2195, 2017.

138. L. Dai, B. Wang, Y. Yuan, S. Han, C. L. I, and Z. Wang, "Non-orthogonal Multiple Access for 5G: Solutions, Challenges, Opportunities, and Future Research Trends," *IEEE Commun. Mag.*, vol. 53, no. 9, pp. 74–81, 2015.

139. Y. Chen et al., "Toward the Standardization of Non-orthogonal Multiple Access for Next Generation Wireless Networks," *IEEE Commun. Mag.*, vol. 56, no. 3, pp. 19–27, 2018.

140. G. Romano, "3GPP Activity Towards IMT-2020 3GPP Roadmap," 3GPP Presentation at the ITU October 2017 5G Workshop, 11 July, Geneva, Switzerland, 2017.

141. 3GPP, *3rd Generation Partnership Project; Technical Specification Group Radio Access Network; Study on Non-orthogonal Multiple Access (NOMA) for NR.* 3GPP TR 38.812, 2018.

142. 3GPP, *Work Item Description: Study on Non-orthogonal Multiple Access for NR.* 3GPP RP-170829, 2017.

143. Z. Ding, Y. Liu, J. Choi, Q. Sun, M. Elkashlan, and H. V. Poor, "Application of Non-orthogonal Multiple Access in LTE and 5G Networks Application of Non-orthogonal Multiple Access in LTE and 5G Networks," *IEEE Commun. Mag.*, vol. 55, no. 2, pp. 185–191, 2017.

144. Z. Ding, F. Adachi, and H. V. Poor, "The Application of MIMO to Non-Orthogonal Multiple Access," *IEEE Trans. Wirel. Commun.*, vol. 15, no. 1, pp. 537–552, 2016.

145. Y. Saito, Y. Kishiyama, A. Benjebbour, T. Nakamura, A. Li, and K. Higuchi, "Non-orthogonal Multiple Access (NOMA) for Cellular Future Radio Access," in *IEEE Vehicular Technology Conference (VTC)*, 2–5 June, Dresden, Germany, 2013.

146. M. Al-Imari, P. Xiao, M. A. Imran, and R. Tafazolli, "Uplink Non-orthogonal Multiple Access for 5G Wireless Networks," in *2014 11th International Symposium on Wireless Communications Systems (ISWCS)*, 26–29 August, Barcelona, Spain, 2014, pp. 781–785.

147. G. L. Stuber, *Principles of Mobile Communication.* New York: Springer, 2011.

148. R. Hoshyar, F. P. Wathan, and R. Tafazolli, "Novel Low-Density Signature for Synchronous CDMA Systems Over AWGN Channel," *IEEE Trans. Signal Process.*, vol. 56, no. 4, pp. 1616–1626, 2008.

149. H. Nikopour and H. Baligh, "Sparse Code Multiple Access," in *IEEE International Symposium on Personal, Indoor and Mobile Radio Communications (PIMRC)*, 8–11 September, London, UK, 2013, pp. 332–336.

150. M. Taherzadeh, H. Nikopour, A. Bayesteh, and H. Baligh, "SCMA Codebook Design," in *IEEE Vehicular Technology Conference (VTC)*, 14–17 September, Vancouver, BC, Canada, 2014.

151. M. Peng, Y. Sun, X. Li, Z. Mao, and C. Wang, "Recent Advances in Cloud Radio Access Networks: System Architectures, Key Techniques, and Open Issues," *IEEE Commun. Surv. Tutorials*, vol. 18, no. 3, pp. 2282–2308, 2016.

152. C. Fan, Y. J. A. Zhang, and X. Yuan, "Advances and Challenges Toward a Scalable Cloud Radio Access Network," *IEEE Commun. Mag.*, vol. 54, no. 6, pp. 29–35, 2016.

153. Z. Yan, M. Peng, and C. Wang, "Economical Energy Efficiency: An Advanced Performance Metric for 5G Systems," *IEEE Wirel. Commun.*, vol. 24, no. 1, pp. 32–37, 2017.

154. Y. Li, T. Jiang, K. Luo, and S. Mao, "Green Heterogeneous Cloud Radio Access Networks: Potential Techniques, Performance Trade-offs, and Challenges," *IEEE Commun. Mag.*, vol. 55, no. 11, pp. 33–39, 2017.

HOMEWORK

12.1 Show that the single-user massive MIMO solutions (12.6) and (12.12) are special cases of the optimal MIMO solutions in Chapter 11.

12.2 Using considerations similar to Section 12.5.1.3, derive the expression of ZF and RZF combiners for uplink massive MIMO. Compare your results with those in [70, sec. 9.2.3].

CHAPTER *13*

CLOSING REMARKS AND FURTHER EXPLORATION

13.1 INTRODUCTION

Digital communication has had a journey of more than a half century. Claude Shannon in 1948 established the information theory (Chapter 2), based on which the search for near-perfect channel coding has taken many decades (Chapter 5). At about the same time, digital modulation and demodulation techniques were researched and developed (Chapters 3 and 4). As the computation power of digital signal processing increases, more and more sophisticated processing algorithms emerged, significantly enhancing communication system performances (Chapters 7–9). As system performances approach channel capacities, a better understanding of the channels became more and more critical. Knowledge of channel properties is especially valuable for wireless communications in urban and other complex environments. Therefore, the study of communication channels is an integral part of digital communication (Chapter 6). From the 1990s, advanced technologies were built on top of the classical modulation and coding schemes. OFDM (Chapter 10) and MIMO (Chapter 11) quickly found their way to the commercial products, in answer to the fast-growing demand for wireless communication capabilities. Innovations are still emerging, with the current state of the art reflected in the 5G cellular systems (Chapter 12).

In this last chapter, we outline a few more areas for further exploration.

13.2 ANALOG CIRCUITRY

The scope of the book is limited to the baseband digital domain. However, even if an engineer works exclusively in the digital domain, there is certain knowledge about the analog circuitry he should know about. The most relevant circuitry knowledge includes:

- Amplifiers. One needs to know the definitions and typical values of some amplifier parameters such as gain, bandwidth, noise figure, output power, and dynamic range. These performance parameters typically represent compromises

Digital Communication for Practicing Engineers, First Edition. Feng Ouyang.
© 2020 by The Institute of Electrical and Electronics Engineers, Inc.
Published 2020 by John Wiley & Sons, Inc.

among themselves and trade-offs with other considerations such as size, power consumption, and cost.

- Filters. One needs to have a general sense of typical analog filter edge steepness, group delay, and time domain spreading.

- Analog-to-digital converter (ADC) and digital-to- analog converter (DAC). The trade-offs among conversion speed, accuracy, and cost should be kept in mind. Such trade-offs affect not only performance expectations but also architectural designs. For example, whether one should put intermediate frequency (IF) operations in digital or analog domain depends on the available ADC/DAC performances. The allocation of echo (self-interference) cancellation gains between analog and digital domains also depend on the number of bits a DAC can provide (see Chapter 8).

- Stability of analog components. Factors such as aging, temperature, mechanical vibration, and so on. affect operational parameters of various analog components, such as the frequency of a clock or the gain of an amplifier. Stability considerations affect digital domain design decisions, such as how often adaptations and calibrations need to be performed and what tolerance we need to assume about analog design parameters.

- Mixer. A mixer converts signals between baseband and passband (Chapter 3). In this book, we assume that the mixers are perfect in functionality. In reality, however, mixers have different architectures and may introduce additional impairments such as phase noise, I/Q mismatch, and direct current (DC) offset. Usually, the mixer performance is not a bottleneck. Therefore, the assumption of a perfect mixer is valid, although it needs to be verified during system design.

There are many books on analog circuitry that provides more detailed considerations on these and other topics. However, analog technology evolves over time, especially in response to more stringent requirements on performance, cost and power consumption imposed by the latest communication systems. The emergence of full-duplex radio (Chapter 8) and increased accuracy requirement in gain calibration may require more communication and coordination between the analog and digital domains. To obtain up-to-date information, one can check out the latest catalogs and application notes of primary analog IC providers. A more effective approach is talking to analog design engineers in your organization.

13.3 SOFTWARE-DEFINED RADIO (SDR)

The software-defined radio (SDR), also known as software radio, is not a technology but a platform. It has a different architecture from the conventional radios. As described in Chapter 3, a typical digital communications radio can be divided into the digital domain and the analog domain.

As shown in Chapter 3, at the transmitter, the signal (waveform) is generated in the digital domain as a sequence of samples, each expressed by some bits. These samples are converted to analog ones by a digital-to-analog converter (DAC). The output of DAC, after proper filtering, forms the analog signal, which is then up-converted from baseband to passband before being amplified and transmitted. The process is reversed at the receiver. Such transceivers are usually designed for a specific waveform. Its analog circuitry (and a part of the digital processing components) is implemented in hardware, optimized for such waveform.

The SDR has a different architecture [1, 2]. For the transmitter, the digital domain processing is implemented in a programmable fashion, either with general-purpose processors or with field–programmable gate arrays (FPGA). The DAC is performed at a rate higher than baseband (typically several hundred MHz or higher). The baseband signal is first converted to an intermediate frequency (IF) while staying in the digital domain. The IF signal is then converted to the analog domain and is up-converted again to the carrier frequency. A reverse process is implemented for the receiver. Such architecture allows for much more flexibility in waveform implementation. The same radio can support multiple waveforms with different modulation schemes and bandwidths. Waveforms can even be changed during operation by loading new software to the radio. Furthermore, the SDR architecture, when following certain standards and conventions, allows for reuse of software over multiple hardware platforms.

The US military views SDR as an avenue toward multiwaveform, easily upgradable and multivendor solution to various communications needs. It published the Software Communications Architecture (SCA) standard [3]. SCA has been used in tactical radios for the military [4, 5]. The National Aeronautics and Space Administration (NASA) also developed an SDR architecture for space communications [6].

Commercially, SDR has found broad application in lab prototyping as well as in products [7–9]. GNU Radio is a popular open-source software package [10] for SDR. GNU Radio is supported by several hardware platforms [2, 11].

Reflecting the state of the arts, the *IEEE Communications Magazine* had special issues on SDR in September 2015 (Volume 53, Issue 9) and January 2016 (Volume 54, Issue 1).

13.4 COGNITIVE RADIO (CR) AND DYNAMIC SPECTRUM ACCESS (DSA)

Cognitive radio (CR) is a very general concept proposed in 1998 [12, 13]. A CR is equipped with various sensors and can change its transmitting and receiving behaviors based on the environment. For example, it may limit its frequency band based on its knowledge of the current location and local regulations. It may also select the power and waveform based on current user applications. A CR is often implemented with the software-defined radio (SDR) architecture described in Section 13.3. Highly intelligent CRs have been envisioned but have not been used in mainstream applications, except for dynamic spectrum access (DSA) [14, 15]. In fact, the terms CR and DSA are often used interchangeably in the literature. Recently, applications of CR beyond DSA have found interest in the 5G cellular technologies [16, 17].

DSA is a spectrum management technique allowing more users to share the spectrum [18].[1] Traditionally, frequency bands (spectrum) are licensed to authorized users, who own the exclusive rights to transmit in these bands. However, the authorized users often do not transmit continuously in this band. In other words, spectrum utilization may be low. DSA addresses such issue by allowing unlicensed users to use the frequency bands, as well.

In the DSA framework, the licensed and unlicensed users are known as the primary and secondary users, respectively. The primary users (PU) use the spectrum in the same way as in traditional spectrum management schemes. Namely, they can transmit at will, under the constraint of location and power stipulated in the license. The secondary users (SU) can transmit only when such transmission does not produce unacceptable interference to the PUs. There are several ways for the SUs to avoid harmfully interfering the PUs.

The first way uses a "whitelist," which is a database documenting spectrum availability to the SUs. An SU is allowed to transmit only at the time, location, and frequency listed on the whitelist. The FCC regulation 47CFR15.711 provides such a method for unlicensed devices to use the spectrum licensed to TV stations (known as TV white space).

The second way is "sensing." In this way, an SU must continuously sense the electromagnetic environment and halt its transmission as soon as it detects transmission from the PU. Sensing is the main focus of DSA research activities [14]. There are some technical challenges to this scheme, including:

- How to detect very weak PU signals. Cyclostationary detection techniques (Chapter 7) are often used [19–22].
- How to defend against "spoofing" attacks, where a malicious transmitter mimics a PU to keep the SU out of the spectrum [23–25]. Also, we need to distinguish between transmissions from PUs and those from other SUs.
- How to fuse sensing results from multiple SUs to form complete and reliable cognition (known as cognitive networks or cooperative sensing) [26–29].

The third is "underlay," where an SU transmits with low power, so that its interference to the PU is acceptable [30, 31]. The ultrawide Band (UWB) method is suitable for underlay transmission because of its low power spectral density (Section 13.5). Underlay is especially attractive when MIMO is used (Chapter 11), so that the transmission from the PU and SU can be spatially separated [32–34]. Such scheme can be extended to the case of massive MIMO (Chapter 12) cognitive radios [35]. One challenge, however, is that unlike in "pure" MIMO cases discussed in Chapter 11, in a DSA situation channel state information is not usually available to the SU transmitter [36, 37].

In a DSA scheme, an SU needs to make complex decisions about its transmitting frequency and power based on channel conditions, the current measurement of

---

[1] For early research results on DSA and CR, see special issue of *IEEE Journal on Selected Areas in Communications*, January 2008 (Volume 26, Issue 1).

electromagnetic radiation, and prior knowledge about the PUs and other operational conditions. Automatic reasoning and learning are often explored to support such decision-making process [26, 38–41]. Making complex decisions is the hallmark of the cognitive radio (CR). Therefore, DSA and CR are closely related.

Another issue with DSA is how multiple SUs share the time and frequency resource available to them. The availability of such resource is usually highly dynamic (depending on the activities of the PU). The SUs may be cooperative (e.g., belong to the same network) or uncooperative. Resource allocation is an interesting research topic. The game theory has found much application in this area [42–44].[2]

Although DSA has been an active research area for more than 10 years, its adaptation is rather slow. The main obstacle is ensuring PU protection against harmful interferences.

The IEEE standard 802.22 is a DSA solution for wireless regional area networks (WRAN) [45]. Such a network has link ranges of tens of kilometers with data rates of tens of Mbps. An 802.22 device, as an SU, shares the frequency band of 54–862 MHz with TV broadcast services. Interference to the PUs is controlled by whitelist database and spectral sensing. Furthermore, 802.22 provides a mechanism to coordinate among multiple SU networks, either by over-the-air beacons or over the IP network.

Another IEEE standard 802.11af was developed in 2014 to support wireless local area network (WLAN) operations in the TV frequency bands [46]. 802.11af was later merged into the 802.11-2016, known as the television very high throughput (TVHT) mode [47, ch. 22]. It uses the database method for coexistence with TV services.

As the demand for cellular services increase, the 4G cellular service is expanding to the unlicensed band. Such technology is known as LTE-unlicensed (LTE-U). LTE-U needs to coexist with services already deployed in these bands, such as the WLAN (IEEE 802.11). DSA techniques have the potential of improving overall performance [48–50]. More works on spectrum sharing and aggregation for cellular systems can be found in the special issues of *IEEE Journal on Selected Areas in Communications*, October and November 2016 (Volume 34, Issues 10 and 11), and January 2017 (Volume 35, Issue 1).

On the military side, the US Defense Advanced Research Projects Agency (DARPA) conducted a field trial, known as "Next Generation Communications" (XG), of DSA technologies in supporting coexistence of multiple mobile radio systems [51, 52]. One of the focuses of the XG program was developing tools for describing and managing the spectrum sharing behavior of cognitive radios. The result is a "policy-based management" approach. DARPA further conducted a program "Wireless Network after Next" (WNaN), which uses cognitive functionalities beyond spectrum sharing to reduce radio cost [51]. The US military has some other efforts to explore the application of cognitive radios [53].

---

[2] For more applications of the game theory in communications, see special issues of *IEEE Journal on Selected Areas in Communications*, February and March of 2017 (Volume 35, Issues 2 and 3).

13.5 ULTRAWIDE BAND (UWB)

Ultrawide band (UWB) has initially been a radar technology, where ultrashort pulses (and thus ultrawide bandwidth) are generated and detected for ranging. The pulse generation and detection are performed in the analog domain with special circuits. Therefore, the baseband design is not burdened by the wide bandwidth. Communication applications of UWB were proposed in 2000 [54]. With UWB, the transmission power is spread over a wide bandwidth, resulting in low power spectral density (PSD). Therefore, communication can be conducted without significant interference to other traditional (narrow-band) systems. Such "underlay" scheme is similar to the nonorthogonal multiple access (NOMA) (Chapter 12). According to the US Federal Communications Commission (FCC) definition, if the bandwidth of a system is larger than 500 MHz or larger than 20% of the mean frequency, it is qualified as a UWB system (47 CFR 15.503).

UWB was a candidate technology for high-rate wireless personal area network (WPAN) under IEEE 802.15.3 standard. However, such effort did not produce an agreement on the final solution. Currently, UWB is a physical layer technology for the low-rate WPAN standard IEEE 802.15.4 [55] for both communications and ranging.

UWB is also a promising technology for the Internet of Things (IoT) applications due to its low power consumption and positioning capabilities [56, 57].

13.6 RELAYING AND COOPERATIVE COMMUNICATIONS

Relaying is an intuitive way to extend the range of communications. The simplest case involves three nodes, the source, the relay, and the destination.

Relaying (also known as cooperative communications), in the context here, is different from multihop networking. In a multihop networking, the "relay" node receives data from the source and retransmits it to the destination. The two paths are independent at the physical layer. In the relaying discussed here, however, the relay may manipulate received signal at the physical layer. The destination may receive signals from both the source and the relay.

The concept of relaying started in the 1970s [58, 59]. In the original configuration, there are three parties to the transmission: the source, the relay, and the destination. In addition to the two channels (source to relay and relay to destination), there is also a (weaker) channel between the source and the destination. The data transmission is divided into two time slots. In time slot 1, the source sends data to the relay and the destination. In slot 2, the relay sends the received data to the destination. At the same time, the source sends additional data directly to the destination.[3] Proper channel

---

[3] The source and the relay can share the slot 2 channel in any way, such as FDMA, TDMA, or superposition.

coding schemes are used to match the data rates with channel capacities (Chapter 5). Such scheme makes use of the third ("direct") channel and achieves better performance than the naive "forwarding" scheme. Various coding/decoding schemes can be used for such vision [60–62].

In practice, instead of using optimum coding schemes, simplified relaying methods are often used. The simplest form is amplify-and-forward (the relay simply amplifies the signal from the source and retransmit it in the next time slot). Other forms include decode-and-forward (the relay recovers received bits and re-modulates them before transmission), and compress-and-forward (the relay codes received bits incompletely, expecting the destination to supplement relay transmission with signals directly from the source) [59].

The basic "three nodes" configuration can be extended in many ways. For example, multiple nodes can cooperate in transmission, where each node helps to relay signals from other nodes, in addition to transmitting its own data [63, 64]. A common relay can be shared by multiple pairs of users [65]. Conversely, there can be multiple relays to be selected by the source–destination pairs based on channel conditions [66]. Additional issues arise when two-way relays are employed [65].

Relaying becomes much more interesting when multi-antenna nodes are used (Chapter 11) [67–69]. One form of cooperative communications is "distributed MIMO," where a cluster of single-antenna transceivers collectively act as a multi-antenna node [70]. In such a configuration, data to be transmitted are first distributed among the nodes within the cluster. These nodes then transmit jointly toward the remote receiver, which can also be such a cluster. In this way, each node can be viewed as a relay for other nodes within the cluster. Another widely studied configuration is using single or multiple relays to achieve space-time diversity. Namely, the same data are transmitted over multiple channels, similar to the space-time coding discussed in Chapter. However, the channel models and transceiver constraints are different [71–73].

Relaying technologies have been proposed for cellular applications [74–76]. The 4G cellular technology LTE-Advanced adopts two types of arrays. Type 1 mimics a separate cell, while Type 2 provides an additional path for downlink data traffic using amplify-and-forward or decode-and-forward methods, without requiring different processing techniques at the mobile users [77–79].

13.7 CODE DIVISION MULTIPLE ACCESS (CDMA)

Code Division Multiple Access (CDMA) is a multiaccess technology used for 2G and 3G cellular systems. It was briefly described in Chapter 12. Although the latest cellular systems use other technologies, CDMA is still studied as a statistical multiplexing technique and for overlay/underlay dynamic spectrum access (Section 13.4) [80–82]. It has also found applications in light communications [83, 84]. For more details on the CDMA techniques and challenges, see [85, ch. 12].

13.8 INTERFERENCE MANAGEMENT

In wireless communications, especially in cellular systems (Chapter 12), spectrum scarcity is a significant limitation to capacity growth. One solution for the cellular systems is increasing frequency reuse (Chapter 12) by reducing cell sizes. With frequency reuse, the bandwidth (thus the total data rate) available to each cell is fixed. When the cell sizes are reduced, the total data rate per area is increased due to frequency reuse.

As a result of reduced cell size, mutual interference among the cells is more significant. In fact, in modern cellular systems, mutual interference levels can far exceed that of the "natural noise" such as thermal noise. Therefore, various techniques are developed to reduce mutual interference, thus increasing system capacity. Interference management techniques include several classes.

13.8.1 Multiuser Detection (MUD)

Multiuser detection (MUD) is the most straightforward interference management method. One form of MUD (successive interference cancellation) was discussed in the context of nonorthogonal multiple access (NOMA) in Chapter 12.

MUD jointly decodes all the signals that are superimposed in the same frequency and time. It is theoretically effective in addressing interference under certain conditions [86–88]. However, the processing complexity is usually high [89–91].

13.8.2 Interference Coordination

All cells in a neighborhood can coordinate their transmission to reduce or avoid mutual interference. For example, cells can avoid allocating the same frequency-time resources to their cell-edge users, while reusing frequency when the users are at the cell centers. Such allocation (known as fractional frequency reuse), when combined with power control, effectively reduces inter-cell interference [92–95]. At the physical layer, multi-antenna beams can be coordinated among the neighboring cells to avoid spatial conflict [96, 97].

13.8.3 Interference Alignment

Interference alignment is a technique for multiple transmitters to coordinate their waveform and coding to reduce mutual interference. Such coordination depends on the channel conditions, but not on the actual data content.

The basic idea of interference alignment is considering the transmitting and receiving signal as a vector in a high-dimensional "signal space." Such space can be spanned by symbols at multiple time instances or transmitted by multiple antennas, for example. Through proper precoding, all undesired signals (i.e., interference) at a given receiver are "aligned" into a subspace, leaving the rest of the signal space available for communication. The dimension of such "interference-free" subspace is known as the degree of freedom (DoF), which determines the number of data streams that can be transmitted without interference. The goal of interference alignment, of course, is designing a precoding scheme to achieve larger DoF per user.

The general theory of interference alignment was developed in the 2000s [98–101]. The general precoding schemes and DoF upper limit studies are mathematically involved. It is not clear whether such schemes can be implemented with reasonable complexity. Furthermore, the assumption that all transmitters have information of all channel states may be unrealistic in many cases.[4] On the other hand, for cellular applications, only a few transmitters (in neighboring cells) are involved. Interference alignment may be more practical in such settings [102–106].

13.9 OTHER MODULATION SCHEMES

This book focuses on phase shift keying (PSK) and quadrature amplitude modulation (QAM) when studying modulation and demodulation techniques (Chapters 3 and 4). These modulation schemes assume that the receiver has (near perfect) information about the channel state. Therefore, amplitude and phase changes introduced by the channel can be compensated for at the receiver. Modulations based on such assumption are known as coherent modulation.

On the other hand, it is possible that the receiver does not have enough channel state information for coherent demodulation. Such a situation could happen when the communication protocol does not support channel estimation, or the channel changes too fast to be estimated with reasonable overhead. In such cases, another modulation class, the noncoherent modulations, can be used. Frequency shift keying (FSK) and its variances are commonly used as a noncoherent modulation, although it can also be demodulated coherently with higher performance. For FSK and other related modulations, see other textbooks such as [107, chs. 3, 4].

This book does not cover analog modulation techniques, either. Such modulations are discussed in [108].

13.10 OPTICAL COMMUNICATIONS

Modern optical communications include three primary forms:

- Fiber optics: highly coherent light traveling through optical fibers to achieve long distance, high data rate communications.
- Free-space optical communication (FSO): light traveling through free space to achieve high data rate communications.
- Visible light communication (VLC): communication with unfocused, often noncoherent light sources. It is typically used for indoor applications.[5]

---

[4] There are many studies on interference alignment without complete channel information.

[5] Some literatures treat VLC as a special case of FSO or combine FSO and VLC as optical wireless communications.

Fiber optics was first commercially used in the 1970s. It now serves as the predominant technology for backbone and regional communication infrastructures, as well as a local area network (LAN) technology. Traditionally, fiber optics technologies are considered a separate area from digital communications at the physical layer. Because fiber optics has vast bandwidth available (leading to very high symbol rates) and the processing speed of electronic units is limited, modulation and demodulation techniques are usually kept simple. However, with the progress of electronics processing power, more sophisticated signal processing techniques are introduced in fiber optics systems [109, 110]. More recent developments on this aspect were captured in a special issue of *IEEE Signal Processing Magazine*, March 2014 (Volume 31, Issue 2).

Free-space optical communication started with military and space applications [111]. They are attractive to commercial communications because of the very high bandwidth available. For example, they can be used to provide high-speed data links between buildings, or between ground and airplanes. However, there are multiple technical challenges such as tracking and acquisition with the very narrow beams. The channel for FSO is also subject to blocking and fading.

Visible light communication uses light emitting diodes (LED) or similar light sources, instead of the lasers typically used for FSO [112]. A popular application of VLC is indoor local area networking (LAN), similar to the familiar WiFi use cases but possibly with higher data rates. In this case, VLC is also referred to as LiFi.

There are numerous researches on channel characteristics and modulation/demodulation techniques for FSO and VLC [113]. Many of them leverage results and techniques from digital communication.

There are several special issues of IEEE journals on light communications, including:

- *IEEE Wireless Communications*, April 2015 (Volume 22, Issue 2)
- *IEEE Journal on Selected Areas in Communications*, September 2015 (Volume 33, Issue 9)
- *IEEE Journal on Selected Areas in Communications*, January 2018 (Volume 36, Issue 1)
- *IEEE Communications Magazine*, February 2018 (Volume 56, Issue 2)

13.11 GREEN COMMUNICATIONS

Green communications mean techniques that reduce system energy consumption, including both transmission energy and that consumed by transmitter and receiver processing units. Green communications are often associated with Internet of Things (IoT), where the available energy to the transceivers is severely limited. Another related topic is energy harvesting, where a transceiver obtains power from ambient radiation [114]. If such radiation is intentionally provided, then it is the case of remote powering or wireless charging [115, 116].

Green communication is a part of the requirements for the 5G cellular systems (Chapter 12) [117]. More green communications studies can be found in the special issues of *IEEE Journal on Selected Areas in Communications*, May 2016 (Volume 34, Issue 5) and December 2016 (Volume 34, Issue 12).

13.12 APPLICATIONS OF ARTIFICIAL INTELLIGENCE (AI)

Application of artificial intelligence (AI) and machine learning (ML) to communications is an emerging and intriguing topic. Cognitive radios are obvious beneficiaries (Section 13.4). *IEEE Journal of Selected Topics in Signal Processing* had a special issue on ML for cognition in February 2018 (Volume 12, Issue 1). AI and ML are applied to other parts of physical layer processing, as well [118–120].

The prevalence of big data and their applications presents unique communication needs, which drive some of the wireless network architecture and operation considerations. Big data, in turn, can also help optimizing wireless network operations. *IEEE Wireless Communications* had a special issue on wireless big data in February 2018 (Volume 25, Issue 1).

13.13 APPLICATION OF GAME THEORY

Game theory deals with multiple parties cooperating or competing with each other, choosing some strategy to maximize certain utility functions [121]. Game theory has found applications in digital signal processing {866} and wireless networks [122].

The most fruitful application of game theory is in cognitive radio and spectrum sharing (Sections 13.4 and 13.8), where multiple parties interact to achieve some optimal allocation of the common resource (spectrum or time) [42, 43, 123–126]. Game theory is also applied to waveform adaptation [127], beamforming [128], network routing [129, 130], electronic attacks {631, 632, 531}, and so on.

IEEE Signal Processing Magazine had a special issue on game theory in September 2009 (Volume 26, Issue 5). Also, *IEEE Journal on Selected Areas in Communications* recently had two special issues on game theory applications to networking, in February and March of 2017 (Volume 35, Issues 2 and 3).

13.14 SECURITY

Security in communications is another issue that this book does not cover in detail. Security concerns have two aspects: we wish our communication is not disrupted by an adversary, and we wish our communication to remain secret to the adversary. As discussed in Chapter 11, as far as the physical layer is concerned, the first concern is mainly anti-jamming (AJ). The second concern is low probability of detection (LPD) and low probability of interception (LPI). One should bear in mind, however, that modern communication attacks and defenses are often cross-layer, holistic techniques. AJ and LPD/LPI are only parts of the integrated strategies.

The classical physical layer security measure is spread-spectrum [131]. One form (direct sequence spread spectrum) is similar to the CDMA technique briefly discussed in Chapter 12, except that the spreading code is kept secret. Another form (frequency hopping) changes the transmit frequency over a broad range, following a secret sequence. In both forms, the receiver must know the secret sequence in order to recover the data content. A typical adversary detector or jammer is disadvantageous without the knowledge of the secrete sequences.

The simplest detection threat is the energy detector (also known as radiometer), which bases its decision on measured RF power [132–136]. If the detector possesses some knowledge about the transmitted signal, more efficient detection methods can be used. For example, cyclic-stationary detection, mentioned in Section 13.4, can be used to detect typical modulated signals. One can also search for the synchronization or reference signal portion of the transmission if the details are known.

The simplest jammer is the noise jammer, which transmits noise-like RF over a selected bandwidth. Depending on hardware limitations, the noise can be limited to a narrow band, or in the form of impulses. More sophisticated jamming attacks can explore particular features and vulnerabilities of the targeted signal. For example, one can focus on the training and synchronization mechanism to "mislead" the target system into sub-optimal states. More sophisticated jamming techniques can be used on specific targets [137–139].

Geolocation of wireless transmitters, especially in a scattering environment such as indoors, is a widely researched topic [140–143]. Geolocation has various applications, including electronic warfare.

Another interesting topic is the security for MIMO systems, where the spatial dimension comes into play. This topic is discussed in Chapter 11.

Beyond traditional jamming and detection, there are various other electronic attacking and defense techniques. Some examples are proposed in [144–146]. The *IEEE Communication Magazine* publishes a special issue on military communications every October, including works on communication security.

13.15 NETWORK CODING

Consider a case where a network transports data from multiple sources to multiple directions. Typically, the data flows are separately transported while sharing the network resources. Network coding is a technique to improve network performance by combining the multiple data flows ("coding") in some paths of the network. To illustrate the concept, consider a simple example [147].

Figure 13.1 shows an example of network coding. The network topology is shown in panel A. The source S transmits data to destinations X and Y (a case of multicast). Nodes T, V, U, and W are routers. In a conventional network shown in panel A, two bits (b_1, b_2) can be transmitted in each time slot, following two

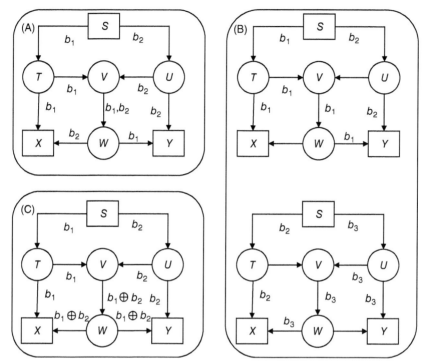

Figure 13.1 Network coding example [147]. A source (S) broadcasts data bits b_i, to two destinations X and Y. T, U, V, and W are intermediate routers. A: Classical network with no bandwidth limitation between V and W. Two bits are transmitted in each time slot. B: The bandwidth between V and W is limited to one bit per slot. Three bits can be transmitted in two time slots. C: Network coding with the same bandwidth limit as B. Two bits can be transmitted per time slot, improving performance comparing to B.

separate paths. The two paths share the same link between V and W, which needs twice the bandwidth as other links.

Panel B shows a case where the link between V and W has the same bandwidth as other links. Namely, it can only transmit one bit per time slot. In this case, this link needs to be used in turn by the two paths. Namely, it transmits b_1 that is fed to Y in the odd time slots (upper diagram in panel B) and b_3 destined for X in the even time slots (lower diagram). Thus, we can transmit three bits (b_1, b_2, b_3) in two time slots. The throughput is lower than the case in Panel A due to the bottleneck link between V and W.

Panel C shows a case with network coding. In this case, two bits (b_1, b_2) are transmitted in each time slot. The bandwidth for the "bottleneck" link is still limited to one bit per time slot. However, it transmits the exclusive or of b_1 and b_2, which is one bit. The same data are passed to X and Y. At X, b_1 is sent through a different path. Knowing b_1, X can recover b_2 from the exclusive or transmitted through the shared path. The same happens at Y. Therefore, with network coding, we can achieve the transmission rate of two bits per time slot.

Similar schemes can be used with multiple sources. For example, b_1 and b_2 can from separate data sources, to be multicast to both X and Y.

For general theories on network coding, works from Yeung, a pioneer of the subject, may serve as a good starting point [147–149]. More overviews and surveys can be found in the literature [150–152]. Network coding can be used to increase diversity against channel disruptions [153–155]. It can also improve system capacity [151, 156, 157]. Network coding has been proposed as a candidate technology for the 5G cellular systems to address the use cases involving complex interconnection among thousands of devices [158].

Although working at the bit level, network coding is closely related to physical layer considerations. A particularly interesting area is combining network coding with relaying (Section 13.6) [159, 160].

IEEE Journal on Selected Areas in Communications had two special issues on network coding: June 2009 (Volume 27, Issue 5) and February 2015 (Volume 33, Issue 2).

13.16 SUMMARY

As shown in this chapter, the foundation of digital communication techniques concerning a pair of transmitter and receiver has been stable for the past few decades. However, continuous innovation is still the norm. Most emerging techniques address various situations where multiple users share the media, with the objective of improving *system* performance regarding throughput and reliability. These new techniques work at various levels (modulation, coding, network configuration), are based on various mathematical treatments, and are fueled by various enabling technology advances such as greatly enhanced processing power, better and cheaper RF circuitry, sensors, and so on. Therefore, the field of digital communications today is more diverse and fast-changing. As an engineer in digital communications, there are ample opportunities for innovation and significant contributions. This book provides a launch pad for the journey.

REFERENCES

1. C. Moy and J. Palicot, "Software Radio: A Catalyst for Wireless Innovation," *IEEE Commun. Mag.*, vol. 53, no. 9, pp. 24–30, 2015.
2. G. Sklivanitis, A. Gannon, S. N. Batalama, and D. A. Pados, "Addressing Next-Generation Wireless Challenges with Commercial Software-Defined Radio Platforms," *IEEE Commun. Mag.*, vol. 54, no. 1, pp. 59–67, 2016.
3. Joint Tactical Networking Center, "Software Communications Architecture Specification," 2015 [Online]. Available at www.public.navy.mil/jtnc/SCA/SCAv4_1_Final/SCA_4.1_ScaSpecification. pdf [Accessed March 5, 2018].
4. Harris, "Harris Falcon III® AN/PRC-152A Wideband Networking Handheld Radio," 2018 [Online]. Available at https://www.harris.com/solution/harris-falcon-iii-anprc-152a-wideband-networking-hand-held-radio [Accessed March 5, 2018].

5. Harris, "Harris Airborne Multi-channel Radio (HAMR)," 2018 [Online]. Available at https://www.harris.com/solution/harris-airborne-multi-channel-radio-hamr [Accessed March 5, 2018].

6. R. C. Reinhart, S. K. Johnson, T. J. Kacpura, C. S. Hall, C. R. Smith, and J. Liebetreu, "Open Architecture Standard for NASA's Software-Defined Space Telecommunications Radio Systems," *Proc. IEEE*, vol. 95, no. 10, pp. 1986–1993, 2007.

7. B. Bloessl, M. Segata, C. Sommer, and F. Dressler, "An IEEE802.11a/g/p OFDM Receiver for GNU Radio," in *Proc. 2013 2nd ACM SIGCOMM Workshop of Software Radio Implementation Forum {(SRIF)}*, 12 August, Hong Kong, China, 2013, pp. 9–16.

8. W. Zhou, G. Villemaud, and T. Risset, "Full Duplex Prototype of OFDM on GNURadio and USRPs," in *IEEE Radio and Wireless Symposium, RWS*, 19–23 January, Newport Beach, CA, 2014.

9. D. C. Tucker and G. A. Tagliarini, "Prototyping with GNU Radio and the USRP—Where to Begin," in *IEEE Southeastcon 2009*, 5–8 March, Atlanta, GA, 2009.

10. GNU Radio [Online]. Available at https://gnuradio.org/ [Accessed March 5, 2018].

11. GNU Radio, "A Quick Guide to Hardware and GNU Radio," 2017 [Online]. Available at https://wiki..org/index.php/Hardware [Accessed March 5, 2018].

12. J. Mitola III, *Cognitive Radio Architecture*. Hoboken, NJ: Wiley, 2006.

13. Y. Zhao, S. Mao, J. Neel, and J. H. Reed, "Performance Evaluation of Cognitive Radios: Metrics, Utility Functions, and Methodology," *Proc. IEEE*, vol. 97, no. 4, pp. 642–659, 2009.

14. M. Amjad, F. Akhtar, M. H. Rehmani, M. Reisslein, and T. Umer, "Full-Duplex Communication in Cognitive Radio Networks: A Survey," *IEEE Commun. Surv. Tutorials*, vol. 19, no. 4, p. 2158, 2017.

15. J. Kim, S. W. Choi, W. Y. Shin, Y. S. Song, and Y. K. Kim, "Group Communication Over LTE: A Radio Access Perspective," *IEEE Commun. Mag.*, vol. 54, no. 4, pp. 16–23, 2016.

16. M. Z. Shakir et al., "Emerging Applications, Services, and Engineering for Cellular Cognitive Systems: Part II [Guest Editorial]," *IEEE Commun. Mag.*, vol. 53, no. 7, pp. 66–68, 2015.

17. M. Z. Shakir et al., "Emerging Applications, Services and Engineering for Cellular Cognitive Systems [Guest Editorial]," *IEEE Commun. Mag.*, vol. 53, no. 5, pp. 32–34, 2015.

18. J. M. Peha, "Sharing Spectrum Through Spectrum Policy Reform and Cognitive Radio," *Proc. IEEE*, vol. 97, no. 4, pp. 708–719, 2009.

19. T. Yucek and H. Arslan, "A Survey of Spectrum Sensing Algorithms for Cognitive Radio Applications," *IEEE Commun. Surv. Tutorials*, vol. 11, no. 1, pp. 116–130, 2009.

20. G. Ganesan, Y. Li, B. Bing, and S. Li, "Spatiotemporal Sensing in Cognitive Radio Networks," *IEEE J. Sel. Areas Commun.*, vol. 26, no. 1, pp. 5–12, 2008.

21. P. D. Sutton, K. E. Nolan, and L. E. Doyle, "Cyclostationary Signatures in Practical Cognitive Radio Applications," *IEEE J. Sel. Areas Commun.*, vol. 26, no. 1, pp. 13–24, 2008.

22. K. Kosmowski, M. Suchanski, J. Pawelec, and M. Kustra, "A Novel OFDM Sensing Method Based on CAF-Max for Hybrid Detectors Architecture," in *2016 International Conference on Military Communications and Information Systems, ICMCIS 2016*, 23–24 May, Brussels, Belgium, 2016.

23. T. Newman and T. Clancy, "Security Threats to Cognitive Radio Signal Classifiers," in *Virginia Tech Wireless Personal Communications Symposium*, 2009. https://scholar.google.com/scholar?cluster=15613493635253173387&hl=en&as_sdt=0,21

24. A. G. Fragkiadakis, E. Z. Tragos, and I. G. Askoxylakis, "A Survey on Security Threats and Detection Techniques in Cognitive Radio Networks," *IEEE Commun. Surv. Tutorials*, vol. 15, no. 1, pp. 428–445, 2013.

25. R. Chen, J. M. Park, and J. H. Reed, "Defense Against Primary User Emulation Attacks in Cognitive Radio Networks," *IEEE J. Sel. Areas Commun.*, vol. 26, no. 1, pp. 25–37, 2008.

26. M. T. Masonta, M. Mzyece, and N. Ntlatlapa, "Spectrum Decision in Cognitive Radio Networks: A Survey," *IEEE Commun. Surv. Tutorials*, vol. 15, no. 3, pp. 1088–1107, 2013.

27. H. Li et al., "Utility-Based Cooperative Spectrum Sensing Scheduling in Cognitive Radio Networks," *IEEE Trans. Veh. Technol.*, vol. 66, no. 1, pp. 645–655, 2013.

28. I. Salah, W. Saad, M. Shokair, and M. Elkordy, "Cooperative Spectrum Sensing and Clustering Schemes in CRN: A Survey," in *13th International Computer Engineering Conference (ICENCO)*, 27–28 December, Cairo, Egypt, 2017.

29. Min Song, Chunsheng Xin, Yanxiao Zhao, and Xiuzhen Cheng, "Dynamic Spectrum Access: from Cognitive Radio to Network Radio," *IEEE Wirel. Commun.*, vol. 19, no. 1, pp. 23–29, 2012.

30. W. Zhang and U. Mitra, "Spectrum Shaping: A New Perspective on Cognitive Radio, Part I: Coexistence with Coded Legacy Transmission," *IEEE Trans. Commun.*, vol. 58, no. 6, pp. 1857–1867, 2010.

31. R. Di Taranto and P. Popovski, "Outage Performance in Cognitive Radio Systems with Opportunistic Interference Cancelation," *IEEE Trans. Wirel. Commun.*, vol. 10, no. 4, pp. 1280–1288, 2011.

32. L. Zhang, R. Zhang, Y. C. Liang, Y. Xin, and S. Cui, "On the Relationship Between the Multi-antenna Secrecy Communications and Cognitive Radio Communications," *IEEE Trans. Commun.*, vol. 58, no. 6, pp. 1877–1886, 2010.

33. V. D. Nguyen, H. V. Nguyen, and O.-S. Shin, "An Efficient Zero-Forcing Precoding Design for Cognitive MIMO Broadcast Channels," *IEEE Commun. Lett.*, vol. 20, no. 8, pp. 1575–1578, 2016.

34. F. S. Al-Qahtani, R. M. Radaydeh, S. Hessien, T. Q. Duong, and H. Alnuweiri, "Underlay Cognitive Multihop MIMO Networks with and Without Receive Interference Cancellation," *IEEE Trans. Commun.*, vol. 65, no. 4, pp. 1477–1493, 2017.

35. H. Xie, B. Wang, F. Gao, and S. Jin, "A Full-Space Spectrum-Sharing Strategy for Massive MIMO Cognitive Radio Systems," *IEEE J. Sel. Areas Commun.*, vol. 34, no. 10, pp. 2537–2549, 2016.

36. F. Gao, R. Zhang, Y.-C. Liang, and X. Wang, "Multi-antenna Cognitive Radio Systems: Environmental Learning and Channel Training," in *IEEE International Conference on Acoustics, Speech, and Signal Processing (ICASSP)*, 19–24 April, Taipei, Taiwan, 2009.

37. M. H. Al-Ali and K. C. Ho, "Transmit Precoding in Underlay MIMO Cognitive Radio with Unavailable or Imperfect Knowledge of Primary Interference Channel," *IEEE Trans. Wirel. Commun.*, vol. 15, no. 8, pp. 5143–5155, 2016.

38. X. Y. Wang, A. Wong, and P.-H. Ho, "Extended Knowledge-Based Reasoning Approach to Spectrum Sensing for Cognitive Radio," *IEEE Trans. Mob. Comput.*, vol. 9, no. 4, pp. 465–478, 2010.

39. M. Bkassiny, Y. Li, and S. K. Jayaweera, "A Survey on Machine-Learning Techniques in Cognitive Radios," *IEEE Commun. Surv. Tutorials*, vol. 15, no. 3, pp. 1136–1159, 2013.

40. L. Gavrilovska, V. Atanasovski, I. Macaluso, and L. A. Dasilva, "Learning and Reasoning in Cognitive Radio Networks," *IEEE Commun. Surv. Tutorials*, vol. 15, no. 4, pp. 1761–1777, 2013.

41. Y. Zhang, W. P. Tay, K. H. Li, M. Esseghir, and D. Gaiti, "Learning Temporal—Spatial Spectrum Reuse," *IEEE Trans. Commun.*, vol. 64, no. 7, pp. 3092–3103, 2016.

42. G. Scutari and D. P. Palomar, "MIMO Cognitive Radio: A Game Theoretical Approach," *IEEE Trans. Signal Process.*, vol. 58, no. 2, pp. 761–780, 2010.

43. Y. Liu and L. Dong, "Spectrum Sharing in MIMO Cognitive Radio Networks Based on Cooperative Game Theory," *IEEE Trans. Wirel. Commun.*, vol. 13, no. 9, pp. 4807–4820, 2014.

44. J. W. Huang and V. Krishnamurthy, "Game Theoretic Issues in Cognitive Radio Systems," *J. Commun.*, vol. 4, no. 10, pp. 790–802, 2009.

45. C. R. Stevenson, G. Chouinard, Z. Lei, W. Hu, S. J. Shellhammer, and W. Caldwell, "IEEE 802.22: The First Cognitive Radio Wireless Regional Area Network Standard," *IEEE Commun. Mag.*, vol. 47, no. 1, pp. 130–138, 2009.

46. A. B. Flores, R. E. Guerra, E. W. Knightly, P. Ecclesine, and S. Pandey, "IEEE 802.11af: A Standard for TV White Space Spectrum Sharing," *IEEE Commun. Mag.*, vol. 51, no. 10, pp. 92–100, 2013.

47. IEEE, *802.11-2016—IEEE Standard for Information Technology—Telecommunications and Information Exchange Between Systems Local and Metropolitan Area Networks–Specific Requirements—Part 11: Wireless LAN Medium Access Control (MAC) and Physical Layer (PHY)*. IEEE, 2016.

48. A. C. Sumathi, M. Akila, and S. Arunkumar, "Study of CR Based U-LTE Co-existence Under Varying Wi-Fi Standards," in *Frontier Computing*, J. C. Hung, N. Y. Yen, and L. Hui, Eds. Singapore: Springer, 2017, pp. 366–375.

49. Z. Wang, H. Shawkat, S. Zhao, and B. Shen, "An LTE-U Coexistence Scheme Based on Cognitive Channel Switching and Adaptive Muting Strategy," in *2017 IEEE 28th Annual International Symposium on Personal, Indoor, and Mobile Radio Communications (PIMRC)*, 8–13 October, Montreal, QC, Canada, 2017, pp. 1–6.

50. A. Bhorkar, C. Ibars, A. Papathanassiou, and P. Zong, "Medium Access Design for LTE in Unlicensed Band," in *IEEE Wireless Communications and Networking Conference Workshops (WCNCW)*, 9–12 March, New Orleans, LA, 2015.

51. P. F. Marshall, "Extending the Reach of Cognitive Radio," *Proc. IEEE*, vol. 97, no. 4, pp. 612–625, 2009.

52. M. McHenry, E. Livsics, T. Nguyen, and N. Majumdar, "XG Dynamic Spectrum Access Field Test Results," *IEEE Commun. Mag.*, vol. 45, no. 6, pp. 51–57, 2007.

53. A. N. Mody et al., "On Making the Current Military Radios Cognitive Without Hardware or Firmware Modifications," *Proc. IEEE Mil. Commun. Conf.*, pp. 2327–2332, 2010.

54. M. Z. Win and R. A. Scholtz, "Ultra-Wide Bandwidth Time-Hopping Spread-Spectrum Impulse Radio for Wireless Multiple-Access Communications," *IEEE Trans. Commun.*, vol. 48, no. 4, pp. 679–691, 2000.

55. E. Karapistoli, F. N. Pavlidou, I. Gragopoulos, and I. Tsetsinas, "An Overview of the IEEE 802.15. 4a Standard," *IEEE Commun. Mag.*, vol. 48, no. 1, pp. 47–53, 2010.

56. B. Lewis, "UWB Is Back ... This Time for IoT Location-Based Services," *Embedded Computing Design*, 2014 [Online]. Available at http://www.embedded-computing.com/embedded-computing-design/uwb-is-back-this-time-for-iot-location-based-services [Accessed April 28, 2018].

57. A. Al-Fuqaha, M. Guizani, and M. Mohammadi, "Internet of Things: A Survey on Enabling Technologies, Protocols, and Applications," *IEEE Commun. Surv. Tutorials*, vol. 17, no. 4, pp. 2347–2376, 2015.

58. T. M. Cover and A. A. El Gamal, "Capacity Theorems for the Relay Channel," *IEEE Trans. Inf. Theory*, vol. 25, no. 5, pp. 572–584, 1979.

59. H. F. Chong and M. Motani, "On Achievable Rates for the General Relay Channel," *IEEE Trans. Inf. Theory*, vol. 57, no. 3, pp. 1249–1266, 2011.

60. M. Karkooti and J. R. Cavallaro, "Distributed Decoding in Cooperative Communications," in *Conference Record of the Forty-First Asilomar Conference on Signals, Systems and Computers*, 4–7 November, Pacific Grove, CA, 2007, pp. 824–828.

61. V. Stankovic, A. Host-Madsen, and Z. Xiong, "Cooperative Diversity for Wireless Ad Hoc Networks," *IEEE Signal Process. Mag.*, vol. 23, no. 5, pp. 37–49, 2006.

62. A. Chakrabarti, E. Erkip, A. Sabharwal, and B. Aazhang, "Code Designs for Cooperative Communication," *IEEE Signal Process. Mag.*, vol. 24, no. 5, pp. 16–26, 2007.

63. A. Sendonaris, E. Erkip, and B. Aazhang, "User Cooperation Diversity—Part I: System Description," *IEEE Trans. Commun.*, vol. 51, no. 11, pp. 1927–1938, 2003.

64. A. Sendonaris, E. Erkip, and B. Aazhang, "User Cooperation Diversity—Part II: Implementation Aspects and Performance Analysis," *IEEE Trans. Commun.*, vol. 51, no. 11, pp. 1939–1948, 2003.

65. M. Chen and A. Yener, "Multiuser Two-Way Relaying: Detection and Interference Management Strategies," *IEEE Trans. Wirel. Commun.*, vol. 8, no. 8, pp. 4296–4305, 2009.

66. D. Chen, M. Haenggi, and J. Laneman, "Distributed Spectrum-Efficient Routing Algorithms in Wireless Networks," *IEEE Trans. Wirel. Commun.*, vol. 7, no. 12, pp. 5297–5305, 2008.

67. C. Rao and B. Hassibi, "Diversity-Multiplexing Gain Trade-Off of a MIMO System with Relays," in *Proceedings of the 2007 IEEE Information Theory Workshop on Information Theory for Wireless Networks*, 1–6 July, Solstrand, Norway, 2007.

68. K. Azarian, H. El Gamal, and P. Schniter, "On the Achievable Diversity Multiplexing Tradeoff in Half-Duplex Cooperative Channel," *IEEE Trans. Inf. Theory*, vol. 51, no. 12, pp. 4152–4172, 2005.

69. B. Wang, J. Zhang, and A. Host-Madsen, "On the Capacity of MIMO Relay Channels," *IEEE Trans. Inf. Theory*, vol. 51, no. 1, pp. 29–43, 2005.

70. A. Ozgur, O. Leveque, and D. Tse, "Spatial Degrees of Freedom of Large Distributed MIMO Systems and Wireless Ad Hoc Networks," *IEEE J. Sel. Areas Commun.*, vol. 31, no. 2, pp. 202–214, 2013.

71. H.-Y. Y. Shen, H. Yang, B. Sikdar, and S. Kalyanaraman, "A Distributed System for Cooperative MIMO Transmissions," in *Proc. IEEE Global Telecommunications Conference, GLOBECOM*, 30 November–4 December, New Orleans, LO, 2008.

72. J. Harshan and B. S. Rajan, "A Non-differential Distributed Space-Time Coding for Partially-Coherent Cooperative Communication," *IEEE Trans. Wirel. Commun.*, vol. 7, no. 11, pp. 4076–4081, 2008.

73. Y. Jing and H. Jafarkhani, "Distributed Differential Space-Time Coding for Wireless Relay Networks," *IEEE Trans. Commun.*, vol. 56, no. 7, pp. 1092–1100, 2008.

74. A. Agustin, O. Muhoz, and J. Vidal, "A Game Theoretic Approach for Cooperative MIMO Schemes with Cellular Reuse of the Relay Slot," in IEEE International Conference on Acoustics, Speech, and Signal Processing, 17–21 May, Montreal, Quebec, Canada, 2004.

75. N. Devroye, N. Mehta, and A. Molisch, "Asymmetric Cooperation Among Wireless Relays with Linear Precoding," *IEEE Trans. Wirel. Commun.*, vol. 7, no. 12, pp. 5420–5430, 2008.

76. J. Park, E. Song, and W. Sung, "Capacity Analysis for Distributed Antenna Systems Using Cooperative Transmission Schemes in Fading Channels," *Wirel. Commun. IEEE Trans.*, vol. 8, no. 2, pp. 586–592, 2009.

77. Y. Yang, H. Hu, J. Xu, and G. Mao, "Relay Technologies for WiMAX and LIE-Advanced Mobile Systems," *IEEE Commun. Mag.*, vol. 47, no. 10, pp. 100–105, 2009.

78. K. Loa et al., "IMT-Advanced Relay Standards," *IEEE Commun. Mag.*, vol. 48, no. 8, pp. 40–48, 2010.

79. 3GPP, *3rd Generation Partnership Project; Technical Specification Group Radio Access Network; Evolved Universal Terrestrial Radio Access (E-UTRA); Further Advancements for E-UTRA Physical Layer Aspects*, 3GPP TR 36.814 v9.2.0.

80. S.-Y. Sun, H.-H. Chen, and W.-X. Meng, "A Survey on Complementary-Coded MIMO CDMA Wireless Communications," *IEEE Commun. Surv. Tutorials*, vol. 17, no. 1, pp. 52–69, 2015.

81. F. Jasbi and D. K. C. So, "Hybrid Overlay/Underlay Cognitive Radio Network with MC-CDMA," *IEEE Trans. Veh. Technol.*, vol. 65, no. 4, pp. 2038–2047, 2015.

82. Z. Liu, Y. L. Guan, and H. H. Chen, "Fractional-Delay-Resilient Receiver Design for Interference-Free MC-CDMA Communications Based on Complete Complementary Codes," *IEEE Trans. Wirel. Commun.*, vol. 14, no. 3, pp. 1226–1236, 2015.

83. Y.-A. Chen, Y. Chang, Y.-C. Tseng, and W.-T. Chen, "A Framework for Simultaneous Message Broadcasting Using CDMA-Based Visible Light Communications," *IEEE Sens. J.*, vol. 15, no. 12, pp. 6819–6827, 2015.

84. M. H. Shoreh, A. Fallahpour, and J. A. Salehi, "Design Concepts and Performance Analysis of Multicarrier CDMA for Indoor Visible Light Communications," *J. Opt. Commun. Netw.*, vol. 7, no. 6, pp. 554–562, 2015.

85. G. L. Stuber, *Principles of Mobile Communication*. New York: Springer, 2011.

86. K. Moshksar, A. Ghasemi, and A. K. Khandani, "An Alternative to Decoding Interference or Treating Interference as Gaussian Noise," *IEEE Trans. Inf. Theory*, vol. 61, no. 1, pp. 305–322, 2015.

87. C. Geng, N. Naderializadeh, S. Avestimehr, and S. A. Jafar, "On the Optimality of Treating Interference as Noise: A Combinatorial Optimization Perspective," *IEEE Trans. Inf. Theory*, vol. 41, no. 4, pp. 1753–1767, 2015.

88. A. Dytso, D. Tuninetti, and N. Devroye, "Interference as Noise: Friend or Foe?," *IEEE Trans. Inf. Theory*, vol. 62, no. 6, pp. 3561–3596, 2016.

89. H. Wang, N. Liu, Z. Li, Z. Pan, and X. You, "Joint MUD Exploitation and ICI Mitigation Based Scheduling with Limited Base Station Cooperation," in *IEEE International Symposium on Personal, Indoor and Mobile Radio Communications, PIMRC*, 9–12 September, Sydney, NSW, Australia, 2012.

90. L. Tadjpour, S. H. Tsai, and C. C. J. Kuo, "Complexity Reduction of Maximum-Likelihood Multiuser Detection (ML-MUD) Receivers with Carrier Interferometry Codes in MC-CDMA," in *IEEE International Conference on Communications (ICC)*, 19–23 May, Beijing, China, 2008.

91. A. I. Canbolat and K. Fukawa, "Joint Interference Suppression and Multiuser Detection Schemes for Multi-Cell Wireless Relay Communications: A Three-Cell Case," *IEEE Trans. Commun.*, vol. 66, no. 4, pp. 1399–1410, 2018.

92. P. Lee, T. Lee, J. Jeong, and J. Shin, "Interference Management in LTE Femtocell Systems Using Fractional Frequency Reuse," in *Advanced Communication Technology (ICACT)*, 7–10 February, Phoenix Park, South Korea, 2010.

93. T. D. Novlan, R. K. Ganti, A. Ghosh, and J. G. Andrews, "Analytical Evaluation of Fractional Frequency Reuse for OFDMA Cellular Networks," *IEEE Trans. Wirel. Commun.*, vol. 10, no. 12, pp. 4294–4305, 2011.

94. T. Novlan, J. G. Andrews, I. Sohn, R. K. Ganti, and A. Ghosh, "Comparison of Fractional Frequency Reuse Approaches in the OFDMA Cellular Downlink," in IEEE Global Telecommunications Conference (GTLOBECOM), 2010.

95. Z. Xie and B. Walke, "Frequency Reuse Techniques for Attaining Both Coverage and High Spectral Efficiency in OFDMA Cellular Systems," in *IEEE Wireless Communications and Networking Conference, WCNC*, 18–21 April, Sydney, NSW, Australia, 2010, pp. 1–6.

96. H. Shirani-Mehr, H. Papadopoulos, S. A. Ramprashad, and G. Caire, "Joint Scheduling and ARQ for MU-MIMO Downlink in the Presence of Inter-Cell Interference," *IEEE Trans. Commun.*, vol. 59, no. 2, pp. 578–589, 2011.

97. J. Jang et al., "Smart Small Cell with Hybrid Beamforming for 5G: Theoretical Feasibility and Prototype Results," *IEEE Wirel. Commun.*, vol. 23, no. 6, pp. 124–131, 2016.

98. G. Bresler, D. Cartwright, and D. Tse, "Feasibility of Interference Alignment for the MIMO Interference Channel," *IEEE Trans. Inf. Theory*, vol. 60, no. 9, pp. 5573–5586, 2014.

99. S. A. Jafar, "Interference Alignment—A New Look at Signal Dimensions in a Communication Network," *Found. Trends Commun. Inf. Theory*, vol. 7, no. 1, pp. 1–134, 2010.

100. V. R. Cadambe and S. A. Jafar, "Interference Alignment and Spatial Degrees of Freedom for the K User Interference Channel," *IEEE Int. Conf. Commun.*, vol. 54, no. 8, pp. 971–975, 2008.

101. C. M. Yetis, T. Gou, S. A. Jafar, and A. H. Kayran, "On Feasibility of Interference Alignment in MIMO Interference Networks," *IEEE Trans. Signal Process.*, vol. 58, no. 9, pp. 4771–4782, 2010.

102. H. Zeng et al., "OFDM-Based Interference Alignment in Single-Antenna Cellular Wireless Networks," *IEEE Trans. Commun.*, vol. 65, no. 10, pp. 4492–4506, 2017.

103. C. Wang, C. Qin, Y. Yao, Y. Li, and W. Wang, "Low Complexity Interference Alignment for mmWave MIMO Channels in Three-Cell Mobile Network," *IEEE J. Sel. Areas Commun.*, vol. 35, no. 7, pp. 1513–1523, 2017.

104. X. Jing, L. Mo, H. Liu, and C. Zhang, "Linear Space-Time Interference Alignment for K-User MIMO Interference Channels," *IEEE Access*, vol. 6, pp. 3085–3095, 2018.

105. H. J. Yang, W. Y. Shin, B. C. Jung, C. Suh, and A. Paulraj, "Opportunistic Downlink Interference Alignment," *IEEE Trans. Wirel. Commun.*, vol. 16, no. 3, pp. 1533–1548, 2017.

106. C. Suh and D. Tse, "Interference Alignment for Cellular Networks," in *46th Annual Allerton Conference on Communication, Control, and Computing*, 23–26 September, Urbana-Champaign, IL, 2008, pp. 1037–1044.

107. J. Proakis and M. Salehi, *Digital Communications*, 5th ed. McGraw-Hill Education, 2007.

108. A. B. Carlson and P. B. Crilly, *Communication Systems*. McGraw-Hill, 2010.

109. G. Katz and D. Sadot, "Wiener Solution of Electrical Equalizer Coefficients in Lightwave Systems," *IEEE Trans. Commun.*, vol. 57, no. 2, pp. 361–364, 2009.

110. W. Shieh and I. Djordjevic, *OFDM for Optical Communications*. Burlington, MA: Academic Press/Elsevier, 2010.

111. H. Kaushal, V. K. Jain, and S. Kar, *Free Space Optical Communication*. Springer (India), 2017.

112. Y. Qiu, S. Chen, H.-H. Chen, and W. Meng, "Visible Light Communications Based on CDMA Technology," *IEEE Wirel. Commun.*, vol. 25, no. 2, pp. 178–185, 2018.

113. V. Janyani, M. Tiwari, G. Singh, and P. Minzioni, *Optical and Wireless Technologies*. Springer, 2018.

114. M. Ku, W. Li, Y. Chen, and K. J. R. Liu, "Advances in Energy Harvesting Communications: Past, Present, and Future Challenges," *IEEE Commun. Surv. Tutorials*, vol. 18, no. 2, pp. 1384–1412, 2016.

115. X. Lu, P. Wang, D. Niyato, D. I. Kim, and Z. Han, "Wireless Charging Technologies: Fundamentals, Standards, and Network Applications," *IEEE Commun. Surv. Tutorials*, vol. 18, no. 2, pp. 1413–1452, 2016.

116. Y. Zeng, B. Clerckx, and R. Zhang, "Communications and Signals Design for Wireless Power Transmission," *IEEE Trans. Commun.*, vol. 65, no. 5, pp. 2264–2290, 2017.

117. Q. Wu, G. Y. Li, W. Chen, D. W. K. Ng, and R. Schober, "An Overview of Sustainable Green 5G Networks," *IEEE Wirel. Commun.*, vol. 24, no. 4, pp. 72–80, 2017.

118. D. Sebastian, S. Cammerer, S. Member, and J. Hoydis, "Deep Learning Based Communication Over the Air," *IEEE J. Sel. Top. Signal Process.*, vol. 12, no. 1, pp. 132–143, 2018.

119. E. Nachmani, E. Marciano, L. Lugosch, W. J. Gross, D. Burshtein, and Y. Be'ery, "Deep Learning Methods for Improved Decoding of Linear Codes," *IEEE J. Sel. Top. Signal Process.*, vol. 12, no. 1, pp. 119–131, 2018.

120. Z. Qin, H. Ye, G. Y. Li, and B.-H. F. Juang, "Deep Learning in Physical Layer Communications," *arXiv Preprint* [Online]. Available at http://arxiv.org/abs/1807.11713 [Accessed September 28, 2018].

121. S. Lasaulce, M. Debbah, and E. Altman, "Methodologies for Analyzing Equilibria in Wireless Games," *IEEE Signal Process. Mag.*, vol. 26, no. 5, pp. 41–52, 2009.

122. D. T. Hoang, X. Lu, D. Niyato, P. Wang, D. I. Kim, and Z. Han, "Applications of Repeated Games in Wireless Networks: A Survey," *IEEE Commun. Surv. Tutorials*, vol. 17, no. 4, pp. 2102–2135, 2015.

123. Z. Al-banna, *"Spectrum Utilization Using Game Theory,"* Brunel University, 2009.

124. G. Scutari, D. P. Palomar, J.-S. Pang, and F. Facchinei, "Flexible Design of Cognitive Radio Wireless Systems," *IEEE Signal Process. Mag.*, vol. 26, no. 5, pp. 107–123, 2009.

125. E. G. Larsson, E. A Jorswieck, L. Johannes, and R. Mochaourab, "Game Theory and the Flat-Fading Gaussian Interference Channel," *IEEE Signal Process. Mag.*, vol. 26, no. 5, pp. 18–27, 2009.

126. N. B. Chang and M. Liu, "Optimal Competitive Algorithms for Opportunistic Spectrum Access," *IEEE J. Sel. Areas Commun.*, vol. 26, no. 7, p. 1183–1192, 2008.

127. S. Buzzi, H. V. Poor, and D. Saturnino, "Noncooperative Waveform Adaptation Games in Multiuser Wireless Communications," *IEEE Signal Process. Mag.*, vol. 26, no. 5, pp. 64–76, 2009.

128. W. Zhong, Y. Xu, and H. Tianfield, "Game-Theoretic Opportunistic Spectrum Sharing Strategy Selection for Cognitive MIMO Multiple Access Channels," *IEEE Trans. Signal Process.*, vol. 59, no. 6, pp. 2745–2759, 2011.

129. Y. Chen and S. Kishore, "A Game-Theoretic Analysis of Decode-and-Forward User Cooperation," *IEEE Trans. Wirel. Commun.*, vol. 7, no. 5, pp. 1941–1951, 2008.

130. R. Kannan and S. S. Iyengar, "Game-Theoretic Models for Reliable Path-Length and Energy-Constrained Routing with Data Aggregation in Wireless Sensor Networks," *IEEE J. Sel. Areas Commun.*, vol. 22, no. 6, pp. 1141–1150, 2004.

131. D. Torrieri, *Principles of Spread-Spectrum Communication Systems*, 4th ed. Springer, 2018.

132. M. Bloch, "Covert Communication over Noisy Channels: A Resolvability Perspective," *IEEE Trans. Inf. Theory*, vol. 62, no. 5, pp. 1–1, 2016.

133. B. He, S. Yan, X. Zhou, and V. K. N. Lau, "On Covert Communication with Noise Uncertainty," *IEEE Commun. Lett.*, vol. 21, no. 4, pp. 941–944, 2017.

134. L. Rugini, P. Banelli, and G. Leus, "Small Sample Size Performance of the Energy Detector," *IEEE Commun. Lett.*, vol. 17, no. 9, pp. 1814–1817, 2013.

135. B. A. Bash, D. Goeckel, D. Towsley, and S. Guha, "Hiding Information in Noise: Fundamental Limits of Covert Wireless Communication," *IEEE Commun. Mag.*, vol. 53, no. 12, pp. 26–31, 2015.

136. S. Lee, R. J. Baxley, M. A. Weitnauer, and B. Walkenhorst, "Achieving Undetectable Communication," *IEEE J. Sel. Top. Signal Process.*, vol. 9, no. 7, pp. 1195–1205, 2015.

137. M. Lichtman, R. P. Jover, M. Labib, R. Rao, V. Marojevic, and J. H. Reed, "LTE/LTE-A Jamming, Spoofing, and Sniffing: Threat Assessment and Mitigation," *IEEE Commun. Mag.*, vol. 54, no. 4, pp. 54–61, 2016.

138. S. Amuru, H. S. Dhillon, and R. M. Buehrer, "On Jamming Against Wireless Networks," *IEEE Trans. Wirel. Commun.*, vol. 16, no. 1, pp. 412–428, 2017.

139. C. Shahriar et al., "PHY-Layer Resiliency in OFDM Communications: A Tutorial," *IEEE Commun. Surv. Tutorials*, vol. 17, no. 1, pp. 292–314, 2015.

140. A. Yassin et al., "Recent Advances in Indoor Localization: A Survey on Theoretical Approaches and Applications," *IEEE Commun. Surv. Tutorials*, vol. 19, no. 2, pp. 1327–1346, 2017.

141. C. Chen, Y. Han, Y. Chen, and K. J. R. Liu, "Indoor Global Positioning System with Centimeter Accuracy Using Wi-Fi," *IEEE Signal Process. Mag.*, vol. 33, no. 6, pp. 128–134, 2016.

142. P. Davidson and R. Piché, "A Survey of Selected Indoor Positioning Methods for Smartphones," *IEEE Commun. Surv. Tutorials*, vol. 19, no. 2, pp. 1347–1370, 2017.

143. Q. D. Vo and P. De, "A Survey of Fingerprint-Based Outdoor Localization," *IEEE Commun. Surv. Tutorials*, vol. 18, no. 1, pp. 491–506, 2016.

144. X. Wu and Z. Yang, "Physical-Layer Authentication for Multi-Carrier Transmission," *IEEE Commun. Lett.*, vol. 19, no. 1, pp. 74–77, 2015.

145. Q. Xu, R. Zheng, W. Saad, and Z. Han, "Device Fingerprinting in Wireless Networks: Challenges and Opportunities," *IEEE Commun. Surv. Tutorials*, vol. 18, no. 1, pp. 94–104, 2016.

146. T. Xu, M. Zhang, and T. Zhou, "Statistical Signal Transmission Technology: A Novel Perspective for 5G Enabled Vehicular Networking," *IEEE Wirel. Commun.*, vol. 24, no. 6, pp. 22–29, 2017.

147. R. W. Yeung, S.-Y. R. Li, N. Cai, and Z. Zhang, "Network Coding Theory Part I: Single Source," *Found. Trends Commun. Inf. Theory*, vol. 2, no. 4, pp. 241–329, 2005.

148. R. W. Yeung, S.-Y. R. Li, N. Cai, and Z. Zhang, "Network Coding Theory Part II: Multiple Source," *Found. Trends Commun. Inf. Theory*, vol. 2, no. 4, pp. 330–381, 2005.

149. M. Tan, R. W. Yeung, S.-T. Ho, and N. Cai, "A Unified Framework for Linear Network Coding," *IEEE Trans. Inf. Theory*, vol. 57, no. 1, pp. 416–423, 2011.

150. R. Bassoli, H. Marques, J. Rodriguez, K. W. Shum, and R. Tafazolli, "Network Coding Theory: A Survey," *IEEE Commun. Surv. Tutorials*, vol. 15, no. 4, pp. 1950–1978, 2013.

151. R. Ahlswede, N. Cai, S.-Y. R. Li, and R. W. Yeung, "Network Information Flow," *IEEE Trans. Inf. Theory*, vol. 46, no. 4, pp. 1204–1216, 2000.

152. M. Médard, F. H. P. Fitzek, M. J. Montpetit, and C. Rosenberg, "Network Coding Mythbusting: Why It Is Not About Butterflies Anymore," *IEEE Commun. Mag.*, vol. 52, no. 7, pp. 177–183, 2014.

153. Y. Chen, S. Kishore, and J. Li, "Wireless Diversity Through Network Coding," in *IEEE Wireless Communications and Networking Conference, 2006. WCNC 2006*, 3–6 April 2006, pp. 1681–1686.

154. M. Xiao and M. Skoglund, "Multiple-User Cooperative Communications Based on Linear Network Coding," *IEEE Trans. Commun.*, vol. 58, no. 12, pp. 3345–3351, 2010.

155. S. Li and A. Ramamoorthy, "Protection Against Link Errors and Failures Using Network Coding," *IEEE Trans. Commun.*, vol. 59, no. 2, pp. 518–528, 2011.

156. J. El-Najjar, H. M. K. Alazemi, and C. Assi, "On the Interplay Between Spatial Reuse and Network Coding in Wireless Networks," *IEEE Trans. Wirel. Commun.*, vol. 10, no. 2, pp. 560–569, 2011.

157. S. H. Lim, Y. H. Kim, A. El Gamal, and S. Y. Chung, "Noisy Network Coding," *IEEE Trans. Inf. Theory*, vol. 57, no. 5, pp. 3132–3152, 2011.

158. P. T. Compta, F. H. P. Fitzek, and D. E. Lucani, "Network Coding Is the 5G Key Enabling Technology: Effects and Strategies to Manage Heterogeneous Packet Lengths," *Trans. Emerg. Telecommun. Technol.*, no. January 2015, pp. 46–55, 2015.

159. U. Niesen and P. Whiting, "The Degrees of Freedom of Compute-and-Forward," *IEEE Trans. Inf. Theory*, vol. 58, no. 8, pp. 5214–5232, 2012.

160. V. Namboodiri, K. Venugopal, and B. S. Rajan, "Physical Layer Network Coding for Two-Way Relaying with QAM," *IEEE Trans. Wirel. Commun.*, vol. 12, no. 10, pp. 5074–5086, 2013.

INDEX

Digital Communication for Practicing Engineers, First Edition. Feng Ouyang.
© 2020 by The Institute of Electrical and Electronics Engineers, Inc.
Published 2020 by John Wiley & Sons, Inc.

IEEE PRESS SERIES ON
DIGITAL AND MOBILE COMMUNICATION

John B. Anderson, *Series Editor*
University of Lund